S0-CAS-690

ORGANIC PHOSPHORUS COMPOUNDS

ORGANIC PHOSPHORUS COMPOUNDS

Volume 3

G. M. KOSOLAPOFF
Auburn University

and

L. MAIER
Monsanto Research S. A.

WILEY-INTERSCIENCE, a Division of John Wiley & Sons, Inc.
New York • London • Sydney • Toronto

Library of Congress Cataloging in Publication Data

Kosolapoff, Gennady M

Organic phosphorous compounds.

1950 ed. published under title: Organophosphorus compounds.
Includes bibliographies.
1. Organophosphorus compounds. I. Maier, L., joint author. II. Title.

QD412.P1K55 1972 547'.07 72-1359
ISBN 0-471-50442-4 (v. 3)

Printed in the United States of America

10 9 8 7 6 5 4 3 2 1

Contents

ORGANIC PHOSPHORUS COMPOUNDS

Chapter 5A. Phosphine Alkylenes and Other Phosphorus Ylids

H. J. BESTMANN and R. ZIMMERMANN

Institut für Organische Chemie der Universität Erlangen-Nürnberg, Germany

A. PHOSPHINE ALKYLENES (PHOSPHONIUM YLIDS)

A.1. Preparation of Phosphine Alkylenes

I. YLIDS FROM PHOSPHONIUM SALTS

Phosphine alkylenes (2) are generally prepared by the interaction of bases and phosphonium salts (1).

$$[R^1R^2CH\text{-}PR^3_3]\ X \underset{+HX}{\overset{-HX}{\rightleftharpoons}} R^1R^2C\text{-}PR^3_3$$

(1) (2)

Using this method, Michaelis and Gimborn[434] obtained in 1894 a phosphonium ylid by treating triphenyl(methoxy-carbonylmethyl)phosphonium chloride with an aqueous solution of sodium hydroxide, without, however, recognizing its structure. The strength of the base necessary to effect the deprotonation depends on the acidity of the phosphonium salt used. As bases one may use ammonia,[472] triethylamine,[190,403,643] pyridine,[190,643] sodium carbonate,[376,435] sodium hydroxide,[341,434] sodium ethoxide,[677] potassium t-butoxide,[400] sodium t-amylate,[388] sodium amide,[675] lithium diethylamide,[289] lithium piperide,[675] sodium hydride,[174,368a] methylsulfinylcarbanion,[172,252] tritylsodium,[167] butyllithium,[167] phenyllithium,[682] sodium,[464] sodium acetylide,[185,188] potassium in HMPT,[112] and, finally, strongly basic phosphine alkylenes.[40]

In the case of phosphonium salts, whose anion is strongly nucleophilic, the latter takes over the role of the base and makes the use of an auxiliary base unnecessary. With phosphonium fluorides Wittig reactions in nonbasic media are therefore possible.[521]

For the successful performance of many reactions of phosphine alkylenes it is important to use lithium-salt-free solutions, since lithium salts form complexes with ylids which interfere considerably with the course of the reaction. Trialkylphosphine alkylenes (alkylene-trialkylphosphorane) form stable complexes with lithium salts which dissociate only at elevated temperature. For this reason the preparation of pure ylids which have three alkyl groups attached to phosphorus has been achieved only recently by desilylating silyl-substituted methylenephosphoranes.[535,682] Alkylene-trialkylphosphoranes are also obtained by treating the corresponding phosphonium chlorides or bromides with sodium amide in boiling tetrahydrofuran.[368a]

Since alkylene-trialkylphosphoranes, on heating or in the presence of proton-active compounds, exchange the hydrogen atoms linked to the α-carbon atoms in such a way that the ligands on the phosphorus atom are interchangeably present as alkyl or alkylene residue,[89,536] an equilibrium mixture of isomeric ylids is sometimes obtained[368a] from phosphonium salts with different alkyl groups.

The tendency of alkyl groups to rearrange into alkylene

groups decreases in the following way, according to Köster:[368a]

$$CH_2\text{-}\overset{CH_3}{\overset{|}{C}}{=}CH_2 \gg CH_3 \gg C_3H_7; \quad C_4H_9 > C_2H_5; \quad CH(CH_3)_2$$

For the preparation of salt-free solutions of ylids the sodium amide method[41,675] (with liquid ammonia as solvent) is well suited.

Schöllkopf[543] has pointed out that elimination reactions may interfere with ylid formation.

The interaction of bases and phosphonium salts that contain activated hydrogen atoms in β-position to the phosphorus atom yields an ylid that may decompose by intermolecular proton transfer into an olefin and triphenylphosphine.[62 106,250,283,288,660] In such cases sodium alcoholate is unsuited as a base, since it favors the Hofmann degradation.[62,286]

Furthermore, when using organolithium compounds as bases, a ligand exchange at the phosphorus atom is possible.[565,568] Also organolithium compounds react with phosphonium salts that contain halogen atoms at the α-carbon attacking both halogen and hydrogen.[368,566-7] The ratio of attack on hydrogen to attack on halogen is dependent on the halogen atom as well as on the base used.

From β-halophosphonium salts and phenyllithium result vinyl-salts to which a second mole base may add.[562]

In general, care must be taken that the phosphonium salt used as a starting material contains, in addition to the α-proton, no other groups which would also react with the base.

In addition to this generally applicable method, the so-called salt method, other special procedures also allow the synthesis of ylids.

II. YLIDS FROM CARBENES (OR CARBENOIDES) AND TRIPHENYLPHOSPHINE

$$Ph_3P + :C{<}^{X}_{Y} \longrightarrow Ph_3P{=}C{<}^{X}_{Y}$$

Phosphine alkylenes with one or two halogen atoms on the α-carbon atom of the C-P double bond are preparable in this way.[563,564,569,597,599,683] Difluoromethylenetriphenylphosphorane,[232,281] caused to react in statu nascendi with carbonyl compounds, yields 1,1-difluoroolefins.

The structure of difluoromethylenetris(dimethylamino)-phosphorane, synthesized by Mark,[422] has been revised. Ramirez and Smith[493] could show that the isolated compound is a phosphorane containing pentacovalent phosphorus.

The mixed fluorochloromethylenephosphorane[154] has also

been prepared according to the carbene method.

Dihalomethylenephosphoranes are formed in the reaction of triphenylphosphine and carbon tetrachloride or bromide.[459,476,482]

$$2Ph_3P + CX_4 \longrightarrow Ph_3P{=}CX_2 + Ph_3PX_2$$

$$X = Cl, Br$$

The mechanism of this reaction has not been clarified completely.

Diazoaliphatics may also serve as a carbene source for the preparation of ylids. However, they must be decomposed in the presence of Cu(I) salts,[684] since generally triphenylphosphine reacts with diazo compounds to form phosphazines (3).[48,610,614-5,677]

$$Ph_3P + \overset{\oplus}{N}{\equiv}\overset{\ominus}{N}\text{-}CR_2 \longrightarrow Ph_3P{=}CR_2 + N_2$$

$$Ph_3P{=}N\text{-}N{=}CR_2$$

(3)

It is not known whether diphenylmethylenetriphenylphosphorane, first obtained by Staudinger in 1919 by pyrolysis of benzophenonetriphenylphosphazine, is formed through cleavage of the phosphazine into triphenylphosphine and diphenyldiazomethane (which would then give diphenylcarbene). The failure to synthesize stable ylids, such as cyclopentadienylidenetriphenylphosphorane,[485] from the corresponding phosphazine seems to disprove this route. Recently, however, Regitz and Liedhegener[500] were successful in obtaining ylids from triphenylphosphine and substituted diazocyclopentadienes in the melt.

Bis(alkylmercapto)carbenes have also been trapped with phosphines with the formation of ylids.[383,555a]

III. YLIDS FROM PHOSPHINES AND COMPOUNDS WITH ACTIVATED C-C MULTIPLE BONDS

$$Ph_3P + \overset{H}{\underset{|}{\overset{|}{C}}}{=}\overset{|}{C}\text{-}R \longrightarrow Ph_3P{=}\underset{|}{C}\text{-}\underset{\underset{H}{|}}{\overset{|}{C}}\text{-}R$$

$$R = CO_2R^1,\ CONH_2,\ C{\equiv}N,\ \text{etc.}$$

In the interaction of acrylic acid derivatives with

triphenylphosphine, ylids are formed that may be trapped with carbonyl compounds.[426,458,518,639] Ylids synthesized in this way can hardly be prepared otherwise because of the easily occurring Hofmann degradation.

Diphenyl-1-phenylvinylphosphine is a special case: the initially formed addition product cyclizes to give cyclic phosphine alkylenes.[518]

$$Ph_2PCPh{=}CH_2 + CH_2{=}CH\text{-}R \longrightarrow Ph_2\overset{\oplus}{P}(CPh{=}CH_2)(CH_2\text{-}\overset{\ominus}{C}H\text{-}R) \longrightarrow$$

$$Ph_2P{=}C(Ph)\text{-}CH_2\text{-}CH(R)\text{-}CH_2 \text{ (ring)}$$

R = CN CO$_2$Me

The following olefin components are also suitable: dibenzoylethylene,[488,490-1] maleic anhydride,[4,319,461,548] maleic imide,[274] and isomaleic imide.[274] Under this heading may also be considered the addition of triphenylphosphine to p-benzoquinone to give a stable ylid.[478,549]

1,2-Dichloroperfluorocycloalkanes react with phosphines to form 1:1 adducts which on hydrolysis yield stable phosphine alkylenes.[618]

$$C_4F_4Cl_2 + PR_3 \longrightarrow [C_4F_4Cl_2(\overset{\oplus}{P}R_3)]^{\ominus} \longrightarrow C_4F_4Cl(\overset{\oplus}{P}R_3) \xrightarrow{H_2O} F_2C_4O_2{=}PR_3$$

Divinylogous phosphine alkylenes may be obtained from conjugated diynes, trialkylphosphines, and activated methylene compounds.[605]

Phosphines and acetylene derivatives react in the mole ratio 2:1 to give stable diphosphoranes.[572,574] Sometimes there is isolated a 1:2[652,653] or a 1:1[323] adduct, which may also have a cyclic structure. Biphosphines react with acetylene-dicarbonyl esters to form cyclic biphosphoranes containing two phosphorus atoms in the ring.[575]

IV. YLIDS BY PYROLYSIS OF PHOSPHONIUM SALTS

The thermal decomposition of triphenyl(α-alkoxycarbonyl-alkyl)phosphonium salts (4) also produces phosphine alkyl-enes.[68] The resulting ylids, if stable, can be isolated.

$$\underset{(4)}{R{-}\underset{\underset{\oplus}{Ph_3P}}{\overset{H}{C}}{-}\underset{O}{\overset{}{C}}{-}\bar{O}{-}R^1\ X^{\ominus}} \longrightarrow R{-}\overset{H}{C}{=}PPh_3 + CO_2 + R^1X$$

They can also be trapped by reaction with various partners, or they can react further intramolecularly.

Miller describes the preparation of trimethylsilyl-methylenetrimethylphosphorane (5) by pyrolysis of the corresponding phosphonium salt (6).[438] In the primary step Me_3SiCl is split off with formation of methylenetrimethyl-phosphorane, which removes a proton from undecomposed (6).

$$\underset{(6)}{2[Me_3SiCH_2\overset{\oplus}{P}Me_3]Cl^{\ominus}} \longrightarrow \underset{(5)}{Me_3Si{-}CH{=}PMe_3} + Me_4PCl + Me_3SiCl$$

V. YLIDS FROM VINYLPHOSPHONIUM SALTS (7)[561,562]

$$\underset{(7)}{[Ph_3PCH{=}CH_2]X} \xrightarrow{RM} Ph_3P{=}CH{-}CH_2R + MX$$

M = metal

If R contains a carbonyl function, an intramolecular Wittig reaction ensues with the ylid grouping. In this way cyclic (also heterocyclic) compounds are obtainable.[551-2,553]

VI. YLIDS FROM DIHALOTRIPHENYLPHOSPHORANES AND REACTIVE METHYLENE COMPOUNDS

Dihalotriphenylphosphoranes react with active methylene compounds in the presence of tertiary amines to yield stable α-disubstituted ylids.[198a,309]

$$Ph_3PX_2 + CH_2R^1R^2 \xrightarrow[-2HX]{NR_3} Ph_3P{=}C\begin{matrix} R^1 \\ R^2 \end{matrix}$$

$$X = Cl, Br$$

VII. YLIDS BY AN ELECTROCHEMICAL METHOD

The electrolysis of phosphonium salts, using a carbon electrode, gives, in the presence of carbonyl compounds, olefins in good yields.[576]

VIII. YLIDS FROM BENZYNE

Only one example of ylid preparation by this method has been described.[560]

$$C_6H_4 \text{ (benzyne)} + Ph_2PCH_3 \longrightarrow C_6H_5\text{-}PPh_2{=}CH_2$$

IX. YLIDS FROM CONVERSION OF OTHER PHOSPHINE ALKYLENES

A great number of complex ylids are obtainable through alkylation,[40,60,100,101,405,408,409] acylation,[37,46,47,111,248] alkoxycarbonylation,[99] and halogenation[259,402,403] of simple phosphine alkylenes if they contain a proton on the α-carbon atom of the alkylidene group. The originally formed phosphonium salt reacts under "transylidation" (Umylidierung)[40] with a second mole of starting ylid. In some cases the second ylid molecule can be replaced by another base.[190,248,604] Silylmethylenephosphoranes (5a) are also obtainable by this transylidation method.[438,534,570]

$$2Ph_3P{=}CH_2 + Me_3SiCl \longrightarrow [Ph_3P\text{-}CH_3]^{\oplus}Cl^{\ominus} + Ph_3P{=}CH\text{-}SiMe_3$$

(5a)

Schmidbaur and Tronich[535] were successful for the first

time in obtaining pure methylenetrialkylphosphoranes (9) by desilylating silyl-substituted phosphine alkylenes (5), using trimethylsilanol (8) or methanol.

$$\underset{(5)}{Me_3Si\text{-}CH{=}PMe_3} + \underset{(8)}{Me_3SiOH} \longrightarrow Me_3Si\text{-}O\text{-}SiMe_3 + \underset{(9)}{Me_3P{=}CH_2}$$

A.2. Reactions of Phosphine Alkylenes

A.2.1. Phosphonium Salts and Phosphine Alkylenes as Corresponding Acid-Base Pairs

Phosphine alkylenes (2), obtainable from phosphonium salts (1) by splitting off HX, are able to add HX, whereby the original phosphonium salt is regenerated. Therefore phosphonium salts may be considered as "Brönsted

$$\underset{(1)}{[R^1R^2CH\text{-}PR^3_3]^{\oplus}X^{\ominus}} \underset{+HX}{\overset{-HX}{\rightleftharpoons}} \underset{(2)}{R^1R^2\overset{\ominus}{C}\text{-}\overset{\oplus}{P}R^3_3}$$

acids," and phosphine alkylenes as the corresponding bases.

The basicity of ylids (2) is also seen in their reaction with water.[167,338,391,543,677,685] Thereby a phosphonium hydroxide (10) is formed by addition of a proton to the carbon atom of the CP double bond (tautomeric structure). The phosphonium hydroxide (10) decomposes by reaction with a second OH^- molecule, through the intermediate formation of (11), with a pentavalent phosphorus atom, into a carbanion (13), which immediately reacts with water to give the hydrocarbon (14) and the phosphine oxide (12)[216,432]

$$\underset{(2)}{R^1R^2C{=}PR^3_3} \xrightarrow{H_2O} \underset{(10)}{[R^1R^2CH\text{-}PR^3_3]^{\oplus}OH^{\ominus}} \rightleftharpoons \underset{(11)}{R^1R^2CH\text{-}\underset{OH}{\underset{|}{P}}R^3_3}$$

$$\xrightarrow{OH^{\ominus}} \underset{(12)}{R^3_3PO} + \underset{(13)}{R^1R^2CH^{\ominus}} \xrightarrow{H_2O} \underset{(14)}{R^1R^2CH_2}$$

Of the four ligands attached to the phosphorus atom in (2), the ligand that is the most electronegative or the best resonance stabilized[216,283] is always the one split off as hydrocarbon.

The basic character of phosphine alkylenes is also indicated by their ability to induce a Cannizzaro reaction[267] and to split off acetic acid from acetates.[268,595,690]

The acid-base strength of phosphonium salts (1) and phosphine alkylenes (2) is decidedly determined by groups R^1 and R^2. Electron-attracting groups enhance the acidity of phosphonium salts (1) and weaken the basicity of ylids (2) as the corresponding bases. The pK_a values of phosphonium salts can be determined potentiometrically.[219,336,337,350,480,600] The basicity of the base that is necessary to split of HX can also serve as a measure of the acid strength of a phosphonium salt. Thus, from aqueous solutions of triphenylmethoxycarbonylmethylphosphonium bromide, which is a strong acid because of the ester group, the corresponding ylid is precipitated by treatment with dilute Na_2CO_3 solutions.[81,340-1] The ylid obtained in this way is stable toward cold water, since the ester group decreases the basicity and thus renders more difficult the attack of a proton on the carbon atom of the ylid. Only on boiling with water does decomposition ensue to give methyl acetate and triphenylphosphine oxide.[81]

Generally an ylid is stabilized by substituents that stabilize the negative charge on the α-carbon atom by inductive and/or mesomeric effects. Organosilyl-substituted phosphine alkylenes, such as ylid (5a), are examples showing the possibility of electron delocalization because of (p→d) π-interaction.[438,534,570]

$$\mathrm{Me_3Si{=}\overset{\ominus}{C}H{-}\overset{\oplus}{P}Ph_3 \longleftrightarrow Me_3Si{-}\overset{\ominus}{C}H{-}\overset{\oplus}{P}Ph_3 \longleftrightarrow Me_3Si{-}CH{=}PPh_3}$$

(5a)

The acid and base character of phosphonium salts (1) and ylids (2), respectively, is also determined by the ligand R^3 attached to the phosphorus atom. The electron-attracting inductive effect of the three phenyl groups in (15a) weakens the C-H bond of the CH_2 group. If the three phenyl groups are replaced by cyclohexyl groups, the +I-effect of these groups strengthens the C-H bonds of the methylene group.

$$\mathrm{[MeO_2C{-}CH_2{-}\overset{\oplus}{P}R^3_3]Br^{\ominus} \longrightarrow MeO_2C{-}CH{=}PR^3_3}$$

(15a, b) (16)

R = Ph or cyclohexyl

a b

Therefore, to deprotonate (15b) a Na_2CO_3 solution is not sufficient; a dilute NaOH solution is necessary.[81] Inversely, methoxycarbonylmethylenetricyclohexylphosphorane is more basic than the corresponding triphenyl compound. In contrast to (16a), compound (16b) is decomposed on contact with cold water.

A.2.2. Transylidation (Umylidierung)

Since phosphonium salts and phosphine alkylenes are related as an acid-base pair to each other, and since groups R^1 and R^2 influence the acid respectively base strength, an acid-base equilibrium is adopted between phosphine alkylenes (17) and phosphonium salt (18).[40] The position

$$R^1\text{-CH=PPh}_3 + [R^2\text{-CH}_2\text{-}\overset{\oplus}{P}\text{Ph}_3]X^{\ominus} \rightleftharpoons [R^1\text{-CH}_2\text{-}\overset{\oplus}{P}\text{Ph}_3]X^{\ominus} + R^2\text{-CH=PPh}_3$$

(17) (18) (19) (20)

of the equilibrium is determined by groups R^1 and R^2: If the basicity of (17) and (20) and the acid character of (18) and (19), respectively, are very different, then formation of the weakest basic phosphine alkylene and of the least acidic phosphonium salt is favored.

In the reaction of methylenetriphenylphosphorane (21) with triphenylphenacylphosphonium bromide (22), triphenylmethylphosphonium bromide (23) and benzoylmethylenetriphenylphosphorane (24) are obtained in nearly 90% yield. This is an example of a reaction for which the term transylidation has been coined.[40] The transylidation renders

$$\text{Ph}_3\text{P=CH}_2 + [\text{Ph-CO-CH}_2\text{-}\overset{\oplus}{P}\text{Ph}_3]\text{Br}^{\ominus} \longrightarrow$$

(21) (22)

$$[\text{Ph}_3\overset{\oplus}{P}\text{CH}_3]\text{Br}^{\ominus} + \text{Ph-CO-CH=PPh}_3$$

(23) (24)

possible a comparison of the strengths of the mesomeric and inductive effects exercised by the substituents on the CH acidity of the phosphonium salts and the basicity of the phosphine alkylenes, respectively. In agreement with the pK_a values of the phosphonium salts $[RCH_2PPh_3]X$, the following series of decreasing acceptor ability for groups R is observed:[336]

$\overset{+}{P}Ph_3 > COPh > COMe > CN > CO_2alkyl > P(S)Ph_2 \approx P(O)Ph_2 > PPh_2 > Ph > H > CH_3$

Treating phosphonium salts $[R^1CH_2PPh_3]X$ with phosphine alkylenes R^2-$CH{=}PPh_3$, in which R^2 stand to the right of R^1 in the above series, results in transylidation.

By reaction of tritium-labeled phosphonium salts it was proved that transylidation occurs also if $R^1 = R^2$ in (17) and (18), or if the inductive and mesomeric effects, respectively, of R^1 and R^2 are only slightly different from each other.[40] The transylidation in the case of $R^1 = R^2$ causes the phenomenon of temperature-dependent PCH spin-spin coupling in the NMR spectrum of phosphine alkylenes.[89,177,536] Traces of acid effect the transformation of a small amount of ylid into the corresponding phosphonium salt. Through a fast, reversible proton exchange both enter into a transylidation equilibrium.

A.2.3. Reaction of Phosphine Alkylenes with Halogen Compounds

The basic character of phosphine alkylenes is seen also in their reactions with methyl iodide.[682] Thus the phosphine alkylene (21) is methylated to the corresponding phosphonium salt (25).

$$\underset{(21)}{Ph_3P{=}CH_2} + CH_3I \longrightarrow \underset{(25)}{[Ph_3\overset{\oplus}{P}CH_2CH_3]I^{\ominus}}$$

Halogen compounds RX with an electron-attracting group convert ylids into phosphonium salts, which, because of their acidity, can be deprotonated in a transylidation reaction by a second mole of starting ylid.[100] In addition to splitting off a proton from the α-position of the salts resulting from the alkylation by a second mole of ylid (transylidation), occasionally β- or γ-elimination is also observed.

A.2.3.1. Phosphine Alkylenes and Alkyl Halides. The alkylation of ylids (17) leads to phosphonium salts (26).

$$\underset{(17)}{Ph_3P{=}CHR^1} + R^2X \longrightarrow \underset{(26)}{[Ph_3P\text{-}CHR^1R^2]^{\oplus}X^{\ominus}} \xrightarrow[-HX]{+B} \underset{(27)}{Ph_3P{=}CR^1R^2}$$

Dehydrohalogenation of the salt (26) with a base produces the substituted ylid (27). The rôle of the base can be taken over by a second mole of ylid. (17)

Alkylation of methoxycarbonylmethylenetriphenylphosphorane (28), followed by hydrolysis of the phosphorane (29) formed by transylidation, allows the formation of carbonic acids of type $R-CH_2-COOH$.[100] The success of the

$$\underset{(28)}{Ph_3P=CH-CO_2Me} + RX \longrightarrow [R-CH(PPh_3^{\oplus})-CO_2Me]X^{\ominus} \xrightarrow{(28)} \underset{(29)}{R-C(=PPh_3)-CO_2Me} + [CH_2(PPh_3^{\oplus})-CO_2Me]X^{\ominus}$$

reaction appears to lie in substituent R, whose inductive effect makes possible a complete transylidation.[100]

Benzylenetriphenylphosphoranes (30) react with benzyl halides (31) in the mole ration 1:1 to give phosphonium salts (32) without the occurrence of a transylidation reaction.[114] 1,2-Diarylethanes are obtainable from the

$$\underset{(30)}{R-C_6H_4-CH^{\ominus}-PPh_3^{\oplus}} + \underset{(31)}{XCH_2-C_6H_4-R^1} \longrightarrow \underset{(32)}{[R-C_6H_4-CH(PPh_3^{\oplus})-CH_2-C_6H_4-R^1]X^{\ominus}}$$

salts (32) by alkaline cleavage or by an electrolytic method.

Acylmethylenetriphenylphosphoranes (33) may be represented by the mesomeric forms (33a) and (33b). These ylids

$$\underset{(33a)}{R-C(=O)-C(R^1)=PPh_3} \longleftrightarrow \underset{(33b)}{R-C(O^{\ominus})=C(R^1)-PPh_3^{\oplus}}$$

exhibit an ambident character and can be alkylated by halogen compounds on the oxygen as well as on the carbon atom.[260,479,490-1,584] As a result either the enol ether-substituted phosphonium salts (35) or ketophosphonium salts (34) are formed.

Alkoxycarbonylalkylidenetriphenylphosphoranes (36) are C-alkylated only by alkyl halides.[100,106] In the reaction of (36) with triethyloxonium fluoroborate (37), however, O-alkylation is observed with the formation of 1-substituted 2-ethoxy-2-alkoxyvinyltriphenylphosphonium tetrafluoro-

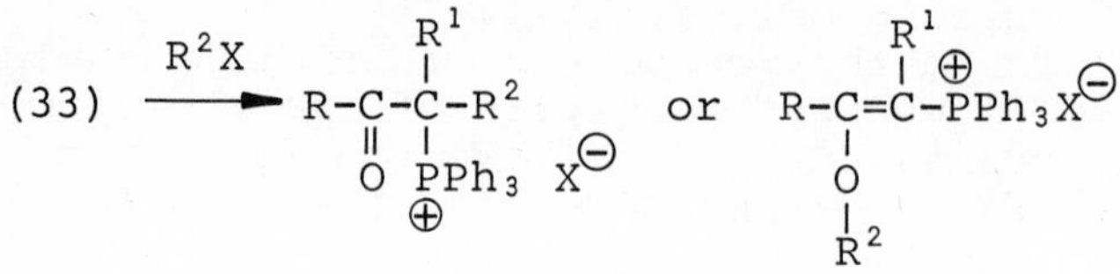

$$(\underline{33}) \xrightarrow{R^2X} R\text{-}\underset{\underset{O}{\|}}{C}\text{-}\underset{\underset{\overset{\oplus}{P}Ph_3}{|}}{\overset{\overset{R^1}{|}}{C}}\text{-}R^2 \; X^{\ominus} \quad \text{or} \quad R\text{-}\underset{\underset{\underset{R^2}{|}}{\underset{O}{|}}}{C}=\overset{\overset{R^1}{|}}{C}\text{-}\overset{\oplus}{P}Ph_3 X^{\ominus}$$

(34) (35)

borates (38a) and (38b), respectively.[97] If R ≠ Et, then

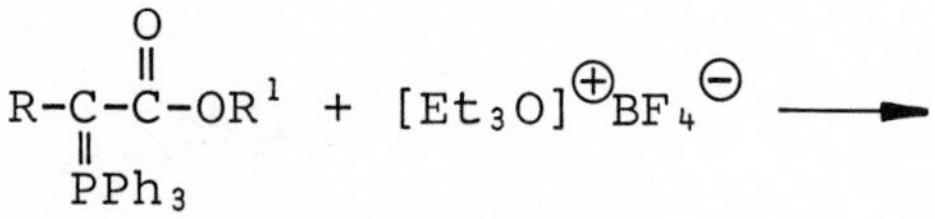

$$R\text{-}\underset{\underset{PPh_3}{\|}}{C}\text{-}\overset{\overset{O}{\|}}{C}\text{-}OR^1 + [Et_3O]^{\oplus}BF_4^{\ominus} \longrightarrow$$

(36) (37)

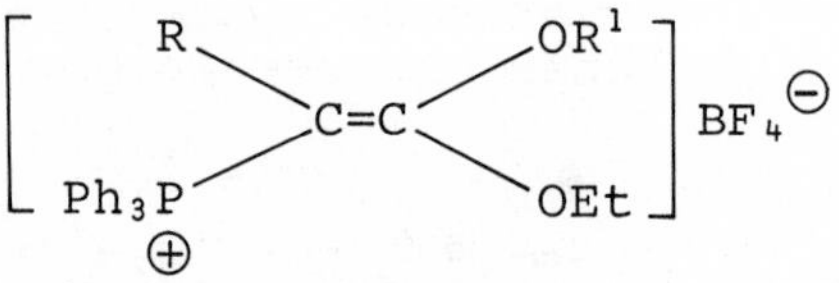

$$\left[\begin{matrix} R \\ \\ Ph_3\overset{\oplus}{P} \end{matrix} \!\!>\! C{=}C \!<\!\! \begin{matrix} OR^1 \\ \\ OEt \end{matrix}\right] BF_4^{\ominus}$$

(38a)

$$\left[\begin{matrix} R \\ \\ Ph_3\overset{\oplus}{P} \end{matrix} \!\!>\! C{=}C \!<\!\! \begin{matrix} OEt \\ \\ OR^1 \end{matrix}\right] BF_4^{\ominus}$$

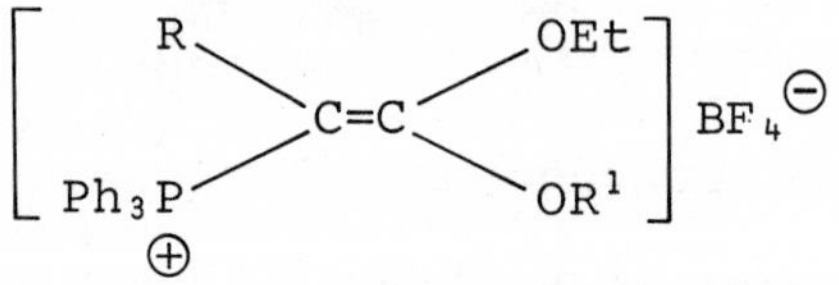

(38b)

a mixture of geometric isomers (38a) and (38b) is formed, as deduced from ^{1}H-NMR investigations. If R = H, treatment of the salt (38) with sodium amide produces the cumulative ylid (39).

$$Ph_3P{=}C{=}C(OEt)_2$$

(39)

A further possibility for alkylating acyl- as well as alkoxycarbonylmethylenetriphenylphosphoranes at the ylid carbon atom consists in the use of Mannich bases as alkylating agents.[619]

Reactions between phosphine alkylenes and halogen compounds can also occur intramolecularly.[60,65,444,520]

$$CH_2-CH_2-Br / CH_2-CH_2-\overset{\oplus}{P}Ph_3 Br^{\ominus} \xrightarrow{\text{base}} CH_2-CH_2-Br / CH_2-CH{=}PPh_3 \longrightarrow CH_2-CH_2 / CH_2-CH-\overset{\oplus}{P}Ph_3 Br^{\ominus}$$

(40) (41) (42)

$$\xrightarrow{\text{base}} CH_2-CH_2 / CH_2-C{=}PPh_3 \xrightarrow[-OPPh_3]{PhCHO} CH_2-CH_2 / CH_2-C{=}CH-Ph$$

(43) (44)

Treatment of the salt (40) with phenyllithium gives the ylid (41), which cyclizes to the salt (42).[444] From salt (42) there is obtainable ylid (43), which, when treated with benzaldehyde, forms benzylidene-cyclobutane (44) in a Wittig reaction.[65,520] Extension of this ring-closure reaction to compounds of type (40) with more than four CH_2 groups failed because these monophosphonium salts could not be prepared.[65,230]

Directed ring-closure reactions are possible with bis-halides whose C-X bonds are equally reactive (these, however, form monophosphonium salts with 1 mole of triphenylphosphine)[64,230] or with those that contain two differently reactive C-X bonds, one of which reacts preferentially with the phosphine.[65]

The ring-closure reaction by intramolecular C-alkylation of phosphine alkylenes is particularly well suited for the formation of polycyclic compounds and opens up a multiplicity of synthetic possibilities.[64,65]

Aromatic compounds of type (45), which are substituted in the nucleus by at least one bromomethyl group and carry the grouping $(CH_2)_nBr$ in the side chain, can be caused to react with 1 mole of triphenylphosphine to give monophosphonium salts (46), which, when treated with bases, yield the corresponding phosphine alkylenes (47). By intramolecular C-alkylation these ylids (47) produce polycyclic phosphonium salts (48), which yield polycyclic olefins when thermally decomposed.

$$Ar\langle^{CH_2-Br}_{(CH_2)_n-CH_2-Br} \xrightarrow{PPh_3} Ar\langle^{CH_2-\overset{\oplus}{P}Ph_3 Br^{\ominus}}_{(CH_2)_n-CH_2-Br} \xrightarrow{\text{base}}$$

(45) (46)

$$\text{Ar}\langle{}^{C(H)=PPh_3}_{(CH_2)_n-CH_2-Br}\ (\underline{47}) \longrightarrow \text{Ar}\langle{}^{CH(\overset{\oplus}{P}Ph_3Br^{\ominus})}_{(CH_2)_n}\rangle CH_2\ (\underline{48}) \xrightarrow{\Delta} \text{Ar}\langle{}^{CH}_{(CH_2)_n}\rangle CH\ (\underline{49})$$

$$(\underline{48}) \xrightarrow{\text{base}} \text{Ar}\langle{}^{C(=PPh_3)}_{(CH_2)_n}\rangle CH_2\ (\underline{51})$$

$$(\underline{48}) \xrightarrow{H_2O/OH^{\ominus}} \text{Ar}\langle{}^{CH_2}_{(CH_2)_n}\rangle CH_2\ (\underline{50})$$

$$(\underline{51}) \xrightarrow{H_2O} (\underline{50}); \quad (\underline{51}) \xrightarrow{R-CHO} \text{Ar}\langle{}^{C(=CHR)}_{(CH_2)_n}\rangle CH_2\ (\underline{52})$$

A second mole of base converts these polycyclic phosphonium salts (48) into the ylids (51). On hydrolysis the latter produce polycyclic hydrocarbons (50) and, in a Wittig reaction, compounds of type (52) with an exocyclic double bond. In this way five-, six-, and seven-membered ring compounds have been prepared.[65] The ring-closure reaction is limited by the fact that the ω-haloalkyltriphenylphosphonium salts are obtainable only with difficulty when the two halogen atoms in the bishalide are separated by an aliphatic or a cycloaliphatic group (Y), respectively. This difficulty can be circumvented, however, by a combination of intramolecular and intermolecular C-alkylation and by using the principle of transylidation.[79]

Methylenetriphenylphosphorane (21) reacts with ω,ω'-dihalides in an intermolecular C-alkylation to give the phosphonium salt (54), which, together with a second mole of (21), is in a transylidation equilibrium with triphenylmethylphosphonium halide (23) and an ω-halosubstituted ylid (55). Ylid (55) is converted into a cyclic phosphonium salt (56) by intramolecular C-alkylation and is therefore constantly removed from the equilibrium, so that the

$$Y\langle{}^{CH_2-X}_{CH_2-X}\ (\underline{53}) + CH_2{=}PPh_3\ (\underline{21}) \longrightarrow Y\langle{}^{CH_2-CH_2-\overset{\oplus}{P}Ph_3\ X^{\ominus}}_{CH_2-X}\ (\underline{54}) \xrightleftharpoons{(\underline{21})}$$

$$Y\begin{matrix}\diagup CH_2-CH=PPh_3 \\ \diagdown CH_2-X\end{matrix} \quad + \quad CH_3-\overset{\oplus}{P}Ph_3 \; X^{\ominus}$$

(55) (23)

$$\downarrow$$

$$Y\begin{matrix}\diagup CH_2 \diagdown \\ \diagdown CH_2 \diagup\end{matrix}CH-\overset{\oplus}{P}Ph_3 \; X^{\ominus} \xrightarrow{\text{base}} Y\begin{matrix}\diagup CH_2 \diagdown \\ \diagdown CH_2 \diagup\end{matrix}C=PPh_3$$

(56) (57)

reaction ensues completely with the formation of (56).

The salts (56) and (23) are precipitated as a mixture. Because of the good solubility of (23) in water, they can easily be separated. In many cases, however, separation of the salts is unnecessary before further reaction.

The ylid (57), obtainable from (56), may be used as the nucleophilic reaction partner in the synthesis of various cyclic derivatives containing the basic cyclic structure of (56). One case is known in which a third mole of (21) takes over the role of the base in the ylid formation from (56).[66]

The method described above allows ring-closure reactions with the formation of four-, five-, six-, and seven-membered rings. The reaction of 1,7-dibromoheptane with the ylid (21), however, does not produce the cyclic phosphonium salt, but rather leads to the linear bissalt. Obviously in this case intermolecular C-alkylation with the second C-halogen bond occurs faster than the intramolecular reaction.

Ring closures of dihalides with ylids that contain only one proton at the ylid carbon atom lead to cyclic phosphonium salts, from which, however, no ylid can any longer be prepared. The triphenylphosphine residue must be eliminated in this case by hydrolysis or pyrolysis.[70]

The reaction of bisalkylidenephosphoranes with bishalomethyl compounds opens up new possibilities for the formation of cyclic compounds.[96]

In the reaction of the bisylid (58) with o-dibromoxylene (59) the bisphosphonium salt (60) is formed. When hydrolyzed with alkali, this yields 9,10,15,16-tetrahydrotribenzo[a,c,g]cyclodecene (61), in addition to triphenylphosphine oxide.

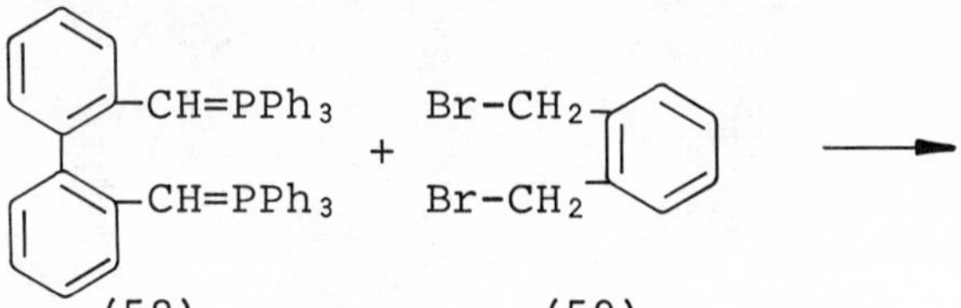

(58) (59)

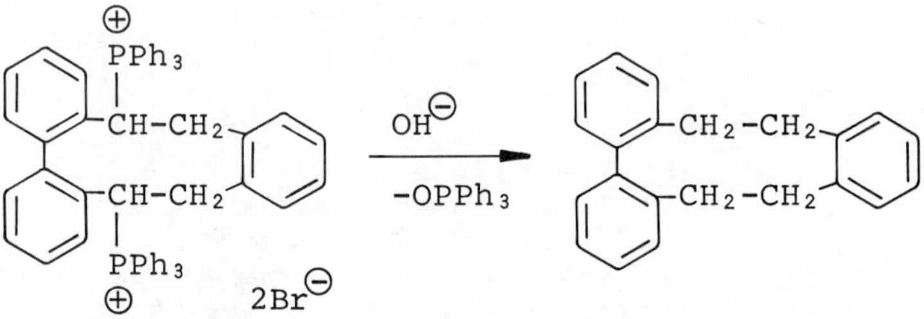

(60) (61)

For further examples see ref. 96.

A.2.3.2. Phosphine Alkylenes and α-Haloketones. From α-bromoketones (62) and methoxycarbonylmethylenetriphenylphosphorane (28) there is formed first the phosphonium salt (63), which, by a transylidation reaction with a second mole of (28), yields the ylid (64) and the salt (65). In ylid (64) the hydrogen atoms in β-position to the phosphorus atom are strongly activated by the neighboring carbonyl group. The ylid decomposes by way of a Hofmann degradation into a β-acylacrylic acid ester (66) and triphenylphosphine.[62,102,106] It could be shown, by using

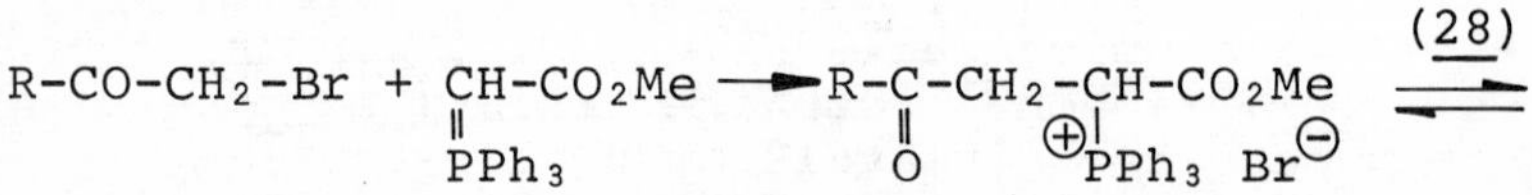

$$R{-}CO{-}CH_2{-}Br + \underset{\displaystyle PPh_3}{\overset{\|}{CH}}{-}CO_2Me \longrightarrow R{-}\underset{\displaystyle O}{\overset{\|}{C}}{-}CH_2{-}\underset{\displaystyle \oplus PPh_3\ Br^{\ominus}}{CH}{-}CO_2Me \xrightleftharpoons{(28)}$$

(62) (28) (63)

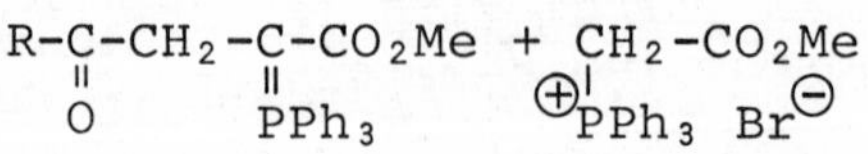

$$R{-}\underset{\displaystyle O}{\overset{\|}{C}}{-}CH_2{-}\underset{\displaystyle PPh_3}{\overset{\|}{C}}{-}CO_2Me + \underset{\displaystyle \oplus PPh_3\ Br^{\ominus}}{CH_2}{-}CO_2Me$$

(64) (65)

$$\downarrow -PPh_3$$

$$R{-}\underset{\displaystyle O}{\overset{\|}{C}}{-}CH{=}CH{-}CO_2Me$$

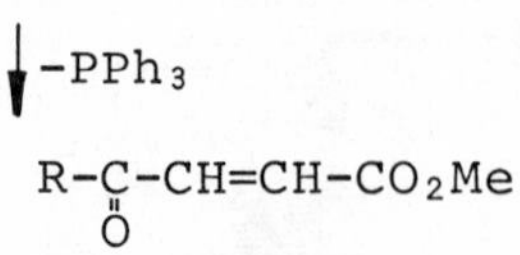

(66)

completely deuteriated compounds, that in the last step of the reaction an intermolecular hydrogen transfer occurs, wherein the rate of the reciprocal β-elimination is determined by the groups attached to the ylid carbon atom.[62]

This procedure makes possible the synthesis of β-acylacrylic acid esters with aliphatic and cycloaliphatic groups, compounds which could previously be prepared only with difficulty. α-Haloketones (the same is true for β-chlorovinyl ketones[701]) and phosphine alkylenes may, however, also undergo a Wittig reaction to give olefins.[143,701]

A.2.3.3. Phosphine Alkylenes and α-Haloacetic Acid Esters. The course of the reaction of phosphine alkylenes with α-haloacetic acid esters is decidedly determined by the halogen atom of the esters.[51,94]

(1) α-Iodo- or α-bromoacetic acid esters (67) form salts (69) with ylids (68), from which, by a transylidation reaction, ylids (70) are produced. The activation of the protons by the ester group in the β-position to the phosphorus atom causes the decomposition of (70) into an α,β-unsaturated carboxylic acid ester (71) and triphenylphosphine.[52]

$$\underset{(68)}{RCH{=}PPh_3} + \underset{(67)}{Br\text{-}CH_2CO_2R^1} \longrightarrow \underset{(69)}{\underset{\oplus PPh_3}{RCH}\text{——}CH_2CO_2R^1 \quad Br^{\ominus}\ (I^{\ominus})} \xrightarrow{(68)}$$

(I)

$$\underset{(70)}{\underset{\|\atop PPh_3}{RC}\text{-}CH_2CO_2R^1} + [RCH_2\text{-}\overset{\oplus}{P}Ph_3]Br^{\ominus}\ (I^{\ominus})$$

$$\downarrow$$

$$\underset{(71)}{RCH{=}CHCO_2R^1} + Ph_3P$$

α-Bromo- and iodocarbonic acid esters and vinylogous halocarbonic acid esters react in the same way.[52] In addition to the intermediate formation of (70) (α-elimination), followed by an intermolecular hydrogen transfer reaction, the direct decomposition of (69) in a β-elimination process is observed.[52]

The reaction of (67) with acylmethylenetriphenylphosphorane leads to the isolable ylids (72).[57] They decompose on heating (150 to 180°) into triphenylphosphine and β-acylacrylic acid ester (74). Since, however, secondary reactions take place at the high decomposition temperature, Hofmann degradation through the phosphonium salt (73) is to be preferred.[57]

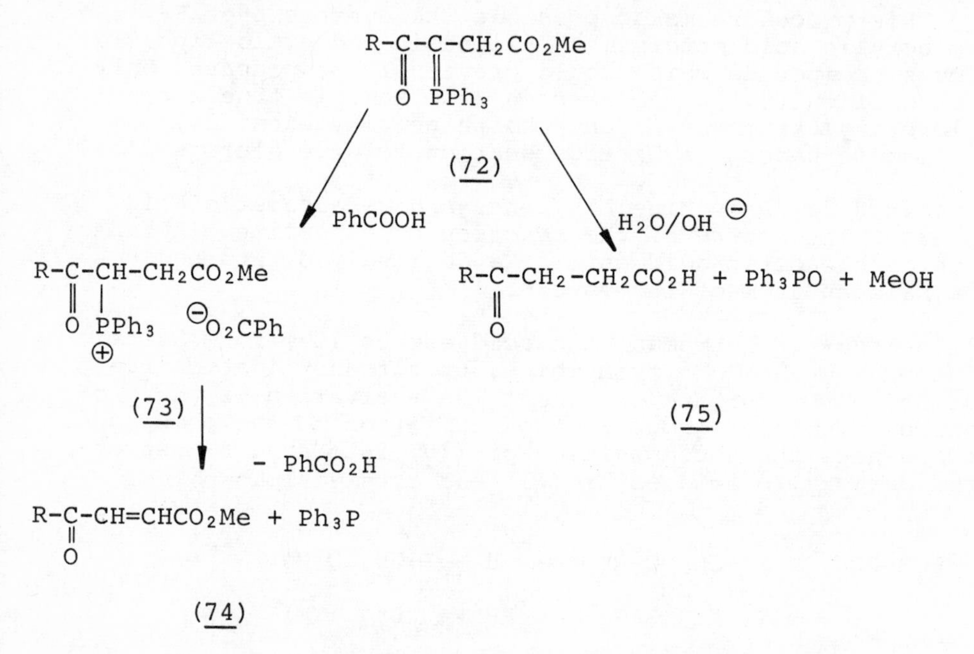

The alkaline hydrolysis of acylmethoxycarbonylmethylenetriphenylphosphorane (72) leads to γ-ketoacids (75).[57]

(2) From the interaction of alkylidenetriphenylphosphoranes (76) and chloroacetic acid methyl ester (77) are produced trans-cyclopropane-1,2,3-tricarbonic acid trimethyl ester (78) and the phosphonium salts (79). Isotopic labeling

$$\underset{(76)}{3\ RR^1C{=}PPh_3} + \underset{(77)}{3\ Cl{-}CH_2CO_2Me} \longrightarrow$$

$$\underset{(78)}{\text{cyclo-}(CH{-}CO_2Me)_3} + \underset{(79)}{3\ [RR^1CH{-}\overset{\oplus}{P}Ph_3]Cl^{\ominus}}$$

indicates that this reaction probably proceeds through the

intermediate formation of fumaric acid.[53,94]

(3) Mono-, di, and trifluoroacetic acid esters (81) and alkylenetriphenylphosphoranes (80) undergo a Wittig reaction at the ester-carbonyl group which results in the formation of enol ethers (82) of fluoroketones.[51]

$$R\text{-}CH{=}PPh_3 + X\text{-}CO_2Et \longrightarrow R\text{-}CH{=}C\begin{matrix} X \\ OEt \end{matrix} + OPPh_3$$

(80) (81) (82)

$$X = CH_2F,\ CHF_2,\ CF_3$$

A.2.3.4. Acylation of Phosphine Alkylenes. Phosphine alkylenes (83) react with acid chlorides (84) according to the following scheme:[37,42]

$$Ph_3P{=}CHR^1 + R^2COCl \longrightarrow [Ph_3\overset{\oplus}{P}\text{-}CHR^1(C{=}O)R^2]Cl^{\ominus} \xrightarrow{(83)}$$

(83) (84) (85)

$$Ph_3P = CR^1(C{=}O)R^2 + [Ph_3PCH_2R^1]^{\oplus}Cl^{\ominus}$$

(86) (87)

The first-formed salt (85) undergoes a transylidation reaction with a second mole of ylid (83) to give an acylated phosphine alkylene (86) and the phosphonium salt (87). The stoichiometric course of the reaction is described by the following scheme:

$$2(83) + (84) \longrightarrow (86) + (87)$$

When optically active acid chlorides are used, this reaction sequence is suitable for the kinetic racemate cleavage of an organophosphorus compound if it contains a chiral phosphorus atom.[113] Since two molecules of racemic ylid (one as substrate, the other as base) react with one mole of optically active acid chloride, it is unnecessary to stop the reaction prematurely, as is otherwise done in kinetic racemate cleavages.[212] Furthermore, the resulting compounds (86) and (87) can easily be separated, since the salt (87) precipitates from the benzene solution.

By using phosphine alkylenes with strongly electron-attracting groups at the ylid carbon atom, the second mole of ylid may be replaced by triethylamine.[248] These reactions are successful, however, only if the amine is a stronger base than the ylid.[42]

Since the preparation of acyl ylids, starting from α-bromoketones, is straightforward in only a few cases,[144,145,290,598,635] the C-acylation of phosphine alkylenes is of preparative importance. Acylation with acid chlorides has the disadvantage that one mole of the ylid used is precipitated as the phosphonium salt (87); even so, compound (83) can be regenerated from it.

An acylation procedure that avoids this disadvantage is the reaction of phosphine alkylenes with esters. In general, salt-free solutions of ylids do not react with ethyl or methyl esters.[37,42,46] Only ω-ethoxycarbonylalkylidenetriphenylphosphoranes (88) are converted by an intramolecular C-alkylation into cyclic ylids (89).[34,315] In

$$(CH_2)_n\begin{cases}CO_2Et \\ CH{=}PPh_3\end{cases} \longrightarrow (CH_2)_n\begin{cases}C{=}O \\ | \\ C{=}PPh_3\end{cases} + EtOH$$

(88) (89)

the presence of lithium salts, however, the alkyl esters react with phosphine alkylenes,[643,685] although the yields of acylated ylids are poor. Better results are obtained in the reaction of salt-free phosphine alkylenes with activated esters, e.g., phenyl esters or thiophenyl esters,[37,42,46] as well as with acylimidazoles.[111] Particularly suitable for the acylation reaction are thiocarbonic acid S-ethyl esters (91).[42,46,47]

$$R^1CH{=}PPh_3 + R^2{-}\underset{\underset{O}{\|}}{C}{-}SEt \longrightarrow$$

(90) (91)

$$R^2{-}\underset{\underset{O}{\|}}{C}{-}CR^1{=}PPh_3 + [R^1CH_2PPh_3]^{\oplus}SEt^{\ominus}$$

(92) (93)

$$[R^1CH_2PPh_3]^{\oplus}SEt^{\ominus} \rightleftharpoons R^1CH{=}PPh_3 + HSEt$$

(93) (90) (94)

This reaction occurs also with transylidation. In the first step the acylated ylid (92) and the phosphonium ethylthiolate (93) are formed. Analogously to the elimination of alcohol from phosphonium alkoxides,[677] (93) splits off ethylmercaptan (94) on heating, whereby the ylid (90) is regenerated and reacts again with the thiol ester (91). In comparison with the acid chloride process, one mole of ylid is saved with this procedure: (90) + (91) → (92) + (94).

Acylalkylenetriphenylphosphoranes (92) are suited for the ketone synthesis. Ketones (95) are obtained by hydrolysis,[42,46,47,479] reduction cleavage with zinc-acetic acid,[42,643] or electrolysis[42,308] of the corresponding ylid. The choice of method depends on the substituents attached to the ylid carbon atom.

$$\underset{(92)}{R^2\text{-}\underset{\underset{PPh_3}{\|}}{C}\text{-}CO\text{-}R^1} \longrightarrow \underset{(95)}{R^2\text{-}CH_2\text{-}CO\text{-}R^1} + OPPh_3 \quad \text{or} \quad PPh_3$$

The formation of α-branched α,β-unsaturated ketones (96) is possible from the Wittig reaction of acylalkylenetriphenylphosphoranes (92) and aldehydes.[37,46,47,479]

$$\underset{(92)}{R^1\text{-}CO\text{-}\overset{\overset{R^2}{|}}{C}{=}PPh_3} + R^3CHO \longrightarrow \underset{(96)}{R^1\text{-}CO\text{-}\overset{\overset{R^2}{|}}{C}{=}CHR^3} + OPPh_3$$

The C-acylation of methoxycarbonylmethylenetriphenylphosphorane with acid chlorides, which occurs with transylidation, makes α-acyl-α-methoxycarbonylmethylenetriphenylphosphoranes (97) readily available. Thermal decom-

$$\underset{(97)}{R\text{-}\underset{\underset{O}{\|}}{C}\text{-}\underset{\underset{PPh_3}{\|}}{C}\text{-}CO_2Me} \xrightarrow{\Delta} \underset{(98)}{R\text{-}C{\equiv}C\text{-}CO_2Me} + OPPh_3$$

position of (97) gives acetylene-carboxylic acid esters (98) (intramolecular Wittig reaction).[166,247,248,401] Bis-acetylene-carboxylic acid esters can also be prepared in this way.[248] Instead of ester groups, other groups may be present (compare, e.g., Ref. 164). In one case the reversal of this reaction was successful, i.e., the addition of triphenylphosphine oxide to dicyanoacetylene to give an acylated ylid.[166a]

$$NC{-}C{\equiv}C{-}CN + Ph_3PO \rightleftharpoons NC{-}\overset{\ominus}{C}{=}C(CN){-}O{-}\overset{\oplus}{P}Ph_3 \rightleftharpoons$$

$$NC{-}C{=}C{-}CN\ (\text{ring } Ph_3P{-}O) \rightleftharpoons Ph_3P{=}C(CN)(COCN)$$

The reaction of ylids (97) with PCl_5 or Vilsmeier reagents also leads to phosphonium salts, which can be converted, with NaOH, into (98) without isolation.[404]

The phosphonium salts (101), obtainable from ethoxycarbonylalkylenetriphenylphosphoranes (100), $R^2 \neq H$, and acid chlorides (99), also undergo a γ-elimination.[67,69]

An additional mole of ylid (100) eliminates from (101) a proton in the γ-position to phosphorus. The betaine

$$\underset{(99)}{R^1CH_2COCl} + \underset{(100)}{Ph_3P{=}C(R^2){-}CO_2Et} \longrightarrow \underset{(101)}{[R^1CH_2{-}C({=}O){-}C(R^2)(\overset{\oplus}{P}Ph_3){-}CO_2Et]Cl^{\ominus}} \quad \xrightarrow{(100)}$$

$$\underset{(102a)}{R^1\overset{\ominus}{C}H{-}C({=}O){-}C(R^2)(\overset{\oplus}{P}Ph_3){-}CO_2Et} + \underset{(104)}{HC(R^2)(\overset{\oplus}{P}Ph_3){-}CO_2Et\ Cl^{\ominus}}$$

$$\underset{(102b)}{R^1CH{=}C(\overset{\ominus}{O}){-}C(R^2)(\overset{\oplus}{P}Ph_3){-}CO_2Et} \xrightarrow{-OPPh_3} \underset{(103)}{R^1CH{=}C{=}C(R^2){-}CO_2Et}$$

formed (102) splits off triphenylphosphine oxide spontaneously and gives allenecarboxylic acid esters (103).

If this reaction is carried out with optically active acid chlorides (99) and a racemic alkylidenephosphorane (100), whose phosphorus atom is the center of chirality, then in the end products the salt (104) and the phosphine oxide are optically active through kinetic racemate cleavage, and the allenecarboxylic acid ester (103) is optically active through partial asymmetric synthesis.[113] Also, if racemic carboxyalkylidenetriphenylphosphoranes, in which the phosphorus atom is achiral but the carbon atom of the alcohol component in the ester group is chiral, are brought into reaction with optically active acid chlorides, a kinetic racemate cleavage again ensues[113] (see also Ref. 632).

Like alkylations, the alcylations of acylalkylenetriphenylphosphoranes can take place on the carbon atom as well as on the oxygen atom.[164,205,388a] Carboxylic acid anhydrides are also suited for the C-acylation of acylalkylenetriphenylphosphoranes.[164]

A.2.3.5. Reactions of Phosphine Alkylenes with Acid Chlorides in the Molar Ration 1:1. Generally, in reactions of phosphine alkylenes with halo compounds, the initially formed phosphonium salt reacts with a second mole of ylid with transylidation or γ-elimination. By working at low temperatures, the γ-elimination can be prevented in some cases.[58] The acylated salts (105) obtainable in this way produce, when electrolyzed, α-branched β-oxocarbonic acid esters (106).[58]

$$R^1CO{-}\overset{\overset{\displaystyle R^2}{|}}{\underset{\underset{\displaystyle \overset{\oplus}{P}Ph_3\ Cl^{\ominus}}{|}}{C}}{-}CO_2Et \xrightarrow{-PPh_3} R^1CO{-}\overset{\overset{\displaystyle R^2}{|}}{C}H{-}CO_2Et$$

(105) (106)

A.2.3.6. Phosphine Alkylenes and Chloroformic Acid Esters. Phosphine alkylenes (107) react with chloroformic acid esters (108) with transylidation to give alkoxycarbonylalkylenetriphenylphosphoranes (109).[99,103] Carbalk-

$$2R^1CH{=}PPh_3 + Cl{-}COOR^2 \longrightarrow R^1\underset{\underset{\displaystyle COOR^2}{|}}{C}{=}PPh_3 + [R^1CH_2\overset{\oplus}{P}Ph_3]Cl^{\ominus}$$

(109) (108) (109)

oxylated phosphine alkylenes (109), readily obtainable in this way, may be used for the synthesis of carboxylic acids.[99,100,103]

Allylenetriphenylphosphorane (110) adds chloroformic acid ester in the γ-position to the phosphorus atom (vinylogous substitution).[103] As an intermediate the salt (111) is formed; this is deprotonated in the γ-position to the phosphorus atom by a second mole of ylid (100). Wittig reactions with the ylid (112) produce polyenecarboxylic acid ester.

$$Ph_3P{=}CH{-}CH{=}CH_2 \longleftrightarrow Ph_3\overset{\oplus}{P}CH{=}CH{-}\overset{\ominus}{C}H_2 \xrightarrow{ClCO_2Me}$$

(110)

$$[Ph_3\overset{\oplus}{P}{-}CH{=}CH{-}CH_2CO_2Me]Cl^{\ominus} \xrightarrow{(110)} Ph_3\overset{\oplus}{P}{-}\overset{\ominus}{C}H{-}CH{=}CHCO_2Me \rightleftharpoons$$

(111) (112a)

$$Ph_3\overset{\oplus}{P}{-}CH{=}CH{-}\overset{\ominus}{C}HCO_2Me$$

(112b)

A.2.3.7. Halogenation of Phosphine Alkylenes. While the halogenation of strongly basic phosphine alkylenes is accompanied by side reactions, stable phosphine alkylenes react with halogen with transylidation.[402,403] If the

$$Ph_3P{=}CH{-}\underset{\underset{O}{\|}}{C}{-}R + X_2 \longrightarrow [Ph_3\overset{\oplus}{P}{-}\underset{X}{\overset{H}{C}}{-}\underset{\underset{O}{\|}}{C}{-}R]X^{\ominus}$$

(113) (114)

$$(113) + (114) \longrightarrow Ph_3P{=}\underset{X}{C}{-}\underset{\underset{O}{\|}}{C}{-}R + [Ph_3\overset{\oplus}{P}{-}CH_2{-}\underset{\underset{O}{\|}}{C}{-}R]X^{\ominus}$$

(115)

halogenation is carried out in the presence of triethylamine, then the second mole of ylid (113) may be saved.[190,403] At -70° the reaction stops at the stage of the salt (114). By treating salt (114) with bases, the ylid (115) can be prepared.[190,604]

For chlorination reactions phenyliododichloride is particularly well suited.[402,403,698] Keto ylids, which contain no proton at the ylid carbon atom, produce salts when treated with phenyliododichloride. In the alkaline hydrolysis these salts yield unsymmetrical α-haloketones.[698]

Cyanogen bromide transfers its nitrile group with transylidation to alkoxycarbonylmethylenetriphenylphosphoranes and cyanomethylenetriphenylphosphorane.[424] Under the same conditions benzoylmethylenetriphenylphosphorane is brominated and cyanogenated.[424,582]

In the presence of triethylamine, cyanogen bromide exclusively brominates alkoxycarbonylmethylenetriphenylphosphorane.[424] Cyanoacid esters are also suitable as cyanogenation agents.[424]

Acylmethylenetriphenylphosphoranes can be iodinated with bromine iodide, BrI.[259] The iodine atom in ylids, obtainable in this way, can be exchanged by the thiocyanide group using KSCN.[259]

The psuedohalogen $(SCN)_2$ (116) reacts with acyl ylids (117), but not with salt formation; rather, a betaine (118) results. Its further behavior depends on the substituents R^1, R^2, and R^3 of the ylid used (117).[696] When $R^1 = R^2 =$

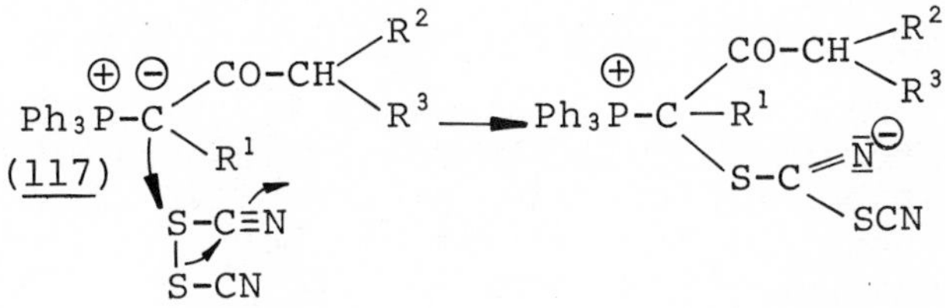

(117) $Ph_3\overset{\oplus}{P}-\overset{\ominus}{C}(R^1)-CO-CH(R^2)(R^3)$ + NC-S-S-CN (116) ⟶ $Ph_3\overset{\oplus}{P}-C(R^1)(CO-CH(R^2)(R^3))-S-C(=\overset{\ominus}{N})-SCN$ (118)

(116) (118)

R^3 = H, transformation to the ylid (119) ensues. If the

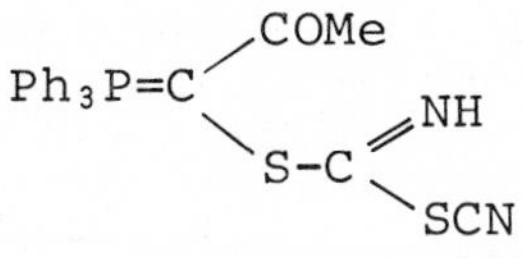

$Ph_3P=C(COMe)-S-C(=NH)-SCN$ $\quad$ Pr(NCS)C=C=CH_2

(119) 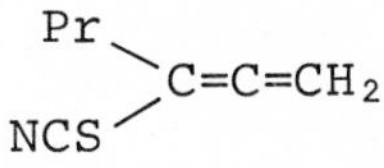 (120)

$R^1(NCS)C=C(NCS)-CH(R^2)(R^3)$

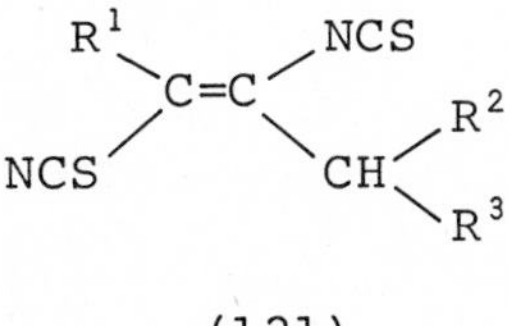

(121)

starting ylid contains no proton at the α-carbon atom, the betaine (118) forms allene rhodanides (120) and 1-thiocyanato-2-isothiocyanatoethylenes (121), respectively.

A.2.3.8. Allenes from Phosphine Alkylenes and 1-Dimethylamino-1-chloro Compounds. Interaction of the 1-dimethylaminochloro compound (122) and the phosphine alkylene (123) yields the salt (124), which is converted by transylidation into the nitrogen-containing ylid (125). This ylid (125) decomposes in an intramolecular β-elimination reaction into the allene (125), dimethylamine, and triphenylphosphine.

$$PhCHClNMe_2 + RCH_2CH{=}PPh_3 \longrightarrow \underset{\overset{\oplus}{Ph_3P}CHCH_2R\ Cl^{\ominus}}{Ph\text{-}CHNMe_2} \xrightarrow{(123)}$$

(122) (123) (124)

$$\underset{\overset{\oplus\ \ominus}{Ph_3P\text{-}C\text{-}CH_2R}}{Ph\text{-}CHNMe_2} \longrightarrow Ph\text{-}CH{=}C{=}CHR + HNMe_2 + PPh_3$$

(125) (126)

The phenyl group in the α-haloamine (122) is necessary for the allene formation. With hydrogen instead of the phenyl group, stable phosphoranes of type (125) are obtained; these can be olefinated with carbonyl compounds.[93]

Benzylidenetriphenylphosphorane [(123), Ph instead of RCH_2], reacts with (122) affording an ylid (125) which is converted by elimination of the dimethylamino anion into a triphenyl-1,2-diphenylvinylphosphonium salt.[93]

A.2.3.9. Reactions with N,N-Dimethyldichloroformamide. The ylid (127) (R = Ph or CO_2Me) and dichloroformamide (128) interact to give the salt (129), which is converted by transylidation to the phosphine alkylene (130). The ylid (130) isomerizes to the phosphonium salt (131), which as an enamine reacts with dilute acids to yield the salt (132).[405] The ylid corresponding to the salt (132) can be caused to react with aldehydes to give α-branched α,β-unsaturated aldehydes.

A.2.3.10. Reaccions with Various Halogen Compounds. The reactions of ylids with halogen compounds of the elements of Groups 4, 5, and 6 lead to the formation of salts, often followed by a transylidation reaction.[445a,468-9,515a,b,534,536a,559,564,570]

Interaction of phosphine alkylenes (133) with dialkyl(aryl)phosphinous chlorides (134) produces phosphine alkylenes of type (135).[335,536a] In this type of reaction, methylenetriphenylphosphorane, which is able to undergo a

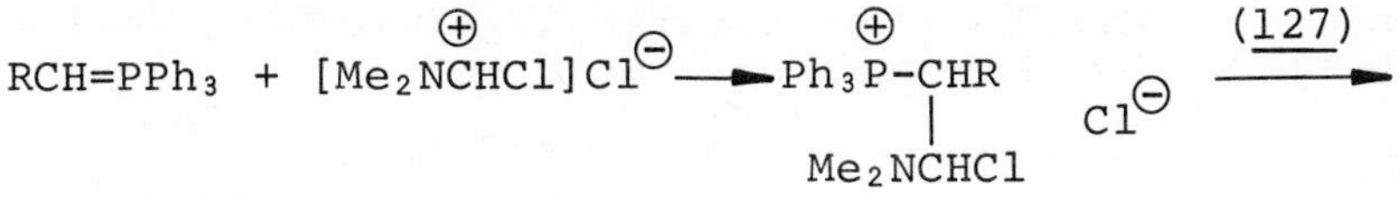

$$\mathrm{RCH{=}PPh_3 + [Me_2N\overset{\oplus}{C}HCl]Cl^{\ominus} \longrightarrow Ph_3\overset{\oplus}{P}{-}CHR(Me_2NCHCl)\ Cl^{\ominus} \xrightarrow{(127)}}$$

(127) (128) (129)

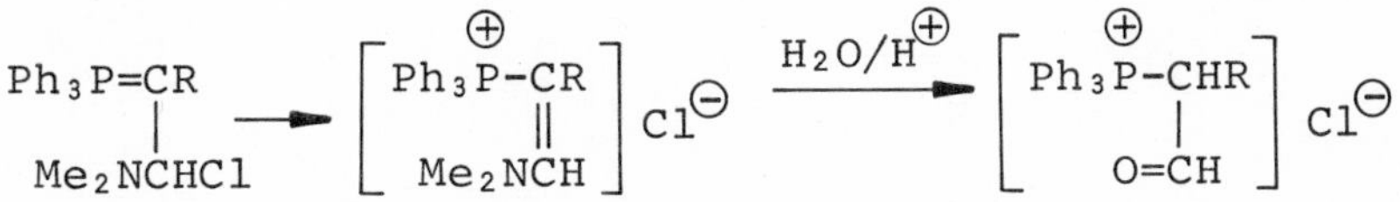

$$\mathrm{Ph_3P{=}CR(Me_2NCHCl) \longrightarrow \left[Ph_3\overset{\oplus}{P}{-}CR{=}CHNMe_2\right]Cl^{\ominus} \xrightarrow{H_2O/H^{\oplus}} \left[Ph_3\overset{\oplus}{P}{-}CHR(CH{=}O)\right]Cl^{\ominus}}$$

(130) (131) (132)

$$\mathrm{R{-}CH{=}PPh_3 + R^1_2PCl \longrightarrow R^1_2P{-}C(R){=}PPh_3 + [RCH_2\overset{\oplus}{P}Ph_3]Cl^{\ominus}}$$

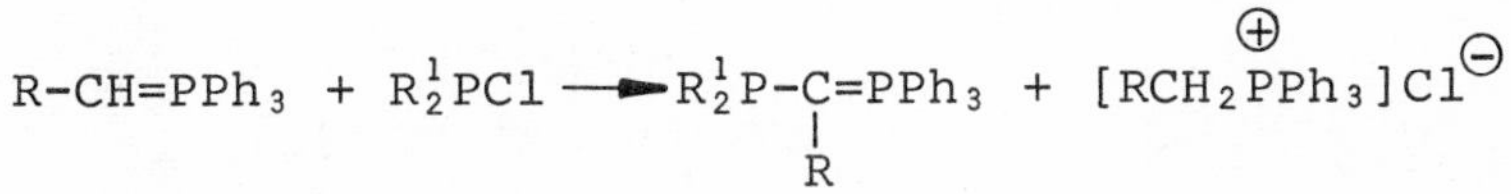

(133) (134) (135)

double proton-exchange reaction because of its $CH_2{=}P$ grouping, gives the ylid (136).

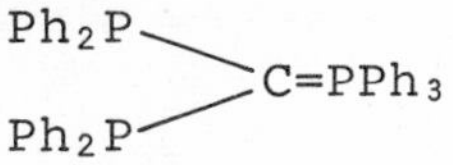

$$\mathrm{(Ph_2P)_2C{=}PPh_3}$$

(136)

Diphenylphosphinic chloride reacts in the same way as diaryl- and dialkylphosphinous chlorides.[336] Acylmethylenephosphoranes are phosphinylated at the oxygen atom.[336]

Triphenylphosphoranylidenemethylphosphonates (137) yield vinylphosphonates (138) with aldehydes in a Wittig reaction.[355] This reaction can also be carried out with

$$\mathrm{Ph_3\overset{\oplus}{P}{-}\overset{\ominus}{C}H{-}P(O)(OR)_2 + R^1CHO \longrightarrow Ph_3PO + R^1{-}CH{=}CH{-}P(O)(OR)_2}$$

(137) (138)

protected nucleosides.[354] Sulfenyl chlorides[445a,515a] and sulfonyl chlorides,[515b] as well as silyl, germyl, and stannyl chlorides,[534] react with phosphine alkylenes which still contain a proton at the α-carbon atom, again first with salt formation, followed by a transylidation reaction. Aryl halides also behave according to the above reaction scheme if they are activated by nitro groups.[462b]

Cyanuric chloride reacts with ylids with transylidation

to give α-(4,6-dichloro-2-s-triazinyl)-α(R-substituted)-methylenetriphenylphosphoranes.[462a]

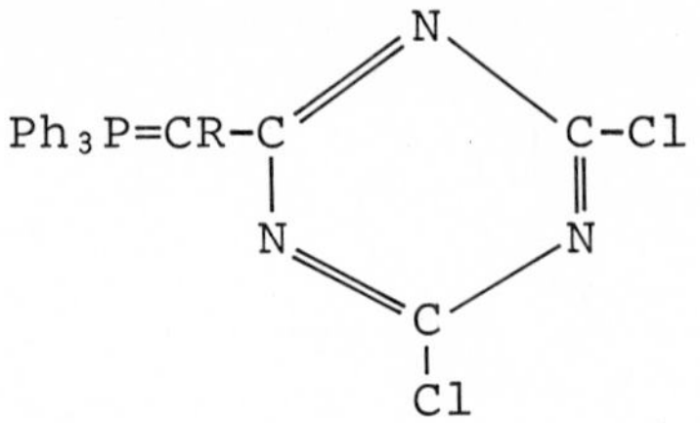

$$Ph_3P{=}CR{-}C_3N_3Cl_2$$

Phosphine alkylenes form salts with Hg(II) chloride, which are, in solution, in equilibrium with the starting components.[450a-451]

$$Ph_3P{=}CR^1R^2 + HgCl_2 \rightleftharpoons [Ph_3\overset{\oplus}{P}{-}CR^1R^2HgCl]Cl^{\ominus}$$

Nitrosyl chloride[362] interacts with ylids to form α-substituted α-hydroxyiminomethyltriphenylphosphonium chlorides (139). These chlorides (139) decompose when

$$Ph_3P{=}CHR + NOCl \longrightarrow [Ph_3\overset{\oplus}{P}{-}C({=}NOH){-}R]Cl^{\ominus}$$

(139)

heated to the corresponding nitriles (140) and phosphine oxide, e.g.:

$$(139) \xrightarrow{\Delta} RCN + Ph_3PO + HCl$$

(140)

Nürrenbach and Pommer[456] also describe the formation of benzonitrile from benzylidenetriphenylphosphorane and nitrosyl chloride. These authors, however, assume that the reaction occurred at the polarized N-O double bond.

Electrophilic carbenes such as dichlorocarbene (142) react with phosphine alkylenes (141) to form olefins (143) and triphenylphosphine.[343,457,674] In the rate-determining

$$Ph_3\overset{\oplus}{P}{-}\overset{\ominus}{C}R^1R^2 + |CCl_2 \longrightarrow Ph_3P{-}CR^1R^2(CCl_2) \longrightarrow Cl_2C{=}CR^1R^2 + PPh_3$$

(141) (142) (143)

step the carbene attacks the negatively charged ylid carbon atom. The reaction is terminated by a fast elimination of triphenylphosphine.[343]

The reaction of phosphine alkylenes (144) with sulfenes (145) (generated from sulfonyl chlorides) produces episulfones (146), their decomposition products, or substituted phosphine alkylenes (147).[344] If the ylid used (144) contains a proton on the α-carbon atom, the intermediate product (148) is stabilized by a proton transfer to give the corresponding ylid.

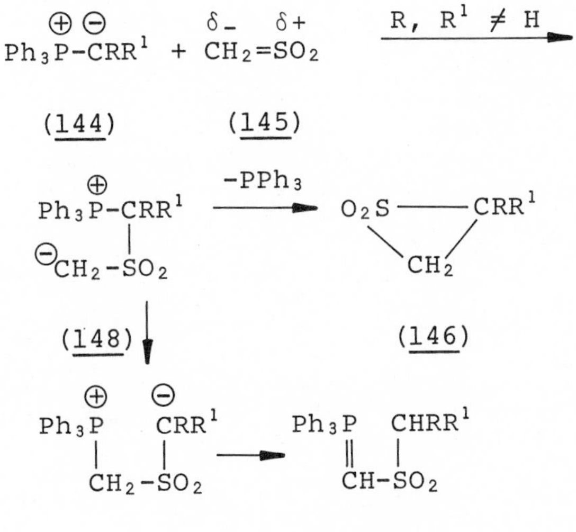

$$\overset{\oplus}{Ph_3P}-\overset{\ominus}{C}RR^1 + \overset{\delta-}{CH_2}=\overset{\delta+}{SO_2} \xrightarrow{R,\ R^1 \neq H}$$

(144) (145)

$$\overset{\oplus}{Ph_3P}-CRR^1(^{\ominus}CH_2-SO_2) \xrightarrow{-PPh_3} O_2S\text{—}CRR^1\text{—}CH_2 \text{ (ring)}$$

(148) (146)

$$\overset{\oplus}{Ph_3P}-CH_2-SO_2-\overset{\ominus}{C}RR^1 \longrightarrow Ph_3P{=}CH-SO_2-CHRR^1$$

(147)

Alkoxymethylenetriphenylphosphoranes react with carboxylic acid anhydrides to form acylated phosphonium salts, which rearrange into acylated ylids by eliminating R^1CO_2H.[164] In the case of chloroacetic anhydride the

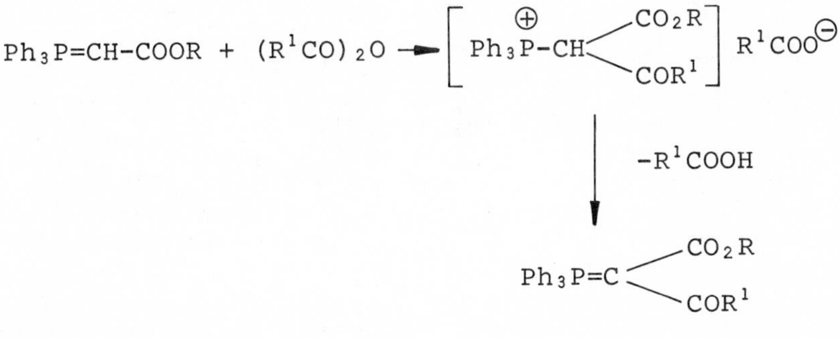

$$Ph_3P{=}CH-COOR + (R^1CO)_2O \longrightarrow \left[\overset{\oplus}{Ph_3P}-CH\begin{matrix}CO_2R\\ COR^1\end{matrix}\right] R^1COO^{\ominus}$$

$$\xrightarrow{-R^1COOH} Ph_3P{=}C\begin{matrix}CO_2R\\ COR^1\end{matrix}$$

resulting ylid cyclizes, with elimination of RCl, to give a cyclic phosphorane.

$$Ph_3P{=}C(COCH_2Cl)(COOR) \longrightarrow Ph_3P{=}C\langle CO{-}CH_2{-}O{-}CO \rangle$$

Aromatic diazonium compounds react with resonance-stabilized phosphine alkylenes to give phosphonium salts, from which phosphorus ylids containing an azo group have been obtained by treatment with a base.[400] Azo ylids are also formed by coupling diazonium salts with cyclopentadienylidene- and indenylidenephosphorane, respectively.[224a]

A.2.4. Reactions of Phosphine Alkylenes with Multiple Bonds

A.2.4.1. Reactions with the Carbonyl Group (Wittig Reaction). The well-known carbonyl olefination reaction, discovered by Wittig with Geissler[676] and Schöllkopf[685] and known today as the Wittig reaction, gave new impetus to phosphorus ylid chemistry.

The ylid carbon atom adds to the carbonyl group to give as an intermediate the betaine (148). In the reaction

$$R^1R^2C^{\ominus}{-}P^{\oplus}Ph_3 + {}^{\oplus}CR^3R^4{-}O^{\ominus} \longrightarrow \underset{(148)}{Ph_3P^{\oplus}{-}CR^1R^2{-}CR^3R^4{-}O^{\ominus}} \longrightarrow \underset{(149)}{Ph_3P{-}CR^1R^2{-}CR^3R^4{-}O\ (\text{ring})} \longrightarrow$$

$$\underset{(150)}{R^1R^2C = CR^3R^4} + Ph_3PO$$

of methylenetriphenylphosphorane with benzaldehyde the intermediate (148) can be isolated as the hydrobromide.[685] In the second step a four-membered ring (149) is formed, which is decomposed into an olefin (150) and phosphine oxide.

Since the betaine (148) formed as an intermediate can exist in two diastereomeric forms (151) and (152), the terminating cis elimination of phosphine oxide may lead to a cis- or a trans-olefin. The reversibility of the addition step has been demonstrated for stable ylids, as well as reactive (in the presence of lithium salts)[219,524] and moderated[357] (for nomenclature see ref. 525) ones. Furthermore, the separation of a diastereomeric pair (151) and (152) (for R = Me, R^1 = Ph) and their specific transformations into a cis- and a trans-olefin, respectively, were achieved.[524,525] Through their decomposition into an aldehyde and a phosphorane betaines (151) and (152) maintain

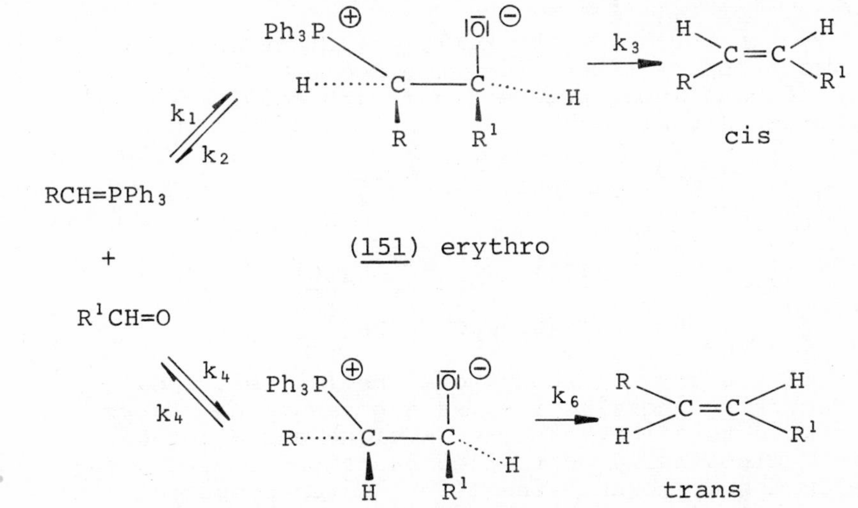

$$RCH{=}PPh_3 + R^1CH{=}O \underset{k_2}{\overset{k_1}{\rightleftharpoons}} \text{(151) erythro} \xrightarrow{k_3} \text{cis-}RCH{=}CHR^1$$

(151) erythro

$$RCH{=}PPh_3 + R^1CH{=}O \underset{k_4}{\overset{k_4}{\rightleftharpoons}} \text{(152) threo} \xrightarrow{k_6} \text{trans-}RCH{=}CHR^1$$

(152) threo

equilibrium with each other.[524,525]

Trippett and Walker[646] were successful in isolating a rearrangement product (154) of the betaine (153) (formed from benzaldehyde and methylenetriphenylphosphorane).

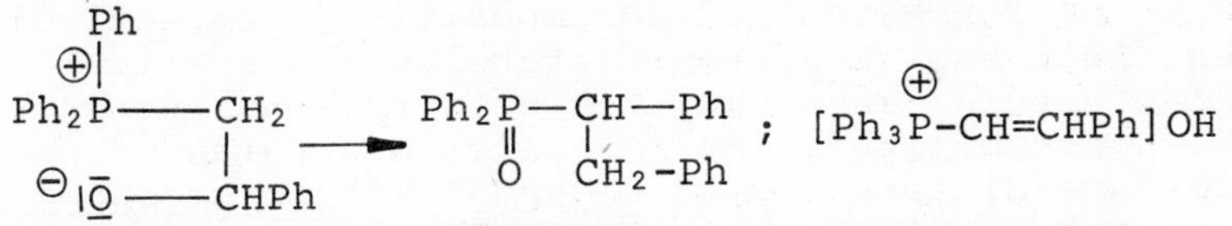

$$Ph_2P^{\oplus}(Ph){-}CH_2{-}CHPh{-}\bar{O}^{\ominus} \longrightarrow Ph_2P(=O){-}CH(Ph){-}CH_2{-}Ph \; ; \; [Ph_3\overset{\oplus}{P}{-}CH{=}CHPh]OH^{\ominus}$$

(153) (154) (155)

Richards and Tebby[508] propose that from (153) triphenyl-β-styrylphosphonium hydroxide (155) is formed first.

The primary addition product (156) of a Wittig reaction could be converted into the β-oxido ylid (157) by treatment with a base at low temperature.[175,176,526] The reactions

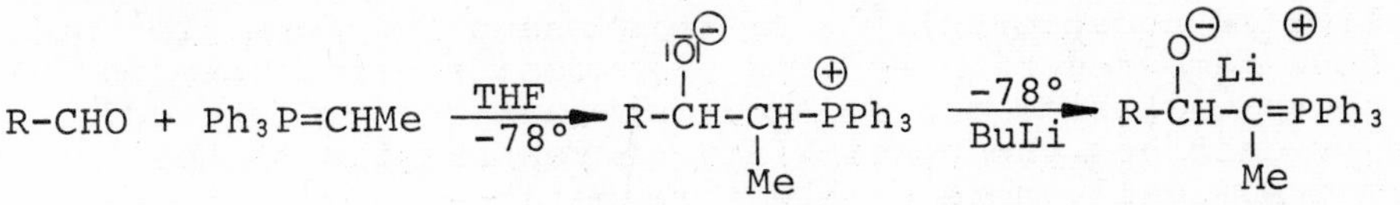

$$R{-}CHO + Ph_3P{=}CHMe \xrightarrow[-78^\circ]{THF} R{-}CH(\bar{O}^{\ominus}){-}CH(Me){-}\overset{\oplus}{P}Ph_3 \xrightarrow[BuLi]{-78^\circ} R{-}CH(O^{\ominus}\,Li^{\oplus}){-}C(Me){=}PPh_3$$

(156) (157)

of (157), a so-called betaine-ylid, with aldehydes or halogenating agents make available new methods for the synthesis of β-hydroxyolefins, haloolefins, ketones, and acetylenes.[175,176,526]

The cyclic course of the Wittig reaction is supported by the isolation of the cyclic intermediate (158) from the reaction of hexafluoroacetone (159) with hexaphenylcarbodiphosphorane (160).

$$Ph_3P{=}C{=}PPh_3 + O{=}C(CF_3)_2 \longrightarrow \begin{array}{l} Ph_3P{-}C{=}PPh_3 \\ \quad | \quad\ | \\ \quad O{-}C(CF_3)_2 \end{array} \longrightarrow$$

(160) (159) (158)

$$Ph_3PO + (CF_3)_2C{=}C{=}PPh_3$$

Most of the known carbonyl olefination reactions are stereochemically nonselective, as a consequence of two opposite effects:[525] phosphine alkylene and aldehyde unite in a kinetically controlled reaction to give preferentially the erythrobetaine (151), which produces mainly the cis-olefin if the betaine formation is made irreversible by using salt-free, strongly basic ylids. Thermodynamically, however, the threobetaine (152) is more stable, and if diastereomeric equilibrium between erythro- and threobetaine (through ylid and aldehyde) is reached quickly, mainly trans-olefins are formed. This is the case with weakly nucleophilic ylids or so-called betaine ylids.

Structural factors and reaction conditions that influence the rate of formation and decomposition, respectively, of betaines (151) and (152) thus determine the steric course of the Wittig reaction. For this reason resonance-stabilized, nonstabilized, and so-called moderated[525] phosphorus ylids are different with regard to their stereochemistry in the Wittig reaction.

Electron-donating groups at the aromatic nucleus of benylidenephosphoranes decrease the rate of reaction.[582b]

In addition to the substituents at the ylid carbon atom the groups attached to phosphorus also determine the cis-trans ratio of the olefins formed. The preferential formation of trans-olefins from methylenetricyclohexylphosphorane is based on the reduced electrophilicity of the phosphorus atom in the primary betaine (as compared to the triphenyl-substituted phosphorus).[81] By this means the formation of the four-membered ring and its decomposition into olefin and phosphine oxide are delayed, so that the primary addition product has the possibility of rearranging to the thermodynamically more stable threobetaine.

In addition to the influence of the substituents on the equilibrium constants k_1, k_2, k_4, and k_5 and the decomposition constants k_3 and k_6, a number of other factors are of importance for the rate and the stereochemical course of the

Wittig reaction. These factors are, e.g., solvents,[29,30,32,316,512,582a,596] inorganic halides,[29,30,32,41,316,523,525] acidic catalysts,[512] reactant ratio,[29,31,36,206] reaction time,[29] and temperature.[127,128,206,272,582a]

The decisive role of the lithium salts in the Wittig reaction is particularly stressed. Nonstabilized phosphorus ylids and their betaines, formed with aldehydes, add lithium halides and thereby lose their original reactivity.[30,523] They can be reactivated by treatment with potassium t-butoxide.[49,523] According to the need, one can therefore "decrease or increase" the rate of the Wittig reaction.[525]

Through suitable choice of the reaction conditions the generally stereochemically unspecific Wittig reaction can be made stereoselective.[29,32,525]

The Wittig reaction occurs with retention at the phosphorus atom.[130,313] The use of optically active benzylidene-methylphenyl-n-propylphosphorane in the Wittig reaction makes possible the partial asymmetric synthesis of substituted benzylidenecycloalkanes (axial chiral).[87] It is also possible to catalyze with optically active acids a partial asymmetric synthesis by the Wittig reaction.[88] The Wittig reaction of cyclic ylids with carbonyl compounds leads to olefins that still contain a phosphine oxide function in the molecule.[323,381,409,518]

Silyl ketones react with ylids to give siloxylalkenes[148a] in addition to the normal Wittig products.

The Wittig reaction becomes increasingly important for the chain lengthening of carbohydrates.[44a,646a,716-722] In view of the great number of review articles concerning the Wittig reaction,[42,347,398,543,638] there is no need for further discussion here of the various preparative possibilities of this carbonyl olefination reaction.

A.2.4.2. Reactions with the NO Group. Whereas the characteristic feature of the Wittig reaction is the formation of a C-C double bond, the reactions of nitroso compounds with phosphine alkylenes lead to the synthesis of a C-N double bond.[186,543,545] In addition the formation of olefins of type $R^1R^2C{=}CR^1R^2$ is also observed. They are

$$Ph_3P{=}CR^1R^2 + O{=}N{-}R^3 \longrightarrow Ph_3P{=}O + R^1R^2C{=}N{-}R^3$$

(161)

apparently produced by reaction of the imine (161), formed as an intermediate, with a second mole of ylid.[108,109]

Interaction of N-methyl-N-nitroso-p-toluenesolfonamide and phosphorus ylids yields the corresponding nitriles (163), probably by decomposition of the first-formed tosylhydrazone (163).[456]

$$RCH{=}N{-}N\begin{matrix}Tos\\Me\end{matrix} \xrightarrow{B} R{-}\overset{\ominus}{C}{=}N{-}N\begin{matrix}Tos\\Me\end{matrix} \longrightarrow R{-}C{\equiv}N + \overset{\ominus}{N}\begin{matrix}Tos\\Me\end{matrix}$$

(162) (163)

Nitrogen oxide reacts with phosphine alkylenes in an unclear way to give a series of products. Included among those isolated are nitriles, olefins, and phosphine oxide.[363]

That the applicability of the Wittig olefination reaction is not limited to carbonyl compounds is shown by the transfer of the reaction to N_2O.[513] Dinitrogen oxide reacts with methylenetriphenylphosphorane to give diazomethane and triphenylphosphine oxide. With excess ylid

$$Ph_3P{=}CH_2 + \underset{\ominus}{\overline{N}}{=}\underset{\oplus}{N}{=}O \longrightarrow \underset{\ominus}{\overline{N}}{=}\underset{\oplus}{N}{=}CH_2 + Ph_3P{=}O$$

diazomethane is deprotonated to the diazomethyl anion, whose hydrolysis yields, depending on the pH, diazomethane or isodiazomethane.

A.2.4.3. Reactions with the C-C Double Bond. Phosphine alkylenes (164) give betaine (166) with olefins containing an activated C-C double bond (165). This intermediate can itself stabilize by various secondary reactions.[104] The reaction path is determined by the residue R^1 of the ylid and the groups R^2 and R^3 of the olefin.

$$\underset{(164)}{Ph_3\overset{\oplus}{P}{-}\overset{\ominus}{C}HR^1} + \underset{(165)}{R^2{-}\overset{\oplus}{C}H{-}\overset{\ominus}{C}H{-}R^3} \longrightarrow \underset{(166)}{R^2{-}CH(CHR^1\overset{\oplus}{P}Ph_3){-}\overset{\ominus}{C}H{-}R^3}$$

path c: $R^2{-}C({=}CHR^1){-}CH_2{-}R^3 + PPh_3$ (169)

path a: $R^2{-}CH{-}CH{-}R^3$ bridged by $C(R^1)(H)$ (cyclopropane) (167)

path b: $R^2{-}CH(C({=}PPh_3)R^1){-}CH_2{-}R^3$ (168)

Path a: If the residue R^1 in the phosphine alkylene (164) is electron donating, the betaine (166) is stabilized by splitting off triphenylphosphine to give the cyclopropane derivative.[21a,104,228,229,431,575a]

Path b: If the residue R^1 in (164) is electron withdrawing, the α-proton in the betaine (166) is loosened. This proton then migrates to the negative part of the molecule with the formation of the ylid (168).[21a,104,575a] This reaction may be considered as a Michael addition of the phosphine alkylene (164) to the olefin (165). A further possibility is observed if the olefin contains several strongly activating groups.[636]

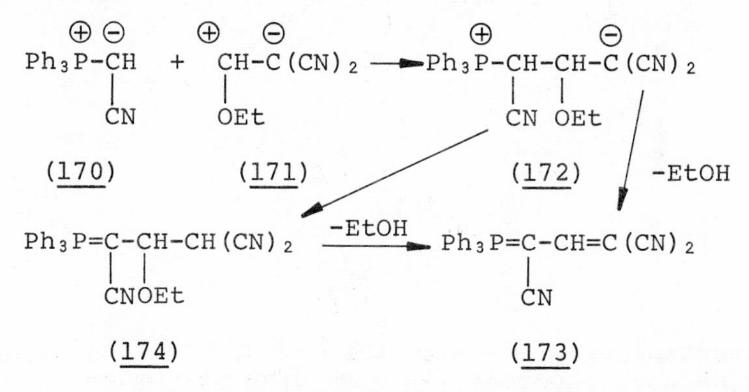

Betaine (172) is formed from cyanomethylenetriphenylphosphorane (170) and the strongly activated olefin (171). According to Trippett, this betaine is stabilized by a direct transformation into (173), whereby alcohol is split off.

However, the possibility cannot be excluded that first a Michael addition occurs to give (174), which is followed by the splitting of alcohol to form (173). Thus the question remains whether there is a possible third method of stabilizing betaines of type (166).

A similar reaction path has been observed in the interaction of cyclopentadienylidenetriphenylphosphorane with cyanoolefins.[510] Here, however, the primary attack of the olefin occurs, not at the ylid carbon atom, but in the α-position to it.

Path c: By a transfer of a proton from the β- into the γ-position, followed by elimination of triphenylphosphine, unsaturated compounds of type (169) are formed. This type of reaction is observed with maleic and fumaric acid esters, respectively, as the olefin component.[73]

McClure[427] reported another type of reaction of ylids with activated C-C double bonds. Acrylonitrile (175), when caused to react with methoxycarbonylmethylenetriphenylphosphorane (176), gives cyanomethylenetriphenylphosphorane (178) and acrylic acid ethyl ester (179). The author proposes a four-membered cyclic intermediate (177), which opens to give the reaction products (178) and (179).

$$CH_2{=}CH{-}CN + Ph_3P{=}CHCO_2Et \rightleftharpoons Ph_3\overset{\oplus}{P}{-}CHCO_2Et{-}CH_2{-}\overset{\ominus}{C}H(CN) \longrightarrow$$

(175) (176)

$$\underset{(177)}{Ph_3P{-}CHCO_2Et,\ CH(CN){-}CH_2 \text{ (ring)}} \rightleftharpoons Ph_3\overset{\oplus}{P}{-}CH(CN){-}CH_2{-}\overset{\ominus}{C}HCO_2Et \longrightarrow$$

$$\underset{(178)}{Ph_3P{=}CHCN} + \underset{(179)}{CHCO_2Et{=}CH_2}$$

The reaction of ylids with the C-C double bond of carbon suboxide also proceeds via a phosphacyclobutane ring.[688a]

Reactions at the C-C double bond are also observed as subsequent reactions when phosphine alkylenes are allowed to react with quinones,[86] α,β-diketones,[90] diphenylcyclopropenone,[36a] and fluorenone.[50]

1,2-Dicarbonyl compounds (180) react with the bisylid (181) in a Wittig reaction to give compound (182). By intramolecular nucleophilic attack of the ylid at the double bond, (182) is converted, with formation of a cyclopropane ring and elimination of triphenylphosphine, into a norcaradiene derivative (183).

In a Wittig reaction, p-quinones (184) and methoxycarbonylmethylenetriphenylphosphorane (185) form a p-quinomethide (186), which adds an additional mole of ylid to give compound (187) (Michael addition).[86]

o-Quinones react with (185) to form o-quinomethides, which also add a second mole of ylid. The betaines produced undergo a Hofmann degradation to yield α-hydroxyarylethylene-dicarboxylic acid esters, which by an intramolecular transesterification are converted into cumaryl-4-carboxylic acid esters.[86]

In a similar way reaction of two moles of benzylidenetriphenylphosphorane with one mole of o-quinone leads to a stilbene derivative, which cyclizes to a dihydrofuran derivative.[86,625] The light-induced reaction of diphenyl-

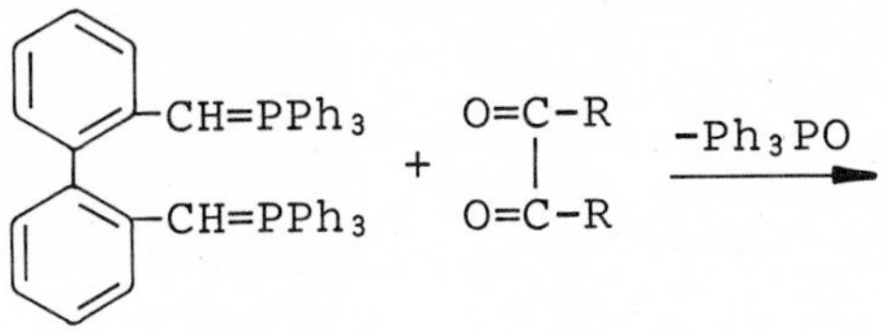

$$(181) + (180) \xrightarrow{-Ph_3PO}$$

(181) (180)

$$\xrightarrow{-PPh_3}$$

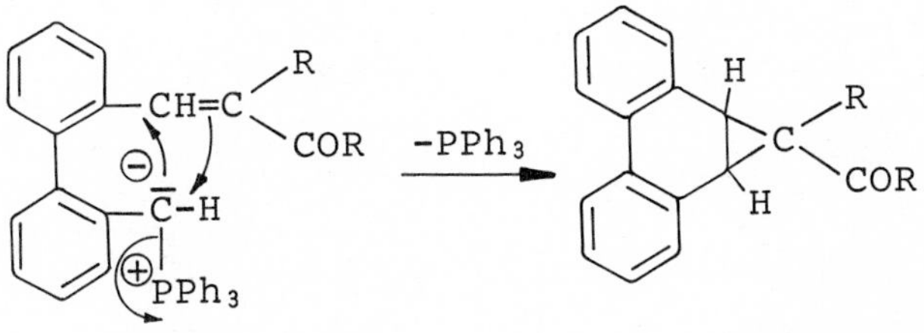

(182) (183)

$$(184) + HC(CO_2Me){=}PPh_3\ (185) \xrightarrow{-OPPh_3} (186) \longrightarrow (187)$$

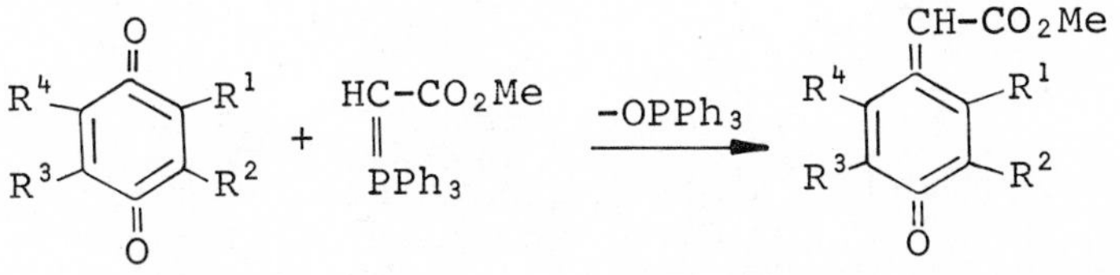

(184) (185) (186)

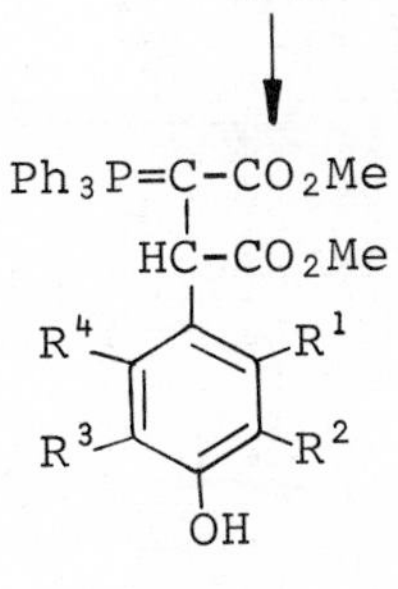

(187)

methylenetriphenylphosphorane with excess 1,1-diphenylethylene to give a cyclopropane ring is the earliest example of a photochemical reaction with phosphine alkyl-

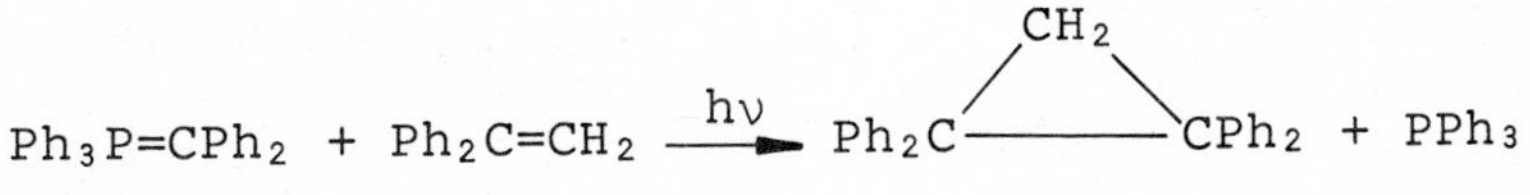

$$Ph_3P{=}CPh_2 + Ph_2C{=}CH_2 \xrightarrow{h\nu} Ph_2C\overset{CH_2}{\frown}CPh_2 + PPh_3$$

enes.[511] Instead of the ylid, the corresponding phosphazine may be used.

Pyran derivatives (184) may be constructed by Michael addition of acylalkylenetriphenylphosphoranes (113) to acylallenes (188).[621a,622] The addition product (189) splits off triphenylphosphine oxide to give the acetylene derivative (190), which in its enol form closes the ring to the pyran (191).

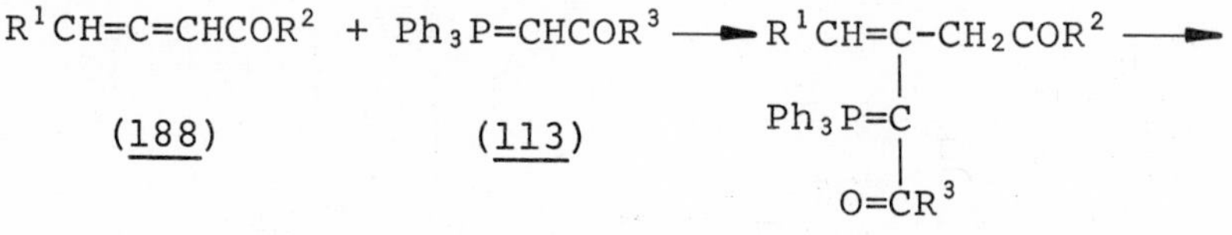

$$R^1CH{=}C{=}CHCOR^2 + Ph_3P{=}CHCOR^3 \longrightarrow R^1CH{=}C(-C(=PPh_3)-COR^3)-CH_2COR^2 \longrightarrow$$

(188) (113) (189)

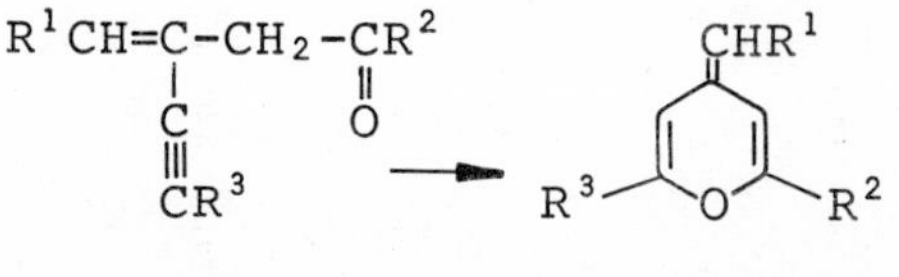

$$R^1CH{=}C(-C{\equiv}CR^3)-CH_2-C(=O)R^2 \longrightarrow$$

(190)

(191): 4-(CHR^1)-4H-pyran with R^3 at C-2 and R^2 at C-6

Diazoketones undergo a Wolff rearrangement under the catalytic influence of acylmethylenetriphenylphorene to yield ketenes. The Wittig reaction of ketenes with acylmethylenetriphenylphosphorane produces acylallenes of type (188), which react with a second mole of ylid in the way shown above.[621a] The ketene can also be produced in several cases from the acid chloride by splitting off HCl by acyl ylids.[621]

Thermolysis of β-ketoacid esters in the presence of acylmethylenetriphenylphosphorane also gives pyran derivatives. By elimination of alcohol from β-keto esters ketoketenes are first formed; then, in a Wittig reaction, these are converted to acylated allenes.[584a,621a] In addition to acylated allenes, acylated acetylenes are also formed in this reaction; these compounds, without reacting with a second mole of ylid, cyclize to give pyrones.

Acylketenimines (188) (R^1-N= instead of R^1-CH=), which are obtained from acylmethylenetriphenylphosphorane and isocyanates,[621a] react in the same way as acylallenes (188).

A.2.4.4. Reaction with the C-N Double Bond.

(1) Reaction with Schiff's Bases

The reaction path of Schiff's bases with phosphine alkylenes depends on whether or not the phosphine alkylene contains a CH_2 group in β-position to the phosphorus atom.[108,109]

Ylids (192) that contain no β-CH_2 group react with Schiff's bases (193) in a Wittig-type reaction through the intermediate formation of a four-membered ring (194) to give as final products olefins (196) and triphenylphosphine phenylimine (195).

$$R^1CH{=}PPh_3 + R^2CH{=}N\text{-}Ph \longrightarrow \begin{array}{l} R^2CH - \overset{\ominus}{N}\text{-}Ph \\ \;| \qquad\quad | \\ R^1CH - \underset{\oplus}{P}Ph_3 \end{array} \longrightarrow$$

(192) (193)

$$\begin{array}{l} R^2CH - N\text{-}Ph \\ \;| \qquad\quad | \\ R^1CH - PPh_3 \end{array} \longrightarrow \begin{array}{l} R^2CH \\ \;\| \\ R^1CH \end{array} + \begin{array}{l} N\text{-}Ph \\ \;\| \\ PPh_3 \end{array}$$

(194) (196) (195)

Ylids (197) that contain a CH_2 group in β-position to the phosphorus atom yield, in the interaction with benzalaniline (198), first a betaine (199), which decomposes on warming to give an allene (201), triphenylphosphine, and aniline.[108,109]

$$RCH_2CH{=}PPh_3 + PhCH{=}NPh \longrightarrow \begin{array}{l} PhCH - \overset{\ominus}{N}Ph \\ \quad\;\; | \\ Ph_3\overset{\oplus}{P}CHCH_2R \end{array} \longrightarrow$$

(197) (198)

(199)

$$\begin{array}{l} PhCH - NHPh \\ \quad | \qquad\quad H \\ \quad | \qquad\quad | \\ Ph_3\overset{\oplus}{P}\text{-}\overset{\ominus}{C} - CR \\ \qquad\qquad\;\; | \\ \qquad\qquad\;\; H \end{array} \longrightarrow PhCH{=}C{=}CHR + Ph_3P + H_2NPh$$

(200) (201)

Probably the isolable betaine (199) is first converted, by a proton transfer, into the ylid (200), which by an intramolecular β-elimination undergoes a Hoffman degradation. The last reaction step is analogous to that of allene formation from 1-dimethylamino-1-chloro compounds and phosphine alkylenes.[107]

(2) Reaction with Isocyanates

Whereas diphenylmethylenetriphenylphosphorane (202) yields diphenylketenimine (204) with phenyl isocyanate (203) in a Wittig reaction,[612] phosphine alkylenes of type (205), which contain a proton on the α-carbon atom, produce a new ylid (206) in this reaction.[516,640]

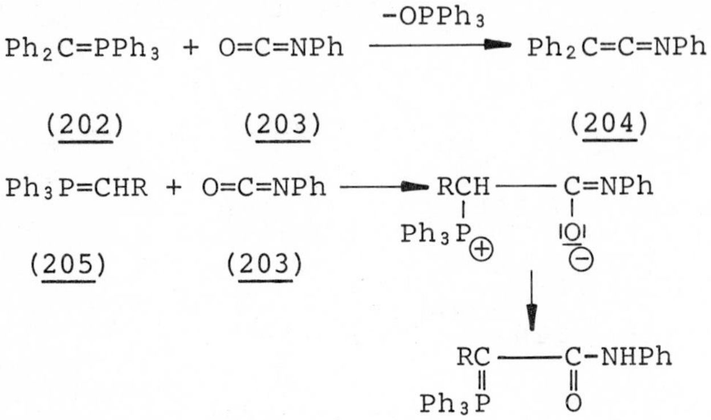

$$Ph_2C{=}PPh_3 + O{=}C{=}NPh \xrightarrow{-OPPh_3} Ph_2C{=}C{=}NPh$$

(202) (203) (204)

$$Ph_3P{=}CHR + O{=}C{=}NPh \longrightarrow \underset{Ph_3\overset{\oplus}{P}}{RCH}-\underset{|\overline{\underline{O}}|^{\ominus}}{C{=}NPh} \longrightarrow \underset{Ph_3P}{\overset{}{RC}}{=}\!-\underset{O}{\overset{}{C}}(=O)-NHPh$$

(205) (203) (206)

For phosphoranes resulting from aromatic isocyanates and acyl ylids NMR investigations indicate a chelate structure (207).[131] At elevated temperature a partial transfer of the proton from nitrogen to the oxygen is observed. Finally (207) decomposes into its components (path a) or into triphenylphosphine oxide and a ketenimine (path b).

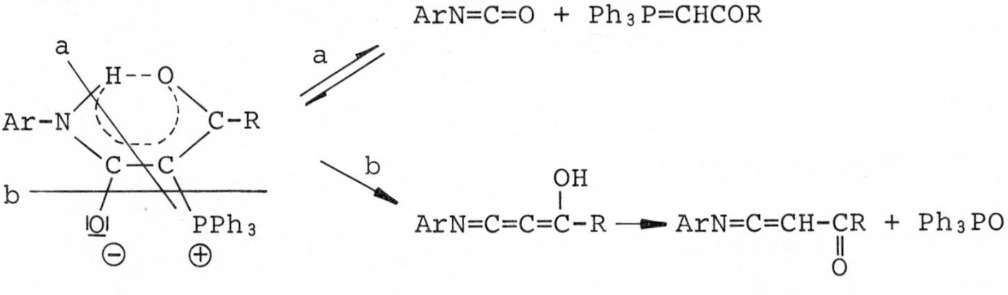

(207) —a⇌ $ArN{=}C{=}O + Ph_3P{=}CHCOR$

(207) —b→ $ArN{=}C{=}C{=}C(OH){-}R \longrightarrow ArN{=}C{=}CH{-}C(=O)R + Ph_3PO$

(207)

A.2.4.5. Reaction with Nitriles. The reaction of phosphoranes with nitriles,[26,129] followed by hydrolysis of the adduct (208) formed as an intermediate, allows the synthesis of ketones (209). The addition of ylids to nitriles is catalyzed by lithium ions.[26]

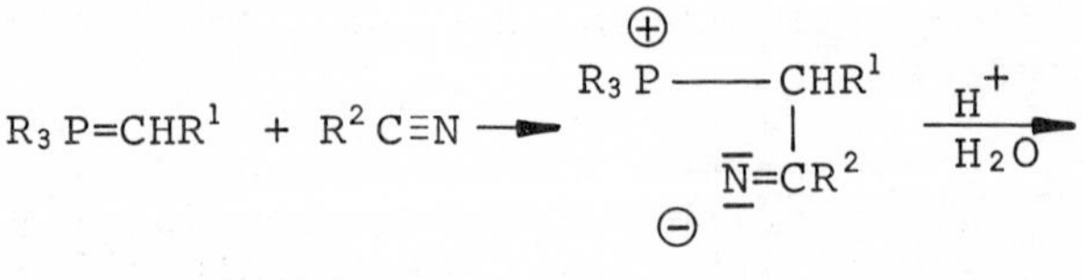

$$R_3P{=}CHR^1 + R^2C{\equiv}N \longrightarrow R_3\overset{\oplus}{P}-CHR^1-\overset{\ominus}{\overline{N}}{=}CR^2 \xrightarrow[H_2O]{H^+}$$

(208)

$$R_3P{=}O + NH_4^{\oplus} + R^2\underset{\underset{O}{\|}}{C}{-}CH_2R^1$$

(209)

Hydrolysis of (208) occurs apparently by two competing reactions. For once a direct attack of water on the phosphorus atom (coupled with inversion) must be assumed, and in addition there exists the possibility of hydrating the imino group, which is followed by an intramolecular attack of oxygen on phosphorus. This last reaction occurs with retention of the configuration on phosphorus. That both reactions occur side by side is supported by studies with optically active ylids and different substituents R^2, respectively.[26]

Ciganek[166] demonstrated that benzylidenetriphenylphosphorane (210) and benzonitrile interact to give the iminophosphorane (214). The mechanism of this reaction is

$$Ph_3P{=}CHPh + PhCN \longrightarrow Ph_3\overset{\oplus}{P}{-}CHPh,\ \overset{\ominus}{\underline{\overline{N}}}{=}CPh \longrightarrow$$

(210) (211) (212)

$$\begin{array}{cc} Ph_3P{-}CHPh \\ | \quad\quad | \\ N{=}CPh \end{array} \longrightarrow \begin{array}{cc} Ph_3P \quad CHPh \\ \| \quad\quad \| \\ N{-}CPh \end{array}$$

(213) (214)

assumed to be similar to that of the reaction of acetylenedicarboxlic esters with ylids. Presumably the primary product (212) cyclizes to the four-membered ring (213), which then opens up to the phosphine imine (214).

Resonance-stabilized ylids, which do not react with benzonitrile, also form iminophosphoranes with activated nitriles (dicyanogen, trifluoroacetonitrile).[166]

A.2.4.6. Reactions with the C-C Triple Bond. The mechanism of the addition of methylenephosphoranes to acetylenes is determined by steric and electronic factors.[149] Phosphoranes that contain an α-proton can undergo a Michael addition[278,279,636] (in protic solvents) or a cycloaddition[95,151,278,279] (in aprotic solvents).

The reaction of ylids (215) with acetylene-dicarboxylic acid esters (216) leads, either through the betaine (217) or by direct cycloaddition, to the phosphacyclobutene (218), which undergoes ring opening to give the stable phosphorane (219).[95]

$$Ph_3\overset{\oplus}{P}-\overset{\ominus}{C}HR + MeO_2CC{\equiv}CCO_2Me \longrightarrow RCH(\overset{\oplus}{P}Ph_3)-C(CO_2Me){=}\overset{\ominus}{C}-CO_2Me \longrightarrow$$

(215) (216) (217)

(218) → (219): $RCH{=}C(CO_2Me)-C(CO_2Me){=}PPh_3$

The interaction of ylids with the C-C triple bond of dehydroaromatics leads to phosphines (222) by rearrangement of the primary adduct (220) and transfer of a residue from phosphorus to the negative center of the intermediate product (221).[692,693]

$R_3P{=}CHR^1$ + benzyne → (220) → (221) → (222)

A similar rearrangement has been observed in the pyrolysis of the betaine (224), obtained from o-hydroxyphenyltriphenylphosphonium iodide (223) by treatment with bases.[71]

(223) $\xrightarrow{B}$ (224) → o-$PPh_2C_6H_4OPh$

Allylidenetriphenylphosphorane reacts with benzyne at the γ-carbon atom to give cinnamoylidenetriphenylphosphorane.[694a]

$$Ph_3P{=}CH{-}CH{=}CH_2 + C_6H_4 \text{(benzyne)} \longrightarrow Ph_3\overset{\oplus}{P}{-}CH{=}CH{-}CH_2{-}C_6H_4^{\ominus} \longrightarrow Ph_3P{=}C{-}CH{=}CH{-}C_6H_5$$

A.2.4.7. Reaction with the C-S Double Bond

(1) Reaction with Carbon Disulfide

The stable ylids diphenylmethylene- and fluorenylidene-triphenylphosphorane react with carbon disulfide to give polymeric or dimeric thioketenes.[547,616]

Salt-free solutions of basic phosphine alkylenes (225) interact with CS_2 in the mole ratio 2:1,[54] giving phosphonium salts (226) of α-(triphenylphosphoranylidene)-dithiocarboxylic acids. Treatment of the salts (226) with alkyl halides (R^1X) results in the formation of α-(triphenylphosphoranylidene)-dithiocarboxylic acid esters (227) and triphenylalkylphosphonium salts (228). The renewed alkylation of the ylids (227), followed by hydrolysis of the resulting (229), allows the synthesis of ketene mercaptals (230).

$$\underset{(225)}{2\ RCH{=}PPh_3} + CS_2 \longrightarrow \underset{(226)}{[RC({=}PPh_3){-}C({=}S){-}\bar{S}|\,]^{\ominus}[RCH_2PPh_3]^{\oplus}} \xrightarrow{R^1X}$$

$$\underset{(227)}{RC(=PPh_3){-}C({=}S){-}SR^1} + \underset{(228)}{[RCH_2\overset{\oplus}{P}Ph_3]X^{\ominus}}$$

$$\xrightarrow{R^2X} \underset{(229)}{\left[RC(-\overset{\oplus}{P}Ph_3){=}C\begin{matrix}SR^1\\SR^2\end{matrix}\right]X^{\ominus}} \xrightarrow[-OPPh_3]{OH^{\ominus}} \underset{(230)}{RCH{=}C\begin{matrix}SR^1\\SR^2\end{matrix}}$$

Interaction of cyanomethylenephosphoranes with CS_2 yields α-dithiocarboxy-α-cyanomethylenetriphenylphosphorane (231). The ylid (231) forms with triethylamine a salt that, when treated with alkyl halides, produces the corresponding S-substituted ylids.[462c]

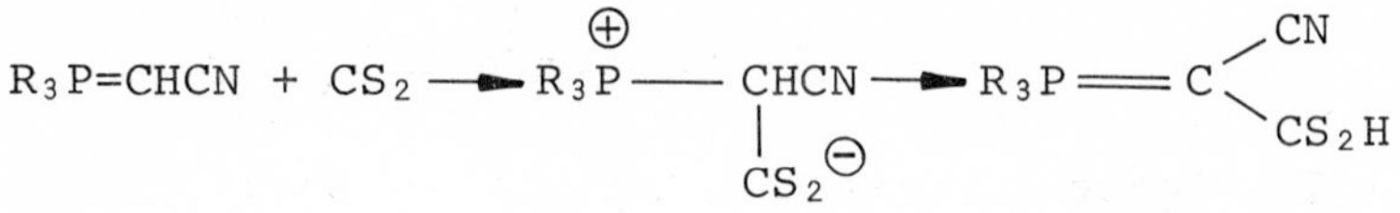

$$R_3P{=}CHCN + CS_2 \longrightarrow R_3\overset{\oplus}{P}{-}\underset{CS_2^{\ominus}}{CHCN} \longrightarrow R_3P{=}C(CN)(CS_2H)$$

(231)

(2) Reaction with Isothiocyanates

Alkylidenetriphenylphosphoranes give with isothiocyanates either betaines (232) (if R exhibits an +I effect)[91] or ylids (233) (for groups with an -I effect),[91,516] whose subsequent reactions make possible the synthesis of thiocarboxylic acid amides (234) of α,β-unsaturated carboxylic acids, and thioimido acid derivatives (235) and (236), as well as substituted vinylphosphonium salts (237).[91]

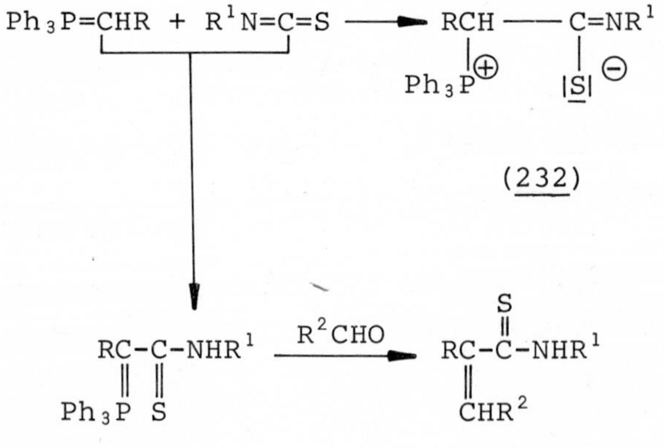

$$Ph_3P{=}CHR + R^1N{=}C{=}S \longrightarrow \underset{Ph_3P^{\oplus}}{RCH}{-}\underset{|\underline{S}|^{\ominus}}{C}{=}NR^1$$

(232)

$$\underset{Ph_3P}{RC}(=)\text{-}\underset{S}{C}(=)\text{-}NHR^1 \xrightarrow{R^2CHO} RC(=CHR^2)\text{-}C(=S)\text{-}NHR^1$$

(233) (234)

$$(\underline{232}) \text{ or } (\underline{233}) \xrightarrow{MeI} \underset{Ph_3\overset{\oplus}{P}}{RC}=\underset{SMe}{C}NHR^1 \quad I^{\ominus}$$

$$\downarrow OMe^{\ominus}$$

$$\underset{(\underline{235})}{RCH_2-\underset{SMe}{C}=NR^1} \xleftarrow{H_2O} \underset{Ph_3P}{RC}-\underset{SMe}{C}=NR^1$$

$$\xrightarrow{R^3CHO} \underset{(\underline{236})}{\underset{CHR^3}{RC}-\underset{SMe}{C}=NR^1} \qquad \xrightarrow{MeJ} \underset{(\underline{237})}{\underset{Ph_3\overset{\oplus}{P}}{RC}=\underset{SMe}{C}-NMeR^1 \quad I^{\ominus}}$$

A.2.4.8. Reactions with the N-N Double Bond. Phosphine alkylenes (239) add to the electron-deficient N-N double bond of azodicarboxylic acid esters (238).

In the case of resonance-stabilized ylids ($R = CO_2R^1$, COMe, COPh) the first-formed betaine (240) is stabilized by a proton transfer to give the new ylid (241)[116] and not, as was previously assumed,[115] by decomposition into a phosphinimine and a Schiff's base.

$$\underset{(\underline{238})}{R^1O_2C-N=N-CO_2R^1} + \underset{(\underline{239})}{RCH=PPh_3} \longrightarrow \underset{(\underline{240})}{R^1O_2C-\underset{RCH\overset{\oplus}{P}Ph_3}{N}-\overset{\ominus}{\bar{N}}-CO_2R^1} \longrightarrow$$

$$\underset{(\underline{241})}{R^1O_2C-\underset{RC=PPh_3}{N}-NH-CO_2R^1}$$

On pyrolysis (241) gives triphenylphosphine and an amidine (242). The latter reacts in the case of methoxycarbonylmethylenetriphenylphosphorane (239) ($R = CO_2Me$) with excess starting ylid (239) in a Wittig-type reaction to give the phosphinimine (244) and the olefin (243).

$$R^1O_2C\text{-}NH\text{-}CR{=}NCO_2R^1 + MeO_2CCH{=}PPh_3 \longrightarrow$$

(242)

$$R^1O_2C\text{-}NH\text{-}CR{=}CHCO_2Me + Ph_3P{=}NCO_2R^1$$

(243) (244)

In the reaction of nonstabilized ylids (239) (R = Et, Ph) with azodicarboxylic esters the primary product (240) spontaneously splits off triphenylphosphine, thus forming an azomethine imine. The split-off triphenylphosphine reacts with a second mole of the azo compound (238) and produces the 1,3-dipole (246).[152] In a subsequent reaction this dipole and the azomethine imine (245) eliminate triphenylphosphine and give the heterocycle (247).

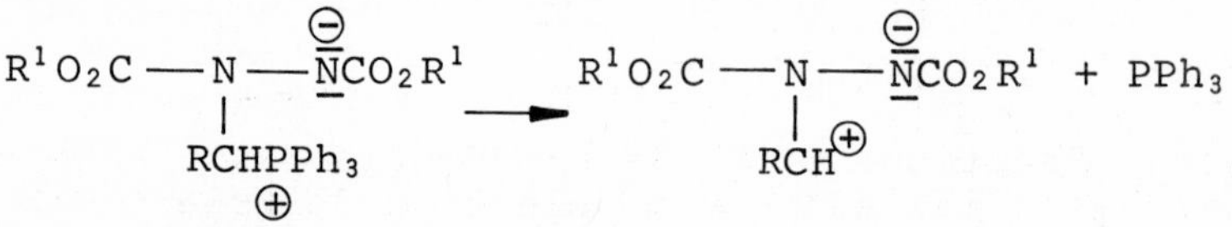

$$R^1O_2C-N(CH(R)\overset{\oplus}{P}Ph_3)-\overset{\ominus}{N}CO_2R^1 \longrightarrow R^1O_2C-N(\overset{\oplus}{C}HR)-\overset{\ominus}{N}CO_2R^1 + PPh_3$$

(240) (245)

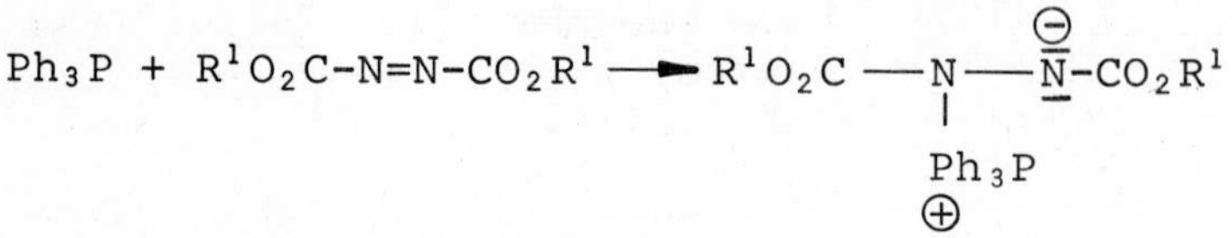

$$Ph_3P + R^1O_2C\text{-}N{=}N\text{-}CO_2R^1 \longrightarrow R^1O_2C-N(\overset{\oplus}{P}Ph_3)-\overset{\ominus}{N}\text{-}CO_2R^1$$

(238) (246)

(246) + (245) ⟶ [six-membered intermediate: C(H)(R) bridging R^1O_2C-N and $N-CO_2R^1$; $R^1O_2C-\overset{\ominus}{N}$; $\overset{\oplus}{N}-PPh_3$; $C(=O)OR^1$] ⟶

[ring: C(H)(R), R^1O_2C-N, $N-CO_2R^1$, R^1O_2C-N, $N-PPh_3$ (⊕), $C(OR^1)(\overline{O}^{\ominus})$] $\xrightarrow{-OPPh_3}$ [ring: C(H)(R), R^1O_2C-N, $N-CO_2R^1$, R^1O_2C-N, N, $C(OR^1)$=N]

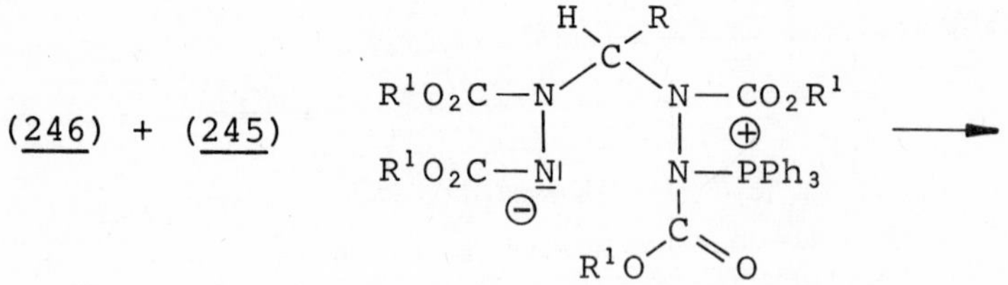

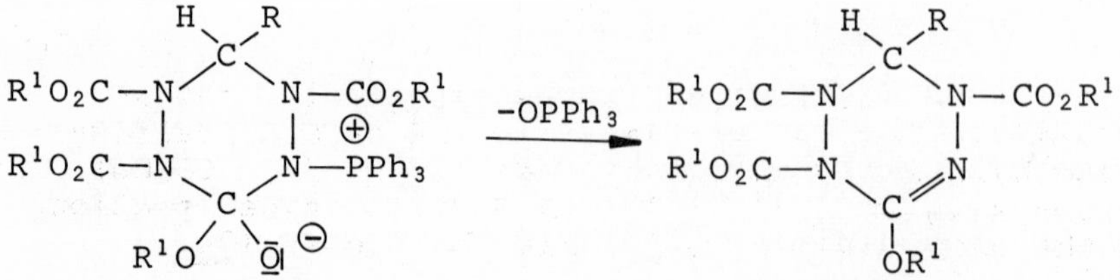

(247)

R^1 = Et; R = Ph, Et

A.2.4.9. Reaction with Oxygen and Other Oxidizing Agents. The action of oxygen on ylids leads to a carbonyl compound (248) and phosphine oxide.[38,83] The carbonyl compound (248) can react with excess starting ylid to give an olefin (249) and an additional mole of phosphine oxide.

$$RR^1C{=}PPh_3 + O_2 \longrightarrow RR^1C{=}O + O{=}PPh_3$$

(248)

$$(248) + RR^1C{=}PPh_3 \longrightarrow RR^1C{=}CR^1R + O{=}PPh_3$$

(249)

For R^1 = H, only the end products, a symmetrical olefin and phosphine oxide, are isolable. When starting from ylids containing no α-proton, however, the ketone formed as an intermediate and phosphine oxide are obtained. As a result of the lower reactivity of the initially formed ketones, as compared to aldehydes, the rate of autoxidation is greater than that of the Wittig reactron, so that tetrasubstituted olefins (249) are not produced. They can be obtained, however, if the introduction of oxygen is interrupted after 1/2 mole has been added, and if the reaction mixture is then heated.[83]

Sulfur reacts with phosphine alkylenes in the same way as does oxygen. In this case triphenylphosphine sulfide is formed in addition to symmetrically substituted ethylenes.[397]

The olefin formation in the autoxidation of phosphine alkylenes makes possible, among other syntheses, the preparation of β-carotene from two molecules of vitamin A.[80] By autoxidation of bis(phosphine alkylenes) (250), cycloolefins (251) have been synthesized.[61]

$$(CH_2)_n\begin{cases} CN{=}PPh_3 \\ CH{=}PPh_3 \end{cases} \xrightarrow{O_2} (CH_2)_n\begin{cases} CH \\ \| \\ CH \end{cases} + 2OPPh_3$$

(250) (251)

This procedure also allows the construction of polycyclic compounds.[61,63] In addition to intramolecular ring closure, intermolecular cyclization reactions, which lead to the formation of macrocycles, may be occurring.[92] The intermediate (252) resulting from the reaction of the bis-ylid (250) with oxygen, may react with excess starting ylid (250) to form a new bis-ylid which then undergoes ring closure to the macrocycle (253).

$$(250) + O_2 \xrightarrow{-OPPh_3} (CH_2)_n\begin{matrix} CHO \\ CH{=}PPh_3 \end{matrix} \xrightarrow{-OPPh_3} (CH_2)_n\begin{matrix} CH \\ \| \\ CH \end{matrix}$$

(252) (251)

$$(250) + (252) \longrightarrow (CH_2)_n\begin{matrix} CH{=}CH{-}(CH_2)_n \\ CH{=}PPh_3 \quad CH{=}PPh_3 \end{matrix} \xrightarrow[-2OPPh_3]{O_2}$$

$$(CH_2)_n\begin{matrix} CH{=}CH \\ CH{=}CH \end{matrix}(CH_2)_n$$

(253)

A valuable supplement to the autoxidation consists in peracid oxidation[191,192,315] and in ozone cleavage[492] of ylids. Stable ylids that do not undergo autoxidation with molecular oxygen may be oxidized with peracids and ozone, respectively. Here the same rules apply as for autoxidation.

As additional oxidizing agents the following are suitable: ethyl nitrite,[695] lead tetraacetate, dibenzoyl peroxide, phenyl iodide-diacetate, and lead dioxide,[700] as well as potassium permanganate and osmium tetraoxide.[697] The oxidation with periodate was particularly successful.[45] By this procedure α,β-dicarbonyl compounds (255) can be synthesized, starting from acylalkylenetriphenylphosphorane (254).

$$\underset{O}{\overset{}{RC}}{-}\underset{PPh_3}{CR^1} \xrightarrow{2NaIO_4} \underset{O}{RC}{-}\underset{O}{CR^1} + Ph_3PO + 2NaIO_3$$

(254) (255)

By treatment with a base, phosphonium periodates (256) are converted into triphenylphosphine oxide, an olefin (259), and iodate. The corresponding ylid (257) is first formed from (256) and the base, half of which is oxidized by the two moles of periodate to the aldehyde (258); this then interacts with the other half of (257) to give the olefin (259) in a Wittig reaction. This new method is particularly applicable for converting into olefins those ylids that react with molecular oxygen only with difficulty. For this reason periodate oxidation complements autoxidation in an ideal way.[45]

$$2[RCH_2PPh_3]^{\oplus}\ IO_4^{\ominus} \xrightarrow[-2H^{\oplus}]{B} OPPh_3 + 2IO_3^{\ominus} + RCHO + RCH{=}PPh_3$$

(256) (258) (257)

$$(257) + (258) \longrightarrow RCH{=}CHR$$

(259)

A.2.5. Reactions of Phosphine Alkylenes with Ring Systems

A.2.5.1. Phosphine Alkylenes and Epoxides. Epoxides (260) react with methylenephosphoranes in different ways, depending on their substitutents,[189,193,236a,429,430,630,691]

While ethoxycarbonylmethylenetriphenylphosphorane interacts with styrene oxide to yield 2-phenylcyclopropanecarbonic ester,[193] the same epoxide reacts with benzylidenetriphenylphosphorane to give cis-1,3-diphenylpropene,[691] and finally with benzylidenemethylethylphenylphosphorane to produce a zwitterion that decomposes into methylethylphenylphosphine and benzylacetophenone.[430] Trippett[638] formulated all three reaction possibilities via an open-chain, zwitterionic adduct (261) and a 1,2-P(V)-oxaphospholidine (262) as intermediates.

$$R^1_3P{=}CHR^2 + R^3CH\overset{O}{\frown}CHR^4 \longrightarrow R^1_3\overset{\oplus}{P}\text{-}CHR^2\text{-}CHR^3\text{-}\underset{}{\overset{|\overline{O}|^{\ominus}}{C}}HR^4$$

(260) (261)

The elimination of R^1_3P from (261), followed by a 1,3-hydrogen transfer to give (265a), is favored if the ring closure to (262) is rendered more difficult by electron-donating groups on the phosphorus atom.

If the group R^2 is able to stabilize a negative charge, cyclopropanes (265b) are formed via the intermediate (263).

The cleavage of (262) at the C-O bond, followed by a hydride transfer (or transfer of R^3 as anion)[236a,691] leads to olefins (265c) with elimination of phosphine oxides via the intermediate (264).

Carbethoxymethylenetributylphosphorane reacts with a series of epoxides to give a,β-unsaturated esters.[236a] With cyclohexene oxide and cycloheptene oxide a ring contraction by one carbon atom results as a consequence of the transfer of R^3 as anion.

Huisgen and Wulf[330] postulate a further decomposition

(261) ⇌ R^1_3P–O–CHR^4–CHR^3–CHR^2 (262) → $R^1_3P^{\oplus}$ O CHR^4 $CHR^2{}^{\ominus}$— CHR^3 (263)

(261) → R^1_3P + R^2CH_2-CHR^3-CR^4(=O) (265a)

(263) —OPR_3→ R^2HC–CHR^4–CHR^3 (265b)

(262) → R^1_3P $\overline{O}|^{\ominus}$ $CHR^4{}^{\oplus}$ H CHR^2—CR^3 (264) —$-R^1_3P{=}O$→ $R^2CH{=}CR^3$-CH_2R^4 (265c)

mechanism of 1,2-P(V)-oxaphospholidines (262). According to these authors, the cycloadduct (266), obtainable from styrene oxide and isopropylidenetriphenylphosphorane, undergoes a P-O heterolysis to give (267), from which the unsaturated alcohol (268) is produced by intramolecular E_2-elimination via a six-membered cyclic intermediate.

H Ph C CH_2 O MeC PPh_3 CH_3 (266) → H Ph C CH_2 $\overline{O}|^{\ominus}$ H MeC—CH_2 $Ph_3P^{\oplus}$ (267) → H Ph C CH_2 OH MeC=CH_2 (268)

Whereas in the case of stable ylids only the end products are obtainable, the isolation of the cyclic intermediate has been achieved with a series of strongly basic ylids.[50,85,330] Thus Huisgen isolated the cycloadducts (266) and (269). On pyrolysis (269) is converted into propiophenone *via* elimination of triphenylphosphine and a 1,3-hydride transfer.

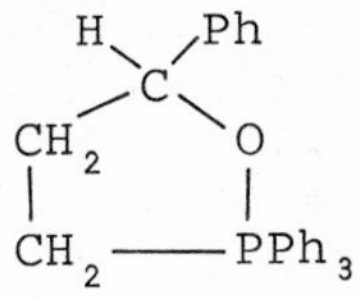

(269)

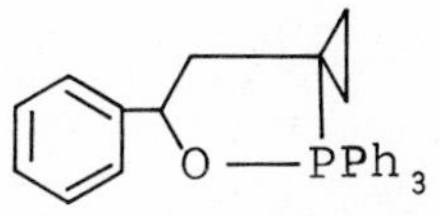

(272)

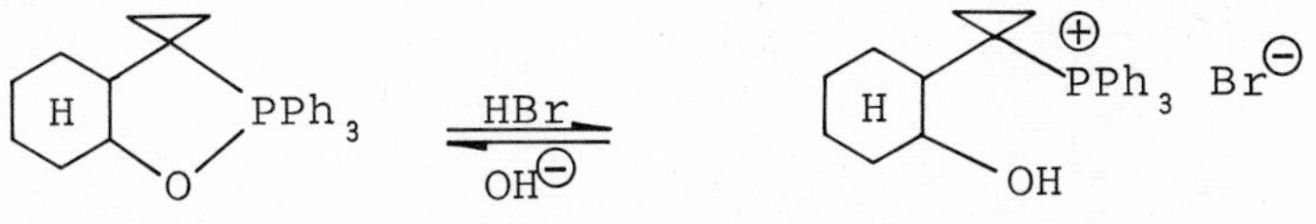

(270) $\underset{OH^{\ominus}}{\overset{HBr}{\rightleftharpoons}}$ (271)

Bestmann et al.[50] obtained, from cyclohexene oxide and cyclopropylidenetriphenylphenylphosphorane, the distillable cycloadduct (270), whose five-membered ring can be opened with HBr at the P-O bond affording (271), which can be closed again with bases. Treatment with methyl iodide causes ring opening to give the corresponding O-methylated phosphonium salt. Interaction of cyclopropylidenetriphenylphosphorane and styrene oxide produced the cycloadduct (272).[50]

A.2.5.2. Ylids and Aziridines. The aziridine ring (273) is also cleaved by phosphine alkylenes (274).[275] Depending on whether or not the ylid used still contains a proton, one obtains a new ylid (275) or, after splitting off phosphine oxide, the pyrrolin (276).

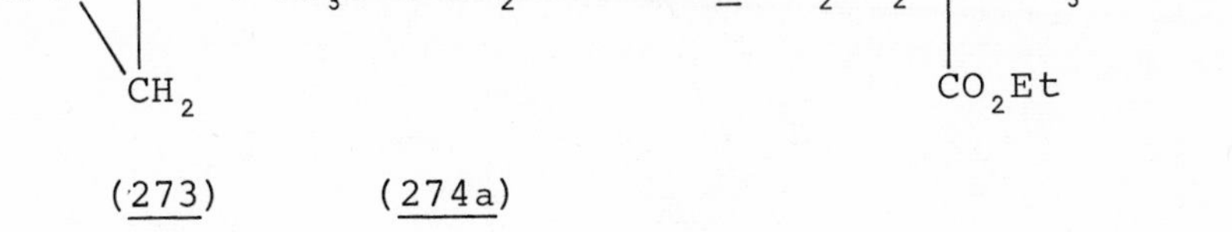

$$\text{R-N}(CH_2)_2 \; (273) + Ph_3P{=}CHCO_2Et \; (274a) \longrightarrow R\text{-}\overset{\ominus}{\underline{N}}\text{-}CH_2CH_2\text{-}\overset{\oplus}{C}HPPh_3(CO_2Et) \longrightarrow$$

$$RNHCH_2CH_2\text{-}C(CO_2Et){=}PPh_3$$

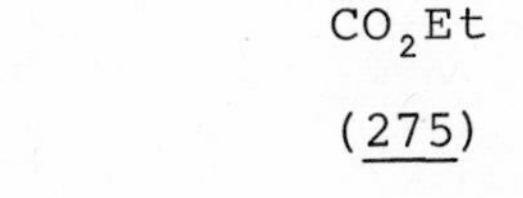

(275)

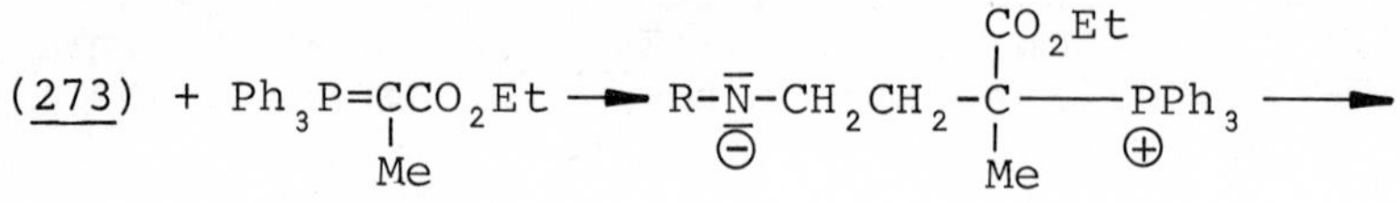

$$(\underline{273}) + Ph_3P{=}\underset{Me}{C}CO_2Et \longrightarrow R{-}\bar{N}^{\ominus}{-}CH_2CH_2{-}\underset{Me}{\overset{CO_2Et}{C}}{-}\overset{\oplus}{P}Ph_3 \longrightarrow$$

(274b)

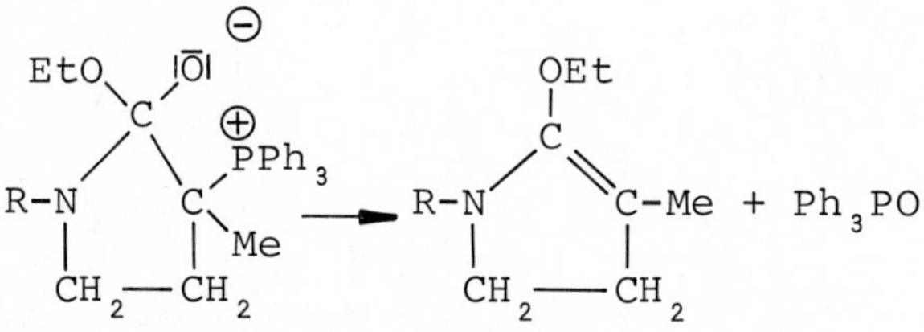

(276)

A.2.5.3. Phosphine Alkylenes and Pyrylium Salts. Pyrylium salts (277) react with phosphine alkylenes (278) to give phosphonium salts (279), which stand in a transylidation equilibrium with a second mole of ylid (278). The resulting new ylid (280) is converted, by an intramolecular Wittig reaction of its tautomeric form (281), into the aromatic compound (282).[407]

In a similar type of reaction methylenetriphenylphosphorane and pyrylium salts produce azulenes.[202]

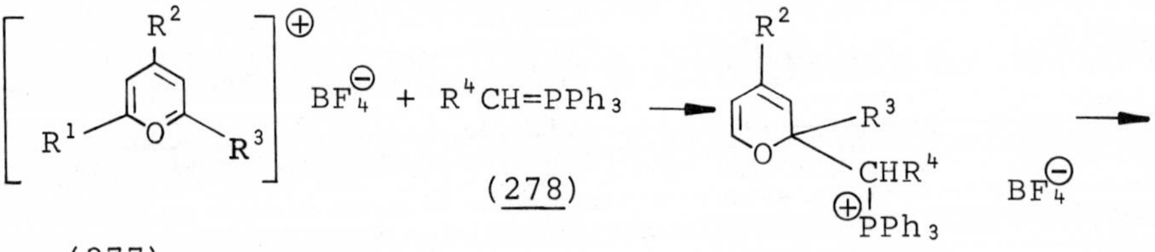

$$(\underline{277})\ BF_4^{\ominus} + R^4CH{=}PPh_3\ (\underline{278}) \longrightarrow (\underline{279})\ BF_4^{\ominus} \longrightarrow$$

(277)

(278)

(279)

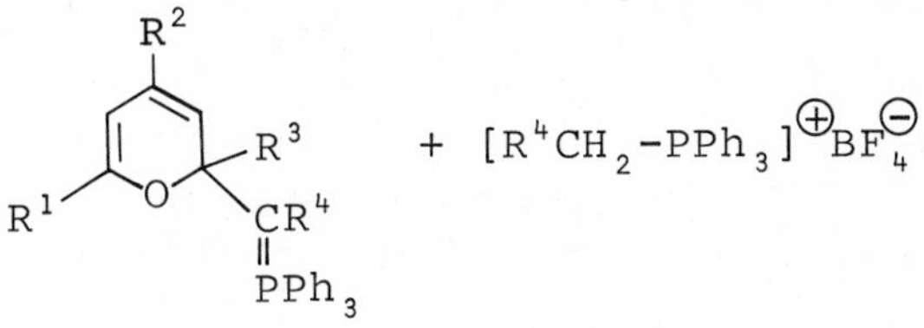

R^2, R^3, R^1, O, CR^4, PPh_3 + $[R^4CH_2-PPh_3]^{\oplus}BF_4^{\ominus}$

(280)

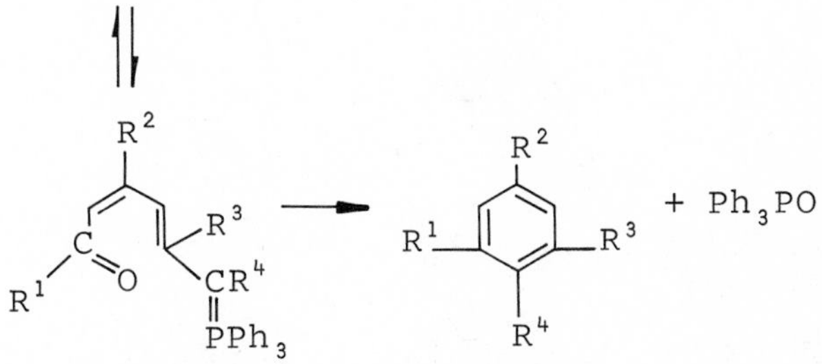

R^2, R^3, C, O, R^1, CR^4, PPh_3 → R^2, R^1, R^3, R^4 + Ph_3PO

(281) (282)

A.2.5.4. Phosphine Alkylenes and Enol Lactones. Treatment of ylids with cyclic enol lactones (283) results in the formation of enolate betaine (284), which, by a proton transfer, is converted into the ketophosphorane (285). By an intramolecular Wittig reaction this phosphorane (285) yields the α,β-unsaturated cyclic ketone (286).[280]

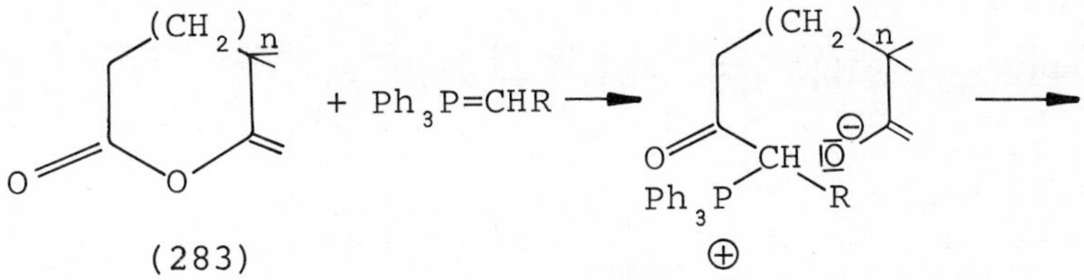

$(CH_2)_n$, O, O + $Ph_3P{=}CHR$ → $(CH_2)_n$, O, CH, $\overline{O}^{\ominus}$, $Ph_3P^{\oplus}$, R →

(283) (284)

$(CH_2)_n$, $^{\ominus}\overline{O}$, C, O, R, $PPh_3^{\oplus}$ → $(CH_2)_n$, O, C, R + Ph_3PO

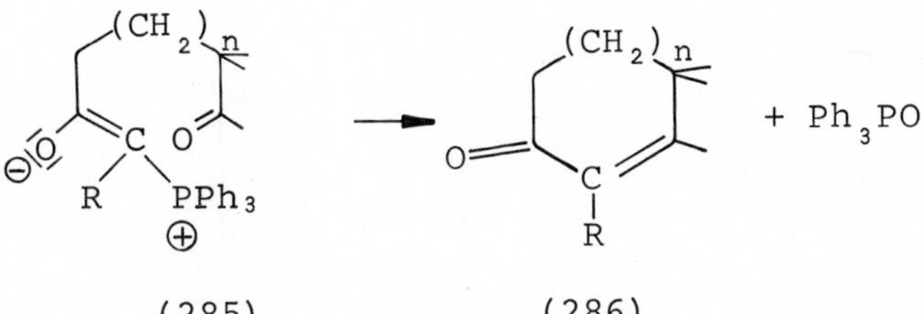

(285) (286)

This procedure is used in the synthesis of steroids.[280]

A.2.6. Reactions of Phosphine Alkylenes with 1,3-Dipolar Compounds

A.2.6.1. Phosphine Alkylenes and Azides. The reaction path of azides with phosphine alkylenes is determined by the group R of the phosphine alkylene, as well as the substituent R^1 of the azide.

Phenylazide (288) reacts with benzylidenetriphenylphosphorane (287) with evolution of nitrogen and formation of benzalaniline and N-phenyltriphenylphosphinimine (291).[289]

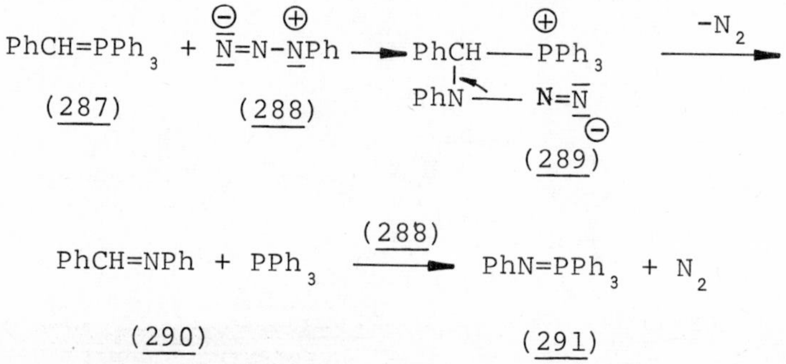

$$PhCH{=}PPh_3 + \overset{\ominus}{\bar{N}}{=}N{-}\overset{\oplus}{\underline{N}}Ph \longrightarrow \begin{matrix} PhCH{-}\overset{\oplus}{P}Ph_3 \\ | \qquad\quad | \\ PhN{-}N{=}\overset{\ominus}{\bar{N}} \end{matrix} \xrightarrow{-N_2}$$

(287) (288) (289)

$$PhCH{=}NPh + PPh_3 \xrightarrow{(288)} PhN{=}PPh_3 + N_2$$

(290) (291)

It is assumed that first a betaine (289) is formed; this decomposes with N_2 evolution into the Schiff's base (290) and triphenylphosphine. The latter reacts then with another mole of azide (288) to form the phosphinimine (291). If the azide (288) is used in substoichiometric quantities, triphenylphosphine is isolable.

The interaction of acylmethylenetriphenylphosphorane (292) and azides (293) leads either to α-diazocarbonyl compounds (295) and a phosphinimine (296) (path a), or, via elimination of phosphine oxide, to 1,5-disubstituted 1,2,3-triazoles (298) (path b).[1,2,271b]

Although Harvey[271b] proposes a stepwise mechanism for the formation of the cyclic intermediates (294) and (297), kinetic investigations by L'Abbé[2] indicate a synchronously occurring cycloaddition of azides to ylids.

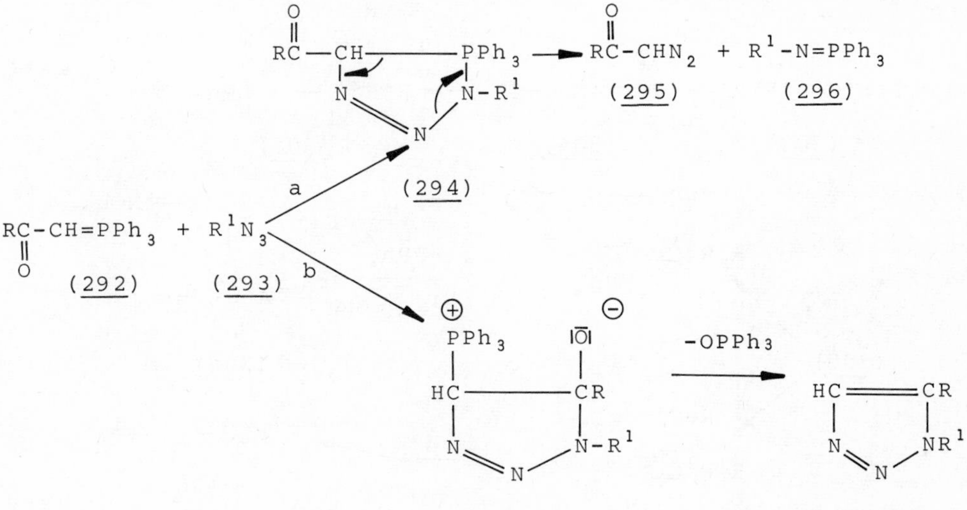

$$RC(=O)\text{-}CH=PPh_3 \ (292) + R^1N_3 \ (293) \xrightarrow{a} (294) \longrightarrow RC(=O)\text{-}CHN_2 \ (295) + R^1\text{-}N=PPh_3 \ (296)$$

$$(292) + (293) \xrightarrow{b} (297) \xrightarrow{-OPPh_3} (298)$$

(292) (293) (294) (295) (296) (297) (298)

A.2.6.2. Ylids and Nitrones. Phospine alkylenes (299) add to nitrones (300) to give a new heterocyclic system, i.e., 1,2,5-P(V)-oxazaphospholidines (301).[689a] It has been proposed that the addition proceeds by a one-step multiple-center process.[329]

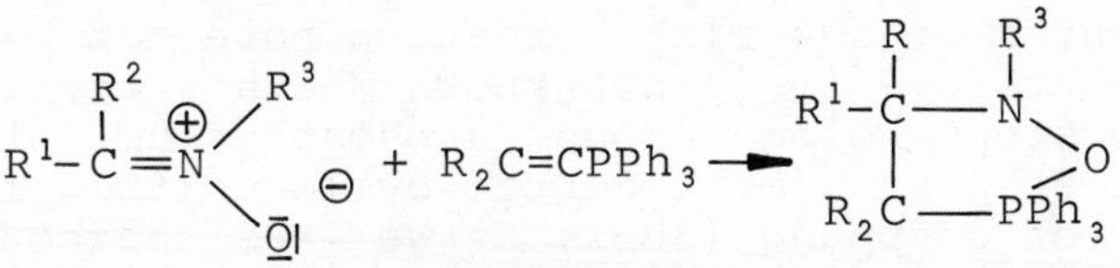

$$R^1\text{-}C(R^2)=\overset{\oplus}{N}(R^3)\text{-}\overset{\ominus}{O} + R_2C=CPPh_3 \longrightarrow (301)$$

(300) (299) (301)

The thermolysis of the cycloadduct from diphenylnitrone and ylids formally occurs by elimination of benzyne, C_6H_4. Substituted phosphine oxides (302) are obtained as reaction products.

$$PhNHCHPhCR_2\overset{O}{\overset{\|}{P}}Ph_2$$

(302)

A.2.6.3. Ylids and Nitrile Oxides. Phosphine alkylenes (303) and nitrile oxides (304) interact to give 4,5-dihydro-1,2,5-P(V)-oxazaphospholenes (305), which decompose thermally.[84,85,325,328,648]

$$R^1R^2C{=}PPh_3 + RC{\equiv}\overset{\oplus}{N}{-}\overset{\ominus}{O} \longrightarrow \text{(305)}$$

(303) (304) (305)

(305): five-membered ring R–C(–N–O–PPh$_3$–CR^{1}R^2–)

(305) → $RC{=}N{-}O^{\ominus}$ / $R^1R^2C{-}\overset{\oplus}{P}Ph_3$ $\xrightarrow[R^1=R^2=Me]{-PPh_3}$ $RC(=NOH){-}C(Me){=}CH_2$ (308)

(305) → $R{-}C{=}N{-}O{-}\overset{\oplus}{P}Ph_3$ / $R^1R^2\overset{\ominus}{C}$ → $R^1R^2C{=}C{=}NR$ (306) + azirine (307)

The mechanics of the ring formation have not been investigated; therefore it is not known whether the reaction occurs in two steps going through a phosphonium intermediate or in one step as a multicenter cycloaddition reaction. The stability of compound (305), as well as the course of its decomposition, is dependent on the substituents R, R^1, and R^2.[85]

Groups R^1 and R^2 with -I and -M effects cause the formation of ketene imines (306), while electron-donating groups R^1 and R^2 effect the formation of azirines (307). In both cases phosphine oxide is also split off. The ketene imines undergo subsequent reactions with excess phosphorane.

If R exhibits a -I and at the same time R^1 and R^2 give a +I effect, then triphenylphosphine and α,β-unsaturated oximes (308) are isolated as subsequent products of the decomposition.[85]

A.2.6.4. Phosphine Alkylenes and Nitrile Imines or Azomethine Ylids. Whereas the addition of nitrones or nitrile oxides[85,328] to phosphine alkylenes produces cyclic products, the reaction of ylids with nitrile imines or azomethine ylids yields open-chain 1:1 adducts.[330]

Interaction of the nitrile imine (309) with alkoxycarbonylmethylenetriphenylphosphorane (310) gives the ylid

(312), which on thermolysis forms the pyrazolone ring (313), apparently going through the intermediate (311).

$$\overset{\oplus}{RC\equiv N}-\overset{\ominus}{\underline{N}}Ph + R^1O_2CCH{=}PPh_3 \longrightarrow$$

(309) (310)

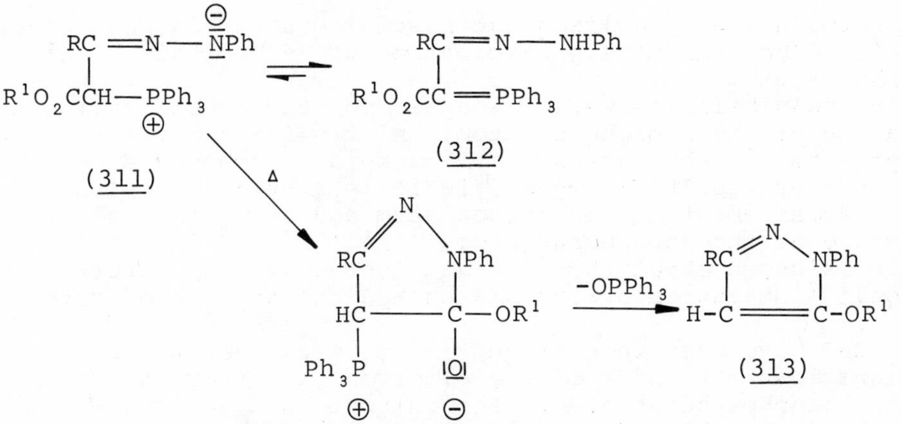

$$\underset{(311)}{\begin{matrix} RC = N - \overset{\ominus}{\underline{N}}Ph \\ | \\ R^1O_2C\underset{\oplus}{H}-PPh_3 \end{matrix}} \rightleftharpoons \underset{(312)}{\begin{matrix} RC = N - NHPh \\ | \\ R^1O_2CC = PPh_3 \end{matrix}}$$

(311) $\xrightarrow{\Delta}$ [ring intermediate: N=RC–CH(PPh_3^⊕)–C(OR^1)(O^⊖)–NPh] $\xrightarrow{-OPPh_3}$ (313) [pyrazole ring: RC=N–NPh–C(OR^1)=C(H)–]

(313)

A.2.6.4. Phosphine Alkylenes and Diazo Compounds. Ylids (314) react with diazo aliphatics (315) to form azines (316) and triphenylphosphine.[399,684]

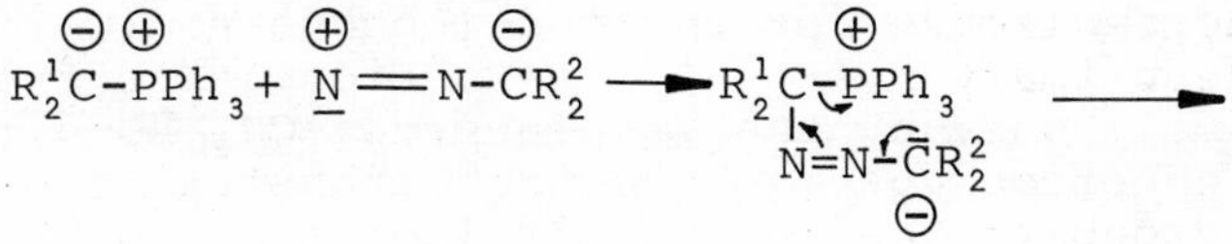

$$R^1_2\overset{\ominus}{C}-\overset{\oplus}{P}Ph_3 + \overset{\oplus}{\underline{N}}{=}N-\overset{\ominus}{C}R^2_2 \longrightarrow R^1_2C-\overset{\oplus}{P}Ph_3 \;(N{=}N-\overset{\ominus}{C}R^2_2) \longrightarrow$$

(314) (315)

$$R^1_2C{=}N-N{=}CR^2_2 + PPh_3 \xrightarrow{(315)} R^2_2C{=}N-N{=}PPh_3$$

(316) (317)

The resulting phosphine reacts with excess diazo compounds tc give the phosphazene (317).

A.3 Physical Properties of Phosphine Alkylenes

Phosphine alkylenes, including the recently synthesized alkylidenetrialkylphosphoranes,[368a,535] are thermodynamically stable compounds when air and humidity are excluded. (Alkylenetriphenylphosphoranes can be decomposed photolytically,[447a] a process wherein primarily a cleavage of the P-Ph bond occurs.) The electronic structure of phosphine alkylenes is generally described as a resonance hybrid:

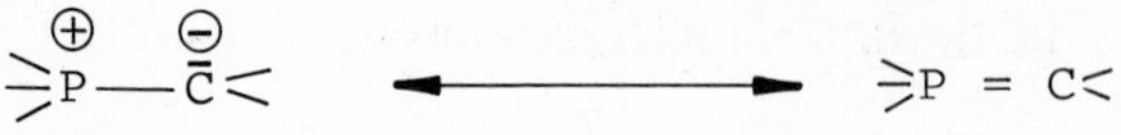

$$\geqslant \overset{\oplus}{P} - \overset{\ominus}{\bar{C}} < \quad \longleftrightarrow \quad \geqslant P = C <$$

(A) (B)

This formulation symbolizes the partial double-bond character of the P-C bond and the participation of 3d-orbitals of the phosphorus atom.

Two possibilities were discussed for the hybridization situation at the phosphorus atom, and two for the ylid carbon atom[347] (phosphorus: sp^3 or sp^3d, carbon: sp^2 or sp^3). Newer results, however, indicate a trigonal, planar ligand arrangement at the carbon atom and a tetrahedral structure at the phosphorus atom.[27,161,417,535,603,617,662]. Little is known about the type and degree of $p\text{-}d_\pi$ interaction.[347] Measurements by ESR on radical cations of type $Ph_3\overset{+}{P}-\dot{C}HR$ have been interpreted[390] to mean that there is no significant transfer of the electron pair from the carbon to the phosphorus atom. According to all the results presently available, phosphine alkylenes are to be considered first of all as stabilized carbanions.

A theoretical treatment of the bonding situation in phosphine alkylenes, using semiempirical molecular orbital calculations, is being carried out by Hoffmann et al.[285] These authors discuss for the model compound (C), in addition to the primary $d_{xz}\text{-}p_x\text{-}\pi$ interaction, a contribution, albeit a small one, by the in-plane $d_{yz}\text{-}CH_2$ bond, comparable to a hyperconjugative electron transfer from the methylene group to the $3d_{yz}$-orbital of the phosphorus atom.

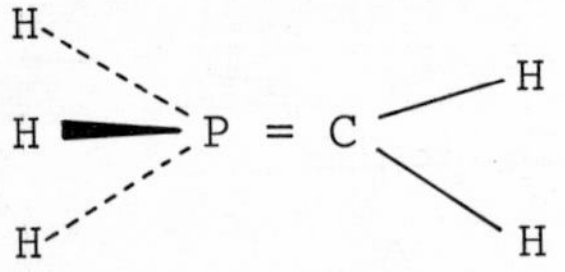

$$H_3P = CH_2 \qquad\qquad Ph_3P = \overset{1}{C}<\begin{matrix}\overset{2}{CH_2}\\ |\\ \underset{3}{CH_2}\end{matrix}$$

(C) 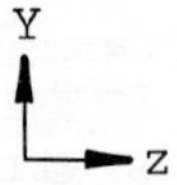 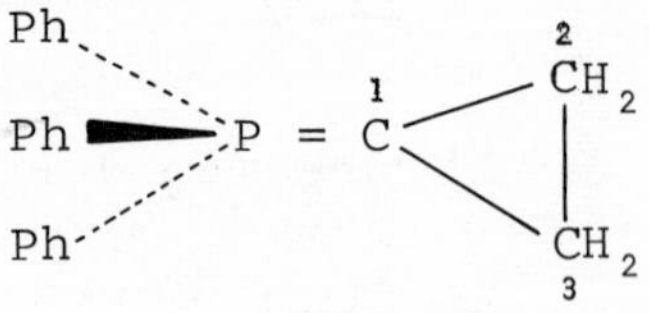 (D)

This electron transfer leads, in the case of cyclopropylidenetriphenylphosphorane, in addition to increased P-C bond strength, to a weakening of the $C_1\text{-}C_2$ and $C_1\text{-}C_3$ bonds and a strengthening of the $C_2\text{-}C_3$ bond.[285a]

The partial double-bond character of the P-C bond is also indicated by the P-C bond length of 1.66 to 1.74 Å.[27,161,181,182,417,603,617,662]. This value lies between the estimated lengths of a double bond (~1.66 Å[467]) and of a single bond (1,863 Å[369]).

X-ray structural investigations[603] of α-ketoylids (E) confirm the participation of mesomeric forms (F), in which the negative charge is localized at the oxygen atom. The C-O bond is considerably longer than a double bond, and the C-C distance of 1.35 to 1.36 Å is considerably shorter than a C-C single bond.

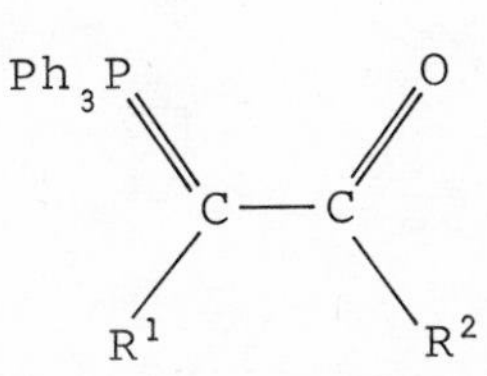

(E) 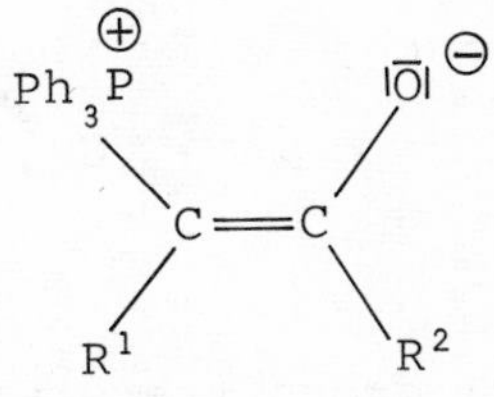 (F)

In the strongly basic methylenetriphenylphorane the P-C (ylid) distance (1.66 Å)[27] is considerably shorter than in resonance-stabilized ylids.[603,617,662] The decreased double-bond character of the P-C bond in the latter is caused by a delocalization of the negative charge. The decreased double-bond character of the CO group in α-ketoylids is also indicated by the carbonyl frequencies in the IR spectra.[4,47,100,163,429,572,574,600,685]

Investigations by ^{1}H-NMR of stable ylids of type (E) point to cis-trans isomerism of the mesomerism-conditional C-C double bond.[72,110,177,178,494,572,574,703] The influences of the substituents of the ylid carbon atom on the isomer ratio and on the coalescence temperature have been investigated extensively in the case of substituted methoxycarbonylmethylenetriphenylphosphorane.[703] Whereas in ester ylids (E, R^2= OR) and formylmethylenetriphenylphosphorane[594] cis-trans isomers are present, in ketoylids (E, R^2= R) one can detect, by NMR spectroscopic investigations,[704] only one form, i.e., the form in which the phosphorus and the keto oxygen are cis to each other.

In addition to the hindered rotation, as well as the occurrence of geometrical isomers, a further phenomenon, dependent on the temperature and solvent, was observed by ^{1}H-NMR spectroscopy. The spin-spin coupling between the phosphorus atom and a proton on the ylid carbon atom (α-proton) is temperature dependent.[89,110,177,494,536]

A fast, reversible proton exchange has been assumed to be the cause of this effect, which is catalyzed by a trace of acid. Methylenetriphenylphosphorane, $Me_3P=CH_2$, exhibits only one proton NMR signal at elevated temperature. In addition to the proved, acid-catalyzed, intermolecular mechanism, a purely intramolecular exchange process is also conceivable, one that eventually goes through the transition state (G) with pentacovalent phosphorus. Theoretical calculations indicate, however, that the intermolecular hydrogen transfer is preferred.[285]

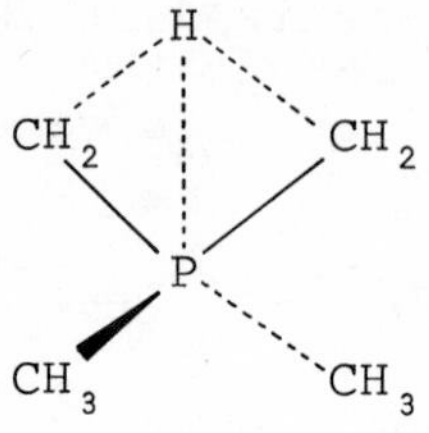

H
CH_2 CH_2
P
CH_3 CH_3

(G)

The silyl group transfer in silyl-substituted phosphorus ylids[531] is formally comparable to the proton exchange.

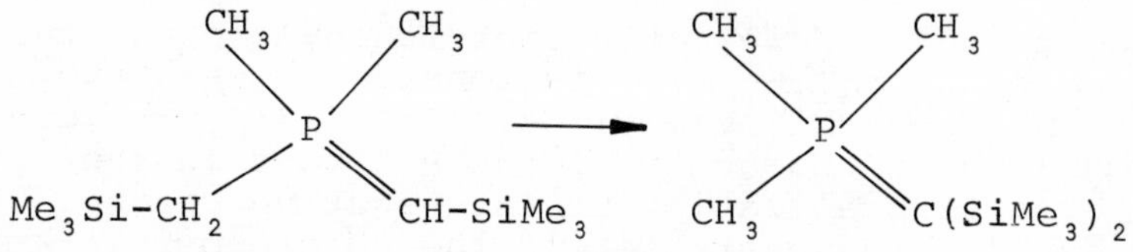

$$(CH_3)_2P(CH_2\text{-}SiMe_3)=CH\text{-}SiMe_3 \longrightarrow (CH_3)_3P=C(SiMe_3)_2$$

The goal, to clarify the electronic situation in phosphine alkylenes by means of ^{31}P-NMR spectroscopy, was not accomplished.

The comparatively small number of known values shows ^{31}P chemical shifts between δ = -6 to -25 ppm (for alkylidenetriphenylphosphoranes)[257,347] and δ = +10.2 to -40.5 ppm (for alkylidenetrialkylphosphoranes).[368a] Ramirez and Smith[493] showed that the phosphorane discovered by Mark[422] and having a value of δ = +66.5 ppm has the structure not of a phosphorus ylid, but rather of a phosphorane containing a pentacovalent phosphorus atom.

Whereas the chemical shifts of resonance-stabilized alkylidenes differ only insignificantly from those of their corresponding phosphonium salts,[347] the corresponding ^{31}P chemical shifts of unstable ylids differ to a larger extent.[257] Investigations by NMR spectroscopy of metallated alkylidenetrialkylphosphoranes (H) suggest a strong bonding relation of the metal atom with the ylid carbon atom, as indicated by formula (I) or (J).[536b]

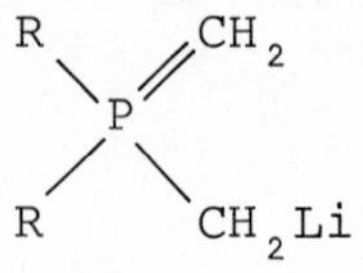

$$R_2P(=CH_2)CH_2Li \qquad R_2P(CH_2)_2Li \qquad R_2P(CH_2\text{-}Li\text{--}CH_2)(CH_2\text{--}Li\text{-}CH_2)PR_2$$

(H) 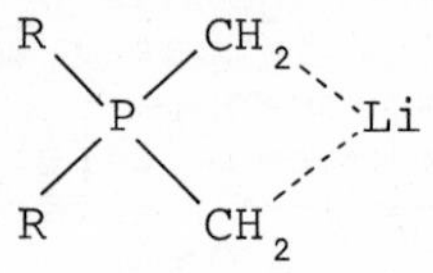 (I) 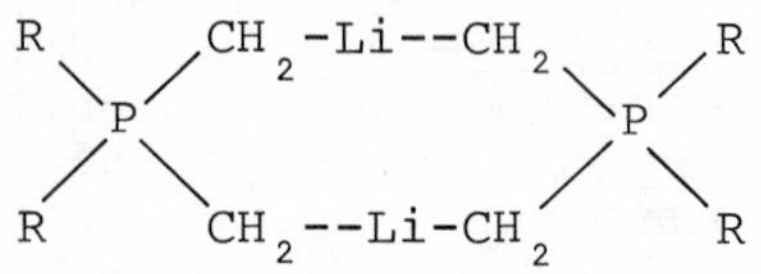 (J)

On the basis of NMR measurements a tetrahedral valence arrangement at the ylid carbon atom, with a high portion of covalent C-Li bond, is assumed for lithium halide adducts of methylenetrialkylphosphoranes.[536b]

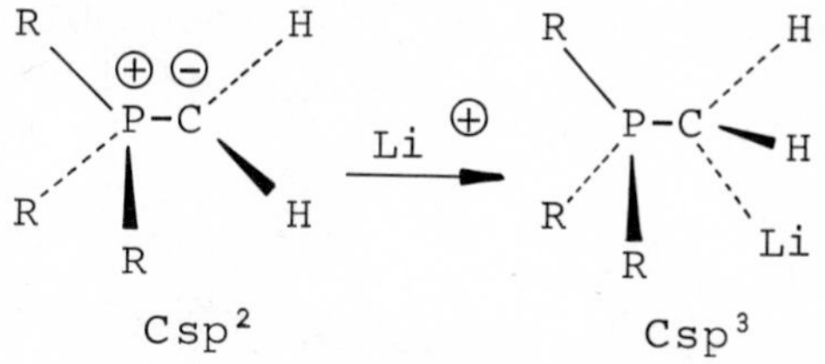

Csp^2 $\xrightarrow{Li^{\oplus}}$ Csp^3

In the IR spectra of alkylenetriarylphosphoranes four characteristic bands are observed (1450 to 1425 cm^{-1}, 1220 to 1100 cm^{-1}, 1010 to 990 cm^{-1}, 730 to 710 cm^{-1}), which are typical for a tetrahedrally configurated phosphorus atom with the PAr_3 grouping. In resonance-stabilized ylids, such as cyclopentadienylidenetriphenylphosphorane, a characteristic absorption in the range of 1230 to 1180 cm^{-1} is observed,[310,629] in addition to the four bands mentioned above.

From IR spectroscopic investigations of isotopically labeled compounds it follows that the bonding order is only a little higher in phosphine alkylenes than in the corresponding quaternary salts.[392] A crude estimate yields a value of 1.3. For the P=C valence frequency of $Ph_3P{=}CH_2$, a value of 900 cm^{-1} has been given (in $[Ph_3PCH_3]I$; P-C is found at 787 cm^{-1}).

Phosphine alkylenes exhibit in the UV-to-visible region two main absorptions:[258,348] one band centered at 265 mμ, which has been ascribed to the benezene absorption, and a second band in the region 340 to 400 mμ. Methylenetriphenylphosphorane exhibits a characteristic absorption at 341 mμ (~3.6 eV).

Alkyl substitution at the ylid carbon atom of unstable ylids effects a bathochromic shift to longer wavelengths.[258] In the case of resonance-stabilized phosphine alkylenes the bathochromic shift increases with increasing charge distribution.[218] Whereas in the case of resonance-stabilized ylids the substituent on the methylene carbon atom is the carrier of the chromophor, the chromophor in nonstabilized ylids is considered to be the P-C bonding system.[258]

The dipole moments of phosphine alkylenes indicate that these compounds are polar molecules.[348,479,483] The basicity of phospine alkylenes may be quantitatively expressed by the pK_a values of their corresponding acids (i.e., phosphonium salts).[336] The available pK_a values are in accord with the basicity scale of ylids[336] established by other criteria (transylidation,[40] strength of the base necessary to deprotonate the phosphonium salt[81,339]). Phenyl-substituted phosphonium salts are distinguished by their greater acid strength in comparison to the ethyl- or cyclohexyl derivatives.[81]

Experiments with deuterium-labeled compounds indicate that electron-induced decomposition of phosphine alkylenes (K) in the mass spectrograph occurs presumably through the formation of phosphafluorenyl ions (L) as intermediates.[168, 671]

$$(D\text{-}C_6H_2D_2)_3P{=}CH_2 \xrightarrow[-D]{-e} \text{[phosphafluorenyl cation, } D_3H_2C_6 \text{ and } CH_2D \text{ on } P^{\oplus}]$$

(K) (L)

Phosphine alkylenes containing a proton on the α-carbon atom, which is made reactive by a neighboring carbonyl group, can be determined quantitatively with methyl magnesium iodide.[205a]

B. OTHER PHOSPHORUS YLIDS

If one considers the phosphine alkylenes as stabilized carbanions, whose stability results from the interaction of the electron pair on the ylid carbon atom with the 3d-orbitals of the phosphorus atom, then this possibility should exist also in regard to other phosphorus-containing carbanions.

I. PHOSPHINE OXIDE CARBANIONS

By treatment with strong bases phosphine oxides (1) can be converted into their anions (2) which afford, with carbonyl compounds (3), an olefin and a phosphinate anion.[287,300,304] Triphenylphosphine oxide is also suitable as a starting material for various phosphine oxide carbanions, since on treatment with lithium alkyls a ligand exchange ensues and the resulting diphenylalkylphosphine oxides are then deprotonated to the corresponding anions.[571]

The mechanism and stereochemistry of olefin formation from phosphine oxide anions (2) and carbonyl compounds (3) are essentially the same as in the Wittig reaction. The carbanion and carbonyl compound react to give isolable betaines (4), which may be present in two diastereomeric

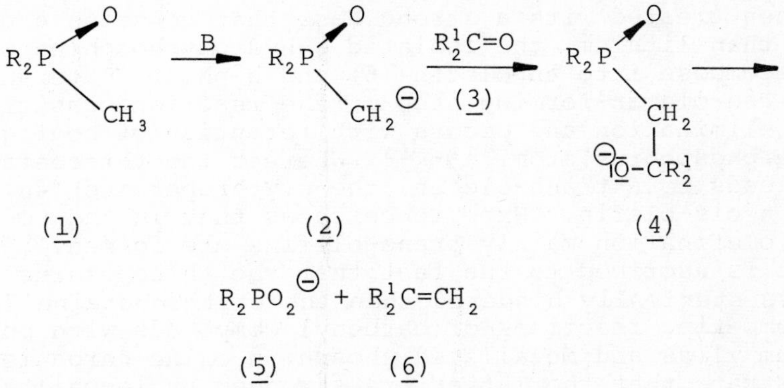

$$R_2P(\rightarrow O)-CH_3 \xrightarrow{B} R_2P(\rightarrow O)-CH_2^{\ominus} \xrightarrow[(3)]{R^1_2C=O} R_2P(\rightarrow O)-CH_2-CR^1_2-\overline{\underline{O}}|^{\ominus} \longrightarrow R_2PO_2^{\ominus} + R^1_2C=CH_2$$

(1) (2) (4) (5) (6)

forms, (4a) and (4b). The diastereomeric pair was separated by Horner and Klink.[306] The betaine formation is a

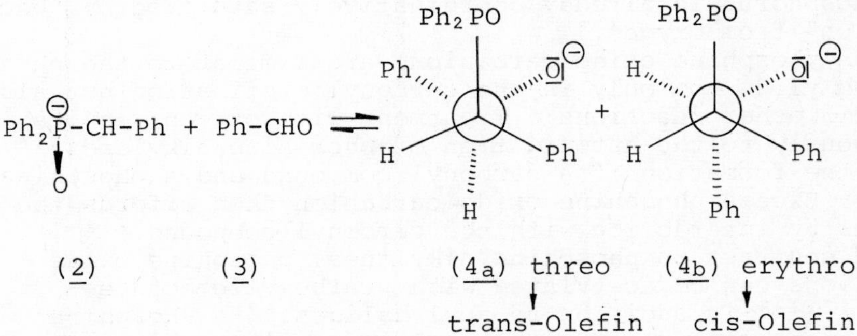

$$Ph_2\overset{\ominus}{P}(\rightarrow O)-CH-Ph + Ph-CHO \rightleftharpoons$$

(2) (3)

(4a) threo → trans-Olefin

(4b) erythro → cis-Olefin

reversible process, so that the two diastereomers, (4a) and (4b), are in equilibrium through the starting products.[306] In contrast to the primary product of the Wittig reaction, the primary adduct (4) easily is interceptable.[301]

The behavior of metallated phosphine oxides is codetermined, to a large extent, by the metal cations of the bases used for the deprotonation. Often the reaction of metallated phosphine oxides with a carbonyl compound stops at the betaine stage (4), particularly if the cation used is lithium. Addition of a proton to (4) affords hydroxyphosphine oxides (7).[301,509]

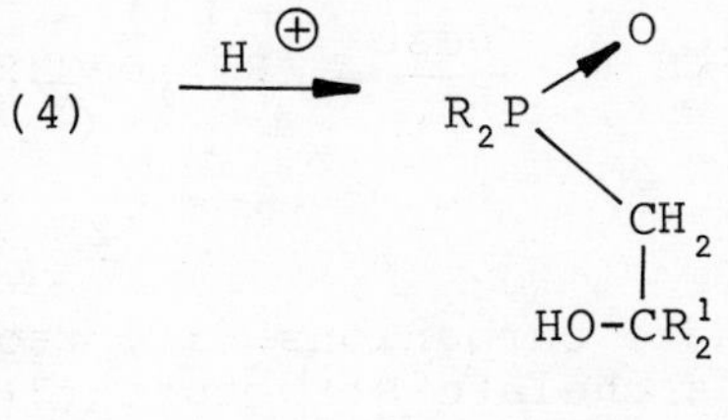

$$(4) \xrightarrow{H^{\oplus}} R_2P(\rightarrow O)-CH_2-CR^1_2-OH$$

(7)

When treated with a strong base that contains a cation other than lithium, the isolated β-hydroxyphosphine oxides (7) decompose into an olefin (6) and a phosphinate anion (5). The olefin-forming step of the reaction consists in a cis elimination and occurs with retention of configuration at the phosphorus atom.[306,313] Whereas the threobetaine (4a) leads to a trans-olefin, the erythrobetaine (4b) produces a cis-olefin. Experience shows that in the PO-activated olefination mainly trans-olefins are formed.[306] This result is ascribed to the fact that the threobetaine (4a) is less sterically hindered than the erythrobetaine (4b).

Competing reactions of carbonyl compounds with phosphonium ylids and metallated phosphine oxide carbanions demonstrate that the latter are stronger nucleophiles.[305] The reason for this is the fact that the negative charge is stabilized to a larger extent by the phosphonium group than by the phosphine oxide residue, because in the latter case the phosphorus is already coordinatively saturated by "back donation" from oxygen.[347]

The phosphine oxide carbanions are similar to the phosphonium ylids not only in the carbonyl olefination but also in their other reactions. Treatment with oxygen results, analogously to the autoxidation of phosphine alkylenes,[38,83] in the formation of a carbonyl compound and a phosphinate anion. Excess phosphine oxide carbanion then affords the olefins by interaction with the carbonyl compound.[302]

In contrast to phosphine alkylenes, phosphine oxide carbanions can be acetylated with carboxylic acid esters,[287,374] in addition to acyl halides.[374] Phosphine oxide carbanions react with epoxides to give cyclopropanes.[303,429] With optically active styrene oxide the reaction occurs at the asymmetric carbon atom with retention of configuration.[631]

II. PHOSPHONATE CARBANIONS

The activation of methylene groups alpha to phosphorus in phosphonates (1) also allows the preparation of phosphorus-containing carbanions (2).[20,21]

$$(RO)_2\overset{O}{\overset{\uparrow}{P}}-CH_2R^1 \xrightarrow{\text{Base}} (RO)_2\overset{O}{\overset{\uparrow}{P}}-\overset{\ominus}{C}HR^1$$

(1) (2)

Acetonylphosphonate carbanions (3), isolable as sodium or zinc salts, have a chelate structure (3) (comparable to that of acetylacetonate), according to IR, UV, and ^{31}P spectra.[179]

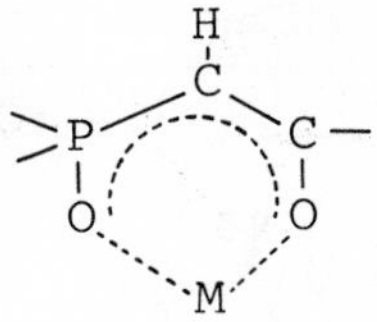

H
C
>P C—
O O
M

(3)

Like phosphonium ylids and phosphine oxide carbanions, the phosphonate carbanions also react with carbonyl compounds to yield olefins and dialkyl phosphates.[300,650] By using substituted phosphonate ester carbanions, the following compounds, among others, have been synthesized in this way: acrylic acid esters,[142,208,372,627,638] α,β-unsaturated ketones,[454] acrylonitrile derivatives,[142 208] bisolefins,[304] α-chlorostilbenes,[708] enol ethers[253] (of α-ketoesters, α-ketocarboxylic acid amides, 1,2-diketones, and phenyl ketones), heterocyclic alkenes,[28,367] vitamin A derivaties,[474] enamines,[262,707] cyclic dienes,[663] 1,2-diketones,[262] and α,β-unsaturated aldehydes.[447,454]

The mechanism and stereochemistry of the reactions of phosphonate ester carbanions are presumably analogous to those of the Wittig reaction. Trans-olefins are formed to a larger extent, however, than in the Wittig reaction.[651] Attempts to increase the portion of cis-olefin have not proved very successful[31,651]. Only Jones and Maisey[356] observed a preferred formation of cis-olefins in the reaction of diethylcyanomethylphosphonate carbanions (4) with a series of alkyl phenyl ketones (5).

Although in the case of unbranched alkyl groups R^2 and of phenyl groups R^1, unsubstituted in the ortho position, predominantly trans-olefins (7a) are formed, the introduction of secondary or tertiary alkyl groups R^2 and/or ortho-substituted phenyl groups R^1 makes possible an increase in the portion of cis-olefin (7b). This result agrees with the concept that the threobetain (6a) is more readily formed than is the sterically more hindered erythrobetaine

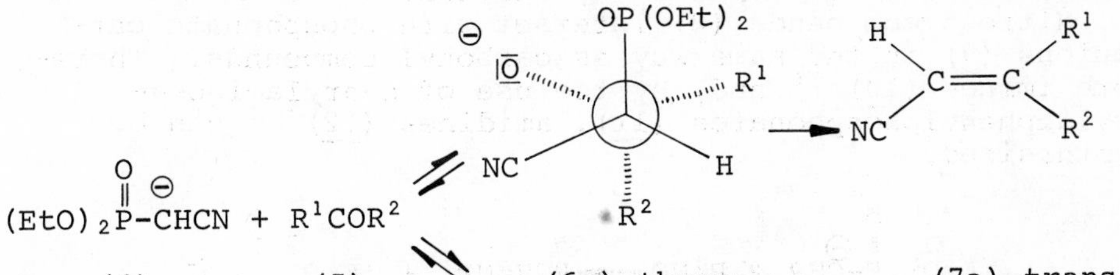

$(EtO)_2\overset{O}{\overset{\|}{P}}-\overset{\ominus}{C}HCN + R^1COR^2$ ⇌ [Newman projection: OP(OEt)$_2$, $\ominus$|O̲, R^1, NC, H, R^2] → $\overset{H}{\underset{NC}{}}C{=}C\overset{R^1}{\underset{R^2}{}}$

(4) (5) (6a) threo (7a) trans

$$\rightleftharpoons \quad \text{(Newman projection: } OP(OEt)_2 \text{ top; } {}^{\ominus}\underline{O}| \text{ and } R^2 \text{ up-left/right; NC and H lower; } R^1 \text{ bottom)} \longrightarrow \underset{NC}{\overset{H}{}}\!\!>C{=}C<\!\!\underset{R^1}{\overset{R^2}{}}$$

(6b) erythro (7b) cis

R^1 = Alkyl; R^2 = Aryl

(6b), and that decomposition of the threobetaine (6a) into the trans-olefin (7a) occurs faster than that of the diastereomeric (6b) into the cis-olefin (7b). This is so because the threobetaine (6a) is, first, sterically less hindered and, second, allows a better conjugative stabilization of the double bond being formed. This stabilization is decreased by ortho substituents in R^1; furthermore, a branching in the alkyl resude R^2 leads to an increased steric hindering in the threobetaine (6a), which effects a decrease in the portion of trans-olefin (7a).

Like the Wittig reaction, the interaction of phosphonate carbanions and carbonyl compounds also makes possible the partial asymmetric synthesis of axial disymmetric systems.[633]

$$(EtO)_2\overset{O}{\overset{\|}{P}}\text{-}CH_2CO_2\text{menthyl} \xrightarrow[(2)\ Me\text{-}C_6H_{10}{=}O]{(1)\ NaH} \text{(H,Me)}C_6H_{10}{=}C(H)CO_2H$$

Phosphonate carbanions and carbonyl compounds may also condense in a kind of Perkin reaction.[475]

Like phosphine oxide carbanions, phosphonate carbanions are more reactive olefin-forming compounds than are phosphonium ylids. The increased nucleophilicity of phosphonate carbanions, as compared to phosphine alkylenes, is due to the fact that the phosphonate group is able to delocalize the negative charge to a lesser extent than the phosphonium residue because of P$\rightleftharpoons$O back donation.[347]

Nitroso compounds (8) interact with phosphonate carbanions (9) in the same way as carbonyl compounds. Therefore imines (10)[396] and, by the use of α-arylamino-α-aryldiphenylphosphonates (11), amidines (12)[709] can be synthesized.

$$(EtO)_2\overset{O}{\overset{\|}{P}}\text{-}\overset{\ominus}{C}HR + R^1NO \longrightarrow RCH{=}NR^1 + (EtO)_2\overset{O}{\overset{\|}{P}}\text{-}\overline{O}|^{\ominus}$$

(9) (8) (10)

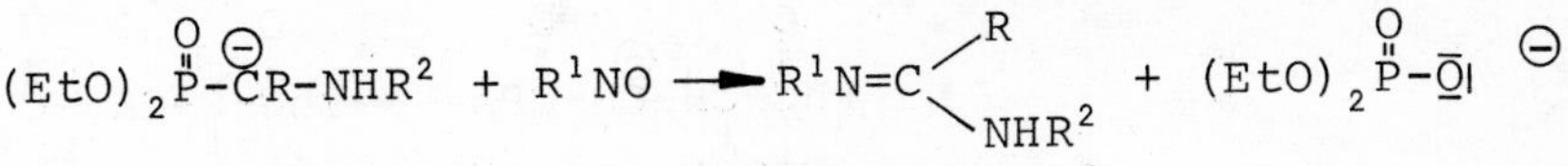

$$(EtO)_2\overset{O}{\overset{\|}{P}}-\overset{\ominus}{C}R-NHR^2 + R^1NO \longrightarrow R^1N{=}C\begin{smallmatrix}R\\NHR^2\end{smallmatrix} + (EtO)_2\overset{O}{\overset{\|}{P}}-\overline{O}|^{\ominus}$$

(11) (8) (12)

Schiff's bases also afford olefins with phosphonate carbanions[364]. Depending on the reaction temperature and solvent used, diethyl p-methylbenzylphosphonate and N-benzylidenaniline afford, in the presence of sodium amide, an erythro- or a threobetaine or a mixture of both. The betaines are in equilibrium through the starting products.

The reactions of phosphonate carbanions are essentially analogous to those of phosphonium ylids. With halogens,[650] alkyl halides,[21,371,475] acyl halides,[374] or aryl cyanates[425] there is observed a halogenation, an alkylation, an acylation, or a cyanogenation, respectively.

Epoxides and phosphonate carbanions produce cyclopropanes.[342,630,650] The mechanism of this reaction is presumably the same as that occurring with phosphine alkylenes.

Like phosphine alkylenes, phosphonate carbanions afford, with enol lactones, cyclic α,β-unsaturated ketones.[280]

Phosphonate carbanions may undergo a Michael addition with activated C-C double bonds.[475]

III. MISCELLANEOUS PHOSPHORUS CONTAINING CARBANIONS

Carbanions are also formed from thiophosphonates (1)[436] or phosphinates (2).[304] Phosphites (3)[486,489] and phos-

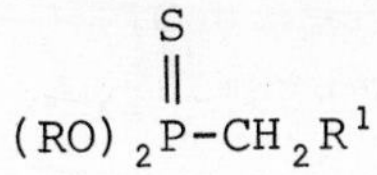

$$(RO)_2\overset{S}{\overset{\|}{P}}-CH_2R^1$$

(1)

$$\begin{smallmatrix}RO\\R^1\end{smallmatrix}\!\!>\overset{O}{\overset{\|}{P}}-CH_2R^2$$

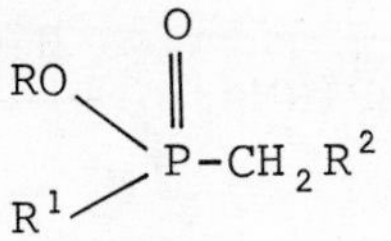

(2)

phinites (4)[487] add to dibenzoylethylene (5). The primary product is stabilized, as in the addition of phosphine to activated double bonds, by a proton transfer to give the corresponding phosphinite anion (6) and phosphitemethylene (7), respectively. The ^{31}P-NMR chemical shifts agree with the proposed structures (6) and (7). Methylenetrialkoxyphosphoranes may also be prepared from thioketones or thiophosgene and trialkyl phosphites.[125,437] The olefin syntheses according to Wittig and Horner-Emmons-Wadsworth can be supplemented to advantage by carbonyl olefination with metallated phosphonic diamides (8).[173]

$$PhC(=O)-CH=CH-C(=O)Ph \; (\underline{5}) \xrightarrow{R_2POR^1 \; (\underline{4})} PhC(=O)-CH(P^{\oplus}R_2OR^1)-\overset{\ominus}{C}H-C(=O)Ph \longrightarrow PhC(=O)-\overset{\ominus}{C}(P^{\oplus}R_2OR^1)-CH_2-C(=O)Ph \; (\underline{6})$$

$$PhC(=O)-CH=CH-C(=O)Ph \; (\underline{5}) \xrightarrow{P(OR)_3 \; (\underline{3})} PhC(=O)-CH(\overset{\oplus}{P}(OR)_3)-\overset{\ominus}{C}H-C(=O)Ph \longrightarrow PhC(=O)-\overset{\ominus}{C}(\overset{\oplus}{P}(OR)_3)-CH_2-C(=O)Ph \; (\underline{7})$$

$$2(RO)_3P + R^1_2C{=}S \longrightarrow (RO)_3P{=}CR^1_2 + (RO)_3P{=}S$$

$$(RO)_3P + CSCl_2 \longrightarrow (RO)_2P(=O)-C(-S-P(O)(OR)_2)(=P(OR)_3)$$

$$R^1R^2C{=}O + R^3R^4C(Li)-P(=O)(NMe_2)_2 \longrightarrow \quad (\underline{8})$$

$$R^1R^2\underset{|\overline{\underline{O}}|^{\ominus}\,Li^{\oplus}}{C}-CR^3R^4-P(=O)(NMe_2)_2 \; (\underline{9}) \xrightarrow{H_2O} R^1R^2C(OH)-CR^3R^4-P(=O)(NMe_2)_2 \; (\underline{10}) \xrightarrow{\Delta}$$

$$R^1R^2C{=}CR^3R^4 \; (\underline{11}) + (Me_2N)_2P(=O)-OH \; (\underline{12})$$

In contrast to the Wittig reaction or the PO-activated olefin synthesis, the primary adduct (9) undergoes no decomposition with olefin formation. However, β-hydroxyphosphonic diamide (10) decomposes when heated into an olefin (11) and phosphoric acid diamide (12). Presumably zwitterions of type (13) are formed as intermediates. They are the result of a proton transfer of the hydroxyl group to the oxygen atom of the P=O group. Compound (13) decomposes

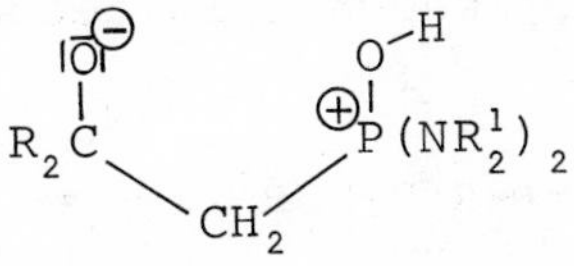

$$\overset{\ominus}{|\overline{O}|}-C(R_2)-CH_2-\overset{\oplus}{P}(OH)(NR^1_2)_2$$

(13)

via cis elimination according to Wittig reaction. Since the isolable β-hydroxyphosphonic diamides (10) can be separated with relative ease into their diastereomeric components, this method allows the stereospecific synthesis of olefins, although the adduct formation is not stereospecific.

A further possibility for the synthesis of stereospecific olefins consists in the stereoselective reduction of β-ketophosphonic amides to diastereomeric β-hydroxyphosphonic amides.[173]

C. PHOSPHINIMINES AND PHOSPHAZINES

Since the appearance of the last review article,[587] the number of publications concerning this class of compounds has been small. The present treatment is limited to tertiary phosphinimines (1) and tertiary phosphazines (2). Phosphonitrilic derivatives are discussed in Chapter 19, and related compounds in Chapter 18.

$R_3P{=}NR^1$ (1) $\qquad$ $R_3P{=}N{-}N{=}CR^1R^2$ (2)

The residues R are connected by a P-C bond to the phosphiniminyl ($P{=}NR^1$) and phosphazinyl group ($P{=}N{-}N{-}CR^1R^2$), respectively.[370,587]

C.1. Phosphinimines

C.1.1. Preparation of Phosphinimines (Iminophosphoranes)

I. FROM TERTIARY PHOSPHINES AND AZIDES

This method of preparation was first described by Staudinger and Meyer.[611] Tertiary phosphines and azides (3) form as an intermediate the so-called phosphazides (4), which decomposes with evolution of nitrogen into phosphinimines.[117,122,137,225,239,296,379,423,433,577,610,611,628,]

[665] The conditions under which this decomposition occurs

$$R_3P + N_3R^1 \longrightarrow R_3P\text{-}N_3R^1 \xrightarrow{-N_2} R_3P{=}NR^1$$

(3) (4) (1)

[sometimes the isolation of the intermediate complex (4) is possible[117,225,296,379,380,611]] depends on the substituents. For the intermediate complexes (4), which have been named Staudinger adducts, structures (4a)[137,225,378,379,445,609,611] and (4b)[296,482,628] have been proposed.

$$\overset{\gamma\ \beta\ \alpha}{R_3P{=}N\text{-}N{=}NR^1} \qquad R_3\overset{\oplus}{P}\text{-}\overset{\alpha}{N}(R^1)\text{-}\overset{\beta}{N}{=}\overset{\gamma}{N}^{\ominus}$$

(4a) (4b)

In the case of the isolable Staudinger adduct from triphenylphosphine and tosylazide, IR spectroscopic studies of ^{15}N-isotopically labeled derivatives confirm the unbranched structure (4a).[134] In the thermal decomposition in benzene, exclusively γ,β-N_2 is split off, according to mass spectroscopic results.

The P=N frequency, in the IR spectrum, of the resulting isotopically labeled iminophosphorane confirms that the nitrogen atom which was originally attached to the azide carbon atom remains in the iminophosphorane. For the decomposition of the Staudinger adduct, Leffler and Temple[378] postulate, as a result of kinetic measurements, a four-membered-ring transition state (5), in which the phosphorus atom is loosely connected to the nitrogen atom attached to the carbon atom. According to the same authors, the phosphazide formation is reversible. This finding could explain

$$R_3P + N_3R^1 \longrightarrow R_3P{=}\overset{\gamma}{N}\text{-}\overset{\beta}{N}{=}\overset{\alpha}{N}R^1\ (4) \longrightarrow [R_3P\cdots\overset{\alpha}{N}R^1;\ R_3P\text{-}\overset{\gamma}{N}{=}\overset{\beta}{N}\text{-}NR^1]\ (5) \longrightarrow R_3P{=}\overset{\alpha}{N}R^1 + \gamma\beta\text{-}N_2$$

(1)

the occurrence of the azide band in the IR spectroscopic

investigation of phosphazides.[628]

The reaction is accompanied in some solvents by a redox side reaction.[225,380]

The Staudinger synthesis of phosphinimine also makes possible the preparation of bisphosphinimines (7) from diphosphines (6) and azides.[282] Hydrazoic acid and trimethyl-

$$Ph_2P-C_6H_4-PPh_2 + 2PhN_3 \xrightarrow{-2N_2} Ph_2P(=NPh)-C_6H_4-PPh_2(=NPh)$$

(6) (7)

phosphine produce trimethylphosphinimine by way of the intermediate trimethylphosphinimidinium azide (8)[528,529]

$$Me_3P + 2HN_3 \longrightarrow N_2 + [Me_3P\text{-}NH_2]^{\oplus}N_3^{\ominus}$$

(8)

$$\downarrow Na$$

$$NaN_3 + \tfrac{1}{2}H_2 + Me_3P{=}NH$$

The azide method is also suitable for the preparation of silyl-, stannyl-, and germyl-substituted aminophosphoranes, starting from phosphines and the corresponding azides.[120-1,124,382,389,537,538,541,665]

II. FROM TERTIARY PHOSPHINES AND AMINE DERIVATIVES

Tertiary phosphines interact with chloramine[9,11,14,588,590] and substituted chloramines (9)[249,590,647,669-70] to yield phosphinimines (1). Phosphonium salts (10) are

$$R_3P + X\text{-}NHR \longrightarrow [R_3\overset{\oplus}{P}\text{-}NHR]X^{\ominus} \xrightarrow{B} R_3P{=}NR$$

(9) (10) (1)

intermediates in this reaction. They are converted to phosphinimines with bases.

For the phosphinimine preparation hydroxylaminosulfonic acid may be used in place of chloramine.[7,8,11] For splitting off HX the following bases are suitable: ammonia,[8] pyridine,[310] triethylamine,[310] sodium amide,[9,14] magnesium hydride,[588] tetramethylguanidine,[265] and lead tetraacetate.[694] The aminophosphonium salts (10) formed as intermediates can also be synthesized by other routes and then be deprotonated.[294,452,453,542,654a]

Sodium salts of chloramines and dichloramine also afford phosphinimine with phosphines, thus making unnecessary the use of a base for the dehydrohalogenation of a phosphonium salt formed as an intermediate.[366,420,577,578]

III. FROM DIHALOTRIPHENYLPHOSPHORANE AND PRIMARY AMINES

The formation of phosphinimines from dihalotriphenylphosphoranes (11) and primary amines (12) constitutes the third generally applicable method for the synthesis of iminophosphoranes (Horner and Oediger[310]). Presumably the

$$\underset{(11)}{R_3PX_2} + \underset{(12)}{H_2NR^1} \longrightarrow [R_3\overset{\oplus}{P}(X)\text{-}NH_2R^1]X^{\ominus} \xrightarrow{-HX} \underset{(13)}{[R_3\overset{\oplus}{P}\text{-}NHR^1]X^{\ominus}} \xrightarrow{-HX} \underset{(1)}{R^1N{=}PR_3}$$

reaction is initiated by attack of the amine on the phosphorus atom, followed by a splitting of HX, repeated twice. The reaction is usually carried out in the presence of base, which binds the split-off hydrogen halide. Interaction of alkylamines and dibromotriphenylphosphorane in the presence of triethylamine stops at the stage of the phosphonium salt (13).[310,711-2] In liquid ammonia salt (13) can, however, be converted to the corresponding phosphinimine (1) by the strong base sodium amide.[249,711-2,714] Dehydrohalogenation to give phosphinimines often occurs also on heating without the presence of an acid binding agent.[249,580]

Hydrazine (14) and dibromotriphenylphosphorane (11) afford N-aminotriphenylphosphinimine (15).[16,714] The use of an excess of dihalophosphorane (11) leads to the bisimine (16).[16] Bisphosphinimines of type (17) are obtainable from aromatic diamines (18) and dihalotriphenylphosphoranes.[310]

$$\underset{(11)}{R_3PBr_2} + \underset{(14)}{N_2H_4} \xrightarrow{NaNH_2} \underset{(15)}{R_3P{=}N\text{-}NH_2}$$

$$2R_3PBr_2 + N_2H_4 \xrightarrow{NaNH_2} R_3P{=}N{-}N{=}PR_3$$

(11) (14) (16)

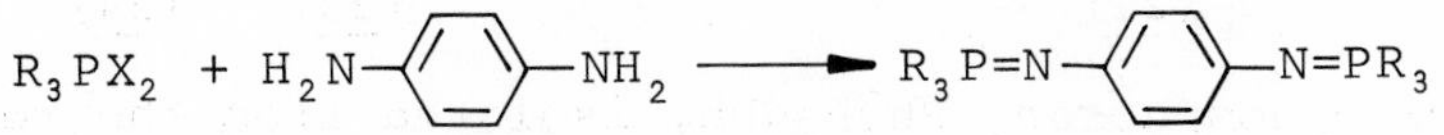

$$R_3PX_2 + H_2N{-}C_6H_4{-}NH_2 \longrightarrow R_3P{=}N{-}C_6H_4{-}N{=}PR_3$$

(11) (18) (17)

Gotsmann and Schwarzmann[249] synthesized phosphinimines from amines, triphenylphosphine, and halogens in one step. Presumably, phosphine and halogen form, as an intermediate, the dihalophosphorane, which then reacts with the amine component in the scheme shown above.

The three generally applicable methods I to III for the synthesis of iminophosphoranes are supplemented by a number of special synthetic procedures.

IV. FROM ACID AMIDES AND PHOSPHORUS PENTACHLORIDE

These compounds afford trichlorophosphinimines (19), whose chlorine groups can be replaced by alkyl or aryl groups using the Grignard procedure.[165,365] As a result phosphinimines of type (20) are formed.

$$Cl_3P{=}N{-}\underset{\underset{O}{\|}}{C}R + 3R^1MgBr \longrightarrow R^1_3P{=}N{-}\underset{\underset{O}{\|}}{C}R$$

(19) (20)

Products resulting from sulfonic acid amides can be treated in the same way. Instead of PCl_5, trichlorodiphenylphosphorane can be used as a starting material.[136,579]

V. FROM PHOSPHINE ALKYLENES

(1) Phosphine alkylenes (21) (with R and R^1 = aryl), which contain no CH_2 group in β-position to phosphorus, and Schiff's bases (22) produce, in a Wittig type of olefination reaction, a phosphinimine (23) and an olefin (24).[42,108,109]

(2) Acylmethylenetriphenylphosphoranes give with azides either α-diazocarbonyl compounds and a phosphinimine or, with elimination of phosphine oxide, 1,5-disubstituted 1,2,3-triazoles.[1,2,271b]

$$R_3P{=}CHR^1 + R^2N{=}CHR^3 \longrightarrow \begin{array}{ccc} R_3P & - & CHR^1 \\ | & & | \\ R^2N & - & CHR^3 \end{array} \longrightarrow \begin{array}{c} R_3P \\ \| \\ NR^2 \end{array} + \begin{array}{c} CHR^1 \\ \| \\ CHR^3 \end{array}$$

(21) (22) (23) (24)

The iminophosphorane $Ph_3P{=}NPh$, isolated from the reaction of phenylazide with benzylidenetriphenylphosphorane, results from the interaction of the triphenylphosphine formed as an intermediate with yet unconsumed azide.[289]

(3) Methylenetriphenylphosphorane (25) and tetrasulfur tetranitride (26) afford a phosphinimine (27) also.[222] The same phosphinimine is also formed in the interaction of

$$R_3P{=}CH_2 + S_4N_4 \longrightarrow N\begin{array}{c} \overset{\oplus}{S}{-}\overset{\ominus}{N} \\ \\ S{-}N \end{array}S{-}N{=}PR_3 + (CH_2S)$$

(25) (26) (27)

triphenylphosphine with S_4N_4.

(4) Iminophosphoranes have been observed as products of the reaction of phosphine alkylenes with nitriles,[166] e.g.:

$$R_3P{=}CHPh + PhCN \longrightarrow \begin{array}{c} R_3\overset{\oplus}{P}{-}CHPh \\ | \\ \overset{\ominus}{\overline{N}}{=}CPh \end{array} \longrightarrow \begin{array}{c} R_3P{-}CHPh \\ | \quad | \\ N{=}CPh \end{array} \longrightarrow \begin{array}{c} R_3P \;\; CHPh \\ \| \quad \| \\ N{-}CPh \end{array}$$

Resonance-stabilized phosphine alkylenes undergo a reaction only with activated nitriles such as $(CN)_2$ or CF_3CN.

VI. FROM α,β-DIBENZOYLHYDROXYLAMINE AND TRIPHENYLPHOSPHINE

α,β-dibenzoylhydroxylamine (28) and triphenylphosphine afford a phosphinimine also.

The reaction is as follows:[657]

$$Ph_3P + PhCOONH{-}COPh \longrightarrow Ph_3P{=}N{-}COPh + PhCO_2H$$

(28)

VII. FROM N-SULFINYLSULFONAMIDES AND TRIPHENYLPHOSPHINE

The reaction is as follows:[557]

$$Ph_3P + RSO_2\text{-}\overset{\ominus\,\oplus}{\underline{N}}\text{-}S{=}O \longrightarrow Ph_3P{=}N\text{-}SO_2R + SO$$

(29) (30)

Compounds of type (30) are also produced from triphenylphosphine oside or sulfide and N-sulfinylsulfonamides (29).[384,557]

$$Ph_3P{=}S + (29) \longrightarrow (30) + S_2O$$

$$Ph_3P{=}O + (29) \longrightarrow (30) + SO_2$$

VIII. FROM AMINOPHOSPHINES

(1) Aminophosphines (31), when treated with activated olefins (32), yield phosphinimines (33).[639]

$$Ph_2P\text{-}NHPh + CH_2{=}CHR \rightleftharpoons Ph_2\overset{\oplus}{P}(\text{—NHR})(\text{—}CH_2\text{—}\overset{\ominus}{C}HR) \longrightarrow RCH_2CH_2\text{-}P(Ph)_2{=}NPh$$

(31) (32) (33)

(2) Tetrahalomethane (34) and aminophosphines (31) give halophosphinimines (36), which can be converted into tertiary phosphinimines (37) by means of Grignard compounds.[322] Phosphonium salts (35) have been postulated as intermediates.

$$Ph_2P\text{-}NHPh + CX_4 \longrightarrow [Ph_2\overset{\oplus}{P}(X)\text{-}NHPh]\overset{\ominus}{C}X_3 \longrightarrow$$

(31) (34) (35)

$$Ph_2P(X){=}NPh \xrightarrow{RMgX} Ph_2P(R){=}NPh$$

(36) (37)

IX. FROM DIFLUORODIAZIRINES

Difluorodiazirine (38) and tertiary phosphines (39) as

well as other tervalent organophosphorus compounds) afford N-cyanophosphinimines (40) and difluorophosphoranes (41).[440]

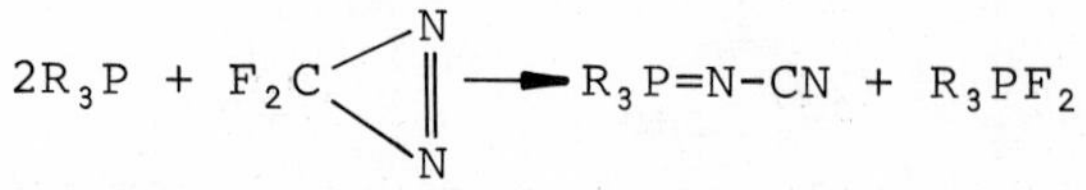

$$2R_3P + F_2C\langle(N{=}N) \longrightarrow R_3P{=}N{-}CN + R_3PF_2$$

(39) (38) (40) (41)

X. FROM QUATERNARY PHOSPHONIUM SALTS

Tetraphenylphosphonium chloride (42) is converted by potassium amide into triphenylphosphinimine (44).[448] The

$$[Ph_4\overset{\oplus}{P}]\overset{\ominus}{Cl} \xrightarrow[-KCl]{KNH_2} [Ph_4\overset{\oplus}{P}]\overset{\ominus}{N}H_2 \xrightarrow{NH_2^{\ominus}} Ph_3P{=}NH + PhH$$

(42) (43) (44)

intermediate formation of tetraphenylphosphonium amide (43) has been assumed.

XI. FROM N-CHLOROIMINOCARBOXYLIC ACID DERIVATIVES AND N-CHLOROARYLAMIDINES

N-Chloroiminocarboxylic acid derivations (45), as well as N-chloroarylamidines, react with triphenylphosphine to form phosphinimines (46).[195,197,198]

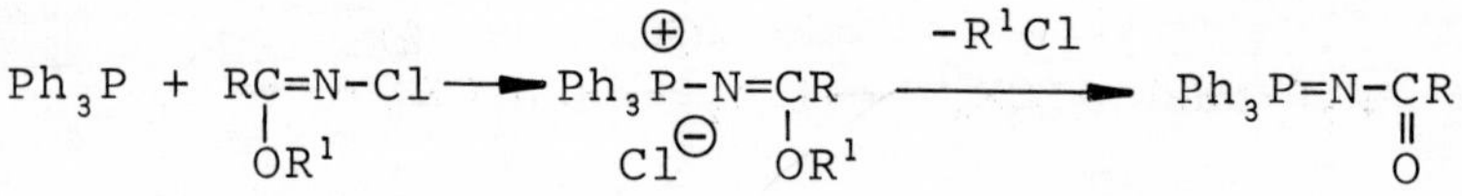

$$Ph_3P + RC(OR^1){=}N{-}Cl \longrightarrow Ph_3\overset{\oplus}{P}{-}N{=}CR(OR^1)\ Cl^{\ominus} \xrightarrow{-R^1Cl} Ph_3P{=}N{-}C({=}O)R$$

(45) (46)

XII. LEAD ACETATE OXIDATION OF AMINOPHOSPHONIUM SALTS

In this method[694] arylaminophosphonium salts are deprotonated to arylphosphinimines by the acetate ion originating from $Pb(OAc)_4$. The liberated ion X^- is able to substitute

$$AcO^{\ominus} + [Ph_3\overset{\oplus}{P}{-}NHAr]X^{\ominus} \longrightarrow AcOH + X^{\ominus} + Ph_3P{=}NAr$$

the N-aryl group in the ortho or para position.

XIII. FROM CONVERSION OF OTHER IMINOPHOSPHORANES

A great number of phosphinimines have also been synthesized by metallation,[496,527,529] desilylation,[120-1] halogenation,[12,13] acylation,[607] tosylation,[11] methanolylis,[465] reaction with Lewis acids,[18,135a,527,537,540,556] or more complex reactions from simpler iminophosphoranes.[23,209,419,440,541,659,694,706]

C.1.2. Reactions of Phosphinimines

The reactions of phosphinimines are often analogous to those of phosphine alkylenes. The mechanisms, however, are generally less well understood.

Iminophosphoranes are basic compounds like methylenephosphoranes. Their reactivity, resulting from the polarity of the P-N bond, is influenced by the substituents on the phosphorus atom and, in particular, on the nitrogen atom. Electron-withdrawing groups on nitrogen, which delocalize the negative charge by inductive and/or mesomeric effects, increase the stability and decrease the reactivity of the corresponding iminophosphorane.

For the generation of iminophosphoranes from their corresponding acids, as in the synthesis of phosphine alkylenes, bases with different strengths are necessary to effect the deprotonation.[8,9,11,14,265-6,310,588,694]

Whereas iminotriphenylphosphorane[10,11,588] and N-alkyliminophosphoranes[611,711-2] are already hydrolyzed in humid air, the cleavage of stabilized iminophosphoranes often succeeds only under drastic conditions.[296,578,611]

The hydrolysis of N-substituted iminophosphoranes leads to a phosphine oxide and a primary amine. N-Trimethylsilyliminophosphorane (47), however, is cleaved with methanolic sulfuric acid at the Si-N bond.[120-1] The acidic hy-

$$\underset{(\underline{47})}{Ph_3P{=}N\text{-}SiMe_3} \xrightarrow[H_2SO_4]{MeOH} Ph_3P{=}NH + MeOSiMe_3$$

drolysis of methylenebis (N-phenyldiphenylphosphinimine) (48) causes cleavage of a P-C bond:[3]

$$\underset{(\underline{48})}{\overset{PhN\qquad\quad NPh}{Ph_2P(=)\text{-}CH_2\text{-}(=)PPh_2}} \xrightarrow[H_2O]{H^{\oplus}} \overset{PhNH\qquad\quad NPh}{Ph_2\overset{\oplus}{P}\text{-}CH_2\text{-}PPh_2(\leftarrow \bar{O}H_2)}$$

$$Ph_2P(O)NHPh + [Ph_2\overset{PhNH}{\overset{|}{P}}=CH_2 \rightleftharpoons Ph_2\overset{PhN}{\overset{\|}{P}}-CH_3] \xrightarrow{H_2O} Ph_2\overset{O}{\overset{\|}{P}}-CH_3 + H_2NPh$$

Hydrolysis of phosphinimines in neutral and basic media probably involves the intermediate product (49) with pentacovalent phosphorus.[312,587] The reaction occurs with

$$R_3P=NR \xrightarrow{H_2O} [R_3\overset{\oplus}{P}-NHR]OH^{\ominus} \rightleftharpoons \underset{(\mathbf{49})}{R_3\underset{OH}{\underset{|}{P}}-NHR} \longrightarrow R_3P=O + H_2NR$$

inversion at the phosphorus atom.[312] Protonation of iminophosphoranes produces the corresponding aminophosphonium salts.[12,296,590,609,710]

$$R_3P=NR \xrightarrow{H^{\oplus}X^{\ominus}} [R_3\overset{\oplus}{P}-NHR]X^{\ominus}$$

Phosphinimines will undergo reaction with a variety of Lewis acids. Coordination compounds have been formed with Cu(II), Co(II), and Ni(II) halides,[15] metal carbonyls,[135a, 284] and mercury and cadmium iodide,[556] as well as with halides and trialkyls of the elements of the third main group,[18,530,540,713] e.g.:

$$R_3P=N-SiR^1_3 + R^2_3M \longrightarrow R_3P=\overset{SiR^1_3}{\overset{|}{\underset{\oplus}{N}}}-\underset{\ominus}{M}R^2_3$$

Treatment of iminophosphoranes, unsubstituted at the nitrogen atom, with lithium, cadmium, zinc, gallium, aluminum, and indium alkyls yields organometal-substituted iminophosphoranes.[496,528-530] Whereas the organometal-

$$Me_3P=NH + LiR \longrightarrow RH + Me_3P=NLi$$

$$Ph_3P=NH + AlR_3 \longrightarrow RH + Ph_3P=NAlR_2$$

substituted iminophosphoranes of zinc and cadmium exhibit a tetrameric structure, the corresponding compounds of aluminum, gallium, and indium exist as dimers.[528-530] N-Lithium triorganophosphinimines are suited for the synthesis of various N-substituted iminophosphoranes[496,527,528] e.g.:

$$Me_2PCl + Me_3P{=}NLi \longrightarrow LiCl + Me_2P{-}N{=}PMe_3$$

C.1.2.1. Iminophosphoranes and Halogen Compounds. Substituted iminophosphoranes (50) are converted with alkyl halides (51) to N-dialkylaminophosphonium salts (52).[16, 296,711-2] In the alkylation of N-unsubstituted triphenyl-

$$R_3P{=}NR + R^1X \longrightarrow [R_3\overset{\oplus}{P}{-}NRR^1]X^{\ominus}$$

(50) (51) (52)

phosphinimine the originally formed N-monoalkylaminophosphonium salt (53) is deprotonated by a second mole of iminophosphorane to give N-alkyliminophosphorane (54).[12,18]

$$Ph_3P{=}NH + RX \longrightarrow [Ph_3\overset{\oplus}{P}{-}NHR]X^{\ominus} \xrightarrow{Ph_3P=NH} Ph_3P{=}NR + [Ph_3\overset{\oplus}{P}{-}NH_2]X^{\ominus}$$

(53) (54)

$$Ph_3P{=}NR \xrightarrow{RX} [Ph_3\overset{\oplus}{P}{-}NR_2]X^{\ominus}$$

(55)

The latter (54) is alkylated again to yield the N-dialkylated salt (55).[12]

The reactions of iminotriphenylphosphorane with acyl halides,[10,11] tosyl chloride,[11] halogens,[12,13] and N,N-carbonylidimidazole[10] proceed in an analogous manner, i.e., with transylidation.

The halogen atom in N-haloiminotriphenylphosphorane (56), obtainable from iminotriphenylphosphorane and halogens, can readily be substituted by other groups,[12 13] e.g.:

$$Ph_3P{=}N{-}Br + Ph_3P \longrightarrow [Ph_3P{-}N{-}PPh_3]^{\oplus}Br^{\ominus}$$

(56)

C.1.2.2. Reactions of Iminophosphoranes with Multiple Bonds

(1) Phosphinimines and Compounds with C-O or C-S Double Bonds

Phosphinimines have been shown to react with a great number of oxygen- or sulfur-containing compounds with elimination of phosphine oxide and phosphine sulfide, respectively.[297-8]

$$
Ph_3P{=}NR \; + \;
\begin{cases}
CO_2 \longrightarrow Ph_3PO + R{-}N{=}C{=}O \\
CS_2 \longrightarrow Ph_3PS + R{-}N{=}C{=}S \\
SO_2 \longrightarrow Ph_3PO + R{-}N{=}S{=}O \\
R^1N{=}C{=}O \longrightarrow Ph_3PO + R^1{-}N{=}C{=}N{-}R \\
R^1_2C{=}C{=}O \longrightarrow Ph_3PO + R^1_2C{=}C{=}NR \\
R^1_2C{=}O \longrightarrow Ph_3PO + R^1_2C{=}NR \\
R^1N{=}C{=}S \longrightarrow Ph_3PS + R^1{-}N{=}C{=}C{-}R
\end{cases}
$$

These reactions may be expressed by the following general equation:[587]

$$Ph_3P{=}NR + R^1_2YZ \longrightarrow Ph_3P{=}Z + R^1_2Y{=}NR$$

$$Y = C,\ N,\ S; \quad Z = O,\ S$$

Reaction partners that are suitable as R^1_2YZ are aldehydes,[19,473,609] ketones,[16,609,702] ketene,[609,612,613] carbon dioxide,[19,433,608,609,611] carbon disulfide,[19,296,609] sulfur dioxide,[609] isocyanates,[158,442,443,609,611] N,N'-carbonyldiimidazoles,[607] and isothiocyanates.[609,611]

The parent iminophosphorane, $Ph_3P{=}NH$, reacts in an analogous manner with C-O and C-S double bonds.[11,12] The mechanism of the reaction of iminophosphoranes with carbonyl compounds is assumed to be similar to that of the Wittig reaction:

$$R_3P{=}NR^1 + R^2_2C{=}O \rightleftharpoons \begin{matrix} R_3\overset{\oplus}{P}{-}NR^1 \\ | \qquad | \\ \ominus\underline{\overline{O}} \quad CR^2_2 \end{matrix} \longrightarrow R_3P{=}O + R^2_2C{=}NR^1$$

The second step of the reaction, namely, the elimination of phosphine oxide, becomes rate determining if the electron density on the phosphorus atom is increased by suitable substituents, so that the driving force of the reaction--the attack of phosphorus on the oxygen with formation of a four-membered cyclic intermediate--is decreased.[351]

(2) Other Reactions Occurring with Elimination of Phosphine Oxide

Phosphinimine and nitrosyl chloride (57) afford a N-nitrosophosphonium salt (58) which decomposes even at -70° into phosphine oxide and a diazonium chloride (59).[312,710]

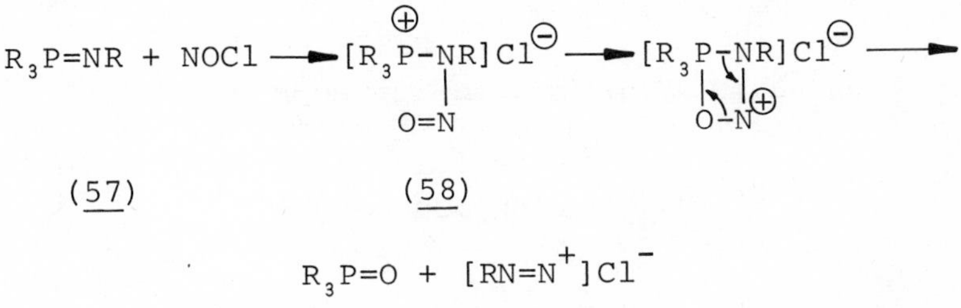

$$R_3P{=}NR + NOCl \longrightarrow [R_3\overset{\oplus}{P}\text{-}NR]Cl^{\ominus} \longrightarrow [R_3P\text{-}NR]Cl^{\ominus} \longrightarrow$$

(57) (58): $[R_3\overset{\oplus}{P}\text{-}N(N{=}O)R]Cl^{\ominus}$; cyclic intermediate with O–N$^{\oplus}$

$$R_3P{=}O + [RN{=}N^+]Cl^-$$

(59)

Similarly to acylated phosphine alkylenes, N-acylaminophosphoranes (60) (and their thio analogs) decompose with the formation of a triple bond into phosphine oxide (and phosphine sulfide) and a nitrile (61).[194,195,296,465,466,609,659]

$$R_3P{=}NCOR^1 \longrightarrow R_3PO + N{\equiv}CR^1$$

(60) (61)

(3) Iminophosphoranes and Triple Bonds

Treatment of iminophosphoranes with acetylene-dicarboxylic ester (62) results in the formation of alkylenephosphoranes (64).[19,149-151] Presumably this reaction involves a four-membered ring (63) with a C-C double bond as an intermediate, as in the case of phosphine alkylenes. Electron-withdrawing groups at the nitrogen atom of the iminophosphorane

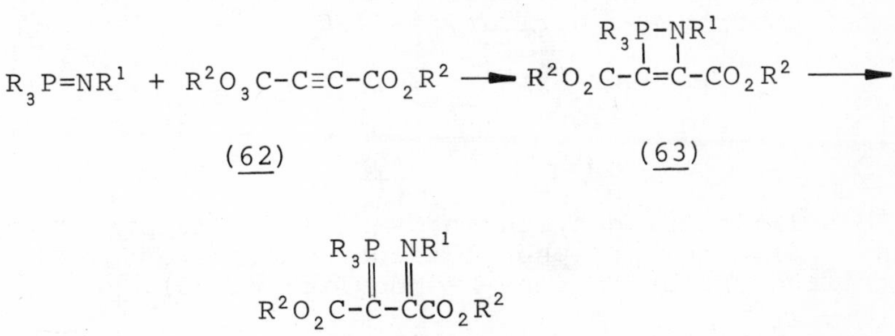

$$R_3P{=}NR^1 + R^2O_3C\text{-}C{\equiv}C\text{-}CO_2R^2 \longrightarrow R^2O_2C\text{-}\underset{|}{C}{=}\underset{|}{C}\text{-}CO_2R^2 \;(\text{ring with } R_3P\text{-}NR^1) \longrightarrow$$

(62) (63)

$$R^2O_2C\text{-}\overset{R_3P}{\overset{\|}{C}}\text{-}\overset{NR^1}{\overset{\|}{C}}CO_2R^2$$

(64)

prevent this reaction.[149]

Iminophosphoranes and activated nitriles interact in a similar way with renewed formation of iminophosphoranes of type (64a).[166]

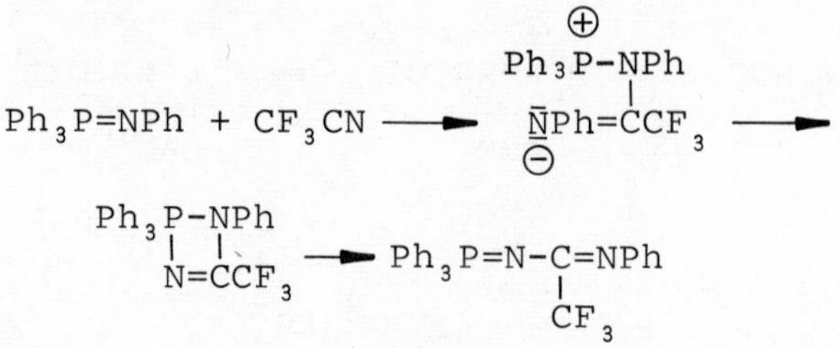

$$Ph_3P{=}NPh + CF_3CN \longrightarrow \overset{\oplus}{Ph_3P}{-}NPh \;|\; \overset{\ominus}{\underline{N}}Ph{=}CCF_3 \longrightarrow$$

$$\begin{matrix} Ph_3P{-}NPh \\ |\quad | \\ N{=}CCF_3 \end{matrix} \longrightarrow Ph_3P{=}N{-}\underset{CF_3}{C}{=}NPh$$

(64a)

An intramolecular cyclization reaction also following this scheme has been observed by Zbiral[694] with 1-thio- or 1-selenocyanato-2-triphenylphosphiniminylnaphthalene (65).

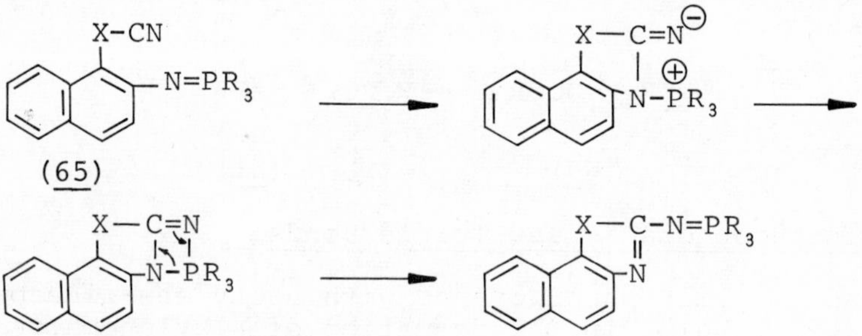

(65)

X = S, Se

C.1.2.3. Miscellaneous Reactions with Iminophosphoranes

Aryllithium converts triarylphosphinimine into penta-arylated phosphorane derivatives (66).[678,681]

$$Ar_3P{=}NR + 2ArLi \longrightarrow Ar_5P + RNLi_2$$

(66)

Phenyliminotriphenylphosphorane (67) and perfluoroisobutene (68) afford triphenyldifluorophosphorane (69) and bis(trifluoromethyl)ketene-N-phenylimine (70).[235]

$$Ph_3P{=}NPh + (CF_3)_2C{=}CF_2 \longrightarrow Ph_3PF_2 + (CF_3)_2C{=}C{=}NPh$$

(67) (68) (69) (70)

Photolysis of N-t-butyliminotriphenylphosphorane causes cleavage of the P=N as well as of the C-N bond.[715]

Zbiral reported a new synthesis of tetrazoles (74) and iminonitriles (77).[699]

Treatment of iminophosphoranes with acid chlorides affords N-acylaminophosphonium salts (71). In absolute aprotic media the nucleophile N_3^- attacks at the carbonyl group of the acetyl group. Elimination of phosphine oxide from (72) produces the imidazide (73), which rearranges to the cyclic isomer (74).

$$R_3P{=}NR^1 + R^2{-}\overset{O}{\overset{\|}{C}}{-}Cl \longrightarrow \overset{\oplus}{R_3P}{-}N(COR^2)(R^1)\ Cl^{\ominus} \xrightarrow{N_3^{\ominus}} \overset{\oplus}{R_3P}{-}N(R^1){-}C(\overset{\ominus}{O})(N_3)R^2 \xrightarrow{-OPR_3}$$

(71) (72)

$$R^1N{=}C(N_3)R^2 \longrightarrow \text{1-}R^1\text{-5-}R^2\text{-tetrazole}$$

(73) (74)

In some cases it is possible to introduce the nucleophile N_3^- directly by acylation with the corresponding acid azide. Acylation of phosphinimines with acyl cyanides (75) yields CN-containing phosphonium betaines (76), which split off phosphine oxide and give imononitriles (77).[699]

$$R_3P{=}NR^1 + R^2COCN \longrightarrow \overset{\oplus}{R_3P}{-}N(R^1){-}C(\overset{\ominus}{O})(CN)R^2 \longrightarrow R_3PO + \begin{matrix}R^2\\NC\end{matrix}\!\!>C{=}NR^1$$

(75) (76) (77)

N-(1-Chloro-2,2-diphenylvinyl)triphenylphosphinimine (78),[465] obtainable from chlorodiphenylacetonitrile and triphenylphosphine, reacts with alcohols, mercaptans, and amines to give acyl-, thioacyl-, and iminophosphinimines, respectively.[465,466]

N-Acylamidotriphenylphosphinimines also react in their enol form.[645a] The corresponding lithium salt is O-acetylated by acid chlorides. α-Acyloxybenzilidenephosphazine, prepared in this way, undergoes an intramolecular Wittig-

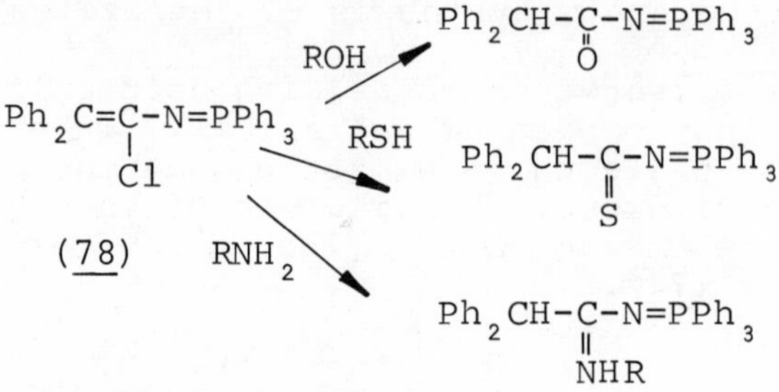

$$Ph_2C{=}C(Cl){-}N{=}PPh_3\ (\underline{78}) \xrightarrow{ROH} Ph_2CH{-}C(=O){-}N{=}PPh_3$$

$$\xrightarrow{RSH} Ph_2CH{-}C(=S){-}N{=}PPh_3$$

$$\xrightarrow{RNH_2} Ph_2CH{-}C(=NHR){-}N{=}PPh_3$$

reaction ring closure to give unsymmetrical 2,5-disubstituted 1,3,4-oxadiazoles.[654a]

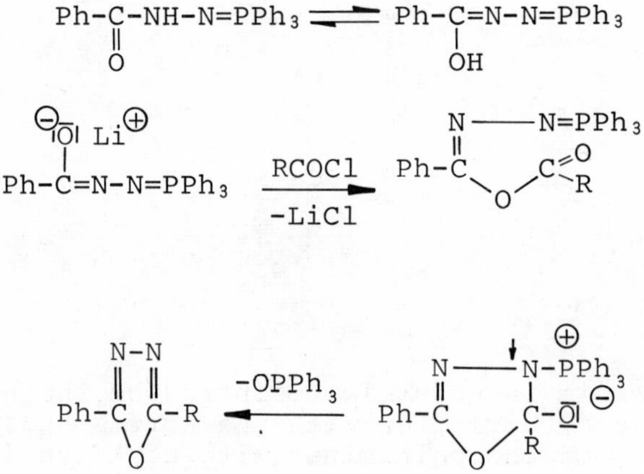

$$Ph{-}C(=O){-}NH{-}N{=}PPh_3 \rightleftharpoons Ph{-}C(OH){=}N{-}N{=}PPh_3$$

$$Ph{-}C(\overset{\ominus}{O}\,Li^{\oplus}){=}N{-}N{=}PPh_3 \xrightarrow[-LiCl]{RCOCl} \text{Ph-C}{=}N{-}N{=}PPh_3 \ldots O{-}C({=}O)R \text{ (ring)}$$

$$\text{N-N=C(R)-O-C(Ph) (1,3,4-oxadiazole)} \xleftarrow{-OPPh_3} \text{Ph-C=N-}\overset{\oplus}{N}{-}PPh_3{-}C(R){-}\overset{\ominus}{O} \text{ (ring via O)}$$

C.1.2.4. Phosphinimines and 1,3-Dipoles. In their dipolar activity iminophosphoranes are inferior to phosphine alkylenes. Whereas in favorable cases the latter allow even the isolation of nitrile oxide or nitrone cycloadducts, with iminophosphoranes only subsequent products are isolable.[326,327] Nitrile oxide (79) and iminophosphorane (80) afford, via a hypothetical 1,2,4,5-P(V)-oxadiazaphole (81), with phosphine oxide elimination, the carbodiimide (82).

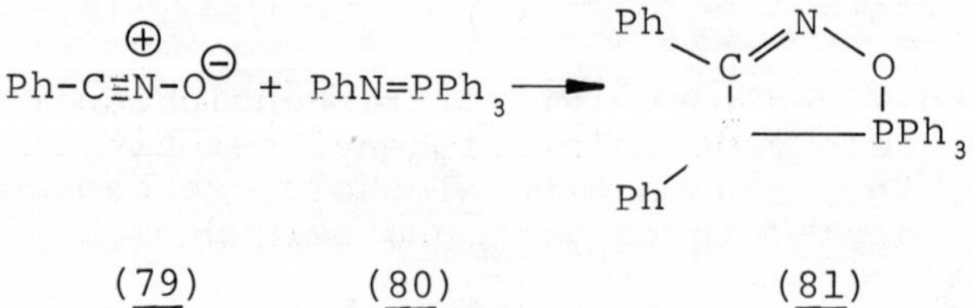

$$Ph{-}C{\equiv}\overset{\oplus}{N}{-}\overset{\ominus}{O} + PhN{=}PPh_3 \longrightarrow \text{Ph-C=N-O-PPh}_3\text{-N(Ph) (ring)}$$

(79) (80) (81)

$$(\underline{81}) \longrightarrow OPPh_3 + PhN{=}C{=}NPh$$

(82)

Fragmentation of the ring (81) is coupled with transfer of a phenyl group. Excess iminophosphorane (80) catalyzes the reaction of carbodiimide (82) with (79) to give 3,4-diphenyloxadiazolone-5-anil (83).

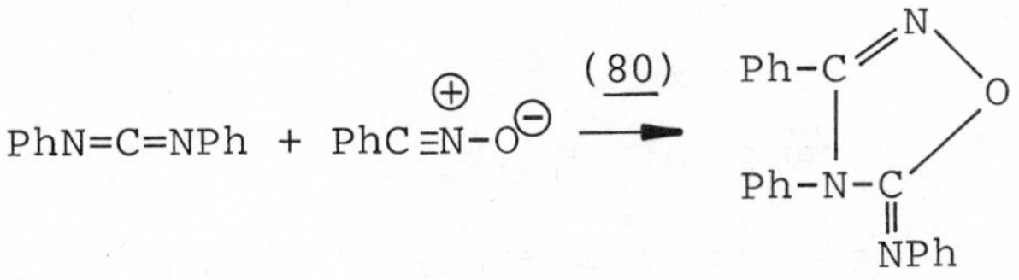

$$PhN{=}C{=}NPh + PhC{\equiv}\overset{\oplus}{N}{-}O^{\ominus} \xrightarrow{(80)} \text{Ph-C=N-O-C(=NPh)-N(Ph) ring}$$

(82) (79) (83)

By the use of suitable substituents in components (79) and (80) the exchange of organic groups between carbodiimide and iminophosphorane via the cycloadduct (87) can be demonstrated[327] to occur as follows:

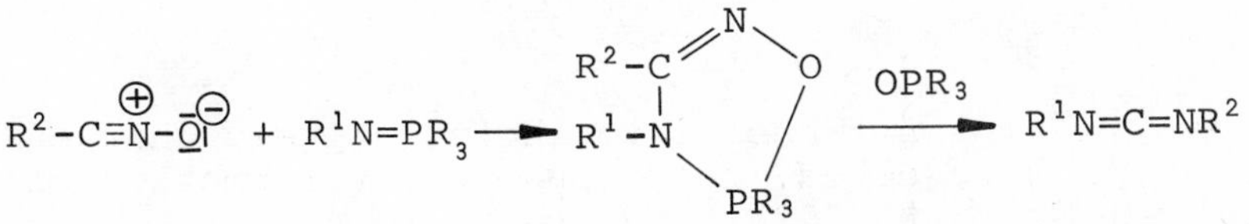

$$R^2{-}C{\equiv}\overset{\oplus}{N}{-}O^{\ominus} + R^1N{=}PR_3 \longrightarrow [R^2{-}C{=}N{-}O{-}PR_3{-}N(R^1)] \xrightarrow{-OPR_3} R^1N{=}C{=}NR^2$$

(84) (85) (86)

$$(85) + (86) \rightleftharpoons [R^1N{=}C{-}N(R^2){-}PR_3{-}N(R^1)] \rightleftharpoons R^1N{=}C{=}NR^1 + R^2N{=}PR_3$$

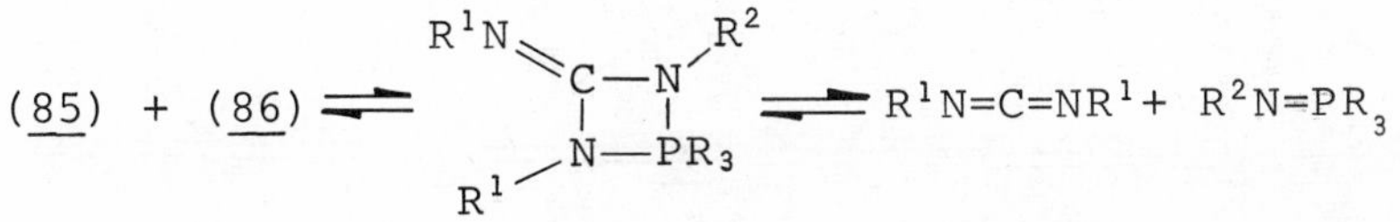

(87) (88) (89)

$R^1 = C_6H_4OMe$, $R^2 = C_6H_4NO_2$ R = Ph

Interaction of (84) and (85) results, not in a mixed carbodiimide (86), but in carbodiimide (88) and an iminophosphorane (89). The exchange via 1,2,3-diazaphosphetane (87) apparently transfers to the iminophosphorane nitrogen the substituent that best stabilizes its anionic partial charge.

Generally nitrones are inactive against iminophosphoranes. Only the reactive N-phenyl-C-benzoylnitrone (90) adds to phenyliminotriethylphosphorane (91) to yield the hypothetical oxadiazaphospholidine (92). The latter is converted, with phosphine oxide elimination and a C→N-benzoyl transfer, into N,N'-diphenyl-N-benzoylformamidine (93).[327]

Nitrile imines (94) and iminophosphoranes (95) do not

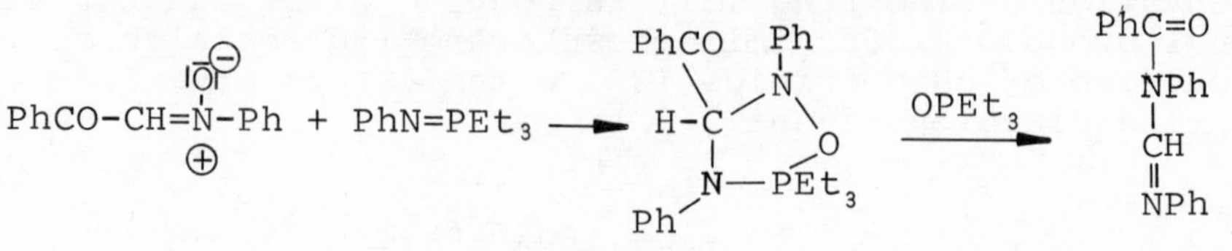

$$PhCO-CH=\overset{\oplus}{N}(O^{\ominus})-Ph + PhN=PEt_3 \longrightarrow \text{(92)} \xrightarrow{OPEt_3} PhC(=O)-NPh-CH=NPh$$

(90) (91) (92) (93)

produce cycloadducts, but instead yield betaines of type (96).

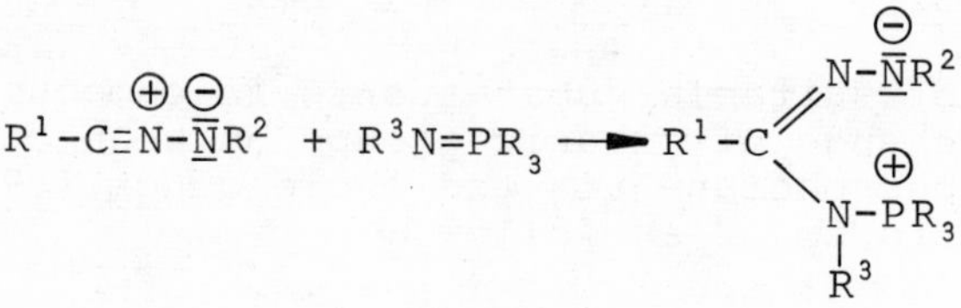

$$R^1-C{\equiv}\overset{\oplus}{N}-\overset{\ominus}{\bar{N}}R^2 + R^3N=PR_3 \longrightarrow R^1-C(=N-\overset{\ominus}{\bar{N}}R^2)-\overset{\oplus}{N}(R^3)-PR_3$$

(94) (95) (96)

For $R^3 = H$, proton transfer to the negatively charged nitrogen atom is observed, resulting in the formation of a new iminophosphorane (97).[327]

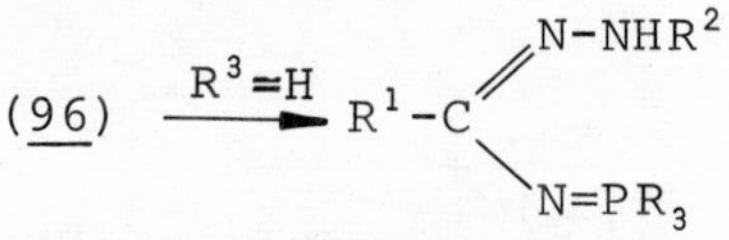

$$(96) \xrightarrow{R^3=H} R^1-C(=N-NHR^2)-N=PR_3$$

(97)

N-Cyanophosphinimines (98) and hydrazoic acid undergo a cycloaddition reaction at the C-N triple bond and produce N-(5-tetrazoyl)phosphinimines (99).[440]

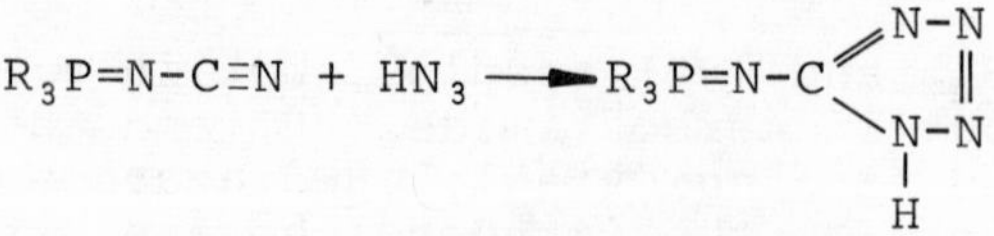

$$R_3P=N-C{\equiv}N + HN_3 \longrightarrow R_3P=N-C\langle(=N-N=N-NH-)\rangle$$

(98) (99)

C.1.3. Physical Properties of Iminophosphoranes

Little is known about the structure and bonding situation in iminophosphoranes, since, in particular, ^{31}P-NMR spectra and X-ray structural analysis data of this class of compounds are scarce.

The hybridization about the phosphorus atom probably is very similar to that in phosphine alkylenes (tetrahedral arrangement of ligands).[347] The nitrogen atom could be sp^2-, sp^3-, and sp-hybridized, corresponding to P-N-R angles of 120°, 109.5°, and 180°,[347] respectively.

The overlapping of vacant 3d-orbitals of the phosphorus atom with the filled hybrid orbitals of nitrogen leads to a p_π-d_π interaction which increases the bond strength as compared to that of a P-N single bond. The assumption of a partial double-bond character of the P-N bond is justified by the high P-N bond strength and the strongly increased valence frequency, as compared to aminophosphonium salts.[392]

Reports concerning the location of the P=N band in the IR are inconsistent.[171] Newer investigations with ^{15}N-labeled phosphinimines indicate that the P-N double bond gives rise to an intensive absorption in the region of 1141 to 1373 cm^{-1} [666-668] This region, however, has real group frequency character as a P=N stretching frequency in only a few cases, since mass effects and, in particular, the coupling with the frequency of substituents linked to nitrogen influence the location of the P=N band in a manner that is difficult to foresee. A further complication exists in the change of the force constant of the P-N double bond as a consequence of dissimilar substituent effects.

Strongly electron-withdrawing groups on the phosphorus increase the bond order, as do groups on nitrogen which are capable of conjugation. In both cases there results a partial positive charge on phosphorus, whose bond-increasing effect may be understood by the assumption of a d-orbital contraction and therewith a better overlap with filled N-hybrid orbitals.[668]

The ^{31}P-NMR data that could clarify the bonding situation in iminophosphoranes are known only for phenyliminotriphenylphosphoranes (0.0 ppm) and trimethylsilyliminotriethylphosphorane (-14.4 ppm).[497,587] Proton resonance spectra have been reported for a larger number of iminophosphoranes.[358,465,530,537,714] The coupling constant P-N-C-H depends on the number and nature of the bonds between phosphorus and hydrogen. Whereas in systems of the type PNCH long-range ^{31}P-H coupling is observed,[358] such an effect could not be seen in compounds with the grouping PNNCH [e.g., in Ph_3P=N-N$(CH_3)_2$].[714]

The dipole moments and NMR and UV spectra of iminophosphoranes point to a considerable π-electron delocalization in the phosphinimine part (formally comparable to a kind of merocyamine mesomerism), corresponding to a strong participation of the mesomeric form (A'). Experimental results, however, provide no evidence for a d_π-p_π interaction in the ground state between the P-aryl and the imine

$$R_3P{=}N{-}C_6H_4{-}NO_2 \longleftrightarrow R_3\overset{\oplus}{P}{-}N{=}C_6H_4{=}NO_2^{\ominus}$$

(A) (A')

part.[211,245,394,395,705]

The blocking of conjugation by P(V) with the coordination number 4 is confirmed by Bock.[133]

The dipole moments in iminophosphoranes are directed from phosphorus to the nitrogen. A reversal of the moments is not observed, even with strongly electronegative substituents on phosphorus.[394]

The polarity of the P-N bond in triphenylphosphinimine is apparently somewhat greater than that of the P-O bond in triphenylphosphine oxide, since it has as high a dipole moment in spite of the lower electronegativity of the nitrogen than oxygen.[471]

The basic strength (characterized by potentiometrically determined pK_a values) of iminophosphoranes decreases with the increasing electronegativity of the substituents on nitrogen and, although to a smaller extent, on phosphorus.[210,211] The π-electron structure of various iminophosphoranes was calculated using the Hückel molecular orbital method.[550]

C.2. Phosphazines (Phosphinazines)

C.2.1. Preparation of Phosphazines

I. FROM DIAZO COMPOUNDS

Staudinger and Meyer[610] synthesized the first phosphinazines (1) from tertiary phosphines and diazoalkanes.[610]

$$R_3P + N_2CR^1R^2 \longrightarrow R_3P{=}N{-}N{=}CR^1R^2$$

(1)

Kinetic results with para-substituted phenyldiphenylphosphines indicate that the rate of phosphinazine formation rises with increasing nucleophilicity of the phosphine used.[243] Treatment of diazoacetic acid azide with triphenylphosphine yielded a phosphazine which, moreover, contained the phosphinimine grouping:[451a]

$$N_2CHCO{-}N_3 + 2Ph_3P \xrightarrow{-N_2} Ph_3P{=}N{-}N{=}CHCO{-}N{=}PPh_3$$

II. FROM HYDRAZONES

Bestmann and Fritsche extended Horner's phosphinimine synthesis from amines and dihalophosphoranes to the preparation of phosphazines.[39,55] Hydrazones (3) and dibromotriphenylphosphoranes (2) produce phosphazines (1) in the presence of a base. The phosphonium salt (4) is formed as an intermediate.[55,585-6]

$$R_3PBr_2 + R^1R^2C{=}N{-}NH_2 \longrightarrow [R_3\overset{\oplus}{P}{-}NH{-}N{=}CR^1R^2]Br^{\ominus} \xrightarrow{\text{base}} R_3P{=}N{-}N{=}CR^1R^2$$

(2) (3) (4) (1)

III. FROM N-AMINOPHOSPHINIMINES

The interaction of triphenylphosphine-N-aminoimine (5) or its hydrobromide with aldehydes or ketones yields phosphazines (1) and water.[654]

$$R_3P{=}N{-}NH_2 + O{=}CR^1R^2 \xrightarrow{-H_2O} R_3P{=}N{-}N{=}CR^1R^2$$

(5) (1)

C.2.2. Reactions of Phosphazines

C.2.2.1. Hydrolysis and Salt Formation. Phosphazines, basic substances whose negative charge is delocalized, may be represented by the following equation:

$$R_3P{=}N{-}N{=}C\Big\langle \longleftrightarrow R_3\overset{\oplus}{P}{-}\overset{\ominus}{\underline{N}}{-}N{=}C\Big\langle \longleftrightarrow R_3\overset{\oplus}{P}{-}N{=}N{-}\overset{\ominus}{\underline{C}}\Big\langle$$

The hydrolytic cleavage[610] probably occurs by an attack of a hydroxyl ion or a water molecule on phosphorus, similarly to the hydrolysis of phosphine alkylenes or phosphinimines. Cyclopentadienylidenetriphenylphosphazine is, however, resistant toward hydrolysis.[485]

The reaction products of the hydrolysis of phosphazines are phosphine oxide and a hydrazone.[48,74,78,498,499,504,610,661] From α-ketophosphazines are thus formed α-ketoaldehyde-al-

$$R^1R^2C{=}N{-}N{=}PPh_3 \xrightarrow[OH^-]{H_2O} R^1R^2C{=}N{-}NH_2 + OPPh_3$$

hydrazones,[48,74] and from acylglyoxyl ester phosphazines the hydrazones of α,β-dioxocarboxylic acid esters,[78] which are easily converted by a Wolf-Kishner reduction into methyl ketones and β-ketocarbonic acids, respectively. β-Dicarbonyl-α-triphenylphosphazines afford α-hydrazono-β-diketones.[499] Particularly suited for the hydrolytic cleavage of α-ketophosphazines is nitrous acid, which leads directly to α-ketoaldehydes.[74]

The products which were named "phosphazine-hydrates" by Staudinger have been shown by Bestmann and Kolm[78] to be actually adducts of phosphine oxide and hydrazones (compare also Ref. 499).

Phosphazines will dissolve with salt formation in dilute mineral acid.[670] Generally the α-nitrogen is protonated.[56,74] With p-benzoquinotriphenylphosphazine, however, a vinylogous protonation on oxygen was observed.[311]

C.2.2.2. Reaction with Alkyl Halides. Alkyl halides when allowed to react with phosphazines also form salts.[610] Evidence that the alkylation takes place at the α-nitrogen was provided by Bestmann and Göthlich,[56] who isolated a phosphine oxide and an alkylhydrazone from the alkaline hydrolysis of the corresponding phosphonium salts.

NMR spectroscopic investigations of the phosphazine-alkyl halide adducts by Singh and Zimmer[585-6] confirm that alkylation takes place on the α-nitrogen. α-Ketotriphenyl-

$$R_3P{=}N{-}N{-}CR^1R^2 + R^3X \longrightarrow [R_3\overset{\oplus}{P}{-}\underset{R^3}{\underset{|}{N}}{-}N{=}CR^1R^2]X^{\ominus}$$

phosphazines and β-diketo-α-triphenylphosphazines form diazoketones and methyltriphenylphosphonium iodide with methyl iodide.[48,56,499] The tricyclohexyl derivatives, however, do not undergo cleavage but rather yield the expected phosphonium salts. The reason for this different behavior is that α-ketotriphenylphosphazines are in solution in an equilibrium with diazoketones and triphenylphosphine, whereby the latter is continuously removed from the equilibrium by salt formation with methyl iodide. This reaction may be used for the purification of diazoketones.[48]

C.2.2.3. Thermolysis of Phosphazines. Phosphazines which were formed from relatively stable diazo compounds decompose on heating reversibly into a tertiary phosphine and the diazo component.[614-5,677,686] Generally, however,

$$R^1R^2C{=}N_2 + PR_3 \rightleftharpoons R^1R^2C{=}N{-}N{=}PR_3$$

the diazo compound is destroyed at the decomposition temperature. In some cases the corresponding azine could be isolated.[610,614-5]

Staudinger and Meyer[610,611] found that benzophenonephosphazine (6) could be pyrolyzed to nitrogen and benzhydrylidenetriphenylphosphorane (7). The applicability of

$$Ph_2C{=}N{-}N{=}PPh_3 \xrightarrow{\Delta} Ph_2C{=}PPh_3 + N_2$$

(6) (7)

this type of phosphine alkylene synthesis is limited, however, to this special case.[684]

Wittig and Schlosser[684] have shown that it is possible to prepare phosphine alkylenes by the thermal decomposition of phosphazines in the presence of Cu(I) salts, which induce the splitting off of nitrogen from the diazo compounds formed as intermediates. Since the resulting phosphine alkylenes may react with undecomposed diazo compound to form ketazines, it is advisable to trap the phosphine alkylenes in situ with carbonyl compounds.[684]

The behavior of phosphazines toward electron bombardment in the mass spectrograph is similar to that observed under pyrolytic conditions.[701a]

C.2.2.4. Wittig-Type Reactions. Phosphazines interact with compounds containing a C-O or a C-S double bond in a manner reminiscent of the Wittig reaction.[39,55,148,608,610,677] The driving force probably is also the formation of the energetically poor P-O or P-S double bond, respectively, e.g.:

$$Ph_2C{=}N{-}N{=}PPh_3 + PhCHO \longrightarrow Ph_2C{=}N{-}\underset{\underset{PhHC-\overline{O}|^{\ominus}}{|}}{N}{-}\overset{\oplus}{P}Ph_3 \xrightarrow{-OPPh_3} Ph_2C{=}N{-}N{=}CHPh$$

Nitrosobenzene (8), which reacts with phosphine alkylenes analogously to carbonyl compounds, affords, with diphenylphosphazine (6), nitrogen, triphenylphosphine oxide, and benzophenone-anil (9).[545] The mechanism of this reaction is unclear.

$$Ph_3P{=}N{-}N{=}CPh_2 + PhNO \longrightarrow N_2 + Ph_3PO + Ph_2C{=}NPh$$

(6) (8) (9)

Interaction of phosphazines (1) with diphenyl ketene (10)[608] leads to Schiff's bases (12) of α-aminodiphenylacetonitrile derivatives.[55] This reaction may have occurred via rearrangement of the mixed azine (11) of the ketene series, which was formed as an intermediate.

$$\underset{(10)}{Ph_2C{=}C{=}O} + \underset{(1)}{Ph_3P{=}N{-}N{=}CR^1R^2} \xrightarrow{-OPPh_3} \underset{(11)}{Ph_2C{=}C{=}N{-}N{=}CR^1R^2}$$

$$\downarrow$$

$$\underset{(12)}{Ph_2C(CN){-}\bar{N}{=}CR^1R^2} \longleftarrow Ph_2\overset{\oplus}{C}(\overset{\ominus}{\bar{C}}{=}N) \leftarrow N{=}CR^1R^2$$

Diphenyl acetyl chloride and phosphazine yields the same products.[44]

C.2.2.5. Phosphazines and Acetylene-Dicarboxylic Acid Esters. Phosphazines (1) react with acetylene-dicarboxylic acid esters (10), as do the substituted phosphinimines,[149,159] to yield phosphine alkylene derivatives (15) via a phosphazacyclobutene intermediate (14).

$$\underset{(1)}{R_3P{=}N{-}N{=}CR^1R^2} + \underset{(13)}{R^3O_2C{-}C{\equiv}C{-}CO_2R^3} \longrightarrow$$

$$\underset{(14)}{\begin{matrix} R_3P & - & N{-}N{=}CR^1R^2 \\ | & & | \\ R^3O_2C{-}C & = & C{-}CO_2R^3 \end{matrix}} \longrightarrow \underset{(15)}{\begin{matrix} R_3P{=}C & - & C{=}N{-}N{=}CR^1R^2 \\ | & & | \\ R^3O_2C & & CO_2R^3 \end{matrix}}$$

Newer results indicate that the monosubstituted triphenylcarboxylidenephosphazine (16) reacts also with acetylene-dicarboxylic acid ester at the α-nitrogen, and not at the γ-carbon atom. The same type of reaction is observed

$$\underset{(16)}{Ph_3P{=}N{-}N{=}C(H)(CO_2R)}$$

in the interaction of phosphazines with activated nitriles.[166] This reaction yielded substituted aminophosphoranes, probably also via a four-membered-ring intermediate.

C.2.2.6. Phosphazines and Metallo-organic Compounds. The type of reaction of phosphazines with metallo-organic compounds RM depends on the kind of phosphazine used and on the reaction temperature.[76,77,677] Whereas at low temperature (-70°) only a substitution at phosphorus occurs, at elevated temperature a decomposition into phosphine, nitrogen, and a homologous metal organyl (17) is observed, e.g.:

$$CH_2{=}N{-}N{=}PPh_3 + RM \longrightarrow Ph_3P + N_2 + RCH_2M$$

(17)

The homologous metal organyl (17) can likewise attack the original phosphazine in a competing reaction, resulting in a repeated chain lengthening of the metallo-organic compound. Acetophenonephosphazine (18) and phenyllithium do not react according to the scheme shown above; instead they yield (after hydrolysis) two isomeric nitrogen-containing phosphine oxides.[76]

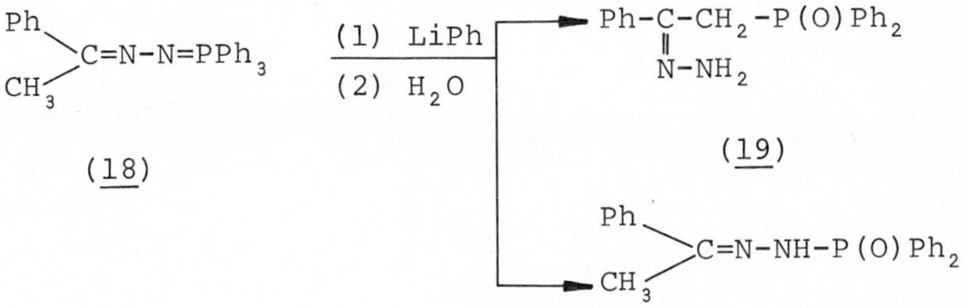

$$(Ph)(CH_3)C{=}N{-}N{=}PPh_3 \xrightarrow[(2)\ H_2O]{(1)\ LiPh} Ph{-}C({=}N{-}NH_2){-}CH_2{-}P(O)Ph_2 + (Ph)(CH_3)C{=}N{-}NH{-}P(O)Ph_2$$

(18) (19) (20)

C.2.3. Application of Phosphazines

Phosphazines may be used for the preparation of methyl ketones,[48] β-ketoacid esters,[78] glyoxals and α,β-dioxo esters[74,75,183,661] (which can be converted into heterocycles such as quinoxalines[74,75,183] and pteridines[661]). Phosphazines are also useful for the characterization of aliphatic diazo compounds[159,199,203,234,264,271,284a,446,447b,451a,499,506,544,544a,624,658] and quinone diazides.[311,498,503,505]

C.2.4. Physical Properties of Phosphazines

The assignment of the various bands in the IR spectra of phosphazines is arduous.[56,204,244]

In aromatic substituted phosphazines a strong triplet is observed in the regions of 490 to 505, 503 to 520, and 513 to 540 cm^{-1}, which apparently is due to the ν_{as} (P-$C_{arom.}$) stretching frequency. The P=N frequency was assumed by Goetz and Juds[244] in the region of 1015 to 1060 cm^{-1}.

Recently Bock et al.[135] were successful in assigning the characteristic valence frequency of the P=N-N=C-system by comparative investigations of ^{15}N-labeled nuclei and methylene deuteriated P-triphenyl-N-methylenephosphaketazines. In view of the ascertained valence frequencies (C=N, 1549; P=N, 1053; and N-N, 847 cm^{-1}), the P=N-N=C bonding system has to be described so that the electron density at the azine carbon atom is increased at the expense of the binding electrons of the P=N system. The P-N double bond is weakened because of an electron deficiency, and the N-C double bond is weakened as a result of excessive negative charge, while the N-N bond hardly profits from this electron density.

Phosphazines exhibit in the UV spectrum an intensive band in the region of 23,000 to 29,000 cm^{-1}. This has been assigned to a π-π^* transition of the phosphazine chromophor, P=N-N=C.[244,586]

The application of a simple electron model allows confirmation of this assignment for fluorenyl-9-phosphazine.[241] A kind of merocyanine mesomerism has been assumed.

According to Goetz and Juds,[244] the considerable bathochromic shift, relative to ketazines, C=N-N=R, is not due to conjugation extending over phosphorus. Rather, it results from a strong lowering of the activation energy in the ground state.

Dipole moment measurements[240,244] on phosphazines and calculations on an electron gas model[242] also point to the fact that P(V) with the coordination number 4 acts as a conjugation barrier. A contrasting view is held by Ried and Appel,[506] who assume, as a result of UV spectroscopic investigations on o-benzoquinone-trisaminophosphazines a continuous conjugation from the amine nitrogen to the quinone oxygen. Largely missing are ^{1}H-NMR spectroscopic investigations. In contrast to the P=N-N-C-H and P-N-N-CH systems, in which no long-range ^{31}P-H coupling is observed, in the phosphazine system P=N-N=CH a spin-spin coupling occurs between phosphorus and hydrogen.[43]

D. LIST OF COMPOUNDS

D.1. Phosphine Alkylenes (Phosphonium Ylids), $X_3P=R$

(Ordered According to the Number of Carbon Atoms in R)

C_1

$Me_3P=CH_2$. I.[535,682,368a] IX [533 535] M. 13-4°, b_{750} 118-20°,[535] ^{1}H-NMR.[535,533,536]
$Me_2(LiCH_2)P=CH_2$. IX. ^{1}H-NMR.[536b]
$Me_2EtP=CH_2$. I.[535] IX.[536b,535] B_{745} 143-5°,[535] ^{1}H-NMR.[535,536,536b]
$Me_2EtP=CH_2 \cdot LiCl$. I. ^{1}H-NMR.[536b]
$MeEt_2P=CH_2$. I. B_{12} 60-2°, ^{1}H-NMR.[536b]
$Et_2(LiCH_2)P=CH_2$. IX. ^{1}H-NMR.[536b]
$Me_2PrP=CH_2$. I. $B_{0.2}$ 25°, ^{1}H-NMR, ^{31}P -4,4.[368a]
$Et_3P=CH_2$. I.[535,368a] IX.[535] B_{12} 80-3°[535], ^{1}H-NMR,[535,536,368a] ^{31}P -23.6.[368a]
$(i\text{-}Pr)_3P=CH_2$. I. $B_{0.001}$ 38°, ^{1}H-NMR, ^{31}P -40,5.[368a]
$Bu_3P=CH_2$. I. $B_{0.001}$ 58°, ^{1}H-NMR, ^{31}P -16,4.[368a]
$Ph_2(CH_2=CH)P=CH_2$. V.[561]
$Ph_3P=CH_2$. I.[368a,568,676,685] II.[684] VIII.[560] M. 96°,[368a] IR,[392] UV,[258] x-ray,[27] ESR,[390] mass spect.,[671] ^{1}H-NMR,[89,110] ^{31}P -20.3.[257]
$(FC_6H_4)_3P=CH_2$. I. ^{19}F-NMR.[353]
$(cyclo\text{-}C_6H_{11})_3P=CH_2$. I.[81,368a] M. 128°.[368a]
$(p\text{-}MeC_6H_4)_3P=CH_2$. I.[688]
$(p\text{-}MeOC_6H_4)_3P=CH_2$. I.[688]
$Ph_3P=CHCl$. I.[569,683] II.[563,683]
$Ph_3P=CHBr$. I.[368,569]
$Ph_3P=CHI$. I.[566-7]
$Bu_3P=CF_2$. II.[233]
$Ph_3P=CF_2$. II.[227,232,281]
$Bu_3P=CCl_2$. II.[599]
$Ph_3P=CCl_2$. II.[599] II.[476,569,599]
$Ph_3P=CBr_2$. I.[599] II.[282,569,599]
$Ph_3P=CClF$. II.[154,599]
$Ph_2(CH_2=CPh)P=CH_2$. I.[518]

C_2

$Et_3P=CH\text{-}Me$. I.[368a,535] IX.[535] B_{12} 86-7°,[535] ^{1}H-NMR,[535,368a,536] ^{31}P -16.9.[368a]
$(i\text{-}Pr)_3P=CH\text{-}Me$. I. $B_{0.001}$ 45.[368a]
$Et_2PrP=CHMe$. I. $B_{0.001}$ 30°, ^{1}H-NMR,^{31}P -14.8.[368a]
$Ph_3P=CH\text{-}Me$. I.[167] IR,[135a] ^{1}H-NMR,[135a] UV,[258] ^{31}P -14.6.[257]
$[Ph_3P=CHMe]_3Mo(CO)_3$. IX. M. 126-40°, IR.[135a]
$Ph_3P=CH\text{-}CN$. I.[522,600] M. 190-2°,[600] ^{1}H-NMR,[110,573] pK_a,[600] ^{31}P -22,6.[600]
$(p\text{-}Me_2NC_6H_4)_3P=CHCN$. I. M. 202-4°.[462c]
$(p\text{-}MeOC_6H_4)_3P=CHCN$. I. M. 118-20°.[462c]
$Ph_3P=CBr\text{-}CN$. I. M. 162.5 - 64.5°, ^{31}P -21.9, pK_a.[600]

$Ph_3P{=}CH{-}CHO$. IX.[606,641,643] M. 186-7°,[643] ^{1}H-NMR,[110,594] ^{31}P -15, -19,[594] IR.[558]
$(p{-}Me_2NC_6H_4){\cdot}Ph_2P{=}CH{-}CHO$. I. M. 210-1°,[644]
$Ph_3P{=}CHOMe$. I.[386,683]
$Ph_3P{=}CHSMe$. I.[683]
$Ph_3\overset{\oplus}{P}{-}\dot{C}HCO_2H$. ERS.[390]
$Ph_3P{=}CHSO_2Me$. I. M. 200-3°, IR, UV.[603]
$Ph_3P{=}CClCHO$. IX. M. 195-7°, IR.[403]
$Ph_3P{=}CBrCHO$. IX. M. 180-1°, IR.[403]
$Ph_3P{=}C{=}C{=}O$. IV.[425b] M. 172-73,5°,[425b] x-ray,[181] IR,[425b] ^{31}P -2.6.[425b]
$Ph_3P{=}C{=}C{=}S$. IV.[425b] M. 224-6°,[425b] x-ray,[182] IR, ^{31}P +7.7[425b]
$Ph_3P{=}CH{-}CONH_2$. I. M. 177-8°.[640]
$Et_3P{=}CCl{-}CONH_2$. I.[603a]

C_3

$Et_3P{=}CHEt$. I. $B_{0.001}$ 30°, ^{1}H-NMR, ^{1}P -14.8.[368a]
$(i{-}Pr)_3P{=}CHEt$. I. $B_{0.001}$ 50°, ^{1}H-NMR, ^{31}P -30.0.[368a]
$Bu_3P{=}CHEt$. I. $B_{0.001}$ 80°,[368a]
$Ph_3P{=}CH{-}Et$. I.[30,167,172,640] ^{31}P -12.2[257] UV.[258]
$(i{-}Pr)_2EtP{=}CMe_2$. I. $B_{0.001}$ 45°.[368a]
$Ph_3P{=}CMe_2$. I.[167] ^{31}P -11.3,[257] UV.[258]
$Ph_3P{=}C(CN)_2$. VI.[309] IX.[424] M. 187-8°,[309] IR,[310,573] mass spect.,[168] UV.[309]
$Ph_3P{=}C(SCH_3)_2$. II.[383]
$Ph_3P{=}\overline{CCH_2CH_2}$. I.[49,388c]
$Ph_3P{=}CH{-}C{\equiv}CH$. I.[213]
$Ph_3P{=}CH{-}CH{=}CH_2$. I.[685]
$Ph_3P{=}CH{-}CH_2{-}CN$. III.[458] IX.[426]
$Ph_3P{=}CH{-}CH_2{-}CH_2{-}OH$. I.[269]
$Ph_3P{=}CH{-}CO{-}Me$. I.[435] I.[479] IX.[47] M. 205-6°,[479] IR,[4,47,479,685] DP,[4] UV,[479] mass spect.,[168] ^{1}H-NMR.[110,704]
$Ph_2MeP{=}CH{-}CO{-}Me$. ^{1}H-NMR.[704]
$(p{-}Me_2NC_6H_4)Ph_2P{=}CH{-}CO{-}Me$. I.[644] M. 176-7°.[644]
$Ph_3P{=}CH{-}CO{-}CH_2Cl$. I.[219,320] M. 179-80°,[320] IR,[320] ^{1}H-NMR.[320]
$Ph_3P{=}CCl{-}CO{-}Me$. IX.[190,403] M. 190-2°,[403] IR.[403]
$Ph_3P{=}CBr{-}CO{-}Me$. IX. M. 163-5°, IR.[403]
$Ph_3P{=}CH{-}CO_2Me$. I.[81,341] IX.[99] M. 164,[341] M. 169-69.5°,[110] IR,[558,600] mass spect.,[168] UV,[341] ^{1}H-NMR,[110,177,494,703] ^{31}P -16.8.[425b]
$Ph_3P{=}CD{-}CO_2Me$. IX. M. 166-9°, ^{1}H-NMR, ^{31}P -17.6.[425b]
$(cyclo{-}C_6H_{11})_3P{=}CH{-}CO_2Me$. I. M. 88-90°,[81]
$MePhNaphtP{=}CH{-}CO_2Me$. I.[113]
$(p{-}MeC_6H_4)_3P{=}CH{-}CO_2Me$. I. M. 141-2°, IR, mass spect., ^{1}H-NMR.[110]
$Ph_3P{=}CCl{-}CO_2Me$. IX. M. 171-3°, UV,[402] IR.[600]
$Ph_3P{=}CBr{-}CO_2Me$. IX. M. 168-9°, UV[402] IR.[600]

$Ph_3P=CI-CO_2Me$. IX. M. 165-7°, UV,[402] IR.[600]
$Ph_3P=C(SO_3Na)-CO_2Me$. IX.[450]
$Ph_3P=CH-CH_2-COOH$. I.[174]
$Ph_3P=CMe-CHO$. IX. M. 220-2°.[641]
$Ph_3P=CH-CS-NHMe$. IX. M. 203°.[91]
$Ph_3P=C(CN)CS_2H$. IX. M. 162-4°.[462c]
$(p\text{-}Me_2NC_6H_4)Ph_2P=C(CN)CS_2H$. IX. M. 136-9°.[462c]
$(p\text{-}MeOC_6H_4)_3P=C(CN)CS_2H$. IX. M. 125-8°.[462c]
$Me_3P=CH-PMe_2$. IX. M. -12 to -10°, b_{12} 80-2°, ^1H-NMR.[536a]
$Me_3P=CH-AsMe_2$. IX. M. -37 to -35°, b_{12} 85-7°, ^1H-NMR.[536a]
$Me_3P=CH-SbMe_2$. IX. M. -26 to -25°, $b_{0.1}$ 40-2°, ^1H-NMR.[536a]

C_4

$Et_3P=CHPr$. I. $B_{0.001}$ 33°, ^1H-NMR, ^{31}P -14.6.[368a]
$Bu_2EtP=CHPr$. I. $B_{0.001}$ 65°, ^1H-NMR, ^{31}P -11.1.[368a]
$Ph_3P=CHPr$. I.[431] UV,[258] ^{31}P -12.6.[257]
(cyclo-Pr)$Ph_2P=CHPr$. I.[49]
$Ph_3P=CMeEt$. I.[257] UV,[258] ^{31}P -10.5.[257]
$Me_3P=CH-CMe=CH_2$. I. $B_{0.1}$ 50°, ^1H-NMR, ^{31}P +10.2.[368a]
$(i\text{-}Pr)_3P=CH-CMe=CH_2$. I. $B_{0.001}$ 72°, ^1H-NMR, ^{31}P -30.0.[368a]
$Ph_3P=CH-CMe=CH_2$. I. M. 115°.[368a]
$Ph_3P=CH-CH=CH-Me$. I.[140]
$Ph_3P=CH-CH_2-CH_2-CH_2$. I.[444,520] IX.[79]
$Ph_3P=CH-(CH_2)_3-Br$. I.[444,520] IX.[79]
$Me_3P=CH-SiMe_3$. I.[534] IV.[438,534] IX.[536] M. -36°,[534,438] b_{11} 66°,[534,438] b_{14} 70-5°,[438] IR,[534,438,535] ^1H-NMR.[438,534,535]
$Me_3P=CH-GeMe_3$. I. M. -33 to -32°, B_{14} 77-80°, ^1H-NMR.[536a]
$(LiCH_2)Me_2P=CH-SiMe_3$. IX.[531]
$Me_2EtP=CH-SiMe_3$. I. B_{14} 83-4°, ^1H-NMR.[535]
$Et_3P=CH-SiMe_3$. I. B_{12} 110-1°, ^1H-NMR.[535]
$(Me_3SiCH_2)Me_2P=CH-SiMe_3$. IX. M. -34 to -35°, $B_{0.01}$ 44.5°, ^1H-NMR.[531]
$(Me_3SiCH_2)_2MeP=CH-SiMe_3$. IX. $B_{0.001}$ 55-6°.[531]
$Ph_3P=CH-SiMe_3$. I. IV.[534] IX.[237,534] M. 76-7°, B_1 150-3°,[535] IR,[534] ^1H-NMR.[534]
$Me_3P=CH-CO_2Et$. I.[642,643]
$Bu_3P=CH-CO_2Et$. I.[127]
$Ph_3P=CH-CO_2Et$. I.[341,434] M. 116-7°,[341] IR,[4,600,601-2] UV,[341,601-2] DP,[4] ^1H-NMR,[177,467] ^{31}P -19.1.[600]
$Ph_3P=CH-CS_2Et$. IX. M. 196°.[54]
$(cyclo\text{-}C_6H_{11})_3P=CH-CO_2Et$. I.[127]
$(n\text{-}C_6H_{13})_3P=CH-CO_2Et$. I.[127]
$(p\text{-}Me_2NC_6H_4)Ph_2P=CH-CO_2Et$. I. M. 141-2°.[644]
$(p\text{-}Me_2NC_6H_4)_2PhP=CH-CO_2Et$. I.[321]
$(p\text{-}MeOC_6H_4)_3P=CH-CO_2Et$. I.[127,512] M. 142°,[512]
$(n\text{-}C_8H_{17})_3P=CH-CO_2Et$. I.[127]
$(n\text{-}C_{10}H_{21})_3P=CH-CO_2Et$. I.[127]

$(p\text{-}Ph\text{-}C_6H_4)_3P{=}CH\text{-}CO_2Et$. I.[127,689]
$(p\text{-}ClC_6H_4)_3P{=}CH\text{-}CO_2Et$. I. M. 162°.[512]
$(NC\text{-}CH_2\text{-}CH_2)Ph_2P{=}CH\text{-}CO_2Et$. III.[639]
$(EtO_2C\text{-}CH_2\text{-}CH_2)Ph_2P{=}CH\text{-}CO_2Et$. III.[639]

(Structure: 2,3,4,5-tetraphenylphosphole ring, P-Ph) $P{=}CH\text{-}CO_2Et$. I.[157,314] M. 161-3°,[157]

$Ph_3P{=}CCl\text{-}CO_2Et$. IX.[190,402,601-2] M. 147.5-148.5°,[601-2] IR,[600,601-2] pK_a,[600] ^{31}P -22.7,[600] UV.[601-2]
$Ph_3P{=}CBr\text{-}CO_2Et$. IX.[190,424,601-2] M. 157-8°,[424] 155-6°,[601-2] IR,[600,601-2] pK_a,[600] UV.[601-2]
$Ph_3P{=}C(CN)COMe$. ^1H-NMR[704]
$Ph_3P{=}C(CN)COCONH_2$. IX. M. 264°, UV, IR.[166a]
$PH_3P{=}C(CN)COCN$. IX. M. 222-3°, UV, IR.[166a]
$Ph_3P{=}C(CN)CO_2Me$. VI.[168,309] IX.[424] M. 217-8°,[424] IR,[100,310,573] mass spect.,[100,168] UV.[309]
$Ph_3P{=}C(CN)CS_2Me$. IX. M. 240-1°.[462c]
$Ph_3P{=}C(CN)COCH_2Cl$. IX. M. 185-6°.[388b]
$Ph_3P{=}C(CN)COCH_2Br$. IX. M. 164°.[388b]
$Ph_3P{=}CClCOEt$. IX. M. 165-7°, IR.[403]
$Ph_3P{=}CBrCOEt$. IX. M. 156-8°, IR.[403]
$Ph_3P{=}C(CHO)CO_2Me$. IX. M. 158-60°, IR.[405]
Ph_3 $P{=}CMeCO_2Me$. IX.[99,100] M. 145°,[99] IR,[558,600] ^1H-NMR.[703]

$Ph_3P{=}C$—CO, O, CH_2-CO (ring). III.[4,274,461] M. 157-9°,[274] IR,[4,274] ^1H-NMR.[274]

$Ph_3P{=}C$—CO, NH, CH_2-CO (ring). III.[274] M. 220°,[274] IR.[274]

$Bu_3P{=}C$—CO, NH, CH_2-CO (ring). III.[274]

$Ph_3P{=}C$—CO, OC—CF_2 (ring). III. M. 228-9°, IR, ^{31}P -4.1, ^{19}F-NMR.[618]

$Ph_3P{=}C$—CO, OC—CCl_2 (ring). IX. M. 224-26.5°, IR.[688a]

$Ph_3P{=}C$—CO, CH_2, CO—O (ring). IX. M. 206-8°, IR, ^1H-NMR.[161a]

$Ph_3P{=}C{=}C(CF_3)_2$. IX. M. 110-55° (+ $OPPh_3$), ^{31}P -24.5, ^{19}F-NMR[126a]
$Ph_3P{=}CMeCHOLi\text{-}Me$. IX.[175]

C_5

$Ph_3P{=}CH\text{-}Bu$. I.[251,523]

$Ph_3P{=}CH{-}i{-}Bu$. I.[334]
$Ph_3P{=}CH{-}t{-}Bu$. I.[570]
$Ph_3P{=}CHEt_2$. I.[247] ^{31}P -10.9.[257]
$Ph_3P{=}CMeCH_2CH{=}CH_2$. V.[561]

$Ph_3P{=}C(CH{=}CH)_2$ (cyclopentadienylidene). I.[483] M. 228-31°,[483] IR,[310,483] UV,[483] DP.[483]

$Ph_3P{=}C(CH_2{-}CH_2)_2$ (cyclopentylidene: $CH_2{-}CH_2$ / $CH_2{-}CH_2$). IX.[79] ^{31}P -4.8.[257]

$Ph_3P{=}C(CH_2{-}CH_2)_2O$. IX.[79]

$Ph_3P{=}C(CH_2{-}CH_2)_2S$. IX.[79]

$Ph_3P{=}C(CH{-}Me)(CH_2)CH_2$ (ring: CH—Me, CH_2, CH_2). IX.[79]

$Ph_3P{=}C{-}CO{-}CH_2{-}CH_2{-}CH_2$ (ring). IX.[34,315] M. 243-5°, IR, UV.[315]

$Ph_3P{=}C{-}CO{-}O{-}CO{-}C(H)Me$ (ring). III. M. 179-81°, IR, ^1H-NMR.[274]

$Ph_3P{=}C(CO{-}CF_2)_2$ (ring: $CO{-}CF_2$ / $CO{-}CF_2$). III. M. 173-4°, ^{31}P -10.2, IR, ^{19}F-NMR.[618]

$Bu_3P{=}C(CO{-}CF_2)_2$ (ring). III. M. 68-9°, ^{31}P -21.7, IR, ^{19}F-NMR.[618]

$BuPh_2P{=}C(CO{-}CF_2)_2$ (ring). III. M. 148-9°, ^{31}P -12.8, IR, ^{19}F-NMR.[618]

$Ph_3P{=}CH{-}C_4H_2O{-}NO_2$ (5-nitro-2-furyl). I.[689b] II.[517a] M. 113°.[689b]
$Ph_3P{=}C(COMe)CO_2Me$. IX. M. 153-5°.[164]
$Ph_3P{=}C(COCH_2Cl)CO_2Me$. IX.[164,388b] M. 138-9°, IR.[164]
$Ph_3P{=}C(COCH_2Br)CO_2Me$. IX.[388b] Oil.
$Ph_3P{=}C(COCCl_3)CO_2Me$. IX. M. 144-6°.[388b]
$Ph_3P{=}C(CO_2Me)CO_2Me$. VI. M. 180-2°, mass spect.[168]
$Ph_3P{=}C(CH_2CN)CO_2Me$. IX. M. 138-9°, IR.[100]
$Ph_3P{=}CEtCO_2Me$. IX.[99,100] M. 125°,[99] ^1H-NMR.[703]
$Ph_3P{=}CMeCO_2Me$. I.[69,339] M. 159-60°,[69] ^1H-NMR,[72,178] mass spect.[168]
$(PhMeNapht)P{=}CMeCO_2Et$. I.[113]
$Ph_3P{=}C(SO_2Me)CO_2Et$. IX. M. 175-7°.[344]
$Ph_3P{=}CMeCS_2Et$. IX. M. 196°.[54]
$Ph_3P{=}CH{-}CH_2{-}CO_2Et$. III.[458]

Ph_3P=CH-CH=CH-CO_2Me. I.[138]
Ph_3P=C(COMe)$_2$. IX. M. 167-9°, IR,[164] ^{1}H-NMR.[704]
Ph_3P=C(COMe)SC(NH)SCN. IX. M. 95-7°, IR.[696]
Ph_3P=CHS-n-Bu. IX.[445a]
Ph_3P=C(SEt)$_2$. II.[383]
Ph_3P=CHO-Bu. I.[674]
Ph_3P=C(CN)$COCO_2$Me. IX. M. 210-1°, UV, IR, ^{1}H-NMR.[166a]
Ph_3P=C(CN)CO_2Et. VI.[168] IX.[424] M. 208-10°,[168] mass spect.[168]
Ph_3P=C(CN)CS_2Et. IX. M. 206-7°,[462c]
Ph_3P=C(CN)(dichlorotriazinyl: N, N, N ring with Cl, Cl). IX. M. 282-3O.[462a]
Ph_3P=CH-Am. I.[257] ^{31}P -12.2,[257] UV.[258]
Ph_3P=CEtPr. I. ^{31}P -10.7.[257]
Ph_3P=C<(CH_2-CH_2)(CH_2-CH_2)>CH_2. I. ^{31}P -6.4,[257] UV.[258]
Ph_3P=C<(CH_2-CH)(CH_2-CH)>CH_2 (CH–CH bridged). IX.[79]
Ph_3P=CH-CMe=$CHCO_2$Me. I.[188]
Ph_3P=CH-(CH=CH)$_2$-Me. I.[140]
Ph_3P=C(CO_2Me)CH_2CO_2Me. IX.[98,100] M. 158-60°,[98] IR.[100]
Ph_3P=CPrCO_2Me. IX.[99,103] M. 105°,[99] ^{1}H-NMR.[703]
Ph_3P=C(CO_2Me)CH_2-CH=CH_2. IX. Oil, IR.[100]
Ph_3P=C(CO_2Me)CH_2-$CHNO_2$-Me. IX. M. 162-4°, IR, UV.[21a]
Ph_3P=C(CO_2Me)CHMeCH_2NO_2. IX. M. 164-6°, IR, UV.[21a]
Ph_3P=C(CO_2Me)(dichlorotriazinyl: N, N, N ring with Cl, Cl). IX. M. 212-3°.[462a]
Ph_3P=C(CO_2Et)COMe. III.[246,309] M. 172-4°,[164] IR,[164] mass spect.,[168] ^{1}H-NMR.[573a]
Ph_3P=C(CO_2Et)$COCH_2$Cl. IX.[164,388b] M. 137-8°, IR.[164]
Ph_3P=C(CO_2Et)$COCH_2$Br. IX.[388b] Oil.
Ph_3P=C(CO_2Et)$COCCl_3$. IX.[164,388b] M. 161-2°,[164] M. 144-5°,[388b] IR.[164]
Ph_3P=C(COMe)CH_2CO_2Me. IX. M. 148-50°, IR, ^{1}H-NMR.[57]
Ph_3P=ClCO-2-thienyl. IX. M. 151-2°.[259]
Ph_3P=C=C(OEt)$_2$. IX. M. 83°, ^{1}H-NMR.[97]
Ph_3P=CH-(CH_2)$_2$-CO_2Et. I.[33,315]
Ph_3P=(5-hydroxy-2-oxo-2H-pyran ring: O, O, OH). III. M. 262-6°, IR, UV.[478]

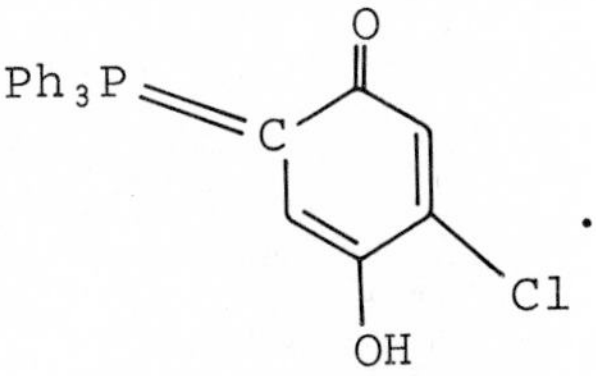

[structure: Ph$_3$P=C ring with C=O, Cl, OH]. I. M. 249-51°, UV, IR.[478]

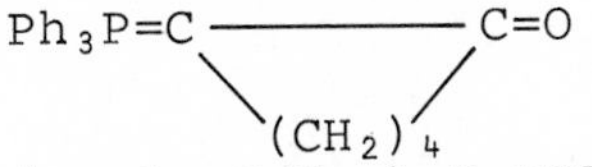

$Ph_3P=C$—C=O with $(CH_2)_4$ ring. IX. M. 245-7°, IR, UV, ^{1}H-NMR.[315]

(cyclo-C_6H_{11})$_3$P=CClCONBu. I.[603a]
$Ph_3P=CH-CO-CH_2-CO_2Et$. I. M. 104-5°, UV.[558a]
$Ph_3P=CH-CO_2$-i-Bu. I. M. 104-5°.[113]
$Ph_3P=CH-SO_2-CHMe-CO_2Et$. IX. M. 129-130°.[344]
$Ph_3P=CMeCH=CHCO_2Me$. I. M. 166-72°.[153]
$Ph_3P=CMeCOPr$. IX. M. 129-31°.[47]
$Ph_3P=C(CN)COPr$. IX. M. 177°,[247] UV.[247]
$Ph_3P=CPrCOMe$. IX. M. 143-5°.[47]
$Ph_3P=C(CN)COCO_2Et$. I. M. 215-6°, UV, IR, ^{1}H-NMR.[166a]
$Ph_3P=C(CN)CH=C(CN)_2$. IX. M. 284-5°, IR,[636] UV, ^{1}H-NMR.[573]
$Ph_3P=C(CN)CS_2CH_2CO_2Me$. IX. M. 118-20°.[462c]

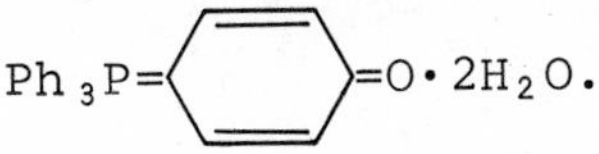

Ph_3P=[cyclohexadienylidene]=$O\cdot 2H_2O$. I. M. 310°.[299]

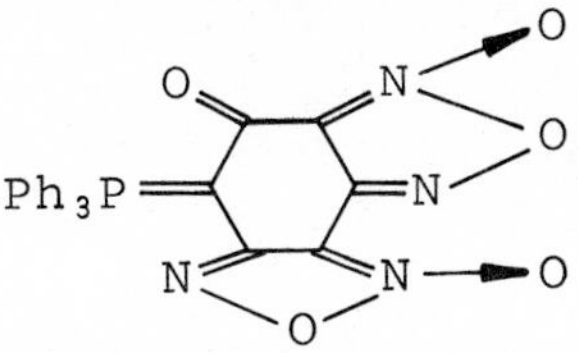

[structure: Ph_3P= ring with O, N-oxide furoxan rings]. X-ray.[22]

C_6

$Ph_3P=C(OMe)CO$-cyclo-Pr. IX. M. 156-9°.[700]
$Ph_3P=C(OMe)CO$-i-Pr. IX. M. 155-9°.[700]
Ph_3P-C-i-$PrCO_2Me$. ^{1}H-NMR.[703]
$Ph_3P=CPrCO_2Me$. IX. M. 105°, IR.[103]
$Ph_3P=CMeSBu$. IX.[445a]
$Me_3P=C(PMe_2)SiMe_3$. IX. M. -8 to -6°, B_{12}122-3°, ^{1}H-NMR.[536a]

C_7

$Ph_3P=CMeAm$. V.[561]
$Ph_3P=CH-(CH=CH)_2-CO_2Me$. IR.[558]
$Ph_3P=C<(CH_2)(CH_2)>C<(CH_2)(CH_2)>CH_2$. IX.[79]
$Ph_3P=C<(CH_2-CH)(CH_2-CH)>(CH_2)_2$. IX.[79]

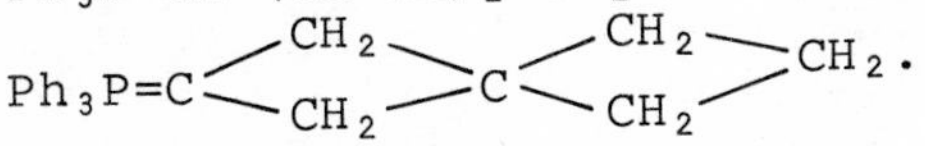

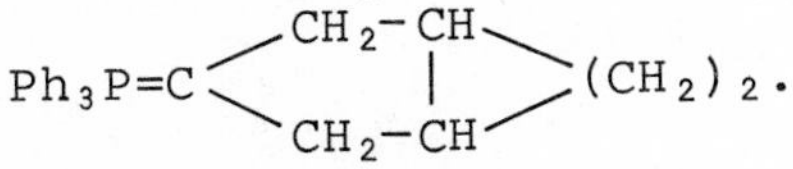

$Ph_3P=C\langle$ CH_2—CH, CH_2, CH_2 — CH $\rangle CH_2$ (bicyclic). IX.[79]

$Ph_3P=C\langle(CH_2)(CH_2)\rangle(CH_2)_4$. IX.[79]

$Ph_3P=C$—C=O, $(CH_2)_5$ (ring). IX. M. 205-8°, IR, UV.[315]

$Ph_3P=CHOPh$. I.[674]
$Ph_3P=CHSPh$. IX.[445a]
$Ph_3P=C(CO_2Me)CHMe-CHMeNO_2$. IX. M. 193.5-95,5°, IR, UV.[21a]
$Ph_3P=C(CN)CS_2Bu$. IX. M. 209-11°.[426c]
$Ph_3P=CH-CH=CH-Bu$. I.[141]
$Ph_3P=C(CN)-NH$-(2-furyl, 5-NO_2). IX.[516]
$Ph_3P=C(CN)$-CO-α-furyl. IX. M. 241-2°.[582]
$Ph_3P=C(SCN)$-CO-α-thienyl. IX. M. 181-2°.[259]
$Ph_3P=C(CN)$-CO-α-thienyl. IX. M. 193-4°.[582]
$Ph_3P=CPrCOEt$. IX. M. 114°, UV.[247]
$Ph_3P=CEtCO$-i-Pr. IX. M. 125-6°.[698]
$Ph_3P=C(OMe)CO$-cyclo-Bu. IX. M. 178-181°.[700]
$Ph_3P=C(OMe)COBu$. IX. M. 131-3°.[700]
$Ph_3P=C(COMe)CH=CHCOMe$. IX. M. 174-6°.[701]
$Ph_3P=C(i-Bu)CO_2Me$. ^1H-NMR.[703]
$Ph_3P=C(COEt)CO_2Et$. IX. M. 123-5°, IR.[164]
$Ph_3P=C(CO_2Et)_2$. VI. M. 106-7°,[309] IR,[310,600] mass spect.,[168] UV.[309]
$Ph_3P=CPrCO_2Et$. IX. M. 89-90°.[69]
$Ph_3P=CH-(CH_2)_3-CO_2Et$. I.[33,315]
$Ph_3P=C(CN)C(CN)=C(CN)_2$. IX. M. 239-40°, IR,[636] UV,[573] ^1H-NMR.[573]
$Ph_3P=CH$-cyclo-C_6H_{11}. I.[333]
$Ph_3P=CH$-1-cyclohexenyl. I.[332]
$Ph_3P=CH-CH_2-N$(pyrrolyl, 2-CHO). V.[553]
$MeEtPhP=CHPh$. I.[26,113,429]
$MePrPhP=CHPh$. I.[87,113]
$Bu_3P=CHPh$. I.[26,349]
$Me_2PhP=CHPh$. I.[357,642,643]
$Ph_2MeP=CHPh$. I.[357,642,643]
$Ph_2(CH=CH)P=CHPh$. IX.[518]
$EtO_2C-(CH_2)_2-Ph_2P=CHPh$. III.[639]
$NC-(CH_2)_2Ph_2P=CHPh$. III.[639]
$H_2N-C(=O)-(CH_2)_2Ph_2P=CHPh$. III.[639]
(cyclo-$C_6H_{11})_3P=CHPh$. I.[81,357]
MePh(α-Napht)P=CH-Ph. I.[113]

$Ph_2(C_6H_4NMe_2$-p)P=CHPh. I.[644]
Ph_3P=CH-Ph. I.[357,685]

O=(cyclohexadienylidene with 2,6-Ph)P(Ph)=CHPh. I.[413]

$Ph_3P=CH-C_6H_4Cl$-P. I.[114]
$Ph_3P=CH-C_6H_4NO_2$-p. I.[251,349] M. 171-2°, IR, UV, ^{31}P -13.[251]
$Ph_3P=CH-C_6H_3(NO_2)_2$-2,4. I. M. 209-10°, IR.[462b]
$(p-ClC_6H_4)_3P=CH-C_6H_4NO_2$-p. I.[349]
$Bu_3P=CH-C_6H_4NO_2$-p. I.[349]
$Me_3P=C(SiMe_3)_2$. I.[534] IX.[534] M. 14-18°, b_1 60-62°,[534] IR,[535] ^1H-NMR.[531,535]
$Me_3P=C(GeMe_3)_2$. IX. B_1 91-5°, ^1H-NMR.[536a]
$Me_3P=C(SnMe_3)_2$. IX. M. -30 to -28°, $b_{0.1}$ 73-6°, ^1H-NMR.[536a]
$Me_3P=C(SiMe_3)GeMe_3$. IX. M. 14-5°, b_1 60-5°, IR, ^1H-NMR.[534]
$Me_3P=C(SiMe_3)SnMe_3$. IX. M. 11-3°, b_1 51-3°, IR, ^1H-NMR.[534]
$(Me_3SiCH_2)Me_2P=C(SiMe_3)_2$. IX. M. -40, $b_{0.001}$ 55°, ^1H-NMR.[531]
$Ph_3P=C(GeMe_3)_2$. IX. M. 145-7°, $b_{0.001}$ 158-61°, IR, ^1H-NMR.[534]
$Ph_3P=C(SnMe_3)_2$. IX. M. 129-30°, b_1 183°, IR, ^1H-NMR.[534]

C_8

$Ph_3P=CH-(CH_2)_6$-Me. I.[456]
$Ph_3P=CH-CH_2$-Ph. V.[561,562]

Ph_3P=CH-CH=(cyclohexylidene). I.[332]

Ph_3P=CH-CH=(2-hydroxycyclohexylidene, HO). I.[270]

Ph_3P=C(COMe)CO-NH-α-furyl. IX.[516]

Ph_3P=C(COMe)CO-NH-(5-nitro-2-furyl, NO_2). IX.[516]

Ph_3P=C(CN)CS-NH-CO-α-furyl. IX.[516]
$Ph_3P=C(CN)-C(CO_2Me)=CHCO_2Me$. IX. M. 229-30°, IR.[636]
Ph_3P=C(CN)SPh. IX. M. 130°.[515a]
$Ph_3P=C(CN)SC_6H_4Br$-p. IX. M. 172°.[515a]
$Ph_3P=C(CN)SC_6H_4Cl$-p. IX. M. 171°.[515a]
$Ph_3P=C(CN)SC_6H_4NO_2$-o. IX. M. 203°.[515a]

Ph_3P=CMeSPh. IX.[445a]
$Ph_3P=C(CN)C_6H_3(NO_2)_2$-2.4. IX. M. 249-50°, IR, UV.[462b]
$Ph_3P=C(CN)C_6H_2(NO_2)_3$-2.4,6. IX. M. 245-6°, IR, UV.[462b]
$Ph_3P=C(CO_2Et)CH_2CO_2Et$. III. M. 104-6°.[288]
$Ph_3P=C(CO_2Et)$COCH=CHMe. IX. M. 146-7°, UV.[247]
$Ph_3P=C(CO_2Et)$COPr. IX.[164,247] M. 132-3°,[164] UV,[247] IR.[164]
$Ph_3P=C(CO_2Et)$CH=CHCOMe. IX. M. 172-5°.[701]
$Ph_3P=C(CO_2$-t-Bu)$COCH_2Cl$. IX. M. 153°.[161a]
$Ph_3P=C(CO_2Me)C(CO_2Me)$=CHMe. IX. M. 174-5°, IR, ^{1}H-NMR.[95]
$Ph_3P=C(CO_2Me)$CO-α-furyl. IX. M. 167-9°.[401]
$Ph_3P=C(CO_2Me)$Am. ^{1}H-NMR.[703]
Ph_3P=C(COPr)Pr. IX.[247,697] M. 97-9°.[697]
Ph_3P=C(COCHClEt)Pr. IX. M. 173-6°.[697]
Ph_3P=C(CO-i-Pr)Pr. IX. M. 139-42°.[697]
Ph_3P=C(CO-cyclo-Pr)Pr. IX. M. 124-6°.[697]
Ph_3P=C(CO-trans-CH=CHMe)Pr. IX. M. 168-70°.[697]
Ph_3P=C(COMe)Am. IX. M. 183-6°.[698]
$Ph_3P=C(COPr)CH_2CO_2Me$. IX.[57]
Ph_3P=CH-CS-NHPh. IX. M. 203°.[91]
Ph_3P=CH-CONHPh. IX. M. 177-8°.[640]
Ph_3P=C(COAm)Me. IX.[47] Oil.
$Ph_3P=C[CON(allyl)_2]Cl$. I.[603a]
$(Cyclopentyl)_3P=C(CONPr_2)Cl$. I.[603a]
$(Butenyl)_3P=C[CONH(CH_2)_6(NH_2)]Cl$. I.[603a]
$Bu_3P=C[CONH(CH_2)_6(NH_2)]Cl$. I.[603a]
$Ph_3P=C[CON(propargyl)_2]I$. I.[603a]
Ph_3P=CMePh. I.[95] 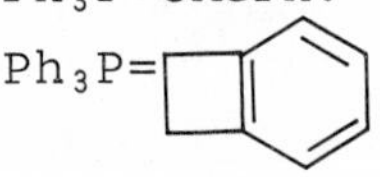
Ph_3P= . I.[132]
$Ph_3P=CH-C_6H_4OMe$-p. I.[114,219,349,361] M. 154-5°.[219]
Ph_3P=CPhCHO. IX. M. 124-6°, IR.[405]
$Ph_3P=CHOC_6H_4Me$-p. I.[674]
$Ph_3P=CH-(CH_2)_4-CO_2Et$. I.[33,315] 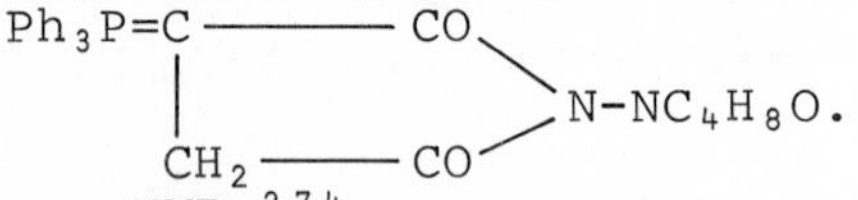
Ph_3P=C——CO
|　　　　　　N-NC_4H_8O. III. M. 132-4°, IR, ^{1}H-NMR.[274]
CH_2——CO
$Ph_3P=CH-SO_2-C_6H_4$-Me-p. I. M. 186-7°, IR, UV,[603] x-ray.[603,662]
$Ph_3P=C(SO_2C_6H_4Me$-p)Br. IX. M. 188-9°, IR.[603]
$Me_nPh_{(3-n)}P$=CH-CO-Ph. I.[642,643] M. 126-7° (n = 2), 123-4° (n = 1).[643]
Ph_3P=CH-COPh. I.[435,479] IX.[47] M. 178-80°,[479] 181-2°,[425] IR,[4,47,392,479,494,600,601-2] UV,[479,601-2] ^{1}H-NMR,[494] mass spect.,[168] ^{31}P -21.6,[600] pK ,[425a,600] DP.[4,479]
$(p-MeC_6H_4)_3P$=CH-COPh. I. M. 177-8, pK_a.[425a]
$(m-MeC_6H_4)_3P$=CH-COPh. I. M. 156-7°, pK_a.[425a]
$(p-MeOC_6H_4)_3P$=CH-COPh. I. M. 142-3°, pK_a.[425a]

$(m\text{-}MeOC_6H_4)_3P{=}CH\text{-}COPh$. I. M. 161-2°, pK_a.[425a]
$Ph_3P{=}CH\text{-}CO\text{-}C_6H_4NO_2\text{-}p$. I.[219] IX.[47] M. 159-61°,[47] IR.[47]
$(p\text{-}ClC_6H_4)_3P{=}CH\text{-}CO\text{-}C_6H_4NO_2\text{-}p$. I. M. 225.5-26°, pK_a.[425a]
$(p\text{-}MeC_6H_4)_3P{=}CH\text{-}CO\text{-}C_6H_4NO_2\text{-}m$. I. M. 197-8°, pK_a.[425a]
$Ph_3P{=}CH\text{-}CO\text{-}C_6H_4Br\text{-}p$. I. M. 200°.[621]
$(p\text{-}MeC_6H_4)_3P{=}CH\text{-}CO\text{-}C_6H_4Cl\text{-}p$. I. M. 169-70°, pK_a.[425a]
$(p\text{-}ClC_6H_4)_3P{=}CH\text{-}CO\text{-}C_6H_4Cl\text{-}p$. I. M. 245-6°, pK_a.[425a]
$Ph_3P{=}C(COPh)HgCl$. I. M. 226-7°.[451]
$Ph_3P{=}C(COPh)Br$. IX.[403,600] M. 167-9°,[403] IR,[403,600,601,602] pK,[600] ^{31}P -19.8,[600] x-ray,[603] UV.[601-2]
$Bu_3P{=}C(COPh)Br$. IX.[600]
$Ph_3P{=}C(COPh)Cl$. IX.[190] M. 156-8°,[601-2] IR,[600,601-2] UV,[601-2] pK_a,[600] x-ray.[603]
$Ph_3P{=}C(COPh)I$. IX.[259,601-2] M. 186-7°,[601-2] M. 156.5-57,5°,[259] IR,[600,601-2] UV,[601-2] x-ray,[603] pK_a,[600] ^{31}P -19.5.[600]
$Ph_3P{=}C(COC_6H_4NO_2\text{-}p)I$. IX. M. 183.5-84.5°.[259]
$Ph_3P{=}C(COC_6H_4Br\text{-}p)I$. IX. M. 180-1°.[259]
$Ph_3P{=}C(COC_6H_4Cl\text{-}p)I$. IX. M. 170-1°.[259]
$Ph_3P{=}C$ (ring: CH_2—CH—CH_2 / O bridge / CH_2—CH—CH_2, closing at CH_2). IX.[79]

C_9

$Ph_3P{=}C$ (ring: CH_2—CH / CH_2—CH, fused with $(CH_2)_4$). IX.[79]

$Ph_3P{=}C$ (ring: CH_2—CH—CH_2—CH=CH—CH_2—CH—CH_2, CH—CH bridged). IX.[79]

$Ph_3P{=}CH\text{-}CH{=}CH\text{-}C_6H_{13}$. I.[141]
$Ph_3P{=}CH\text{-}CH{=}CH\text{-}Ph$. I.[38,439,456]
$(cyclo\text{-}C_6H_{11})_3P{=}CH\text{-}CH{=}CH\text{-}Ph$. I.[81]
$Ph_3P{=}CH\text{-}C_6H_3(2\text{-}OMe, 5\text{-}CHO)$. I.[428]
$Ph_3P{=}CH\text{-}CH_2\text{-}COPh$. I.[255,256]
$Ph_3P{=}CH\text{-}CHMe\text{-}Ph$. V.[561]
$Ph_3P{=}C(CH_2Ph)Me$. V.[561]
$Ph_3P{=}C(CHOLi\text{-}n\text{-}C_6H_{13})Me$. IX.[175,176]
$Ph_3P{=}C(CHOLi\text{-}Ph)Me$. IX.[175]
$Ph_3P{=}CH\text{-}C_6H_4CO_2Me\text{-}p$. I.[156]
$Ph_3P{=}C(CO_2Me)Ph$. IX.[99,103] M. 155°,[99] mass spect.,[168] ^1H-NMR.[703]
$Ph_3P{=}C(CO_2Me)C_6H_4NO_2\text{-}p$. ^1H-NMR.[703]
$Ph_3P{=}C(CO_2Me)C_6H_3(NO_2)_2\text{-}2,4$. IX. M. 230-1°, IR, UV.[462b]
$Ph_3P{=}C(CO_2Me)C_6H_2(NO_2)_3\text{-}2,4,6$. IX. M. 192-3°, IR, UV.[462b]
$Ph_3P{=}C(CO_2Me)C_6H_4Br\text{-}p$. ^1H-NMR[703]
$Ph_3P{=}C(CO_2Me)CH(C_4H_9O)\text{-}CH_2NO_2$. IX. M. 184.5-85.5°, IR, UV.[21a]

$Ph_3P{=}C(CO_2Me)$cyclo-C_6H_{11}. IX.[99,103] Oil.
$Ph_3P{=}C(CO_2Me)N{=}N{-}Ph$. IX. M. 197-8°, UV.[400]
$Ph_3P{=}C(CO_2Me)N{=}N{-}C_6H_4NO_2$-p. IX. M. 196-8°, UV.[400]
$Ph_3P{=}C(COMe)CO$-3-pyridyl. IX. M. 212-6°.[700]
$Ph_3P{=}C(COMe)CO$-2-furylvinyl. IX. M. 201-6°.[700]
$Ph_3P{=}C(COMe)CH{=}CHCO$-i-Pr. IX. M. 146-8°.[701]
$Ph_3P{=}C(COMe)Ph$. I.[246,247] IX.[47,642] M. 171°.[247]
$Ph_3P{=}C(CO_2Et)CO$-NH-α-furyl. IX.[516]

$Ph_3P{=}C(CO_2Et)CO$-NH-(5-nitro-2-furyl, structure with NO_2). IX.[516]

$Ph_3P{=}C(CO_2Et)CH{=}CHCOEt$. IX. M. 170-1°.[701]
$Ph_3P{=}C(CHO)CO$-NHPh. IX. M. 230-1°.[640]
$Ph_3P{=}CH{-}CO{-}CH_2Ph$. IX. 147-8°,[47] ^{1}H-NMR,[704] IR.[47]
$Ph_3P{=}CH{-}CO{-}C_6H_4OMe$-p. I. M. 154-5°.[219]
$(p\text{-}ClC_6H_4)_3P{=}CH{-}CO{-}C_6H_4OMe$-p. I.[349]
$(p\text{-}ClC_6H_4)_3P{=}CH{-}CO{-}C_6H_4Me$-p. I. M. 236.5-38°, pK_a,[425a]
$(p\text{-}MeC_6H_4)_3P{=}CH{-}CO{-}C_6H_4Me$-p. I. M. 169.5-70.5°, pK_a.[425a]
$(p\text{-}MeOC_6H_4)_3P{=}CH{-}CO{-}C_6H_4Me$-p. I. M. 178-9°, pK_a.[425a]
$Ph_3P{=}CH{-}CO{-}CMe_2{-}CO$-i-Pr. I.[517]
$Ph_3P{=}CH{-}SO_2{-}CHMe{-}CO_2Ph$. IX. M. 153°.[344]
$Ph_3P{=}C(COPh)Me$. IX.[46,47] M. 170-2°, IR.[47]
$Ph_3P{=}C(COPh)SO_2Me$. IX. M. 193-4°.[344]
$Ph_3P{=}C(COPh)CN$. IX.[247,424,582] M. 208°, UV.[247]
$Ph_3P{=}C(COPh)SCN$. IX. M. 177-8°.[259]
$Ph_3P{=}C(COC_6H_4Me\text{-}p)I$. IX. M. 173-4°.[259]
$Ph_3P{=}C(COC_6H_4OMe\text{-}p)I$. IX. M. 177-8°.[259]
$Ph_3P{=}C(COC_6H_4NO_2\text{-}p)I$. IX. M. 167-8°.[259]
$Ph_3P{=}C(COC_6H_4Br\text{-}p)SCN$. IX. M. 164-5°.[259]
$Ph_3P{=}C(COC_6H_4Cl\text{-}p)SCN$. IX. M. 167-8°.[259]
$Ph_3P{=}C(COC_6H_4NO_2\text{-}p)SCN$. IX. M. 170-1°.[259]
$Ph_3P{=}C(COC_6H_4NO_2\text{-}p)CN$. IX. M. 220-1°.[582]
$Ph_3P{=}C(COC_6H_4Cl\text{-}p)CN$. IX. M. 209-10°.[582]
$Ph_3P{=}C(CN)CO$-NH-CH=CH-α-furyl. IX.[516]
$Ph_3P{=}C(CN)CO$-NHPh. IX. M. 205-6°, IR.[640]
$Ph_3P{=}CH{-}C(SMe)NPh$. IX. M. 162°.[91]
$Ph_3P{=}C(CN)SC_6H_4Me$-p. IX. M. 170°.[515a]
$Ph_3P{=}C(CS_2Me)Ph$. IX. M. 244°.[54]
$Ph_3P{=}C(CS{-}NHMe)Ph$. IX. M. 153°.[91]
$Ph_3P{=}C(CO$-cyclo-Pr)Bu. IX. M. 172°.[697]
$Ph_3P{=}C(CO_2$-cyclo-$C_6H_{11})Me$. I. M. 133.5-34.5°.[344]
$Ph_3P{=}CH(CH_2)_2{-}O{-}C_6H_4CHO$-o. I.[555]
(cyclo-$C_6H_{11})_3P{=}CH{-}CH{=}CHPh$. I.[81]
$Ph_3P{=}C(SO_2Ph)SO_2CH{=}CH_2$. VI. M. 216°.[198a]
$Ph_3P{=}C$(indenylidene structure). I. M. 218-20°, UV.[224a]

C_{10}

$Ph_3P=CH-CH=CMe-(CH_2)_2-CH=CMe$. I.[340-1]
$Ph_3P=CH-(CH_2)_2-(C\equiv C)_2-Pr$. I.[139]
$Ph_3P=CH-(CH_2)_2-COPh$. I.[256]
$Ph_3P=CH-(CH_2)_7-CO_2Me$. I.[268]
$Ph_3P=CH$-4-bornyl. I.[180]
$Ph_3P=CH-CO_2-CHMePh$. I. M. 113-4°.[113]
$Ph_3P=CH-CO-CH=CH-Ph$. I. M. 99-102°,[320] ^{1}H-NMR.[704]
$Ph_3P=CH-CO-(CH_2)_2-Ph$. IX. M. 148-50°,[47] ^{1}H-NMR,[704,110] IR.[47]

Me Me
$Ph_3P=CH$ (2,6,6-trimethylcyclohex-1-enyl ring) . I.[166,340]
Me

$Ph_3P=CH-SO_2-CHMe-CO_2$-cyclo-C_6H_{11}. IX. M. 154-5°.[344]
$Ph_3P=C(CO_2Me)CH_2Ph$. IX. M. 185-7°,[98] IR.[100]
$Ph_3P=C(CO_2Me)C(CO_2Me)=CHPh$. IX. M. 133-7°, IR, ^{1}H-NMR.[95]
$Ph_3P=C(CO_2Me)CS-NHPh$. IX. M. 175°.[91]
$Ph_3P=C(CO_2Et)SPh$. IX.[515,515a] M. 199°.[515a]
$Ph_3P=C(CO_2Et)SC_6H_4NO_2$-p. IX.[515,515a] M. 194°.[515a]
$Ph_3P=C(CO_2Et)SC_6H_4Cl$-p. IX.[515,515a] M. 199°.[515a]
$Ph_3P=C(CO_2Et)SC_6H_4Br$-p. IX.[515,515a] M. 214°.[515a]
$Ph_3P=C(CO_2Et)SeAr$. IX. M. 184-6°(Ar = Ph), M. 183-5° [Ar = 2,4-$C_6H_3(NO_2)_2$][468]
$Ph_3P=C(CO_2Et)SO_2Ph$. IX. M. 199°.[515b]
$Ph_3P=C(CO_2Et)SO_2C_6H_4Cl$-p. IX. M. 199°.[515b]
$Ph_3P=C(CN)CH=C(CO_2Et)_2$. IX. M. 183-4°, IR.[636]
$Ph_3P=C(CN)COC_6H_4Me$-p. IX. M. 218-9°.[582]
$Ph_3P=C(CN)COC_6H_4OMe$-p. IX. M. 225-6°.[582]
$Ph_3P=C(CHMePh)Me$. V.[561]
$Ph_3P=C(CO_2Et)CS-NH-CO$-α-furyl. IX.[516]
$Ph_3P=C(CO_2Et)CO-(CH=CH)_2-Me$. IX. M. 158-9°, UV.[247]
$Ph_3P=C(CH=CHMe)CH=CH-CO$-i-Pr. IX. M. 160-2°.[701]
$Ph_3P=C(CO_2Et)CH=CH-CO$-i-Pr. IX. M. 138-40°,[701]
$Ph_3P=C(CO_2Et)Ph$. ^{1}H-NMR.[178]
$Ph_3P=C(C(SMe)NMe)Ph$. IX. M. 138°.[91]
$Ph_3P=C(CS_2Et)$cyclo-C_6H_{11}. IX. M. 202°.[54]
$Ph_3P=C(COC_6H_4OMe\text{-}p)SCN$. IX. M. 158-9°.[259]
$Ph_3P=C(COC_6H_4Me\text{-}p)SCN$. IX. M. 171-2°.[259]
$Ph_3P=C(CS_2CH_2Ph)Me$. IX. M. 155°.[54]
$Ph_3P=C(C(SMe)NPh)Me$. IX. M. 148°.[91]
$Ph_3P=C(CO_2CH_2Ph)Me$. I. M. 105-6°.[344]
$Ph_3P=C(COMe)COPh$. IX. M. 172-3°, IR.[164]
$Ph_3P=C(COMe)COC_6H_4NO_2$-p. IX. M. 190-1°, IR.[164]
$Ph_3P=C(COPh)CO_2Me$. IX. M. 133-5°.[47]
$Ph_3P=C(COC_6H_4Cl\text{-}o)CO_2Me$. IX. M. 171-3°.[401]
$Ph_3P=C(COC_6H_4Cl\text{-}p)CO_2Me$. IX. M. 147-9°.[401]
$Ph_3P=C(COC_6H_4NO_2\text{-}m)CO_2Me$. IX. M. 86-8°.[401]
$Ph_3P=C(COPh)COCH_2Cl$. IX. M. 190°.[388a]
$Ph_3P=C(COC_6H_4Cl\text{-}p)COCH_2Cl$. IX. M. 210°.[388a]
$Ph_3P=C(COC_6H_4Cl\text{-}p)COCCl_3$. IX. M. 160°.[388a]

$Bu_3P=C(CO)CH_2CO$ NPh (cyclic). III.[274]

$Ph_3P=C(CO)CH_2CO$ NPh (cyclic). III. M. 176.5-78.5°, IR, ^{1}H-NMR.[274]

$Ph_3P=C$(cyclopentadienyl)$C(CN)=C(CN)_2$. IX. M. 222-4°, IR, UV.[510]

$Ph_3P=C(CO_2Me)C_6H_4OMe$-p. ^{1}H-NMR.[703]
$Ph_3P=C(COMe)CH_2Ph$. IX. M. 183-5°.[698]
$Ph_3P=C(CN)CS_2CH_2Ph$. IX. M. 207-8°.[462c]
$Ph_3P=C(CN)CS_2-CH_2C_6H_4NO_2$-p. IX. M. 216-8°.[462c]

$Ph_3P=C$(dichlorotriazinyl)$C_6H_3(NO_2)_2$-2,4. IX. M. 264-5°.[462a]

$Ph_3P=C$(dichlorotriazinyl)$C_6H_4NO_2$-p. IX. M. 247°.[462a]

$Ph_3P=C(CO-N(methallyl)_2)Cl$. I.[603a]
$Ph_3P= C(COMe)CONHPh$. IX. M. 191-2°, IR.[640]

C_{11}

$Ph_3P=CH-(CH_2)_3-COPh$. I.[118-9]
$Ph_3P=CH-SO_2-CHMe-CO_2CH_2Ph$. IX. M. 60-2°.[344]
$Ph_3P=CH(CH_2)_7CO_2Et$. I.[33]
$Ph_3P=C(CO_2Me)C(SMe)NPh$. IX. M. 185°.[91]
$Ph_3P=C(CO_2Me)CHPh-CH_2NO_2$. IX. M. 180-1°, IR, UV.[21a]
$Ph_3P=C(CO_2Me)C_6H_4NMe_2$-p. ^{1}H-NMR.[703]
$Ph_3P=C(CO_2Et)COPh$. IX.[164,247] M. 142-3°,[247] UV,[247] IR.[164]
$Me_2PhP=C(CO_2Et)COPh$. IX. M. 108-9°, UV.[247]
$Ph_3P=C(CO_2Et)SC_6H_4Me$-p. IX. M. 205°.[515a]
$Ph_3P=C(CO_2Et)SO_2C_6H_4Me$-p. IX.[468-9,515b] M. 205-10°, IR.[468-9]
$Ph_3P=C(COC_6H_4Me\text{-p})CO_2Me$. IX. M. 182-82.5°.[401]
$Ph_3P=C(COC_6H_4OMe\text{-o})CO_2Me$. IX. M. 181-3°.[401]
$Ph_3P=C(COC_6H_4OMe\text{-m})CO_2Me$. IX. M. 157.5-59.5°.[401]
$Ph_3P=C(COCH_2Ph)CO_2Me$. IX. M. 147-9°.[401]
$Ph_3P=C(CO_2Et)CH_2C_6H_4OH$-p. IX. M. 203-6°.[619]
$Ph_3P=C(CO_2Et)CO-NH-CH=CH-\alpha$-furyl. IX.[516]
$Ph_3P=C(CO_2Et)CONHPh$. IX.[275,640] M. 189-90°.[640]
$Ph_3P=C(CO_2-2\text{-octyl})Me$. I.[632]

$Ph_3P{=}C(CO_2(CH_2)_2Ph)Me$. I. M. 130-2°.[344]
$Ph_3P{=}C(COMe)COC_6H_4OMe$-p. IX.[164,388a] M. 195-6°,[388a] IR.[164]
$Ph_3P{=}C(COMe)COC_6H_4Me$-p. IX. M. 210-1°.[388a]
$Ph_3P{=}C(COCH_2Cl)COC_6H_4Me$-p. IX. M. 140°.[388a]
$Ph_3P{=}C(COCH_2Cl)COC_6H_4OMe$-p. IX. M. 150°.[388a]
$Ph_3P{=}C(COCCl_3)COC_6H_4Me$-p. IX. M. 158°.[388a]
$Ph_3P{=}C(COCCl_3)COC_6H_4OMe$-p. IX. M. 174°.[388a]
$Ph_3P{=}C(CO\text{-cyclo-}C_6H_{11})Pr$. IX. M. 164-6°.[698]
$Ph_3P{=}C(CO\text{-cyclo-}C_6H_{11})CH_2CO_2Me$. IX. M. 167-8°, ^{1}H-NMR.[57]
$Ph_3P{=}C(COPh)CH_2CO_2Me$. IX. M. 177-8°, ^{1}H-NMR.[57]
$Ph_3P{=}C(COCH{=}CH\text{-}Ph)OMe$. IX. M. 201-5°.[700]
$Ph_3P{=}C(COCH{=}CHPh)Me$. IX. M. 205-8°.[47]
$Ph_3P{=}C(COCH_2\text{-}CH_2\text{-}Ph)Me$. IX. M. 164-6°.[47]

```
Ph3P=C ——CO
     |      >NPh.
   H-C ——CO
     |
     Me
```
III. M. 184-6°, IR, ^{1}H-NMR.[274]

$Ph_3P{=}CH$-2-naphthyl. I.[36a]
$Ph_3P{=}CH(CH{=}CH)_2C_6H_{13}$-n. I.[140]

C_{12}

$Ph_3P{=}CH\text{-}(CH_2)_4\text{-}COPh$. I.[256]
$Ph_3P{=}CH\text{-}CH_2\text{-}C(CO_2Et)_2\text{-}CH_2\text{-}COMe$. I.[554]
$Ph_3P{=}CH\text{-}SO_2\text{-}CHMe\text{-}CO_2(CH_2)_2Ph$. IX. M. 154-5°.[344]
$Ph_3P{=}C(CO_2Me)CH_2\text{-}CH{=}CHPh$. IX. Oil, IR.[100]
$Ph_3P{=}C(CO_2Me)COCH{=}CHPh$. IX. M. 184-84.5°.[401]

```
         CO2Me
        /
Ph3P=C
        \
         CH(CO2Me)-C6R4OH-p
```
. IX. M. 160°(R = H)[86] M. 139°(R = Cl).[86]

$Ph_3P{=}C(CO_2Me)CHPh\text{-}CHMeNO_2$. IX. M. 204-7°, IR, UV.[21a]
$Ph_3P{=}C(CO_2Et)CH_2\text{-}CH_2Ph$. IX. M. 162-3°.[69]
$Ph_3P{=}C(CO_2Et)SO_2C_6H_4NHCOMe$-p. IX. M. 240°.[515b]
$Ph_3P{=}C(COC_6H_4OMe\text{-p})CH_2CO_2Me$. IX. M. 150-3°, ^{1}H-NMR.[57]
$Ph_3P{=}C(CO\text{-}\beta\text{-}C_{10}H_7)I$. IX. M. 183-4°.[259]
$(Allyl)_3P{=}C(CO\text{-}N(pentenyl)_2)I$. I.[603a]
$Ph_3P{=}C(COEt)COC_6H_4OMe$-p. IX. M. 176-7°.[388a]
$Ph_3P{=}CH$-CO-1-adamantyl. I. M. 208-9°.[377a]
$Ph_3P{=}C(SPh)CO\text{-}\alpha$-furyl. IX.[515,515a] M. 192°.[515a]
$Ph_3P{=}C(SC_6H_4Cl\text{-p})\text{-}CO\text{-}\alpha$-furyl. IX.[515,515a] M. 188°.[515a]
$Ph_3P{=}C(SC_6H_4Br\text{-p})CO\text{-}\alpha$-furyl. IX.[515,515a] M. 188°.[515a]
$Ph_3P{=}C(SC_6H_4NO_2\text{-o})CO\text{-}\alpha$-furyl. IX.[515,515a] M. 231°.[515a]
$Ph_3P{=}CPh\text{-}CH{=}CH\text{-}CO\text{-}Et$. IX. M. 216-8°.[701]
$Ph_3P{=}CSPh\text{-}CH{=}CH\text{-}CO\text{-}Et$. IX. M. 188-9°.[701]
$Ph_3P{=}C(CS_2Et)CH_2\text{-}CH_2Ph$. IX. M. 196°.[54]

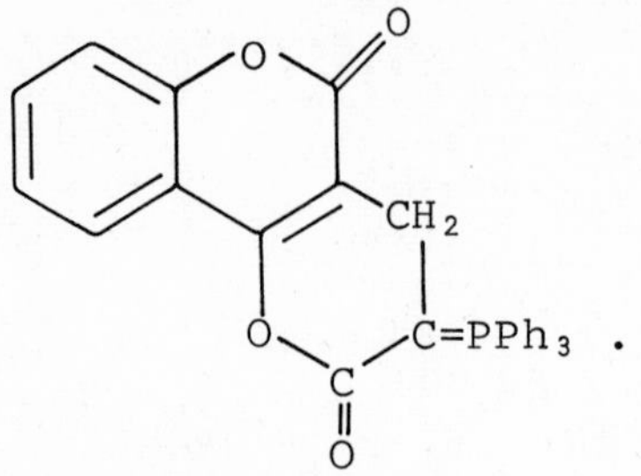

O O
CH_2
O $C=PPh_3$.
C
O

IX. M. 208-11°.[619]

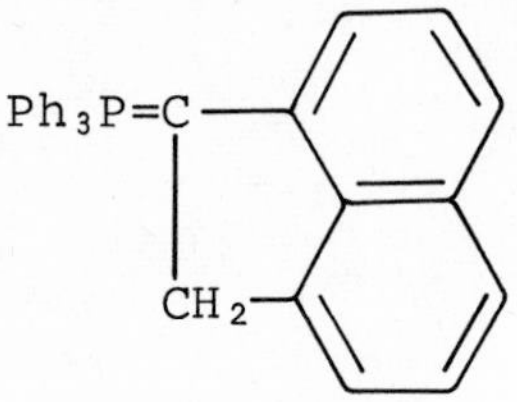

$Ph_3P=C$
CH_2

. IX.[60]

C_{13}

$R_3P=$

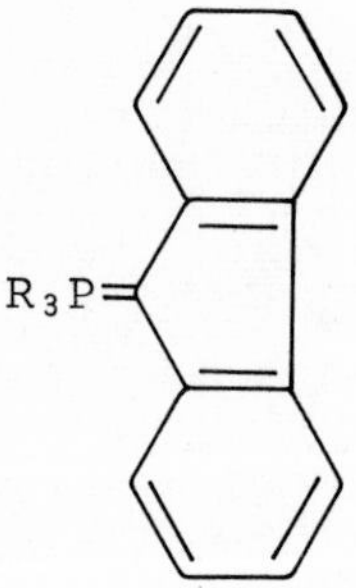

R = Me. I.[680]
= Bu. I.[350]
= Ph. I.[218,348,472] M. 289-91°,[348] IR,[310] DP,[348] UV.[218,348]
= FC_6H_4. I. M. 254°(p-F), M. 194° (m-F), ^{19}F-NMR.[353]

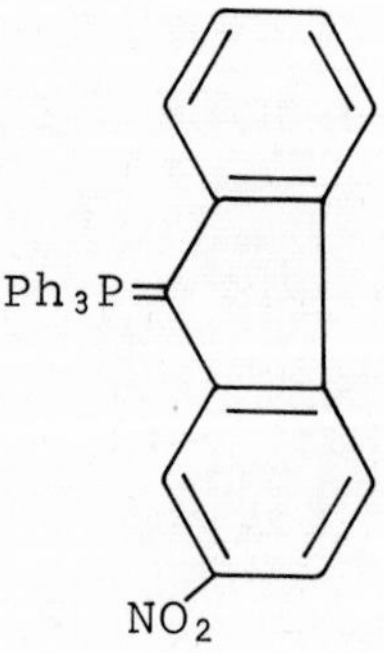

$Ph_3P=$

. I.[221]

NO_2

[Structure: 2,7-dibromofluorenylidene-triphenylphosphorane, $Ph_3P=$] . I. M. 294-5°.[546]

$Ph_3P=C(N=NPh)C_6H_4NO_2$-p. IX. M. 146-7°, UV.[400]
$Ph_3P=C(N=NC_6H_4NO_2$-p)[$(CH=CH)_2CO_2Me$]. IX. UV.[400]
$Ph_3P=CH-C_6H_4-SO_2Ph$-p. I.[583]
$Ph_3P=CH-C_6H_4-SO_2C_6H_4Br$-p,p. I.[583]
$Ph_3P=CH(CH_2)_3CO_2Et$. I.[33]
$Ph_3P=CH-POPh_2$. IX. M. 157-8°, IR, ^{1}H-NMR.[480-1]
$Ph_3P=CH-PO(OPh)_2$. I. IX. M. 149-50°.[355]
$Ph_3P=CPh_2$. II.[611] I.[167,307] M. 172°.[307]
$Ph_3P=C(C_6H_4NO_2$-p)Ph. I. M. 230°, UV.[207]
$Ph_3P=C(N=N-Ph)_2$. IX. M. 139-41°, UV.[406]
$Ph_3P=C(OPh)_2$. I.[674]
$Ph_3P=C(SO_2Ph)_2$. VI.[309] I.[291] M. 267-9°.[309]
(n-Bu)$_3P=C(SPh)_2$. II. M. 79-81°, ^{1}H-NMR, UV.[555a]
$Ph_3P=C(SPh)_2$. II.[555a] IX.[445a] M. 170-2°,[555a] M. 164-5°,[445a] ^{1}H-NMR, UV.[555a]
$Ph_3P=C(COPh)COCH_2CO_2Et$. I. M. 115°, UV, IR.[558a]
$Ph_3P=C(COC_6H_4R$-p)CO-NH-α-furyl (R = H, Br). IX.[516]

$Ph_3P=C(COC_6H_4R$-p)CO-NH-[5-nitro-2-furyl] (R = H, Br). IX.[516]

$Ph_3P=C(COC_6H_4NO_2$-p)CO-2-thienyl. IX. M. 165°.[205]
$Ph_3P=C(CO_2Me)C(CO_2Me)=CHPh$. IX. M. 182-4°, IR.[95]
$Ph_3P=C(CO_2Me)C(NPh)-CH_2CO_2Me$. IX. M. 204-5°, IR, ^{1}H-NMR, ^{31}P -17.[328]
$Ph_3P=C(CO_2Et)CH_2CH_2-NH-COC_6H_4NO_2$-p. IX. M. 191-2°, IR.[275]

$Ph_3P=C(CO_2Et)CH_2$-[3-indolyl] . IX. M. 187-9°.[619]

$Ph_3P=C(CO_2Et)CO-CH=CHPh$. IX. M. 118-20°, IR.[164]
$Ph_3P=C(CO_2Et)CH_2CH_2-NHSO_2C_6H_4Me$-p. IX. M. 184-6°.[275]
$Ph_3P=C(CO_2$-menthyl)Me. I.[632]
$Ph_3P=C(COCH_2-CH_2Ph)CH_2CO_2Me$. IX. M. 130-1°, ^{1}H-NMR.[57]
$Ph_3P=C(COCH_2CH_2Ph)Pr$. IX. M. 147-9°.[47]
$Ph_3P=C(SC_6H_4Me$-p)CO-α-furyl. IX. M. 227°.[515a]

$Ph_3P=C(SC_6H_4Cl\text{-}p)C_6H_4NO_2\text{-}p$. IX.[515a]
$Ph_3P=C(CH_2Ph)SO_2CHMe\text{-}CO_2Et$. IX. M. 65-8°.[344]
$Ph_3P=CPh\text{-}CH=CH\text{-}CO\text{-}i\text{-}Pr$. IX. M. 227-9°.[701]
$Ph_3P=C(SPh)\text{-}CH=CH\text{-}CO\text{-}i\text{-}Pr$. IX. M. 235-7°.[701]

C_{14}

$Ph_3P=CH\text{-}SO_2$-fluorenyl. IX. M. 208-10°.[344]

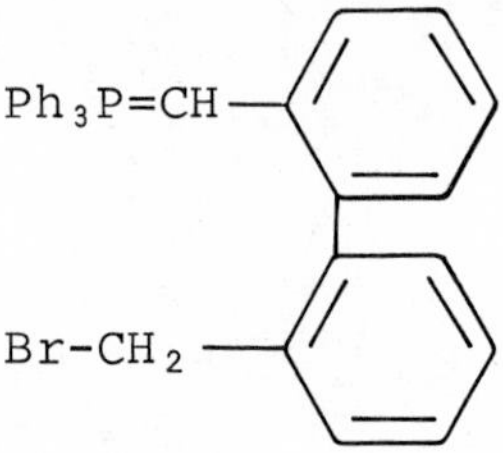

[Structure: $Ph_3P=CH$– and $Br\text{-}CH_2$– substituted biphenyl] . I.[64]

[Structure: $Ph_3P=C$ and CH_2 bridged biphenyl] . IX.[64]

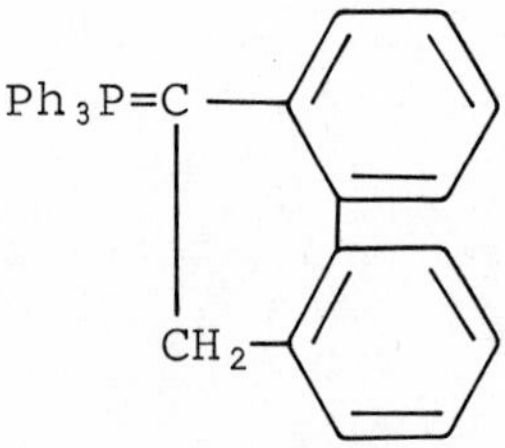

$Ph_3P=C(N=NC_6H_4\text{-}p)_2$. IX.[406] UV.[406]
$Ph_3P=C(COPh)N=NPh$. IX. M. 160-2°.[400]
$Ph_3P=C(COPh)Ph$. I.[640] IX.[47,642,643] M. 192-4°.[47]
$Ph_3P=C(COC_6F_5)Ph$. IX. M. 245-7°.[217]
$Ph_3P=C(COC_6F_5)C_6F_5$. IX. M. 245-6°.[217]
$Ph_3P=C(CH_2Ph)\text{-}CH=CH\text{-}CO\text{-}i\text{-}Pr$. IX. M. 185-8°.[701]
$Ph_3P=C(COPh)SPh$. IX. M. 205°.[515a]
$Ph_3P=C(COPh)SC_6H_4Cl\text{-}p$. IX. M. 177°.[515a]
$Ph_3P=C(COPh)SC_6H_4Br\text{-}p$. IX. M. 181°.[515a]
$Ph_3P=C(COPh)C(CO_2Me)=CHCO_2Me$. IX. M. 164-6°, IR, UV, ^1H-NMR.[278]
$Ph_3P=C(CO_2Me)C(CO_2Me)=CHCOPh$. IX. M. 230-1°, IR, UV, ^1H-NMR.[278]
$Ph_3P=C(CO_2Me)CH(CO_2Me)CH_2COPh$. IX.[104,278] M. 192-3°,[278] 187°,[104] IR, UV, ^1H-NMR.[278]
$Ph_3P=C(CO_2Et)CH_2CHMe\text{-}NHCOC_6H_4NO_2\text{-}p$. IX. M. 183-4°, IR.[275]

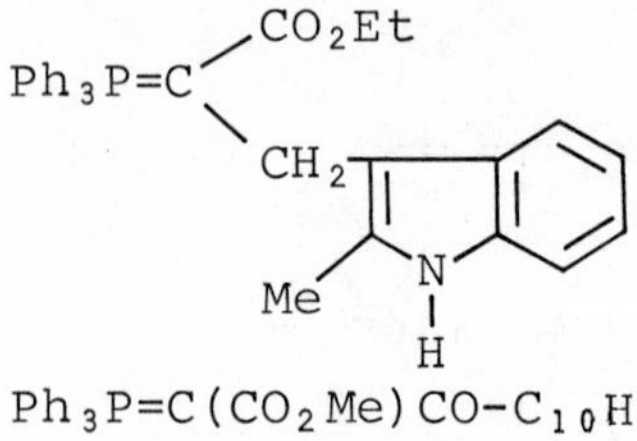

[Structure: $Ph_3P=C(CO_2Et)CH_2$-(2-methylindol-3-yl)] . IX. M. 181.5-83°.[619]

$Ph_3P=C(CO_2Me)CO\text{-}C_{10}H_7\text{-}1$. IX. M. 186.5-87.5°.[401]

$Ph_3P=C(CO_2Me)C(CO_2Me)=CMePh$. IX. M. 213-6°, IR.[95]
$Ph_3P=C=CPh_2$. I.[237]
$Ph_3P=C(CO(CH_2)_3Ph)Pr$. IX. M. 142°.[697]
$Ph_3P=C(CSNHPh)Ph$. IX. M. 140°.[91]
$Ph_3P=CH-CO-NPh_2$. I. M. 180°, IR, UV,[601-2] IR, pK_a.[600]
$Bu_3P=C(CONPh_2)Cl$. I.[603a]
$Ph_3P=CClCONPh_2$. I.[601-2,603a] IX.[601-2] M. 170°, IR, UV.[601-2]
$Et_3P=CClCON(hexynyl)_2$. I.[603a]
$Pr_3P=CClCON(cyclo-C_6H_{11})_2$. I.[603a]
$Me_3P=CBrCON(cyclo-C_6H_{11})_2$. I.[603a]
$Ph_3P=CBrCONPh_2$. IR, pK_a.[600]
$MePh_2P=CPhCH(OH)Ph$. I.[357]

C_{15}

$Ph_3P=C(COPh)_2$. IX.[164,388a] M. 191-2°,[164] IR.[164]
$Ph_3P=C(COPh)COC_6H_4NO_2$-p. IX. M. 170-1°.[205]
$Ph_3P=C(COPh)COC_6H_4Cl$-p. IX. M. 166-7°.[388a]
$Ph_3P=C(COC_6H_4Cl\text{-}p)_2$. IX. M. 165-6°.[205]
$Ph_3P=C(COC_6H_4NO_2\text{-}p)_2$. IX. M. 100°.[205]
$Ph_3P=C(COPh)SC_6H_4Me$-p. IX. M. 213°.[515a]
$Ph_3P=C(COPh)CONHPh$. IX. M. 189°.[640]
$Ph_3P=CHCONHPh$. ^1H-NMR.[704]
$Ph_3P=C(COPh)CH_2C_6H_4OH$-p. IX. M. 259-60.5°.[619]
$Ph_3P=C(CONHPh)_2$. IX. M. 172-3°.[640]

CO_2Me
$Ph_3P=C$. IX. M. 213-6°, IR.[95] 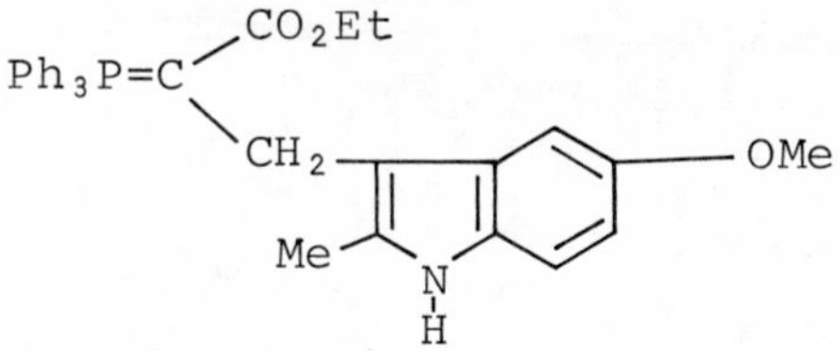
$C(CO_2Me)=CH-CH=CHPh$

CO_2Et
$Ph_3P=C$
CH_2 – OMe
Me N H . IX. M. 190-3°.[619]

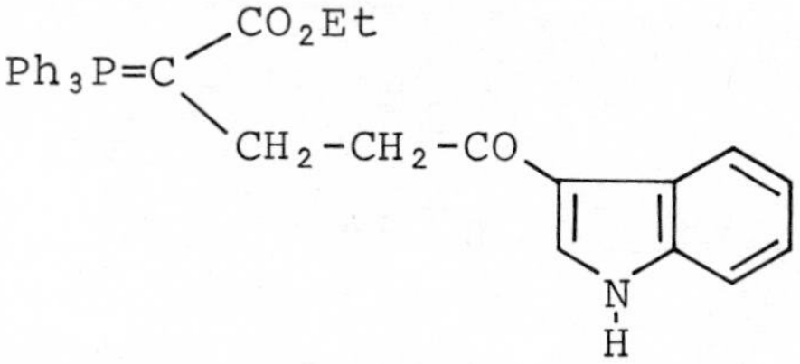

CO_2Et
$Ph_3P=C$
CH_2-CH_2-CO . IX. M. 189-92°.[619]
N H

$Ph_3P=CPhCS_2CH_2Ph$. IX. M. 216°.[54]
$Ph_3P=CPhC(NPh)SMe$. IX. M. 181°.[91]
$Ph_3P=CH-CH=fluorenyl$. IX. M. 204-6°, UV.[218]

Ph_3P=CH-CH=CMe-CH=CH(2,6,6-trimethylcyclohex-1-enyl). I.[185,474]

$Ph_3P=C(N=NC_6H_4OMe\text{-}o)_2$. IX. UV.[406]

Ph_3P=(cyclopentadienylidene, 2-CPh=C(CN)$_2$). IX.[510]

Ph_3P=(cyclopentadienylidene, 2-CH(C$_6$H$_4$R-p)-CH(CN)$_2$) (R = OMe, NO_2, H). IX.[510]

Ph_3P=(fluorenylidene, 2-NMe$_2$). I.[221]

Ph_3P=C(indenylidene, 3-N=N-$C_6H_4NO_2$-p), IX. M. 140-3°, UV.[224a]

Ph_3P=C(indenylidene, 3-N=N-Ph). IX.[177a,224a] M. 210-2°, UV.[224a]

$Ph_3P=CPhCOCH_2Ph$. IX. M. 192-3°, IR, ^{1}H-NMR.[325]

C_{16}

$Bu_3P=C(COPh)CH_2COPh$. III.[488,490-1] M. 97°, IR, ^{1}H-NMR, ^{31}P -21.3.[490-1]

$Et_2PhP=C(COPh)CH_2COPh$. III. M. 121°, IR, ^{1}H-NMR, ^{31}P -21.1.[490-1]

$Ph_3P=C(COPh)CH_2COPh$. III.[488,490-1] M. 122°,[490-1] IR, ^1H-NMR, ^{31}P -16.9.[490-1]
$Ph_3P=C(COPh)COC_6H_4OMe$-p. IX. M. 182-3°.[388a]
$Ph_3P=C(COPh)COC_6H_4Me$-p. IX. M. 172-3°.[388a]
$Ph_3P=C(COC_6H_4Me\text{-}p)COC_6H_4NO_2$-p. IX. M. 170-70.5°.[205]
$Ph_3P=C(SPh)CH=CHCOPh$. IX. M. 218-20°.[701]
$Ph_3P=C$ (ring: CO—N=C(-OPh)—N=C(OPh)—). IX. M. 280-1°.[424]

$Ph_3P=C$ (cyclopentadienylidene) Me-C(Ph)-$CH(CN)_2$. IX.[510]
$Ph_3P=C(CONHPh)_2$. IX. M. 172-3°.[640]
$Ph_3P=C$—C=O / O=C—fluorenyl. IX. M. 207-11°, IR.[688a]
$Ph_3P=C$—C=O / O=C—C(Ph)—Ph . IX. M. 221-22°, IR.[688a]
$Bu_3P=C(CH_2\text{-morph})\text{-}CH=CH\text{—}C(CH_2\text{-morph})=C(CN)_2$. III. M. 200°.[605]
$Bu_3P=CPhCH=CHCPh=NCN$. III. M. 136°.[605]
$Ph_3P=C(CO_2Me)CHPhCHPhNO_2$. IX. M. 173-4°, IR, UV.[21a]
$Ph_3P=CPhCH=CHCPh=C(CN)_2$. III. M. 150°.[605]
$Ph_3P=C(COPh)CH_2$-(3-indolyl, N-H) . IX. M. 268-72°.[619]

$Ph_3P=C$ (2,5-diphenylcyclopentadienylidene) . II. M. 234-5°, UV.[500]
$Ph_3P=CH\text{-}CH=CH\text{-}CH=$fluorenyl. IX. UV.[218]
$Ph_3P=C(N=N\text{-}C_6H_4NMe_2\text{-}p)_2$. IX. UV.[406]

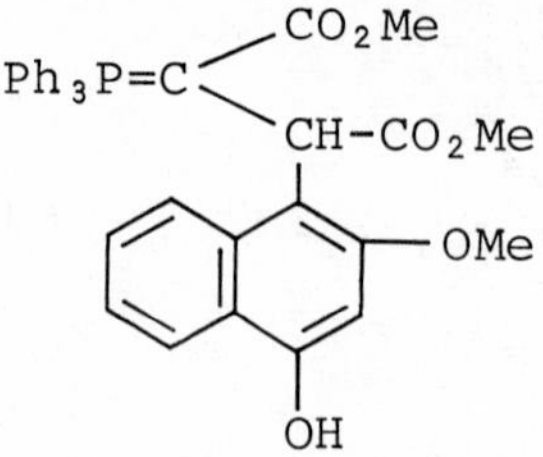

CO_2Me
$Ph_3P=C$
$CH-CO_2Me$ IX. M. 184°.[86]
OMe
OH

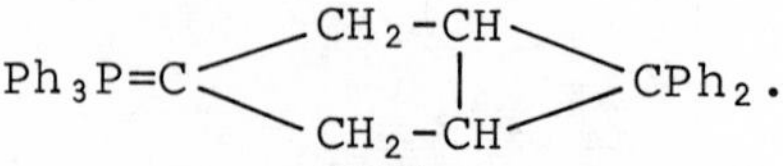

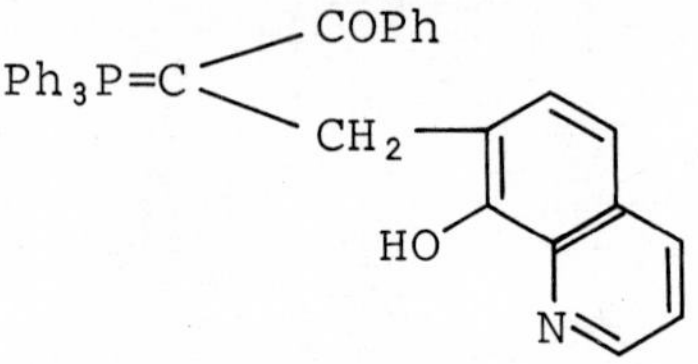

CH_2-CH
$Ph_3P=C$ CPh_2. IX.[79]
CH_2-CH

COPh
$Ph_3P=C$
CH_2 . IX. M. 190-92.5°.[619] 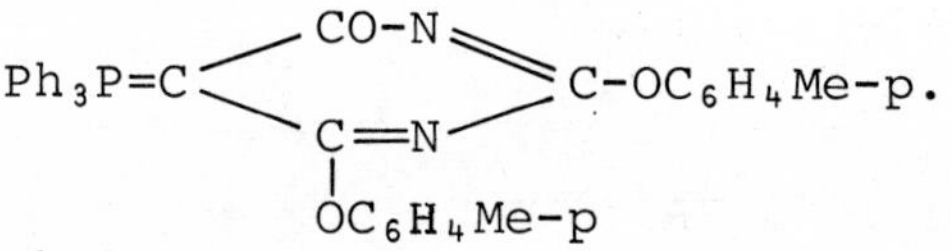
HO
N

CO-N
$Ph_3P=C$ $C-OC_6H_4Me-p$. IX. M. 276-7°.[424]
C=N
OC_6H_4Me-p

$Ph_3P=CHCO_2C_{16}H_{33}-n$. I. M. 90-1°, UV.[340-1]

C_{19}

$Ph_3P=CMeCO_2C_{16}H_{33}-n$. I. M. 68-9°, UV.[340-1]
$Bu_3P=CPhCH=CHCPh=C(CN)_2$. III.[605]
$Ph_3P=C(CO_2Me)C(CO_2Me)=CPh_2$. IX. M. 257°, UV.[150]

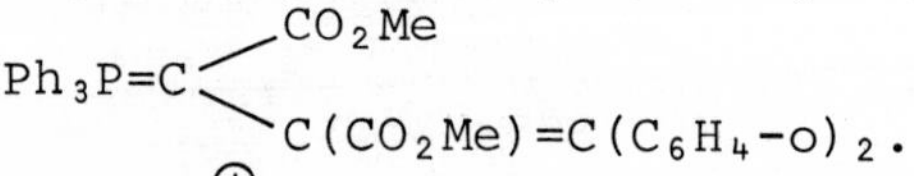

CO_2Me
$Ph_3P=C$ IX. M. 211°, UV.[150]
$C(CO_2Me)=C(C_6H_4-o)_2$.

$Ph_3P=CH\overset{\oplus}{P}Ph_3Br^{\ominus}$. I.[261,480-1] M. 272-4°,[261] IR.[480-1]

$Ph_3P=CBr\overset{\oplus}{P}Ph_3\overset{\ominus}{Br}$. IX. M. 278-9°, IR.[480-1]

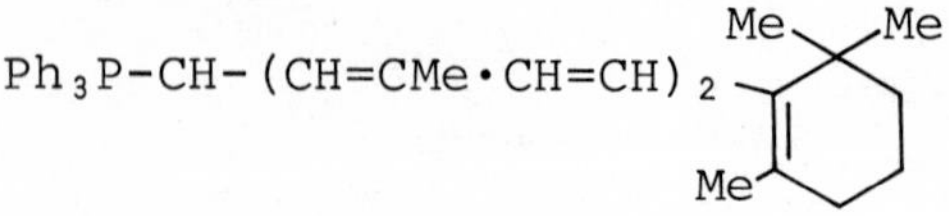

Me Me
$Ph_3P-CH-(CH=CMe\cdot CH=CH)_2$. I.[80]
Me

C_{21}

$Ph_3P=CPhC(NPh)CH_2Ph$. IX. M. 209-10°, IR, ^{1}H-NMR, ^{31}P -11.[325]

C_{22}

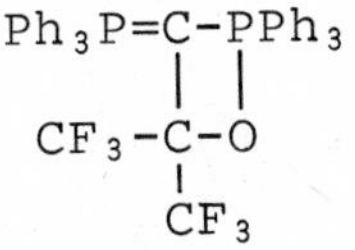

$Ph_3P{=}C{-}PPh_3$ / $CF_3{-}C{-}O$ / CF_3 (ring). IX.[126,126a,161] M. 157-8°,[126a] x-ray,[161] ^{19}F-NMR, ^{31}P -7,3, +54.[126]

C_{24}

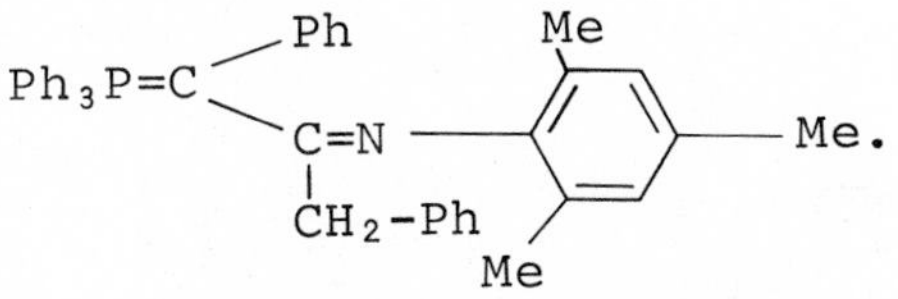

$Ph_3P{=}C(Ph)C(CH_2{-}Ph){=}N{-}C_6H_2Me_3$ (2,4,6-trimethylphenyl). IX. M. 194-5°, IR, ^{1}H-NMR, ^{31}P -11.5.[325]

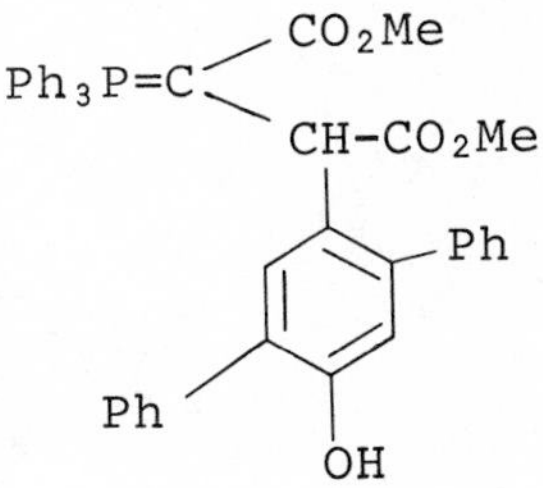

$Ph_3P{=}C(CO_2Me)CH(CO_2Me){-}C_6H_2(Ph)_2(OH)$. IX. M. 233°.[86]

C_{29}

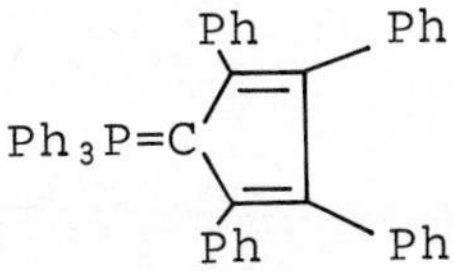

$Ph_3P{=}C$ (tetraphenylcyclopentadienylidene: Ph, Ph, Ph, Ph) . II. M. 306-8°, UV.[500]

$Ph_3P{=}C(CH{=}C(C_6H_4{-}o)_2)_2$. IX. M. 173-5°, UV.[218]

$Bu_3P{=}C(Ph)CH{=}CH{-}CPh{=}C(C_6H_4{-}o)_2$. III. M. 215°.[605]

C_{30}

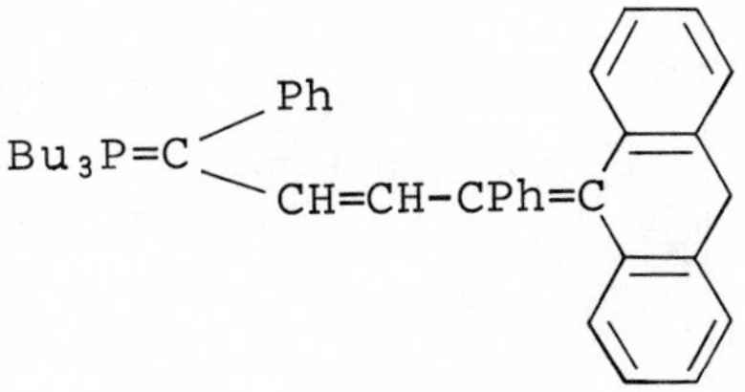

$Bu_3P{=}C(Ph)CH{=}CH{-}CPh{=}C$ (anthracen-9(10H)-ylidene) . III. M. 210°.[605]

$Ph_3P{=}CH{-}CH{=}C(CH{=}C(C_6H_4{-}o)_2)_2$. I. M. 218-20°, UV.[218]

D.1.1. Cyclic Phosphonium Ylids

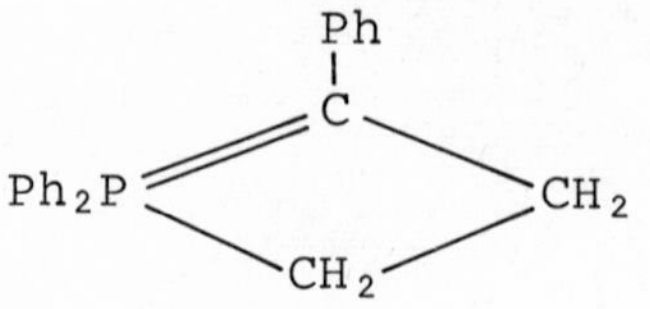

$Ph_2P{=}C(Ph){-}CH_2{-}CH_2{-}$ (ring). IX.[518]

$Ph_2P{=}CH{-}CH_2{-}CH(Ph){-}$ (ring). IX.[518]

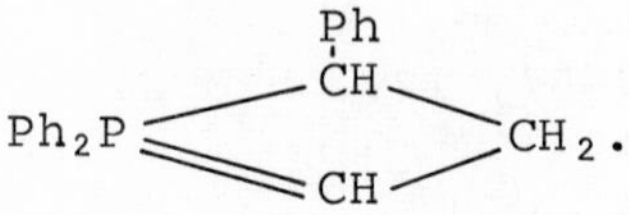

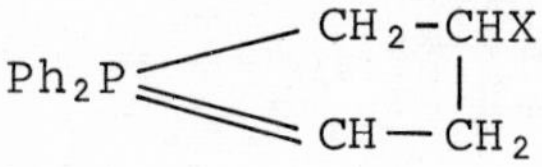

$Ph_2P{=}CH{-}CH_2{-}CHX{-}CH_2{-}$ (ring) (X = CN, CO_2Me). III.[518]

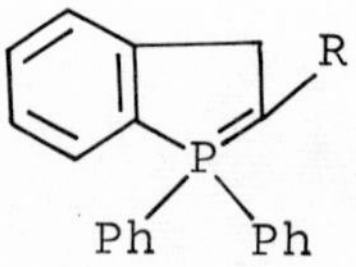

R (R=H, Ph, CO_2Me). I.[409]

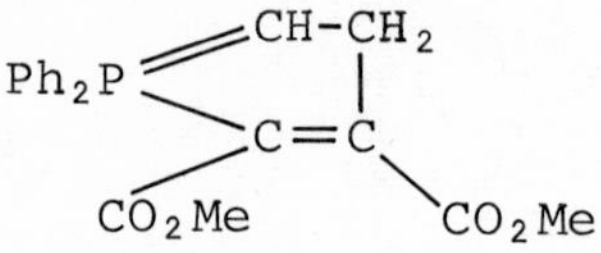

Ph Ph

$Ph_2P{=}CH{-}CH_2{-}C(CO_2Me){=}C(CO_2Me){-}$ (ring). III.[323]

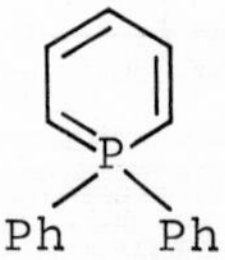

. I.[410] UV.[410]

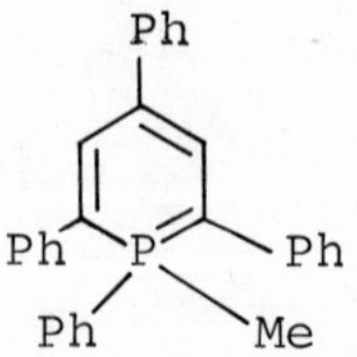

Ph Ph

Ph

Ph P Ph . I.[414,415] M. 169-70°, IR, UV, ^{1}H-NMR.[415]

Ph Me

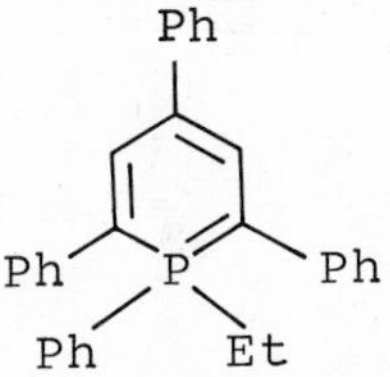

Ph

Ph P Ph

Ph Et

. I. M. 151-2°, IR, UV, ^{1}H-NMR.[415]

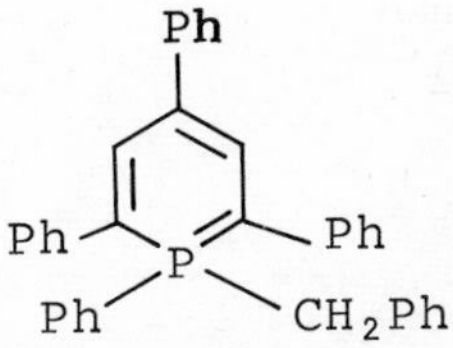

Ph

Ph P Ph

Ph CH_2Ph

. I. M. 201-3°, IR, UV, ^{1}H-NMR.[415]

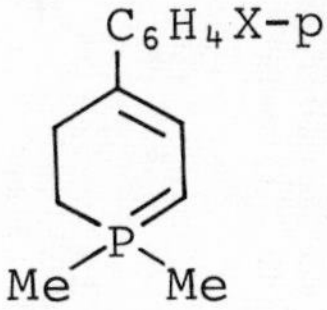

C_6H_4X-p

Me P Me

(X = H, OMe, Me, F). I.[381]

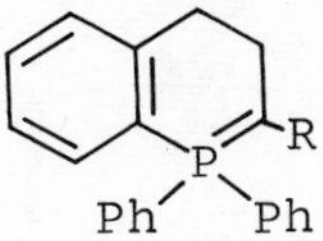

R

Ph P Ph

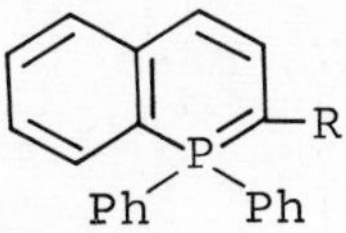

(R = Ph, H). I.[408]

P—R

Ph Ph

(R = Ph, H). I.[408]

CO_2Me CO_2Me

Ph P CO_2Me

Ph C

Ar CO_2Me

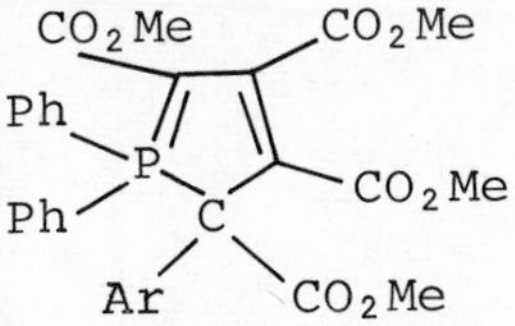

. III. M. 253-5° (Ar = Ph), M. 228-30° (Ar = p-$CH_3C_6H_4$), IR.[652]

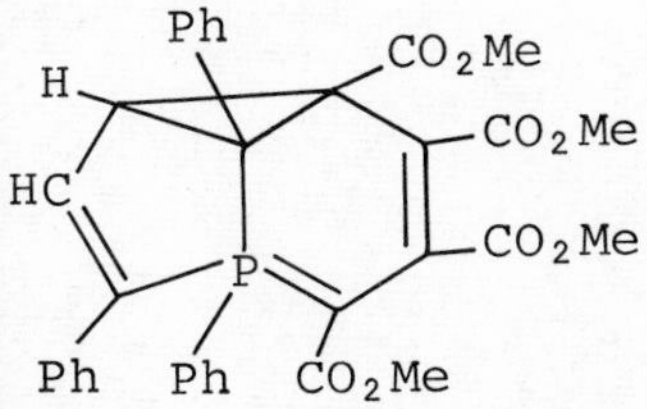

Ph CO_2Me

H CO_2Me

HC CO_2Me

P

Ph Ph CO_2Me

. III. M. 180-2°, IR, ^{1}H-NMR, ^{31}P -37,[653] UV.[324]

D.1.2. Bisphosphonium Ylids, $R_3P=X=PR_3$ (Ordered According to the Number of Carbon Atoms in X)

C_1

$Ph_3P=C=PPh_3$. I. M. 208-10°, IR, UV.[480-1]

C_2

$Ph_3P=CH-O-CH=PPh_3$. I.[200,201]
$Ph_3P=CH-S=CH=PPh_3$. I.[200,236]

C_3

$Ph_3P=CH-CH_2-CH=PPh_3$. I.[675]

C_4

$Ph_3P=CMe-CH_2-CH=PPh_3$. V. IX.[561]
$Ph_3P=CH-CH=CH-CH=PPh_3$. I.[276,277,474]
$Ph_3P=CH-(CH_2)_2-CH=PPh_3$. I.[59,444,675]
$Ph_3P=C(CN)COCH=PPh_3$. I. M. 300°.[388b]

C_5

$Ph_3P=C(CO_2Me)COCH=PPh_3$. I. M. 200°.[388b]
$Ph_3P=CH-(CH_2)_3-CH=PPh_3$. I.[59]

C_6

$MePh_2P=C(CO_2Me)C(CO_2Me)=PPh_2Me$. III. M. 215-7°, IR, ^{1}H-NMR.[572]
$Ph_3P=C(CO_2Me)C(CO_2Me)=PPh_3$. III. M. 220-2°, IR, mass spect., ^{1}H-NMR.[572]
$Ph_3P=CH-CO_2-(CH_2)_2-CO_2-CH=PPh_3$. I.[224]
$Ph_3P=CH-(CH_2)_4-CH=PPh_3$. I.[59]

C_7

$Ph_3P=CH-(CH_2)_5-CH=PPh_3$. I.[59]

C_8

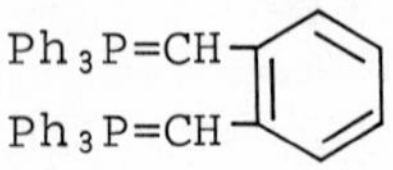

$o-C_6H_4(CH=PPh_3)_2$ (Ph3P=CH– groups on adjacent positions of a benzene ring). I.[254]

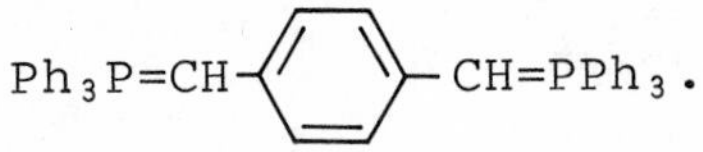

$Ph_3P=CH-C_6H_4-CH=PPh_3$. I.[428]

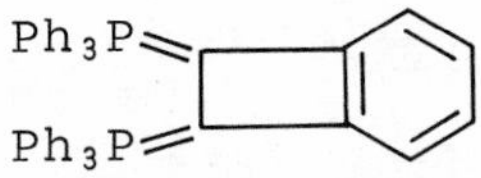

$Ph_3P=$ / $Ph_3P=$ (benzocyclobutene-1,2-diylidene) . I.[132]

$Ph_3P=C(-CO-)-CH_2-CO-N-N(-CO-C(=PPh_3)-CO-CH_2)$. III. M. 350°, ^{1}H-NMR, IR.[274]

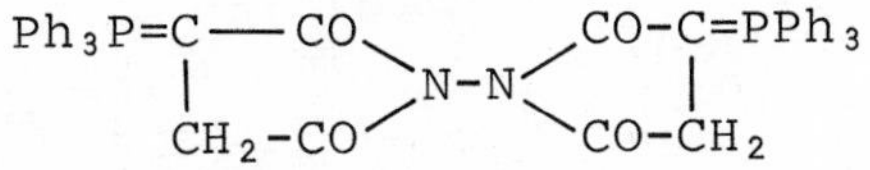

C_9

$Ph_3P=CH-(CH_2)_7-CH=PPh$. IX.[79]

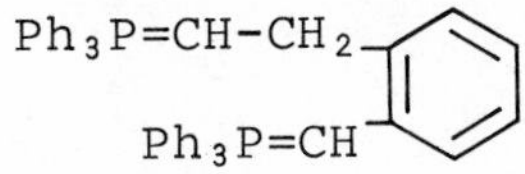

$Ph_3P=CH-CH_2-C_6H_4-CH=PPh_3$. I.[63]

C_{10}

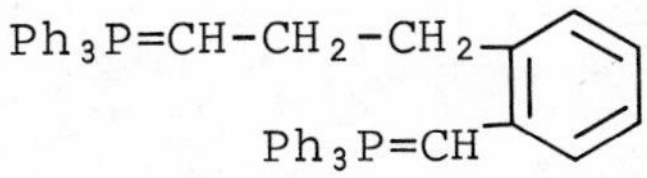

$Ph_3P=CH-CH_2-CH_2-C_6H_4-CH=PPh_3$. I.[63]

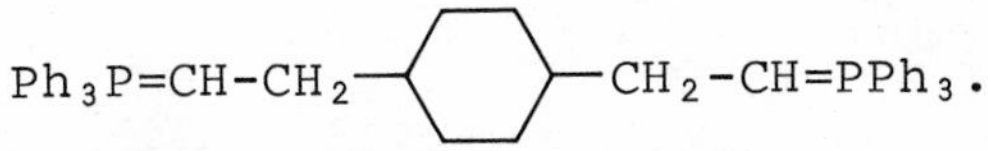

$Ph_3P=CH-CH_2-C_6H_{10}-CH_2-CH=PPh_3$. IX.[79]

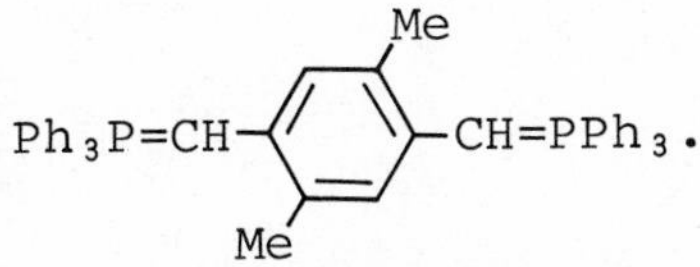

$Ph_3P=CH-C_6H_2(Me)_2-CH=PPh_3$. I.[156]

$Ph_3P=CH-C(Me)=CH-CH$
$Ph_3P=CH-C(Me)=CH-CH$. I.[626]

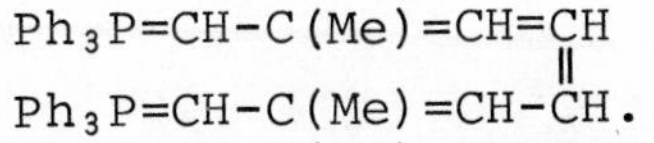

$Ph_3P=CH-C(Me)=CH-C\equiv C-CH=C(Me)-CH=PPh_3$. I.[626]

C_{12}

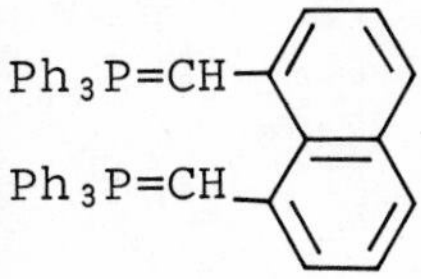

$Ph_3P=CH-$ / $Ph_3P=CH-$ (1,8-naphthalenediyl) . I.[63,82]

$Ph_3P{=}C(CN){-}C(CN){=}C(CN)$
. III.[573] M. 237-9°,[497] IR,
$Ph\ P{=}C(CN){-}C(CN){=}C(CN)$
UV,[497,573] mass spect.[573]

$Ph_3P{=}C(CO_2Me)C(CO_2Me)$
‖ . III. M. 266°, ^{31}P -22.9
$Ph_3P{=}C(CO_2Me)C(CO_2Me)$
(-28.9),[573a] ^1H-NMR,[573a] UV,[573,573a] mass spect.[573a]

C_{14}

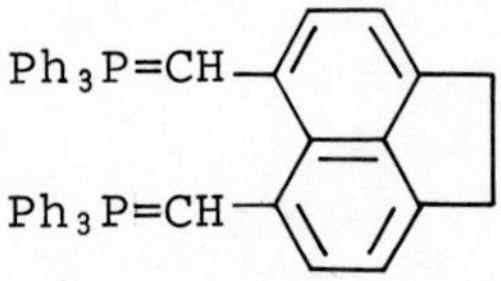

$Ph_3P{=}CH$ (acenaphthene, bis-substituted)
$Ph_3P{=}CH$
. I.[45]

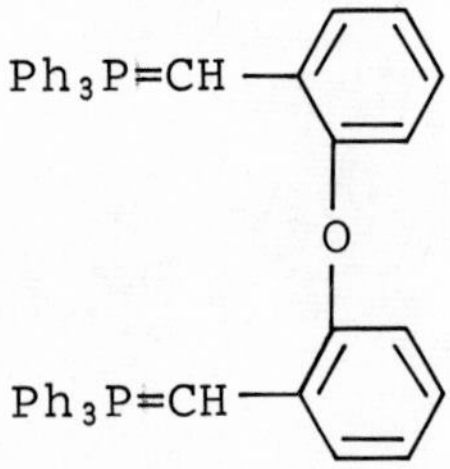

$Ph_3P{=}CH$–C_6H_4–O–C_6H_4–$CH{=}PPh_3$
. I.[45,63]

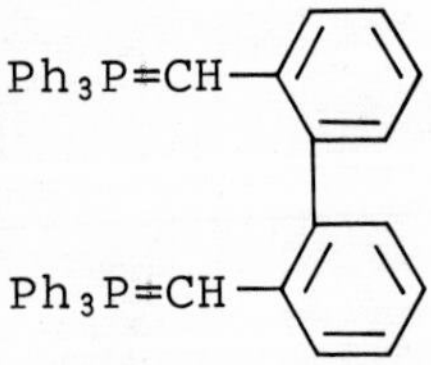

$Ph_3P{=}CH$–C_6H_4–C_6H_4–$CH{=}PPh_3$
. I.[59,90]

C_{15}

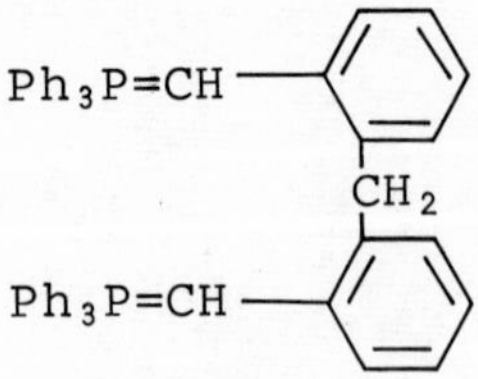

$Ph_3P{=}CH$–C_6H_4–CH_2–C_6H_4–$CH{=}PPh_3$
. I.[63]

C_{16}

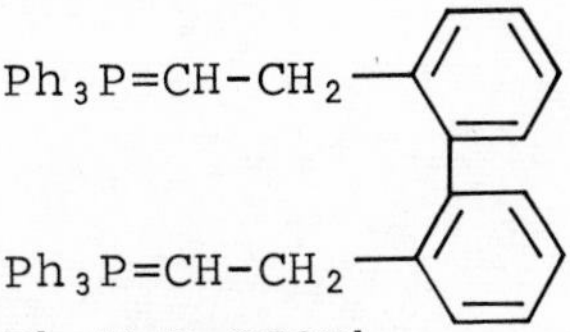

$Ph_3P{=}CH{-}CH_2$-(2,2'-biphenyl)-$CH_2{-}CH{=}PPh_3$. I.[63]

$Ph_2MeP{=}C(COPh){-}C(COPh){=}PMePh_2$. III. M. 188-9°, IR, ^{1}H-NMR.[574]

$Ph_3P{=}C(COPh){-}C(COPh){=}PPh_3$. . III. M. 199-200°, IR, ^{1}H-NMR.[574]

C_{18}

$MePh_2P{=}C(COC_6H_4Me\text{-}p){-}C(COC_6H_4Me\text{-}p){=}PPh_2Me$. . III. M. 201-2°, IR, ^{1}H-NMR.[574]

$Ph_3P{=}C(COC_6H_4Me\text{-}p){-}C(COC_6H_4Me\text{-}p){=}PPh_3$. III. M. 179-80°, IR, ^{1}H-NMR.[574]

C_{22}

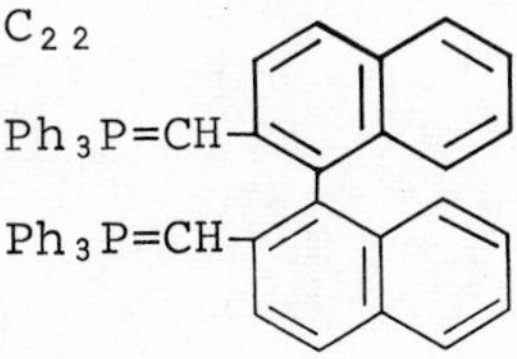

$Ph_3P{=}CH$-(1,1'-binaphthyl-2,2'-diyl)-$CH{=}PPh_3$. I.[45]

C_{24}

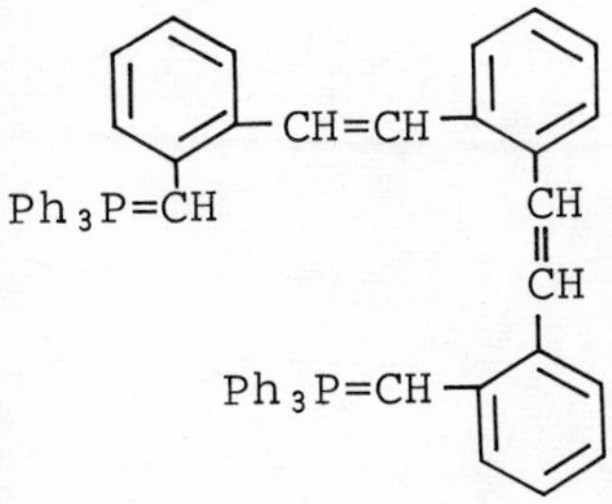

$Ph_3P{=}CH$-C_6H_4-CH=CH-C_6H_4-CH=CH-C_6H_4-$CH{=}PPh_3$ (all ortho) . I.[254]

D.1.3. Cyclic Bisphosphonium Ylids

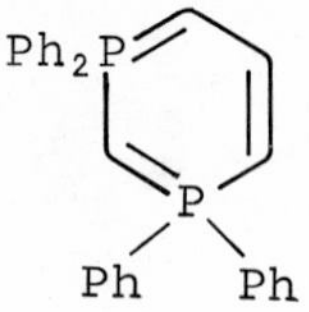

[Structure: 1,1,4,4-tetraphenyl-1,4-diphosphorin] . I.[411]

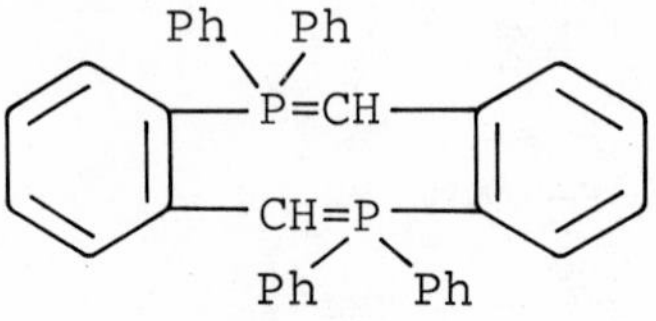

[Structure: dibenzo ring with Ph₂P=CH and CH=PPh₂ bridges] . I.[411]

[Structure: Ph₂P=CH–PPh₂=C(CO₂Me)–CH(CO₂Me) ring] . III. M. 176-7°, IR, mass spect, ^{1}H-NMR.[575]

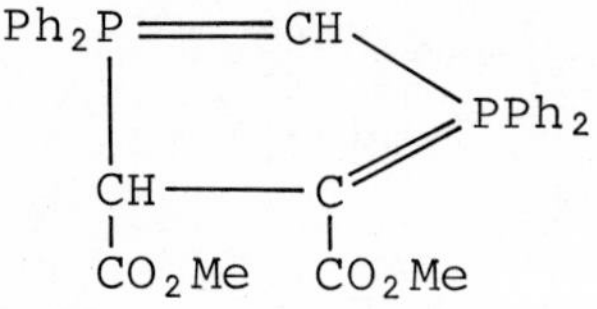

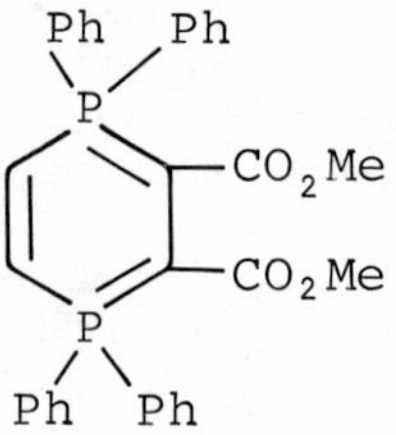

[Structure: 1,1,4,4-tetraphenyl-2,3-bis(CO₂Me)-1,4-diphosphorin] . III. IR, mass spect.[575]

D.2. Other Phosphorus Ylids (Ordered According to the Number of Carbon Atoms in R)

C_1

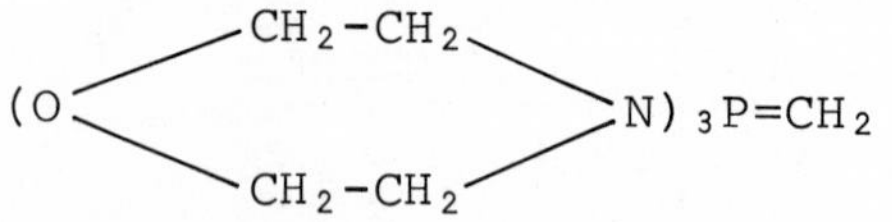

$(O(CH_2-CH_2)_2N)_3P=CH_2$. I.[688]

$(CH_2(CH_2-CH_2)_2N)_3P=CH_2$. I.[688]

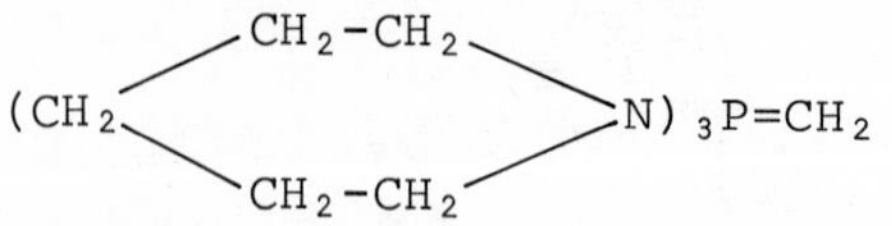

C_3

$(Me_2N)_3P=CH-CO_2Me$. I.[460]
$(MeO)_3P=C(CF_3)_2$. $B_{0.35}$ 61-2°.[437]
$(EtO)_3P=C(CF_3)_2$. $B_{0.25}$ 74-5°.[437]
$(i\text{-}PrO)_3P=C(CF_3)_2$. $B_{0.5}$ 75-6°.[437]
$(n\text{-}C_{12}H_{25}O)_3P=CCCF_3)_2$.[437]

$(Me_2N)_2PhP{=}CH{-}CO_2Et$. I. IR.[321]

C_7

$(Me_2N)_3P{=}CH{-}Ph$. I.[460]
$(Me_2N)_2PhP{=}CH{-}Ph$. I.[321]

C_{13}

$(MeO)_3P{=}C(o{-}C_6H_4)_2$. M. 105-7°.[437]

C_{16}

$Ph_2(EtO)P{=}C(COPh)CH_2COPh$. III.[487,490-1] M. 124°,[490-1] IR, ^1H-NMR, ^{31}P -54.2.[490-1]

C_{16}

$(MeO)_3P{=}C(COPh)CH_2COPh$. III.[486,489,490-1] IR, ^1H-NMR,[489,490-1] ^{31}P -56.2.[489,490-1]
$(Me_2N)_3P{=}C(COPh)CH_2COPh$. III. M. 137°, IR, ^1H-NMR, ^{31}P -63.2.[490-1]

D.3. Phosphinimines

D.3.1. Tertiary Phosphinimines, $R_3P{=}N{-}X$ (Ordered According to the Number of Carbon Atoms in X)

C_0

$Me_3P{=}NH$. I.[528,529] M. 59-60°, b_1 70°, IR, ^1H-NMR.[529]
$Me_3P{=}NLi$. XIII.[527-529] IR.[496]
$Et_3P{=}NH$. XIII. B_{11} 94°.[120-1]
$Et_3P{=}NLi$. XIII. M. 134-7°.[527]
$Pr_3P{=}NH$. XIII. B_{11} 129°,[120-1] DP.[394]
$Bu_3P{=}NH$. XIII.[120-1] $B_{0.1}$ 104°,[120-1] ^{31}P -31.[495]
$Bu_3P{=}NLi$. XIII. IR, ^{31}P -10.[496]
$Et_2PhP{=}NH$. II. M. 26.5°.[14]
$EtPh_2P{=}NH$. II. M. 74-5°.[14]
$(o{-}MeC_6H_4)_3P{=}NH$. II. M. 118°.[14]
$(m{-}MeC_6H_4)_3P{=}NH$. II. M. 117°.[14]
$(p{-}MeC_6H_4)_3P{=}NH$. II. M. 135-7°.[14]
$(C_{12}H_{25})Me_2P{=}NH$. II. IR, ^1H-NMR, ^{31}P -18.[495]
$(C_{12}H_{25})Me_2P{=}NLi$. XIII. IR, ^{31}P +10.[496]
$Ph_3P{=}NH$ I.[14] II.[8,10,11,588] X.[488] XIII.[122] M. 128°,[448] DP.[394]
$Ph_3P{=}NLi$. XIII. M. 214-9°.[527]
$Ph_3P{=}NH{\cdot}BH_3$. XIII.[18]
$Ph_3P{=}NH{\cdot}BPh_3$. XIII. M. 212°.[18]
$Ph_3P{=}NH{\cdot}BF_3$. XIII. M. 198-203°.[18]
$(Ph_3P{=}NH)_2{\cdot}Mo(CO)_4$. XIII. M. 127-40°, IR.[135a]

$(Ph_3P{=}NH)_2{\cdot}W(CO)_4$. XIII. M. 170-4°, IR.[135a]
$[(Ph_3P{=}NH)_2{\cdot}Mo(CO)_3]_2$. XIII. M. 235-45°, IR.[135a]
$(Ph_3P{=}NH)_2{\cdot}CoCl_2$. XIII. M. 217°.[15]
$(Ph_3P{=}NH)_2{\cdot}CuCl_2$. XIII. M. 193°.[15]
$(Ph_3P{=}NH)_4[V(CO)_6]_2$. XIII.[284]
$(Ph_3P{=}NH)_4{\cdot}NiI_2{\cdot}CH_3CN$. XIII. M. 226°.[15]
$(o\text{-}MeC_6H_4)_3P{=}NH$. II. M. 118°.[14]
$(p\text{-}MeC_6H_4)_3P{=}NH$. II. M. 135-7°.[14]
$(m\text{-}MeC_6H_4)_3P{=}NH$. II. M. 117°.[14]
$(p\text{-}CF_3C_6H_4)_3P{=}NH$. DP.[394]
$Ph_3P{=}N\text{-}Cl$. XIII. M. 179-80°.[13]
$Ph_3P{=}N\text{-}Br$. XIII.[12,13] M. 170-2°.[13]
$Ph_3P{=}N\text{-}I$. XIII. M. 174-5°.[13]
$[Me_3P{=}N\text{-}AlBr_2]_2$. XIII. M. 291-4°, IR, ^1H-NMR.[540]
$Ph_3P{=}N\text{-}POCl_2$. XIII. M. 186°.[5]
$Ph_3P{=}N\text{-}POBr_2$. XIII. M. 182-4°.[5]
$Ph_3P{=}N\text{-}PBr_4$. XIII.[5]
$Ph_3P{=}N\text{-}PBrCl_3$. XIII.[5]
$Ph_3P{=}N\text{-}SO_2NH_2$. XIII. M. 198°.[12]
$Ph_3P{=}N\text{-}N_2^{\oplus}SbCl_5N_3^{\ominus}$. I. M. 90-4°.[146]
$Ph_3P{=}N\text{-}N_2^{\oplus}SbCl_6^{\ominus}$. I. M. 166-9°.[146]
$Bu_3P{=}N\text{-}N_2^{\oplus}SbCl_6^{\ominus}$. I. M. 155-9°.[146]

```
          ⊖ ⊕
          N-S
         //   \\
Ph3P=N-S        N   .  V.  M.  180°.[222]
         \     /
          N=S
```

```
            Cl
            |
        N-P—NH2
      //     \\
Ph3P=N-P       N     .  III.  M.  172°.[359]
      |  \    /
      Cl  N=P
            | \
            Cl NH2
```

```
          Cl  Cl
            \/
         N-P
       //    \\
Ph3P=N-P       N    .  III.  M.  214-5°.[359]
      |  \    /
      Cl  N=P
            /\
          Cl  Cl
```

C_1

$Bu_3P{=}N\text{-}CN$. IX. $B_{0.155}$ 63°, IR.[440]
$(Et_2N\text{-}CH_2)_3P{=}N\text{-}CN$. IX. M. 90-2°, IR.[440]
$Ph_3P{=}N\text{-}CN$. IX. M. 194-6°, IR.[440]
$Et_3P{=}N\text{-}Me$. I. B_{11} 94-6°.[609]
$Ph_3P{=}N\text{-}Me$. I.[609,668,711-2] M. 67°,[711-2] IR,[668] UV,[711-2] ^1H-NMR.[358]

Ph_3P=N-CF_3. I.[418] XII.[419] M. 162°.[418]
Ph_3P=N-SO_2Me. I.[295] IV.[668] M. 192°,[295] IR.[668]
Pr_3P=N-CO-NH_2. DP.[394]
Ph_3P=N-CO-NH_2. DP.[394]
(p-$CF_3C_6H_4$)$_3$P=N-CO-NH . DP.[394]
Ph_3P=N-C(=NNO_2)NH_2. I. M. 178.5-79.5°.[455]
(Me_3P=N-ZnMe)$_4$. XIII. M. 350°, IR, ^{1}H-NMR.[529]
(Et_3P=N-ZnMe)$_4$. XIII. M. 292°, IR, ^{1}H-NMR.[529]
(Me_3P=N-CdMe)$_4$. XIII. M. 230°, IR, ^{1}H-NMR.[529]
(Et_3P=N-CdMe)$_4$. XIII. M. 214-8°, IR, ^{1}H-NMR.[529]

```
           N-N
          //  ||
Bu3P=N-C      ||   .  XIII. M.  169-70°.[440]
          \   ||
           N-N
           |
           H
```

```
           N-N
          //  ||
Ph3P=N-C      ||   .  XIII. M.  225-6°.[440]
          \   ||
           N-N
           |
           H
```

C_2

Me_3P=N-Et. I. B_{10} 56°, D(P-C).[165]
Et_3P=N-Et. I. B_{11} 93.5°.[609]
(i-Am)$_3$P=N-Et. I. $B_{0.03}$ 119°.[609]
Ph_3P=N-Et. I.[609] II.[160] III.[711-12] XIII.[12] M. 96°,[12,160,711-2] D(P-C).[165]
Ph_3P=N-NMe_2. ^{1}H-NMR.[358]
Ph_3P=N-COMe. XI.[195] M. 163-5°.[195]
Ph_3P=N-CO-$CHCl_2$. XIII.[465] M. 136-9°,[465] IR.[465]
Pr_3P=N-CO-CCl_3. DP.[394]
Ph_3P=N-CO-CCl_3. DP.[394]
(p-$CF_3C_6H_4$)$_3$P=N-CO-CCl_3. DP.[394]
Ph_3P=N-CO_2Me. II. M. 136-7°.[578]
(C_7H_{15})$_2$PhP=N-CO_2Me. XI. M. 25-30°, IR.[197]
($C_{10}H_{21}$)$_2$PhCH=CHP=N-CO_2Me. XI. M. 66-8°, IR.[197]
Ph_3P=N-C(=NCN)NH_2. I. M. 222-3°.[455]
Ph_3P=N-SO_2Et. I. M. 163-4°.[295]
Ph_3P=N-SO_2NMe_2. IV. M. 156-8°, IR.[441]
(m-MeC_6H_4)$_3$P=N-SO_2NMe_2. IV. M. 159°, IR.[441]
(p-MeC_6H_4)$_3$P=N-SO_2NMe_2. IV. M. 196°, IR.[441]
[Me_3P=N-$AlMe_2$]$_2$. XIII.[528,529,540] M. 129-31°,[540] IR, ^{1}H-NMR.[540]
[Et_3P=N-$AlMe_2$]$_2$. XIII.[529] M. 138-40°,[529] IR, ^{1}H-NMR.[540]
[Ph_3P=N-$AlMe_2$]$_2$. XIII.[529,530] M. 268-72°, IR, ^{1}H-NMR.[529]
[Et_3P=N-$GaMe_2$]$_2$. XIII. M. 100-3°, IR, ^{1}H-NMR.[529]
[Ph_3P=N-$GaMe_2$]$_2$. XIII.[529,530] M. 238-42°, IR, ^{1}H-NMR.[529]
[Et_3P=N-$InMe_2$]$_2$. XIII. M. 83-5°, IR, ^{1}H-NMR.[529]

$[Ph_3P{=}N{-}InMe_2]_2$. XIII.[529,530] M. 230-2°, IR, ^{1}H-NMR.[529]
$Me_3P{=}N{-}GeMe_2N_3$. I. M. 87-9°, IR, ^{1}H-NMR.[541]
$Me_3P{=}N{-}SnMe_2N_3$. I. M. 238-42°, IR.[541]
$Et_3P{=}N{-}SnMe_2N_3$. I. M. 218-24°, IR, ^{1}H-NMR.[541]
$Ph_3P{=}N{-}SnMe_2N_3$. I. M. 225-30°, IR.[541]
$Me_3P{=}N{-}PMe_2$. XIII. M. 8.5-10°, IR, ^{1}H-NMR.[529]
$Me_3P{=}N{-}AsMe_2$. XIII. B_{12}96-9°, IR, ^{1}H-NMR.[529]
$Et_3P{=}N{-}AsMe_2$. XIII. B_{12} 114-5°, IR, ^{1}H-NMR.[529]

C_3

$Ph_3P{=}N{-}Pr$. III. M. 112-4°.[711-2]
$Ph_3P{=}N{-}i{-}Pr$. III. M. 126-7°.[711-2]
$Pr_3P{=}N{-}CO_2Et$. DP.[394]
$Ph_3P{=}N{-}CO_2Et$. I. M. 136-7°,[668] DP,[394] IR.[668]
$(C_6H_{11})_2PhP{=}N{-}CO_2Et$. XI. Oil, IR.[197]
$(C_7H_{15})_2PhP{=}N{-}CO_2Et$. XI. Oil, IR.[197]
$(C_8H_{17})_2PhP{=}N{-}CO_2Et$. XI. M. 28-34°, IR.[197]
$(C_9H_{19})_2PhP{=}N{-}CO_2Et$. XI. M. 25-32°, IR.[197]
$(C_{10}H_{21})_2PhCH{=}CH{-}P{=}N{-}CO_2Et$. XI. M. 71-3°, IR.[197]
$(p{-}CF_3C_6H_4)_3P{=}N{-}CO_2Et$. DP.[394]
$Ph_3P{=}NC({=}NCN)NHMe$. I. M. 229-30°.[455]
$Me_3P{=}N{-}SiMe_3$. I.[538] B_{11} 57°,[527] IR, ^{1}H-NMR.[527,538]
$Me_3P{=}N{-}SiMe_3{\cdot}GaCl_3$. XIII.[527,540] IR, ^{1}H-NMR.[527]
$Me_3P{=}N{-}SiMe_3{\cdot}InCl_3$. XIII.[527,540] IR, ^{1}H-NMR.[527]
$Me_3P{=}N{-}SiMe_3{\cdot}InBr_3$. XIII.[527,540] IR, ^{1}H-NMR.[527]
$Me_3P{=}N{-}SiMe_3{\cdot}AlX_3$. (X = Cl, Br, I). XIII.[527,540] IR, ^{1}H-NMR.[527]
$Me_3P{=}N{-}SiMe_3{\cdot}AlMe_3$. XIII. M. 79-80°.[537]
$Me_3P{=}N{-}SiMe_3{\cdot}GaMe_3$. XIII.[537,538] M. 32-4°,[537] IR, ^{1}H-NMR.[537,538]
$Me_3P{=}N{-}SiMe_3{\cdot}InMe_3$. XIII.[537,538] M. 43-4°,[537] IR, ^{1}H-NMR.[537,538]
$Me_3P{=}N{-}SiMe_3{\cdot}AlEt_3$. XIII.[537,538] M. 28-30°,[538] IR, ^{1}H-NMR.[537,538]
$Et_3P{=}N{-}SiMe_3$. I. B_{11} 89.5°.[120-1]
$Et_3P{=}N{-}SiMe_3{\cdot}AlMe_3$. XIII.[537,538] M. 169°,[537] IR, ^{1}H-NMR.[538]
$Et_3P{=}N{-}SiMe_3{\cdot}InMe_3$. XIII.[537,538] IR, ^{1}H-NMR.[538]
$Et_3P{=}N{-}SiMe_3{\cdot}GaMe_3$. XIII.[537,538] M. 114°,[537] IR, ^{1}H-NMR.[538]
$Et_3P{=}N{-}SiMe_3{\cdot}AlEt_3$. XIII.[537,538] M. 57-9°,[538] IR, ^{1}H-NMR.[538]
$Pr_3P{=}N{-}SiMe_3$. I. B_{11} 119°.[120-1]
$Bu_3P{=}N{-}SiMe_3$. I. B_{11} 149°.[120-1]
$C_{12}H_{25}Me_2P{=}N{-}SiMe_3$. I. B. 95-100°, IR, ^{1}H-NMR, ^{31}P -6.[495]
$Ph_3P{=}N{-}SiMe_3$. I.[123,538] XIII.[527] M. 76-77°,[123] IR, ^{1}H-NMR.[527]
$Ph_3P{=}N{-}SiMe_3{\cdot}AlMe_3$. XIII. M. 126-30°, IR, ^{1}H-NMR.[538]
$Ph_3P{=}N{-}SiMe_3{\cdot}GaMe_3$. XIII. M. 87-9°, IR, ^{1}H-NMR.[538]

[Structure: Ph$_2$P=C–N(SiMe$_3$)–MR$_2$ fused to benzene ring] (M=Al, Ga). XIII. M. 222-4°(Al), ^{1}H-NMR, IR,[539] M. 202-4°(Ga), ^{1}H-NMR, IR.[539]

$Ph_3P=N-SiMe_3 \cdot AlPh_3$. XIII. M. 202-12°, IR, ^{1}H-NMR.[539]
$Ph_3P=N-SiMe_3 \cdot GaPh_3$. XIII. M. 175-6°, IR, ^{1}H-NMR.[539]
$Me_3P=N-GeMe_3$. XIII.[527,541] B_{12} 69-70°,[541] IR,[527] ^{1}H-NMR.[527,541]
$Et_3P=N-GeMe_3$. XIII.[527,541] $B_{0.5}$ 55-7°,[541] IR, ^{1}H-NMR.[527]
$Ph_3P=N-GeMe_3$. XIII. $B_{0.1}$ 164°, IR, ^{1}H-NMR.[527]
$Me_3P=N-SnMe_3$. XIII. B_{11} 88-9°, IR, ^{1}H-NMR.[527]
$Et_3P=N-SnMe_3$. XIII. $B_{0.5}$ 73-4°, IR, ^{1}H-NMR.[527]
$Ph_3P=N-SnMe_3$. XIII. $B_{0.5}$ 172-6°, IR, ^{1}H-NMR.[527]

$Ph_3P=N$–[4,6-diamino-1,3,5-triazin-2-yl]. I.[377]

$Et_2PhP=N$–[4,6-diamino-1,3,5-triazin-2-yl]. I.[377]

C_4

$Ph_3P=N$-i-Bu. III. M. 79-81°.[711-2]
$Ph_3P=N$-t-Bu. III. M. 146-8°, UV, ^{1}H-NMR.[711-2]
$PhCH_2Ph_2P=N$-t-Bu. II.[265-6,593] M. 92.5-93°.[265-6]
$Ph_3P=N-C(CF_3)_3$. XIII. M. 146-47.5°.[209]
$Ph_3P=N-CH_2-CO_2Et$. I.[609]

$Ph_3P=N-CO-N$[imidazol-1-yl]. XIII. M. 200-1°.[607]

$Ph_3P=N-SO_2-NEt_2$. IV. M. 127°, IR.[441]
$(p-MeC_6H_4)_3P=N-SO_2NEt_2$. IV. M. 160°, IR.[441]

$Ph_3P=N-SO_2-N$[morpholino]O. IV. M. 181-2°, IR.[441]

$(m-MeC_6H_4)_3P=N-SO_2-N$[morpholino]O. IV. M. 124-5°, IR.[441]

$(p\text{-}MeC_6H_4)_3P{=}N\text{-}SO_2\text{-}N$ (morpholino ring) O. IV. M. 129°, IR.[441]

$Ph_3P{=}NC({=}NCN)NMe_2$. I. M. 159-60°.
$Ph_3P{=}N\text{-}C(CF_3){=}CHCN$. V. M. 162-6°, ^{1}H-NMR.[166]
$Ph_3P{=}N\text{-}C(CN){=}CHCN$. V. M. 228-30°, IR, ^{1}H-NMR, UV, ^{31}P -13.7.[166]
$Et_3P{=}N\text{-}AlEt_2$. XIII. M. 104-7°, $B_{0.02}$ 195-8°, IR, ^{1}H-NMR.[529]
$Ph_3P{=}NP(O)(OEt)_2$. I.[656]
$Ph_3P{=}N\text{-}N{=}N\text{-}SO_2NEt_2$. I.[226]

C_5

$Ph_3P{=}N\text{-}C(CONH_2){=}CHCO_2Me$. V. M. 163-4°, UV, IR, ^{1}H-NMR.[166]
$Ph_3P{=}N\text{-}C(CF_3){=}CHCO_2Me$. V. M. 114-5°, ^{1}H-NMR.[166]
$Ph_3P{=}N\text{-}C(CN){=}CHCO_2Me$. V. M. 173-4°, UV, IR, ^{1}H-NMR.[166]
$Ph_3P{=}N\text{-}C(CN){=}CBrCO_2Me$. V. M. 158-9°, UV, IR, ^{1}H-NMR.[166]
$Ph_3P{=}N\text{-}C(CN){=}C(CN)_2$. XIII. M. 192-4°, UV.[706]
$Me_3P{=}N\text{-}GeMe_2\text{-}O\text{-}SiMe_3$. XIII. M. -12 to -10°, IR, ^{1}H-NMR.[541]

$Ph_3P{=}N$-(4,6-dimethoxy-1,3,5-triazin-2-yl: ring N, N, N; OMe, OMe). XIII.[377]

$Ph_3P{=}N$-(5-iodo-2-pyridyl: ring N)-I. XII. M. 208°.[694]

C_6

$Et_3P{=}N\text{-}Ph$. I. $B_{0.08}$ 116°.[609]
$Pr_3P{=}N\text{-}Ph$. DP.[394]
$PhEt_2P{=}N\text{-}Ph$. I. M. 69-70°,[609] pK_a.[210]
$Bu_3P{=}N\text{-}Ph$. I. $B_{0.01}$ 89-94°, IR.[668]
$Ph_2EtP{=}N\text{-}Ph$. I. $B_{0.01}$ 165-70°, IR.[668]
$Am_3P{=}N\text{-}Ph$. I. $B_{0.04}$ 161°.[609]
$Ph_3P{=}N\text{-}Ph$. I.[378,611,668] III.[310] V.[108] VIII.[322] M. 131-2°,[668] ^{31}P ±0.0,[497] IR,[310,667,668] UV,[310] DP.[394]
$Ph_3P{=}N\text{-}Ph\cdot CdI_2$. XIII.[556]
$Ph_3P{=}N\text{-}Ph\cdot CoCl_2$. XIII.[556]
$Ph_3P{=}N\text{-}Ph\cdot HgI_2$. XIII. M. 178°.[556]
$[Ph_3P{=}N\text{-}Ph]_2\cdot NiCl_2$. XIII.[556]
$(p\text{-}CF_3C_6H_4)_3P{=}N\text{-}Ph$. I. M. 138-9°,[705] DP.[394]
$(p\text{-}Me_2NC_6H_4)_3P{=}N\text{-}Ph$. I. M. 146-8°.[705]
$(p\text{-}CF_3C_6H_4)_2PhP{=}N\text{-}Ph$. I. M. 100-2°.[705]
$(Ph_2P{=}CH_2)Ph_2P{=}N\text{-}Ph$. I. M. 96-97.5°.[238]
$[(p\text{-}O_2NC_6H_4)Ph_2P\text{-}CH_2]Ph_2P{=}N\text{-}Ph$. I. M. 197.5-98.5°.[238]
$(NC\text{-}CH_2\text{-}CH_2)Ph_2P{=}N\text{-}Ph$. VIII.[639]
$(H_2NCOCH_2\text{-}CH_2)Ph_2P{=}N\text{-}Ph$. VIII.[639]

(9-phenyl-9-phosphafluorene structure) P=N-Ph . I.[679] M. 194-5°.[679]

$PhEt_2P=NC_6H_4Cl$-p. pK_a.[210]
$Ph_3P=N-C_6H_4Cl$-p. III. M. 118-20°, IR, UV.[310]
$Ph_3P=N-C_6H_4Cl$-o. III. M. 137-8°.[310]
$Ph_3P=N-C_6H_4Cl$-m. III.[310] I.[378] M. 118-9°.[310]
$Ph_3P=N-C_6H_3Cl_2$-2.4. I. M. 125-7°.[706]
$PhEt_2P=N-C_6H_4Br$-p. pK_a.[210]
$Ph_3P=N-C_6H_4Br$-m. I. M. 107-9°.[706]
$Ph_3P=N-C_6H_3Br_2$-2.4. XII. M. 176-80°.[694]
$Ph_3P=N-C_6H_4I$-p. XII. M. 156-7°.[694]
$Ph_3P=N-C_6H_4F$-m. I. M. 141-2°.[378]
$Ph_3P=N-C_6H_4F$-p. I. M. 140.5-42°.[378]
$PhEt_2P=N-C_6H_4NH_2$-p. pK_a.[210]
$PhEt_2P=N-C_6H_4NO_2$-p. pK_a, UV.[211]
(p-$MeOC_6H_4$)$Et_2P=N-C_6H_4NO_2$-p. pK_a, UV.[211]
(p-$Me_2NC_6H_4$)$Et_2P=N-C_6H_4NO_2$-p. pK_a, UV.[211]
(p-MeC_6H_4)$Et_2P=N-C_6H_4NO_2$-p. pK_a, UV.[211]
(p-ClC_6H_4)$Et_2P=N-C_6H_4NO_2$-p. pK_a, UV.[211]
(p-$MeO_2CC_6H_4$)$Et_2P=N-C_6H_4NO_2$-p. pK_a, UV.[211]
$Pr_3P=N-C_6H_4NO_2$-p. I. M. 92-4°, DP, UV, ^{1}H-NMR.[245]
$PhPr_2P=N-C_6H_4NO_2$-p. I. M. 84°, DP, UV, ^{1}H-NMR.[245]
(p-ClC_6H_4)$Pr_2P=N-C_6H_4NO_2$-p. I. M. 90-2°, DP, UV, ^{1}H-NMR.[245]
(p-BrC_6H_4)$Pr_2P=N-C_6H_4NO_2$-p. I. M. 91°, DP, UV, ^{1}H-NMR.[245]
(p-$Me_2NC_6H_4$)$Pr_2P=N-C_6H_4NO_2$-p. I. M. 115°, DP, UV, ^{1}H-NMR.[245]
$Ph_3P=N-C_6H_4NO_2$-o. M. 151-2°, UV.[310]
$Ph_3P=N-C_6H_4NO_2$-m. I. M. 138-38.5°,[378] UV.[310]
$Ph_3P=N-C_6H_4NO_2$-p. I.[378] III.[245,668] M. 156-8°,[245] M. 160-2°,[668] ^{1}H-NMR,[245] DP,[245,394] IR,[135,310] UV.[245,310]
(p-ClC_6H_4)$_3P=N-C_6H_4NO_2$-p. III. M. 206°, DP, UV, ^{1}H-NMR.[245]
(p-$MeOC_6H_4$)$_3P=N-C_6H_4NO_2$-p. III. M. 133°, DP, UV, ^{1}H-NMR.[245]
(p-$Me_2NC_6H_4$)$_3P=N-C_6H_4NO_2$-p. I. M. 220-1°, DP, UV, ^{1}H-NMR.[245]
(p-$CF_3C_6H_4$)$_3P=N-C_6H_4NO_2$-p. I. M. 212-3°,[705] DP.[394]
(p-$CF_3C_6H_4$)$Ph_2P=N-C_6H_4NO_2$-p. I. M. 161-3°.[705]
(p-$Me_2NC_6H_4$)$Ph_2P=N-C_6H_4NO_2$-p. I. M. 156-8°.[705]

$(4\text{-}Me_2N\text{-}3,5\text{-}Me_2C_6H_2)_3P{=}N\text{-}C_6H_4NO_2$-p. I. M. 250°.[245]

$Ph_3P{=}N\text{-}C_6H_3(NO_2)_2$-2,4. III. M. 200-1°.[310]
$Ph_3P{=}N\text{-}C_6H_2(NO_2)_3$-2,4,6. III. M. 183-4°.[310]
$Me_3P{=}N\text{-}SO_2Ph$. VII. M. 111-2°.[384]
$Et_3P{=}N\text{-}SO_2Ph$. VII. M. 100-1°.[384]
$Pr_3P{=}N\text{-}SO_2Ph$. DP.[394]
$Bu_3P{=}N\text{-}SO_2Ph$. I.[226] VII.[384] M. 42-3°.[384]
$Me_2PhP{=}N\text{-}SO_2Ph$. II. M. 116-8°.[620]
$Et_2PhP{=}N\text{-}SO_2Ph$. II. M. 95-6°.[620]
$EtPh(PhCH_2)P{=}N\text{-}SO_2Ph$. II. M. 107-8°.[581]
$Ph_3P{=}N\text{-}SO_2Ph$. III.[310 580] VII.[384] M. 157-8°,[384] DP.[394]
$(PhCH_2)_3P{=}N\text{-}SO_2Ph$. VII. M. 172-3°.[384]
$(C_6H_{13})_2PhP{=}N\text{-}SO_2Ph$. II. M. 44-5°.[215]
$(C_7H_{15})_2PhP{=}N\text{-}SO_2Ph$. II. M. 38-40°.[215]
$(C_8H_{17})_2PhP{=}N\text{-}SO_2Ph$. II. M. 45-6°.[215]
$(C_9H_{19})_2PhP{=}N\text{-}SO_2Ph$. II. M. 48°.[215]
$(C_{10}H_{21})_2PhP{=}N\text{-}SO_2Ph$. II.[215] M. 47°.[197]
$(p\text{-}MeC_6H_4)Ph_2P{=}N\text{-}SO_2Ph$. I. II. III. M. 133-4°.[577]
$(p\text{-}MeOC_6H_4)(p\text{-}BrC_6H_4)PhP{=}N\text{-}SO_2Ph$. I. II. III. M. 140-1°.[577]
$(p\text{-}MeC_6H_4)(p\text{-}BrC_6H_4)PhP{=}N\text{-}SO_2Ph$. I. II. III. M. 141-2°.[577]
$(p\text{-}MeOC_6H_4)_2PhP{=}N\text{-}SO_2Ph$. I. II. III. M. 94-5°.[577]
$(p\text{-}MeC_6H_4)_2PhP{=}N\text{-}SO_2Ph$. I. II. III. M. 128-30°.[577]
$(p\text{-}CF_3C_6H_4)_3P{=}N\text{-}SO_2Ph$. DP.[394]
$(p\text{-}MeC_6H_4)_2PhP{=}N\text{-}SO_2C_6H_4Cl$-p. I. II. III. M. 123-5°.[577]
$(p\text{-}MeOC_6H_4)_2PhP{=}N\text{-}SO_2\text{-}C_6H_4Cl$-p. I. II. III. M. 143-4°.[577]
$(p\text{-}MeC_6H_4)Ph_2P{=}N\text{-}SO_2\text{-}C_6H_4Cl$-p. I. II. III. M. 169-70°.[577]
$(p\text{-}MeC_6H_4)(p\text{-}BrC_6H_4)PhP{=}N\text{-}SO_2\text{-}C_6H_4Cl$-p. I. II. III. M. 163-5°.[577]
$(p\text{-}MeOC_6H_4)(p\text{-}BrC_6H_4)PhP{=}N\text{-}SO_2\text{-}C_6H_4Cl$-p. I. II. III. M. 120-1°.[577]
$(p\text{-}MeC_6H_4)Ph_2P{=}N\text{-}SO_2C_6H_4NO_2$-p. I. II. III. M. 149-50°.[577]
$(p\text{-}MeOC_6H_4)_2PhP{=}N\text{-}SO_2C_6H_4NO_2$-p. I. II. III. M. 168-9°.[577]
$(p\text{-}MeC_6H_4)(p\text{-}BrC_6H_4)PhP{=}N\text{-}SO_2\text{-}C_6H_4NO_2$-p. I. II. III. M. 138-40°.[577]
$(p\text{-}MeOC_6H_4)(p\text{-}BrC_6H_4)PhP{=}N\text{-}SO_2\text{-}C_6H_4NO_2$-p. I. II. III. M. 122-3°.[577]
$Ph_3P{=}N\text{-}SO_2C_6H_4NO_2$-o. III. M. 185-6°.[580]
$Ph_3P{=}N\text{-}SO_2NPr_2$. IV. M. 178-9°, IR.[441]
$(p\text{-}MeC_6H_4)_3P{=}N\text{-}SO_2NPr$. IV. M. 175°, IR.[441]
$Ph_3P{=}N\text{-}P(O)N_3Ph$. I. M. 143-5°, IR.[23]

Ph_3P=N-P(S)N_3Ph. I. IR.[23]
Ph_3P=N-P(O)OHPh. XIII. M. 145-6°, IR.[23]
Bu_3P=N-$SnEt_3$. I. B. 101-3°.[389]
Ph_3P=N-$SnEt_3$. I. B. 180-2°, M. 49-52°, ^{1}H-NMR.[389]
Ph_3P=N-C(CN)=N-t-Bu. V. M. 150-1°, ^{1}H-NMR.[166]
Ph_3P=N-C(CN)=CMe-CO_2Me. V. M. 178-9°, ^{1}H-NMR.[166]
Ph_3P=N-C(CF_3)=CMe-CO_2Me. V. M. 88-90°, IR, UV, ^{1}H-NMR, ^{19}F-NMR.[166]
Ph_3P=N-C(Me)=CH-CO_2Et. I. M. 135-135.5°, IR, UV, ^{1}H-NMR.[271a]
Ph_3P=N-N=N-SO_2Ph. I. M. 94-5°.[226]

C_7

Ph_3P=N-N=N-$SO_2C_6H_4$Me-p. I.[226]
Et_2PhP=N-C_6H_4Me-p. pK_a.[210]
Ph_3P=N-C_6H_4Me-p. I.[378,611] III.[310] M. 135-6°,[378] IR.[310]
Ph_3P=N-C_6H_4Me-o. III. M. 129-30°.[310]
Ph_3P=N-C_6H_4Me-m. I.[378] III.[310] M. 111.5-112.5°.[378]
Et_2PhP=N-C_6H_4OMe-p. pK_a.[210]
Ph_3P=N-C_6H_4OMe-p. I.[378] III.[310] M. 118°, IR.[310]
Ph_3P=N-C_6H_4OMe-m. I. M. 125-125.5°.[378]
Et_2PhP=N-C_6H_4CN-p. pK_a.[210]
Ph_3P=N-C_6H_4CN-p. I.[378] III.[310] M. 194-95.5°,[378] IR, UV.[310]
Ph_3P=N-C_6H_4CN-m. I. M. 150-2°.[378]
Ph_3P=N-$C_6H_4CF_3$-p. III. M. 113-4°,[705] UV.[395]
(p-$Me_2NC_6H_4$)Ph_2P=N-$C_6H_4CF_3$-p. I. M. 150-1°,[705] UV.[395]
(p-$CF_3C_6H_4$)Ph_2P=N-$C_6H_4CF_3$-p. I. M. 129-31°,[705] UV.[395]
(p-$CF_3C_6H_4$)$_3$P=N-$C_6H_4CF_3$-p. I. M. 152-4°,[705] UV.[395]
Ph_3P=N-$C_6H_4CO_2$H-o. I. M. 215-6°.[378]
Et_3P=N-COPh. I. 62.5-63°.[609]
Bu_3P=N-COPh. I. Oil,[668] IR.[668]
$PhEt_2P$=N-COPh. I. M. 73-4°.[609]
Ph_3P=N-COPh. I.[609,668] IV.[194] VI.[657] XI.[195] M. 196-7°,[668] IR.[135,668]
Pr_3P=N-CO-C_6H_4Br-p. DP.[394]
Ph_3P=N-CO-C_6H_4Br-p. I. M. 140-2°,[196] IR,[196] DP.[394]
(p-$CF_3C_6H_4$)$_3$P=N-CO-C_6H_4Br-p. DP.[394]
Ph_3P=N-CO-C_6H_4Cl-p. IV. M. 54-6°.[194]
Pr_3P=N-CO-$C_6H_4NO_2$-m. DP.[394]
Ph_3P=N-CO-$C_6H_4NO_2$-m. I. M. 149-51°,[196] DP,[394] IR.[196]
(p-$CF_3C_6H_4$)$_3$P=N-CO-$C_6H_4NO_2$-m. DP.[394]
Ph_3P=N-CO-$C_6H_4NO_2$-p. I.[196] X.[195] M. 200-1°.[196]
(C_8H_{17})$_2$PhP=N-CO_2Ph. XI. Oil, IR.[197]
($C_{10}H_{21}$)$_2$(PhCH=CH)P=N-CO_2Ph. X. M. 72-4°, IR.[197]
Et_3P=N-$SO_2C_6H_4$Me-p. II. M. 119°.[420]
Pr_3P=N-$SO_2C_6H_4$Me-p. II. M. 66°.[420]
Bu_3P=N-$SO_2C_6H_4$Me-p. II.[420] I.[668] M. 54°,[420] $B_{0.01}$ 130-4°.[668]
Me_2PhP=N-$SO_2C_6H_4$Me-p. I. M. 105-6°.[620]

$Me_2PhP{=}N{-}SO_2C_6H_4Me$-o. II. M. 97-8°.[620]
$Et_2PhP{=}N{-}SO_2C_6H_4Me$-p. II.[420,620] M. 82°.[420]
$Et_2PhP{=}N{-}SO_2C_6H_4Me$-o. II. M. 125-7°.[620]
$EtPh_2P{=}N{-}SO_2C_6H_4Me$-p. I. 116-7°, IR.[668]
$Ph_3P{=}N{-}SO_2C_6H_4Me$-p. II.[668] III.[310 580] IV.[365] XII.[11]
M. 193-4°,[668] IR.[135,264,668]
$Ph_3P{=}N{-}SO_2C_6H_4Me$-o. III.[580] IV.[365] M. 168-9°.[580]
$(2{-}C_5H_4N)_3P{=}N{-}SO_2C_6H_4Me$-p. II. M. 177°.[421]
$Et(PhCH_2)PhP{=}N{-}SO_2C_6H_4Me$-p. I. II. M. 102-3°[581]
$Et(PhCH_2)PhP{=}N{-}SO_2C_6H_4Me$-o. I. II. M. 125-6°.[581]

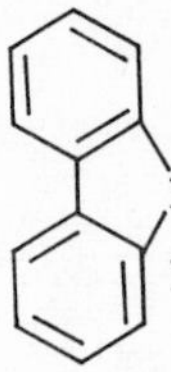

$P{=}N{-}SO_2C_6H_4Me$-p. II. M. 176-7°.[681]
Ph

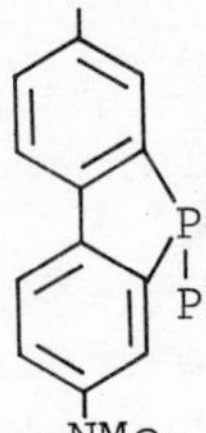

NMe_2
$P{=}N{-}SO_2C_6H_4Me$-p. II.[681]
Ph
NMe_2

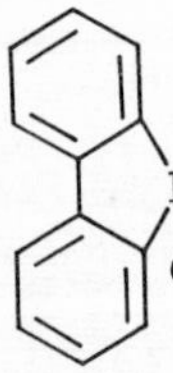

$P{=}N{-}SO_2C_6H_4Me$-p. II. M. 198-200°.[681]
$C_6H_4NMe_2$-p.

$(p{-}MeOC_6H_4)_2PhP{=}N{-}SO_2C_6H_4Me$-p. I. II. III. M.
134-5°.[577]
$(p{-}MeC_6H_4)Ph_2P{=}N{-}SO_2C_6H_4Me$-p. I. II. III. M. 143-4°.[577]
$(p{-}MeC_6H_4)_2PhP{=}N{-}SO_2C_6H_4Me$-p. I. II. III. M. 142-3°.[577]
$(p{-}MeC_6H_4)Et_2P{=}N{-}SO_2C_6H_4Me$-p. II. M. 120°.[420]
$(o{-}ClC_6H_4)_3P{=}N{-}SO_2C_6H_4Me$-p. II. M. 235-6°.[420]
$(p{-}ClC_6H_4)_3P{=}N{-}SO_2C_6H_4Me$-p. II. M. 232°.[420]
$(o{-}MeC_6H_4)_3P{=}N{-}SO_2C_6H_4Me$-p. II. M. 188°.[420]
$(m{-}MeC_6H_4)_3P{=}N{-}SO_2C_6H_4Me$-p. II. M. 98° (hydrate).[420]
$(p{-}MeC_6H_4)_3P{=}N{-}SO_2C_6H_4Me$-p. II. M. 174°.[420]
$(o{-}MeOC_6H_4)_2P{=}N{-}SO_2C_6H_4Me$-p. II. M. 273-4°.[420]
$(m{-}MeOC_6H_4)_3P{=}N{-}SO_2C_6H_4Me$-p. II. M. 112°.[420]

$(p-MeOC_6H_4)_3P=N-SO_2C_6H_4Me-p$. II. M. 155°.[420]

$(p-MeC_6H_4)(p-BrC_6H_4)PhP=N-SO_2C_6H_4Me-p$. I. II. III. M. 144-5°.[577]

$(p-MeOC_6H_4)(p-BrC_6H_4)PhP=N-SO_2C_6H_4Me-p$. I. II. III.[577]

$Et_2PhP=N-SO_2C_6H_4Me-o$. II. M. 125-7°.[620]

$Me_2PhP=N-SO_2C_6H_4Me-o$. II. M. 97-8°.[620]

$Ph_3P=NC(=NPh)NH_2$. I. M. 194-5°.[455]

$Ph_3P=NC(=NH)C_6H_4NO_2-p$. XI. M. 165-6°.[198]

$Ph_3P=NC(=NH)C_6H_4Br-p$. XI. M. 170-1°.[198]

$Ph_3P=NC(=NH)Ph$. XI. M. 70-80°.[198]

$Ph_3P=N-NH-CO-Ph$. III. M. 202-3°, IR.[654a]

$Ph_3P=N(=NSO_2C_6H_4NO_2-p)NH_2$. I. M. 232-3°.[455]

$Ph_3P=N-C(CF_3)=CCO_2CH_2CH_2CH_2$. V. M. 165-9°.[166]

$Ph_3P=N-C(CN)=CCO_2CH_2CH_2CH_2$. V. M. 218-20°.[166]

$Ph_3P=N-P(S)(OMe)Ph$. XIII. M. 121-3°.[23]

$Ph_3P=N-P(O)(OMe)Ph$. XIII. M. 133-5°.[23]

C_8

$Et_2PhP=N-C_6H_4CO_2Me-p$. pK_a.[210]

$Ph_3P=N-C_6H_3Me_2-3,5$. I. M. 130-1°.[611]

$Ph_3P=N-C_6H_2Me_2-3,5,NO_2-4$. III. M. 195-8°.[245]

$Ph_3P=N-C_6H_4NMe_2-p$. I.[378,705] III.[310] M. 146-8°.[705]

$(p-CF_3C_6H_4)Ph_2P=N-C_6H_4NMe_2-p$. III. M. 146-8°.[705]

$(p-CF_3C_6H_4)_3P=N-C_6H_4NMe_2-p$. III. M. 136-7°.[705]

$(XYZ)P=N-C_6H_4NMe_2-p$. UV.[395]

$(XYZ)P=N-CO-C_6H_4Me-p$. UV.[395]

$(XYZ)P=N-CO-C_6H_4CF_3-p$. UV.[395]

$Ph_3P=N-CO-NH-CO-Ph$. I. M. 153-4°.[449]

$Ph_3P=N-CO-NH-COC_6H_4Cl-p$. I. M. 175-6°.[449]

CN

CN

$Ph_3P=N-CO-$ [ring] $-CN$. XIII. M. 215-7°,[659] IR.[659]

O

$Ph_3P=N-CO-NH-SO_2C_6H_4Me-p$. I. M. 146-46.5°.[449]

$Ph_3P=N-C(=NCN)NHPh$. I. M. 200-2°.[455]

$Ph_3P=N-SO_2NBu_2$. IV. M. 149°, IR.[441]

$(p-MeC_6H_4)_3P=N-SO_2NBu_2$. IV. M. 155°, IR.[441]

$Ph_3P=N-C(CF_3)=N-Ph$. V. M. 122-4°.[166]

$Ph_3P=N-C(CN)=N-Ph$. V. M. 197-8°,[166] IR, UV.[166]

$Ph_3P=N-P(O)PhNMe_2$. I. M. 126.5-29°.[23]

$Ph_3P=NN_3P_3(NH_2)(NMe_2)_4$. III. M. 181°.[359]

$Ph_3P=N-N=-SO_2-C_8H_{17}$. I.[226]

NHEt

$Ph_3P=N-$ [ring: N, N, N] . I.[377]

NHPr

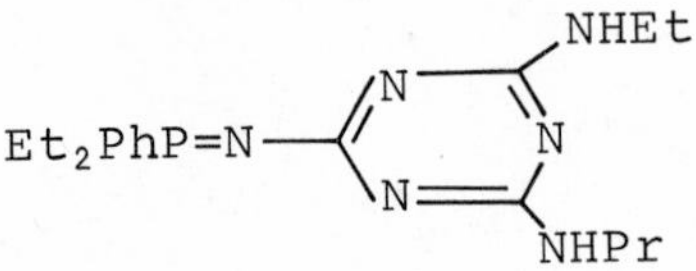

$Et_2PhP=N-$[2-(4-NHEt-6-NHPr-1,3,5-triazinyl)] . I.[377]

C_9

$Ph_3P=N-C_6H_4CO_2Et$-p. III. M. 135-6°, IR.[310]
$Ph_3P=N-CO-NH-CO-C_6H_4Me$-p. I. M. 180-1°.[449]
$Ph_3P=N-CO-NH-CO-C_6H_4OMe$-o. I. M. 59-60°.[449]
$Ph_3P=N-CO-NH-CO-C_6H_4OMe$-m. I. M. 170-1°.[449]

C_{10}

$Ph_3P=N-C_{10}H_7$-1. I. M. 141-3°.[609]
$Ph_3P=N-C_{10}H_7$-2. III. M. 142-3°.[310]
$Ph_3P=N-C_6H_4$-t-Bu-p. I. M. 171-3°.[378]
$Ph_3P=N-C_6H_4$-t-Bu-m. I. M. 127.5-29°.[378]
$Ph_3P=N-C_6H_4NEt_2$-p. III. M. 138-40°.[668]
$Ph_3P=N-(1-I-C_{10}H_6)-2$. XII. M. 143°.[694]

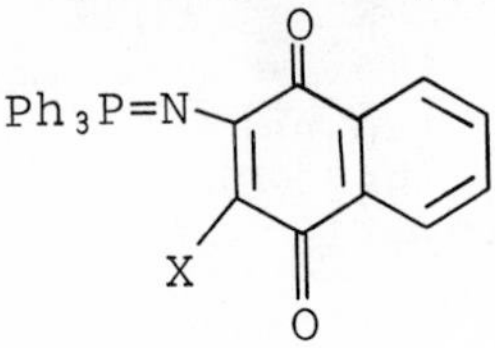

$Ph_3P=N-$[2-(3-X-1,4-naphthoquinonyl)] (X = H, NH_2, N_3). I. M. 176.5-77° (X = H), M. 188-90° (X = NH_2), M. 122.5-23.5° (X = N_3), UV.[445]

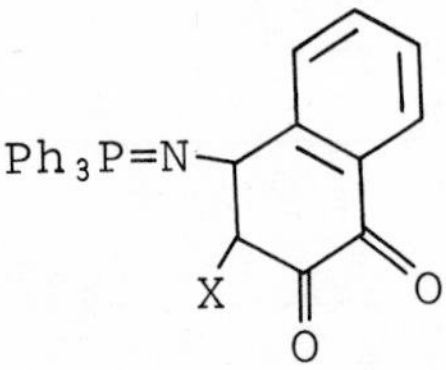

$Ph_3P=N-$[4-(3-X-1,2-naphthoquinonyl)] (X = H, Cl). I. M. 259-60° (X = H),[455] M. 220-1° (X = Cl), UV.[445]

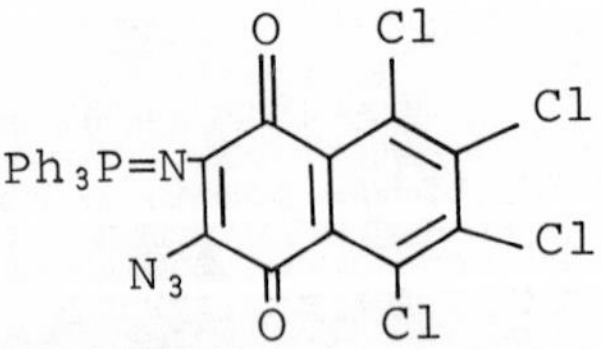

$Ph_3P=N-$[2-(3-N_3-5,6,7,8-tetrachloro-1,4-naphthoquinonyl)] . I. M. 131.5-33.5°, UV.[445]

$Me_2PhP=N-SO_2C_{10}H_7$-1. II. M. 84-6°.[620]
$Me_2PhP=N-SO_2C_{10}H_7$-2. II. M. 167-8°.[620]
$Et_2PhP=N-SO_2C_{10}H_7$-2. II. M. 86-8°.[620]
$Et_2PhP=N-SO_2C_{10}H_7$-1. II. M. 150-1°.[620]
$Et(PhCH_2)PhP=N-SO_2C_{10}H_7$-1. II. M. 105-6°.[581]
$Ph_3P=N-SO_2C_{10}H_7$-1. III.[580] IV.[365] M. 201-2°.[580]
$Ph_3P=N-SO_2C_{10}H_7$-2. IV.[365] M. 162-4°.[365]
$(p-MeC_6H_4)Ph_2P=N-SO_2C_{10}H_7$-2. I. II. III.[577] Glass.[577]

$(p\text{-}MeOC_6H_4)_2PhP{=}N\text{-}SO_2C_{10}H_7\text{-}2$. I. II. III.[577] M. 135-6°.[577]
$(p\text{-}MeC_6H_4)_3P{=}N\text{-}SO_2C_{10}H_7\text{-}2$. I. II. III.[577] M. 155-6°.[577]
$Ph_3P{=}N\text{-}C(CF_3){=}CH\text{-}CO_2Ph$. V.[166] M. 163-5°,[166] ^{1}H-NMR.[166]
$Ph_3P{=}N\text{-}C(CN){=}CH\text{-}CO_2Ph$. V.[166] M. 198-200°,[166] ^{1}H-NMR.[166]
$Ph_3P{=}N\text{-}P(O)PhNEt_2$. XIII.[23] M. 119-20.5°.[23]
$Ph_3P{=}NN_3P_3(NMe_2)_5$. III.[359] M. 150°.[359]

C_{11}

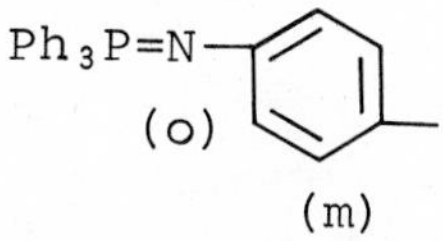

$Ph_3P{=}N$–(o)–C_6H_4–(m)–$C(CN){=}C(CN)_2$. XIII.[706] M. 241-3°, M. 180-2°(m-Br), M. 171-3° (m-Cl), M. 241-3°(o-Cl), UV.[706]

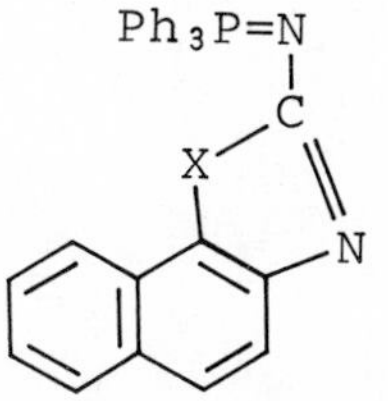

$Ph_3P{=}N$–C (naphtho-fused X–C=N ring) (X = S, Se). XIII. M. 195°(X = S), M. 211° (X = Se).[694]

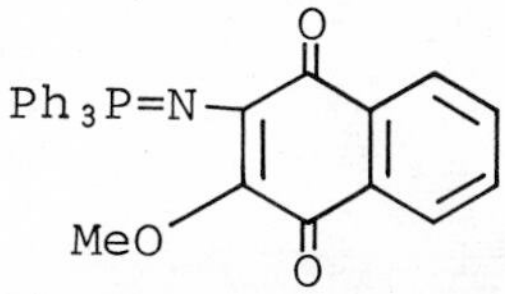

$Ph_3P{=}N$–, MeO– (1,4-naphthoquinone). I. M. 185°, UV.[445]

$Ph_3P{=}N$-β-D-triacetylxylyl. I. M. 128°.[433]

C_{12}

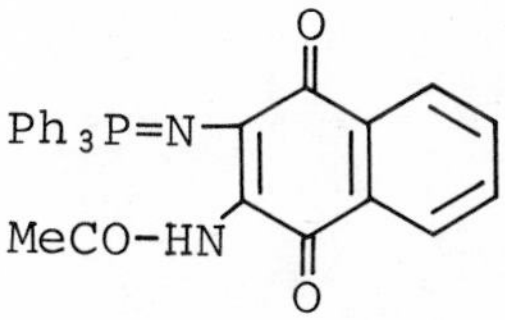

$Ph_3P{=}N$–, MeCO-HN– (1,4-naphthoquinone). I. M. 173-4°, UV.[445]

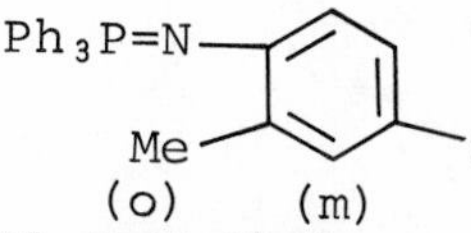

$Ph_3P{=}N$–(o)Me–C_6H_3–(m)–$C(CN){=}C(CN)_2$. XIII. M. 242-4° (o-Me), M. 180-1°(m-Me), UV.[706]

$Ph_3P{=}N\text{-}SbPh_2$. I. M. 123.5-24°, IR.[502]
$(NCC_2H_4)_3P{=}N\text{-}P(O)Ph_2$. I. M. 193-4°.[25]
$Me_2PhP{=}N\text{-}P(O)Ph_2$. I. M. 159.5-61°.[25]
$Bu_3P{=}N\text{-}P(O)Ph_2$. I. M. 66-7°.[25]
$BuPh_2P{=}N\text{-}P(O)Ph_2$. I.[656]

$Ph_3P=N-P(O)Ph_2$. I.[24,25,668] M. 170-1°,[24] M. 152-3°,[668] IR.[668]
$(p-MeC_6H_4)_3P=N-P(O)Ph_2$. I.[656]
$(p-ClC_6H_4)_3P=N-P(O)Ph_2$. I. M. 135-40°.[24]
$Ph_3P=N-P(O)(C_6H_4Cl-p)_2$. I.[24,25] M. 205-6°.[24]
$Ph_3P=N-P(O)Ph(NHPh)$. XIII. M. 208-10°.[23]
$Ph_3P=NP(S)Ph_2$. I.[23,25] M. 179-79.5°.[25]

C_{13}

$Ph_3P=N-CO-NPh_2$. I. M. 178-80°, IR.[379]

C_{14}

$Ph_3P=N-CO-CHPh_2$. XIII.[465,466] M. 151-53.5°, IR, ^{1}H-NMR.[465]
$Ph_3P=N-CPh=CHPh$. V. M. 157-57.5°, IR, UV, ^{1}H-NMR, ^{31}P +0.3.[166]
$Ph_3P=N-CCl=CPh_2$. –[465]

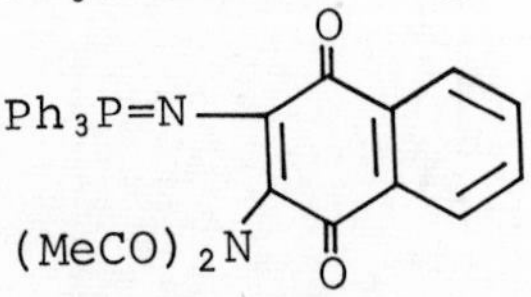

$Ph_3P=N$–(2-naphthoquinonyl, 3-$N(COMe)_2$) . I. M. 208-208.6°, UV.[445]

$Ph_3P=N-P(O)(C_6H_4Me-p)_2$. I.[24,25] M. 176-8°.[24]
$Ph_3P=N$-β-D-tetraacetylglycosyl. I. M. 136°.[433]
$Ph_3P=N$-β-D-tetraacetylgalaktosyl. I. M. 129°.[433]

C_{15}

$Ph_3P=N-C(CF_3)=C(o-C_6H_4)_2$. V. M. 253-4°.[166]
$Ph_3P=N-C(CN)=C(o-C_6H_4)_2$. V. M. 272-3°.[166]
$Ph_3P=N-C(CF_3)=N-N=CPh_2$. V. M. 223-4°, IR, UV, ^{19}F-NMR.[166]
$Ph_3P=N-C(CN)=N-N=CPh_2$. V. M. 219-21°.[166]

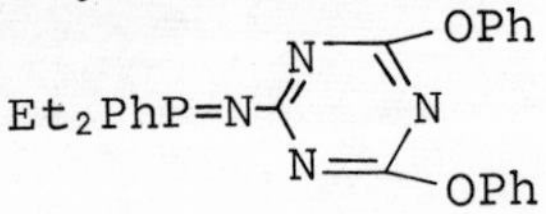

$Et_2PhP=N$–(4,6-di-OPh-1,3,5-triazin-2-yl) . I.[377]

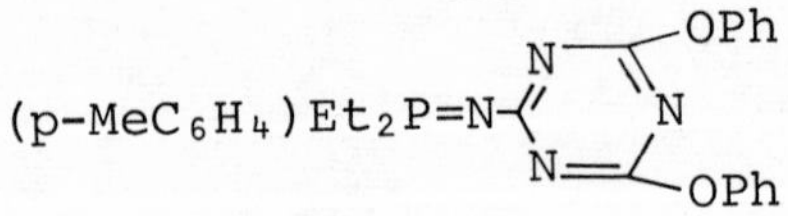

$(p-MeC_6H_4)Et_2P=N$–(4,6-di-OPh-1,3,5-triazin-2-yl) . I.[377]

$PhPr_2P=N$–(4,6-di-OPh-1,3,5-triazin-2-yl) . I.[377]

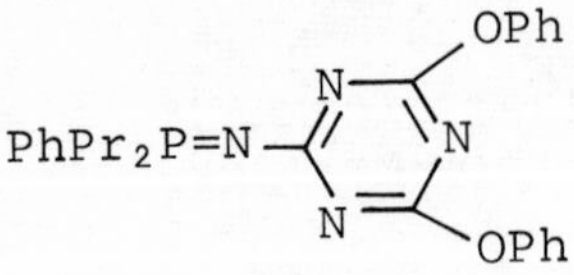

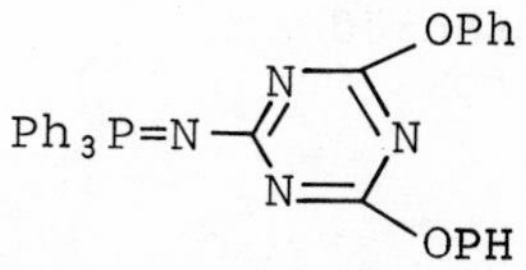

$Ph_3P{=}N$–(2,4-di-OPh-1,3,5-triazin-6-yl) [triazine ring bearing OPh and OPH substituents]. I.[377]

C_{16}

$(p\text{-}MeOC_6H_4)_2PhP{=}N\text{-}SO_2\text{-}C_6H_4(C_{10}H_7\text{-}2)\text{-}p$. I. II. III.[577] M. 135-6°.[577]

C_{18}

$Bu_3P{=}N\text{-}SnPh_3$. I.[382]
$Ph_3P{=}N\text{-}SnPh_3$. I.[382]
$(n\text{-}Octyl)_3P{=}N\text{-}SnPh_3$. I.[382]
$Ph_3P{=}N\text{-}SiPh_3$. I.[501,628] I.[655,664] M. 216°,[501] I .[628]
$Ph_3P{=}N\text{-}GePh_3$. I.[501,628] M. 192-3°,[501] IR.[628]
$Ph_3P{=}N\text{-}\overset{\oplus}{P}Ph_3\overset{\ominus}{X}$. XIII.[6,12,17,170] M. 269-71° (Cl),[6] M. 232° (NO_3),[12] M. 214-6° (N_3),[170] M. 256° (Br),[12] M. 264-6° (ClO_4),[170] M. 203.5-204° (I_3).[17]
$Ph_3P{=}N\text{-}\overset{\oplus}{As}Ph_3\overset{\ominus}{Br}$. XIII.[12] M. 226°.[12]
$Ph_3P{=}N\text{-}\overset{\oplus}{P}(OPh)_3\overset{\ominus}{Br}$. XIII.[13]
$Ph_3P{=}N\text{-}CPh(o\text{-}C_6H_4)_2$. I. M. 223-4°.[379]

C_{19}

$Ph_3P{=}N\text{-}N{=}N\text{-}CPh_3$. I.[379]
$Ph_3P{=}N\text{-}CPh_3$. XIII. M. 228-9°.[18]

C_{20}

$Ph_3P{=}N\text{-}C(CHPh_2){=}NC_6H_3Cl_2\text{-}3,4$. XIII. M. 144-6°, IR, ^1H-NMR.[466]
$Ph_3P{=}N\text{-}C(CHPh_2){=}NC_6H_4NO_2\text{-}p$. XIII. M. 187-9°, IR, ^1H-NMR.[466]
$Ph_3P{=}N\text{-}C(CHPh_2){=}NPh$. XIII. M. 185-7°, IR, ^1H-NMR.[466]

C_{22}

$Ph_3P{=}N\text{-}C(CHPh_2){=}NC_6H_4NMe_2\text{-}p$. XIII. M. 168-73°, IR, ^1H-NMR.[466]

C_{26}

$Ph_3P{=}N$-β-D-heptaacetylcellobiosyl. I. M. 90-5°.[433]

D.3.2. Polyphosphinimines, $(R_3P{=}N{-})_nA$ (Ordered According to the Number of Carbon Atoms in A)

n = 2

C_0

$Ph_3P{=}N{-}N{=}PPh_3$. III. M. 184°, UV.[16]
$Ph_3P{=}N{-}SO_2{-}N{=}PPh_3$. IV.[441,649] XIII.[5] M. 245-6°,[441] IR.[441]

C_2

$Me_3P{=}N{-}GeMe_2{-}N{=}PMe_3$. I. M. 9-10°, IR, ^{1}H-NMR.[541]

C_3

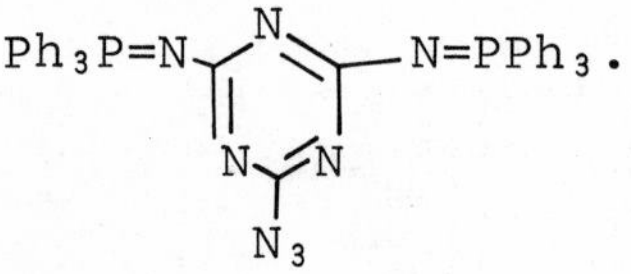

$Ph_3P{=}N$–(triazine ring with N_3)–$N{=}PPh_3$. I. M. 243°.[360]

C_6

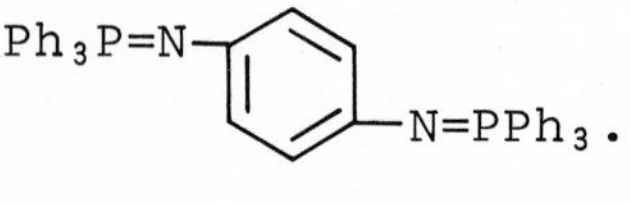

$Ph_3P{=}N{-}C_6H_4{-}N{=}PPh_3$ (para). I.[282] III.[310] M. 255-7°.[282]

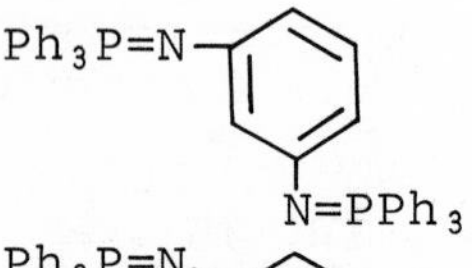

$Ph_3P{=}N{-}C_6H_4{-}N{=}PPh_3$ (meta). I. M. 213-4°.[706]

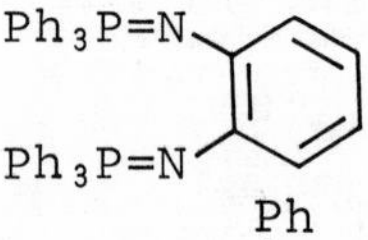

$Ph_3P{=}N{-}C_6H_4{-}N{=}PPh_3$ (ortho). III. M. 206°.[310]

$Ph_3P{=}N{-}P(Ph)(O(S)){-}N{=}PPh_3$. I. M. 192-3° (O), M. 195-7° (S).[23]

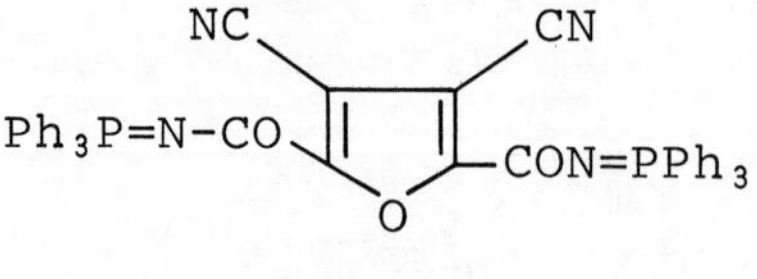

$Ph_3P{=}N{-}CO$–(3,4-dicyanofuran-2,5-diyl)–$CON{=}PPh_3$. I. M. 283-5°.[659]

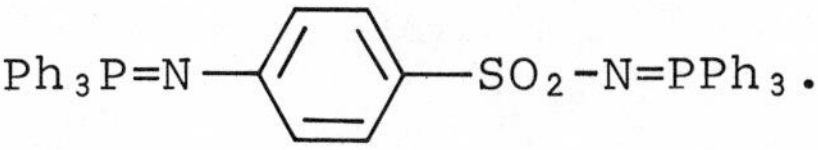

$Ph_3P{=}N{-}C_6H_4{-}SO_2{-}N{=}PPh_3$. III.[310]

C_{10}

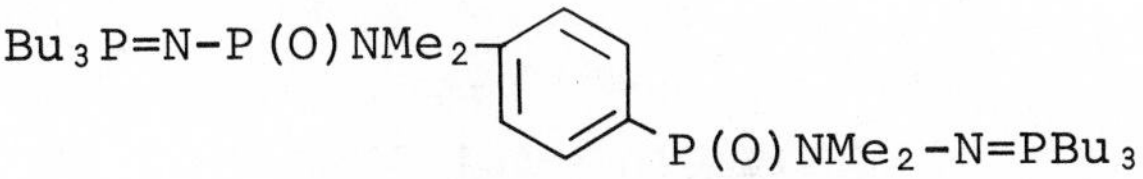

$Bu_3P{=}N{-}P(O)NMe_2{-}C_6H_4{-}P(O)NMe_2{-}N{=}PBu_3$. I.[656]

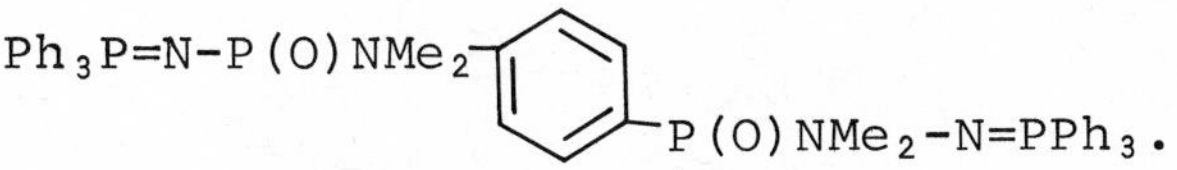

$Ph_3P{=}N{-}P(O)NMe_2{-}C_6H_4{-}P(O)NMe_2{-}N{=}PPh_3$. I.[656]

2,3-bis($Ph_3P{=}N$)-5-X-1,4-naphthoquinone (X = NH_2, NO_2). I. M. 243-5°(NH_2), 287-9°(NO_2), UV.[445]

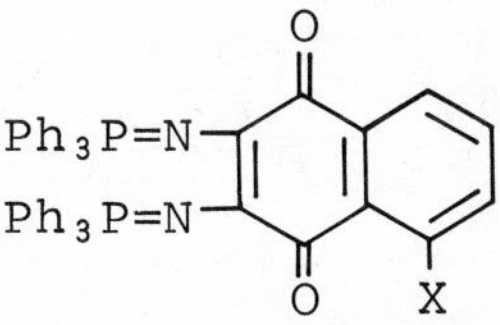

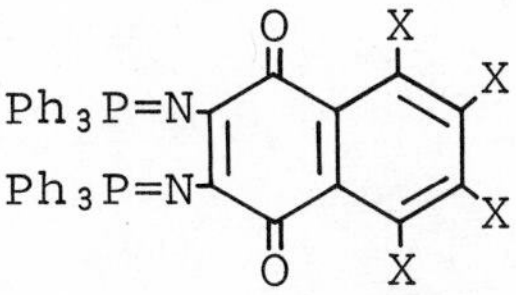

2,3-bis($Ph_3P{=}N$)-5,6,7,8-tetra-X-1,4-naphthoquinone (X = H, Cl). I. M. 244-5°(H), 251.5-252.5°(Cl), UV.[445]

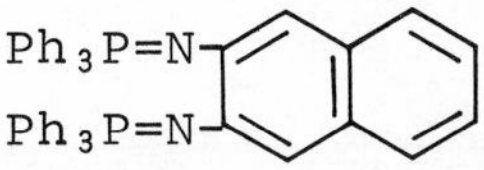

2,3-bis($Ph_3P{=}N$)naphthalene. III. M. 226-8°.[310]

C_{12}

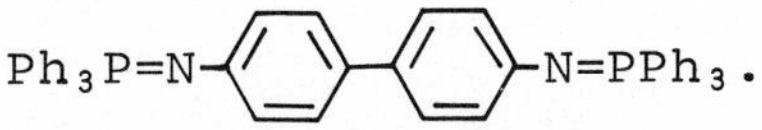

$Ph_3P{=}N{-}C_6H_4{-}C_6H_4{-}N{=}PPh_3$. III. M. 269-70°.[310]

$Ph_3P{=}N{-}SiPh_2{-}N{=}PPh_3$. I.[501,655,664] M. 190-1°.[664]

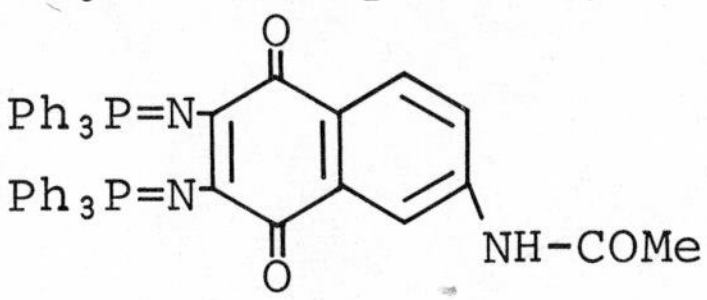

2,3-bis($Ph_3P{=}N$)-6-(NH-COMe)-1,4-naphthoquinone. I. M. 284-6°, UV.[445]

C_{14}

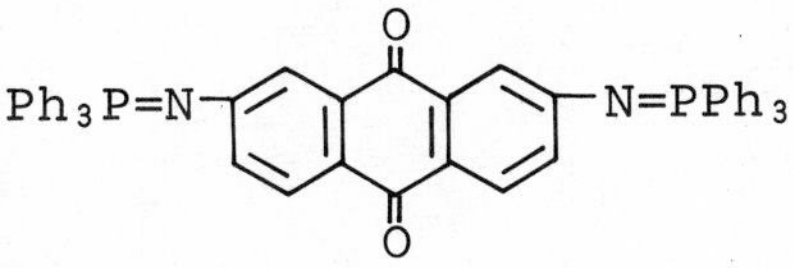

2,7-bis($Ph_3P{=}N$)anthraquinone ($Ph_3P{=}N$ ... $N{=}PPh_3$). I. M. 280°.[147]

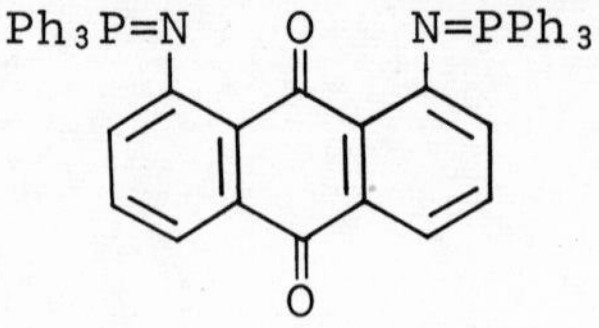

$Ph_3P{=}N$ O $N{=}PPh_3$ (1,8-bis(triphenylphosphoranylideneamino)anthraquinone) . I.[147]

C_{16}

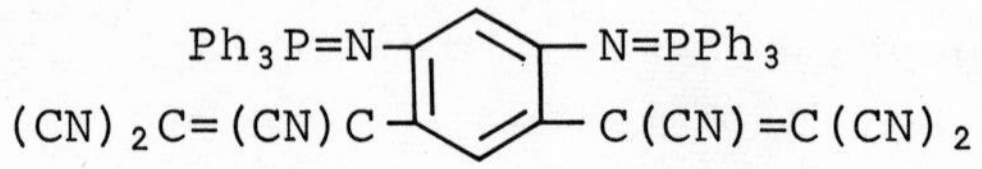

$Ph_3P{=}N$–C_6H_2–$N{=}PPh_3$; $(CN)_2C{=}(CN)C$–, –$C(CN){=}C(CN)_2$. XIII. M. 361-3°, UV.[706]

C_{30}

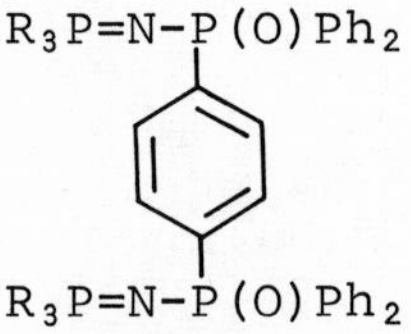

$R_3P{=}N{-}P(O)Ph_2$ –C_6H_4– $R_3P{=}N{-}P(O)Ph_2$ (R = Me, Bu, Ph). I.[656]

n = 3

C_0

$(Ph_3P{=}N{-})_3P{=}O$. I.[656]

C_3

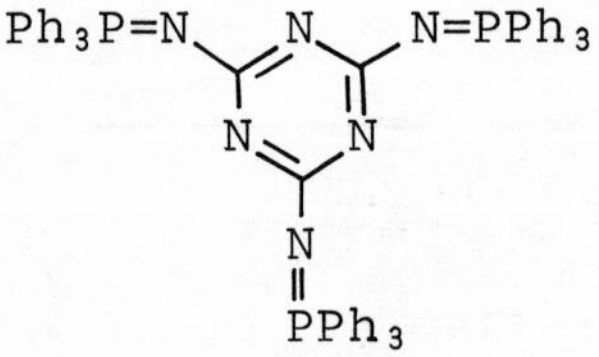

$Ph_3P{=}N$, $N{=}PPh_3$, $N{=}PPh_3$ (triazine) . III.[249] I.[360] M. 240°.[249]

C_6

$(Ph_3P{=}N{-})_3SiPh$. I. M. 225-226°.[501]
$(Ph_3P{=}N{-})_3SnPh$. I.[655]

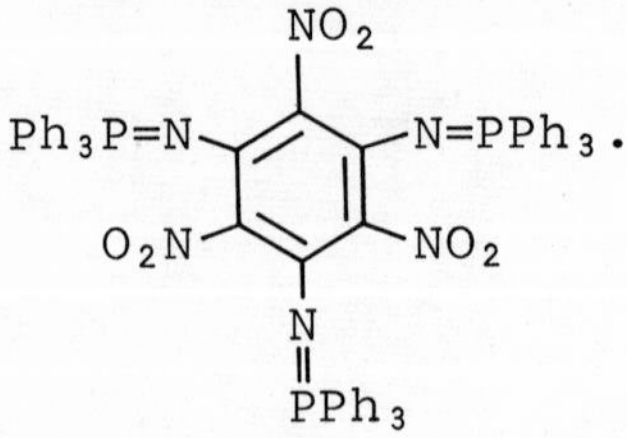

NO_2, $Ph_3P{=}N$, $N{=}PPh_3$, O_2N, NO_2, $N{=}PPh_3$ (benzene) . X-ray.[155]

n = 4

$(Ph_3P=N-)_2P=O$ – C_6H_4 – $(Ph_3P=N-)_2P=O$ (p-phenylene). I.[656]

D.3.3. Bisphosphinimines, R-N=P-A-P=N-R (Derived from Bisphosphines)

$CH_2(-PPh_2=NPh)_2$. I.[3,238] M. 160-2°,[238] IR, ^{1}H-NMR.[3]
$CH_2(-PPh_2=NC_6H_4NO_2$-p$)_2$. I. M. 274-5°.[238]
$(CH_2)_2(-PPh_2=NC_6H_4NO_2$-p$)_2$. I. M. 234-234.5°.[238]
$(CH_2)_2(-PPh_2=N-P(O)Ph_2)_2$. I. M. 262-4°.[25]
$(CH_2)_2(-PPh_2=N-P(S)Ph_2)_2$. I. M. 292-4°.[25]

$Ph_2P=N-Ph$ – C_6H_4 – $Ph_2P=N-Ph$ (p-phenylene). I. M. 118-20°.[282]

$Ph_2P=N-P(O)Ph_2$ – C_6H_4 – $Ph_2P=N-P(O)Ph_2$ (p-phenylene). I. M. 232-4°.[25]

$Ph_2P=N=P(O)Ph_2$ – C_6H_4–C_6H_4 – $Ph_2P=N-P(O)Ph_2$ (4,4'-biphenylene). I. M. 241-3°.[25]

$[=PPh_2-C_6H_4-PPh_2=N-C_6H_4-N=]_n$. I.[282]

D.4. Phosphazines, $R_3P=N-N=A$ (Ordered According to the Number of Carbon Atoms in A)

C_1

$Ph_3P=N-N=CH_2$. I.[135,677] M. 145-6°,[677] IR.[135]

$(p\text{-}PhC_6H_4)_3P{=}N{-}N{=}CH_2$. I. M. 230°.[686]

C_3

$Ph_3P{=}N{-}N{=}CMe_2$. II.[55] III.[654] M. 111°.[55]
$Ph_3P{=}N{-}N{=}C(CF_3)_2$. I.[234]
$Ph_3P{=}N{-}N{=}CMeCHO$. I. M. 129°.[499a]
$(Cyclo.C_6H_{11})_3P{=}N{-}N{=}CHCOCF_3$. I. M. 155-6°.[56]
$Ph_3P{=}N{-}N{=}CHCOMe$. I. M. 141°.[48]

C_4

$Ph_3P{=}N{-}N{=}CH\text{-}i\text{-}Pr$. III.[654]
$Ph_3P{=}N{-}N{=}CHCO_2Et$. I. M. 113-4°.[610]
$Ph_3P{=}N{-}N{=}CEtCHO$. I. M. 134°.[499a]
$Ph_3P{=}N{-}N{=}CBrCO_2Et$. I. M. 117-9°.[544a]
$Ph_3P{=}N{-}N{=}NICO_2Et$. I. M. 109-12°.[544a]

C_5

$Ph_3P{=}N{-}N{=}C(CO_2Me)_2$. I. M. 92°.[614-5]
$Ph_3P{=}N{-}N{=}CHCOCO_2Et$. I. M. 112°.[48]
$Ph_3P{=}N{-}N{=}C(CO_2Me)_2$. I. M. 88°.[499]
$Ph_3P{=}N{-}N{=}C(COMe)_2$. I. M. 98°.[499]
$Ph_3P{=}N{-}N{=}CPrCHO$. I. M. 125°.[499a]
$Ph_3P{=}N{-}N{=}C(CH)_4$. I. M. 151-3°.[56]
$Ph_3P{=}N{-}N{=}C(CCl)_4$. I. M. 142-3°.[367a]
$Ph_3P{=}N{-}N{=}CH\text{-}\alpha\text{-}furyl$. II. M. 129-30°, UV.[585-6]

$Ph_3P{=}N{-}N{=}CH$–(5-nitro-2-furyl: furan ring, O, NO_2). I.[517a]

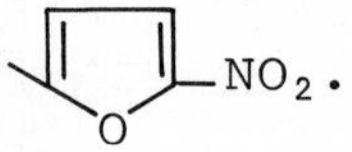

C_6

$Ph_3P{=}N{-}N{=}\overline{C(CH_2)_5}$. III.[654]
$Ph_3P{=}N{-}N{=}C(COMe)CO_2Et$. I. M. 90°,[615] M. 96-7°.[499]

$Ph_3P{=}N{-}N{=}C$(cyclohexadienone ring)$=O$. I. M. 134-5°.[311]

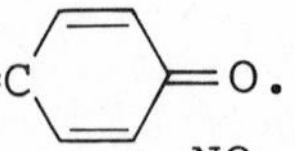

$Ph_3P{=}N{-}N{=}$(cyclohexadienone ring with NO_2)$=O$. I. M. 135-6°.[311]

$Ph_3P{=}N{-}N{=}$(cyclohexadienone ring with Cl, Cl)$=O$. I. M. 183°.[504]

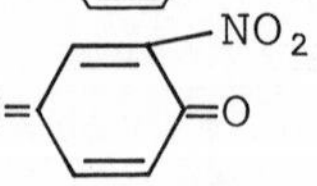

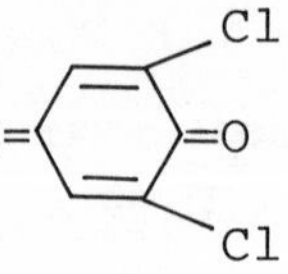

$Ph_3P=N-N=$ (ring: Br, =O, Br) . I. M. 119-23°.[311]

$Ph_3P=N-N=$ (ring: O, NO_2, Cl) . I. M. 119°.[505]

$Ph_3P=N-N=$ (ring: O, NO_2, NO_2) . I.[503,614-5] M. 187°.[614-5]

C_7

$Ph_3P=N-N=$ (ring: O, CO_2H, NO_2) . I.[505]

$Ph_3P=N-N=$ (ring: O, $CO_2^{\ominus}$, NO_2) $Ph_3\overset{\oplus}{P}H$. I. M. 133°.[505]

$Ph_3P=N-N=$ (ring: CO_2H, =O) . I.[311,503,504] M. 173-4°.[504]

$Ph_3P=N-N=$ (ring: CO_2H, =O, NO_2) . I. M. 205-7°.[504]

$Ph_3P=N-N=CHPh$. II.[55,585-6] III.[654] M. 129-30°.[585-6]

$Ph_3P=N-N=CHC_6H_4Cl$-p. II. M. 158.5°, UV.[585-6]

$Ph_3P=N-N=CHC_6H_4NO_2$-p. II.[585-6] III.[654] M. 162°, UV.[585-6]

$Ph_3P=N-N=CHC_6H_4NO_2$-o. I. M. 142-3°.[507]

$(p\text{-}MeOC_6H_4)_3P=N-N=CHC_6H_4NO_2$-o. I. M. 165-6°.[507]

$Ph_3P=N-N=CPh(OH)$. IR.[654a]

$Ph_3P=N-N=C(CO_2Et)_2$. M. 128°,[615] M. 82°.[499]

$Ph_3P=N-N=CH-CH=CH$ (furan ring, O) NO_2. I.[517a]

C_8

$Ph_3P{=}N{-}N{=}CPhMe$. II.[56] III.[654] M. 133-5°.[56]
$Ph_3P{=}N{-}N{=}CHCOPh$. I.[48,399] M. 119-20°.[48]
$(Cyclo{-}C_6H_{11})_3P{=}N{-}N{=}CHCOPh$. I. M. 152-3°.[56]
$Ph_3P{=}N{-}N{=}CHCOC_6H_4NO_2{-}p$. I. M. 139-40°.[48]
$(cyclo{-}C_6H_{11})_3P{=}N{-}N{=}CHCOC_6H_4NO_2{-}p$. I. M. 164-5°.[56]

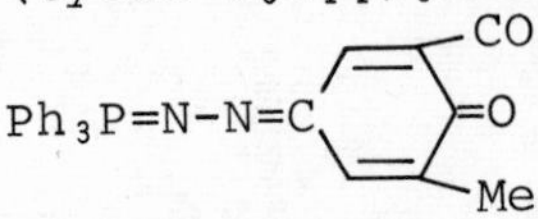

$Ph_3P{=}N{-}N{=}C$ (ring: CO_2H, $=O$, Me) . I. M. 180-1°.[504]

$Ph_3P{=}N{-}N{=}C(COMe)CO_2{-}t{-}Bu$. I. M. 123°.[499]
$Ph_3P{=}N{-}N{=}CH{-}CH_2{-}SO_2Ph$. I. M. 144-5°.[624]

C_9

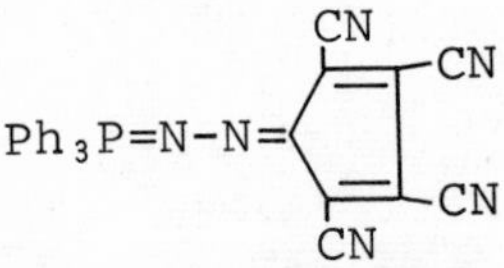

$Ph_3P{=}N{-}N{=}$ (ring: CN, CN, CN, CN) . I. M. 212-4°, IR, UV.[657a]

$Ph_3P{=}N{-}N{=}C(CH_2Ph)CHO$. I. M. 124°.[499a]
$Ph_3P{=}N{-}N{=}CH{-}CH_2{-}SO_2{-}C_6H_4Me{-}p$. I. M. 147-8°.[624]
$Ph_3P{=}N{-}N{=}CH{-}CO{-}C_6H_4OMe{-}p$. I. M. 124°.[48]
$Ph_3P{=}N{-}N{=}CH{-}COCH_2Ph$. I. M. 132-3°.[48]
$Ph_3P{=}N{-}N{=}CMeCOPh$. I. M. 142-4°.[56]

C_{10}

$Ph_3P{=}N{-}N{=}C(COPh)CO_2Me$. I.[399 615] M. 136-8°.[399]
$Ph_3P{=}N{-}N{=}C(COPh)COMe$. I.[499 615] M. 128°.[615]
$Ph_3P{=}N{-}N{=}C(COMe)COC_6H_4Cl{-}p$. I. M. 119-20°.[499]
$Ph_3P{=}N{-}N{=}C(COMe)COC_6H_4Br{-}p$. I. M. 110°.[499]
$Ph_3P{=}N{-}N{=}C(COMe)COC_6H_4NO_2{-}o$. I. M. 76°.[499]
$Ph_3P{=}N{-}N{=}C(COMe)COC_6H_4NO_2{-}p$. I. M. 131-2°.[499]
$Ph_3P{=}N{-}N{=}CH{-}CO(CHOCOMe)_2$ (with CO_2Me on CH) . I. M. 134°.[48]
$Ph_3P{=}N{-}N{=}CHCOC_6H_3(OMe)_2{-}3,4$. I. M. 75°.[48]

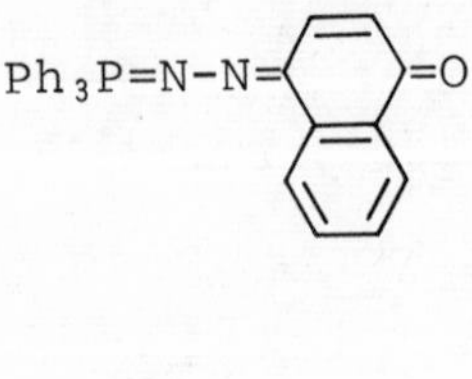

$Ph_3P{=}N{-}N{=}$ (naphthoquinone ring, $=O$) . I. M. 148-9°.[311]

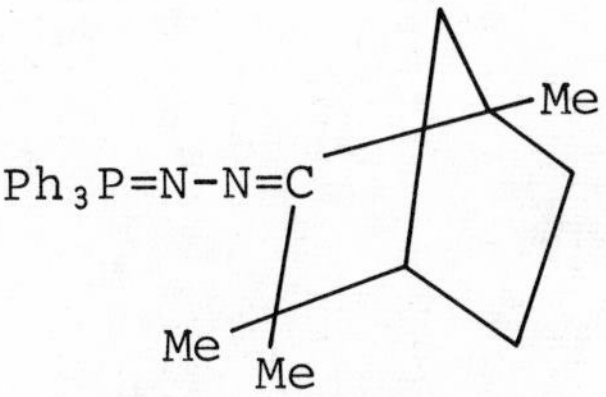

$Ph_3P{=}N{-}N{=}C$ (Fenchonylidene). II. M. 144-5°, UV.[585-6]

$Ph_3P{=}N{-}N{=}C$ (Camphorylidene). II. M. 161-2°, UV.[585-6]

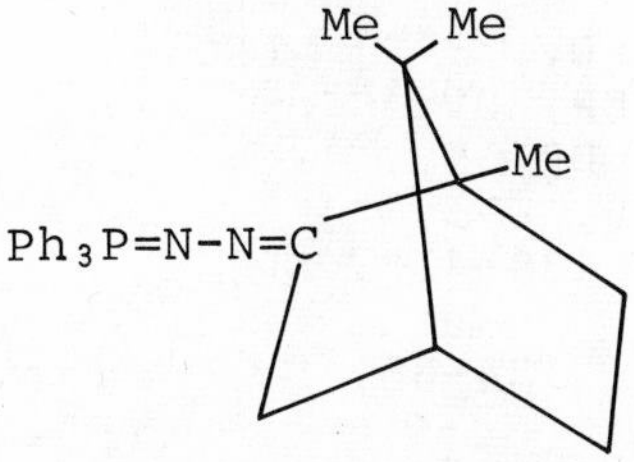

$Ph_3P{=}N{-}N{=}C(COMe)SO_2C_6H_4OMe$-p. I. M. 155-8°, IR.[284a]

C_{11}

$Ph_3P{=}N{-}N{=}CHCOCH(Ph)OCOMe$. I. M. 154-5°.[48]

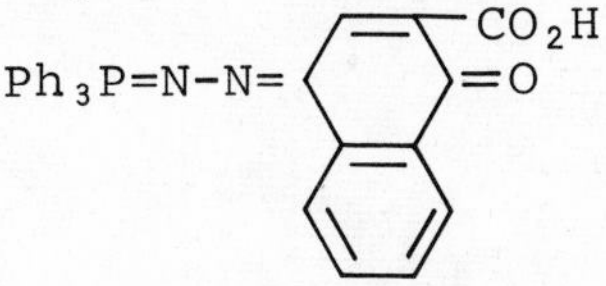

$Ph_3P{=}N{-}N{=}$(3-carboxy-4-oxo-1,4-dihydronaphthylidene, CO_2H, =O). I.[503,504] M. 177°.[503]

$Ph_3P{=}N{-}N{=}C(COMe)COC_6H_4OMe$-p. I. M. 107°.[499]
$Ph_3P{=}N{-}N{=}C(COMe)COC_6H_4Me$-p. I. M. 108-9°.[499]

C_{12}

$Ph_3P{=}N{-}N{=}CH{-}COC_{10}H_7$-1. I. M. 130-2°.[48]
$Ph_3P{=}N{-}N{=}CHCOC_{10}H_7$-2. I. M. 132-3°.[56]
$Ph_3P{=}N{-}N{=}C(CO_2Me)COCH{=}CHPh$. I. M. 174°.[614-5]
$Ph_3P{=}N{-}N{=}CH{-}COC_6H(OMe)_4$-2,3,5,6. I. M. 162°.[519]

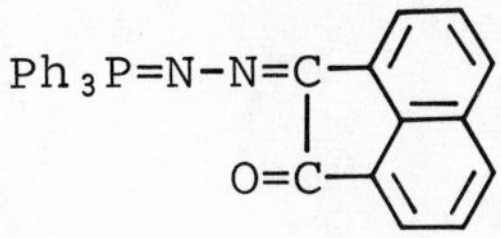

$Ph_3P{=}N{-}N{=}C$ (acenaphthenone, $O{=}C$). I. M. 109-11°.[503]

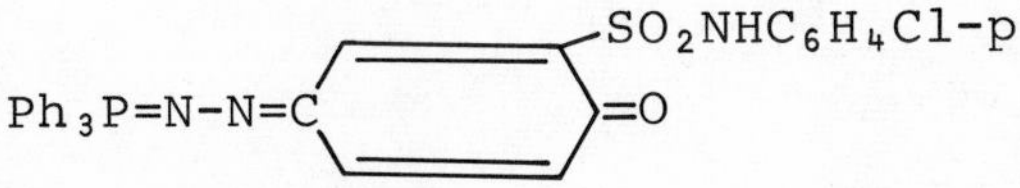

$Ph_3P{=}N{-}N{=}C$ (cyclohexadienone, $SO_2NHC_6H_4Cl$-p, =O). I. M. 194-5°.[311]

$Ph_3P{=}N{-}N{=}C$ (ring with SO_2NHPh and $=O$) . I. M. 188-9°.[311]

C_{13}

$Et_3P{=}N{-}N{=}CPh_2$. I.[610]
$Et_2PhP{=}N{-}N{=}CPh_2$. I. M. 113°.[610]
$Ph_3P{=}N{-}N{=}CPh_2$. II.[55,585-6] M. 173°,[55] UV,[585-6] DP.[244]
$(p\text{-}MeOC_6H_4)_3P{=}N{-}N{=}CPh_2$. I. M. 140-1°, DP, UV.[244]
$(p\text{-}Me_2NC_6H_4)_3P{=}N{-}N{=}CPh_2$. I. M. 170-1°, DP, UV.[244]
$Ph_3P{=}N{-}N{=}CPhC_6H_4NO_2\text{-}p$. I. M. 114-6°, DP, UV.[244]
$(p\text{-}ClC_6H_4)Ph_2P{=}N{-}N{=}CPhC_6H_4NO_2\text{-}p$. I. M. 139-41°, DP, UV.[244]
$(p\text{-}MeOC_6H_4)Ph_2P{=}N{-}N{=}CPhC_6H_4NO_2\text{-}p$. I. M. 154-6°, UV, DP.[244]
$(p\text{-}Me_2NC_6H_4)Ph_2P{=}N{-}N{=}CPhC_6H_4NO_2\text{-}p$. I. M. 176-7°, UV, DP.[244]
$(p\text{-}Me_2NC_6H_4)Ph_2P{=}N{-}N{=}CPhC_6H_4Cl\text{-}p$. I. M. 141-2°, UV, DP.[244]
$Ph_3P{=}N{-}N{=}C(C_6H_4Cl\text{-}p)_2$. I. M. 137-9°, UV, DP.[244]
$(p\text{-}Me_2NC_6H_4)Ph_2P{=}N{-}N{=}C(C_6H_4Cl\text{-}p)_2$. I. M. 155-6°, UV, DP.[244]
$Et_3P{=}N{-}N{=}C(C_6H_4\text{-}o)_2$. I. M. 160°.[610]
$Me_2PhP{=}N{-}N{=}C(C_6H_4\text{-}o)_2$. I. M. 129-31°, DP.[244]
$Bu_3P{=}N{-}N{=}C(C_6H_4\text{-}o)_2$. I. M. 82-3°, UV, DP.[244]
$Et_2PhP{=}N{-}N{=}C(C_6H_4\text{-}o)_2$. I. M. 115°.[610]
$Am_3P{=}N{-}N{=}C(C_6H_4\text{-}o)_2$. I. M. 55.5-56.5°, UV, DP.[244]
$Ph_3P{=}N{-}N{=}C(C_6H_4\text{-}o)_2$. I.[243,244,610] II.[55,585] III.[654] M. 209-10°,[610] DP,[244] UV.[244,585-6]
$(p\text{-}ClC_6H_4)Ph_2P{=}N{-}N{=}C(C_6H_4\text{-}o)_2$. I. M. 176°,[243] DP, UV.[244]
$(p\text{-}BrC_6H_4)Ph_2P{=}N{-}N{=}C(C_6H_4\text{-}o)_2$. I. M. 176-7°,[243] UV, DP.[244]
$(p\text{-}MeOC_6H_4)Ph_2P{=}N{-}N{=}C(C_6H_4\text{-}o)_2$. I. M. 177°,[243] UV, DP.[244]
$(p\text{-}Me_2NC_6H_4)Ph_2P{=}N{-}N{=}C(C_6H_4\text{-}o)_2$. I. M. 185°,[243] UV, DP.[244]
$(p\text{-}ClC_6H_4)_3P{=}N{-}N{=}C(C_6H_4\text{-}o)_2$. I. M. 191-3°, UV, DP.[244]

$Ph_3P{=}N{-}N{=}C$ (dibenzo ring with SO_2) . I. M. 208-9°.[498]

$Ph_3P{=}N{-}N{=}C$(cyclohexadienone, =O, CONHPh) . I. M. 174°.[504]

$Ph_3P{=}N{-}N{=}CHP(O)Ph_2$. I.[375]

$Ph_3P{=}N{-}N{=}C$(benz[e]indane-1,3-dione) . I. M. 119°.[550a]

C_{14}

$Ph_3P{=}N{-}N{=}$(anthrone, =O) . I.[159,498,507] M. 170-1°.[498]
165-6°.[507]
218-20.[159]

$(p\text{-}MeC_6H_4)_3P{=}N{-}N{=}$(anthrone, =O) . I. M. 149-51°.[507]

$Ph_3P{=}N{-}N{=}$(anthrone, =O, Cl) . I.[507]

$(p\text{-MeC}_6H_4)_3P{=}N{-}N{=}$(4-chloro-10-oxo-9(10H)-anthrylidene). I. M. 156-7°.[507]

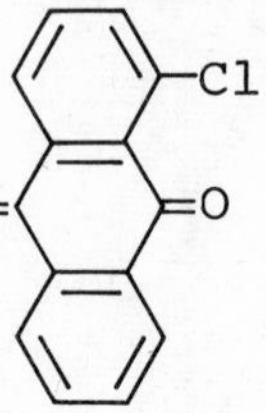

$Ph_3P{=}N{-}N{=}$(10-oxo-9(10H)-phenanthrylidene) $\cdot ZnCl_2$. I.[503,504] M. 202°.[503]

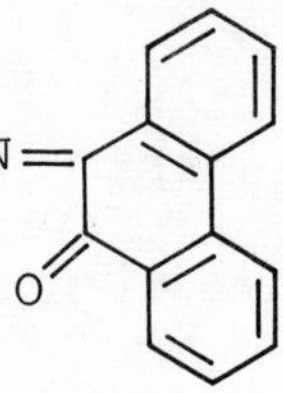

$Ph_3P{=}N{-}N{=}CH[CHOCOMe]_3CH_2OCOMe$. I. M. 138°.[48]

$Ph_3P{=}N{-}N{=}CPhCOPh$. I.[614-5] II.[55] M. 163°,[55] 115-7°.[614-5]
$(\text{Cyclo-}C_6H_{11})_3P{=}N{-}N{=}CPhCOPh$. I. M. 171-2°.[56]
$Ph_3P{=}N{-}N{=}C(COMe)COC_{10}H_7\text{-}2$. I. M. 127-8°.[499]
$Ph_3P{=}N{-}N{=}C(COPh)SO_2C_6H_4NO_2\text{-}p$. I. M. 129.5-31°, IR, UV.[284a]

C_{15}

$Ph_3P{=}N{-}N{=}CHP(O)(CH_2Ph)_2$. I.[375]
$Ph_3P{=}N{-}N{=}C(COPh)_2$. I. M. 137°.[499]
$Ph_3P{=}N{-}N{=}C(COPh)COC_6H_4NO_2\text{-}o$. I. M. 139-40°.[499]
$Ph_3P{=}N{-}N{=}C(COPh)COC_6H_4NO_2\text{-}p$. I. M. 147°.[499]
$Ph_3P{=}N{-}N{=}C(COPh)COC_6H_4NH_2\text{-}p$. I. M. 143-4°.[499]
$Ph_3P{=}N{-}N{=}CPhCH_2SO_2C_6H_4Me\text{-}p$. I. M. 151-2°.[624]
$Ph_3P{=}N{-}N{=}CH\text{-}9\text{-anthryl}$. I. M. 155-7°.[447b]

C_{16}

$Ph_3P{=}N{-}N{=}$(cyclohexadienone ring bearing $SO_2NHC_{10}H_7\text{-}2$ and $=O$). I.[503,504] M. 175°.[503]

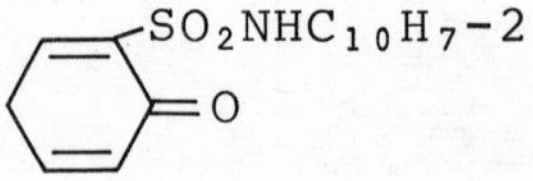

$Ph_3P{=}N{-}N{=}$(cyclohexadienone ring bearing $SO_2NHC_{10}H_7\text{-}1$ and $=O$). I. M. 182-3°.[311]

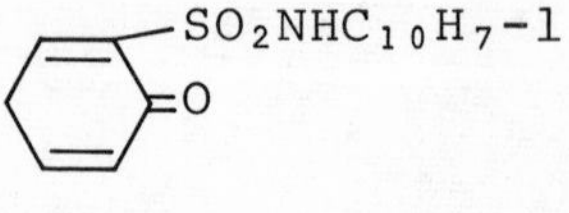

$Ph_3P{=}N{-}N{=}$[structure: 1,2-naphthoquinone-2-hydrazono, 5-SO_2NHPh] . I.[503,504] M. 164-5°.[503]

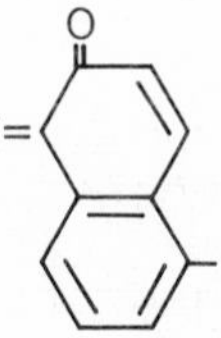

C_{17}

$Ph_3P{=}N{-}N{=}(COPh)COC_6H_4NHCOMe$-p. I. M. 140-1°.[499]

C_{18}

$Ph_3P{=}N{-}N{=}$[structure: cyclohexadienone ring with $SO_2NHC_6H_4Ph$-p and =O] . I. M. 193°.[504]

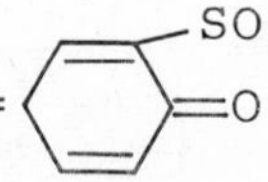

$Ph_3P{=}N{-}N{=}$[structure: cyclohexadienone ring with $SO_2NHC_6H_4OPh$-p and =O] . I.[311,503,504] M. 187°.[503]

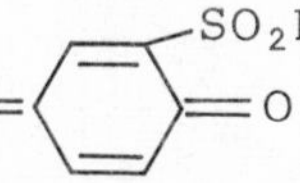

$Ph_3P{=}N{-}N{=}$[structure: chrysene quinone hydrazone, O] $\cdot ZnCl_2$. I.[503,504] M. 204-5°.[503]

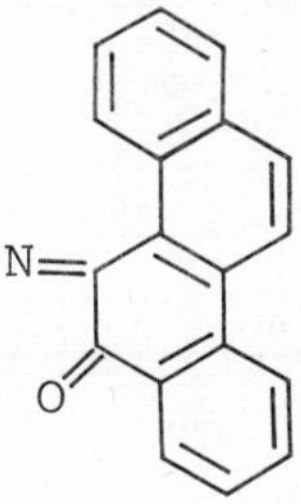

C_{20}

$Ph_3P{=}N{-}N{=}$[structure: 1,2-naphthoquinone-2-hydrazono, O] $SO_3C_6H_4$-t-Bu-p. I.[503,504] M. 130-1°.[503]

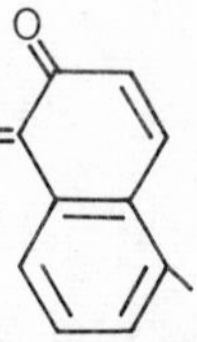

$Ph_3P{=}N{-}N{=}CHCON{=}PPh_3$. I. M. 150-1°.[451a]

$(Ph_3P{=}N{-}N{=}CH{-}CO)_2$. I. M. 123°.[43]

(received April 14, 1971)

REFERENCES

1. L'Abbé, G., and H. J. Bestmann, Tetrahedron Letters, 1969, 63.
2. L'Abbé, G., P. Ykman, and G. Smets, Tetrahedron, 25, 5421 (1969).
3. Aguiar, A. M., H. J. Aguiar, and T. G. Archibald, Tetrahedron Letters, 1966, 3187.
4. Aksnes, G., Acta Chem. Scand., 15, 692 (1961).
5. Appel, R., and G. Büchler, Z. Anorg. Chem., 320, 3 (1963).
6. Appel, R., and G. Büchler, Z. Naturforsch., 17b, 422 (1962).
7. Appel, R., W. Büchner, and E. Guth, Liebigs Ann. Chem., 618, 53 (1958).
8. Appel, R., and E. Guth, Z. Naturforsch., 15b, 57 (1960).
9. Appel, R., and A. Hauss, Angew. Chem., 71, 626 (1959).
10. Appel, R., and A. Hauss, Angew. Chem., 71, 626 (1959).
11. Appel, R., and A. Hauss, Chem. Ber., 93, 405 (1960).
12. Appel, R., and A. Hauss, Z. Anorg. Allgem. Chem., 311, 290 (1961).
13. Appel, R., A. Hauss, and G. Büchler, Z. Naturforsch., 16b, 405 (1961).
14. Appel, R., G. Köhnlein, and R. Schöllhorn, Chem. Ber., 98, 1355 (1965).
15. Appel, R., and R. Schaaf, Z. Naturforsch., 16b, 405 (1961).
16. Appel, R., and R. Schöllhorn, Angew. Chem., 76, 991 (1964).
17. Appel, R., and G. Siegemund, Z. Anorg. Allgem. Chem., 361, 203 (1968).
18. Appel, R., and F. Vogt, Chem. Ber., 95, 2225 (1962).
19. Appel, R., and G. Siegemund, Z. Anorg. Allgem. Chem., 363, 183 (1968).
20. Arbuzov, A. E., and A. A. Dunin, Ber. Deut. Chem. Ges., 60, 291 (1927).
21. Arbuzov, A. E., and A. I. Razumov, J. Russ. Phys. Chem. Soc., 61, 623 (1929).
21a. Asunskis, J., and H. Shechter, J. Org. Chem., 33, 1164 (1968).
22. Bailey, A. S., T. S. Cameron, J. M. Evans, and C. K. Prout, Chem. Commun., 1966, 664.
23. Baldwin, R. A., J. Org. Chem., 30, 3866 (1965).
24. Baldwin, R. A., and R. M. Washburn, J. Am. Chem. Soc., 83, 4466 (1961).
25. Baldwin, R. A., and R. M. Washburn, J. Org. Chem., 30, 3860 (1965).
26. Barhardt, R. G., and W. E. McEwen, J. Am. Chem. Soc., 89, 7009 (1967).

27. Bart, J. C. J., J. Chem. Soc., B, 1969, 350; Angew, Chem., 80, 696 (1968).
28. Bednarek, P., R. Bodalski, J. Michalski, and S. Musierowicz, Bull. Acad. Polon. Sci., 11, 507 (1963).
29. Bergelson, L. D., L. I. Barsukov, and M. M. Shemyakin, Tetrahedron, 23, 2709 (1967).
30. Bergelson, L. D., and M. M. Shemyakin, Tetrahedron, 19, 149 (1963).
31. Bergelson, L. D., and M. M. Shemyakin, Pure Appl. Chem., 9, 271 (1964).
32. Bergelson, L. D., and M. M. Shemyakin, Angew. Chem., 76, 113 (1964).
33. Bergelson, L. D., V. A. Vaver, and M. M. Shemyakin, Izvest. Akad. Nauk SSSR, 1779 (1960); C. A., 55, 14294e (1961).
34. Bergelson, L. D., V. A. Vaver, L. I. Barsukov, and M. M. Shemyakin, Izvest. Akad. Nauk SSSR. 1134 (1963); Bull. Acad. Sci. USSR, 1037 (1963); C. A., 59, 8607d (1963).
35. Bergelson, L. D., V. A. Vaver, L. I. Barsukov, and M. M. Shemyakin, Izvest. Akad. Nauk SSSR. 1053 (1963); Bull. Acad. USSR, 957 (1963); C. A., 59, 8783 (1963).
36. Bergelson, L. D., V. A. Vaver, L. I. Barsukov, and M. M. Shemyakin, Tetrahedron Letters, 1964, 2669.
36a. Bergmann, E. D., and I. Agranat, J. Chem. Soc., 1968, 1621.
37. Bestmann, H. J., Tetrahedron Letters, 1960, No. 4, 7.
38. Bestmann, H. J., Angew. Chem., 72, 34 (1960).
39. Bestmann, H. J., Angew. Chem., 72, 326 (1960).
40. Bestmann, H. J., Chem. Ber., 95, 58 (1962).
41. Bestmann, H. J., Angew. Chem., 77, 612 (1965).
42. Bestmann, H. J., Angew. Chem., 77, 651 (1965).
43. Bestmann, H. J., unpublished results.
44. Bestmann, H. J., unpublished results.
44a. Bestmann, H. J., and J. Angerer, Tetrahedron Letters, 1969, 3665.
45. Bestmann, H. J., R. Armsen, and H. Wagner, Chem. Ber., 102, 2259 (1969).
46. Bestmann, H. J., and B. Arnason, Tetrahedron Letters, 1961, 455.
47. Bestmann, H. J., and B. Arnason, Chem. Ber., 95, 1513 (1962).
48. Bestmann, H. J., H. Buckschewski, and H. Leube, Chem. Ber., 92, 1345 (1959).
49. Bestmann, H. J., and T. Denzel, Tetrahedron Letters, 1966, 3591.
50. Bestmann, H. J., T. Denzel, and R. Kunstmann, Tetrahedron Letters, 1968, 2895.
51. Bestmann, H. J., H. Dornauer, and K. Rostock, Chem. Ber., 103, 2001 (1970).
52. Bestmann, H. J., H. Dornauer, and K. Rostock, Chem. Ber., 103, 685 (1970).

53. Bestmann, H. J., H. Dornauer, and K. Rostock, Liebigs Ann. Chem., 735, 520 (1970).
54. Bestmann, H. J., R. Engler, and H. Hartung, Angew. Chem., 78, 1100 (1966).
55. Bestmann, H. J., and H. Fritzsche, Chem. Ber., 94, 2477 (1961).
56. Bestmann, H. J., and L. Göthlich, Liebigs Ann. Chem., 655, 1 (1962).
57. Bestmann, H. J., G. Graf, and H. Hartung, Liebigs Ann. Chem., 706, 68 (1967).
58. Bestmann, H. J., G. Graf, S. Kolewa, and E. Vilsmeier, Chem. Ber., 103, 2794 (1970).
59. Bestmann, H. J., H. Häberlein, and O. Kratzer, Angew. Chem., 76, 226 (1964).
60. Bestmann, H. J., and H. Häberlein, Z. Naturforsch. 17b, 787 (1962).
61. Bestmann, H. J., H. Häberlein, and O. Kratzer, Angew. Chem., 76, 226 (1964).
62. Bestmann, H. J., H. Häberlein, and I. Pils, Tetrahedron, 20,2079 (1964).
63. Bestmann, H. J., H. Häberlein, H. Wagner, and O. Kratzer, Chem. Ber. 99, 2848 (1966).
64. Bestmann, H. J., H. Häberlein, and W. Eisele, Chem. Ber., 99, 28 (1966).
65. Bestmann, H. J., H. Härtel, and H. Häberlein, Liebigs Ann. Chem., 718, 33 (1968).
66. Bestmann, H. J., and H. Heid, Angew. Chem., 83, 329 (1971).
67. Bestmann, H. J., and H. Hartung, Angew. Chem., 75, 297 (1963).
68. Bestmann, H. J., H. Hartung, and I. Pils, Angew. Chem. 77, 1011 (1965).
69. Bestmann, H. J., and H. Hartung, Chem. Ber., 99, 1198 (1966).
70. Bestmann, H. J., and G. Hofmann, unpublished results.
71. Bestmann, H. J., and G. Hofmann, Liebigs Ann. Chem., 716, 98 (1968).
72. Bestmann, H. J., G. Joachim, I. Lengyel, J. F. M. Oth, R. Merenyi, and J. Weitkamp, Tetrahedron Letters, 1966, 3355.
73. Bestmann, H. J., and G. Joachim, unpublished results.
74. Bestmann, H. J., O. Klein, L. Gothlich, and H. Buckschewski, Chem. Ber., 96, 2259 (1963).
75. Bestmann, H. J., and O. Klein, Liebigs Ann. Chem., 676, 97 (1964).
76. Bestmann, H. J., and O. Klein, Tetrahedron Letters, 1966, 6181.
77. Bestmann, H. J., and O. Klein, unpublished results.
78. Bestmann, H. J., and H. Kolm, Chem. Ber., 96, 1948 (1963).
79. Bestmann, H. J., and E. Kranz, Angew. Chem., 79, 95 (1967); Chem. Ber., 102, 1802 (1969).

80. Bestmann, H. J., and O. Kratzer, Angew. Chem., 73, 757 (1961).
81. Bestmann, H. J., and O. Kratzer, Chem. Ber., 95, 1894 (1962).
82. Bestmann, H. J., and O. Kratzer, Angew. Chem., 74, 494 (1962).
83. Bestmann, H. J., and O. Kratzer, Chem. Ber., 96, 1899 (1963).
84. Bestmann, H. J., and R. Kunstmann, Angew. Chem., 78, 1059 (1966).
85. Bestmann, H. J., and R. Kunstmann, Chem. Ber., 102, 1816 (1969).
86. Bestmann, H. J., and H. J. Lang, Tetrahedron Letters, 1969, 2101.
87. Bestmann, H. J., and J. Lienert, Angew. Chem., 81, 751 (1969).
88. Bestmann, H. J., and J. Lienert, Chem. Ztg., 94, 487 (1970).
89. Bestmann, H. J., H. Liberda, and J. P. Snyder, J. Am. Chem. Soc., 90, 2963 (1968).
90. Bestmann, H. J., and H. Morper, Angew. Chem., 79, 578 (1967).
91. Bestmann, H. J., and S. Pfohl, Angew. Chem., 81, 750 (1969).
92. Bestmann, H. J., and H. Pfüller, Angew. Chem., 84, 528 (1972).
93. Bestmann, H. J., and J. Popp, unpublished results.
94. Bestmann, H. J., K. Rostock, and H. Dornauer, Angew. Chem., 78, 335 (1966).
95. Bestmann, H. J., and O. Rothe, Angew. Chem., 76, 569 (1964).
96. Bestmann, H. J., and D. Ruppert, Angew. Chem., 80, 668 (1968).
97. Bestmann, H. J., R. Saalfrank, and I. P. Snyder, Angew. Chem., 81, 227 (1969).
98. Bestmann, H. J., and H. Schulz, Tetrahedron Letters, 1960, No. 4, 5.
99. Bestmann, H. J., and H. Schulz, Angew. Chem., 73, 27 (1961).
100. Bestmann, H. J., and H. Schulz, Chem. Ber., 95, 2921 (1962).
101. Bestmann, H. J., and H. Schulz, Tetrahedron Letters, 1960, No. 4, 5.
102. Bestmann, H. J., and H. Schulz, Angew. Chem., 73, 620 (1961).
103. Bestmann, H. J., and H. Schulz, Liebigs Ann. Chem., 674, 11 (1964).
104. Bestmann, H. J., and F. Seng, Angew. Chem., 74, 154 (1962).
105. Bestmann, H. J., and F. Seng, unpublished results.
106. Bestmann, H. J., F. Seng, and H. Schulz, Chem. Ber., 96, 465 (1963).

107. Bestmann, H. J., and F. Seng, Angew. Chem., 75, 1117 (1963).
108. Bestmann, H. J., and F. Seng, Angew. Chem., 75, 475 (1963).
109. Bestmann, H. J., and F. Seng, Tetrahedron, 21, 1373 (1965).
110. Bestmann, H. J., and P. J. Snyder, J. Am. Chem. Soc., 89, 3936 (1967).
111. Bestmann, H. J., N. Sommer, and H. A. Staab, Angew. Chem., 74, 293 (1962).
112. Bestmann, H. J., and W. Stransky, unpublished results.
113. Bestmann, H. J., and I. Tömösközi, Tetrahedron, 24, 3299 (1968).
114. Bestmann, H. J., E. Vilsmeier, and G. Graf, Liebigs Ann. Chem., 704, 109 (1967).
115. Bestmann, H. J., and R. Zimmermann, Chem. Ber., 101, 2185 (1968).
116. Bestmann, H. J., and R. Zimmermann, unpublished results.
117. Bergmann, E., and H. A. Wolf, Chem. Ber., 63, 1176 (1930).
118. Bieber, T. I., and E. H. Eismann, J. Org. Chem., 27, 678 (1962).
119. Bieber, T. I., and E. H. Eismann, J. Org. Chem., 27, 678 (1962).
120. Birkofer, L., and S. M. Kim, Chem. Ber., 97, 2100 (1964).
121. Birkofer, L., and S. M. Kim, Chem. Ber., 97, 2100 (1964).
122. Birkofer, L., A. Ritter, and S. M. Kim, Chem. Ber., 96, 3099 (1963).
123. Birkofer, L., A. Ritter, and P. Richter, Chem. Ber., 96, 2750 (1963).
124. Birkofer, L., and A. Ritter, Angew. Chem., 77, 415 (1965).
125. Birum, G. H., U.S. Pat. 3,058,876; C.A., 58, 7975h (1963).
126. Birum, G. H., and C. N. Matthews, Chem. Commun., 1967, 137.
126a. Birum, G. H., and C. N. Matthews, J. Org. Chem., 32, 3554 (1967).
127. Bissing, D. E., J. Org. Chem., 30, 1296 (1965).
128. Bissing, D. E., and A. J. Speziale, J. Am. Chem. Soc., 87, 2683 (1965).
129. Bladé-Font, A., W. E. McEwen, and C. A. Vander Werf, J. Am. Chem. Soc., 82, 2646 (1960).
130. Bladé-Font, A., C. A. Vander Werf, and W. E. McEwen, J. Am. Chem. Soc., 82, 2396 (1960).
131. Blanchard, M. L., H. Strzelecka, G. J. Martin, and M. Simalty, Bull. Soc. Chim. France, 1967, 2677.
132. Blomquist, A. T., and V. J. Hruby, J. Am. Chem. Soc., 86, 5041 (1964).

133. Bock, H., Angew. Chem., 77, 469 (1965).
134. Bock, H., and M. Schmöller, Chem. Ber., 102, 38 (1969).
135. Bock, H., and M. Schmöller, and H. T. Dieck, Chem. Ber., 102, 1363 (1969).
135a. Bock, H., and H. T. Dieck, Z. Naturforsch., 21, 739 (1966).
136. Bock, H., and W. Wiegräbe, Angew. Chem., 74, 327 (1962).
137. Bock, H., and W. Wiegräbe, Angew. Chem., 75, 789 (1963).
138. Bohlmann, F., Chem. Ber., 90, 1519 (1957).
139. Bohlmann, F., and E. Inhoffen, Chem. Ber., 89, 1276 (1956).
140. Bohlmann, F., and H. J. Mannhardt, Chem. Ber., 89, 1307 (1956).
141. Bohlmann, F., and H. G. Viehe, Chem. Ber., 88, 1245, 1347 (1955).
142. Bose, A. K., and R. T. Dahill, J. Org. Chem., 30, 505 (1965).
143. Bose, A. K., R. T. Dahill, and R. M. Ramer, Tetrahedron Letters, 1966, 6263.
144. Borowitz, I. J., and L. I. Grossmann, Tetrahedron Letters, 1962, 471.
145. Borowitz, I. J., and R. Kirkhaus, J. Am. Chem. Soc., 85, 2183 (1963).
146. Bott, K., Angew. Chem., 77, 683 (1965).
147. Brass, K., and F. Albrecht, Chem. Ber., 61, 983 (1928).
148. Braunholtz, W., J. Chem. Soc., 121, 300 (1922).
148a. Brook, A. G., and S. A. Fieldhouse, J. Organometal. Chem., 10, 235 (1967).
149. Brown, G. W., J. Chem. Soc., 1967, 2018.
150. Brown, G. W., R. C. Cookson, and I. D. R. Stevens, Tetrahedron Letters, 1964, 1263.
151. Brown, G. W., R. C. Cookson, I. D. R. Stevens, T. C. W. Mak, and J. Trotter, Proc. Chem. Soc., 1964, 87.
152. Brunn, E., and R. Huishen, Angew. Chem., 81, 534 (1969).
153. Buchta, E., and F. Andree, Chem. Ber., 92, 3111 (1959).
154. Burton, D. J., and H. C. Krutzsch, Tetrahedron Letters, 1968, 71.
155. Camerson, T. S., and C. K. Prout, Chem. Commun., 1967, 455.
156. Campbell, T. W., and R. N. McDonald, J. Org. Chem., 24, 1246 (1959).
157. Campbell, I. G. M., R. C. Cookson, M. B. Hocking, and A. N. Hughes, J. Chem. Soc., 1965, 2184.
158. Campbell, T. W., J. J. Monagle, and V. S. Foldi, J. Am. Chem. Soc., 84, 3673 (1962).

159. Cauquis, G., G. Reverdy, and M. Rastoldo, Compt. Rend., 260 (8), 2259 (1965); C.A., 63, 553e (1965).
160. Chaplin, E. J., and F. G. Mann, Nature, 135, 686 (1934).
161. Chioccola, G., and J. J. Daly, J. Chem. Soc., A, 1968, 568.
161a. Chopard, P. A., Helv. Chim. Acta, 50, 1016 (1967).
162. Chopard, P. A., and R. F. Hudson, Z. Naturforsch., 18b, 509 (1963).
163. Chopard, P. A., and G. Salvadori, Gazz. Chim. Ital., 73, 668 (1963).
164. Chopard, P. A., R. J. G. Searle, and F. H. Devitt, J. Org. Chem., 30, 1015 (1965).
165. Claydon, A. P., P-A. Fowell, and C. T. Mortimer, J. Chem. Soc., 1960, 3284.
166. Ciganek, E., J. Org. Chem., 35, 3631 (1970).
166a. Ciganek, E., J. Org. Chem., 35, 1725 (1970).
167. Coffman, D. D., and C. S. Marvel, J. Am. Chem. Soc., 51, 3496 (1929).
168. Cooks, R. G., R. S. Ward and D. H. Williams, Tetrahedron, 24, 3289 (1968).
169. Cookson, E. A., and P. C. Croft, Angew. Chem., 76, 755 (1964).
170. Cookson, R. C., and A. N. Hughes, J. Chem. Soc., 1963, 6061.
171. Corbridge, D. E. C., Topics in Phosphorus Chemistry, Vol. 6, Wiley, New York, p. 297.
172. Corey, E. J., and M. Chaykovsky, J. Am. Chem. Soc., 84, 866 (1962).
173. Corey, E. J., and G. T. Kwiatkowsky, J. Am. Chem. Soc., 90, 6816 (1968).
174. Corey, H. S., J. R. D. McCormick, and W. E. Swensen, J. Am. Chem. Soc., 86, 1884 (1964).
175. Corey, E. J., J. I. Shulman, and H. Yamamoto, Tetrahedron Letters, 1970, 447.
176. Corey, E. J., and H. Yamamoto, J. Am. Chem. Soc., 92, 226 (1970).
177. Crews, P., J. Am. Chem. Soc., 90, 2961 (1968).
177a. Crofts, P. C., and M. P. Williamson, J. Chem. Soc., 1967, 1093.
178. Crouse, D. M., A. T. Wehlman, and E. E. Schweizer, Chem. Commun., 1968, 866.
179. Cotton, F. A., and R. A. Schunn, J. Am. Chem. Soc., 85, 2394 (1963).
180. Dallacker, F., K. Ulrichs, and M. Lipp, Liebigs Ann. Chem., 667, 50 (1963).
181. Daly, J. J., and P. J. Wheatley, J. Chem. Soc., A, 1966, 1703.
182. Daly, J. J., J. Chem. Soc., A, 1967, 1913.
183. DBP 1,141,273 Farbwerke Hoechst, Erf. H. J. Bestmann; C.A., 58, 10128a (1963).

184. DBP 1,158,971 Farbf. Bayer, Erf. Malz and E. Roos; C.A., 60, 9289d (1964).
185. DBP 954,247, BASF, Erf. G. Wittig and H. Pommer; C. A., 53, 2279e (1959).
186. DBP 1,048,568, BASF, Erf. G. Wittig, U. Schöllhöpf, and H. Pommer; C.A., 55, 4576h (1961).
187. DBP 1,046,046, BASF, Erf. W. Sarnecki and H. Pommer; Angew. Chem., 72, 811 (1960).
188. DBP 1,026,745, BASF, Erf. H. Pommer, G. Wittig, and W. Sarnecki; C.A., 54, 11074f (1960).
189. Denney, D. B., and M. J. Boskin, J. Am. Chem. Soc., 81, 6330 (1959).
190. Denney, D. B., and S. T. Ross, J. Org. Chem., 27, 998 (1962).
191. Denney, D. B., and L. C. Smith, J. Am. Chem. Soc., 82, 2396 (1962).
192. Denney, D. B., L. C. Smith, J. Song, J. Rossi, and C. D. Hall, J. Org. Chem., 28, 778 (1963).
193. Denney, D. B., J. J. Vill, and M. J. Boskin, J. Am. Chem. Soc., 84, 3944 (1962).
194. Derkach, G. I., E. S. Gubnitskaya, and A. V. Kirsanov, Zh. Obshch. Khim., 31, 3679 (1961); C.A., 57, 9876 (1962).
195. Derkach, G. I., E. S. Gubnitskaya, V. A. Shokol, and A. V. Kirsanov, Zh. Obshch, Khim., 32, 1874 (1962); C.A., 58, 6857 (1963).
196. Derkach, G. I., and E. S. Gubnitskaya, Zh. Obshch. Khim., 34, 604 (1964); C.A., 60, 13268 (1964).
197. Derkach, G. I., G. K. Fedorova, and E. S. Gubnitskaya, Zh. Obshch. Khim., 33, 1017 (1963); C.A., 59, 8783 (1963).
198. Derkach, G. I., and A. V. Kirsanov, Zh. Obhsch. Khim., 32, 2245 (1962); C.A., 58, 9126 (1963).
198a. Diefenbach, H., H. Ringsdorf, and R. E. Wilhelms, J. Polymer Sci., Part B, 5, 1039 (1967); C.A., 67, 117390b (1967).
199. Dieckmann, J., J. Org. Chem., 28, 2933 (1963).
200. Dimroth, K., H. Follmann, and G. Pohl, Chem. Ber., 99, 642 (1966).
201. Dimroth, K., and G. Pohl, Angew. Chem., 73, 436 (1961).
202. Dimroth, K., K. H. Wolf, and H. Wache, Angew. Chem., 75, 860 (1963).
203. Disselkötter, H., Angew. Chem., 76, 431 (1964).
204. Dokumentation der Molekülspektroskopie, Verlag Chemie, Weinheim, Nr. 14116-141433.
205. Dombrowski, A. W., V. N. Listvan, A. A. Grigorenko, and M. I. Shewtschuk, Zh. Obshch. Khim., 36, 1421 (1966); C.A., 66, 11004h (1967).
205a. Dombrovski, A. V., and V. N. Listvan, Zh. Obshch. Khim., 37, 2273 (1967); C.A., 69, 36228u (1968).

206. Drefahl, G., D. Lorenz, and G. Schmitt, J. Prakt. Chem., 23, 143 (1964).
207. Drefahl, G., G. Plotner, and R. Scholz, Z. Chem. 1, 93 (1961).
208. Drefahl, G., K. Pensold, and H. Schick, Chem. Ber., 97, 2011 (1964).
209. Dyatkin, B. L., E. P. Mochalina, Yu. S. Konstantinov, SR.Sterlin, and I. L. Knunyants, Izv. Akad. Nauk SSSR, Ser. Khim., 1967 (10), 2297; C.A., 68, 77632u (1968).
210. Edelman, T. G., and B. I. Stepanov, Zh. Obshch. Khim., 37 (4), 963 (1967); C.A., 67, 111875s (1967).
211. Edelman, T. G., and B. I. Stepanov, Zh. Obshch. Khim., 38 (1), 195 (1968); C.A., 69, 58748r (1968).
212. Eliel, E. L., Sterochemie der Kohlenstoffverbindungen, Verlag Chemie, 1966, p. 135.
213. Eiter, K., and H. Oediger, Liebigs Ann. Chem., 682, 62 (1965).
214. Elix, J. A., M. V. Sargent, and F. Sondheimer, J. Am. Chem. Soc., 92, 973 (1970).
215. Fedorova, G. K., and G. A. Lanchuk, Zh. Obshch. Khim., 34, 511 (1964); C. A., 60, 12048 (1964).
216. Fenton, G. W., and C. K. Ingold, J. Chem. Soc., 1929 2342.
217. Filler, R., and E. W. Heffern, J. Org. Chem., 32, 3249 (1967).
218. Fischer, H., and H. Fischer, Chem. Ber., 99, 658 (1966).
219. Flizar, S., R. F. Hudson, and G. Salvatori, Helv. Chim. Acta, 46, 1580 (1963).
220. Flizar, S., R. F. Hudson, and G. Salvatori, Helv. Chim. Acta, 47, 159 (1964).
221. Fletcher, T. L., M. J. Namkung, J. R. Price, and S. K. Schaefer, J. Med. Chem., 8, 347 (1965).
222. Fluck, E., M. Becke-Goehring, and G. Dehoust, Z. Anorg. Allgem. Chem., 312, 60 (1961).
223. Fluck, E., Topics in Phosphorus Chemistry, Vol. 4, Wiley, New York, 1967, pp. 373,374.
224. Ford, J. A., jr., and C. V. Wilson, J. Org. Chem., 26, 1433 (1961).
224a. Ford, J. A., jr., Tetrahedron Letters, 1968, 815.
225. Franz, J. E., and C. Osuch, Tetrahedron Letters, 1963, 841.
226. Franz, J. E., and C. Osuch, U.S. Pat. 3,282,895; C.A. 66, 28888g (1967).
227. Franzen, V., Angew. Chem., 72, 566 (1960).
228. Freeman, J. P., Chem. Ind., 1959, 1254.
229. Freeman, J. P., J. Org. Chem., 31, 538 (1966).
230. Friedrich, K., and H. G. Henning, Chem. Ber., 92, 2756 (1959).
231. Fuqua, S. A., W. G. Duncan, and R. M. Silberstein,

Tetrahedron Letters, 1964, 1461.
232. Fuqua, S. A., W. G. Duncan, and R. M. Silberstein, J. Org. Chem., 30, 1027 (1965).
233. Fuqua, S. A., W. G. Duncan, and R. M. Silberstein, J. Org. Chem., 30, 2543 (1965).
234. Gale, D. M., W. J. Middleton, and G. G. Krespan, J. Am. Chem. Soc., 87, 657 (1965).
235. Gambaryan, N. P., E. M. Rokhlin, Y. V. Zeifman, and I. L. Knunyants, Izv. Akad. Nauk SSSR, Ser. Khim., 1965, 749; C.A., 63, 2913 (1965).
236. Garett, P. J., and K. P. C. Vollhardt, Chem. Commun., 1970, 109.
236a. Gerkin, R. M., and B. Rickborn, J. Am. Chem. Soc., 89, 5850 (1967).
237. Gilman, H., and R. A. Tomasi, J. Org. Chem., 27, 3647 (1962).
238. Gilyarov, V. A., V. Yu. Kovtun, and M. I. Kabachnik, Izv. Akad. Nauk SSSR, Ser. Khim., 1967, 1159; C.A., 68, 21626c (1968).
239. Goedler, J., and H. Ullmann, Chem. Ber., 94, 1067 (1961).
240. Goetz, H., Angew. Chem., 77, 1021 (1965).
241. Goetz, H., Liebigs Ann. Chem., 698, 18 (1967).
242. Goetz, H., Liebigs Ann. Chem., 701, 1021 (1965).
243. Goetz, H., and H. Juds, Liebigs Ann. Chem., 678, 1 (1964).
244. Goetz, H., and H. Juds, Liebigs Ann. Chem., 698, 1 (1966).
245. Goetz, H., and D. Probst, Liebigs Ann. Chem., 715, 1 (1968).
246. Gough, S. T. D., and S. Trippett, Proc. Chem. Soc., 1961, 302.
247. Gough, S. T. D., and S. Trippett, J. Chem. Soc., 1962, 2333.
248. Gough, S. T. D., and S. Trippett, J. Chem. Soc., 1964, 543.
249. Gotsmann, G., and M. Schwarzmann, Liebigs Ann. Chem., 729, 106 (1969).
250. Grayson, M., P. T. Keough, and G. A. Johnson, J. Am. Chem. Soc., 81, 4803 (1959).
251. Grayson, M., and P. T. Keough, J. Am. Chem. Soc., 82, 3919 (1960).
252. Greanwald, R., M. Chaykovsky, and E. J. Corey, J. Org. Chem., 28, 1128 (1963).
253. Grell, W., and H. Machleidt, Liebigs Ann. Chem., 699, 53 (1966).
254. Griffin, C. E., K. R. Martin, and B. E. Douglas, J. Org. Chem., 27, 1627 (1962).
255. Griffin, C. E., and G. Witschard, J. Org. Chem., 27, 3334 (1962).
256. Griffin, C. E., and G. Witschard, J. Org. Chem., 29,

1001 (1964).
257. Grim, S. O., W. McFarlane, and T. J. Marks, Chem. Commun., 1967, 1191.
258. Grim, S. O., and J. H. Ambrus, J. Org. Chem., 33, 2993 (1968).
259. Grigorenko, A. A., M. I. Shevchuk, and A. V. Dombrovsky, Zh. Obshch. Khim., 36 (6), 1121 (1966); C.A., 65, 12230 (1966).
260. Grigorenko, A. A., M. J. Shevchuk, and A. W. Dombrovsky, Zh. Obshch. Khim., 36, 506 (1966); C.A., 65, 737g (1966).
261. Grisley, D. W., J. C. Alm, and C. N. Matthews, Tetrahedron, 21, 5 (1965).
262. Gross, H., and W. Bürger, J. Prakt. Chem., 311, 395 (1969).
263. Gross, H., and B. Costisella, Angew. Chem., 80, 364 (1968).
264. Grünanger, P., P. V. Finzi, and C. Scotti, Chem. Ber., 98, 623 (1965).
265. Grushkin, B., Fr. Pat. 1,345,811; C.A., 60, 12055 (1964).
266. Grushkin, B., Fr. Pat. 1,345,811; C.A., 60, 12055 (1964).
267. Sen Gupta, A. K., Tetrahedron Letters, 1968, 5205.
268. Sen Gupta, A. K., and D. A. Mitchard, Tetrahedron Letters, 1968, 5207.
269. Hands, A. R., and A. J. H. Mercer, J. Chem. Soc., 1967, 1099, and 1968, 2448.
270. Harrison, I. T., and B. Lythgoe, J. Chem. Soc., 1958, 843.
271. Hartzler, H. D., J. Am. Chem. Soc., 86, 2174 (1964).
271a. Harvey, G. R., and K. W. Ratts, J. Org. Chem., 31, 3907 (1966).
271b. Harvey, G. R., J. Org. Chem., 31, 1587 (1966).
272. Hauser, C. F., T. W. Brooks, M. L. Miles, R. A. Raymond, and G. B. Butler, J. Org. Chem., 28, 372 (1963).
273. Hawthorne, M. F., J. Am. Chem. Soc., 83, 367 (1961).
274. Hedaya, E., and S. Theodoropulos, Tetrahedron, 24, 2241 (1968).
275. Heine, H. W., G. B. Lowrie, and K. C. Irving, J. Org. Chem., 35, 444 (1970).
276. Heitman, H., U. K. Pandit, and H. O. Huisman, Tetrahedron Letters, 1963, 915.
277. Heitman, H., J. H. S. Wieland, and H. O. Huisman, Koninkl. Ned. Akad. Wetenschap., Proc., Ser. B, 64, 165 (1961); C.A., 55, 17562 (1961).
278. Hendrikson, J. B., C. Hall, R. Rees, and J. F. Templeton, J. Org. Chem., 30, 3312 (1965).
279. Hendrikson, J. B., R. Rees, and J. F. Templeton, J. Am. Chem. Soc., 86, 107 (1964).

280. Henrik, C. A., E. Böhme, J. A. Edwards, and J. H. Fried, J. Am. Chem. Soc., 90, 5926 (1968).
281. Herkes, F. E., and D. J. Burton, J. Org. Chem., 32, 1311 (1967).
282. Herring, D. L., J. Org. Chem., 26, 3998 (1961).
283. Hey, L., and C. K. Ingold, J. Chem. Soc., 1933, 531.
284. Hieber, W., E. Winter, and E. Schubert, Chem. Ber., 95, 3070 (1962).
284a. Hodson, D., G. Holt, and D. K. Wall, J. Chem. Soc., 1968, 2201.
285. Hoffman, R., D. B. Boyd, and S. Z. Goldberg, J. Am. Chem. Soc., 92, 3929 (1970).
285a. Hoffman, R., and D. B. Boyd, J. Am. Chem. Soc., 93, 1064 (1971).
286. Hofmann, A. W., Liebigs Ann. Chem., 78, 253 (1851), and 79, 11 (1851).
287. Hoffman, H., Angew. Chem., 71, 379 (1959).
288. Hoffman, H., Chem. Ber., 94, 1331 (1961).
289. Hoffmann, H., Chem. Ber., 95, 2563 (1962).
290. Hoffman, H., and H. J. Diehr, Tetrahedron Letters, 1962, 583.
291. Hoffman, H., and H. Forster, Tetrahedron Letters, 1963, 1547.
292. Hoffman, H., L. Horner, H. G. Wippel, and D. Michael, Chem. Ber., 95, 523 (1962).
293. Hoffman, H., L. Horner, and G. Hassel, Chem. Ber., 91, 58 (1958).
294. Hoffman, H., and D. Michael, Chem. Ber., 95, 528 (1962).
295. Horner, L., and A. Christmann, Chem. Ber., 96, 388 (1963).
296. Horner, L., and A. Gross, Liebigs Ann. Chem., 591, 117 (1955).
297. Horner, L., and H. Hoffmann, Angew. Chem., 68, 473 (1956).
298. Horner, L., and H. Hoffmann, Angew. Chem., 68, 473 (1956).
299. Horner, L., H. Hoffmann, H. G. Wippel, and G. Hassel, Chem. Ber., 91, 52 (1958).
300. Horner, L., H. Hoffmann, and H. G. Wippel, Chem. Ber., 91, 61 (1958).
301. Horner, L., H. Hoffmann, H. G. Wippel, and G. Klahre, Chem. Ber., 92, 2499 (1959).
302. Horner, L., H. Hoffmann, G. Klahre, V. G. Toscano, and H. Ertel, Chem. Ber., 94, 1987 (1961).
303. Horner, L., H. Hoffmann, and V. G. Toscano, Chem. Ber., 95, 536 (1962).
304. Horner, L., H. Hoffmann, W. Klink, H. Ertel, and V. G. Toscano, Chem. Ber., 95, 581 (1962).
305. Horner, L., W. Klink, and H. Hoffmann, Chem. Ber., 96, 3133 (1963).
306. Horner, L., and W. Klink, Tetrahedron Letters, 1964,

2467.
307. Horner, L., and E. Lingnau, Liebigs Ann. Chem., 591, 135 (1955).
308. Horner, L., and A. Mentrup, Liebigs Ann. Chem., 646, 65 (1961).
309. Horner, L., and H. Oediger, Chem. Ber., 91, 437 (1958).
310. Horner, L., and H. Oediger, Liebigs Ann. Chem., 627, 142 (1956).
311. Horner, L., and H. G. Schmelzer, Chem. Ber., 94, 1326 (1961).
312. Horner, L., and H. Winkler, Tetrahedron Letters, 1964, 175.
313. Horner, L., and H. Winkler, Tetrahedron Letters, 1964, 3265.
314. Hocking, M. B., Can. J. Chem., 44, 1581 (1966).
315. House, H. O., and H. Babad, J. Org. Chem., 28, 90 (1963).
316. House, H. O., V. K. Jones, and G. A. Frank, J. Org. Chem., 29, 3327 (1964).
317. House, H. O., Modern Synthetic Reactions, Benjamin, New York, 1965, p. 245.
318. Hudson, R. F., Pure Appl. Chem., 9, 371 (1964).
319. Hudson, R. F., and P. A. Chopard, Helv. Chim. Acta, 46, 2178 (1963).
320. Hudson, R. F., and P. A. Chopard, J. Org. Chem., 28, 2446 (1963).
321. Hudson, R. F., P. A. Chopard, and G. Salvadori, Helv. Chim. Acta, 47, 632 (1964).
321a. Hudson, R. F., and P. A. Chopard, U.S. Pat. 3,294,820; C.A., 66, 104827d (1967).
322. Hudson, R. F., R. G. Searle, and F. H. Dewitt, J. Chem. Soc., 1966, 1001.
323. Hughes, A. N., Chem. Ind., 1969, 138.
324. Hughes, A. N., and S. Uaboonkul, Tetrahedron, 24, 3437 (1968).
325. Huisgen, R., and J. Wulf, Tetrahedron Letters, 1967, 917.
326. Huisgen, R., and J. Wulf, Tetrahedron Letters, 1967, 921.
327. Huisgen, R., and J. Wulf, Chem. Ber., 102, 1848 (1969).
328. Huisgen, R., and J. Wulf, Chem. Ber., 102, 1833 (1969).
329. Huisgen, R., and J. Wulf, Chem. Ber., 102, 746 (1969).
330. Huisgen, R., and J. Wulf, Chem. Ber., 102, 1841 (1969).
331. Hullar, T. L., Tetrahedron Letters, 1967, 4921.
332. Inhoffen, H. H., K. Bruckner, G. F. Domagk, and H. M. Erdmann, Chem. Ber., 88, 1415 (1955).
333. Inhoffen, H. H., and K. Irmacher, Chem. Ber., 89,

1833 (1956).
334. Inhoffen, H. H., H. Burkhardt, and G. Quinkert, Chem. Ber., 92, 1564 (1959).
335. Issleib, K., and R. Lindner, Liebigs Ann. Chem., 699, 40 (1966).
336. Issleib, K., and R. Lindner, Liebigs Ann. Chem., 707, 112, 120 (1967).
337. Issleib, K., and R. Lindner, Liebigs Ann. Chem., 713, 12 (1968).
338. Issler, H., Dissertation, E. T. H., Zürich, 1924.
339. Isler, O., H. Gutmann, M. Montavon, R. Rüegg, G. Ryser, and P. Zeller, Helv. Chim. Acta, 40, 1242 (1957).
340. Isler, O., G. Gutmann, H. Lindlar, M. Montavon, R. Rüegg, G. Ryser, and P. Zeller, Helv. Chim. Acta, 39, 463 (1956).
341. Isler, O., G. Gutmann, H. Lindlar, M. Montavon, R. Rüegg, G. Ryser, and P. Zeller, Helv. Chim. Acta, 39, 463 (1956).
342. Inouye, Y., T. Sugita, and H. M. Walborsky, Tetrahedron, 20, 1695 (1964).
343. Ito, Y., M. Okano, and R. Oda, Tetrahedron, 22, 2615 (1966).
344. Ito, Y., M. Okano, and R. Oda, Tetrahedron, 23, 2137 (1967).
345. Jaffe, H. H., Chem. Rev., 53, 191 (1953).
346. Jayawant, M. D., "Reactions of Alkyl and Cycloalkylaminotriphenylphosphonium Halides," Ph.D. Thesis, University of Cincinnati, Ohio, 1967.
347. Johnson, A. W., Ylid Chemistry, Academic Press, New York, 1966.
348. Johnson, A. W., J. Org. Chem., 24, 282 (1959).
349. Johnson, A. W., and V. L. Kylingstad, J. Org. Chem., 31, 334 (1966).
350. Johnson, A. W., and R. B. LaCount, Tetrahedron, 9, 130 (1960).
351. Johnson, A. W., and S. C. K. Wong, 151st Meeting of the American Chemical Society, Pittsburgh, Pa., March 1966, Abstract of Papers.
352. Johnson, A. W., and S. C. K. Wong, Can. J. Chem., 44, 2793 (1966).
353. Johnson, A. W., and H. Lloyd Jones, J. Am. Chem. Soc., 90, 5232 (1968).
354. Jones, G. H., and J. G. Moffat, J. Am. Chem. Soc., 90, 5337 (1968).
355. Jones, G. H., E. K. Hamamura, and J. G. Moffat, Tetrahedron Letters, 1968, 5731.
356. Jones, G., and R. F. Maisey, Chem. Commun., 1968, 543.
357. Jones, M. E., and S. Trippett, J. Chem. Soc., 1966, 1090.

358. Kaplan, F., G. Singh, and H. Zimmer, J. Phys. Chem., 67, 2509 (1963).
359. Keat, R., M. C. Miller, and R. A. Shaw, Proc. Chem. Soc., 1964, 137.
360. Kesting, W., J. Prakt. Chem., 105 (2), 242 (1923).
361. Ketcham, R., D. Jambatkar, and L. Martinelli, J. Org. Chem., 27, 4666 (1962).
362. Akiba, Kin-Ya, C. Eguchi, and N. Inamoto, Bull. Chem. Soc. Japan, 40 (12), 2983 (1967); C.A., 68, 78370u (1968).
363. Akiba, Kin-Ya, M. Imanari, and N. Inamoto, Chem. Commun., 1969, 166.
364. Kirilov, M., and J. Petrova, Tetrahedron Letters, 1970, 2129.
365. Kirsanov, A. V., and Z. D. Nekrasova, Zh. Obshch. Khim., 26, 903 (1956); C.A., 50, 14631 (1956).
366. Kirsanov, A. V., A. S. Shtepanek, and V. I. Shewchenko, Dopovidi Akad. Nauk Ukr. RSR, No. 1, 63 (1962); C.A., 57, 11229a (1962).
367. Knott, E. B., Chem. Soc., 1965, 3793.
367a. Knutson, D., U.S. Pat. 3,251,830; C.A., 65, 2149d (1968).
368. Köbrich, G., Angew. Chem., 74, 33 (1962).
368a. Köster, R., D. Simić, and M. A. Grassberger, Liebigs Ann. Chem., 739, 211 (1970).
369. Kojima, T., E. L. Breig, and C. C. Lim, J. Chem. Phys., 35, 2139 (1961).
370. Kosolapoff, G. M., Organophosphorus Compounds, Wiley, New York, 1950, p. 297.
371. Kosolapoff, G. M., J. Am. Chem. Soc., 75, 1500 (1953).
372. Kovalev, B. G., L. A. Yanovskaya, and V. F. Kucherov, Izv. Akad. Nauk SSSR, 1876 (1962); C.A., 58, 9148d (1963).
373. Kraus, H. L., and H. Jung, Z. Naturforsch., 16b, 624 (1961).
374. Kreutzkamp, N., Chem. Ber., 88, 195 (1955).
375. Kreutzkamp, N., E. Schmidt-Samoa, and A. K. Hersberg, Angew. Chem., 77, 1138 (1965).
376. Kröhnke, F., Chem. Ber., 83, 291 (1950).
377. Kuchar, V. P., M. I. Bukovsky, T. N. Kaseva, V. S. Palejcuk, A. A. Petrasenko, and S. N. Solodusenkov, Zh. Obshch. Khim., 40 (102), 1226 (1970).
377a. Kuchar, M., Collection Czech. Chem. Commun., 33, 880 (1968); C.A., 68, 77830g (1968).
378. Leffler, J. E., and R. D. Temple, J. Am. Chem. Soc., 89, 5235 (1967).
379. Leffler, J. E., U. Honsberg, Y. Tsuno, and I. Forsblad, J. Org. Chem., 26, 4810 (1961).
380. Leffler, J. E., and Y. Tsuno, J. Org. Chem., 28, 902 (1963).
381. Lednicer, D., J. Org. Chem., 35, 2307 (1970).

382. Lehn, W. L., J. Inorg. Chem., 6, 1061 (1967).
383. Lemal, D. M., and E. H. Banitt, Tetrahedron Letters, 1964, 245.
384. Levchenko, E. S., and B. N. Ugarov, Zh. Obshch. Khim., 35, 2080 (1965); C.A., 64, 6683 (1966).
385. Levisalles, J., Bull. Soc. Chim. France, 1958, 1021.
386. Levine, S. G., J. Am. Chem. Soc., 80, 6150 (1958).
387. Lichtenstadt, L., and W. Samuel, Liebigs Ann. Chem., 449, 213 (1926).
388. Limasset, J. C., Bull. Soc. Chim., 1967, 1936.
388a. Listvan, V. N., and A. V. Dombrovskii, Zh. Obshch. Khim., 39, 185 (1969); C.A., 70, 96870c (1969).
388b. Listvan, V. N., and A. V. Dombrovskii, Zh. Obshch. Khim., 38, 601 (1968); C.A., 69, 43979t (1968).
388c. Longone, D. T., and R. R. Doyle, Chem. Commun., 1967, 300.
389. Lorberth, J., H. Krapf, and H. Nöth, Chem. Ber., 100, 3511 (1967).
390. Lucken, E. A. C., and C. Mazeline, J. Chem. Soc., A, 1966, 1074 and 1967, 439.
391. Lüscher, G., Dissertation, E. T. H., Zürich, 1922.
392. Lüttke, W., and K. Wilhelm, Angew. Chem., 77, 867 (1965).
393. Lund, L. G., N. L. Paddock, J. E. Proctor, and H. T. Searle, J. Chem. Soc., 1960, 2542.
394. Lutskii, A. E., L. I. Samarai, L. A. Kochergina, A. V. Shepel, Z. A. Sheochenko, G. I. Derkach, E. S. Kozlov, and B. S. Drach, Zh. Obshch. Khim, 37 (9), 2042 (1967); C.A., 68, 34387a (1968).
395. Lutskii, A. E., Z. A. Sheochenko, L. I. Samarai, and A. M. Pinchuk, Zh. Obshch. Khim, 37 (9), 2034 (1967); C.A., 68, 34386z (1968).
396. Maasen, J. A., Th. A. J. W. Wajer, and Th. J. de Boer, Rec. Trav. Chim., 88, 5 (1969).
397. Mägerlein, H., and G. Meyer, Chim. Ber., 103, 2995 (1970).
398. Maercker, A., Org. Reactions, 14, 270 (1965).
399. Märkl, G., Tetrahedron Letters, 1961, 811.
400. Märkl, G., Tetrahedron Letters, 1961, 807.
401. Märkl, G., Chem. Ber., 94, 3005 (1961).
402. Märkl, G., Chem. Ber., 94, 2996 (1961).
403. Märkl, G., Chem. Ber., 95, 3003 (1962).
404. Märkl, G., Angew. Chem. 74, 217 (1962).
405. Märkl, G., Tetrahedron Letters, 1962, 1027.
406. Märkl, G., Z. Naturforsch., 17b, 782 (1962).
407. Märkl, G., Angew. Chem., 74, 696 (1962).
408. Märkl, G., Angew. Chem., 75, 168 (1963).
409. Märkl, G., Z. Naturforsch., 18b, 84 (1963).
410. Märkl, G., Angew. Chem., 75, 669 (1963).
411. Märkl, G., Angew. Chem., 75, 1121 (1963).
412. Märkl, G., Angew. Chem., 78, 907 (1966).

413. Märkl, G., and H. Olbrich, Angew. Chem., 78, 598 (1966).
414. Märkl, G., F. Lieb, and A. Merz, Angew. Chem., 79, 59 (1967).
415. Märkl, G., and A. Merz, Tetrahedron Letters, 1968, 3611.
416. Maier, L., U.S. Pat. 3,188,294 (1965); C.A., 63, 13318 (1965).
417. Mak, T. C. W., and J. Trotter, Acta Cryst., 18, 81 (1965); C.A., 62, 4725c (1965).
418. Makarov, S. P., A. Ya. Yakubovich, A. S. Filatov, M. A. Englin, and T. Ya. Nikiforova, Zh. Obshch. Khim., 38 (4), 709 (1968); C.A., 69, 18506d (1968).
419. Makarov, S. P., V. A. Shpanskii, V. A. Ginsburg, A. I. Shchekotikhin, A. S. Filatov, L. L. Martynova, I. V. Pavlovskaya, A. F. Golovaneva, and A. Y. Yakubovich, Dokl. Akad. Nauk SSSR, 142, 596 (1962); C.A., 57, 4528 (1962).
420. Mann, F. G., and E. J. Chaplin, J. Chem. Soc., 1937, 527.
421. Mann, F. G., and J. Watson, J. Org. Chem., 13, 502 (1948).
422. Mark, V., Tetrahedron Letters, 1964, 3139.
423. Marsh, F. D., and M. E. Hermes, J. Am. Chem. Soc., 86, 4506 (1964).
424. Martin, D., and H. J. Niclas, Chem. Ber., 100, 187 (1967).
425. Masriera, M., Anales Soc. Espan. Fis. Quim., 21, 418 (1923); C.A., 18, 2139 (1924).
425a. Mastryukova, T. A., T. A. Melenteva, E. P. Lure, and M. I. Kabadnik, Dokl. Akad. Nauk. SSSR, 172 (3), 611 (1967); C.A., 66, 94613q (1967).
425b. Mathews, C. N., and G. H. Birum, Tetrahedron Letters, 1966, 5707.
426. McClure, J. D., Tetrahedron Letters, 1967, 2401.
427. McClure, J. D., Tetrahedron Letters, 1967, 2407.
428. McDonald, R. N., and T. W. Campbell, J. Am. Chem. Soc., 82, 4669 (1960).
429. McEwen, W. E., and A. P. Wolf, J. Am. Chem. Soc., 84, 676 (1962).
430. McEwen, W. E., A. Bladé-Font, and C. A. Vander Werf, J. Am. Chem. Soc., 84, 677 (1962).
431. Mechoulam, R., and F. Sondheimer, J. Am. Chem. Soc., 80, 4386 (1958).
432. Meisenheimer, J., J. Caspar, M. Horing, W. Lauter, L. Lichtenstadt, and W. Samuel, Liebigs Ann. Chem., 449, 213 (1926).
433. Messmer, A., I. Pinter, and F. Szegö, Angew. Chem., 76, 227 (1964).
434. Michaelis, A., and H. V. Gimborn, Ber. Deut. Chem. Ges., 27, 272 (1894).

435. Michaelis, A., and E. Kohler, Ber. Deut. Chem. Ges., 32, 1566 (1899).
436. Michalski, J., and S. Musierowicz, Tetrahedron Letters, 1964, 1187.
437. Middleton, W. J., U.S. Pat. 3,067,233; C.A., 58, 11402 (1963).
438. Miller, N. E., J. Am. Chem. Soc., 87, 390 (1965); Inorg. Chem., 4, 1458 (1965).
439. Misumi, S., and M. Nakagawa, Bull. Chem. Soc. Japan. 36, 399 (1963).
440. Mitsch, R. A., J. Am. Chem. Soc., 89 6297 (1967).
441. Moeller, T., U. A. Vandi, J. Org. Chem., 27, 3511 (1962).
442. Monagle, J. J., J. Org. Chem., 27, 3851 (1962).
443. Monagle, J. J., T. W. Campbell, and H. F. McShane, jr., J. Am. Chem. Soc., 84, 4288 (1962).
444. Mondon, A., Liebigs Ann. Chem., 603, 115 (1957).
445. Mosby, W. L., and M. L. Silva, J. Chem. Soc., 1965, 1003.
445a. Mukaiyama, T., S. Fukuyama, and T. Kumamoto, Tetrahedron Letters, 1968, 3787.
446. Murray, R. W., and A. M. Trozzolo, J. Org. Chem., 29, 1268 (1964).
447. Nagata, W., and Y. Hayase, J. Chem. Soc., 1969, 460.
447a. Nagav, Y., K. Shima, and H. Sakurai, Kogyo Kagaku Zasshi, 72, 236 (1969); C.A., 70, 114372y (1969).
447b. Nakaya, T., T. Tamomoto, and M. Imoto, Bull. Chem. Soc. Japan, 40 (3), 691 (1967); C.A., 67, 81978h (1967).
448. Nast, R., and K. Käb, Liebigs Ann. Chem., 706, 75 (1967).
449. Neidlein, R., Angew. Chem., 78, 333 (1966).
450. Nesmeyanov, N. A., S. T. Zhuzhlikova, and O. A. Reutov, Dokl. Akad. Nauk. SSSR, 151, 856 (1963); C.A., 59, 12838a (1963).
450a. Nesmeyanov, N. A., S. T. Bermann, C. D. Ashkinadze, L. A. Kazityna, and O. A. Reutov, Zh. Obshch. Khim., 4 (10), 1685 (1968); C.S., 70, 19463v (1969).
450b. Nesmeyanov, N. A., V. M. Novikov, and O. A. Reutov, Zh. Org. Khim., 2 (6), 942 (1966); C.A., 65, 15420h (1966).
451. Nesmeyanov, N. A., V. M. Novikov, and O. A. Reutov, Izv. Akad. Nauk SSSR, Ser. Khim., 1964, 772; C. A., 61, 3143e (1964).
451a. Neunhoeffer, H., G. Cuny, and W. K. Franke, Liebigs Ann. Chem., 713, 96 (1968).
452. Nielsen, R. P., J. F. Vincent, and H. H. Sisler, Inorg. Chem., 2, 760 (1963).
453. Nöth, H., L. Meinel, and H. Madersteig, Angew. Chem., 77, 734 (1965).
454. Normant, H., and G. Sturtz, Compt. Rend., 256, 1800 (1963).

455. Norris, W. P., and R. A. Henry, J. Org. Chem., 29, 650 (1964).
456. Nürrenbach, A., and H. Pommer, Liebigs Ann. Chem., 721, 34 (1969).
457. Oda, R., Y. Ito, and M. Okano, Tetrahedron Letters, 1964, 7.
458. Oda, R., T. Kawabata, and S. Tanimoto, Tetrahedron Letters, 1964, 1653.
459. Oda, R., T. Kawabata, and S. Tanimoto, J. Am. Chem. Soc., 84, 1312 (1962).
460. Oediger, H., and K. Eiter, Ann., 682, 58 (1965).
461. Osuch, C., J. E. Franz, and F. B. Zienty, J. Org. Chem., 29, 3721 (1964).
462. Paddock, N. L., Quart. Rev., 18, 168 (1964).
462a. Pappas, J. J., and E. Gancher, J. Heterocycl. Chem., 5, 123 (1968).
462b. Pappas, J. J., and E. Gancher, J. Org. Chem., 31, 1287 (1966).
462c. Pappas, J. J., and E. Gancher, J. Org. Chem., 31, 3877 (1966); U.S. Pat. 3,394,166; C.A., 69, 77484t (1968).
463. Parrick, J., Can. J. Chem., 42, 190 (1964).
464. Partentjev L. N., and A. A. Shamshurin, Zh. Obshch. Khim., 9, 865 (1939); C.A., 34, 392 (1940).
465. Partos, R. D., and A. J. Speziale, J. Am. Chem. Soc., 87, 5068 (1965).
466. Partos, R. D., and K. W. Ratts, J. Am. Chem. Soc., 88, 4996 (1966).
467. Pauling, L., Nature of the Chemical Bond, 3rd ed., Cornell University Press, Ithaca, N. Y., 1960, p. 224.
468. Petragnini, N., and M. de M. Campos, Chem. Ind. (London), 1964, 1461.
469. Petragnini, N., and M. de M. Campos, Chem. Ind. (London), 1964, 1461.
470. Petrov, K. A., V. A. Parshina, B. A. Orlov, and G. M. Tsypima, Zh. Obshch. Khim., 32, 4017 (1962); C.A., 59, 657c (1963).
471. Phillips, G. M., J. S. Hunter, and L. E. Sutton, J. Chem. Soc., 1945, 146.
472. Pinck, L., and G. E. Hilbert, J. Am. Chem. Soc., 69, 723 (1947).
473. Plieninger, H., and D. Brück, Tetrahedron Letters, 1968, 4371.
474. Pommer, H., Angew. Chem., 72, 911 (1960).
475. Pudovik, A. N., and N. M. Lebedeva, Dokl. Akad. Nauk SSSR, 90, 799 (1953); C.A., 50, 2429e (1956).
476. Rabinowitz, R., and R. Marcus, J. Am. Chem. Soc., 84, 1312 (1962).
477. Ramirez, F., O. P. Madan, and C. P. Smith, Tetrahedron, 22, 567 (1966).
478. Ramirez, F., and S. Dershowitz, J. Am. Chem. Soc.,

78, 5614 (1956).
479. Ramirez, F., and S. Dershowitz, J. Org. Chem., 22, 41 (1957).
480. Ramirez, F., N. B. Desai, B. Hansen, and N. McKelvie, J. Am. Chem. Soc., 83, 3539 (1961).
481. Ramirez, F., N. B. Desai, B. Hansen, and N. McKelvie, J. Am. Chem. Soc., 83, 3539 (1961).
482. Ramirez, F., N. Desai, and N. B. McKelvie, J. Am. Chem. Soc., 84, 1745 (1962).
483. Ramirez, F., and S. Levy, J. Am. Chem. Soc., 79, 67 (1957).
484. Ramirez, F., and S. Levy, J. Am. Chem. Soc., 79, 6167 (1957).
485. Ramirez, F., and S. Levy, J. Org. Chem., 23, 2036 (1958).
486. Ramirez, F., and O. P. Madan, 148th Meeting, American Chemical Society, 1964, Abstracts of Papers, p. 136.
487. Ramirez, F., O. P. Madan, and C. P. Smith, J. Am. Chem. Soc., 86, 5339 (1964).
488. Ramirez, F., O. P. Madan, and C. P. Smith, Tetrahedron Letters, 1965, 201.
489. Ramirez, F., O. P. Madan, and C. P. Smith, J. Org. Chem., 30, 2284 (1965).
490. Ramirez, F., O. P. Madan, and C. P. Smith, Tetrahedron, 22, 567 (1966).
491. Ramirez, F., O. P. Madan, and C. P. Smith, Tetrahedron, 22, 567 (1966).
492. Ramirez, F., R. B. Mitra, and N. B. Desai, J. Am. Chem. Soc., 82, 5763 (1960).
493. Ramirez, F., and C. P. Smith, Tetrahedron Letters, 1966, 3651.
494. Randall, F. J., and A. W. Johnson, Tetrahedron Letters, 1968, 2841.
495. Rave, T. W., and H. R. Hays, J. Org. Chem., 31, 2894 (1966).
496. Rave, T. W., J. Org. Chem., 32, 3461 (1967).
497. Reddy, G. S., and C. D. Weis, J. Org. Chem., 28, 1822 (1963).
498. Regitz, M., Chem. Ber., 97, 2742 (1964).
499. Regitz, M., Chem. Ber., 99, 3128 (1966).
499a. Regitz, M., and F. Menz, Chem. Ber., 101, 2622 (1968).
500. Regitz, M., and A. Liedhegener, Tetrahedron, 23, 2701 (1967).
501. Reichle, W. T., Inorg. Chem., 3, 402 (1964).
502. Reichle, W. T., J. Organometal. Chem., 13, 529 (1968).
503. Ried, W., and H. Appel, Z. Naturforsch., 15b, 684 (1960).
504. Ried, W., and H. Appel, Liebigs Ann. Chem., 646, 82 (1961).
505. Ried, W., and H. Appel, Liebigs Ann. Chem., 678, 127 (1964).

506. Ried, W., and H. Appel, Liebigs Ann. Chem., 679, 56 (1964).
507. Ried, W., and H. Ritz, Liebigs Ann. Chem., 691, 50 (1966).
508. Richards, E. M., and J. C. Tebby, Chem. Commun., 1969, 494.
509. Richards, J. J., and C. V. Banks, J. Org. Chem., 28, 123 (1963).
510. Rigby, C. W., E. Lord, and C. D. Hall, Chem. Commun., 1967, 714.
511. Ritter, A., and B. Kim, Tetrahedron Letters, 1968, 3449.
512. Rüchardt, C., P. Panse, and S. Eichler, Chem. Ber., 100, 1144 (1967).
513. Rundle, W., and P. Kästner, Liebigs Ann. Chem., 686, 88 (1965).
514. Ryser, G., and P. Zeller, Helv. Chim. Acta, 40, 1242 (1957).
515. Saikachi, H., and S. Nakamura, Yakugaku Zasshi, 89 (10), 1446 (1969).
515a. Saikachi, H., and S. Nakamura, Yakugaku Zasshi, 88 (6), 715 (1968); C.A., 69, 106824m (1968).
515b. Saikachi, H., and S. Nakamura, Yakugaku Zasshi, 88 (8), 1039 (1968); C.A., 70, 11435s (1969).
516. Saikachi, H., and K. Takai, Yakugaku Zasshi, 89 (10), 1401 (1969).
517. Saunders, M., and G. Burchmann, Tetrahedron Letters, 1959, No. 1, 8.
517a. Sasaki, T., S. Eguchi, and A. Kojina, Bull. Chem. Soc. Japan, 41 (7), 1658 (1968); C.A., 69, 106360g (1968).
518. Savage, M. P., and S. Trippett, J. Chem. Soc., 1968, 591.
519. Schäfer, W., and R. Leute, Chem. Ber., 99, 1632 (1966).
520. Scherer, K. V., jr., and R. S. Lunt, J. Org. Chem., 30, 3215 (1965).
521. Schiemenz, G. P., J. Becher, and J. Stöckigt, Chem. Ber., 103, 2077 (1970).
522. Schiemenz, G. P., and H. Engelhard, Chem. Ber., 94, 578 (1961).
523. Schlosser, M., and K. F. Christmann, Angew. Chem., 76, 683 (1964).
524. Schlosser, M., and K. F. Christmann, Angew. Chem., 77, 682 (1965).
525. Schlosser, M., and K. F. Christmann, Liebigs Ann. Chem., 708, 1 (1967).
526. Schlosser, M., and K. F. Christmann, Synthesis, 1, 38 (1969).
527. Schmidbaur, H., and G. Jonas, Chem. Ber., 100, 1120 (1967).

528. Schmidbaur, H., and G. Jonas, Angew. Chem., 79, 413 (1967).
529. Schmidbaur, H., and G. Jonas, Chem. Ber., 101, 1271 (1968).
530. Schmidbaur, H., G. Kuhn, and U. Krüger, Angew. Chem., 77, 866 (1965).
531. Schmidbaur, H., and W. Malisch, Chem. Ber., 102, 83 (1969).
532. Schmidbaur, H., and W. Malisch, Chem. Ber., 103, 97 (1970).
533. Schmidbaur, H., and W. Tronich, Angew. Chem., 79, 412 (1967).
534. Schmidbaur, H., and W. Tronich, Chem. Ber., 100, 1032 (1967).
535. Schmidbaur, H., and W. Tronich, Chem. Ber., 101, 595 (1968).
536. Schmidbaur, H., and W. Tronich, Chem. Ber., 101, 604 (1968).
536a. Schmidbaur, H., and W. Tronich, Chem. Ber., 101, 3545 (1968).
536b. Schmidbaur, H., and W. Tronich, Chem. Ber., 101, 3556 (1968).
537. Schmidbaur, H., and W. Wolfsberger, Angew. Chem., 78, 306 (1966).
538. Schmidbaur, H., and W. Wolfsberger, Chem. Ber., 100, 1000 (1967).
539. Schmidbaur, H., and W. Wolfsberger, Chem. Ber., 100, 1016 (1967).
540. Schmidbaur, H., W. Wolfsberger, and H. Kröner, Chem. Ber., 100, 1023 (1967).
541. Schmidbaur, H., and W. Wolfsberger, Chem. Ber., 101, 1664 (1968).
542. Schmidtpeter, A., B. Wolf, and K. Dull, Angew. Chem., 77, 737 (1965).
543. Schöllkopf, U., Angew. Chem., 71, 260 (1959).
544. Schöllkopf, U., and H. Schäfer, Angew. Chem., 77, 379 (1965).
544a. Schöllkopf, U., F. Gerhart, M. Reetz, H. Frasnelli, and H. Schumacher, Liebigs Ann. Chem., 716, 204 (1968).
545. Schönberg, A., and K. H. Brosowski, Chem. Ber., 92, 2602 (1959).
546. Schönberg, A., K. H. Brosowski, and E. Singer, Chem. Ber., 95, 2144 (1962).
547. Schönberg, A., E. Frese, and K. H. Brosowski, Chem. Ber., 95, 3077 (1962).
548. Schönberg, A., and A. F. Ismail, J. Chem. Soc., 1940, 1374.
549. Schönberg, A., and R. Michaelis, Chem. Ber., 69, 1080 (1936).
550. Schuster, P., Monatsh. Chem., 98, 1310 (1967).
550a. Schwall, H., and M. Regitz, Chem. Ber., 101, 2633

(1968).
551. Schweizer, E. E., J. Am. Chem. Soc., 86, 2744 (1964).
552. Schweizer, E. E., J. Am. Chem. Soc., 86, 2744 (1964).
553. Schweizer, E. E., and K. K. Light, J. Am. Chem. Soc., 86, 2963 (1964).
554. Schweizer, E. E., and G. J. O'Neill, J. Org. Chem., 30, 2082 (1965).
555. Schweizer, E. E., and R. Schepers, Tetrahedron Letters, 1963, 979.
555a. Seebach, D., Angew. Chem., 79, 469 (1967).
556. Seidel, W., Angew. Chem., 77, 809 (1965).
557. Senning, A., Angew. Chem., 77, 379 (1965).
558. Senyawina, L. B., E. V. Dyatlovitskaya, Yu. N. Sheinker, and L. D. Bergelson, Izv. Akad. Nauk SSSR, Ser. Khim., 1964 (11), 1979; C.A., 62, 9942e (1965).
558a. Serratosa, F., and E. Sole, Anales Real Soc. Espan. Fis. Quim., Ser. B, 62, 431 (1966); C.A., 66, 2623f (1967).
559. Seyferth, D., and K. A. Brandle, J. Am. Chem. Soc., 83, 2055 (1961).
560. Seyferth, D., and J. M. Burlitch, J. Org. Chem., 28, 2463 (1963).
561. Seyferth, D., and J. S. Fogel, J. Organometal. Chem., 6, 205 (1966).
562. Seyferth, D., J. S. Fogel, and J. K. Heeren, J. Am. Chem. Soc., 86, 307 (1964).
563. Seyferth, D., S. O. Grim, and T. O. Read, J. Am. Chem. Soc., 82, 1510 (1960), and 83, 1617 (1961).
564. Seyferth, D., and S. O. Grim, J. Am. Chem. Soc., 83, 1610 (1961).
565. Seyferth, D., M. A. Eisert, and J. K. Heeren, J. Organometal. Chem., 2, 101 (1964).
566. Seyferth, D., J. K. Heeren, and S. O. Grim, J. Org. Chem., 26, 4783 (1961).
567. Seyferth, D., J. K. Heeren, and S. O. Grim, J. Org. Chem., 26, 4783 (1961).
568. Seyferth, D., J. K. Heeren, and W. B. Hughes, jr., J. Am. Chem. Soc., 84, 1764 (1962), 87, 2847,3467 (1965).
569. Seyferth, D., J. K. Heeren, and G. Singh, J. Organometal. Chem., 5, 267 (1966).
570. Seyferth, D., and G. Singh, J. Am. Chem. Soc., 87, 4156 (1965).
571. Seyferth, D., D. E. Welch, and J. K. Heeren, J. Am. Chem. Soc., 86, 1100 (1964).
572. Shaw, M. A., J. C. Tebby, R. S. Ward, and D. H. Williams, J. Chem. Soc., 1967, 2442.
573. Shaw, M. A., J. C. Tebby, R. S. Ward, and D. H. Williams, J. Chem. Soc., 1968, 1609.
573a. Shaw, M. A., J. C. Tebby, J. Ranayne, and D. H. Williams, J. Chem. Soc., 1967, 944.
574. Shaw, M. A., and J. C. Tebby, J. Chem. Soc., 1970, 5.

575. Shaw, M. A., and J. C. Tebby, J. Chem. Soc., 1970, 504.
575a. Shechter, H., U.S. Clearing House Fed. Sci. Tech. Inform. 1968, AD 668343; C.A., 70, 47547u (1969).
576. Shono, T., and M. Mitani, J. Am. Chem. Soc., 90, 2728 (1968).
577. Shevchenko, V. I., A. M. Pinchuk, and A. V. Kirsanov, Zh. Obshch. Khim., 35, 1488 (1965); C.A., 63, 14899g (1965).
578. Shevchenko, V. I., A. S. Shtepanek, and A. V. Kirsanov, Zh. Obshch. Khim., 32, 2595 (1962); C.A., 58, 9126 (1963).
579. Shevchenko, V. I., A. S. Shtepanek, and A. M. Pinchuk, Zh. Obshch. Khim., 30, 1566 (1960); C.A., 55, 1490g (1961).
580. Shevchenko, V. I., V. T. Stratienko, and A. M. Pinchuk, Zh. Obshch. Khim., 34, 3954 (1964); C.A., 62, 9167 (1965).
581. Shevchenko, V. I., V. T. Stratienko, and A. M. Pinchuk, Zh. Obshch. Khim., 35, 1487 (1965); C.A., 63, 14899 (1965).
582. Shewtshuk, M. I., A. A. Grigorenko, and A. V. Dombrowski, Zh. Obshch. Khim., 35, 2216 (1965); C.A., 64, 11243c (1966).
582a. Shubina, L. V., L. Ya. Malkes, V. N. Dmitrieva, and V. D. Bezuglyi, Zh. Obshch. Khim., 37, 437 (1967); C.A., 67, 43251h (1967).
582b. Shubina, L. V., L. Ya. Malkes, B. A. Zadorozhnyi, and I. K. Ishchenko, Zh. Obshch. Khim., 36, 1991 (1966); C.A., 66, 75780a (1967).
583. Simalty-Siemiatycki, M., J. Caretto, and F. Malbec, Bull. Soc. Chim. France, 125 (1962).
584. Simalty-Siemiatycki, M., and H. Strzelecka, Compt. Rend., 250, 3489 (1960).
584a. Simalty, M., H. Strzelacka, and M. Dupré, Compt. Rend., 265, 1284 (1967), and 266, 1306 (1968).
585. Singh, G., and H. Zimmer, J. Org. Chem., 30, 417 (1965).
586. Singh, G., and H. Zimmer, J. Org. Chem., 30, 417 (1965).
587. Singh, G., and H. Zimmer, Organometal. Chem. Rev., 2, 279 (1967).
588. Sisler, H. H., H. S. Ahuja, and N. L. Smith, J. Org. Chem., 26, 1819 (1961).
589. Sisler, H. H., A. Sarkis, H. S. Ahuja, R. J. Drago, and N. L. Smith, J. Am. Chem. Soc., 81, 2982 (1959).
590. Sisler, H. H., A. Sarkis, H. S. Ahuja, R. J. Drago, and N. L. Smith, J. Am. Chem. Soc., 81, 2982 (1959).
591. Sisler, H. H., and N. L. Smith, J. Org. Chem., 26, 4733 (1961).
592. Sisler, H. H., and N. L. Smith, J. Org. Chem., 26,

611 (1961).
593. Smith, N. L., J. Chem. Eng. Data, 8 (3), 461 (1963); C.A., 59, 8785 (1963).
594. Snyder, J. P., and H. J. Bestmann, Tetrahedron Letters, 1970, 3317.
595. Sondheimer, F., and R. Mechoulan, J. Am. Chem. Soc., 80, 3087 (1958).
596. Speziale, A. J., and D. E. Bissing, J. Am. Chem. Soc., 85, 1888,3878 (1963).
597. Speziale, A. J., G. J. Marco, and K. W. Ratts, J. Am. Chem. Soc., 82, 1260 (1960).
598. Speziale, A. J., and R. D. Partos, J. Am. Chem. Soc., 85, 3312 (1963).
599. Speziale, A. J., and K. W. Ratts, J. Am. Chem. Soc., 82, 1260 (1960), and 84, 854 (1962).
600. Speziale, A. J., and K. W. Ratts, J. Am. Chem. Soc., 85, 2790 (1963).
601. Speziale, A. J., and K. W. Ratts, J. Org. Chem., 28, 465 (1963).
602. Speziale, A. J., and K. W. Ratts, J. Org. Chem., 28, 465 (1963).
603. Speziale, A. J., and K. W. Ratts, J. Am. Chem. Soc., 87, 5603 (1965).
603a. Speziale, A. J., and K. W. Ratts, U.S. Pat. 3,325,542; C.A., 67, 108753v (1967).
604. Speziale, A. J., and C. C. Tung, J. Org. Chem., 28, 1353 (1963).
605. Sprenger, H. E., and W. Ziegenbein, Angew. Chem., 75, 1011 (1965).
605a. Sprenger, H. E., Ger. Pat. 1,221,221; C.A., 65, 20164b (1966).
606. Staab, H. A., and N. Sommer, Angew. Chem., 74, 294 (1962).
607. Staab, H. A., Angew. Chem., 74, 420 (1962).
608. Staudinger, H., and W. Braunholtz, Helv. Chim. Acta, 4, 897 (1921).
609. Staudinger, H., and E. Hauser, Helv. Chim. Acta, 4, 861 (1921).
610. Staudinger, H., and J. Meyer, Helv. Chim. Acta, 2, 619 (1919).
611. Staudinger, H., and J. Meyer, Helv. Chim. Acta, 2, 635 (1919).
612. Staudinger, H., and J. Meyer, Chem. Ber., 53, 72 (1920).
613. Staudinger, H., and E. Hauser, Helv. Chim. Acta, 4, 887 (1921).
614. Staudinger, H., and G. Lüscher, Helv. Chim. Acta, 5, 75 (1922).
615. Staudinger, H., and G. Lüscher, Helv. Chim. Acta, 5, 75 (1922).
616. Staudinger, H., G. Rathsam, and F. Kjelsberg, Helv.

Chim. Acta, 3, 853 (1920).
617. Stephens, F. S., J. Chem. Soc., 1965, 5640, 5658.
618. Stockel, R. F., F. Megson, and M. T. Beachem, J. Org. Chem., 33, 4395 (1968).
619. Strandtman, M., M. P. Cohen, C. Puchalski, and J. Shavel, jr., J. Org. Chem., 33, 4306 (1968).
620. Stratienko, V. T., and V. T. Shevchenko, Zh. Obshch. Khim., 34, 1463 (1964); C.A., 61, 5686 (1964).
621. Strzelecka, H., M. S. Siemiatycki, and C. Prevost, Compt. Rend., 254, 696 (1962), and 255, 731 (1962).
621a. Strzelecka, H., Ann. Chim., 1, 201 (1966).
622. Strzelecka, H., M. Simalty-Siemiatycki, and Ch. Prévost, Compt. Rend., 257, 926 (1963).
623. Strzelecka, H., M. Simalty-Siemiatycki, and Ch. Prévost, Compt. Rend., 258, 6167 (1964).
624. Strating, J., J. Heeres, and A. M. van Leusen, Rec. Trav. Chim. Pays-Bas, 85, 1061 (1966).
625. Sullivan, W. W., D. Ullman, and H. Shechter, Tetrahedron Letters, 1969, 457.
626. Surmatis, J. D., and A. Ofner, J. Org. Chem., 26, 1171 (1961).
627. Takahaschi, H., K. Fujiwara, and M. Ota, Bull. Chem. Soc. Japan, 35, 1498 (1962).
628. Thayer, J. S., and R. West, Inorg. Chem., 3, 406 (1964).
629. Thomas, L. C., and R. A. Chittenden, Spectrochim. Acta, 21, 1905 (1965); C.A., 64, 168f (1966).
630. Tomösközi, I., Angew. Chem., 75, 294 (1963); Tetrahedron, 19, 1969 (1963).
631. Tomösközi, I., Chem. Ind. (London), 1965, 689.
632. Tomösközi, I., U. H. J. Bestmann, Tetrahedron Letters, 1964, 1293.
633. Tomösközi, I., and G. Janzo, Chem. Ind. (London), 1962, 2085.
634. Trippett, S., Advances in Organic Chemistry, Vol I, Inter-Science, New York, 1960, p. 83.
635. Trippett, S., J. Chem. Soc., 1962, 2337.
636. Trippett, S., J. Chem. Soc., 1962, 4733.
637. Trippett, S., Organophosphorus Compounds, Special Lectures, International Symposium, Heidelberg, 1964, Butterworths, London, 1964, p. 255.
638. Trippett, S., Quart. Rev., 17, 406 (1963).
639. Trippett, S., Chem. Commun., 1966, 468.
640. Trippett, S., and D. M. Walker, J. Chem. Soc., 1959, 3874.
641. Trippett, S., and D. M. Walker, Chem. Ind. (London), 202 (1960).
642. Trippett, S., and D. M. Walker, Chem. Ind. (London), 933 (1960).
643. Trippett, S., and D. M. Walker, J. Chem. Soc., 1961, 1266.

644. Trippett, S., and D. M. Walker, J. Chem. Soc., 1961, 2130.
645. Trippett, S., and D. M. Walker, Chem. Ind. (London), 990 (1961).
646. Trippett, S., and B. J. Walker, Chem. Commun., 1965, 106.
646a. Tronchet, J. M. J., E. Doelker, and Br. Bachler, Helv. Chim. Acta, 52, 308, 817 (1969).
647. Tsolis, A. K., W. E. McEwen, and C. A. Vander Werf, Tetrahedron Letters, 1964, 3217.
648. Umani-Ronchi, A., M. A. Campora, G. Gaudiano, and A. Selva, Chem. Ind., 49, 388 (1967).
649. Vandi, A. F., and T. Moeller, Chem. Ind. (London), 1962, 221.
650. Wadsworth, W. S., and W. D. Emmons, J. Am. Chem. Soc., 83, 1733 (1961).
651. Wadsworth, D. H., O. E. Schupp, E. J. Seuss, and J. A. Ford, jr., J. Org. Chem., 30, 680 (1965).
652. Waite, N. E., J. C. Tebby, R. S. Ward, and D. H. Williams, J. Chem. Soc., 1969, 1100.
653. Waite, N. E., and J. C. Tebby, J. Chem. Soc., 1970, 386.
654. Walker, C. C., and H. Shechter, Tetrahedron Letters, 1965, 1447.
654a. Walker, C. C., and H. Shechter, J. Am. Chem. Soc., 90, 5626 (1968).
655. Washburn, R. M., and R. A. Baldwin, U.S. Pat. 3,112,331; C.A., 60, 5554 (1964).
656. Washburn, R. M., and R. A. Baldwin, U.S. Pat. 3,189,564; C.A., 63, 9991 (1965).
657. Wassermann, H. H., and R. C. Koch, Chem. Ind. (London), 1956, 1014.
657a. Webster, O. W., J. Am. Chem. Soc., 88, 4055 (1966).
658. Weil, Th., and M. Cais, J. Org. Chem., 28, 2472 (1963).
659. Weis, C. D., J. Org. Chem., 27, 3514 (1962).
660. Welcher, R. P., and N. E. Day, J. Org. Chem., 27, 1824 (1962).
661. Weygand, F., and H. J. Bestmann, Angew. Chem., 72, 535 (1960).
662. Wheatley, P. J., J. Chem. Soc., 1965, 5785.
663. Withlock, H. W., J. Org. Chem., 29, 3129 (1964).
664. Wiberg, N., F. Raschig, and R. Sustmann, Angew. Chem., 74, 716 (1962).
665. Wiberg, N., K. H. Schmid, and W. C. Loo, Angew. Chem., 77, 1042 (1965).
666. Wiegräbe, W., and H. Bock, Angew. Chem., 77, 1042 (1965).
667. Wiegräbe, W., H. Bock, and W. Lüttke, Chem. Ber., 99, 3737 (1966).
668. Wiegräbe, W., and H. Bock, Chem. Ber., 101, 1414 (1968).

669. Wiesboeck, R. A., J. Org. Chem., 30, 3161 (1965).
670. Wiesboeck, R. A., J. Org. Chem., 30, 3161 (1965).
671. Williams, D. H., R. S. Ward, and R. G. Cooks, J. Am. Chem. Soc., 90, 966 (1968).
672. Witschard, G., and C. E. Griffin, Spectrochim. Acta, 19, 1905 (1963); C.A., 59, 13482e (1963).
673. Wittig, G., Pure Appl. Chem., 9, 245 (1964).
674. Wittig, G., and W. Böll, Chem. Ber., 95, 2526, 2514 (1962).
675. Wittig, G., H. Eggers, and P. Duffner, Liebigs Ann. Chem., 619, 10 (1958).
676. Wittig, G., and G. Geissler, Liebigs Ann. Chem., 580, 44 (1953).
677. Wittig, G., and W. Haag, Chem. Ber., 88, 1654 (1955).
678. Wittig, G., and D. Hellwinkel, Angew. Chem., 74, 76 (1962).
679. Wittig, G., and E. Kochendörfer, Chem. Ber., 97, 741 (1964).
680. Wittig, G., and H. Laib, Liebigs Ann. Chem., 580, 57 (1953).
681. Wittig, G., and A. Maercker, Chem. Ber., 97, 747 (1964).
682. Wittig, G., and M. Rieber, Liebigs Ann. Chem., 562, 177 (1949).
683. Wittig, G., and M. Schlosser, Angew. Chem., 72, 324 (1960); Chem. Ber., 94, 1373 (1961).
684. Wittig, G., and M. Schlosser, Tetrahedron, 18, 1023 (1962).
685. Wittig, G., and H. Schöllkopf, Chem. Ber., 87, 1318 (1954).
686. Wittig, G., and K. Schwarzenbach, Liebigs Ann. Chem., 650, 1 (1961).
687. Wittig, G., H. D. Weigmann, and M. Schlosser, Chem. Ber., 94, 676 (1961).
688a. Van Woerden, H. F., H. Cerfontain, and C. F. van Valkenburg, Rec. Trav. Chim. Pays-Bas, 88, 158 (1969).
689. Worrall, D. E., J. Am. Chem. Soc., 52, 2933 (1930).
689a. Wulf, J., and R. Huisgen, Angew. Chem., 79, 472 (1967).
689b. Yoshina, S., A. Tananka, and K. Yamamoto, Yakugaku Zasshi, 88 (1), 65 (1968); C.A., 69, 27134h (1968).
690. Zbiral, E., Monatsh. Chem., 91, 1144 (1960).
691. Zbiral, E., Monatsh. Chem., 94, 78 (1963).
692. Zbiral, E., Monatsh. Chem., 95, 1759 (1964).
693. Zbiral, E., Tetrahedron Letters, 1964, 3963.
694. Zbiral, E., Tetrahedron Letters, 1966, 2005.
694a. Zbiral, E., Monatsh. Chem., 98, 916 (1967).
695. Zbiral, E., and L. Fenz, Monatsh. Chem., 96, 1983 (1965); C.A., 64, 19659b.
696. Zbiral, E., and H. Hengstenberger, Liebigs Ann. Chem., 721, 121 (1969).

697. Zbiral, E., and M. Rasberger, Tetrahedron, 24, 2419 (1968).
698. Zbiral, E., and M. Rasberger, Tetrahedron, 25, 1871 (1969).
699. Zbiral, E., and J. Stroh, Liebigs Ann. Chem., 725, 29 (1969).
700. Zbiral, E., and E. Werner, Monatsh. Chem., 97, 1797 (1966).
701. Zbiral, E., and E. Werner, Angew. Chem., 79, 899 (1967).
701a. Zeeh, B., and R. Beutler, Org. Mass Spectrom., 1968, 1 (6), 791.
702. Zeifmann, Y. V., N. P. Gambaryan, and I. L., Knunyants, Izv. Akad. SSSR, Ser. Khim., 3, 450 (1965); C.A., 63, 482b (1965).
703. Zeliger, H. I., J. P. Snyder, and H. J. Bestmann, Tetrahedron Letters, 1969, 2199.
704. Zeliger, H. I., J. P. Snyder, and H. J. Bestmann, Tetrahedron Letters, 1970, 3313.
705. Zhmurova, I. N., and A. V. Kirsanov, Zh. Obshch. Khim., 56 (7), 1248 (1966); C.A., 16826g (1966).
706. Zhmurova, I. N., and R. I. Yurchenko, Zh. Obshch. Khim., 38 (3), 613 (1968); C.A., 69, 107589a (1968).
707. Zimmer, H., and P. J. Bercz, Liebigs Ann. Chem., 686, 107 (1965).
708. Zimmer, H., P. J. Bercz, O. J. Haltenieks, and M. W. Moore, J. Am. Chem. Soc., 87, 2777 (1965).
709. Zimmer, H., P. J. Bercz, and G. E. Heuer, Tetrahedron Letters, 1968, 171.
710. Zimmer, H., and G. Singh, Angew. Chem., 75, 574 (1963).
711. Zimmer, H., and G. Singh, J. Org. Chem., 28, 483 (1963).
712. Zimmer, H., and G. Singh, J. Org. Chem., 28, 483 (1963).
713. Zimmer, H., and G. Singh, J. Org. Chem., 29, 3412 (1964).
714. Zimmer, H., and G. Singh, J. Org. Chem., 29, 1579 (1964).
715. Zimmer, H., and M. Yayavant, Tetrahedron Letters, 5061 (1966).
716. Zhdanov, Yu. A., Yu. E. Alekseev, and G. N. Dorofenko Zh. Obshch. Khim., 36 (10), 1742 (1966); C.A., 66, 76244r (1967).
717. Zhdanov, Yu. A., and V. G. Alekseeva, Zh. Obshch. Khim., 37 (6), 1408 (1967); C.A., 68, 22161j (1968).
718. Zhdanov, Yu. A., Yu. E. Alekseev, and G. N. Dorofenko, Zh. Obshch. Khim., 37, 98, 2635 (1967); C.A., 66, 95321e, and 70, 11905v (1969).
719. Zhdanov, Yu. A., and V. A. Alekseeva, Zh. Obshch. Khim., 38 (9), 1951 (1968); C.A., 70, 29212b (1969).

720. Zhdanov, Yu. A., G. N. Dorofenko, G. A. Korol' chenko, and A. E. Ozolin, Zh. Obshch. Khim., 36 (2), 492 (1966); C.A., 65, 2337e (1966).
721. Zhdanov, Yu. A., and L. A. Uzlova, Zh. Obshch. Khim., 36 (7), 1211 (1966); C.A., 65, 18670f (1966).
722. Zhdanov, Yu. A., L. A. Uzlova, G. N. Dorofenko, and G. J. Kravchenko, Zh. Obshch. Khim., 36 (6), 1025 (1966); C.A., 65, 12273h (1966).

Chapter 5B. Penta- and Hexaorganophosphorus Compounds

DIETER HELLWINKEL

Organisch-Chemisches Institut der Universität Heidelberg, Deutschland

A. GENERAL SYNTHETIC ROUTES

Compounds with five ligands around the phosphorus atom are generally called phosphoranes. They can be derived formally from the various "ortho" acids of phosphorus with the general formula $R_nP(OH)_{5-n}$, where n = 0 to 5, and from the different P(V) hydrids abstracted in the formula R_nPH_{5-n}, where n = 0 to 5. A schematic basic formulation, which comprises all kinds of compounds with pentacoordinated phosphorus, is as follows:

$$R_lP(Y)_nH_m \quad \text{where} \quad l + n + m = 5, \quad \text{and} \quad l, n, m = 0 \text{ to } 5.$$

Most of the types of compounds discussed in this chapter were totally unknown in 1950, when Kosolapoff's Organophosphorus Compounds, on which this book is based, was published. Those compounds can be arranged into the following general groups. A more detailed classification is, of course, helpful, and will be used at appropriate places, particularly in the list of compounds in Section D.

1. PY_5. 2. RPY_4. 3. R_2PY_3. 4. R_3PY_2. Here Y_n includes any combination of groups like the halogens, -OR, -SR, and $-NR_2$. Excepted are, as purely inorganic compounds, the phosphorus pentahalides and saltlike structures such as $[(RO)_4P^+]$, Cl, Br, I^-, $[R_3(RO)P^+]Cl$, Br, I^-, and their R_2N analogs, which fall within the scope of Chapter 4.

5. R_4PY. Here Y is a group of the type OR. When Y is a halogen, these compounds are tetracoordinated phosphonium salts, which belong in Chapter 4.

6. R_5P. 7. HR_nPY_{4-n}. Here Y_n includes any combination of substituents like F, -OR, and $-NR_2$. This set represents all compounds that contain one hydrogen bonded directly to the pentacoordinated phosphorus center.

In Section A.8 are discussed compounds with six ligands around a central phosphorus atom, with the exception of the inorganic hexahalophosphate complexes.

It may be recalled that many of the substances discussed in this chapter--mainly the chloro-, bromo-, and iodophosphoranes--can dissociate under certain conditions into ionic forms with tetracoordinated phosphorus, or even exist exclusively in a saltlike structure. The close genetic relationship of these zwitter compounds to the true pentacoordinated species, however, demands their inclusion in this chapter.

A.1. Compounds of Type PY_5

I. REACTIONS OF FLUORINE DERIVATIVES OF PENTACOVALENT PHOSPHORUS WITH AMINES, THIOLS, OR ALCOHOLS AND DERIVATIVES THEREOF

Phosphorus pentafluoride reacts with primary or secondary amines at higher temperatures or, better, with organometal(oid) derivatives of amines, alcohols, and thiols at room temperature and below to form noncyclic or cyclic products with one or more fluorine atoms substituted by -OR, -SR, or $-NR_2$ groups.[590] In most cases primary adducts are formed as intermediates at low temperatures; these decompose to the products on warming.[79b]

$$2PF_5 + HNR_2 \longrightarrow R_2NPF_4 + [H_2NR_2]^{\oplus}[PF_6]^{\ominus}$$ [82]

$$6PF_5 + 2RNH_2 + 4R'_3N \longrightarrow (F_3PNR)_2 + 4[R'_3NH]^{\oplus}[PF_6]^{\ominus}$$ [223,233]

Without tertiary amines, complex reactions[233] occur. Sterically hindered primary amines react in a simpler way to form noncyclic phosphoranes: $2RNH_2 + 2PF_5 \longrightarrow RNHPF_4 + [RNH_3]^{\oplus}[PF_6]^{\ominus}$.[233]

$$PF_5 + R_2N\text{-}SiR'_3 \longrightarrow R_2NPF_4 + R'_3SiF$$ [115,578]

$$2PF_5 + 2MeN(SiMe_3)_2 \longrightarrow (F_3PNMe)_2 + 4Me_3SiF$$ [114,115,582]

$$PF_5 + (Me_2N)_2S \longrightarrow Me_2NPF_4 + \text{Solids}$$ [79b]

$$PF_5 + (MeO)_3P \longrightarrow MeOPF_4 + FP(OMe)_2$$ [81,108]

$$PF_5 + MeOLi \longrightarrow MeOPF_4 + LiF$$ [108]

$$PF_5 + (Me_2N)_3P \longrightarrow Me_2NPF_4 + FP(NMe_2)_2$$ [80]

$$Me_2NPF_4 + PhO\text{-}SiMe_3 \longrightarrow (Me_2N)(PhO)PF_3 + FSiMe_3$$ [455]

$$R_2NPF_4 + 2R'NH_2 \longrightarrow R_2N(R'NH)PF_3 + R'NH_2 \cdot HF$$ [148]

$$PF_5 + Me_2N\text{-}PMe_2 \longrightarrow Me_2NPF_4 + \text{Solids}$$ [80]

$$PF_5 + RS\text{-}SiMe_3 \longrightarrow RSPF_4 + FSiMe_3$$ [456,456a]

Similar reactions can be carried out with trifluorodichlorophosphorane, in which case, as a rule, only the chlorine atoms are substituted.[51,54,339,340]

II. EXCHANGE REACTIONS IN CHLORINE DERIVATIVES OF PENTAVALENT PHOSPHORUS WITH INORGANIC FLUORIDES

In these syntheses, chlorine derivatives of pentavalent phosphorus are caused to react with fluorides of silver,[37,52] arsenic,[135] antimony,[135,350,592a] or lead.[135] For example:

$$\mathrm{O{=}C(NMe)_2PCl_3 \xrightarrow{SbF_3} O{=}C(NMe)_2PF_3}$$

III. REACTIONS OF PHOSPHORUS PENTACHLORIDE WITH COMPOUNDS CONTAINING ACTIVE HYDROGEN

Phosphorus pentachloride has been used in many condensation reactions with phenols,[101,347,418,499,500] phenolic esters,[75,213,218] amines,[37,95,221,222,347,463,696-698,700,701,703] amides,[449,643] oxamide,[55,55a] sulfonamides,[52] hydrazides,[152] semicarbazides,[53,274] and nitriles.[333] Generally, in these reactions, two to five chlorines are substituted by the O- or N-containing organic radicals. Some typical examples follow.

$$\mathrm{2PCl_5 + 3\,C_6H_4(OH)_2 \xrightarrow{-6HCl} (C_6H_4O_2)_2PCl + C_6H_4O_2PCl_3} \quad [500]$$

$$\mathrm{PCl_5 + C_6H_4O_2CHOEt \longrightarrow C_6H_4O_2PCl_3 + Cl_2CHOEt} \quad [213]$$

$$\mathrm{PCl_5 + 2RHN{-}C(O){-}NHR' \longrightarrow O{=}C(NR)(NR')PCl_3 + [RHN{-}\overset{\oplus}{C}(Cl){-}NHR']Cl^{\ominus} + H_2O} \quad [643]$$

$$\mathrm{PCl_5 + 2Me\overset{\oplus}{N}H_3Cl^{\ominus} \longrightarrow Cl_3P(NMe)_2PCl_3 + 6HCl} \quad [49]$$

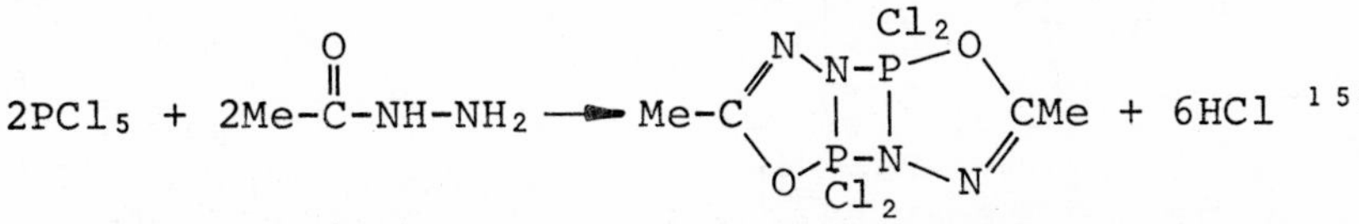

$$2PCl_5 + 2Me\text{-}C(=O)\text{-}NH\text{-}NH_2 \longrightarrow Me\text{-}C\langle\text{cyclic } N\text{-}N\text{-}PCl_2\text{-}O \ldots\rangle CMe + 6HCl$$ [152]

$$Cl_3P(\text{-}NMe\text{-})_2PCl_3 + 2MeNHSO_2NHMe \xrightarrow{-4HCl} O_2S(NMe)_2PCl(NMe)_2PCl(NMe)_2SO_2$$ [51]

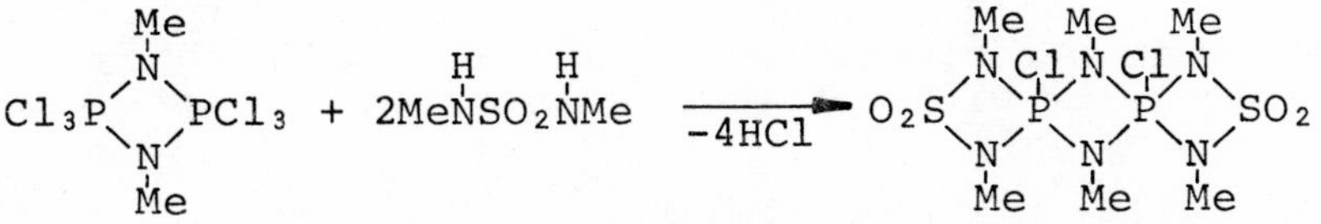

In addition to the monocyclic compounds, the reaction of primary amine hydrochlorides with phosphorus pentachloride can lead to small amounts of related tricyclic compounds.[46,49] It is interesting to note that amines with $K_b = 10^{-14}$ to 10^{-19} react to monomeric phosphine imine derivatives (see Chapter 18), whereas amines with $K_b = 10^{-9}$ to 10^{-13} give the dimeric, cyclic products.[700] Amines that are sterically hindered in the α-position also form only monomeric products.[222,696]

All of the products synthesized by the procedures of this section behave in principle like phosphorus pentachloride itself and can undergo further substitutions by the same kind of reagents, leading in many cases to polycyclic compounds.[46,50,51,173]

It is now established that in the reactions of salicyclic acid with phosphorus pentachloride, contrary to former views, pentacoordinate phosphorus compounds are formed, at the most, as unstable intermediates.[31,465-469]

IV. ADDITION OF HALOGENS TO PHOSPHITES AND AMINO-SUBSTITUTED ANALOGS

The simplest way of obtaining various substituted phosphorus pentahalides is via the reaction of phosphorus derivatives of lower valence with the halogens.[347] Thus aryl phosphites, when allowed to react with chlorine,[100,174,215,499,564] bromine,[100,216,231,564] iodine,[100,176,231,564] and the mixed halogens[100] at lower temperatures, give phosphorane derivatives. In most cases, however, the primary

$$P(OPh)_3 + X_2 \longrightarrow X_2P(OPh)_3$$ [100,564]

adducts are rather unstable and easily undergo very complex disproportionation reactions to form ionic structures of the following general formulation:

$$[(PhO)_nPX_{4-n}]^{\oplus} [(PhO)_mPX_{6-m}]^{\ominus}$$

X = Cl, Br[564,564a]

No adequate data are available for these products; the identifications were made by hydrolysis and alcoholysis experiments.[101,564,564a] The corresponding iodo compounds have simpler ionic structures, which tend to add further iodine to yield salts with polyiodide anions.[176,231,564]

$$(PhO)_3P + nI_2 \longrightarrow [(PhO)_3PI]^{\oplus}[I_{n-1}]^{\ominus}$$

Halophosphoranes with diorganylamino groups are synthesized by similar procedures. Appropriate amino derivatives of trivalent phosphorus are reacted with chlorine[37,103,150,349,403,573] or bromine.[150,450-452]

$$R_2NPF_2 + X_2 \longrightarrow R_2NPF_2X_2$$

X = Cl, Br

V. ADDITION OF HALOAMINES TO PHOSPHORUS TRIHALIDES

Fluorine-containing phosphorus trihalides add chloroamines, which are substituted with trifluoromethyl groups.[156]

$$PF_2Cl + (CF_3)_2NCl \longrightarrow (CF_3)_2NPF_2Cl_2$$

VI. REACTIONS OF PHOSPHITES AND AMINOSUBSTITUTED ANALOGS WITH HALOGENATING AGENTS

In the case of fluorine, halogenations of this type can be carried out with sulfur tetrafluoride,[79b] trifluoroacetone, hexafluoroacetone,[508,535] and difluoroazirine.[415] Corresponding chlorinations are possible using sulfuryl chloride, sulfur dichloride, disulfur dichloride,[472] arylsulfur chloride,[460] iron trichloride,[691] copper chloride,[103] phosphorus pentachloride,[699] and tetrachlorophosphoranes.[481]

$$(Me_2N)_3P \xrightarrow{SF_4} (Me_2N)_3PF_2$$

$$4(PhO)_3P + SO_2Cl_2 \longrightarrow 2(PhO)_3PO + (PhO)_3PS + (PhO)_3PCl_2$$

$$Me_2NPF_2 + 2CuCl_2 \longrightarrow Me_2NPF_2Cl_2 + 2CuCl$$

VII. REACTIONS OF PEROXIDES AND OZONE WITH TRIALKYL PHOSPHITES

Trimethyl phosphite[123] and triethyl phosphite[126,127] react at room temperature over weeks with dimethyl peroxide and diethyl peroxide, respectively, to form the corresponding pentacoordinated derivatives. The products were not

$$(RO)_3P + RO\text{-}OR \longrightarrow (RO)_5P$$

obtained in pure form, but could be characterized by spectroscopic data and degradation reactions.

Analogous reactions are possible with cyclic phosphites[123,126] and corresponding aminosubstituted phosphorus compounds.[118] Phosphonites, phosphinites,[118] and phosphines[118,119,279] also react with dialkyl peroxides, yielding the corresponding phosphoranes, at least as unstable intermediates.[116,128] Similar reactions with other peroxy compounds, such as acyl peroxides, also lead to unstable intermediates with, presumably, phosphorane structure;[121,122,124,207,650] these decompose very easily to the corresponding oxophosphorus compounds. In these cases, however, ionic structures seem to play a predominant role.

In this section belong also the reactions of phosphites with ozone. Triphenyl phosphite adds ozone at -70° in methylene dichloride, yielding probably an unstable four-membered ring adduct that easily decomposes around -35° to triphenyl phosphate and singlet oxygen.[436,437,652]

$$(PhO)_3P + O_3 \longrightarrow (PhO)_3P\!\!<\!\!\begin{smallmatrix}O\\O\\O\end{smallmatrix}\!\!> \longrightarrow (PhO)_3PO + {}^1O_2$$

Similar reactions were carried out with trialkyl phosphites, but in these cases no indications of cyclic phosphoranes could be found.[628]

VIII. ADDITION OF PHOSPHITES AND AMINOSUBSTITUTED ANALOGS TO α-DICARBONYL DERIVATIVES AND SIMILAR 1,3-UNSATURATED SYSTEMS

This is the most extensively used reaction for obtaining pentaalkoxy derivatives and corresponding amino or thio compounds of phosphorus.[484-488] In general, trialkyl phosphites at room temperature react exothermically with suitable 1,3-unsaturated systems[70,332,355,478,501,502] in solvents such as methylene dichloride, diethyl ether, and benzene.

The less reactive triaryl phosphites react at higher temperatures, eventually without solvent, whereas the more reactive tris(dialkylamino)phosphines sometimes undergo addition reactions, even at -70°. The following are typical examples.

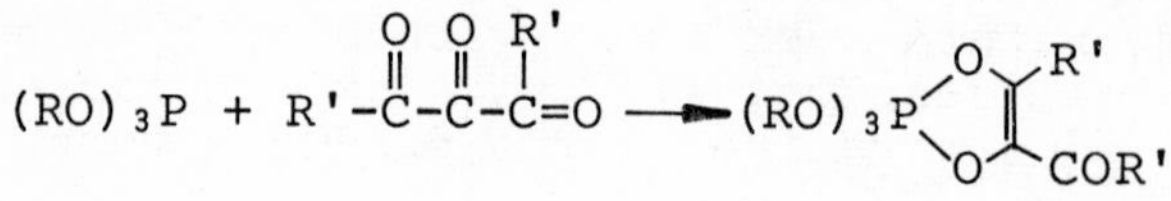

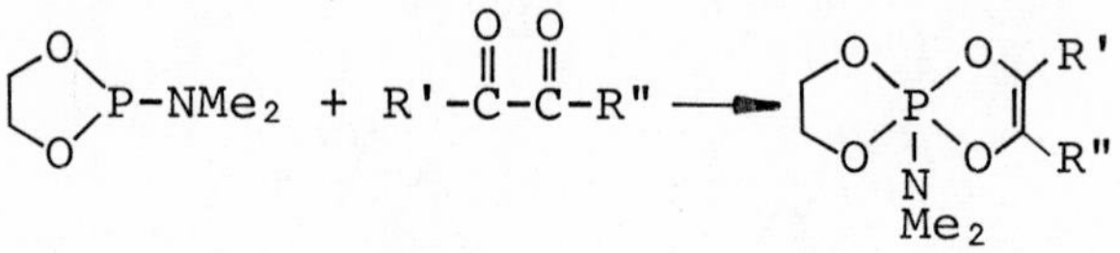

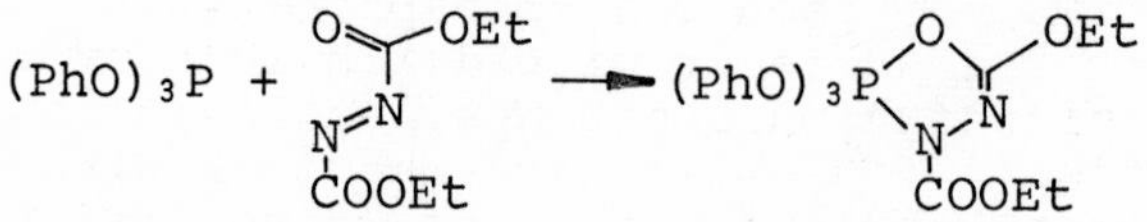

$$(RO)_3P + R'-\overset{O}{\overset{\|}{C}}-\overset{O}{\overset{\|}{C}}-\overset{R'}{\overset{|}{C}}=O \longrightarrow (RO)_3P\overset{O-C(R')}{\underset{O-C(COR')}{<}}$$

$$\text{[cyclic } P\text{-}NMe_2\text{]} + R'-\overset{O}{\overset{\|}{C}}-\overset{O}{\overset{\|}{C}}-R'' \longrightarrow \text{[spiro } P(NMe_2)\text{, } R', R''\text{]}$$

$$(PhO)_3P + EtOOC-N=N-COOEt \longrightarrow (PhO)_3P\overset{O-C(OEt)}{\underset{N(COOEt)-N}{<}} \quad [193]$$

Similar additions are possible with phosphinites,[534] phosphonites,[70,534,574,660] phosphonamidites,[356] and tertiary phosphines.[280,537] In the last case equilibrium between pentacovalent and betainic forms has been observed.

The adducts formed with tris(dimethylamino)phosphine also are open dipolar ions[522] or can, at best, exist in an equilibrium mixture of pentacovalent and ionic forms.[89,93,523]

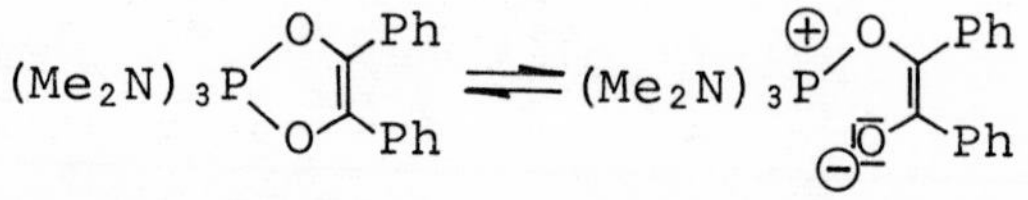

$$(Me_2N)_3P\overset{O-C(Ph)}{\underset{O-C(Ph)}{<}} \rightleftharpoons (Me_2N)_3\overset{\oplus}{P}-O-C(Ph)=C(Ph)-O^{\ominus}$$

The following general principles seem to be valid in these 1,4-addition reactions:[486,523]

1. $(RO)_3P$ compounds have a greater tendency than $(R_2N)_3P$ species to form pentacovalent derivatives (electronegativity difference!).
2. Cyclic tris(dialkylamino)phosphines show a greater tendency to form P(V) derivatives than acyclic analogs (release of crowding).
3. Delocalization of a negative charge on carbon (ester groups) tends to favor an open dipole, but conjugation of the C-C double bond with aromatic rings again favors the pentavalent structures.

IX. CONDENSATIONS OF 2,2,2-TRISUBSTITUTED 2,2-DIHYDRO-1,3,2-DIOXAPHOSPHOLENES WITH CARBONYL COMPOUNDS

When the phospholene compounds prepared as described in Section A.1.VIII are treated further with a variety of carbonyl compounds under mild conditions, new phosphoranes are formed that contain the 2,2-dihydro-1,3,2-dioxaphospholane skeleton.[484,485,488] Some typical examples illustrate the principles of these reactions.

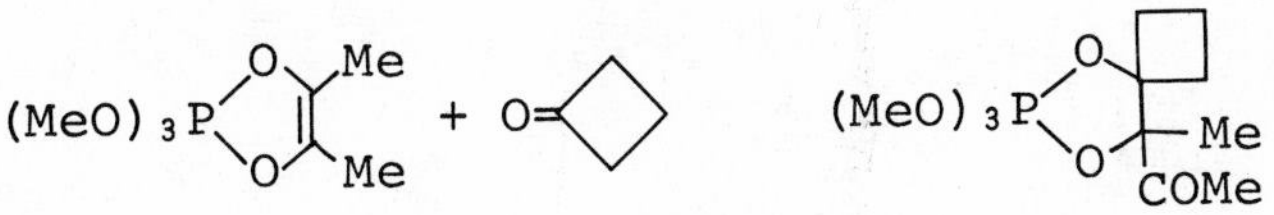

[527]

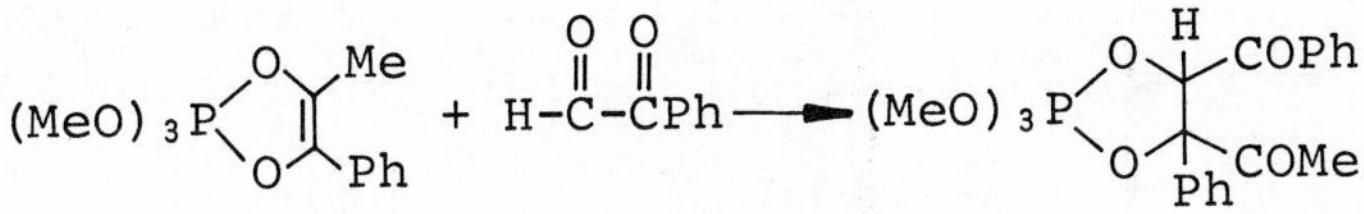

$(MeO)_3P$ (O, Me, O, Me) + O=(cyclobutanone) → $(MeO)_3P$ (O, O; cyclobutane; Me; COMe)

[526]

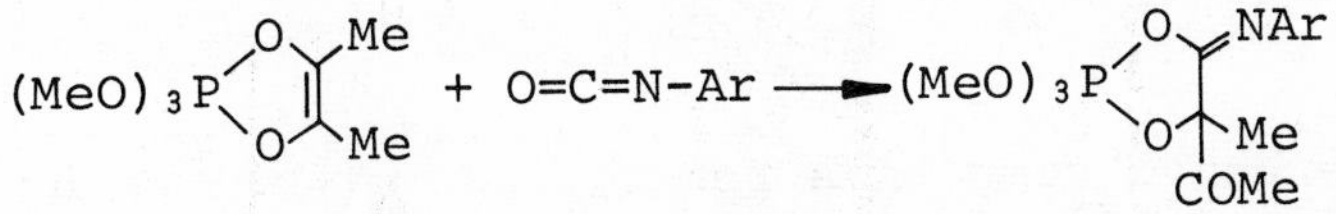

$(MeO)_3P$ (O, Me, O, Ph) + H–C(=O)–C(=O)Ph → $(MeO)_3P$ (O, H, COPh; O, Ph, COMe)

[498]

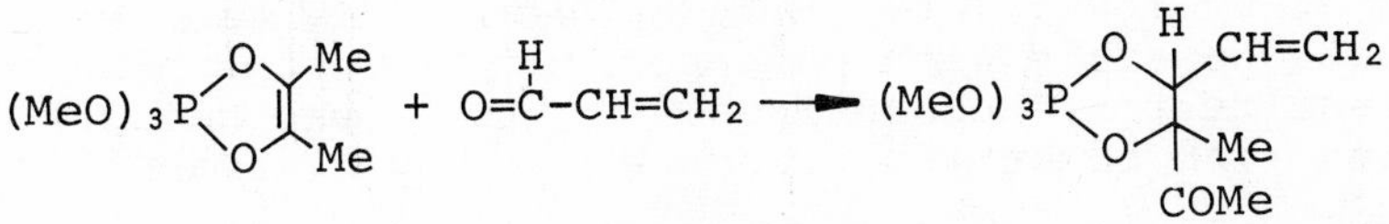

$(MeO)_3P$ (O, Me, O, Me) + O=C=N–Ar → $(MeO)_3P$ (O, =NAr; O, Me, COMe)

[510]

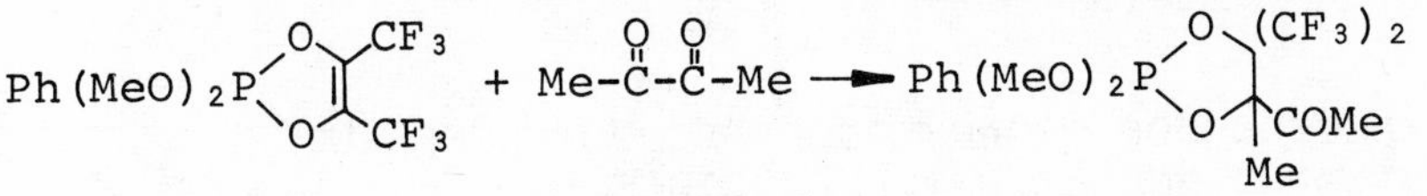

$(MeO)_3P$ (O, Me, O, Me) + O=C(H)–CH=CH_2 → $(MeO)_3P$ (O, H, CH=CH_2; O, Me, COMe)

[488]

$Ph(MeO)_2P$ (O, CF_3, O, CF_3) + Me–C(=O)–C(=O)–Me → $Ph(MeO)_2P$ (O, $(CF_3)_2$; O, COMe, Me)

In many of such condensations, mixtures of cis-trans or of meso-racemic forms are obtained. From these, one isomer can usually be separated in pure form by fractional crystallization.

X. REACTIONS OF TRIALKYL PHOSPHITES AND SIMILAR AMINO-SUBSTITUTED COMPOUNDS WITH MONOFUNCTIONAL CARBONYL COMPOUNDS

In these syntheses 2 equivalents of a carbonyl compound react with 1 equivalent of the phosphorus compound, yielding 2,2,2-trisubstituted 2,2-dihydro-1,3,2-dioxaphospholanes.[484,485] Aldehydes, ketones, α-ketoesters, and α-ketonitriles have been used in this way. Hexafluoroacetone is especially reactive, leading in general to very vigorous reactions.[508,560] Again, mixtures of meso and racemic products are obtained, from which one isomer can usually be crystallized in pure form.

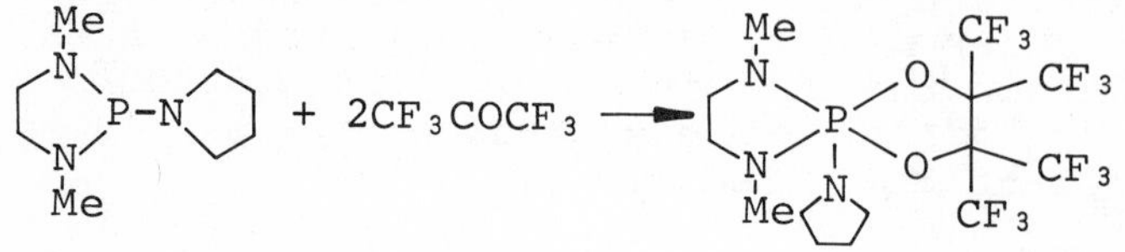

Me Me CF_3 [508]
N N O CF_3
P-N + $2CF_3COCF_3$ ⟶ P
N N O CF_3
Me Me N CF_3

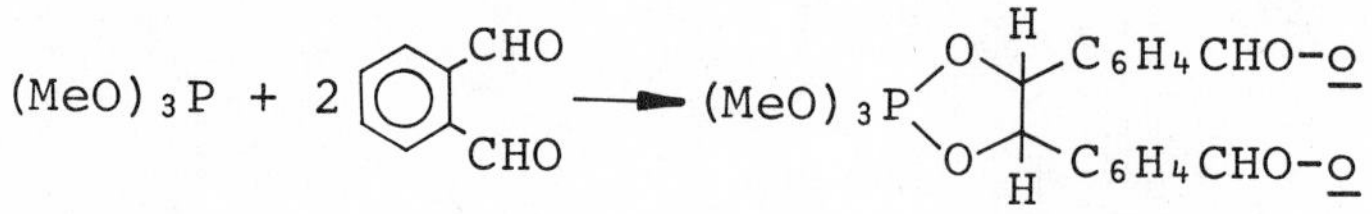

H
$(MeO)_3P$ + 2 [CHO / CHO] ⟶ $(MeO)_3P$ O C_6H_4CHO-o [492]
O C_6H_4CHO-o
H

Meso + racemic

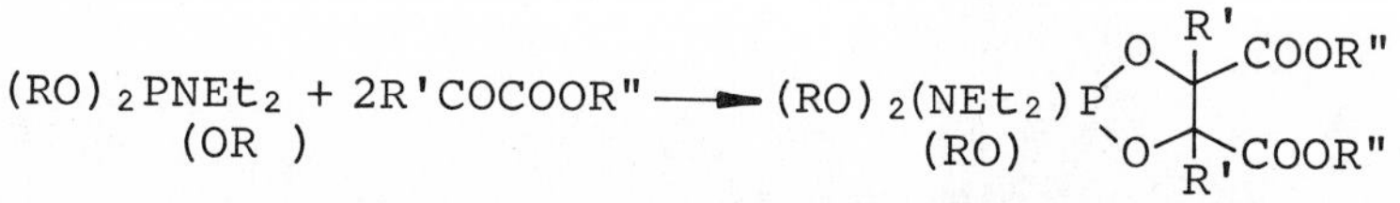

R' [433,475]
O COOR"
$(RO)_2PNEt_2$ + 2R'COCOOR" ⟶ $(RO)_2(NEt_2)P$
(OR) (RO) O COOR"
R'

Difunctional carbonyl compounds react first of all with only one function.[439,492,531] The reactivity order of the phosphites is $(i\text{-}PrO)_3P > (EtO)_3P > (MeO)_3P$.[533] These reactions are also possible with phosphonites,[474] phosphinites,[201,534] and tertiary phosphines,[71,537,621] the adducts of the last reaction sometimes being reversibly cleavable.[71]

A.2. Compounds of Type RPY_4

XI. REACTIONS OF ORGANYLDICHLOROPHOSPHINES WITH ANTIMONY AND ARSENIC FLUORIDES

Dichlorophosphines react easily with antimony pentafluoride[105,590,616] or with a mixture of SbF_5 and SbF_3,[590,616] to form organyltetrafluorophosphoranes. When only the

$$3RPCl_2 + 3SbF_5 \longrightarrow 5RPF_4 + SbF_3 + 2SbCl_3$$

trifluorides of antimony or arsenic are used, the tetrafluorophosphoranes are formed in a redox reaction.[144,343,483,574a,576,590,681] It is clear that a difluorophosphine can also undergo such a redox reaction.[585] Similar reac-

$$3RPX_2 + 4MF_3 \longrightarrow 3RPF_4 + 2M + 2MX_3$$

$$X = Cl, F; \quad M = Sb, As$$

tions occur, but more smoothly, with substituted chloro- and fluorophosphines.[143,145,371,586]

$$3RPCl(NR'_2) + 3MF_3 \longrightarrow 3RPF_3(NR'_2) + 2M + MCl_3$$

In cases where the dichlorophosphines contain strong electronegative groups, especially perfluoroalkyl groups, the redox reaction shown above does not occur.[576,590]

XII. REACTIONS OF TETRACHLOROPHOSPHORANES WITH FLUORINATING AGENTS

Tetrachlorophosphoranes and their addition complexes with aluminum chloride can be transformed to tetrafluorophosphoranes[590] by reaction with antimony trifluoride,[210,343,398,681,682] AsF_3, KF,[343] and HF,[99,343,345] respectively.

$$RPCl_4 \xrightarrow[\text{agent}]{\text{fluorinating}} RPF_4$$

Mixed fluorochlorophosphoranes were also subjected to this reaction.[170,443,444]

XIII. REACTIONS OF PHOSPHONYL AND THIOPHOSPHONYL COMPOUNDS WITH INORGANIC FLUORIDES

Phosphonyl and thiophosphonyl halides react with sulfur tetrafluoride[617,618] or antimony trifluoride[324] to form tetrafluorophosphoranes. Reaction with SF_4 is also possible in the case of phosphonic acids.[617,618]

$$RP(=O(S))(Y)_2 \xrightarrow{SF_4,\, SbF_3} RPF_4$$

$$Y = F,\ Cl,\ OH$$

XIV. DISPROPORTIONATION REACTIONS OF ORGANYLDIFLUOROPHOSPHINES

Several examples are known in which difluorophosphines disproportionate at room temperature or thermally to yield tetrafluorophosphoranes and cyclopolyphosphines.[7,171,369,596]

$$2RPF_2 \longrightarrow \frac{1}{n}(RP)_n + RPF_4$$

Analogous disproportionations have been observed with diorganylfluorophosphines[595] (see Chapter 2).

XV. REACTIONS OF PHOSPHORUS PENTAFLUORIDE WITH ORGANOMETALLIC COMPOUNDS

Although reactions of this kind appear very general, only a few examples are actually known. Thus tetraorganyltin derivatives can be used in reactions with phosphorus pentafluoride to form organyltetrafluorophosphoranes.[601,631]

$$R_4Sn + PF_5 \longrightarrow RPF_4 + R_3SnF$$

XVI. REACTIONS OF PHOSPHORUS PENTACHLORIDE WITH OLEFINS

Phosphorus pentachloride adds easily to many olefinic substances, in the sense $Cl\text{-}PCl_4$, to form chlorosubstituted aliphatic or olefinic compounds of type $RPCl_4$.[11,346] Acetylenes react similarly.[8,15,346] The reaction is usually conducted in an inert solvent like benzene. Simple monoolefins react according to the Markovnikov rule,[348,709] which is in disagreement with the statements of two patents.[677,678] In many cases the initial addition products are not obtained, the yield consisting only of the olefinic tetrachlorophosphoranes resulting from dehydrochlorination[14,692] of the initial adducts. When excess PCl_5 is used, complexes of the tetrachlorophosphoranes with 1 equivalent of PCl_5 are formed.[8,162,321] Simple diolefins also undergo these addition reactions with one double bond, forming halogenated alkenyltetrachlorophosphoranes.[9,10] The general course of these reactions is as follows:

$$R_2C{=}CH_2 + PCl_5 \longrightarrow R_2C(Cl)\text{-}CH_2\text{-}PCl_4 \xrightarrow{-HCl} R_2C{=}CHPCl_4$$

Enol ethers[14,16,561,602,638] or appropriate precursors thereof[563] react in a similar way to form primary adducts, which on warming are dehydrochlorinated to organyloxyalkenyltetrachlorophosphoranes or their PCl_5 adducts. In

$$R'OC(H){=}C(R)(H) + PCl_5 \longrightarrow R'OC(Cl)(H)\text{-}C(R)(H)\text{-}PCl_4 \xrightarrow{-HCl} R'OC(H){=}C(R)\text{-}PCl_4$$

exactly the same way phosphorus pentachloride adds to enol

esters.[390-394]

Most of the moisture-sensitive compounds obtained by the syntheses described in this section were not isolated in pure form or even adequately identified. Rather, they were characterized by unequivocal chemical reactions.

XVII. ADDITION OF ORGANIC HALIDES TO PHOSPHORUS TRICHLORIDE

Reactions of this type occur readily when the organic halides phosphorus trichloride or phosphorus tribromide[220] and aluminum chloride (bromide) are mixed without solvents or with a solvent like methylene chloride.[329] Thus alkyl chlorides form alkyltrichlorophosphonium tetrachloroaluminate complexes,[97,99,262,329] which can be cleaved to the tetrachlorophosphoranes by diethylphthalate.[458,557] In some cases complexes with the $Al_2Cl_7^-$ anion are formed.[570]

When alkyl bromides and iodides are used, mixed alkyltrihalophosphonium complexes are obtained.[220]

$$RX + PCl_3 + AlCl_3 \longrightarrow [RPCl_2X]^{\oplus}[AlCl_4]^{\ominus}$$

$$X = Cl, Br, I$$

An activated chloride like tricyanomethyl chloride reacts at 0° with PCl_3 without the aid of aluminum chloride.[630]

XVIII. ADDITION OF HALOGENS TO DIHALOPHOSPHINES

Halogens react exothermically with halophosphines in appropriate solvents such as carbon tetrachloride or other chlorohydrocarbons.[346] When alkyldichlorophosphines are treated with chlorine, after the initial formation of the tetrachlorophosphorane[41,635,688] successive chlorination of the alkyl group can occur.[481,687,689,690,706]

$$MePCl_2 \xrightarrow{Cl_2} MePCl_4 \xrightarrow{Cl_2} ClCH_2PCl_4 \xrightarrow{Cl_2} Cl_2CHPCl_4 \xrightarrow{Cl_2} Cl_3PCl_4$$

Fluoroalkyldichlorophosphines can be chlorinated without such complications;[105,208,210] the same is true of aryldichlorophosphines[117,471,682,683] and t-alkyldichlorophosphines.[43] Difluorophosphines have also been converted to mixed phosphoranes when caused to react with chlorine.[136,170,443,444] Reactions of dihalophosphines with ClF[209] and with bromine[86,136,402,443,667] were also successful. In an interesting reaction, nitrogen trichloride was used as chlorinating agent to form phenyltetrachlorophosphorane.[175]

In this section belong also the reactions of biphosphines and cyclopolyphosphines with chlorine,[399] bromine, and iodine,[168a,169a,298,300] which give tetrahalophosphoranes (see Chapter 2).

$$(RP)_4 + 8X_2 \longrightarrow 4RPX_4$$

XIX. REACTIONS OF PHOSPHONYL AND THIOPHOSPHONYL COMPOUNDS WITH CHLORINE OR PHOSPHORUS PENTACHLORIDE

Arylphosphonyl halides react with phosphorus pentachloride to give the PCl_5 adducts of the corresponding tetrachlorophosphoranes.[162,705] Alkylphosphonyl halides initially also form the tetrachlorophosphoranes, but then undergo further chlorination in the α-position.[163]

$$\underset{\displaystyle O}{\overset{}{RP\!\!\stackrel{\|}{}}}Cl_2 + 2PCl_5 \longrightarrow RPCl_4{\cdot}PCl_5 + POCl_3$$

Arylthiophosphonyl dichlorides are converted to the corresponding tetrachlorophosphoranes by chlorine in carbon tetrachloride.[381a]

XX. REACTIONS OF TETRAHALOPHOSPHORANES WITH AMINES AND METAL(OID) DERIVATIVES OF AMINES, ALCOHOLS, AND THIOLS

Tetrafluorophosphoranes react easily with primary[145,312,371,593] or secondary amines[137,145,371,555,589,593] to yield trifluorophosphoranes containing amino groups. Introduction of a second amino group is also possible.[137] As by-products, derivatives of hexavalent phosphorus are often obtained[589,593] (Section A.8.LII). When chlorotrifluorophosphoranes are used, only the chlorine atom is substituted by a dialkylamino group.[305,371,579] The general formulation is as follows:

$$RPF_3X + 2HNR'_2 \longrightarrow RPF_3NR'_2 + [R'_2NH_2]^{\oplus}X^{\ominus}$$

$$X = F, Cl$$

At slightly elevated temperatures silylated secondary amines are also suitable agents for transforming tetrafluorophosphoranes to the corresponding trifluoroaminophosphoranes[578,589] (see Section A.1.I). The introduction of organyloxy and organylthio groups is possible with silyl ethers, silyl thioethers,[146,454,456,456a] and sodium thiolates.[311]

$$RPF_4 + (R'Z)SiMe_3 \longrightarrow RPF_3(ZR') + Me_3SiF$$

$$Z = O,\ S,\ NR''$$

Disilylated primary amines produce under comparable conditions cyclic products with the 1,3,2,4-diazadiphosphetidine structure.[581,582,588] Such cyclic compounds are

$$2RPF_4 + (Me_3Si)_2NMe \longrightarrow RF_2P\langle(NMe)_2\rangle PF_2R + 4Me_3SiF$$

also obtained from tetrachlorophosphoranes and primary amines[275,695,702] in accordance with similar reactions of phosphorus pentachloride (Section A.1.III).

XXI. REACTIONS OF TRIVALENT PHOSPHORUS COMPOUNDS WITH 1,3-UNSATURATED SYSTEMS

Phosphites and similar aminosubstituted phosphorus compounds can react with a great variety of α,β-unsaturated carbonyl compounds. The general aspects of such addition reactions have been summarized.[544] Most of these reactions were carried out with compounds containing an acrylic skeleton. Thus acrolein[29,319] reacts exothermically with trialkyl phosphites to give distillable 1,2-oxaphospholenes.[29] Simple vinylic ketones[201,367,647] react at slightly elevated temperatures according to the same principle. The most thoroughly studied reactions of this type, however, are those with olefins bearing two geminal acyl groups on the double bond.[20,22,25,201,514] Ethylideneacetoacetic esters give similar addition products.[19,20,22] Corresponding reactions with acrylic acids have been reported,[317,320] but no stable adducts were obtained. In this case probably only open betainic structures are formed.[318,357-359,365]

All these reactions can be shown in general form by the following formula:

$$(RO)_3P + O{=}C(R'')\text{–}C(R''')\text{=}CR'_2 \longrightarrow (RO)_3P\langle O\text{–}C(R'')\text{=}C(R''')\text{–}CR'_2\rangle$$

$R'' = H, HO,$ alkyl, aryl; $R''' = H,$ alkyl, aryl, $-\overset{O}{\overset{\|}{C}}R$, $-\overset{O}{\overset{\|}{C}}-OR$

XXII. REACTIONS OF TRIALKYL PHOSPHITES WITH ALDEHYDES

In contrast to the reactions of certain monocarbonyl compounds with trialkyl phosphites (see Section A.1.X) starting with $\geqslant P \rightarrow O{=}C\lessdot$ attack, several aldehydes are able to react under primary $\geqslant P \rightarrow \overset{|}{\underset{|}{C}}{=}O$ condensation, yielding not 1,3,2- but 1,4,2-dioxaphospholanes.[361,521,530] From some aldehydes that are substituted with electron-withdrawing groups one obtains at low temperatures initial carbon attack, leading to 1,4,2-dioxaphospholanes, and at higher temperatures[488,530] oxygen attack, leading to 1,3,2-dioxaphospholanes (Section A.1.X). Naturally, cis-trans mixtures are always formed.

$$(RO)_3P\langle O-CH(R')-CH(R')-O\rangle \xleftarrow{80°} (RO)_3P + 2R'\text{-CHO} \rightleftharpoons (RO)_3P\langle O-CH(R')-O-CH(R')\rangle$$

Corresponding adducts are obtained with phosphonites.[488]

XXIIIA. INSERTION OF TRIALKYL PHOSPHITES INTO C-F BONDS

A general-appearing but until now rarely used reaction is the insertion of trialkyl phosphites into a C-F bond of perfluoroolefins.[334]

$$(CF_3)_2C{=}CF_2 + (EtO)_3P \longrightarrow (CF_3)_2C{=}CF\text{-}PF(OEt)_3$$

XXIIIB. INSERTION OF CARBONYL GROUPS INTO P-H BONDS

Spirophosphoranes with P-H bonds of the type described in Section A.7.LI react with aldehydes with insertion of the carbonyl group into the P-H bond to form tetraoxyphosphoranes with one P-C bond.[91]

A similar insertion into the P-H bond is possible with certain fluorohydridophosphoranes at slightly elevated temperatures.[305a]

$$\text{(bicyclic spirophosphorane with P–H)} + RCH{=}O \rightarrow \text{(spirophosphorane with P–CH(OH)R)}$$

$$ArPF_3H + RCHO \longrightarrow ArPF_3(CHOHR)$$

A.3. Compounds of Type R_2PY_3

XXIV. REACTIONS OF CHLOROPHOSPHINES OR TRICHLOROPHOSPHORANES WITH FLUORINATING AGENTS

The reactions of diorganylmonochlorophosphines with antimony trifluoride[343,576] or antimony pentafluoride[105] need no further comment because they are fully comparable to the analogous reactions given in Section A.2.XI.

$$3R_2PCl + 3SbF_3 \longrightarrow 3R_2PF_3 + SbCl_3 + 2Sb$$

The same is true for the halogen-exchange reactions of diorganyltrihalophosphoranes (or their $AlCl_3$ adducts) with arsenic (antimony) trifluoride,[170,398,444] hydrogen fluoride,[343] or sulfur tetrafluoride,[398] forming trifluorophosphoranes (Section A.2.XII). Syntheses of trifluoro-

$$R_2PCl_3(\cdot AlCl_3) \xrightarrow{\text{fluorinating agent}} R_2PF_3$$

phosphoranes from diorganylhydridophosphines and SF_4 have also been reported.[205,583,590]

XXV. REACTIONS OF PHOSPHINYL AND THIOPHOSPHINYL COMPOUNDS WITH INORGANIC FLUORIDES

Like the corresponding phosphonyl and thiophosphonyl compounds (Section A.2.XIII), the phosphinyl species undergo fluorinations to trifluorophosphoranes when caused to react with antimony trifluoride[577,587] or with sulfur tetrafluoride.[172,337,618]

$$R_2P(=O(S))Y \xrightarrow{SbF_3;\ SF_4} R_2PF_3$$

$$Y = -OH,\ -P(=S)R_2$$

XXVI. REACTIONS OF ORGANYLTETRACHLOROPHOSPHORANES WITH UNSATURATED SYSTEMS

Alkenylorganyltrichlorophosphoranes are obtained when organyltetrachlorophosphoranes or phosphorus pentachloride is combined with olefins[165,167,167a] at higher temperatures. Enol ethers behave analogously.[12,164] The products of these reactions are very hygroscopic and were characterized, for the most part, by subsequent chemical reactions.

$$RPCl_4 + H_2C{=}C\langle^{R'(OR')}_{H} \longrightarrow R\left(\,^{R'}_{H}\rangle C{=}CH{-}\right)PCl_3 + HCl$$

When acetylenes are used, addition occurs and β-chloroalkenylorganyltrichlorophosphoranes are the final products.[166-167a]

$$RPCl_4 + Ph{-}C{\equiv}CH \longrightarrow R(PhClC{=}CH{-})PCl_3$$

XXVII. ADDITION OF ALKYL HALIDES TO ORGANYLDICHLOROPHOSPHINES

Like phosphorus trichloride (Section A.2.XVII), dichlorophosphines add exothermically to alkyl halides in the presence of aluminum chloride as catalyst.[220,306,308,570] Again, very moisture-sensitive complexes with $AlCl_3$ are obtained.

$$RPCl_2 + R'Cl + AlCl_3 \longrightarrow [RR'\overset{\oplus}{P}Cl_2][Al\overset{\ominus}{C}l_4]$$

A reaction sequence probably involving an addition of methyl bromide to methyldibromophosphine has also been reported.[401]

XXVIII. REACTIONS OF HALOGENS WITH DERIVATIVES OF TRIVALENT PHOSPHORUS

The most widely used method of preparation for trihalophosphoranes is the addition of halogens to phosphorus derivatives of lower valence.[346] Thus dialkyl-,[372] diperfluoroalkyl-,[154,157,211,444] alkylperfluoroalkyl-,[212] aralkyl-,[684] and diaryl-[117,224] chlorophosphines or fluorophosphines[444] are easily transformed to trichlorophosphoranes when caused to react with chlorine under mild conditions in such solvents as carbon tetrachloride, hexane, or benzene. Diorganylhydridophosphines can also be chlorinated to trichlorophosphoranes.[84]

Similar chlorinations are possible when aminosubstituted[5,626] or phosphinosubstituted[6] diorganylphosphines are used. For example:

$$R_2P\text{-}PR_2 \xrightarrow{Cl_2} R_2Cl_2P\text{-}PR_2 \xrightarrow{2Cl_2} 2R_2PCl_3$$

Corresponding syntheses of tribromophosphoranes by bromination of diorganylbromophosphines and similar systems are also known.[86,300,354] Syntheses of triiodophosphoranes according to these schemes, however, are rarely reported,[300] one interesting example being the iodination of tris(dicyclohexylphosphino)vanadium to dicyclohexyltriiodophosphorane in tetrahydrofuran.[302]

XXIX. REACTIONS OF PHOSPHINYL AND THIOPHOSPHINYL COMPOUNDS WITH VARIOUS HALOGENATING AGENTS

When phosphinic acids[277,608] or phosphinic chlorides[608,684] are treated with excess phosphorus pentachloride, adducts of the corresponding trichlorophosphoranes with PCl_5 are formed at elevated temperatures. A similar reaction with diphenylphosphinic amide is reported to give aminodiphenyldichlorophosphorane, obviously possessing an ionic structure.[47,48]

In the case of thiophosphinic chlorides and dithiophosphinic acids, treatment with chlorine in benzene or carbon tetrachloride is sufficient for the transformation to trichlorophosphoranes.[107,260,278,372] A patent claims that phosphinic and thiophosphinic compounds are transformed to trihalophosphoranes by use of carbon tetrachloride (bromide) as halogenating agent in the presence of complex-forming metal halides.[32] Dithiotetraorganylbiphosphines have also been used in similar preparations of trichloro- or tribromophosphoranes.[41,353,354,559]

All the reactions of this section can be expressed by the following scheme:

$$RR'P(=O(S))Y \xrightarrow{Cl_2;\ PCl_5;\ CCl_4} RR'PCl_3$$

$$Y = Cl,\ OH,\ SH,\ -\overset{S}{\overset{\|}{P}}R_2;\quad (OR,\ SR,\ -O\overset{S}{\overset{\|}{P}}R_2,\ -S\overset{S}{\overset{\|}{P}}R_2)^{32}$$

XXX. CLEAVAGE OF ONE ORGANYL GROUP FROM TRIORGANYL-PHOSPHORUS COMPOUNDS

Two examples are known in which an organyl group can be split off from a triorganylphosphine[351] or a triorganylphosphine oxide,[178] respectively, by reaction with chlorine or phosphorus pentachloride.

$$(ClCH_2)_3P \xrightarrow{Cl_2} (Cl_3C)_2PCl_3 \xleftarrow{PCl_5} (ClCH_2)_3P{=}O$$

A similar cleavage occurs when tetramethylbiphosphine disulfide is treated with chlorine under UV irradiation.[559]

$$Me_2(S)PP(S)Me_2 \xrightarrow{Cl_2,\ UV} MePCl_4$$

XXXI. SUBSTITUTION OF HALOGENS IN TRIHALOPHOSPHORANES

Fluorine atoms in trifluorophosphoranes can be replaced by organyloxy or organylthio groups on reaction with silyl ethers[131a] or silyl thioethers.[454,456a] Similar substitutions occur when the trifluorophosphoranes are treated with amines[312] or silylamines.[590] Trichlorophosphoranes also undergo such a substitution when treated with amines[48,352] or vinyl ethers.[633] All these methods are counterparts of syntheses already described in Section A.1.I and A.2.XX, and no additional examples are required.

XXXII. 1,4-ADDITION OF DERIVATIVES OF TRIVALENT PHOSPHORUS TO DIENES

1,1,1-Trisubstituted phospholene compounds are accessible by reactions of suitable derivatives of trivalent phosphorus with 1,3-dienes.[478] Thus phosphorus trihalides[27,28,236] and chloro-,[545] bromo-, and fluorophosphites[159,543] at room temperature easily undergo 1,4-addition with 1,3-dienes. The same reaction with phosphites, however, needs higher temperatures (about 100°).[542] The following general formulation is valid:

$$(RO)_nPX_{3-n} + \text{1,3-diene} \longrightarrow (RO)_nX_{3-n}P\text{(phospholene ring)}$$

X = F, Cl, Br; n = 0, 3, 2

The products are often very sensitive toward moisture

and therefore are characterized only by means of chemical reactions.

XXXIII. ADDITION OF DERIVATIVES OF PHOSPHONOUS ACIDS TO 2,3-UNSATURATED CARBONYL COMPOUNDS

Phosphonites in solvents like hexane or benzene add readily at room temperature to α,β-unsaturated carbonyl systems.[201,359,528] As a rule, a mixture of stereoisomers is formed. Corresponding results are obtained with phosphinites.[528]

$$RP(OR')_2 + CH_2{=}C(R'')CH{=}O \longrightarrow R(R'O)_2P\text{(cyclic: }P{-}O{-}CH{=}C(R''){-}CH_2{-}P)$$

$$R'' = -\overset{O}{\overset{\|}{C}}Me,\ -\overset{O}{\overset{\|}{C}}-OMe$$

Dichlorophosphines have also been used in similar reactions with acrylic acid derivatives, but no stable primary products have been isolated or characterized.[327]

A.4. Compounds of Type R_3PY_2

XXXIV. HALOGENATION OF TERTIARY PHOSPHINES

The most direct synthetic method for the preparation of triorganyldihalophosphoranes is the addition of halogens to triorganylphosphines.[346] These reactions are usually conducted in such solvents as benzene, carbon tetrachloride, nitrobenzene, or acetonitrile at room temperature or at lower temperatures. Thus chlorinations of triarylphosphines[117,155,282,663] and trialkylphosphines[57,295,663] occur without difficulty (sometimes with concomitant chlorination of the alkyl groups[687]), as well as brominations[4,56,66,86,282,295,299,663] and iodinations.[56,66,169,202,284,295,299,301]

$$R_3P + X_2 \longrightarrow R_3PX_2$$

For the preparation of triorganyldifluorophosphoranes direct fluorination is not used. Reactions with fluorinating agents,[590] such as sulfur tetrafluoride,[172,335,398] fluoroamines,[35] tetrafluorohydrazine,[172,183] and even special fluorocarbon compounds,[412,413,415] are found to be more appropriate.

Such indirect halogenation reactions[621a] were also employed successfully in chlorinations with phosphorus pentachloride,[562] phosphorus trichloride,[179,619] dichloro- and chlorophosphines,[179,619] and similar arsenic[313] and selenium compounds.[459] Acid chlorides like phosgen[17] and the carbon tetrahalides[482,503] are also useful for the transformation of tertiary phosphines to the corresponding phosphoranes.

Compounds containing an activated halogen, such as α-bromoketones,[96,262a,287] β-dibromoesters,[642] and α,α-dibromodisulfones,[263] form dibromotriphenylphosphorane when treated with triphenylphosphine, but in these cases a pure product is seldom obtained. The same is true for the reaction of triphenylphosphine with meso-stilbene dibromide.[77] Reactions of this type have been reviewed.[414]

XXXV. ADDITION OF ALKYL HALIDES TO DIORGANYLHALOPHOSPHINES[346]

Alkyl chlorides,[194,264,640] alkyl bromides,[194] and alkyl iodides[264,297] add to diorganylhalophosphines to form dihalophosphoranes, which are often characterized only by subsequent chemical reactions. Sometimes the reactions of alkyl chlorides are conducted in the presence of aluminum chloride,[308] to form the corresponding complexes containing the $AlCl_4^-$ anion. Similar moisture-sensitive and insufficiently characterized adducts are formed with α-chloroethers,[634,636,637] α-chlorothioethers[641] α-, β-, or γ-chloroesters, and similar compounds.[195] The general formulation is as follows:

$$R_2PX + X\text{-}\underset{R'}{\overset{H}{C}}\text{-}Y \longrightarrow R_2\underset{X}{\overset{X}{P}}\text{-}\underset{R'}{\overset{H}{C}}Y$$

X = Cl, Br, I; Y = alkyl, aryl, -OR", -SR", $-(CH_2)_{0,1,2}$-COOR

One case of a corresponding addition of benzoyl fluoride is known.[79a]

Very interesting phosphorane-like substances are obtained on reaction of chlorophosphines with carbon tetrachloride, titanium tetrachloride, sulfur dichloride, and disulfur dichloride.[194] No details of the structures, however, are known.

$$Me_2PCl + TiCl_4 \longrightarrow Ti(PMe_2Cl_2)_4$$

XXXVI. REACTIONS OF PHOSPHINE OXIDES, PHOSPHINE SULFIDES, AND PHOSPHINE IMINES WITH HALOGENATING AGENTS

Triorganyldifluorophosphoranes can be obtained by reaction of the corresponding phosphine oxides and phosphine sulfides with conventional fluorinating agents such as antimony (arsenic) trifluoride[134,170,577,591] and sulfur tetrafluoride[618] at elevated temperatures. But an unusual fluorinating substance such as perfluoroisobutene[182] has also been found appropriate in a corresponding reaction with triphenylphosphine imine (see Chapter 5A).

$$2R_3PO + SF_4 \longrightarrow 2R_3PF_2 + SO_2$$

$$3R_3PS + 2MF_3 \longrightarrow 3R_3PF_2 + M_2S_3 \text{ (As, Sb)}$$

$$Ph_3PNPh + (CF_3)_2C{=}CF_2 \longrightarrow Ph_3PF_2 + (CF_3)C{=}C{=}NPh$$

Similar preparations of dichloro- (and dibromo-) phosphoranes are possible using phosgen,[18,680] thionyl chloride,[202,235,416,615] phosphorus pentachloride[192,346,416] (and pentabromide[202]), arsenic and antimony pentahalides,[416] chlorosulfonic acid,[416] and chlorine.[155]

XXXVII. ADDITION OF DIHALOPHOSPHINES AND PHOSPHONITES TO DIENES[478]

Organyldichloro- and dibromophosphines add to dienes with formation of unstable, moisture-sensitive 1,1-dihalophospholene derivatives,[24,34,407,410,479] which usually are not isolated but are used directly in hydrolysis or reduction reactions. In general, solvents are not necessary for these reactions. Analogous additions occur with cyclic phosphonites,[541,661] the reactions being somewhat slower than in the case of the dihalophosphines.

$$RPY_2 + \text{(diene)} \longrightarrow RY_2P\text{(phospholene ring)}$$

$$Y = Cl, Br, OR$$

XXXVIII. REACTIONS OF PHOSPHORUS PENTACHLORIDE, OR TETRACHLOROPHOSPHORANES WITH DIAZOMETHANE

Diazomethane reacts with phosphorus pentachloride[688] or with tetrachlorophosphoranes[192,683] to form intermediate dichlorophosphoranes, which were not isolated but were hydrolyzed to the corresponding phosphine oxides (see Chapter 6).

$$RPCl_4 + 2CH_2N_2 \longrightarrow R(ClCH_2)_2PCl_2 + 2N_2$$

A.5. Compounds of Type R_4PY

XXXIX. CLEAVAGE OF PENTAARYLPHOSPHORANES WITH PHENOL

One example is known in which a pentaphenylphosphorane undergoes cleavage to a stable tetraphenylorganyloxyphosphorane on reaction with phenol in pyridine at room temperature.[548]

$$Ph_5P + PhOH \longrightarrow Ph_4P\text{-}OPh + PhH$$

XL. RING CLOSURE OF PHOSPHONIUM SALTS CONTAINING A γ- OR δ-HYDROXYLATED CHAIN

Phosphonium salts with a carbon chain or a mixed carbon-heteroatom chain[184,406] that contains an actual[227,228,593a] or a potential[110] hydroxyl group in the γ- or δ-position undergo ring closure when treated with bases. Two examples illustrate this type of reaction. Some of these reactions can be reversed with acids.

$$Ph_3\overset{\oplus}{P}\text{-}CH_2\text{-}C(Ph)\text{=}N\text{-}OH \xrightarrow{\text{base}} Ph_3P\text{(}CH_2\text{-}C(Ph)\text{=}N\text{-}O\text{)}$$

$$Bu_3\overset{\oplus}{P}\text{-}CH_2CH_2CH_2C(\text{=}O)CH_3 \xrightarrow{\text{base}} Bu_3P\text{(}CH_2CH_2CH\text{=}C(CH_3)\text{-}O\text{)} + Bu_3P\text{(}CH_2CH_2CH_2C(\text{=}CH_2)\text{-}O\text{)}$$

XLI. ADDITION OF KETONES TO PHOSPHINE ALKYLENES

One case is reported of the addition of a phosphine alkylene to hexafluoroacetone, yielding a 1,2-oxaphosphetane ring system.[71,72]

$$Ph_3P\text{=}C\text{=}PPh_3 + (CF_3)_2CO \longrightarrow Ph_3P\text{(}O\text{-}C(CF_3)_2\text{-}C(\text{=}PPh_3)\text{)}$$

XLII. ADDITION OF 1,3-DIPOLES TO PHOSPHINE ALKYLENES

Although reactions of this kind became known only recently, their general applicability is obvious. Thus phosphine alkylenes have been combined with nitrones,[290] nitrile oxides,[64,291,644] and epoxides,[63,679] the lasts being considered as latent or potential 1,3-dipoles. The general formulation[290] gives some idea of the versatility of this synthesis.

$$R_3P{=}CR'_2 + \overset{\oplus}{A}\text{=}B{-}\overset{\ominus}{D} \longrightarrow R_3P\langle D{-}B{=}A{-}CR'_2\rangle$$

A.6. Compounds of Type R_5P

Pentaorganylphosphoranes have been reported to be formed in various addition reactions of tertiary phosphines to olefinic[556] and acetylenic[254,255,289,315,556] substances according to the following general scheme:

$$R_3P + (\text{diyne}) \longrightarrow R_3P(\text{cyclic diene})$$

Exact structural studies in recent times have shown, however, that in nearly all of these syntheses <u>no</u> pentacoordinated species are the final products,[204a,254a,603,604,648,649] phosphoranes being only intermediates.

XLIII. REACTIONS OF TETRAARYLPHOSPHONIUM SALTS WITH METAL ORGANYLS

The first synthesis of a compound of this class was achieved in 1948, just at the time when Kosolapoff's Organophosphorus Compounds was written. It involved the reaction of tetraphenylphosphonium iodide with phenyllithium.[675,676] The successful extension of this method was reported soon afterward.[551,670]

$$[Ph_4\overset{\oplus}{P}]\overset{\ominus}{X} + RLi \xrightarrow[-LiX]{} Ph_4PR \qquad R = \text{aryl, trityl}$$

Corresponding reactions of tetraphenyl- or triphenylmethylphosphonium salts with alkyllithium[599,600] and alkenyllithium[598] compounds are reported to give pentacovalent phosphorus species only as unstable intermediates. In the reaction of tetraphenylphosphonium chloride with several alkinylpotassium derivatives, unstable phosphoranes, which decompose at temperatures higher than -20°, were isolated.[440] Very recently, stable alkylarylphosphoranes[325a] and even an unexpectedly stable pentaalkylphosphorane!,[642a] all containing the phosphahomocubane system, could be synthesized by similar procedures.

Stable spirocyclic tetraarylalkyl- and pentaarylphosphoranes can be obtained when the stabilizing bis-2,2'-biphenylylene phosphorus skeleton is introduced.[238a] Thus bis-2,2'-biphenylylenephosphonium salts react easily with magnesium or lithium organic compounds to give the desired phosphoranes.[237,241,245,247,250] Even the reaction with lithium aluminum hydride[240] or sodium borohydride[246,668] leads to a stable compound, bis-2,2'-biphenylylene-phosphorane, which represents the parent compound of this series of spirocyclic phosphoranes.

R = H, Me; R' = H, alkyl, aryl

All these reactions were carried out in solvents like diethyl ether or tetrahydrofuran under a nitrogen atmosphere at room temperature or at slightly elevated temperatures.

XLIV. REACTIONS OF PHOSPHINE IMINES AND RELATED COMPOUNDS WITH LITHIUM ORGANYLS

Simple cyclic and spirocyclic pentaarylphosphoranes are also accessible by the reaction of methylated phosphine imines with lithium organyls in ethereal solvents.[672,673] This reaction can be simplified by using phosphine tosylimines.[239,671,674]

$$[Ar_3P{=}NMePh]^{\oplus}Br^{\ominus} \xrightarrow{2Ar'Li} Ar_3PAr'_2 \xleftarrow{2Ar'Li} Tos{-}N{=}PAr_3$$

A direct synthesis of pentaarylphosphoranes is possible when the tosylimine method is slightly varied and phosphite tosylimines are used as starting materials.[571,572]

$$(RO)_3P{=}NTos + 5ArLi \longrightarrow Ar_5P$$

XLV. EXCHANGE REACTIONS WITH SPIROCYCLIC PHOSPHORANES

Spirocyclic phosphoranes containing the bis-2,2'-biphenylylene phosphorus skeleton undergo exchange reactions of the singly bonded ligand when treated with lithium organyls[246,572] in ethereal solvents.

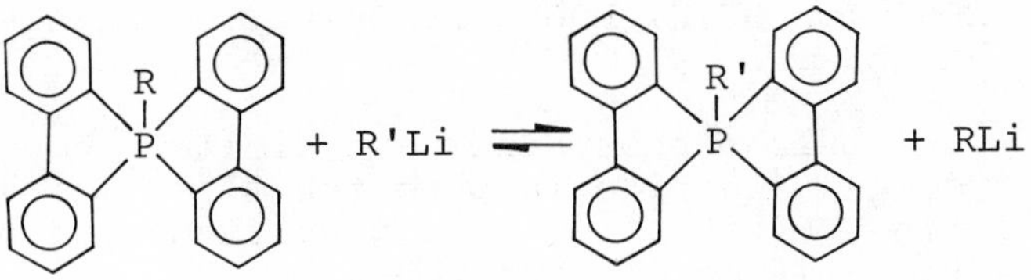

R = H, aryl, alkyl; R' = alkyl, aryl

XLVI. CLEAVAGE OF SPIROCYCLIC HEXAARYL-ATE COMPLEXES OF PHOSPHORUS

Tris-2,2'-biphenylylene phosphates react readily with hydrochloric acid[237,241-243] in ethereal or alcoholic solvents or with bromine in methylene chloride,[252,668] yielding spirocyclic phosphoranes.

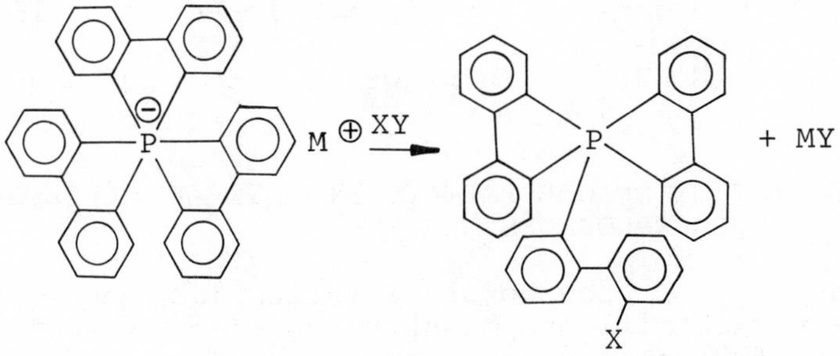

X = H, Br; Y = Cl, Br

A.7. Compounds of Type HR_nPY_{4-n}

XLVII. REDUCTION OF TETRAFLUOROPHOSPHORANES

Alkyltetrafluorophosphoranes are reducible by trialkylstannanes to alkyltrifluorohydridophosphoranes,[197] which are stable at lower temperatures.

$$RPF_4 + R'_3SnH \longrightarrow RPF_3H + R'_3SnF$$

XLVIII. ADDITION OF HYDROGEN FLUORIDE TO DIHALOPHOSPHINES

Dichloro- or difluorophosphines and similar compounds readily add hydrogen fluoride in pure form[145,594] or in latent form as in KHF_2[136,304,307] to give the corresponding organylhydridotrifluorophosphoranes.

$$RPCl_2 + KHF_2 \longrightarrow 2KCl + HF + RPF_3H$$

XLIX. ADDITION OF ALCOHOLS AND AMINES TO DIFLUOROPHOSPHINES

Difluorophosphines can be transformed to alkylhydridoalkoxydifluorophosphoranes by treatment with the corresponding alcohols at low temperatures[136,138,142] in the presence of catalytic amounts of amines. Analogous addition reactions are possible with primary and secondary amines.[139-143] The general formulation is as follows:

$$RPF_2 + ROH; \quad (R'NH_2) \longrightarrow RHPF_2(OR'); \quad RHPF_2(NHR')$$

L. SUBSTITUTION OF ONE HALOGEN ATOM IN ALKYLHYDRIDOTRIFLUOROPHOSPHORANES

As in the case of phosphorus pentafluoride (see Section A.1.I) alkylhydridotrifluorophosphoranes react with sodium alcoholates to form alkoxylated compounds.[136]

$$RHPF_3 + NaOR' \longrightarrow RHPF_2(OR') + NaF$$

LI. REACTIONS OF PHOSPHITES AND ANALOGOUS AMINOSUBSUBSTITUTED PHOSPHORUS COMPOUNDS WITH GLYCOLS AND AMINO ALCOHOLS

When, in structures of general type HR_nPY_{4-n}, n is taken as zero and Y = OR, compounds of type $HP(OR)_4$ are obtained; their syntheses are described in this section. β-Glycols and β-amino alcohols are able to react with phosphites,[184a,206,441] chlorophosphites,[131,328,441,566] phosphorus trichloride,[441,558] and phosphorous acid[92] to yield spirocyclic phosphoranes containing one P-H bond. Analogous products are formed when tris(dimethylamino)phosphine is treated with β-glycols[87,568] or β-amino alcohols.[87,90,558,567,568] Naturally, exactly the same is true for mixed aminooxyphosphines.[567,568] The general aspects of these reactions are shown in the formulas; as a rule mixtures of diastereomers are formed.

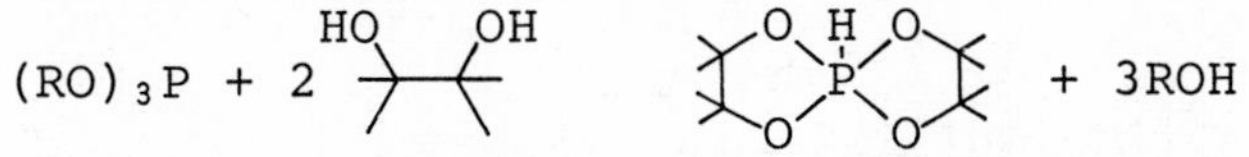

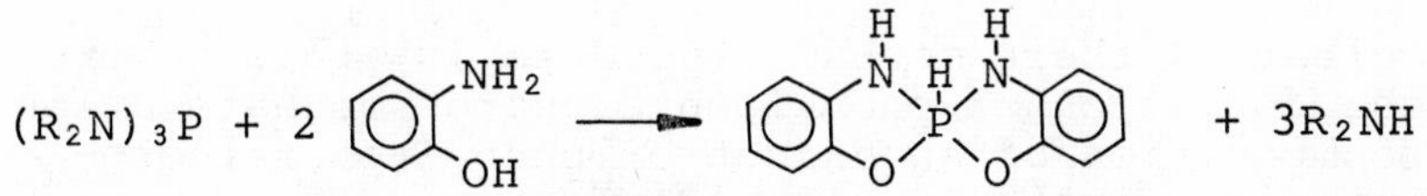

$(RO)_3P$ + 2 HO OH → O H O P O O + 3ROH

$(R_2N)_3P$ + 2 NH_2 OH ⟶ H H N H N P O O + $3R_2NH$

When such reactions were made for the first time, open structures such as the following:

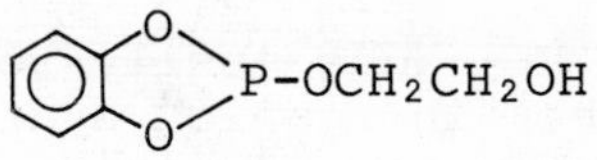

O O P-OCH_2CH_2OH

were formulated,[87,131,328,441] but later investigations have shown that actually spirocyclic species are present.[568] Nevertheless, equilibria between spirocyclic and open structrues are possible when the ring nitrogens are substituted.[566,568]

A.8. Compounds with Hexavalent Phosphorus

It is clearly foreseeable that compounds of this kind will show in the future a diversity similar to that characterizing the pentaorganylphosphorus compounds. However, because not so many organic compounds containing hexavalent phosphorus are known as yet, their treatment in a single section is still possible.

LII. ADDITION OF FLUORIDE IONS TO FLUOROPHOSPHORANES

Several examples are known in which direct[94a,144,147,593] or indirect addition (following disproportionation reactions[94,455,555,579,586,589,593]) of fluoride ions to fluorophosphoranes occurs. A typical reaction of the second type is as follows:

$$RPF_4 + R'_2NH \longrightarrow RPF_3(NR'_2) \longrightarrow [RPF_5]^{\ominus}[RPF(NR'_2)_2]^{\oplus}$$

Similar transfers of fluoride ions must be envisaged in reactions of phosphorus pentafluoride with tin organic compounds[314] and in reactions of tetrafluorophosphoranes with several aminosilanes.[592] All these reactions are not

$$PF_5 + Me_3SnCF_3 \longrightarrow Me_3SnCF_3 \cdot PF_5 \longrightarrow Me_3SnF + CF_3PF_4 \longrightarrow$$

$$[Me_3Sn]^{\oplus}[F_3CPF_5]^{\ominus}$$

entirely clear in their mechanistic details; this is true also of the formation of phenylpentafluorophosphate anion in the decomposition of an adduct of phenyltetrafluorophosphorane with dimethyl sulfoxide.[424]

LIII. ADDITION OF HF_2^- TO PHOSPHINES

At room temperature HF_2^- adds easily to phosphorus trifluoride,[448] yielding the "inorganic" hydridopentafluorophosphate anion, and to fluorophosphines, forming organic hexacoordinated phosphorus species.[447]

$$(CF_3)_nPF_{3-n} + HF_2^{\ominus} \longrightarrow (CF_3)_nPHF_{5-n}^{\ominus}$$

LIV. REACTIONS OF PHOSPHORUS PENTAHALIDES WITH DIFUNCTIONAL BASES

Phosphorus pentachloride (and pentafluoride) reacts with certain difunctional bases such as acetylacetone,[83] tropolone,[428] dimethylurea,[50,378] and phenanthroline[44,654] with substitution of some or of all halogen atoms. In all cases inner onium-ate complexes are formed. The following is a typical reaction:

$$2PCl_5 + MeHNC(O)NHMe \longrightarrow Cl_4P^{\ominus}(NMe)_2CCl^{\oplus} + POCl_3 + 2HCl$$

LV. REACTIONS OF PHOSPHONITRILIC COMPOUNDS WITH CATECHOL AND ORTHO-AMINOPHENOL

Interesting and unexpected reactions occur when a variety of phosphonitrilic compounds are treated with catachol[1] and o-aminophenol.[2] It is now totally clear that the catechol derivative is hexacoordinated and not pentacoordinated.[1a] Recent investigations have shown that at least the triscatecholate anion can exist also in solution (see Section A.8.LVI).

$$(NPCl_2)_3 + C_6H_4(OH)_2 \xrightarrow{Et_3N} [(C_6H_4O_2)_3P^{\ominus}]\ Et_3NH^{\oplus}$$

? ⇌

$$(C_6H_4O_2)_2P{-}O{-}C_6H_4{-}O\cdots H\cdots NEt_3$$

LVI. REACTION OF PHENOLATES WITH PHOSPHORUS PENTA-CHLORIDE AND SIMILAR PHOSPHORUS COMPOUNDS

When phosphorus pentachloride [or chlorobis(o-phenylene-dioxy)phosphorane] is caused to react with dilithio- or disodiocatecholate, the tris(o-phenylene-dioxy)phosphate anion is formed. It can be isolated in the form of onium-ate complexes.[251] Similar ate complexes are obtained when bis-2,2'-biphenylylenephosphonium iodide is treated with dilithiocatecholate or dilithio-1,8-naphthalinedio-late.[251] All these reactions were conducted in tetrahydro-furan with exclusion of oxygen. In the case of the tris-catecholate complex, the preparation can also be carried out in alcoholic or even alcoholic aqueous solvents!

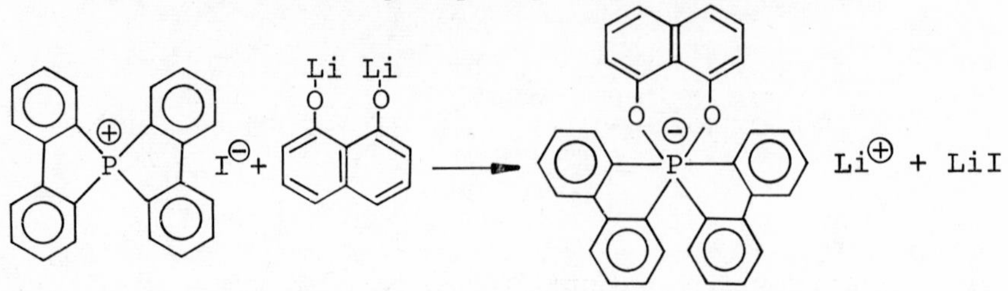

LVII. REACTIONS WITH ARYLLITHIUM COMPOUNDS LEADING TO SPIROCYCLIC HEXAARYL PHOSPHATES

Bidentade lithium organyls like 2,2'-dilithiobiphenyl react easily with phosphorus pentachloride[237] or even better, with spirocyclic phosphonium salts[241-244,251] to give tris-2,2'-biphenylylene phosphate complexes. These reactions were conducted in ethereal solvents under exclusion of oxygen.

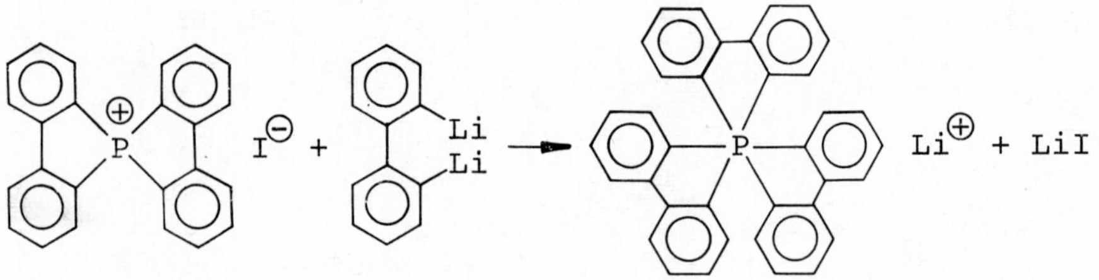

Complexes so formed are stable against water but can readily be cleaved to pentaarylphosphoranes (Section A.6. XLVI) by the action of aqueous mineral acids.

LVIII. ADDUCTS OF PENTAVALENT PHOSPHORUS HALIDES WITH MONOFUNCTIONAL BASES

Generally, phosphorus pentahalides can react with a great variety of organic bases to form adducts that can be considered as inner onium-ate complexes and thus as derivatives of hexavalent phosphorus. But because for most of

$$PX_5 + :B \longrightarrow X_5\overset{\ominus}{P}-\overset{\oplus}{B}$$

these adducts sufficient structural data are not available, only some key references to this subject will be given. The general aspects of the topic were reviewed some years ago,[653] and a great number of PF_5 adducts were described in an even earlier report.[423] Phosphorus pentafluoride forms adducts with acetonitrile,[341,623,662] tertiary amines,[341,623] and several derivatives of trivalent phosphorus.[80] Pyridine also forms adducts with phosphorus pentachloride[44,45,272,662] and mixed fluorochlorophosphoranes.[271,272] When phosphorus pentachloride is allowed to react with triphenylphosphine oxide, a very moisture-sensitive product is isolated which, on the basis of ^{31}P-NMR data, can be viewed as an inner onium-ate complex containing a hexacoordinated and a tetracoordinated phosphorus atom.[68]

$$Ph_3P{=}O + PCl_5 \longrightarrow Ph_3\overset{\oplus}{P}\text{-}O\text{-}\overset{\ominus}{P}Cl_5$$

B. BASIC CHEMISTRY

B.1. Fluorophosphoranes

The chemistry of the fluorophosphoranes has been reviewed several times in great detail.[434,583,584,590] Fluorophosphoranes, as PF_5, are Lewis acids, and their reactivity depends on the factors influencing the acceptor properties. Thus inductive and steric effects are responsible for the decrease of acceptor character in the following sense:[434]

$$PF_5 > ArPF_4 > AlkPF_4 >> R_2PF_3 \sim R_3PF_2$$

Most of the known fluorophosphoranes are liquids at room temperature and can be distilled, at least under reduced pressure. More complex species, like the fluorophosphoranes containing the 1,3,2,4-diazadiphosphetidine ring, are solids that usually can be sublimed or distilled under reduced pressure. The great majority of the fluorophosphoranes have at least moderate thermal stability. Exceptions are some organyloxy- and organylthiosubstituted derivatives,[81,108,146,456a] which decompose readily to form derivatives of tetracoordinated phosphorus, e.g.,

$$RPF_3OR' \xrightarrow[\text{temp.}]{\text{room}} RP(=O)F_2; \quad RP(=O)(F)OR'$$

In this connection it is interesting that some trifluoromethyl-substituted fluorophosphoranes undergo spontaneous reorganization at room temperature with reversible elimination of difluorocarbene, CF_2.[397,398]

All fluorophosphoranes, with a few exceptions,[82,582,588] are more or less moisture sensitive but can be handled in glass apparatus in a dry atmosphere. Slow attack on glass occurs when moisture is not excluded. In Teflon or stainless steel containers, fluorophosphoranes can be stored indefinitely. The hydrolysis of fluorophosphoranes leads finally to the corresponding acids, but intermediate steps can be isolated.

$$R_nPF_{5-n} \xrightarrow[-HF]{H_2O} R_nP(O)F_{3-n} \xrightarrow[-HF]{H_2O} R_nP(O)(OH)_{3-n} \quad (n = 1 \text{ to } 3)$$

The following qualitative order of the hydrolytic stability of fluorophosphoranes has been given:[590]

$$R_3PF_2 > R_2ArPF_2 > Ar_3PF_2 \sim Ar_2PF_3 > RArPF_3 \sim R_2PF_3 > RPF_4 > PF_5 > ArPF_4$$

Comparable results have been reported for reactions with alcohols, in which alkyl fluorides are formed at higher temperatures in an autoclave.[335-337]

The most important chemical reactions of fluorophosphoranes of any kind are their interconversions by amines or organometal(oid) derivatives of amines, alcohols, and thiols to other phosphorane derivatives. All these reactions were given in Section A (A.1.I, A.2.XX, A.3.XXXI, and A.8.LII) and need no further comment. Some special cases, however, must be mentioned.

Some mixed halofluorophosphoranes undergo extensive intermolecular ligand exchanges even at 0°,[149] whereas normally such disproportionations occur only at more elevated temperatures. In Sections A.2.XX and A.3.XXXI it was shown that fluorophosphoranes react with silyl ethers to give new organyloxy-substituted phosphoranes. But such reactions can also lead to degradations, forming derivatives of tetracoordinated phosphorus.[170,576,580,590]

$$R_nPF_{5-n} + R'_3Si\text{-}OR'' \longrightarrow R_nP(O)F_{3-n} + R'_3SiF + R''F$$

Phosphinosubstituted triorganylsilanes, on the other hand, reduce tetrafluorophosphoranes to difluorophosphines.[435] A good method for the conversion of carboxylic acid derivatives into acyl fluorides consists in the reaction of acid anhydrides with tetrafluorophosphoranes.[395,576]

$$RPF_4 + (R'CO)_2O \longrightarrow RPOF_2 + 2R'COF$$

Reactions of fluorophosphoranes with sulfur and numerous metal sulfides have also been reported.[343]

$$RPF_4,\ R_2PF_3 \xrightarrow{\text{S or sulfides}} RPSF_2,\ R_2PSF$$

Fluorophosphoranes containing a P-H bond show some

special reactions in connection with the acidity of this hydrogen. For example, with strong bases like ammonia[136] or tertiary amines,[138,140,142] dehydrofluorination is possible.

$$RPHF_2(OR') \xrightarrow[-HF]{NR''_3} RPF(OR')$$

A further reaction that is typical of hydridofluorophosphoranes is easy chlorination to chlorofluorophosphoranes.[145,304,305] Like the other fluorophosphoranes, the hydridofluorophosphoranes react with sulfur. In this case, however, HF is eliminated and thiophosphonic difluorides[304] are formed.

In regard to the practical value of such species, a great variety of fluorophosphoranes have been reported to be useful as cocatalysts in the anionic polymerization of lactams with strong bases.[79,588]

B.2. Other Halophosphoranes

The general stability of these phosphoranes is not high, and therefore often the more stable complexes with $AlCl_3$ or PCl_5 are prepared. Many of the reactions discussed in this section are valid for the free phosphoranes and the complexes as well. Thermal decomposition of the phosphoranes generally leads to the elimination of RX with the formation of halophosphines, this reaction often being of practical value.[277,346,408]

As a class, chlorophosphoranes and similar bromo and iodo compounds are, with few exceptions, very moisture-sensitive substances and are therefore easily hydrolyzed to the corresponding oxo derivatives, $RPO(OH)_2$, R_2POOH, and R_3PO. Other reactions common to all kinds of chlorophosphoranes are the reactions with SO_2 and H_2S whereby two halogen atoms are replaced by a doubly bonded oxygen or sulfur atom, respectively. These reactions are often used for identification of the halophosphoranes.[13,14,411,470] Reactions with P_2O_5,[348,678] sulfur,[306,342] P_2S_5,[651,678] metal sulfides,[306,342] thiols,[322,323] epoxides, and episulfides[220,325] lead also to the substitution of two halogens by one oxygen or sulfur atom. In many cases, not the halophosphoranes themselves, but rather their addition complexes with phosphorus pentachloride or aluminum trichloride are used.

$$R_nPX_{5-n} \xrightarrow{\text{O or S donor}} R_nP(O,S)X_{3-n}$$

Reactions of various chloro- (bromo-, iodo-) phosphoranes with alcohols generally lead to alkyl halides and the corresponding oxophosphorus compounds.[283,293,418] A kinetic study suggests the following mechanism for these reactions:[664]

$$R_3PX_2 + R'OH \xrightarrow[MeCN]{fast} [R_3\overset{\oplus}{P}\text{-}O\text{-}R'][X\text{-}HX]^{\ominus} \xrightarrow{slow}$$

$$R_3P{=}O + R'X + HX$$

At higher temperatures even phenols can be transformed to aryl halides by reaction with appropriate halophosphoranes.[265] Use has also been made of such reactions in transformations of certain sugar derivatives to halogenosugars by organyloxybromo- and organyloxyiodophosphoranes.[161,338] In the steroid series, dibromophosphoranes react in a fourfold manner: exchange of OH against Br, H_2O elimination, addition of Br_2 to double bonds, and α-bromination of ketones.[384-386] Dehydration with dihalophosphoranes has also been observed in other systems.[283]

Various other oxygenated organic compounds,[214a] such as phosphonic and phosphinic acids,[256] esters,[214,216] carboxylic acids,[65,196,217,703] anhydrides,[27,196] ethers,[4,639,693] ketones,[213,283,310] aldehydes,[283] and corresponding ortho derivatives thereof,[294] react with all types of chloro- (bromo-, iodo-) phosphoranes, according to the same principles. Triphenoxydichlorophosphorane has also been used in the polymerization of tetrahydrofuran[691] and some diphenols.[686]

Iminophosphorus compounds are formed when halophosphoranes are caused to react with ammonia,[573,645] amines,[221] amides,[605,606,609] and urethanes.[610] The reaction characteristics are shown in the following schematic equation:

$$Y_nPX_{5-n} + R'NH_2 \xrightarrow{-2HX} R'N{=}PY_nX_{3-n} \quad (n = 0 \text{ to } 3)$$

Y = R, RO, et al.; R' = alkyl, aryl, R-C(=O)–, RSO_2-, R-O-C(=O)–

Trichlorophosphoranes derived from some four-membered heterocyclic phosphorus compounds react with isocyanates with ring enlargement and formation of new cyclic phosphoranes.[275,373-377,380] The chemistry of such cyclophosphoranes has been reviewed.[46,173] The following is a typical example:

$$Cl_3P\langle N(Me)\rangle_2C{=}O \;+\; MeN{=}\overset{\oplus}{C}{-}\overset{\ominus}{O} \longrightarrow Cl_3P\langle N(Me){-}C(O)\rangle_2NMe$$

Reduction of halophosphoranes to the corresponding phosphines is possible by means of many metal(oid)s.[219,220,259,416] In particular, red and white phosphorus,[11,344] antimony,[321,458] aluminum,[32,308,344] magnesium,[480] the alkali metals,[276] and carbon[32] have been used for this purpose. Simple and complex hydrides,[219,278] as well as lithium and magnesium organic compounds,[125,407,694] are also suitable for the reduction of halophosphoranes.

Finally, some dichloro- and dibromophosphoranes have been reported to show herbicidal[37] and cytostatic[611-614] activity.

B.3. Oxyphosphoranes

The vast majority of oxyphosphoranes are represented by mono- and spirocyclic compounds containing 1,3,2-dioxaphospholene or 1,3,2-dioxaphospholane rings and their nitrogen analogs.[484-488] Oxyphosphoranes of all types are more or less sensitive to moisture with only a few remarkable exceptions.[21,158] In general, the stability of oxyphosphoranes increases from the simple noncyclic derivatives through the monocyclic to the spirocyclic species,[500] the same being true of compounds in which one or more oxy groups are replaced by amino groups.[508,522]

Until now, of the simple pentaoxy derivatives only pentamethoyx-, pentaethoxy-, and pentaphenoxyphosphorane have been investigated. The alkyl derivatives very easily alkylate water, alcohols, enols, and carboxylic acids with formation of trialkyl phosphates and alcohols, ethers, and esters.[129] Corresponding reactions are observed with similar oxyphosphoranes containing fewer than five oxy groups.[118] The more stable pentaphenoxyphosphorane reacts with HCl to give triphenoxydichlorophosphorane and phenol.[499]

Most of the 1,3,2-dioxaphospholenes are thermally stable under nitrogen up to about 120°.[502] At higher temperatures cleavage to the starting components can occur;[59,360] then an attack of the phosphite on the carbon atom of one carbonyl group leads to a polar intermediate, which undergoes an Arbuzov-like rearrangement.[364]

When the benzil adduct of triethyl phosphite was heated to 215°,* triethyl phosphate and diphenylacetylene were the final products.[431] It was shown that an intermediate in

*In the presence of an excess phosphite.

$$(RO)_2\overset{\oplus}{P}(OR)-C(R')(O^{\ominus})-C(R')=O \longrightarrow (RO)_2P(=O)-C(R')(OR)-C(R')=O$$

this reaction is diphenylketene, which reacts with further phosphite to give rearrangement to the acetylene.[432] The metal salt-catalyzed thermolysis of the same phospholenes showed that very probably benzoylphenylcarbene is formed as the first intermediate, which can be trapped in the presence of alcohols.[429]

$$(EtO)_3P\langle O_2C_2Ar_2\rangle \xrightarrow{\Delta} (EtO)_3PO + Ar-C(=O)-\ddot{C}-Ar$$

rearrangement → $O{=}C{=}CAr_2$

dimerization → $ArC(=O)-C(Ar){=}C(Ar)-C(=O)Ar$

ROH → $ArCH(OR)-C(=O)-Ar$

Phospholenes of this kind are sensitive toward oxygen[516] and react readily with ozone.[490,504,517] The products formed are phosphates and α-dicarbonyl compounds, carboxylic acid anhydrides, or diacyl peroxides.

Three possibilities exist for the hydrolysis of simple dioxaphospholenes. After the initial substitution of one alkoxy group by a hydroxy group, either methanol is split off with formation of a cyclic phosphate, or the ring is cleaved and a 2-ketophosphate is formed. At higher temperatures, hydrolysis to trialkyl phosphate and an enediol can also occur. In all of these reactions intramolecular rearrangements called pseudorotation (see Section C) play an important role, in order to place the leaving groups in a convenient position.[515,622] The same stereochemical phenomena and similar intermediates[494] are discussed in connection with the hydrolysis reactions of certain cyclic phosphate esters.[78,656] An equivalent hydrolysis scheme holds for 2,2,2-tris(dimethylamino)-2,2-dihydro-1,3,2-dioxaphospholenes.[89]

2-Ketophosphates are the only products obtained on reaction of dioxaphospholenes with HCl.[491] With bromine, either 1-halo-2-ketophosphates and alkyl bromide are formed or total cleavage of the triorganyl phosphate entity occurs with formation of an α,α-dibromoketone.[538]

A corresponding reaction has been observed with Nbromophthalimide.[461] When the 1:1 adduct of trimethyl phosphite and biacetyl is photolyzed in the presence of bromotrichloromethane, in a radical reaction a 1-trichloromethylated 2-ketophosphate and methyl bromide are formed.[58] Other photolytic reactions with simple 1,3,2-dioxaphospholenes in inert solvents result in the formation of phosphites, dicarbonyl compounds, and phosphates;[59] in ketone solvents at -75° addition to the double bond occurs with formation of unstable oxetane intermediates, which then rearrange to 1,3,2-dioxaphospholanes.[61]

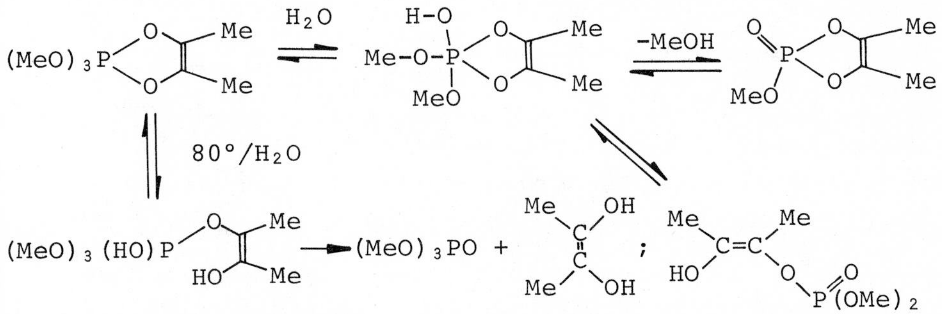

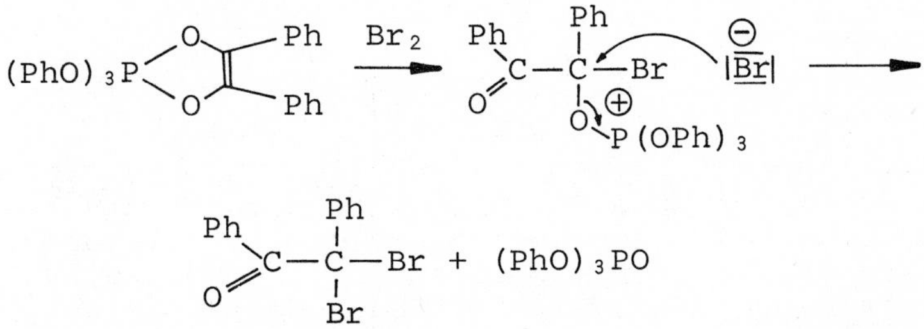

In reactions with acyl chlorides or anhydrides, O-acylation to acylated enediol phosphates[417,507] or C-acylation to phosphate esters of 2-hydroxy-1,3-diketones[489] is possible.[204b] When cumulated systems such as acyl or aryl isocyanates are reacted with dioxaphospholenes, phosphate is eliminated and 2-oxazoline-4-ones[540] or 4-oxazoline-2-ones[429] are formed. In the latter reaction, which is copper catalyzed, obviously the intermediate is benzoylphenylcarbene, which adds to the isocyanate 1,3-dipole. When aryl isocyanates are used in an uncatalyzed reaction, normal C=O condensation (see Section A.1.IX) occurs,[497,498] as it does with carbon suboxide, C_2O_3.[513] The dioxapholanes formed here in the first step easily undergo further transformations with the cumulated substrates. Reaction of the trimethyl phosphite biacetyl adduct with CS_2 in complex

$$(MeO)_3P(O_2C_2(Me)(COMe)(=NPh)) + PhN{=}C{=}O \longrightarrow (MeO)_3PO + \text{MeOC(Me)-imidazolidinedione (N,N'-diphenyl)}$$

sequences leads finally to sulfur heterocycles of the following kind:[330]

$$(MeO)_3P(O_2C_2Me_2) \xrightarrow{CS_2} (Me)(MeOC)C{=}C\langle(S)_n,(S)_m\rangle C{=}C(Me)(COMe)$$

$$(n,\ m = 1,\ 2)$$

The most important reactions of such 1,3,2-dioxaphospholenes are condensation reactions with a variety of carbonyl compounds, as described in Section A. Although the mechanistic details of these reactions are not fully understood,[526] it has been shown that the reactivity in this rather general C-C condensation reaction increases from monoketones through aldehydes to dicarbonyl derivatives. Electron-withdrawing substituents also enhance the reactivity.[527]

1,3,2-Dioxaphospholanes are also formed when phosphites and other derivatives of trivalent phosphorus are treated with 2 moles of a monofunctional carbonyl compound, such as ketones, aldehydes, or esters. Most of these syntheses can be explained by attack of the phosphorus on the oxygen of the carbonyl group, but there exists compelling evidence that equilibria between P-O and P-C primary adducts also play an important role.[404,474,486,488,509]

$$-\overset{|}{\underset{|}{C}}-\overset{\oplus}{P}\lessgtr \;\; \xleftarrow{:P\lessgtr} \;\; >C{=}O \;\; \xrightarrow{\gtrless P:} \;\; -\overset{\ominus}{C}:-O-\overset{\oplus}{P}\lessgtr$$

$$\downarrow\ {>}C{=}O \qquad\qquad\qquad\qquad \downarrow\ O{=}C{<}$$

$$\text{1,3,2-dioxaphospholane (P–O–C–C–O ring)} \qquad\qquad \text{1,3,2-dioxaphospholane (O–C–C–O–P ring)}$$

A condensation reaction involving an unisolated 1,3,2-dioxaphospholane intermediate of the above type has been used in the partial asymmetric synthesis of 2,3-dimethyltartaric acid from trimethyl phosphite and (-)-menthyl pyruvate.[433]

The 2:1 adducts of hexafluoroacetone with various trivalent phosphorus compounds have been investigated in more detail. It has been shown that the stabilities of the 1,3,2-dioxaphospholane species decrease in the following order,[537] which, incidentally, seems to be of rather general significance:

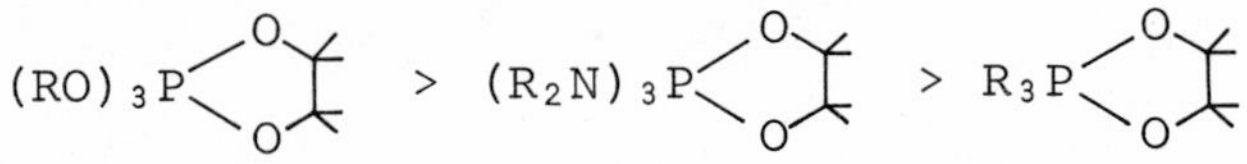

$(RO)_3P$ > $(R_2N)_3P$ > R_3P

Thus the phosphite adducts of hexafluoroacetone decompose only when heated above 170°,[560,666] but some phosphine adducts undergo very interesting rearrangements to 1,2-oxaphosphetane derivatives[72] when heated only to 80 to 100°. It is concluded that a cleavage of the primarily formed 1,3,2-dioxaphospholane back to the starting materials is involved here.[536] These products are of the same type as

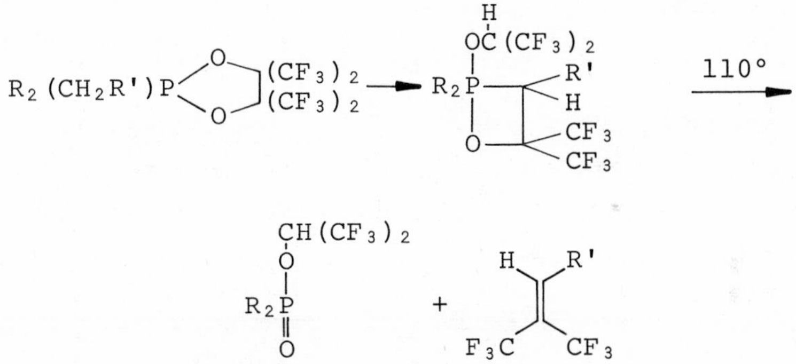

$R_2(CH_2R')P$ $(CF_3)_2$ $(CF_3)_2$ → $OC(CF_3)_2$ H R' H R_2P CF_3 CF_3 —110°→ $CH(CF_3)_2$ O R_2P O + H R' F_3C CF_3

the intermediate formulated in the Wittig reaction and similar ones (Chapter 5A) and can therefore be thermolyzed to olefins and phosphinates.

A rearrangement of another type occurs when 2:1 fluorenone-phosphite adducts are heated above 80°, the corresponding phosphates and biphenylylenedihydrophenanthrone being formed.[76,473,533] When the 2:1 adducts of aromatic

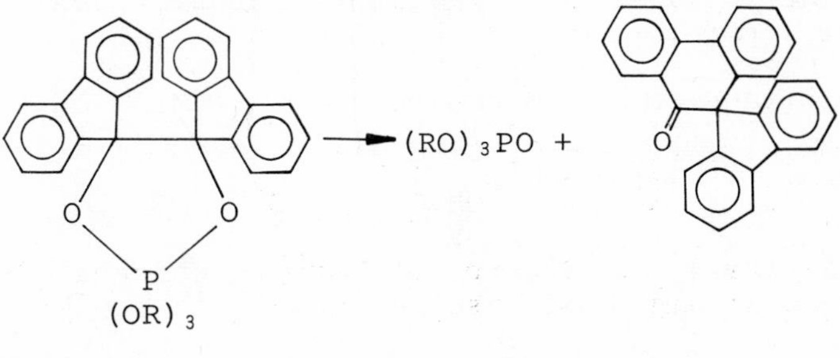

O O P $(OR)_3$ → $(RO)_3PO$ + O

aldehydes and phosphites are thermolyzed, epoxides are formed.[126,430,493] The same is true for similar adducts of tris(dimethylamino)phosphine.[404,509] In each case the trans epoxide is formed predominantly.

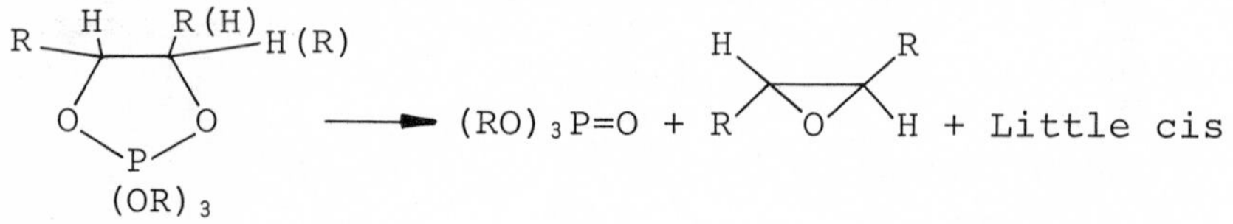

R H R(H) H(R)

O O

P

$(OR)_3$

$\longrightarrow$ $(RO)_3P{=}O$ + H R R O H + Little cis

The hydrolysis of 2,2,2-trisubstituted-2,2-dihydro-1,3,2-dioxaphospholanes again involves different mechanistic pathways. Thus with 1 mole of water formation of cyclic phosphates can be realized,[519] whereas excess water leads to the formation of a phosphate and the corresponding diole.[474,526] This corresponds to similar reactions of dioxaphospholenes (see above). An interesting special example can be given, in which no cyclic phosphate triester could be detected.[496]

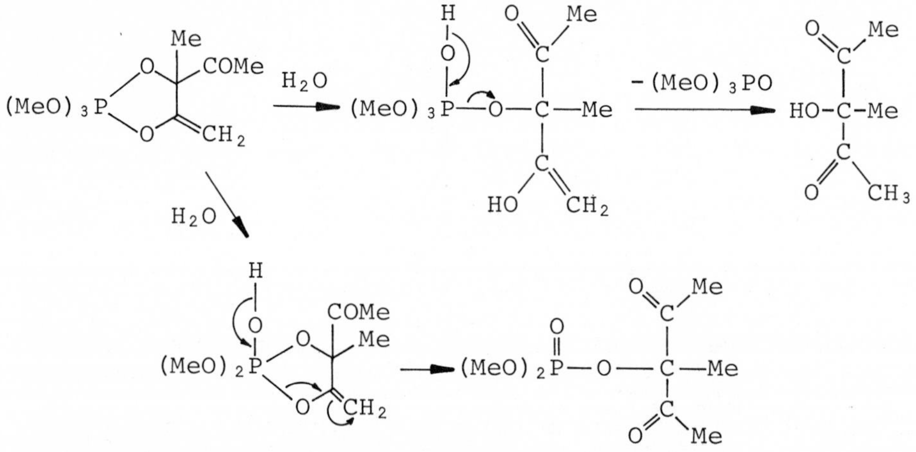

Me O COMe $(MeO)_3P$ O CH_2 $\xrightarrow{H_2O}$ H O $(MeO)_3P$—O—C—Me O C Me C HO CH_2 $\xrightarrow{-(MeO)_3PO}$ O C Me HO-C-Me C O CH_3

H_2O

H O $(MeO)_2P$ O COMe Me O CH_2 $\longrightarrow$ $(MeO)_2P$(=O)—O—C—Me O C Me C O Me

In these hydrolysis reactions, polyoxygenated compounds related to carbohydrates are obtained. Obviously the following synthetic principles have been formally realized here via dioxaphospholenes → phospholanes:

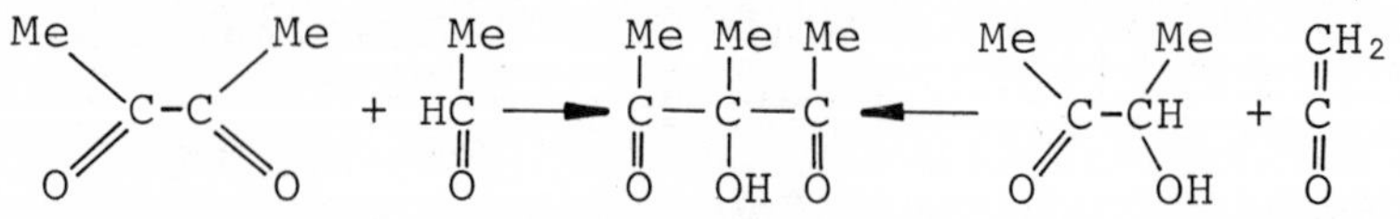

Me Me C-C O O + Me HC O $\longrightarrow$ Me Me Me C—C—C O OH O $\longleftarrow$ Me Me C-CH O OH + CH_2 C O

Two additional hydrolytic pathways in which a cleavage of a C-C bond is involved must be shown.[511]

Finally, it has been observed that some thermal alcoholysis reactions include a rearrangement step,[493] e.g.:

$$MeCOMe + MeC(=O)-C_6H_4-OH$$

Oxyphosphoranes containing an 1,2-oxaphospholene ring and similar ring systems with one C-P bond are easily hydrolyzed to the corresponding phosphonates.[19,26,521]

$$(RO)_3P(C_3H_4O) + H_2O \longrightarrow (RO)_2P(=O)-CH_2-CH_2-CHO$$

$$(RO)_3P\text{(1,3,2-dioxaphospholane, R'/H substituents)} + H_2O \longrightarrow (RO)_2P(=O)\text{-}CH(R')OH + R'CHO$$

Not much is known, however, about the chemistry of cyclic monooxyphosphoranes derived from tertiary phosphines. Only some thermolysis reactions have been reported.[228,593a] Thus 2,2,2-triphenyl-2,2-dihydro-1,2-oxaphospholane behaves at 90° as an ylid and reacts with aldehydes in a normal Wittig reaction.[228]

$$Ph_3P\text{(oxaphospholane)} \rightleftharpoons Ph_3\overset{\oplus}{P}\text{-}\overset{\ominus}{C}H\text{-}CH_2CH_2OH \xrightarrow[-Ph_3PO]{RCHO} RHC{=}CHCH_2CH_2OH$$

The 1,2,5-oxazaphosphol-2-enes obtained by 1,3-dipolar addition reactions[184,291] are thermally decomposed in three directions, forming azirines, ketene imines, or unsaturated oximes. The following generalizations have been made:[64]

$$\text{oxazaphospholene } (R, R^1, R^2, PPh_3) \xrightarrow{\Delta} \begin{cases} R^1, R^2\ (-)I,\ (-)M^* \longrightarrow R^1R^2C{=}C{=}NR \\ R^1, R^2\ (+)I,\ (+)M \longrightarrow \text{azirine } (R; R^1, R^2) \\ R(-)I;\ R^1R^2\ (+)I \longrightarrow R\text{-}C({=}NOH)\text{-}C(Me){=}CH_2 \end{cases}$$

The chemistry of the spirocyclic tetraoxyphosphoranes (and their nitrogen analogs) containing one P-H bond is determined by the fact that the pentavalent species are in equilibrium with ring-open phosphites. Intermolecular isomerizations occur when unsymmetrical phosphoranes of this type are heated.[566] These reactions are based on the substitution of one alcoholic group for another in the phosphites, as was shown independently.[184a] It may be

$$\text{spiro-}HP(OCH_2CH_2O)_2 \rightleftharpoons \text{(1,3,2-dioxaphospholane)}P\text{-}OCH_2CH_2OH \xrightarrow{R'OH} \text{(1,3,2-dioxaphospholane)}POR' + HOCH_2CH_2OH$$

*I = inductive effect, M = mesomeric effect.

recalled that in earlier reports all these compounds were formulated as ring-open phosphites.[87,131,441]

One report of a cyclic oxyphosphorane with <u>two</u> P-H bonds has appeared, but no details for this very interesting compound are given.[442]

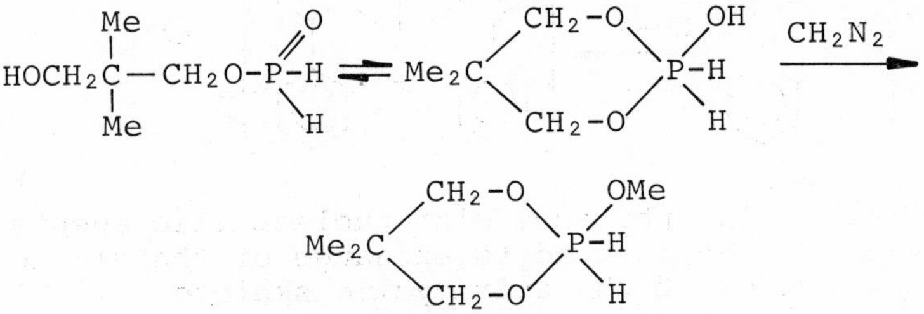

$$HOCH_2C(Me)_2-CH_2O-P(=O)(H)-H \rightleftharpoons Me_2C(CH_2-O)_2P(OH)(H)-H \xrightarrow{CH_2N_2} Me_2C(CH_2-O)_2P(OMe)(H)-H$$

B.4. Pentaorganylphosphoranes

The chemistry of the pentaarylphosphoranes has been reviewed in part several times.[62,248,288,669] In this series of compounds the same stability rules that prevail for the oxyphosphoranes are valid: the general stability increases from the noncyclic through the monocyclic to the spirocyclic species. Pentaphenylphosphorane decomposes at about 130° to benzene, biphenyl, triphenylphosphine, and 2,2'-biphenylylene-phenylphosphine.[670] In such solvents as benzene or pyridine this decomposition occurs at lower temperatures.[547] More recently it has been shown, particularly with the aid of ^{14}C- and deuterium-labeled derivatives of pentaphenylphosphorane, that the thermolysis, as well as the normal and light-catalyzed reactions with several halogenated hydrocarbons, are radical reactions and that the splitting of all five phenyl groups is equivalent.[546,549-554] The equivalence of the phenyl splitting was also demonstrated for some heterolytic reactions of pentaphenylphosphorane.[552] Because of the radical decompositions of Ph_5P it is used as a synergist in the polymerization of styrene-polybromo compounds.[370]

Spirocyclic pentaaryl- and tetraarylalkylphosphoranes isomerize to interesting nine-membered cyclic phosphines when heated above their melting points[237,674] (see also Chapter 1). In electrophilic reactions with mineral acids,

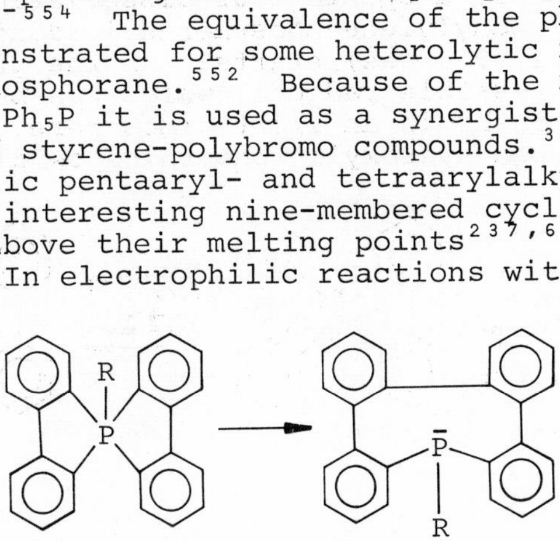

[Structure: spirocyclic bis(2,2'-biphenylylene)phosphorane with R on P → nine-membered cyclic phosphine with $\bar{P}$–R]

halogens, or metal salts, spirophosphoranes of this type are cleaved at one ring, monocyclic phosphonium salts being formed.[247,674] On the other hand, reactions of

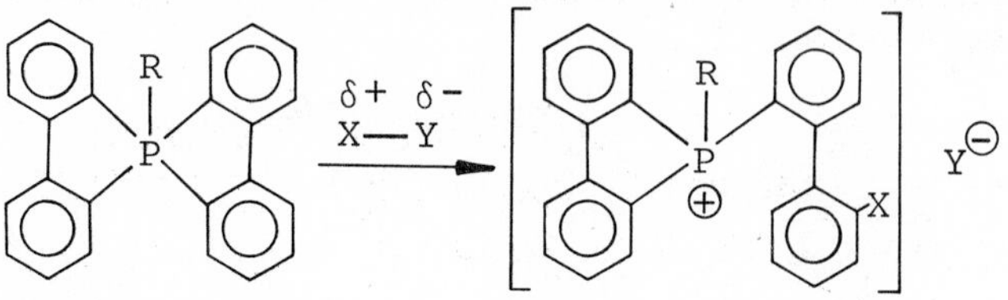

these spirocyclic phosphoranes with nucleophilic agents such as lithium organyls lead to exchange of the single group with retention of the spirocyclic skeleton.[246,572]

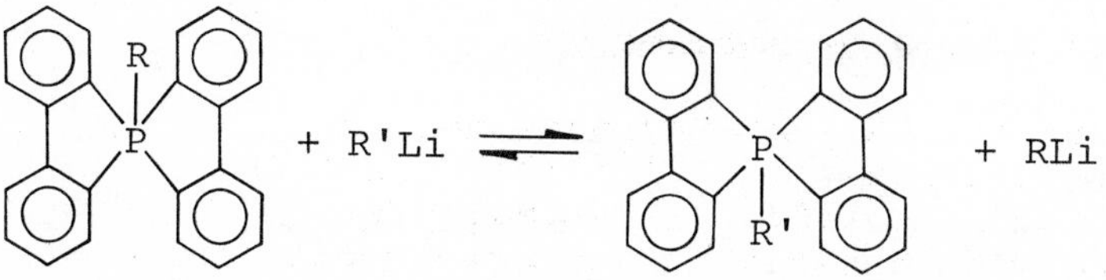

R = H, alkyl, aryl; R' = alkyl, aryl

Pentaphenylphosphorane also can undergo such nucleophilic displacement reactions,[572] but at a much slower rate.[109]

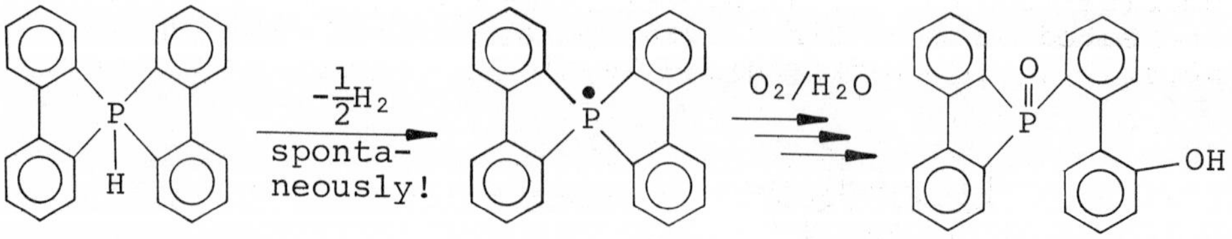

The prototype of the spirocyclic phosphoranes discussed in this section, bis-2,2'-biphenylylene-hydridophosphorane, needs special attention because of its unique chemical properties,[246] all of which are connected with the rather labile P-H bond. Thus, when the hydridophosphorane is dissolved in inert solvents or even in acetone, almost instantaneously a spontaneous dissociation to the violet bis-2,2'-biphenylylene-phosphoranyl radical (ESR doublet with a_P = 17.9 gauss) starts! In the presence of air, the relatively stable radical, which can be observed over a period of months in an inert atmosphere, is transformed to a ring-open hydroxylic phosphine oxide. When air is excluded, the radical dimerizes to a compound containing a triply and a quintuply coordinated phosphorus atom.[246,253]

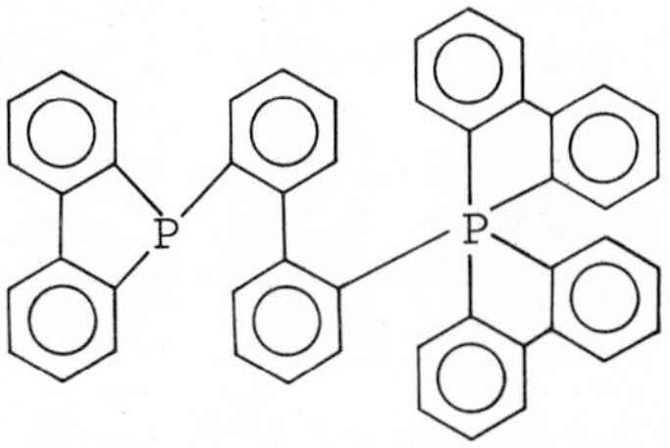

Furthermore, in addition to the radical splitting of the P-H bond of this phosphorane, it is possible to abstract the hydrogen as hydride ion (with strong electrophiles such as protons) or as proton (with strong and bulky nucleophiles such as t-butyllithium). The blue-green

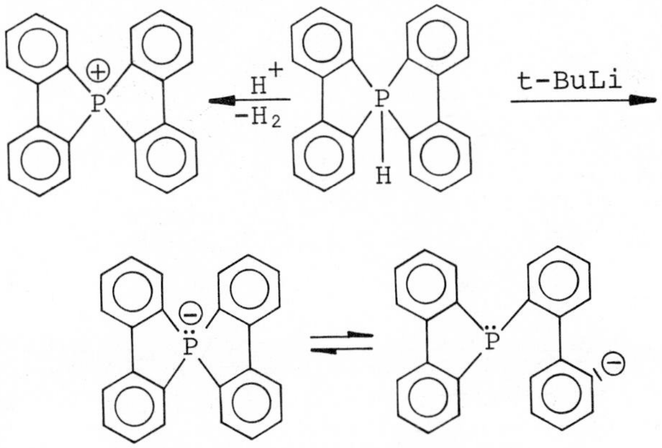

H^+ / $-H_2$; t-BuLi

(in THF) bis-2,2'-biphenylylene-phosphoranyl anion is very probably in equilibrium with a ring-open carbanionic phosphine, which finally can be protonated to 2,2'-biphenylylene-2-biphenylylphosphine.[246]

Corresponding noncyclic tetraorganyl derivatives of PH_5 are not known. They have sometimes been discussed, however, as intermediates in reduction reactions with tetracoordinated phosphorus structures.[104,204,256a,281]

B.5. Hexavalent Phosphorus Compounds

As already stated in Section A, organic compounds of hexacoordinated phosphorus are still so rare and heterogeneous that not much is known about their chemistry. As salts, they are usually solid substances, but a few exceptions are known in the fluorophosphate series, where only oily[144] and sometimes distillable[593] products have been obtained. A general similarity exists in the thermic and hydrolytic stability of hexavalent and corresponding pentavalent phosphorus compounds. Thus chlorine-containing phosphate complexes are very sensitive toward moisture, whereas some fluorophosphate complexes show remarkable hydrolytic stability when big cations and perfluoroalkyl groups are present.[94a] Hexaoxyphosphates of the type tris-(o-phenylenedioxy)phosphate also can be handled in cold aqueous solvent mixtures;[251] on heating, however, tris-(o-hydroxyphenyl)-phosphate is formed.[1]

The hexaaryl ate complexes of the tris-2,2'-biphenylylene phosphate type are completely stable toward water and can be degraded to pentavalent phosphorus species only by strong electrophiles such as protons[237,241-243] or bromine.[252,668] The bromine-substituted phosphorane obtained in the latter reaction reverts to the hexacoordinated ate complex when reacted with butyllithium. The thermal

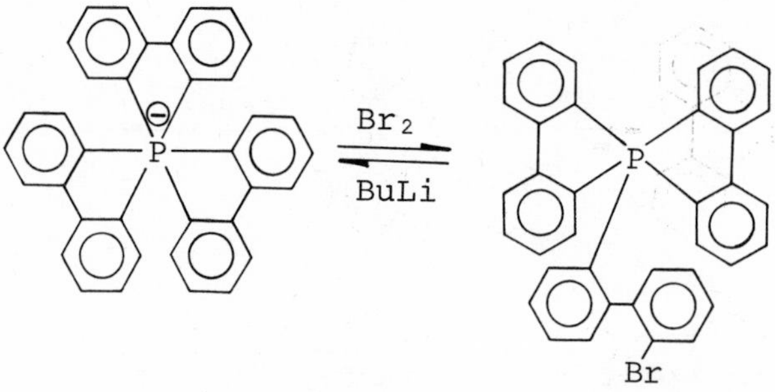

$\xrightarrow[\text{BuLi}]{Br_2}$ (Br)

stability of these ate complexes is very high, the potassium salt, e.g., decomposing only at temperatures above 300°!

C. GENERAL PHYSICAL PROPERTIES OF PENTA- AND HEXAORGANYLPHOSPHORUS COMPOUNDS

Nearly all physicochemical investigations undertaken in this field are closely involved with the structural and stereochemical aspects of phosphoranes. It is necessary, therefore, to give a short introduction into the principal features of coordination number five.[39,286,383,427,590]

The simplest and most widely used picture of the valence bonding situation in pentacoordinated structures of the main group elements is based on the presence of hybrid $sp^3d_{z^2}$-orbitals. When the d_{z^2}-orbital is involved, a trigonal-bipyramidal conformation results; in the case of combination with the $d_{x^2-y^2}$-orbital, a square-pyramidal or a tetragonal-pyramidal conformation is obtained.[328a,389a] At first sight the rather diffuse d-orbitals of the neutral phosphorus atom seem not to allow effective hybridization with the more compact s- and p-orbitals. However, it has been shown that under the influence of suitable electronegative ligands the d-orbitals are contracted, so that effective overlap with the s- and p-orbitals can occur.[106a,405a]

Because of the high promotion energies of main group element d-electrons, attempts are being made to introduce bonding models that avoid the inclusion of d-orbitals. These concepts range from partially delocalized models, in which the axial bonds of a trigonal-bipyramidal structure are relatively weak, and polar delocalized three-center, four-electron bonds formed from a doubly occupied central p-orbital and the σ-orbitals from two ligands,[563a] to fully delocalized models based only on s- and p-orbitals, where no preconception of hybridization or directed valence is necessary.[38] The electronic structure of the hypothetical PH_5 has been calculated without the inclusion of d-orbitals; a trigonal-bipyramidal configuration with equatorial and axial P-H bond energies of 59 and 44 kcal/mole, respectively, was obtained.[296] In an interesting extension of the three-center bond concept an attempt was made recently to handle, in one simple, unifying bonding theory,[438] all kinds of "hypervalent molecules," i.e., molecules in which a main group central atom utilizes more electron pairs than are allowed by the conventional Lewis-Langmuir theory.

Another theory that avoids any a priori fixation of the bond orbital system is the "valence shell electron pair repulsion" or VSEPR theory.[190,191] According to this theory, the stereochemistry of molecular polyhedra is determined mainly by the fact that the electron pairs around a central atom tend to occupy positions in which their mutual distances are maximized. This follows directly from the Pauli exclusion principle. Again, the trigonal-bipyramidal arrangement of five ligands around a central atom with longer axial than equatorial bonds is found to be the most stable one. It is clear that a purely electrostatic treatment of an AB_5 polyhedron must lead to the same results. Such a calculation has shown, moreover, that a tetragonal pyramid with an axial-equatorial bond angle of about 104° is only slightly less stable than a normal trigonal bipyramid.[694a]

In summary, the following conclusions can be drawn from the various stereochemical considerations:

1. The most stable arrangement of five ligands around a phosphorus center should be given by the trigonal-bipyramidal model.

2. The axial bonds in these trigonal bipyramids are weaker, more polarizable, and longer than the equatorial bonds and therefore are the preferred bonds for electronegative substituents.

3. A tetragonal pyramid with an axial-equatorial bond angle of about 104° is only slightly less stable than a trigonal bipyramid.

As point 3 suggests, the trigonal-bipyramidal bond configuration should not be an absolutely stable one. Indeed, all pentacoordinated phosphorus derivatives investigated up to now have trigonal-bipyramidal ground states, but nearly exclusively can undergo, more or less easily, intramolecular ligand reorganizations. These processes, which have been called "pseudorotations,"[62a] "pseudoinversions,"[248,250] and "polyhedral"[419] or "polytopal"[421] rearrangements, can be represented by six different mechanisms.[421,422] But it has been shown theoretically,[457,625] as well as by calculations based on IR and Raman data,[268-270] that the "Berry mechanism"[62a] involving a tetragonal-pyramidal transition state is the most probable one for such ligand reorganizations. Very recently, a new but energetically less favorable mechanism, the "Turnstile Rotation," which includes a complex combination of angle compressions, ligand pair tilts and rotations of a pair against a trio of ligands, has been proposed for these interconversions.[540a]

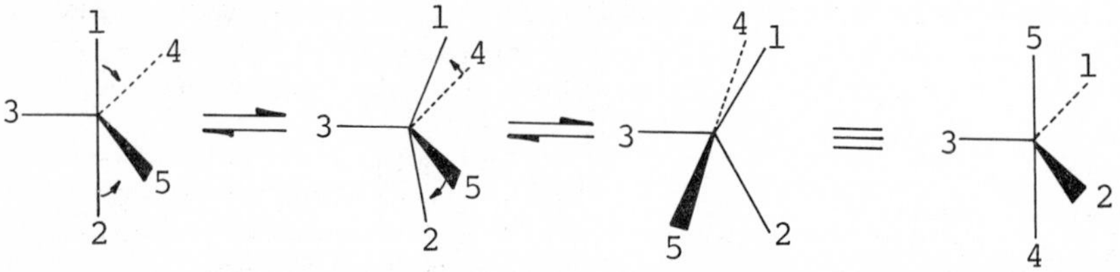

In the case of five different ligands around the central atom 20 stereoisomers are possible when the trigonal-bipyramidal configuration is valid. The manifold interconversions of isomers which must then be considered for such species have been rationalized in topoligical, nontopological, and matrix representations.[33,111,112,151,185,186,188,381,419-422]

C.1. Fluorophosphoranes

The vast majority of fluorophosphoranes are pentacovalent species for which the principles of pentacoordinate stereochemistry must be considered.[434,583,584,590] Only a very few examples are known in which fluorophosphoranes of formula $F_3P(OR)_2$ are in reality ionic structures like $[(RO)_4P^+][PF_6^-]$.[340] Electron diffraction studies of $MePF_4$ and Me_2PF_3 [40] have shown that the alkyl groups occupy equatorial sites of trigonal bipyramids and that the axial P-F bonds are ∿0.8 Å longer than the equatorial P-F bonds. The electron diffraction[3] and crystal structure[106] investigations of fluorophosphoranes containing the 1,3,2,4-diazadiphosphetidine ring have also established trigonal-bipyramidal configurations with the planar rings in axial-equatorial positions and with longer axial than equatorial P-F bonds.

The vibrational spectra of fluorophosphoranes in general confirm the structural evidence gained by other methods. Detailed IR studies of the methylfluorophosphoranes, Me_nPF_{5-n}, have been reported.[133,134] In the case of CF_3PF_4 the vibrational spectra suggest a C_{2v} structure with an equatorial CF_3 group,[210] but the microwave spectrum points to a C_{3v} structure with an axial CF_3 group.[102] For the diazadiphosphetidine system mentioned above, however, the IR and Raman data show C_{2h} symmetry[132,685] in accordance with the diffraction results.

Generally, P-F stretching frequencies of fluorophosphoranes are considerably dependent on the different states of aggregation and on the electronegativity of the other substituents. In the methylfluorophosphorane series, e.g., the axial F-P-F stretching frequencies increase from 500 to 596 cm^{-1} [ν(F-P-F) symm.] and from 690 to 843 cm^{-1} [ν(F-P-F) asymm.] respectively, in going from the difluoro- to the tetrafluorophosphoranes. Successive substitution of the methyl groups in trimethyldifluorophosphorane by phenyl or perfluoroalkyl groups also leads to an increase in the P-F stretching frequencies.[590] A summary of the IR data for various tetrafluorophosphoranes gives asymmetric axial F-P-F stretching modes in the region of 865 ± 30 cm^{-1} and equatorial F-P-F stretching frequencies at 905 ± 30 cm^{-1} (B_1) and 985 ± 40 cm^{-1} (A_1), respectively.[79b]

The greatest amount of structural information on fluorophosphoranes has been gained by ^{19}F-NMR spectroscopy.[590] The ^{19}F spectra are in most cases very characteristic and/or complex because of the various coupling modes between ^{1}H, ^{31}P, and the differently situated ^{19}F nuclei. The simplest spectra, obtained with difluorophosphoranes in which the fluorines are in identical axial positions show a basic doublet. Most δ_F values are here found between +5 and +75 ppm (relative to CCl_3F).[415] The corresponding P-F coupling constants vary considerably with the electronegativities of the other ligands, from 542 Hz in Me_3PF_2 to 1109 Hz in $CCl_3PBr_2F_2$ (see the list of compounds in Section D). Some organylthiodifluorophosphoranes are known in which at -80° the rotation of the thio group is inhibited, thus allowing differentiation of the two unequal axial fluorines.[456a]

The ^{19}F resonance spectra of trifluorophosphoranes

usually show two identical axial fluorine positions and one equatorial fluorine signal. The cyclic tetramethylenetrifluorophosphorane reveals these spectral characteristics only at -100°, whereas at room temperature fast pseudorotation equilibrates the ^{19}F nuclei.[426] In perfluoroalkyl-substituted trifluorophosphoranes only one kind of ^{19}F is observed, and it is not known whether this is due to a fast pseudorotation process or to a trigonal-bipyramidal configuration with three equatorial fluorines.[590] Similar uncertainties exist in regard to the ^{19}F-NMR spectra of fluorophosphoranes containing the diazadiphosphetidine ring, where also only one ^{19}F signal, indicating fast intramolecular ligand reorganizations, could be observed.[233] Again some examples are known in which the two axial fluorines of organylthio- or organylaminotrifluorophosphoranes become different because a freezing of the rotation of a thio or amino group, respectively, is possible.[145,148,371,456a] In general, for trifluorophosphoranes the equatorial δ_F values (+67 to +94 ppm) are considerable higher than the axial δ_F values (+3 to +44 ppm). Usually the axial P-F coupling constants (<900 Hz) are 100 to 200 Hz smaller than those for the equatorial positions (>900 Hz).

In all tetrafluorophosphoranes of type RPF_4 equivalent fluorine atoms are found, even at very low temperatures. This would indicate a tetragonal-pyramidal (C_{4v}) structure, but in the light of the vibrational-spectroscopic and electron-diffraction results a trigonal-bipyramidal model either possessing unmeasurably small δ ^{19}F differences or undergoing pseudorotation by a very fast rate, also at low temperatures, is more probable.[105,170,590] Certain diorganylaminosubstituted tetrafluorophosphoranes, however, reveal at low temperatures two kinds of fluorine signals possessing the chemical shift and coupling characteristics of equatorial and axial fluorine atoms, respectively, in trigonal bipyramids.[624] A line shape analysis has proved that such phosphoranes undergo ligand reorganizations according to the Berry mechanism.[659] In a highly substituted phenylaminotetrafluorophosphorane three kinds of fluorines are observed. This suggests the presence of a trigonal-bipyramidal structure in which the substituted amino group does not rotate freely.[233] The same observations were made in the low-temperature ^{19}F-NMR spectra of some organylthiotetrafluorophosphoranes.[456,456a]

In summary, it can be concluded from the ^{19}F-NMR investigations that the electronegative fluorine ligands always tend to occupy axial positions in trigonal-bipyramidal conformations. Two examples are known, however, in which the steric restrictions in a spirocyclic monofluorophosphorane very probably force the fluorine atom into an equatorial site,[52,131a] as indicated by the high P-F coupling constants (>1000 Hz). A recent study by the heteronuclear INDOR technique has shown that the P-F coupling constants in fluorophosphoranes have negative signs relative to $^{1}J_{HC} > 0$.[135a]

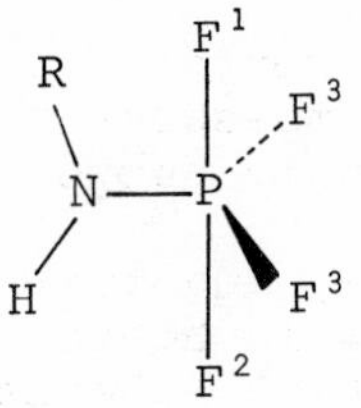

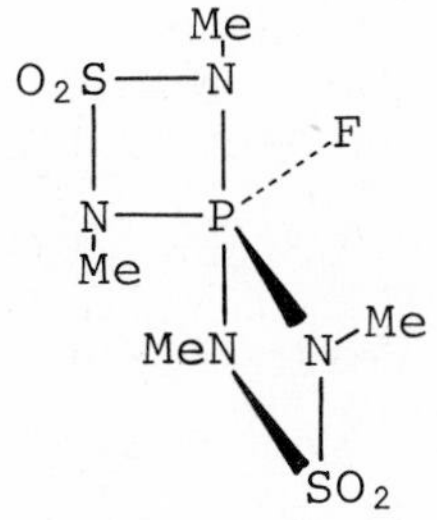

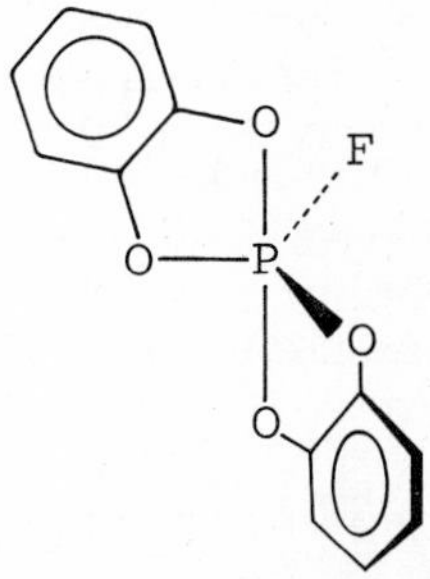

The ^{19}F-NMR resonance shifts have also been used to obtain information concerning d_{π}-p_{π} interactions in tetrafluorophosphoranes. Indeed, the use of one d-orbital in σ-bonding seems not to inhibit the π-bonding ability of another one.[483] Speculations in the same direction have been made for some perfluoroalkenylfluorophosphoranes.[105] On the other hand, in the diazadiphosphetidine-fluorophosphoranes little d_{π}-p_{π} interaction of the $\overset{\ominus\,\oplus}{-P=N-}$ / $\underset{\oplus\,\ominus}{-N=P-}$ type seems to occur, as was concluded from an IR study.[582,590]

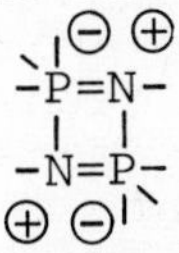

The ^{31}P chemical shifts of fluorophosphoranes are not very specific.[405] Very roughly, they fall in a range of about +50 ± 20 ppm (relative to 85% H_3PO_4), but some remarkable exceptions are known. For example, the simple dialkyltrifluorophosphoranes and the tetramethylenetrifluorophosphorane show even negative values for the phosphorus chemical shift![590]

A preliminary systematic study of the mass spectra of fluorophosphoranes can be summarized as follows: tetrafluorophosphoranes and difluorophosphoranes give weak or no parent ions, whereas for trifluorophosphoranes the parent ions are always present.[73]

C.2. Other Halophosphoranes

Chlorophosphoranes and corresponding bromo and iodo derivatives show a much greater tendency to exist in ionic forms than the fluoro analogs. These species of the general formula $[R_nPX_{4-n}^{\oplus}]X^{\ominus}$ are especially susceptible to nucleophilic attack, a fact which explains the great moisture sensitivity of almost all chlorophosphoranes. In general, halophosphoranes of this type are sparingly soluble in nonpolar solvents but are easily soluble in such polar solvents as methylene chloride, chloroform, nitromethane, nitrobenzene, and dimethylformamide. This again indicates the possibility of dissociation, at least under certain conditions. Furthermore, electric conductivity measurements in solvents like DMF or acetonitrile strongly support this view.[153,353,354,564a]

For triphenyldichloro-,[230] triphenyldibromo-, and triphenyldiiodophosphorane[66] conductometric studies in acetonitrile have established the following dissociation schemes:

$$2Ph_3PCl_2 \longrightarrow [Ph_3PCl]^{\oplus} + [Ph_3PCl_3]^{\ominus}$$

$$Ph_3PBr_2(I_2) \longrightarrow [Ph_3PBr(I)]^{\oplus} + Br^{\ominus}(I^{\ominus})$$

The bromo and iodo compounds add additional halogen to form highly conducting salts with tetrahalide anions.[66] Cryoscopic molecular weight determinations have indicated that triphenyltrichlorophosphorane exists in haloform solution in the form of an octahedral dimer bridged by two chlorines.[30]

Infrared and Raman investigations were conducted with the various methylchlorophosphoranes, R_nPCl_{5-n}.[41] According to these measurements, the crystalline, high-melting substances are chlorophosphonium salts in the solid state. Methyltetrachlorophosphorane is ionic in the solid state but is a pentacoordinate species with C_{2v} symmetry in nonpolar solvents.[43] The readiness of chloro-, bromo-, and iodophosphoranes to form ionic structures is also expressed in their reactions with phosphorus pentachloride and/or other metal(oid) halides to yield halophosphonium salts containing complex anions.[220,299,388,570] These complexes are soluble in polar solvents and show electric conductivity.

A stabilization of the pentacovalent state of chlorophosphoranes seems to occur when trihaloalkyl substituents are present. Bis(trichloromethyl)trichlorophosphorane[178] and diphenyl(trifluoromethyl)dibromo(diiodo)phosphorane[56] are even stable toward water! (Trifluoromethyl)tetrachlorophosphorane[208] and bis(trifluoromethyl)trichlorophosphorane[211] are solids that, according to their vibrational spectra, have trigonal-bipyramidal structures with axial CF_3 groups.

The most conclusive studies with real pentacoordinate chlorophosphoranes were made with compounds based on the diazadiphosphetidine ring. X-ray structure determinations of some trichlorophosphoranes of this type show the trigonal-bipyramidal ligand arrangement around the phosphorus atoms, the planar rings holding axial-equatorial positions, and the axial P-Cl distance being about 0.1 Å longer than the equatorial ones.[258,261,655] The differences between the lengths of the axial and equatorial P-N bonds are somewhat larger. The relatively short P-N distances indicate some p_π-d_π character in these bonds.[258,261] The same conclusion is reached from the P-N bond energy of the following phosphorane, which is greater (74.3 kcal/mole) than the normal P-N single-bond energy.[177]

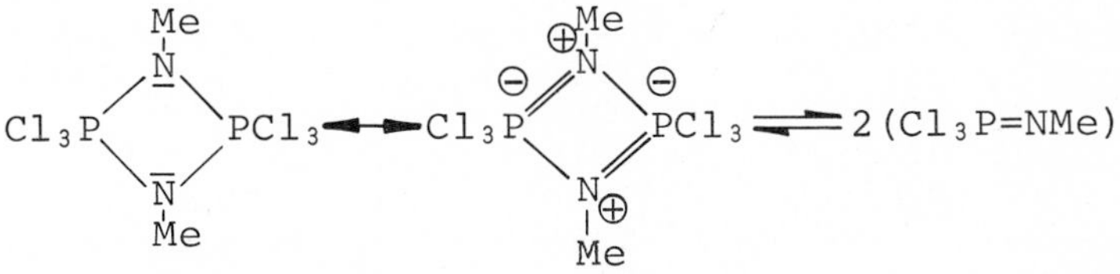

$$Cl_3P\langle N(Me) \rangle_2 PCl_3 \longleftrightarrow Cl_3\overset{\ominus}{P}\langle \overset{\oplus}{N}(Me) \rangle_2 \overset{\ominus}{P}Cl_3 \rightleftharpoons 2(Cl_3P{=}NMe)$$

The extent of p_π-d_π interaction in such compounds seems also to play an important role in determining whether the cyclic pentacoordinated structures or the corresponding monomeric phosphine imine structures are more stable. Generally, diazadiphosphetidines formed from amines of high basicity (K_b = 10^{-9} to 10^{-10}) are cyclic dimeric species, whereas products originating from weaker amines (K_b = 10^{-10} to 10^{-13}) are dimeric in the crystalline state but monomeric in boiling benzene. With bases of K_b = 10^{-14} to 10^{-19}, only monomeric phosphines can be obtained.[222,695,700,704] On the other hand, when one or more chlorine atoms in these cyclic phosphoranes are substituted by less electronegative groups, a considerable increase in the tendency to form the monomeric phosphine imines is again observed, because now the partial negative charge on the phosphorus becomes less favorable.[221,702,707] In the IR spectra, a band around 850 cm^{-1} is said to be characteristic for these diazadiphosphetidine systems.[95,222]

The ^{31}P chemical shifts of chlorophosphoranes are always positive but are spread over a wide range from +9 to +86 ppm.[375,379,405] The cyclic phosphoranes of the diazadiphosphetidine type generally absorb at +60 ± 25 ppm, whereas the simple phenylchlorophosphoranes show smaller ^{31}P values: $PhPCl_4$, +43 ppm; Ph_2PCl_3, +26 ppm; Ph_3PCl_2, +11 ppm (nitrobenzene). In the latter series, increasing amounts of ionic forms in equilibrium are probably responsible for the decrease of the ^{31}P shift.[117] This theory has been tested in more polar solvents like acetonitrile or tetrachloroethane, where indeed very negative ^{31}P resonances indicate the presence of chlorophosphonium ions.[375,665] These negative ^{31}P shifts of about -50 ppm and lower are always observed when a formal chlorophosphorane is in reality a chlorophosphonium salt.[379] Only a few ^{31}P values of bromophosphoranes are known, but these lie at considerably higher field strengths than those of corresponding chlorophosphoranes;[150] (o-phenylene-dioxy)tribromophosphorane, e.g., exhibits a ^{31}P shift of +189 ppm![174]

C.3. Oxyphosphoranes

The physiochemical investigations in this field have concentrated nearly exclusively on the stereochemical phenomena. The pentacoavlent nature of the simple oxyphosphoranes obtained from derivatives of trivalent phosphorus and dialkyl peroxides is proved by their high positive ^{31}P resonances. The pentaalkoxyphosphoranes[123,126] and pentaaroxyphosphoranes[499] have δ ^{31}P values higher than +70 ppm, whereas oxyphosphoranes derived from tertiary phosphines, phosphinites, and phosphonites absorb between +38 and +55 ppm.[118,119] All these compounds show in the proton NMR spectrum only one kind of ligand, even at low temperatures. This is indicative of the occurrence of fast pseudorotation processes. In the case of diphenyltriethoxyphosphorane, however, at -60° two signal groups for the ethoxy ligands are observed, corresponding to the axial and equatorial positions in a frozen trigonal bipyramid.[118]

For the various reactions of triphenylphosphine with hydroperoxides,[121] peresters,[122] and acyl peroxides[124,207] (and also of phosphites with methyl toluenesulfonate[120]) experiments with ^{18}O-labeled substrates have shown that here ion pairs are the most important species and pentacoordinated structures are, at most, very unstable intermediates. The following is a typical example:

$$\left[Ph_3\overset{\oplus}{P}-O\overset{\overset{O}{\|}}{C}R\right]\left[\ominus\, O\overset{\overset{O}{\|}}{C}-R'\right] \rightleftharpoons \left(Ph_3P\begin{matrix} \diagup O\overset{\overset{O}{\|}}{C}R \\ \diagdown O\underset{\underset{O}{\|}}{C}R \end{matrix}\right)$$

The most comprehensive stereochemical studies in the oxyphosphorane series, however, were made with cyclic compounds of the 1,3,2-dioxaphospholene and 1,3,2-dioxaphospholane type. This subject has been reviewed several times in considerable detail.[484-488,656] X-ray diffraction studies with the two allotropic forms of the phenanthrenechinone-triisopropyl phosphite adduct have established a slightly distorted trigonal-bipyramidal structure with the 1,3,2-dioxaphospholene ring in axial-equatorial positions.[225,226,620] The axial P-O bonds of the ring are about 0.1 Å longer than the equatorial ones. The exocyclic P-O bonds are a little shorter than their ring counterparts; this is taken as an indication of more prominent p_π-d_π interaction of the phosphorus with the exocyclic oxygens.

In solution, the ^{1}H-NMR spectra of a great variety of 2,2,2-trimethoxy-2,2-dihydro-1,3,2-dioxaphospholenes show, even at -90°, only one kind of methoxy signal as a doublet with J_{HCP} = 13 Hz. Obviously, rapid positional exchange of axial and equatorial MeO groups in the trigonal bipyramids cannot be inhibited, even at these low temperatures.[486,488]

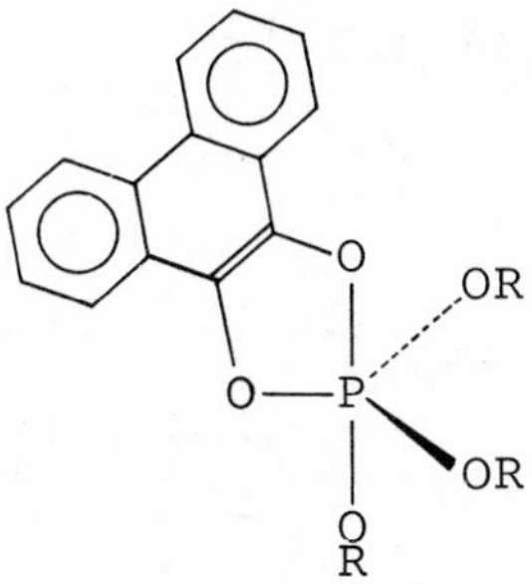

In the IR spectra of such phospholenes the P-O-alkyl stretching vibrations are found in the region from 1087 to 1064 cm^{-1}. Similar triphenyl phosphite adducts show the P-O-aryl stretch at 1215 to 1200 cm^{-1}. The olefinic double bond usually is found at 1664 to 1645 cm^{-1}.[502,523,525] The dipole moments of dioxaphospholenes of this type are all between 2 and 4D.[364,501] The mean atomic refraction of phosphorus in phospholenes (R_p = 3.95) is close to the atomic refraction of phosphorus in phosphates.[362]

In Section A it was shown that, depending on the nature of the substituents at phosphorus in 1,3,2-dioxaphospholenes, all kinds of transitions between true pentavalent and open dipolar forms are possible.[486,523] This can easily be examined by ^{31}P-NMR spectroscopy.

Generally 2,2,2-trialkoxy-2,2-dihydro-1,3,2-dioxaphospholenes show ^{31}P chemical shifts of +44 to +62 ppm.[525] Introduction of a second 1,3,2-dioxaphosphol(a,e)ne ring, as in the spiro derivatives, reduces the ^{31}P shift by about 20 ppm.[518]

When, as in the case of some tris(dialkylamino) derivatives, open dipolar ions are present, strongly negative ^{31}P values of about -30 to -40 ppm are observed. Corresponding derivatives with true pentacovalence of phosphorus absorb above +30 ppm.[522] When equilibria between the two extremes are occurring, solvent-dependent ^{31}P shifts in an intermediate range are registered.[523] The occurrence of dipolar ions is also indicated by their orange color and by the very strong enolate band at 1493 to 1449 cm^{-1} in the IR spectrum.[523] Similar equilibria are discussed for the adducts of phenanthrene-chinone with several tertiary phosphines.[537]

Inhibition of pseudorotation can also not be observed in phosphoranes containing the 1,3,2-dioxaphospholane system. 2,2,2-Trimethoxy derivatives of this type always show one signal for the MeO groups. 2-Organyl-2,2-dimethoxy

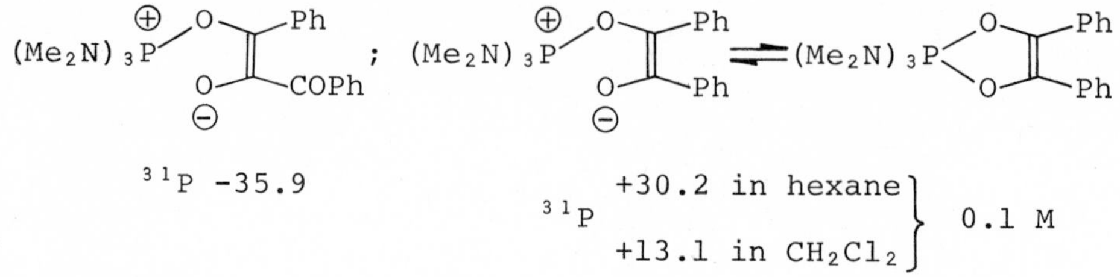

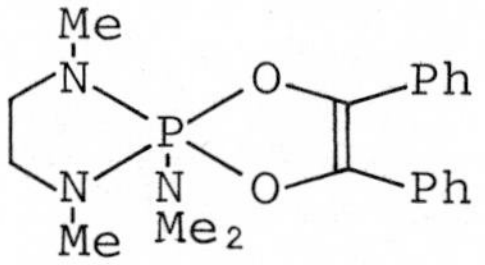

^{31}P -35.9

^{31}P +30.2 in hexane, +13.1 in CH_2Cl_2 (0.1 M)

^{31}P +36.9

derivatives reveal one MeO signal when the phospholane ring (not the molecule as a whole) has a plane of symmetry, and two MeO signals if there is none. Detailed stereochemical analyses have shown that this is consistent with fast-occurring pseudorotations by the Berry mechanism, leading to full [$(MeO)_3$] or partial [$R(MeO)_2$] equilibration of the axial and equatorial methoxy signals.

Intermediates and/or transition states in which the phospholane rings occupy two equatorial positions are forbidden.[488] Furthermore, all these phospholanes can exist in diastereomeric forms, depending on the arrangement of the substituents (when different) at the phospholane ring.

R = OMe, alkyl, aryl;
R^1, R^2, R^3, R^4 = H, aryl, alkyl, COR, COOR

The exact structures are derived chiefly from ^1H-NMR spectroscopy. Especially when phenyl substituents are present, hydrogens or methyl groups cis to the phenyl group are found to absorb at higher field than in the trans case.[492,509,527]

In contrast to the phosphoranes described above, cyclic oxyphosphoranes of the 1,4,2-dioxaphospholane type with one C-P bond in the ring allow inhibition of pseudorotation at -140°. This is proved by the fact that in suitable substituted species the single methoxy signal observed at room temperature splits, on cooling below -110°, into three different signals, belonging to the two equatorial sites and the axial position of a trigonal bipyramid. The ring occupies equatorial-axial positions with the electronegative oxygen in the axial position.[530]

A similar situation is found for a 1,4,2-dioxaphospholane system with two exocyclic methoxy groups and one exocyclic phenyl group. For the cis compound, pseudorotation slows down at -42°, where the trigonal-bipyramidal configuration shown below becomes stable with respect to the NMR time scale. For the corresponding trans product, however, pseudorotation remains effective, even at lower temperatures.[488]

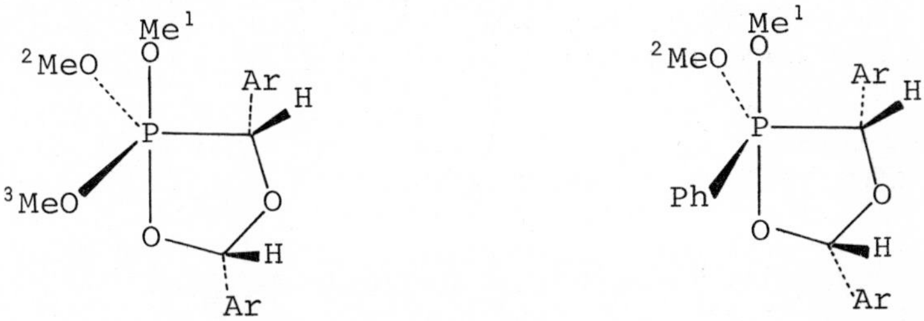

In phosphoranes containing the 1,2-oxaphosphol-4-ene ring, inhibition of pseudorotation at still higher temperatures is possible.[199-201,486,488,528] For the products obtained by the reaction of 3-benzylidene-2,4-pentanedione and dimethyl phenyl phosphonite, at temperatures below 0° two "frozen" diastereomeric trigonal bipyramids are found, each possessing two different MeO signals. Coalescence of the methoxy signals occurs at

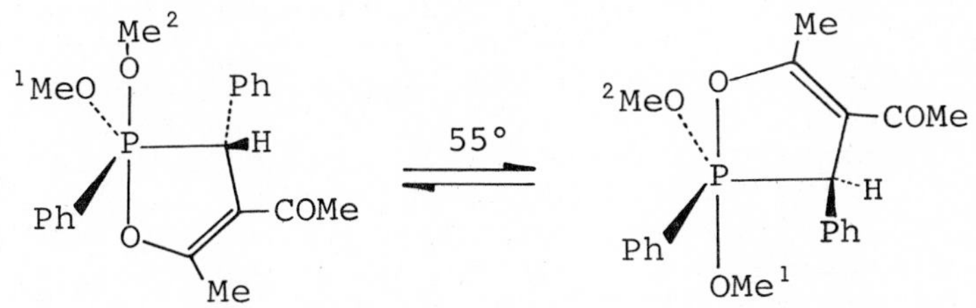

^{31}P +16.7 $\quad$ 55° $\quad$ ^{31}P +13.3

+55°, indicating full pseudorotation between the two diastereomeric trigonal bipyramids. Finally, at temperatures above +110°, reversible bond rupture to ionic species is detected in the proton resonance spectra of this compound and similar ones.

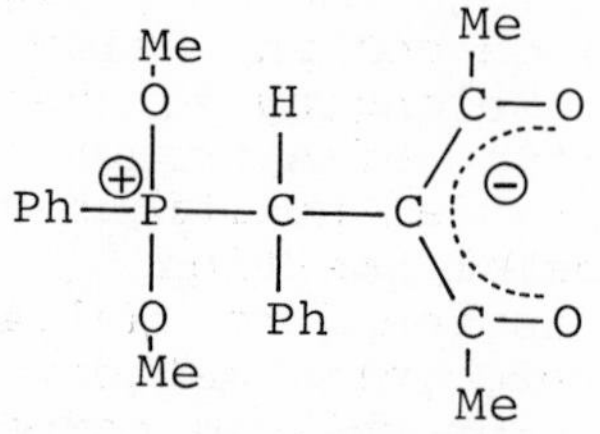

For a corresponding compound made from methyl diphenyl phosphinite and the same unsaturated ketone, the ^{1}H-NMR spectrum shows a totally stable trigonal bipyramid even at +25°, because pseudorotation would force a phenyl group and the ring carbon into the unfavorable axial positions.[528] Finally, with tertiary phosphines only dipolar ions are obtained, because the ring-closed species require at least one unfavorable apical carbon atom.[488,529]

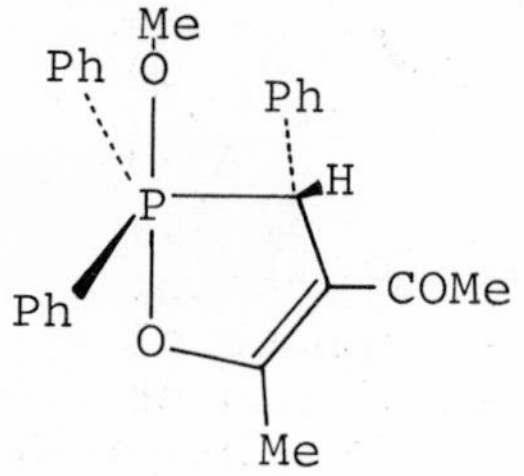

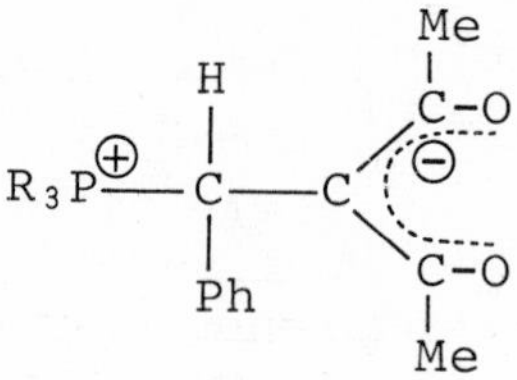

In the ^{1}H-NMR spectra of all these cyclic phosphoranes equatorial methoxy groups are generally found at lower field and with higher H-C-P coupling constants than axial MeO groups.[530]

Kinetic studies of the above processes have shown that the free energies of activation associated with placing a carbon group in apical positions in a trigonal bipyramid vary from 10 to 17 kcal/mole. Bulky substituents on the carbon groups further increase the energy of transition states when these groups are axial. The free energy of activation for ring opening is about 20 kcal/mole. At this level of energy, the presumable occurrence of trigonal-bipyramidal intermediates with a ring in equatorial-equatorial positions has also been discussed.[198]

The same rules that determine the limitations of intramolecular ligand reorganizations in the phospholane series are valid for phosphoranes containing the 1,2-oxaphosphetane system.[488,536] Generally such phosphoranes exist in diastereomeric trigonal-bipyramidal configurations with the electronegative oxygen groups in axial positions. The considerable rigidity of these systems is shown by the ^{1}H-NMR spectra, which are practically unchanged between -60° and +80°.

This strict inhibition of ligand reorganization by pseudorotation can be explained by the fact that each pseudorotation step would bring two quite electropositive carbon atoms into axial positions and/or would include intermediates with the four-membered ring in an equatorial-equatorial arrangement. Under certain conditions, however, equilibration of such diastereomeric species can be observed. This is believed to involve a bond rupture-recombination mechanism.[488,536]

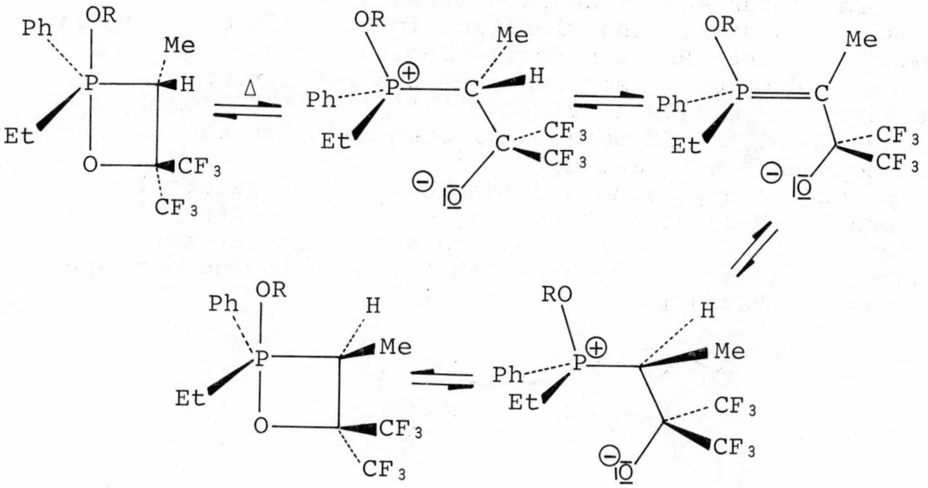

In summary, the following rules can be derived from stereochemical studies with oxyphosphoranes. The more electronegative oxygen groups always tend to occupy apical positions in the trigonal-bipyramidal configurations. Four- and five-membered rings occupy axial-equatorial positions. When these requirements are met, stable species are obtained which can undergo more or less rapid ligand reorganizations by pseudorotation, provided that the intermediate trigonal-bipyramidal species also obey the foregoing rules. When, on the other hand, these restrictions are violated, increasing and sometimes insurmountable barriers to pseudorotation occur. In extreme cases no pentacoordinate structure is possible at all.

The great majority of the oxyphosphoranes described in this section have ^{31}P chemical shifts in a relatively wide range around +50 ± 30 ppm. The dependence of these values on parameters such as ligand electronegativities and ring structures is not very characteristic. The differences in the δ ^{31}P values of diastereomers are between 1 and 5 ppm. Exceptionally low and even negative values in the range from -11 to +11 ppm are found for the ^{31}P chemical shifts of some 1,3,2-dioxaphospholanes made from tertiary phosphines and hexafluoroacetone.[488,537] In general, however, negative values of the phosphorus shift indicate the presence of open dipolar ions containing a phosphonium center.

C.4. Phosphoranes Containing a P-H Bond

Compounds of this type are known in the fluorophosphorane, the spirocyclic tetraoxyphosphorane, and the spirocyclic tetraarylphosphorane series. In the IR spectrum, hydridophosphoranes can easily be identified by the medium-intensity P-H stretching vibration in the 2400 cm^{-1} region. Compounds of type RPF_3H normally absorb between 2430 and 2470 cm^{-1}; $RPF_2(OR')H$ species show the P-H stretch somewhat lower, from 2400 to 2430 cm^{-1}.[136,143,145] Complete assignments of the IR bands have been made for the inorganic compounds HPF_4 and H_2PF_3.[632]

Spirotetra(oxy,amino)hydridophosphoranes generally have broad P-H bands at 2380 ± 30 cm^{-1}.[565] Systematic studies of the fundamental N-H transition in several compounds of this kind have shown that the following conjugation must be taken into consideration:[409]

Extremely low P-H values of about 2080 to 2100 cm^{-1} are found in the IR spectra of the bis(2,2'-biphenylylene)-hydridophosphoranes,[246,668] indicating very weak P-H bonds.

In the proton NMR spectrum, all hydridophosphoranes show the H-P hydrogen with chemical shifts lower than 6.5 ppm.[285a] In the fluorohydridophosphorane series these shifts are found between 7 and 7.5 ppm. The corresponding coupling constants increase with the sum of the electronegativities of the other substituents.[136]

	Me_2PF_2H	$RPF_2(NHR')H$	$RPF_2(OR')H$
J_{HP} (Hz)	733 [594]	740-780	850-860

H_2PF_3	$RPHF_3$	HPF_4
825, 865 [632]	900	1030, 1115 [632]

In the spirotetra(oxy,amino)hydridophosphorane series the ^1H-(P) chemical shift and the H-P couplings show similar variations with the electronegativity of the other ligands. When the oxygens of the rings are substituted by the less electronegative nitrogens, the δ ^{1}H values are shifted to higher field strength and the H-P coupling constants decrease. The opposite is true when phenyl rings are annexed.[565,566]

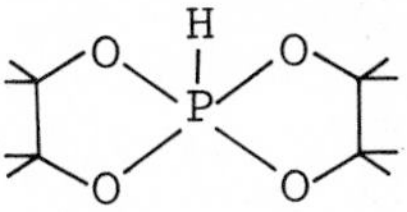

δ H = 7.1 ppm δ H = 7.13 to 6.66 δ H = 6.76 to 6.4

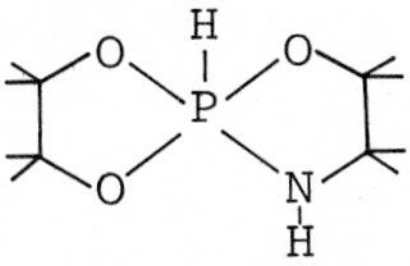

J_{HP} = 829 to 802 Hz J_{HP} = 800 to 760 J_{HP} = 770 to 730

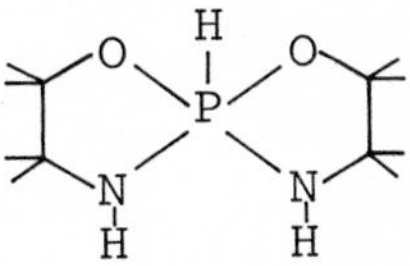

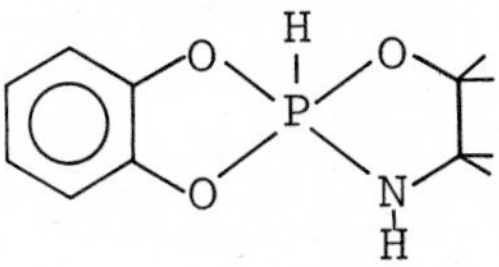

δ H = 7.73 to 7.91 δ H = 8.62

J_{HP} = 818 to 848 J_{HP} = 836

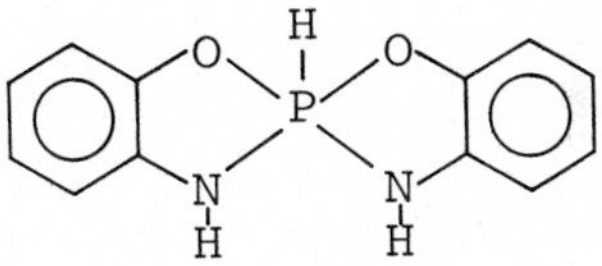

The bis(2,2'-biphenylylene)hydridophosphoranes again have exceptional properties in this respect. They show very low δ H-(P) values of about 9.33 ppm and small H-P coupling constants of about 480 Hz.[246,668] A compilation of the ^{1}H data of hydridophosphorus compounds was published recently.[285a]

The ^{31}P resonances of fluorohydridophosphoranes are generally found in the range of +20 to +50 ppm. Most of the phosphorus chemical shifts of spirotetra(oxy,amino)-hydridophosphoranes are between +35 and +55 ppm. In a few cases in which the ring nitrogens are substituted, however, δ ^{31}P values between +60 and +100 ppm are found.[566] The hydridophosphoranes with the bis-2,2'-biphenylylene phosphorus skeleton also show very high ^{31}P resonances of about +110 ppm.[246,668]

The stereochemical features of the various kinds of hydridophosphoranes are essentially the same as those of their analogs without H-P bonds. The microwave spectrum of HPF_4, which is the prototype of all hydridophosphoranes, has established the assumed trigonal-bipyramidal C_{2v} structure with longer axial than equatorial P-F bonds and an equatorial hydrogen.[462] According to the vibrational spectra, H_2PF_3 is a C_{2v} trigonal bipyramid in the gas phase, but an octahedral dimer with two fluorine bridges in the liquid and solid states.[202a,273] Hydridofluorophosphoranes of type RPF_3H show in the ^{19}F-NMR spectrum two different ^{19}F signals whose F-P couplings are consistent with axial ($J < 900$ Hz) and equatorial ($J > 900$ Hz) positions.[145,425] However, $MePF_3H$ reveals these spectral characteristics only at -60°, when pseudorotation is inhibited.[197,594] In phosphoranes of type $RPF_2(OR')H$ the low J_{FP} values of about 750 Hz also indicate axial fluorine positions.[136] The axial F-P coupling constants are still lower in compounds of type $RPF_2(NHR')H$ (~620 Hz)[139,143] and in Me_2PF_2H (535 Hz)[595] because less electronegative groups are present (Chapter C.1.).

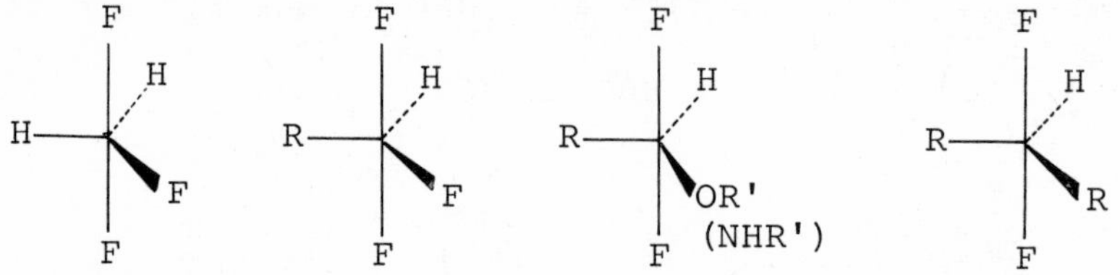

In the spirotetra(oxy,amino)hydridophosphoranes the rings occupy axial-equatorial positions in trigonal-bipyramidal configurations.[565] Many positional isomers, diastereomers, and the corresponding enantiomers are possible for these species, depending on the substituents of the rings. Unfavorable steric interactions often prevent the formation of certain diastereomers.[168]

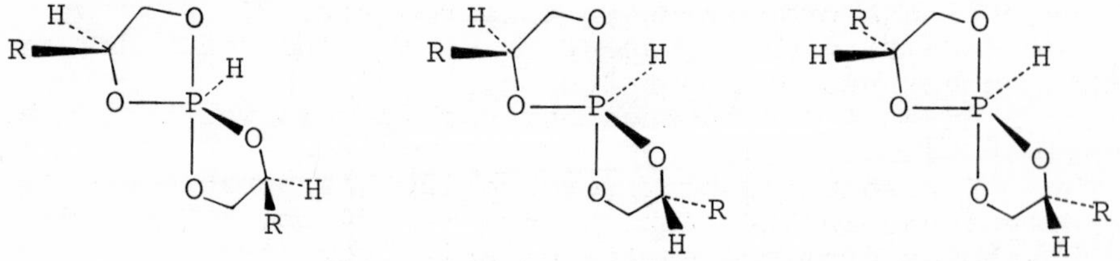

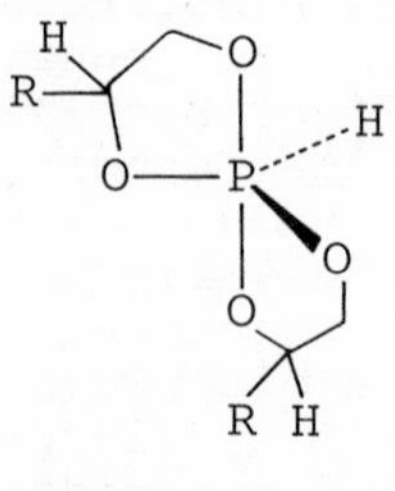

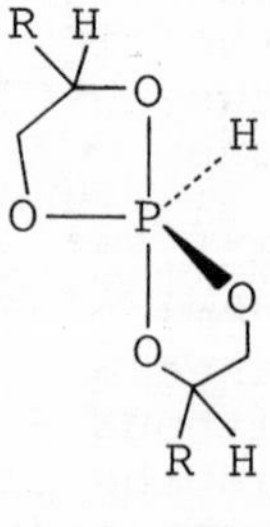

Three diastereomers Three diastereomers

Similarly, if nitrogen atoms are present in the rings, the situation becomes simpler because the electronegative oxygens tend to hold onto the axial positions and therefore some positional isomers do not form. These molecules show no pseudorotation at room temperature because this would imply intermediate trigonal bipyramids with the nitrogen(s) in unfavorable apical position(s).[565] In contrast, spirocyclic tetraoxyphosphoranes undergo intramolecular ligand reorganization by the pseudorotation mechanism even at -70°,[285] as proved by the fact that in the following compound only two different methyl signals can be detected, whereas a rigid trigonal bipyramid would demand four discernible methyl groups:

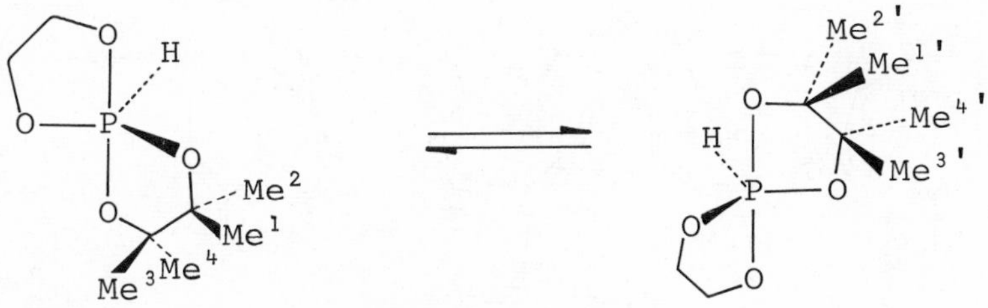

C.5. Pentaorganylphosphoranes

The vibrational spectra of pentaphenylphosphorane initially were incorrectly interpreted in terms of a tetragonal-pyramidal structure[113] but later were shown to be indicative of a trigonal-bipyramidal configuration.[42,203] This is consistent with the results of x-ray structure determinations, which demonstrate a distorted trigonal-bipyramidal configuration with longer axial (1.987 Å) than equatorial (1.850 Å) P-C bonds.[671] The proton NMR spectrum (60 MHz) of penta-p-tolylphosphorane shows one methyl signal, even at -60°. It is not known whether this is an accidental coincidence or an expression of rapidly occurring pseudorotation.[239]

The situation becomes clearer in the case of spirocyclic pentaorganyl phosphoranes of the bis-2,2'-biphenylylene-phosphorane type, where at lower temperatures "axial" and "equatorial" methyl groups of the chiral trigonal-bipyramidal configurations can be differentiated. At room temperature or at slightly elevated temperatures pseudorotation through a tetragonal-pyramidal transition state leads to equilibration of the methyl signals. The free energies of activation for the intramolecular ligand reorganizations depend on the bulkiness of the singly bonded ligand.[245]

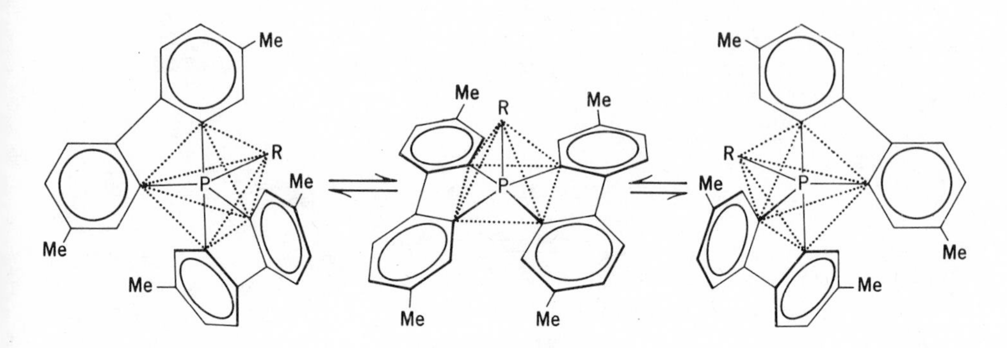

Fig. 1

R =	2-Np	Me	Et	Bz	2-Biphenylyl	1-Np
$\Delta F^\ddagger$ (T_c, °C)	11.9 (-48)	12.5 (-26)	13.6 (-4)	12.0 (-36)	15.7 (+20)	17.1 (+54)

R =	o-i-Pr-Ph	8-Me_2N-1-Np	8-MeO-1-Np
$\Delta F^\ddagger$	20 (E_a) [658]	> 21 [250]	> 23 [250]

In the spirophosphoranes containing the 8-dimethylamino-1-naphthyl and the 8-methoxy-1-naphthyl group, respectively, not only is the trigonal-bipyramidal configuration of the phosphorus fixed but also the nitrogen pyramid of the dimethylamino group seems to be unable to undergo inversion and/or rotation.[250]

Because of the rapid pseudorotation in most spirocyclic phosphoranes of this type, it is clear why no stable, optically active trigonal-bipyramidal species could be obtained.[241,243,674] However, provided that in the tetragonal-pyramidal transition state involved in these pseudorotations some element of chirality could be retained, optically active species should be obtainable. This possibility was realized by the stereoselective cleavage of a suitably substituted, optically active tris-2,2'-biphenylylene-phosphate complex where an optically active phosphorane with $[\alpha]_{578}^{24} = \pm 94 \pm 1°$, $[M]_{578}^{24} = \pm 472 \pm 5°$, could be obtained. This is actually a mixture of two rapidly equilibrating diastereomeric trigonal bipyramidal species.[242]

The stereochemistry of spirocyclic pentaorganylphosphoranes has been reviewed several times.[62,180,248,669]

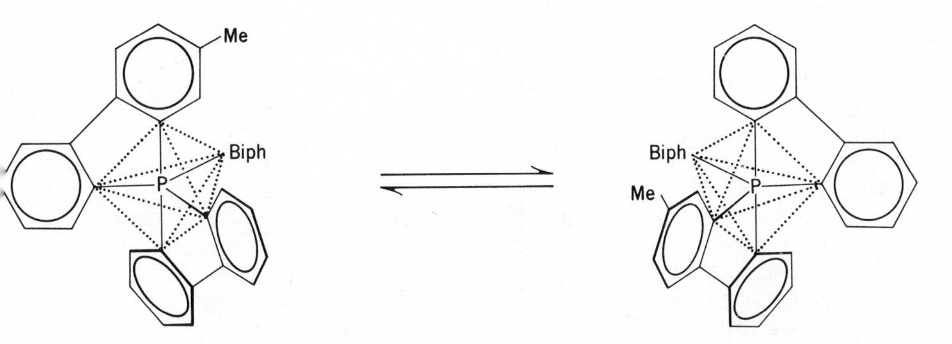

Fig. 2

The ^{31}P chemical shifts of pentaorganylphosphoranes are found in a rather narrow range around +87 ±10 ppm.[246,405]

C.6. Compounds with Hexavalent Phosphorus

Complexes with anions of type RPF_5^- have octahedral structures, as shown by the very characteristic ^{19}F-NMR pattern, consisting of a doublet of doublets for the equatorial fluorines and of a doublet of quintuplets for the axial fluorine.[424,592] Again, in general the equatorial F-P coupling constants are larger ($J > 800$ Hz) than the axial ones ($J < 730$ Hz). The prototype of this group of compounds, the hydridopentafluorophosphate anion, HPF_5^-, reveals corresponding ^{19}F-NMR data but also a characteristic proton NMR signal at 5.4 ppm with $J_{HP} = 955$ Hz.[448] Some adducts of phosphorus pentafluoride with bases have also real hexacovalent octahedral structures, as shown by the comparable ^{19}F-NMR patterns.[423,623]

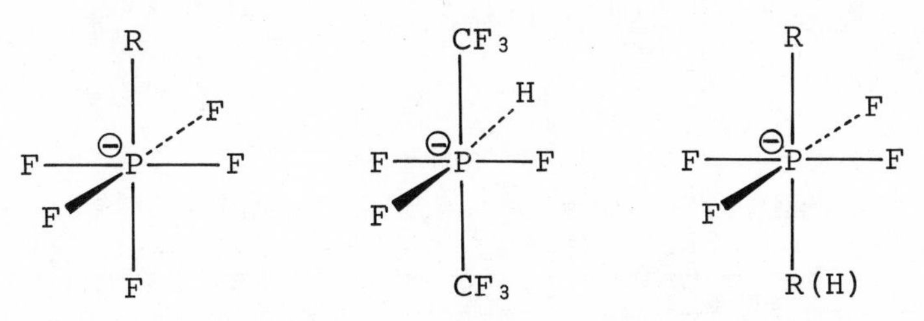

In anions like $R_2PF_4^-$ or $RHPF_4^-$, trans configuration with identical equatorial fluorine atoms is observed ($J_{FP} = 858$

to 898 Hz).[72a,94,314] Only one compound of type $R_2HPF_3^-$ is known. In this case one equatorial fluorine has J_{FP} = 834 Hz, and the other two have J_{FP} = 725 Hz. Again the hydrogen is more strongly shielded (δ H = -5.0 ppm) than in corresponding pentavalent phosphorus compounds.[447] The stable adducts of chlorofluorophosphoranes with pyridine[272] seem also to have octahedral structures but with much greater F-P couplings (ca. 1000 Hz).[271] The ^{31}P chemical shifts of fluorophosphates are much higher than those of corresponding pentacoordinated compounds. The δ ^{31}P values for $AlkPF_5^-$ compounds are found at ~+125 ppm; for $ArPF_5^-$, at ~+136 ppm.[592] The adduct of acetonitrile with PF_5 absorbs at +146 ppm, exactly like the PF_6^- anion itself.[662] The P-F stretching frequencies of fluorophosphate complexes are found in the 620 to 830 cm^{-1} region.[94,94a]

Hexacoordinated chlorophosphates are known only in the form of inner onium-ate complexes obtained by the reaction of phosphorus pentachloride with suitable mono- or bidentate bases. Molecular weight determinations, conductivity measurements, and vibrational spectroscopical studies have demonstrated the hexacovalent nature of the complexes with phenanthroline and pyridine. The P-Cl stretching vibrations are found in the 445 to 568 cm^{-1} range.[44,654] Whereas for these two adducts the cis configuration of the two nitrogen ligands is assumed, the corresponding complexes with THF and tetrahydrothiophene are believed to have the trans configuration.[44] An x-ray structural determination with the compound obtained from PCl_5 and dimethylurea has established a distorted octahedral geometry with a planar diazaphosphetidine ring.[708] The P-N bonds are longer than a single bond, and the N-C (ring) bonds are considerably shorter than N-C single bonds. This is in accordance with a negatively charged phosphorus and with delocalization of the positive charge of the ring carbon.

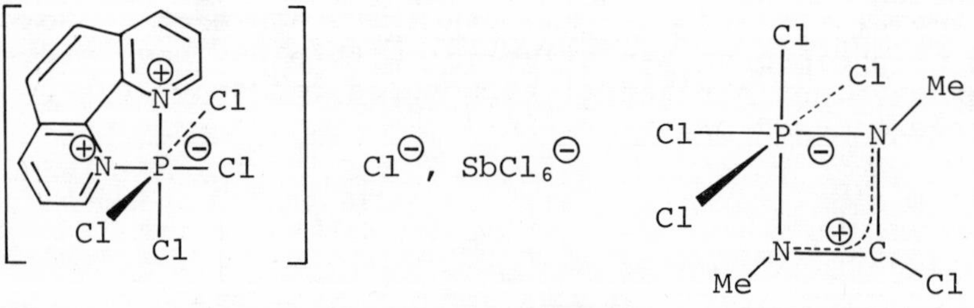

$Cl^{\ominus}$, $SbCl_6^{\ominus}$

The ^{31}P chemical shifts of such chlorophosphates are found at very high field strengths. Chlorophosphates containing diazaphosphetidine rings usually have signals around +200 ppm,[50,378] whereas the adducts of PCl_5 with pyridine[662] or tertiary phosphine oxides absorb at even higher strengths (ca. +300 ppm).[68]

Only a few hexacoordinate phosphate complexes with oxy

groups are known.[1,2,251] For ate complexes with o-phenylene-dioxy groups a close relationship can be shown between the ^{31}P resonance data and the number of o-phenylene-dioxy groups present.[251] In going from the tris-2,2'-biphenylylene-phosphate anion to the tris-(o-phenylene-dioxy)phosphate anion each catechol substituent lowers the δ ^{31}P value by about 30 ppm. The latter complex has, therefore, a relatively low δ ^{31}P value of only +82 ppm, which lies in the normal range for pentacoordinate phosphorus species! In the IR spectra of such complexes the P-O stretchings are found at 802 to 825 cm^{-1}, and the aryl-O stretchings at 1210 to 1223 cm^{-1}.

+181 +147 +106 +82

The octahedral structure of the tris-2,2'-biphenylylene-phosphates was elucidated by their separation into optical isomers with very high molecular rotations: $[M]^{25}_{578}$ $> 10{,}000°$.[241-243] By investigating the circular dichroism and electronic absorption spectra, the (-)tris-2,2'-biphenylylene-phosphate complex was shown to have the **M**(C_3) configuration of a left-handed screw, as shown in the picture.[249a] Whereas this complex has considerable optical

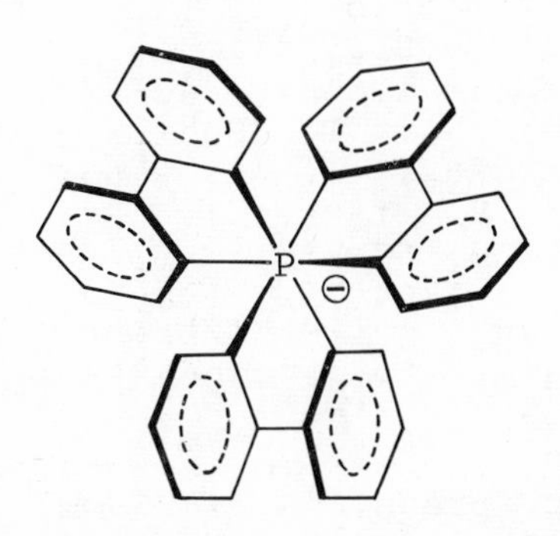

stability, the optically active tris(4,4'-dimethyl-2,2'-biphenylylene)-phosphate anion easily racemizes when warmed in methanol-acetone.[243] It is not clear whether intramolecular ligand reorganization occurs by the trigonal or by the rhombic-twist mechanism.[187,189] Finally, it was shown by means of ^{32}P labeled bis(2,2'-biphenylylene)-^{32}P-phosphonium-tris(2,2'-biphenylylene)phosphate that the onium-ate structure is stable up to 130° in dimethylformamide; no syncoordination to a double phosphorane could be detected.[244]

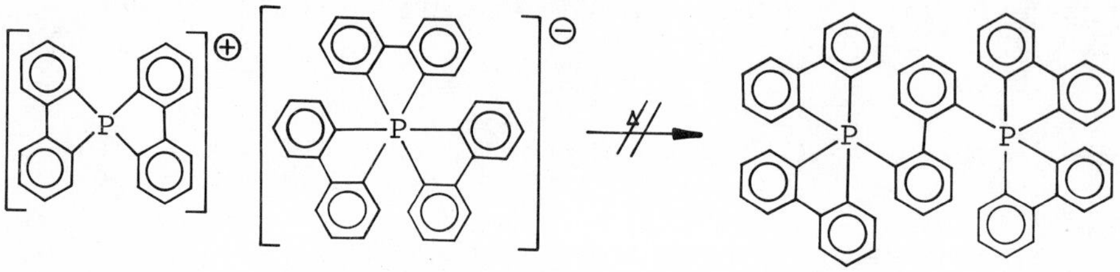

D. LIST OF COMPOUNDS

D.1. Compounds of Type PY_5

TYPE: $RZPX_4$

$MeOPF_4$. I. Not pure, ^1H, ^{19}F[81,108] mass spect., ^{31}P +79,[81] ^{31}P +65.[108]

$MeSPF_4$. I. Liquid, volatile at -45°,[79b] dec. 0°, ^{31}P +34.2,[456a] mass spect., IR, ^1H,[79b] ^{19}F shows three kinds of F at -100°.[456,456a]

$EtSPF_4$. I. Liquid, dec. 0°, ^{31}P +34,[456a] ^{19}F shows 3 kinds of F at -70°.[456,456a]

$PhSPF_4$. I. B_{40} 75-76°,[456a] b_{10} 58°,[456] 3 kinds of ^{19}F at -80°,[456,456a] ^{31}P +38.3.[456a]

$RHNPF_4$. R = 2,4-$Me_2C_6H_3$: I. $B_{0.5}$ 42-44°, m. 32-34°, IR.[233] R = 2,6-$Me_2C_6H_3$: I. Subl.$_{0.5}$ 60°, m. 60-62°, IR.[233] R = 2,6-$Et_2C_6H_3$: I. $B_{0.5}$ 63°, m. 15-17°, IR, 3 kinds of ^{19}F.[233] R = i-Pr: I. B_{750} 37°, IR.[233]

Me_2NPF_4. I. B. 60°,[82] b. 63.7°,[115] m. -80°,[115] m. < -78°,[82] log P_{mm} = -1764.75/T + 8.12,[115] log P_{mm} = -1936/T + 8.69,[82] IR, ^1H,[79b,82,115] ^{31}P +69.7,[590] ^{19}F,[82,115] 2 kinds of ^{19}F at -90°,[590] line shape analysis,[659] mass spect.[73]

Et_2NPF_4. I. B. 99-100°,[578,589] b_{150} 45-47°, n_D^{20} 1.3440, d_4^{20} 1.2368,[149] vap. press.$_{19°}$ = 25 mm,[82] m. -73°,[82] ^{31}P +70.2,[578] ^{19}F,[82,149,578] 2 kinds of ^{19}F at -85°,[425,590] mass spect.[73]

Pr_2NPF_4. I. Liquid, Vap. press.$_{19°}$ = 25 mm, ^{19}F, J_{FP}

856 Hz.[82]

Ph_2NPF_4. I. Solid, subl.$_{0.5}$ 80°,[590] 2 kinds of ^{19}F at -72°,[624] mass spect.[73]

$(CF_3)_2NPF_3Cl$. V. B. 44°, $\log_{10} P_{cm} = -1380/T + 6.23$, IR, 2 kinds of ^{19}F at -70°.[156]

Et_2NPF_3Br. (?) From disprop. of $Et_2NPF_2Br_2$, not isolated, 2 kinds of ^{19}F.[149]

$Me_2NPF_2Cl_2$. IV.[103,150] VI.[103] B. 144°, $\log P_{mm} = -2022/T + 77.2$,[103] n_D^{20} 1.4210, ^{31}P +44,[150] ^1H, IR,[103] ^{19}F.[103,150]

$(CF_3)_2NPF_2Cl_2$. V. B. 81°, $\log_{10} P_{cm} = -1720/T + 6.74$, IR, ^{19}F.[156]

$Et_2NPF_2Cl_2$. IV. D_4^{20} 1.2855, n_D^{20} 1.4280, ^{31}P +42, ^{19}F.[150]

$Me_2NPF_2Br_2$. IV. D_4^{20} 1.9134, n_D^{20} 1.4785, ^{31}P +103, ^{19}F.[150]

$Et_2NPF_2Br_2$. IV. D_4^{20} 1.7531, n_D^{20} 1.4760, ^{31}P +108, ^{19}F.[150]

$PhOPCl_4$. IV. Unstable hygroscopic needles.[347]

$4\text{-}MeC_6H_4OPCl_4$. IV. Very unstable solid.[347]

(2-t-Bu-C$_6$H$_4$)C(Cl)(OPCl$_4$)(2-Cl-6-t-Bu-C$_6$H$_3$) [structure: t-Bu, Cl, C—$OPCl_4$, Cl, t-Bu]. From di-t-butylbenzophenone and PCl_5, yellow solid.[627]

Me_2NPCl_4: $\cdot PCl_5$. III. Dec. 242-244°.[347]

Et_2NPCl_4: $\cdot PCl_5$. III. Dec. 232-233°.[347]

Pr_2NPCl_4. IV. Cryst., $\cdot PCl_5$: III. Dec. 220°.[347]

$i\text{-}Bu_2NPCl_4$. IV. Cryst., $\cdot PCl_5$: III. Dec. 168°.[347]

$MePhNPCl_4$. IV. Cryst.[347]

$H_2C{=}C{=}NPCl_4$: $\cdot PCl_5$. III. Dec. 70°, from MeCN and PCl_5.[333]

$ClHC{=}C{=}NPCl_4$: $\cdot PCl_5$. III. From ClH_2CN and PCl_5, dec. 140°.[333]

$NC\text{-}C(R)(R')\text{-}N(R'')PCl_4$. III. R, R', R" = Me: m. 98-100°. R, R" = Me, R' = Et: m. 70-73°. R, R' = Me, R" = Bu: m. 63-66°. R, R' = $-(CH_2)_5-$, R" = Me: m. 79-82°.[463]

$PhOPCl_2Br_2$. IV. Very unstable solid.[347]

Me_2NPBr_4. IV. Yellow cryst., dec. 114-116°.[451]

TYPE: $(RZ)_2PX_3$

$(MeO)_2PF_3$ = $[(MeO)_4P^+]PF_6^-$. I. M. 108°, ^{19}F, ^{31}P (solution) 0, +147, ^1H.[340]

$(EtO)_2PF_3$ = $[(EtO)_4P^+]PF_6^-$. I. M. 58°, ^1H, ^{19}F, ^{31}P (solution) +5, +147, ^{31}P (solid) +8, +148.[340]

$(PhO)_2PF_3$. I. M. 98°, ^{31}P (solid) +76, in solution = $[(PhO)_4P^+]PF_6^-$, ^{31}P +24, +147, ^{19}F.[340]

$(BuHN)(Me_2N)PF_3$. I. B_{10} 70-73°, d^{20} 1.1200, n_D^{20} 1.403, 3 kinds of ^{19}F.[148]

$(i\text{-}PrHN)(Me_2N)PF_3$. I. B_{20} 50-52°, d^{20} 1.1405, n_D^{20} 1.3970, 3 kinds of ^{19}F.[148]

$(MeHN)(Me_2N)PF_3$. I. B_{25} 41-43°, d^{20} 1.1750, n_D^{20} 1.4070,

3 kinds of ^{19}F.[148]

$(BuHN)(Et_2N)PF_3$. I. B_{13} 84-86°, n_D^{20} 1.4102, 3 kinds of ^{19}F.[148]

$(Me_2N)_2PF_3$. I. B_{19} 43°,[590] m. -22°, IR, ^1H,[82] 2 kinds of ^{19}F,[82,590] mass spect.[73]

$(Me_2N)(Et_2N)PF_3$. Two kinds of ^{19}F.[590]

$(Et_2N)_2PF_3$. I. M. 5°,[82] b_{14} 79°,[578,589] b_{18} 84-86°, n_D^{25} 1.4049,[589] ^{31}P +63.7,[578] ^{31}P +64.5,[593] ^1H,[593] 2 kinds of ^{19}F,[82,425,578,593] mass spect.[73]

$(Pr_2N)_2PF_3$. I. B_4 19°.[590]

$(Me_2N)_2PFCl_2$. IV. n_D^{20} 1.4970, ^{31}P -40 (!?), ^{19}F.[150]

$(Me_2N)_2PFBr_2$. IV. n_D^{20} 1.5540, ^{31}P -32 (!?), ^{19}F.[150]

$(Me_2N)_2PBr_3 = [(Me_2N)_2PBr_2]^+Br^-$. IV. Yellow cryst., dec. 207°.[450,451]

Me
N
OC PF$_3$. II. B_{20} 39-40°, ^{19}F, ^{31}P +55.8.[592a]
N
Me

Me
N
O_2S PF$_3$. I. M. 54°,[51] subl.$_{0.1}$ 50°, ^{31}P +76.8,[54,379] ^1H, ^{19}F.[54]
N
Me

$(MeNPF_3)_2$. I.[114,115,582,588] II.[350] B. 87.9°, log P_{mm} = -2021.24/T + 8.46782,[115] b. 88-90°,[350] b. 88°,[223] b_{757} 91.6°,[582] b_{175} 54°,[588] m. -10.3°,[115] m. -8.3°,[582] m. -11°,[223] n_D^{25} 1.3323, d_4^{25} 1.532,[588] ^1H,[115,446] ^{19}F,[114,234] ^{31}P +69.5,[590] ^{31}P +71.5,[582] μ 0 D,[582] IR and Raman show C_{2h} trig. bipyr.,[132,685] elect. diff. establishes trig. bipyr. struct. with $P-F_{a,e}$ 1.605, 1.570 Å, $P-N_{a,e}$ 1.735, 1.595 Å, F_aPF_e 88.5°, F_ePF_e 103.9°, N_ePF_e 127.5°, N_ePF_a 99.1°, N_aPN_e 77.9°. The ring is planar.[3]

$(RNPF_3)_2$. II.

R	b. (mm)	n_D^{20}	d_4^{20}
Me	89-90°	1.3388	1.5554
Et	123-124°	1.3551	1.4030
Pr	162-163°	1.3750	1.3312
n-Bu	81-82° (12)	1.3854	1.2580
i-Bu	71-73° (12)	1.3848	1.2573

All products are very moisture sensitive.[135]

$(RNPF_3)_2$. I. IR, ^{19}F (1 kind).[233]

R	m.	subl. (mm)
Ph	198-205°	110-130° (0.4)
$2,4-Me_2C_6H_3$	164-168°	140-150° (0.4)
$2,6-Me_2C_6H_3$	255-257°	130° (0.05)
$2,6-Et_2C_6H_3$	151-152°	125° (0.025)
n-Pr	4-7°	b_4 41°, b_{21} 70-72°
t-Bu	70-72°	70° (4)

$(PhNPFCl_2)_2$. I. M. 186°.[51]
$(PhO)_2PCl_3$. IV. Viscous oil,[215] unstable cryst.[347]
$(PhO)_2PClBr_2$. IV. Unstable orange solid.[347]

(o-phenylenedioxy)PX_3. $X_3 = Cl_3$: III.[213,347,500] IV.[174,215] M. 48-54°,[218] m. 61-62°,[213,347,500] $b_{0.5}$ 95°,[500] b_{10} 128-135°,[215] b_{12} 127-131°,[218] b_{13} 132-135°,[213] ^{31}P +26 to +26.4.[174,316,500] $X_3 = Cl_2Br$: IV. ^{31}P +75.8. $X_3 = ClBr_2$: IV. ^{31}P +131.[174] $X_3 = Br_3$: IV. Yellow cryst., m. 67-70°,[216] ^{31}P +189.[174]

(4-Me-o-phenylenedioxy)PCl_3. IV. B_{11} 158.[347]

$(ClC_2H_4O)(Et_2N)PCl_3$. IV. No data, but reaction with SO_2 yields $>P(=O)Cl$.[403]

$MeN{=}C\langle N(Me)\cdot HCl,\ O\rangle PCl_3$. From MeNCO and PCl_5, dec. 183°, ^{31}P -50, shows ionic struct.: $>\overset{+}{P}Cl_2]Cl^-$.[374]

$O{=}C\langle N(R),\ N(R')\rangle PCl_3$. III. R, R' = Me: $b_{1.5}$ 78-79°, IR, ^1H,[643] ^{31}P +60.[375] R, R' = n-Bu: $b_{0.3}$ 105-108°, IR, ^1H. R = Ph, R' = n-Bu: Oil, IR, ^1H.[643]

$(MeNPCl_3)_2$. III. M. 160°,[95] m. 173-176°,[52,696] m. 174-179°,[449] m. 177-178°,[377] $subl_{0.1}$ 140°,[449] ^1H,[632a] ^{31}P +78.2,[172a,222] ^{31}P +78,[377] P-N bond energy 74.3,[177] IR, Raman,[95] x-ray structure: $P\text{-}Cl_{a,e}$ 2.152, 2.018-2.022 Å, $P\text{-}N_{a,e}$ 1.776, 1.629 Å, Cl_ePN_e 124.6-125.2°, Cl_aPN_e 93.8°, Cl_ePCl_e 109.8°, N_ePN_a 81.7°.[258] Another x-ray determination gives similar values,[261] mass spect.[449]
$(ClH_2CNPCl_3)_2$. Photochemical chlorination of $(MeNPCl_3)_2$, m. 134-135.°[135]
$(ClOCCH_2NPCl_3)_2$. III. Dec. 149-151°, forms adduct with benzene.[698]
$(RNPCl_3)_2$. III. IR, ^1H, cryst. from CCl_4.[222] R, m., ^{31}P: Et, 122-124°, +78.8 (+78.5[375]); Pr, 118-120°, +78.7; Bu, 76-79°, +79.3; (i-Bu, m. 101-104°[135]); Am, 69-71°, +78.4; Hex, 52-55°, +79.1; Hep, 34-35°, +79.1; Oct, 22-24°, +79.8; Non, 34-36°, +79.8; Dec, 48-49°, +78.6; i-Pr, 71-73°, —; β-MeBu 37-39°, +77.9; i-Am, 74-77°, +78.7.
$(RNPCl_3)_2$. III.[696] R, m.: Me, 174-176°; Et, 119-122°; n-Pr, 112-115°; n-Bu, 75-78°; n-Am, 70-72°. Similar: III.[697] R, m.: $Me_3CCH_2CH_2$-, 135-137°, $PhCH_2$-, 167-169°; $PhCH_2CH_2$-, 173-176°; $ClCH_2CH_2$-, 145-147°.

$(PhNPCl_3)_2$. III. M. 180-182°,[221,700] IR,[221] ^{31}P +80.2.[375] Used as anti-oxidant in lubricating oil.[387]

$(ArNPCl_3)_2$. III. The following products are dimeric under all conditions: Ar, m.: 3,5-$Me_2C_6H_3$-, 177-179°; p-ClC_6H_4-, 181-183°; p-MeC_6H_4-, 198-200°; p-$MeOC_6H_4$-, 196-198°; p-$EtOC_6H_4$-, 185-187°; p-BrC_6H_4-, 184-186°; o-MeC_6H_4-, 124-126°. The following substances are partly monomeric in boiling benzene: m-MeC_6H_4-, 154-155°; m-ClC_6H_4-, 129-131°; m-BrC_6H_4-, 125-126°. The following are totally monomeric in boiling benzene: o-ClC_6H_4-, 127-128°; o-BrC_6H_4-, 121-123°; 2,4-$Cl_2C_6H_3$-, 116-118°; 2,4-$Br_2C_6H_3$-, 114-115°; 3,5-$Cl_2C_6H_3$-, 136-138°; m-$NO_2C_6H_4$-, 142-143°; p-$NO_2C_6H_4$-, 140-141°. Monomeric even in cold benzene: o-$NO_2C_6H_4$-, 109-111°. All products are dimeric in the solid state.[700]

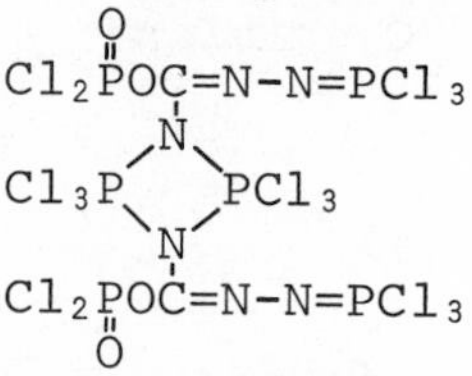

$Cl_2P(O)OC=N-N=PCl_3$ / N / Cl_3P PCl_3 / N / $Cl_2P(O)OC=N-N=PCl_3$. III. From semicarbazide and PCl_5, hygroscopic cryst.,[53] ^{31}P +78.8, -7.6, -14.5,[379] resp. (-15.2[53]), IR.[53]

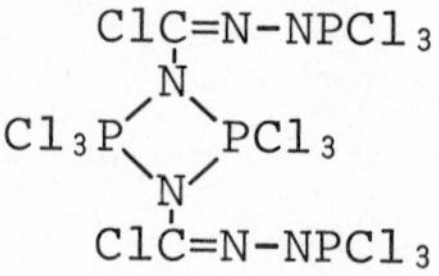

$ClC=N-NPCl_3$ / N / Cl_3P PCl_3 / N / $ClC=N-NPCl_3$. From above product at 60°, hygroscopic cryst., IR,[53] ^{31}P +78.8, -15.2.[53,379]

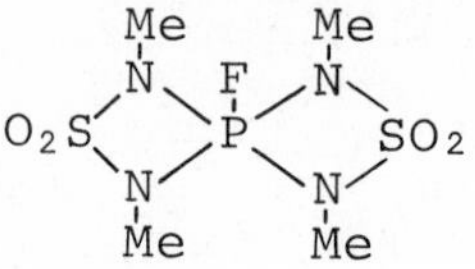

Me Me / N F N / O_2S P SO_2 / N N / Me Me . I.[54] II.[52] M. 184-188°,[51,52] m. 180-182°,[54] ^{19}F,[52,54] ^1H,[54] IR,[52] ^{31}P +84.6,[379] ^{31}P +85,[54] ^{31}P +86.6.[52]

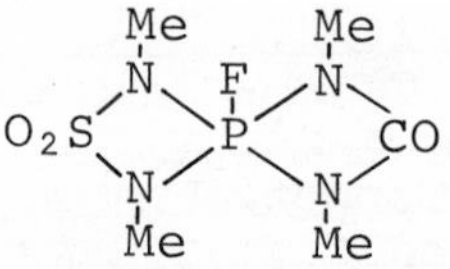

Me Me / N F N / O_2S P CO / N N / Me Me . I. M. 117°,[51] m. 116-118°, ^{19}F, ^1H,[54] ^{31}P +67.[54,379]

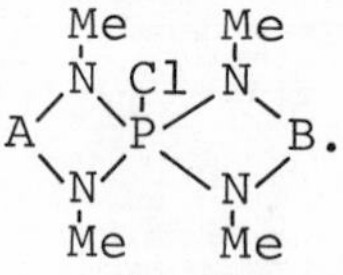

Me Me / N Cl N / A P B / N N / Me Me . A, B = >CO: III. ^{31}P +59.[50,379] A = >CO, B = >SO_2: III. M. 128°, ^1H, ^{31}P +70.5.[50] A, B = >SO_2: III. M. 171°,[51] 170-171.5°, ^1H, IR,[52] ^{31}P +86,[52,379] $\cdot TiCl_4$: yellow cryst. dec. 186-189°; $\cdot 2SbCl_5$: olive cryst., dec. 147°; $\cdot 2BCl_3$: white cryst., dec. > 90°.[52] A, B = SO_2, Ph for Me: m. 188°.[51,52]

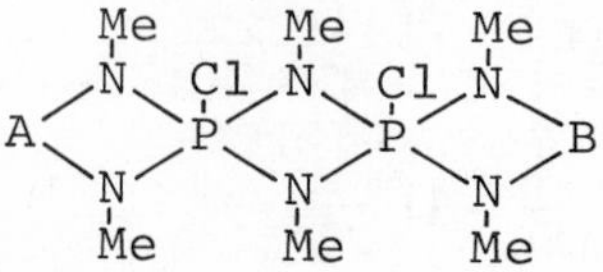

Me Me Me / N Cl N Cl N / A P P B / N N N / Me Me Me . A, B = >CO: III. Dec. 178-180°, ^1H, IR, mass spect.,[50] ^{31}P +67

$(HCCl_3)$,[379] +67 (CH_2Cl_2), +69 (benzene).[50] A, B = $>SO_2$: III. M. 221°,[51] 220-221°, IR,[52] ^{31}P +81.2,[379] +82.1.[52] A, B = $>PCl_3$: III. Dec. 395°, IR,[49] ^{31}P +74.5.[375] X-ray determination:[655] trig. bipyr. with outer P: $P-Cl_{a,e}$ 2.16, 2.05 Å; $P-N_{a,e}$ 1.71, 1.66 Å; N_ePCl_e 123.8, 126.7°; Cl_ePCl_e 109.4°; N_ePN_a 80.7°; all other a,e angles are between 89.3 and 93.6°. Inner P: $P-Cl_e$ 2.10 Å; $P-N_{a,e}$ 1.77, 1.74, 1.67, 1.64 Å; N_ePN_e 125.3°; N_ePCl_e 116.8, 117.8°; N_ePN_a 80.0, 79.0°; all other a,e angles are between 89.4 and 98.1°. A, B = $>P(=O)Cl$: from above with SO_2, dec. 185-220°, IR,[49] ^{31}P +73 (-4).[375] A, B = $>P(=S)Cl$: from above and H_2S, dec. 230°, IR,[49] ^{31}P +68 (-55.3).[375] A, B = $>P(=S)NEt_2$: ^{31}P +68 (-55.3).[375]

[Structure: 1,3-dimethyl-4,5-dioxo-imidazolidine ring with PCl_3] PCl_3. III. ^{31}P +69[55,55a] (not isolated, only presumably).

[Structure: ring R, R', Cl, N-R", PCl_3] PCl_3. III.[463] R, R', R" = Me: prisms, m. 105-106°; $\cdot PCl_5$, dec. 213-214°. R, R" = Me, R' = Et: needles, m. 84-86°, $\cdot PCl_5$, dec. 205-207°. R, R' = Me, R" = Et: needles, m. 70-72°, $\cdot PCl_5$, dec. 190-191°. R, R' = Me, R" = Bu: prisms, m. 98-100°, $\cdot PCl_5$, dec. 175-176°. R, R' = $-(CH_2)_5-$, R" = Bu: needles, m. 141-142°, $\cdot PCl_5$, dec. 159-160°.

[Structure: Cl-C=N-N(Ph), N(Ph) ring with PCl_3] PCl_3. III. Very hygroscopic, dec. 156°,[274] ^{31}P +77.[274,379]

[Structure: bicyclic MeN, C=O, N, SO_2, N⊖, C⊕, N-Me, PCl_3] PCl_3. From $OC(NMe)_2PCl_3$ and $(OCN)_2SO_2$, dec. 132°, ^1H, ^{31}P -44.5[380] = ionic structure!

[Structure: MeN, two C=O, two N-Me ring with PCl_3] PCl_3. From above four membered ring and MeNCO, elect. cond., ^{31}P -50, = ionic structure.[373]

TYPE: $(RZ)_3PX_2$

$(MeO)_3PF_2$. VI. Not isolated, ^{19}F, J_{FP} 706 Hz, ^{1}H.[415]
$(PhO)_3PF_2$. VI. No details, ^{19}F, J_{FP} 721 Hz.[415]
$(Me_2N)_2(MeO)PF_2$. VI. B_4 39-40°, n_D^{23} 1.4069, ^{19}F, IR.[415]
$(BuS)_3PF_2$. VI. $B_{0.05}$ 123°, n_D^{24} 1.5232, ^{19}F, IR.[415]
$(Me_2N)_3PF_2$. VI. B_3 35°, n_D^{25} 1.4190, d^{25} 1.0604, mass spect., IR, ^{31}P +65.7,[535] ^{1}H, ^{19}F.[415,535]

Me–N, N–Me ring; PF_2–N(piperidino). VI. Not isolated, ^{31}P +50.6, J_{PF} 796 Hz.[508]

Me–N, N–Me ring; PF_2–NMe_2. VI. Not isolated, ^{31}P +47.3, J_{PF} 798 Hz.[508]

$(—N—\underset{F_2}{P}—)_n$ (benzo-fused N,N). I. No data, polymeric substance.[232]

$(PhO)_3PCl_2$. IV. M. 70-80°,[472] 76-78°,[699] 80-81°,[100] liquid,[347,460] ^{31}P +22.8,[499] $\cdot PCl_5$ = $[(PhO)_3PCl]^+[PCl_6]^-$: III. Light yellow cryst., m. 124-128°.[75]
$(2\text{-}MeC_6H_4O)_3PCl_2$. III. Liquid, dec. to o-chlorotoluene at 180°.[347]
$(3\text{-}MeC_6H_4O)_3PCl_2$. III. IV. Liquid, dec. to m-chlorotoluene.[347]
$(4\text{-}MeC_6H_4O)_3PCl_2$. III. IV. Liquid, dec. to p-chlorotoluene.[347]
$(1\text{-}NpO)_3PCl_2$. IV. Green oil.[347]
$(2\text{-}NpO)_3PCl_2$. IV. Liquid, dec. to 2-chloronaphthalene at 310°.[347]
$[2\text{-}(1,6\text{-}Br_2C_{10}H_5)O]_3PCl_2$. III. Hygroscopic liquid.[347]
$(2,6\text{-}Me_2C_6H_3O)_3PCl_2$ = $[(RO)_3PCl_3]^-[(RO)_3PCl]^+$. IV. Needles, m. 124-125°.[564]
$(PhO)_3PBr_2$. IV. Reddish plates,[347] reaction with THF;[693] $\cdot Br_2$: Yellow solid, m. 105-106°.[231]
$(1\text{-}NpO)_3PBr_2$. IV. Red liquid.[347]
$(2\text{-}NpO)_3PBr_3$. IV. Dark liquid.[347]
$(Me_3Si)_3PBr_2$. IV. No details, no data.[597]
$(PhO)_3PI_2$. IV. Brown rhombs, m. 68-69°,[564] conduct. titr. shows $[(PhO)_3PI]^+I^-$;[231] $\cdot I_2$: Dark red prisms[176] or plates, m. 76-78°;[564] $\cdot 3I_2$, $\cdot 4I_2$: Green prisms, m. 89-90° these are salts with the anions I_3^-, I_7^- and I_9^-.[176]
$(2,6\text{-}Me_2C_6H_3O)_3PI_2$. IV. Black rhombs, m. 90-93°.[564]

$(R_2N)_3PX^1X^2$.

R	X^1	X^2	m.	Synth.[37]
Pr	Br	Br	115°	IV
Pr	I	I	186-187° dec.	IV
Pr	Cl	I	155-158°	Metathesis
Pr	Br	I	160° dec.	Metathesis
Allyl	Cl	I	67-70°	Metathesis
Pr	F	I	138-140°	II

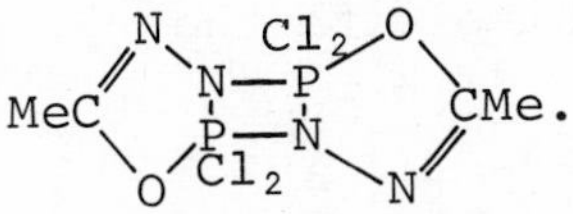

[structure]. III. Dec. 150°, $subl_{0.01}$ 120°, needles, ^1H, ^{31}P +52.7.[152]

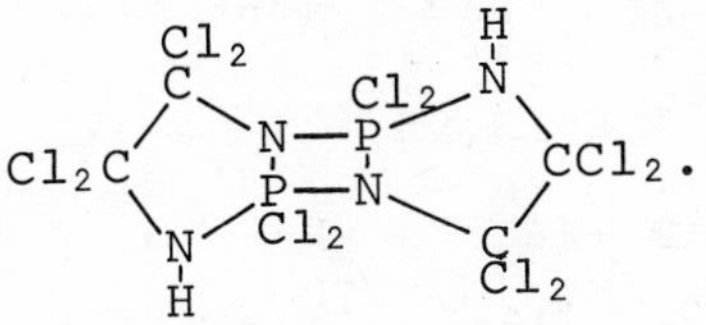

[structure]. III. Dec. 80°, IR,[55a] ^{31}P +55.[55,55a,379]

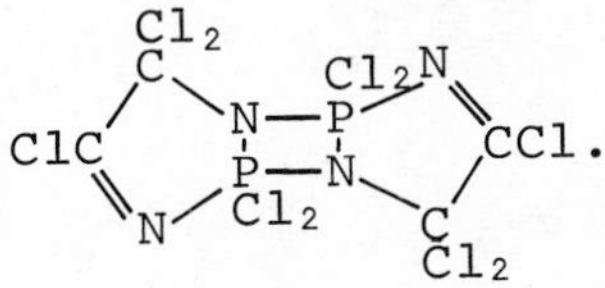

[structure]. III. Dec. > 85°, IR,[55a] ^{31}P +48.[55,55a,379]

TYPE: $(RZ)_4PX$

$(PhO)_4PF$?. I. From $[PCl_4]^+PF_6^-$, ^{31}P +75 (solid).[339]

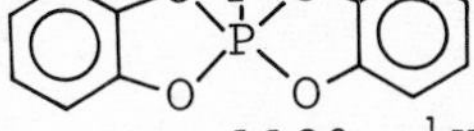

[structure]. I.[131a] Thermolysis of [structure],[160] m. 112°, ^1H, IR,[160] ^{19}F, J_{FP} 1018 Hz.[131a]

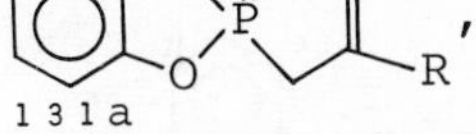

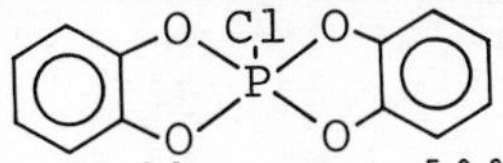

[structure]. III. M. 164-174°, b_{12} 195-202°,[218] m. 166-168°, b_{11} 194°,[347] m. 174-175°, ^{31}P +9.4,[500] ^{31}P +9.7.[174]

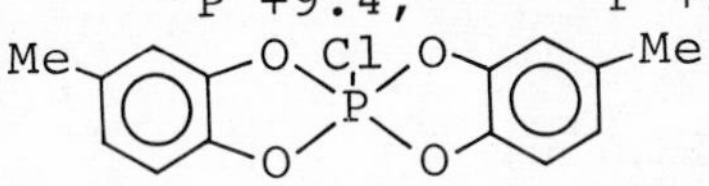

[structure]. III. Hygroscopic solid.[347]

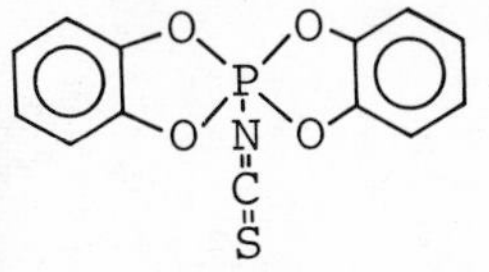

[structure]. No details, $b_{0.05}$ 120°, IR shows ν (NCS) at 1970 cm^{-1}.[36]

TYPE: $(RZ)_5P$

$(MeO)_5P$. VII. Not isolated, ^1H, only 1 kind of Me signal at -3.44 ppm, J_{HCOP} 12 Hz,[123] ^{31}P +8.9?.[405]

$(EtO)_5P$. VII. Not pure, ^1H,[126,127] ^{31}P +70.9,[127] reactions.[129]

$(PhO)_5P$. III. M. 103-104°,[499] 80-90°,[699] 46-52°,[347] IR, ^{1}H, ^{31}P +85.7.[499]

$(PhO)_3P<O_3$ (ring). VII. Identified only at low temp., reactions,[436,437,628] ^{31}P +63.[628] Decomposes to singlet oxygen at ∿-35°.

$(MeO)_3P$(O–CR=CR'–O). All the following products are obtained by reaction of trimethyl phosphite with the appropriate dicarbonyl compound. VIII. R, R' = H: $b_{0.05}$ 36-37°, IR, ^{1}H, ^{31}P +44.2.[507] R = H, R' = Ph: $b_{0.2/0.3}$ 113-122°, n_D^{25} 1.5308, IR, ^{1}H, ^{31}P +45.4.[525] R, R' = Me: $b_{0.2}$ 54-58°,[70] $b_{0.5}$ 45-47°, n_D^{25} 1.4387,[502,532] n_D^{20} 1.4391,[501] d^{25} 1.666,[532] IR,[363,364,502,532] Raman,[363,502] UV,[363,364] μ 2.55 D,[364] R_P 3.95,[362] ^{1}H,[502,505] ^{31}P +53,[502,532] ^{31}P +48.9.[518] R = Me, R' = Et: $b_{0.5}$ 59-62°, n_D^{25} 1.4391, IR, ^{31}P +46.[502] R = Me, R' = Ph: $b_{0.2}$ 117-119°, n_D^{25} 1.5286, IR, ^{1}H, ^{31}P +49.5.[526] R, R' = Ph: m. 49-50°, IR, ^{31}P +53,[501,502] ^{31}P +49.5.[518] R = Ph, R' = PhCO: yellow glass, IR, ^{1}H, ^{31}P +49.3.[525]

$(MeO)_3P$(O,O-tetrachlorocatechol). VIII. M. 65-66°, IR, ^{1}H, ^{31}P +46.2.[491]

$(MeO)_3P$(O,O-1,2-naphthalenediol). VIII. B_1 148-150°,[491] b_4 188-190°, n_D^{20} 1.5809, d_0^{20} 1.2944,[331] IR, ^{1}H, ^{31}P +45.5.[491]

$(MeO)_3P$(O,O-9,10-phenanthrenediol). VIII. M. 74°,[501,534] m. 74-75°,[502] IR,[501,502] ^{31}P +49,[501,502] ^{31}P +44.7.[518,534,537]

$(MeO)_3P$(O,O-pyrenediol). VIII. M. 86-87°, IR, ^{1}H, ^{31}P +44.5.[491]

$(MeO)_3P$(O,O-pyrenetetrol-O,O)$P(OMe)_3$. VIII. M. 206-211°, IR, ^{1}H, ^{31}P +44.3.[491]

$(EtO)_3P$(O–CR=CR'–O). From triethyl phosphite and the appropriate dicarbonyl compound. VIII. R = H, R' = Ph: ^{31}P +50.4.[405] R = Me, R' = Ph: $b_{0.5}$ 122-128°,

n_D^{25} 1.5110,[70] ^{31}P +47.4.[405] R, R' = Me: $b_{0.05}$ 62°,[534] $b_{0.2}$ 56-57°,[70] b_{10} 104-105°, d_0^{20} 1.0825, n_D^{20} 1.4290,[355,366] n_D^{25} 1.4326,[70] IR, UV, Raman,[363,364] μ 3.98 D,[364] R_P 4.01,[362] ^1H,[364] ^{31}P +51.3,[534] ^{31}P +52.1.[405] R = Me, R' = Et: $b_{0.2}$ 63-64°, n_D^{25} 1.4334.[70] R = Me, R' = Am: $b_{0.25}$ 96.5.[70] R, R' = Ph: n_D^{25} 1.5604,[70] ^{31}P +52.2.[534] R, R' = o-ClC_6H_4: ^{31}P +49.[405]

$(EtO)_3P$(cyclohexene-1,2-dioxy). VIII. B_6 115-117°, n_D^{20} 1.4570, d_0^{20} 1.1150,[368] R_P 3.94.[362]

$(EtO)_3P$(o-phenylenedioxy). VII. ^{31}P +50.[126]

$(EtO)_3P$(1,2-naphthylenedioxy). VII.[126] VIII.[331] $B_{7\cdot10^{-5}}$ block 110-115°, ^1H, ^{31}P +45;[126] b_6 180-182°, n_D^{20} 1.5540, d_0^{20} 1.2144.[331]

$(EtO)_3P$(9,10-phenanthrylenedioxy). VIII. M. 74°,[534] ^{19}F,[537] ^{31}P +47.1.[534,537]

$(ClCH_2CH_2O)_3P$(OCR=CRO). VIII. R = Me: n_D^{25} 1.4818,[70] ^{31}P +51.8.[405] R = Ph: n_D^{25} 1.5769,[70] ^{31}P +47.5.[405]

$(PrO)_3P$(OCMe=CMeO). VIII. B_{10} 127-130°, n_D^{20} 1.4305, d_0^{20} 1.0256,[355,366] IR, μ 3.78 D,[364] IR, Raman,[363] R_P 3.92.[362]

$(PrO)_3P$(cyclohexene-1,2-dioxy). VIII. B_{10} 144-146°, n_D^{20} 1.4580, d_0^{20} 1.0656.[368]

$(PrO)_3P$(1,2-naphthylenedioxy). VIII. B_3 184-186°, n_D^{20} 1.5450, d_0^{20} 1.1724.[331]

$(i\text{-}PrO)_3P$(OCR=CR'O). From tri-i-propyl phosphite and α-diketone: VIII. R, R' = Me: $b_{0.5}$ 82-83°, n_D^{25} 1.4260,[502] $b_{0.2}$ 59-61°, n_D^{25} 1.4249,[70] IR, ^{31}P +51.[502] R, R' = Ph: m. 52-53°, IR, ^{31}P +53.8.[502] R = Ph, R' = PhCO: yellow prisms, m. 46-48°, IR, ^1H, UV, ^{31}P +54.9.[525]

$(i\text{-}PrO)_3P$(9,10-phenanthrylenedioxy). VIII. M. 105-106°,[502] m. 107-108°,[226] ^{31}P +48.9,[525] ^{31}P +48.6 to +49.2 in different solvents,[226] IR, ^1H,[226] x-ray determination shows 2 slightly distorted trig. bipyr. forms. Orthorhombic:[225,226] $P-O_{ring,a,e}$ 1.76, 1.62 Å, $P-O_{i-Pr,a}$ 1.63 Å, $P-O_{i-Pr,e}$ 1.58-1.60 Å, $O_{ring,a}PO_{ring,e}$ 89°, all other O_ePO_a angles 85-97°, all O_ePO_e angles between 119 and 121°. Monoclinic:[620] $P-O_{ring,a,e}$ 1.751, 1.633 Å, $P-O_{i-Pr,a}$ 1.638 Å, $P-O_{i-Pr,e}$ 1.574-1.588 Å, $O_{ring,a}PO_{ring,e}$ 89.3°, all other O_ePO_a angles 88.6-91.3°, all O_ePO_e angles between 117.2 and 125.4°.

(BuO)$_3$P(O–CMe=CMe–O). VIII. B_{10} 138-140°, n_D^{20} 1.4335, d_0^{20} 0.9977.[366]

(BuO)$_3$P(1,2-naphthalenedioxy). VIII. B_3 204-206°, n_D^{20} 1.5320, d_0^{20} 1.1135.[331]

(BuO)$_3$P(9,10-phenanthrenedioxy). ^{31}P +47.3.[405]

(HexO)$_3$P(O–CMe=CMe–O). VIII. n_D^{25} 1.4449.[70]

[Me(CH$_2$)$_3$CH(Et)–CH$_2$O]$_3$P(O–CMe=CMe–O). ^{31}P +53.5.[405]

(BzO)$_3$P(O–CMe=CMe–O). From (MeO)$_3$P(O–CMe=CMe–O) and benzyl alcohol,[539] oil, IR, ^1H, ^{31}P +51.3

(BzO)$_2$(MeO)P(O–CMe=CMe–O). $B_{5\cdot10^{-6}}$ 96-100°, IR, ^1H, ^{31}P +50.6.

(BzO)(MeO)$_2$P(O–CMe=CMe–O). $B_{0.05}$ 96-97°, IR, ^1H, ^{31}P +49.8.

(PhO)$_3$P(O–CR=CR–O). VIII. R = Me: not isolated, n_D^{25} 1.5628,[70] ^{31}P +64.7,[534] ^{31}P +67.5.[405] R = Ph: m. 99-100°, IR,[501,502] ^{31}P +62.5.[502,518]

(PhO)$_3$P(1,2-benzenedioxy). III. Plates, m. 95-107°?,[347] m. 86-87°, IR, ^{31}P +60.8.[500]

(PhO$_3$P(tetrachloro-1,2-benzenedioxy). VIII. M. 85-87°, ^{31}P +60.0.[500]

$(PhO)_3P$ [phenanthrene-fused dioxaphospholene]. VIII. M. 147-148°, IR,[501,502] ^{31}P +58.6.[502,518]

$(EtO)_3P$ [$O-CH_2-CH_2-O$ ring]. VII. ^{31}P +52.[126]

$(EtO)_3P$ [$O-CHPh-CH_2-O$ ring]. VII. ^{31}P +53.[126]

$(MeO)_3P$ [$O-CHR-CHR-O$ ring]. R = C_6F_5: X. Meso: ^1H, ^{31}P +50.4; racem: ^1H, ^{31}P +52.3.[488] R = o-$NO_2C_6H_4$-: X. Meso: m. 120-121°, IR, ^1H, ^{31}P +50.1; racem: not isolated, ^1H.[495] R = p-$NO_2C_6H_4$-: X. Meso: m. 138-139°, IR, ^1H, ^{31}P +49.6; racem: m. 131-132°, IR, ^1H, ^{31}P +50.2.[495,(492)] R = o-$OHCC_6H_4$-: X. Meso: m. 123-124°, IR, ^1H, ^{31}P +49.9; racem: IR, ^1H, ^{31}P +50.1.[492] R = PhCO-: IX. Meso: m. 68-70°, IR, ^1H, ^{31}P +46.9.[525]

[Structure: $(MeO)_3P$ dioxaphospholane bearing H, H, o-CHO-phenyl and o-phenylene linked to a second $(MeO)_3P$ dioxaphospholane ring with H, Me(COMe), COMe(Me)] + diastereomer (in the second dioxaphospholane ring). IX. M. 139-140°, IR, ^1H, ^{31}P +50.1.[492]

[Structure: $(MeO)_3P$ dioxaphospholane bearing two aryl groups each linked to a $(MeO)_3P$ dioxaphospholane ring with H, Me, COMe]. IX. Mixture of diastereomers, m. 54-56°, IR, ^1H, ^{31}P +49.9.[492]

$(EtO)_3P$ [$O-CHR-CHR-O$ ring]. R = Me: VII. ^{31}P +56.[126] R = o-$NO_2C_6H_4$-: X. Meso: m. 133-134°, IR, ^1H, ^{31}P +52.8; racem: not isolated, ^1H.[495] R = p-$NO_2C_6H_4$-: X. Meso: m. 138-140°, IR, ^1H, ^{31}P +52.4; racem: not isolated, ^1H, ^{31}P +53.4.[495]

$(RO)_3P$ [$O-CH(COOBu)-CH(COOBu)-O$ ring]. X. Viscous liquids, one δ ^{31}P value is given: +50.[474]

R	b. (mm)	d_4^{20}	n_D^{20}
Me	158-159° (1.5)	1.1591	1.4465
Et	149-153° (0.5)	1.0941	1.4400
Pr	160-164° (1.0)	1.0745	1.4430
i-Pr	146-149° (1.0)	1.0584	1.4360
Bu	179-182° (1.0)	1.0368	1.4420
i-Bu	167.5-169° (0.5)	1.0352	1.4425

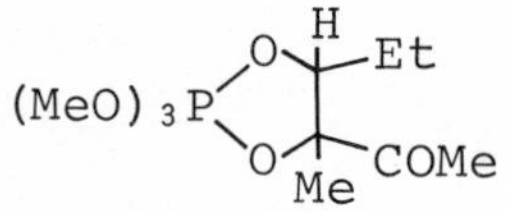

$(MeO)_3P$ (O–CH(Et)–C(Me)(COMe)–O). IX. $B_{0.2}$ 64-66°, n_D^{25} 1.4361,[524] $b_{1.5}$ 88-89°, n_D^{25} 1.4356,[520] ^{31}P +50,[520] ^1H, ^{31}P +51.3, other isomer: ^{31}P +48.9.[405,524]

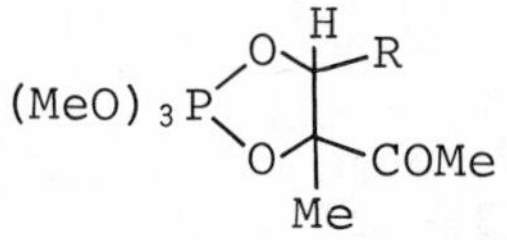

$(MeO)_3P$ (O–CH(R)–C(Me)(COMe)–O). IX. All the following products are made by reaction of the trimethyl phosphite biacetyl adduct with an appropriate aldehyde.

R	b. (mm)	n_D^{25}	IR	^1H	^{31}P	Ref.
Vinyl	96° (0.2)	1.4501	+	+	+50.9	510
n-Pr	72-73° (0.2)	1.4382	+	+	+51.2	524
Propenyl	104° (0.2)	1.4512	+	+	+51.0	512
Hex	96-101° (0.2)	1.4418	+	+	+51.3	524
Ph	110-112° (0.1)	1.4940	+	+	+51.5	524, (520)
$o-OHCC_6H_4$	m. 70-81 (mixt. of diaster.)		+	+	+49.6, +51.1	492
$p-OHCC_6H_4$	m. 74-78°		+	+	+51.2	493
	trans MeH/ product not isol.		-	+	-	
PhCO	Thick oil (mixt.)	1.5081	+	+	+50	525

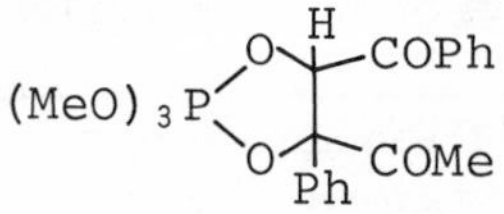

$(MeO)_3P$ (O–CH(COPh)–C(Ph)(COMe)–O). IX. Mixture of diastereomers, IR, ^1H, ^{31}P +49.4.[526]

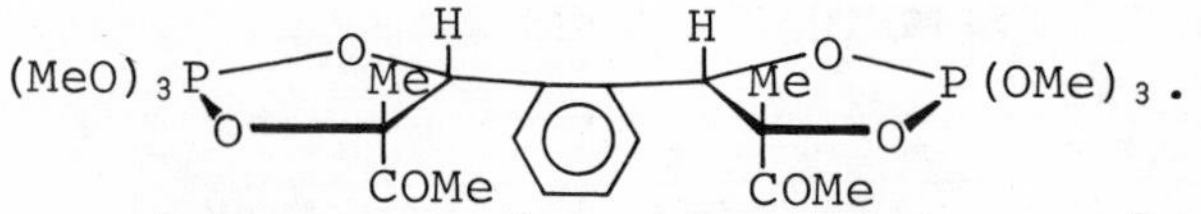

$(MeO)_3P$(O–C(Me)(COMe)–CH–O)–C_6H_4–(O–CH–C(Me)(COMe)–O)$P(OMe)_3$. IX. Cis/cis Me/H: m. 95-100°, IR, ^1H, ^{31}P +49.3. cis/trans Me/H: m. 82-90°, IR, ^1H, ^{31}P +49.5. trans/trans Me/H: not pure, IR, ^1H.[492]

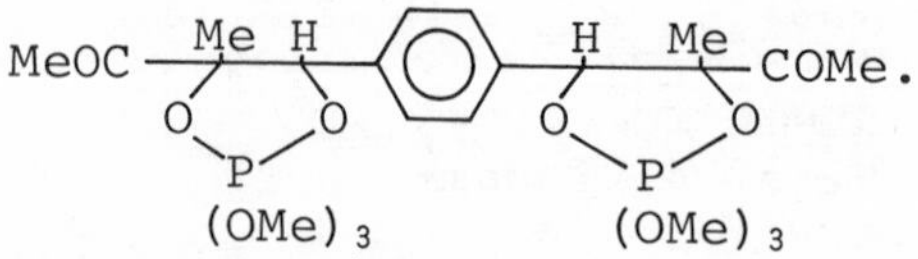

MeOC–C(Me)(O)–CH(O)[$P(OMe)_3$]–C_6H_4–CH(O)–C(Me)(O)[$P(OMe)_3$]–COMe. IX. Cis/cis Me/H: m. 154-155°, IR, ^1H, ^{31}P +51.1.[493]

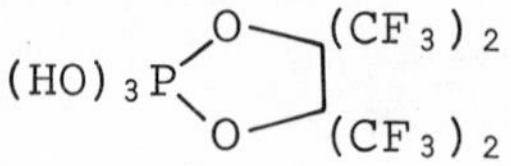

$(HO)_3P$(O–$C(CF_3)_2$–$C(CF_3)_2$–O) !?. From $(EtO)_3P$(O–O) or $(PrO)_2(NEt_2)P$(O–O) and conc. H_2SO_4!, m. 61-62°, no further data.[158,181]

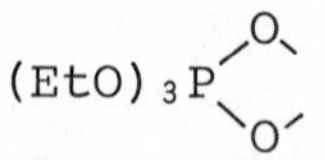

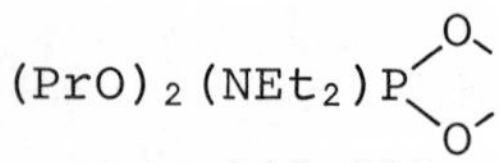

$(MeO)_3P$ dioxaphospholane with $(CF_3)_2$, $(CF_3)_2$. X. $B_{0.5}$ 34°, ^{31}P +50.1,[534] ^{19}F.[537]

$(MeO)_3P$ dioxaphospholane with Biph, Biph. X. M. 162-164°, ^{1}H, ^{31}P +46.7.[533]

$(MeO)_3P$ dioxaphospholane with Me, Me, oxetane ring, CD_3, CD_3. Photochemical addition of $(CD_3)_2CO$ to dioxaphospholene at -75°, not isolated, ^{1}H (-50°).[61]

$(MeO)_3P$ dioxaphospholane with R, R', Me, COMe. IX. From dioxaphospholene and suitable carbonyl compound. R, R' = Me: $b_{0.1}$ 46-47°, m. 26-28°, IR, ^{1}H.[60] R, R' = CF_3: $b_{0.05}$ 60-62°, n_D^{25} 1.3911, IR, ^{1}H, ^{31}P +52.6.[527] R = Ph, R' = Me: $b_{0.07}$ 110°, n_D^{25} 1.4963, IR, ^{1}H, ^{31}P +54.2.[527] R = Ph, R' = CF_3: not isolated, mixture, $b_{0.1}$ 110-112°, n_D^{25} 1.4673, IR, ^{1}H, cis Ph/Me: ^{31}P +52.5, trans Ph/Me: ^{31}P +49.6,[527] 82%/18%. R = Me, R' = p-$NO_2C_6H_4$-: m. 91-92°, IR, ^{1}H, ^{31}P +52.4, trans isomer not isolated, ^{1}H, ^{31}P +54.0.[527] R, R' = -$(CH_2)_3$-: $b_{0.05}$ 62-65°, n_D^{25} 1.4486, IR, ^{1}H, ^{31}P +53.1.[527] R, R' = -$(CH_2)_5$-: $b_{0.05}$ 90-92°, n_D^{25} 1.4641, IR, ^{1}H, ^{31}P +54.3.[527]

$(MeO)_3P$ dioxaphospholane spiro-cyclohexadienone, Me, COMe. IX. Dec. 65°, IR, UV, ^{1}H, ^{31}P +50.3,[511] cryst.

$(MeO)_3P$ dioxaphospholane spiro-naphthalenone, Me, COMe. IX. Dec. rapidly, IR, ^{1}H, ^{31}P +50.[511]

$(MeO)_3P$ dioxaphospholane with Ph, Ph, Ph, COPh. IX. ^{31}P +48.0.[484]

$(MeO)_3P$ dioxaphospholane with R, COR', Me, COMe. IX. From dioxaphospholene and suitable α-dicarbonyl compound. R = Me, R' = Me: m. 31-32°, IR, ^{31}P +51,[532] ^{31}P +54.8,[534] ^{1}H,[505,506,532] racem: liquid,[506] ^{1}H.[505,506] R = Ph, R' = Me: m. 91-93°, IR, ^{1}H, ^{31}P +54.7; trans form not isolated, ^{1}H, ^{31}P +52.4.[526] R = Me, R' = Ph: not isolated, ^{1}H, ^{31}P +54.6, trans form not isolated, ^{1}H.[526] R = Ph, R' = Ph: m. 108-110°, IR, ^{1}H, ^{31}P +54.7; trans form ^{31}P +52.7.[526]

$(MeO)_3P$ [dioxaphospholane fused to acenaphthenone; substituents Me, COMe]. IX. M. 95-96°, IR.[532]

$(MeO)_3P$ [dioxaphospholane, spiro-bis(acenaphthenone)]. X. Cryst. eventually open struct., IR.[531]

$(MeO)_3P$ [dioxaphospholane; COOMe, Me; Me, COMe]. IX. Struct. only presumed, m. 33-35°, IR, ^1H, ^{31}P ∿+50.[520]

$(MeO)_3P$ [dioxaphospholane; Me, COOR; Me, COOR]. X. R = Me: liquid mixture of 2 diastereomers, $b_{0.1}$ 105-110°, n_D^{25} 1.4470,[505] b_2 129-130.5°, n_D^{20} 1.4485, d_4^{20} 1.2620,[477] IR, ^1H, ^{31}P +50,[505] ^{31}P racem: +50.4, meso: +52.5.[405] R = Et: $b_{3.5}$ 141.5-143°, n_D^{20} 1.4440, d_4^{20} 1.2002.[477]

$(MeO)_3P$ [dioxaphospholane; $(COOEt)_2$, $(COOEt)_2$]. X. $B_{0.5}$ 150-152°, n_D^{20} 1.4480, d_4^{20} 1.2494, ^{31}P +60.[474]

$(EtO)_3P$ [dioxaphospholane; Me_2, Me_2]. VII. ^{31}P +60.[126]

$(EtO)_3P$ [dioxaphospholane; $(CF_3)_2$, $(CF_3)_2$]. X. $B_{0.2}$ 33,[534] $b_{0.5}$ 61-62°, m. -8 to -9°, n_D^{20} 1.3555,[181] b_1 63-64°,[560] ^{19}F, ^{31}P +53.2.[534,537]

$(EtO)_3P$ [dioxaphospholane; Biph, Biph]. X. M. 143-145°, ^1H, ^{31}P +49.3.[533]

$(EtO)_3P$ [dioxaphospholane; 4.4'-Br -Biph, 4.4'-Br -Biph]. X. M. 204-206°, ^1H, ^{31}P +48.[76]

$(EtO)_3P$ [dioxaphospholane; R, CN; R, CN]. X. R, m.: Ph, 117°; p-MeC_6H_4-, 84°; p-Cl-C_6H_4-, 111-113°; p-$NO_2C_6H_4$-, 131° (dec.).[430]

$(EtO)_3P$ [dioxaphospholane; Me, COOR; Me, COOR]. X.

R	b. (mm)	n_D^{20}	d_4^{20}	Ref.
Me	132.5-133° (1.5)	1.4420	1.1671	477
Et	132-133° (1.0)	1.4410	1.1301	477

$(RO)_3P$ [spiro-dioxaphospholane fused with two oxindole rings, N–R']. X. Crystalline substances,[439] a few IR and ^1H data.

R:	Me	Et	i-Pr	Me	Et
R':	H	H	H	Me	Me
m.:	122-124	115-117	103-105	145-147	138-140

R:	i-Pr	Me	Et	i-Pr
R':	Me	COMe	COMe	COMe
m.:	143-145	143-145	125-127	158-160

$(EtO)_3P$ [O–C$(COOEt)_2$–C$(COOEt)_2$–O]. X. $B_{0.5}$ 162-165°, n_D^{20} 1.4430, d_4^{20} 1.1750.[474]

$(i\text{-}PrO)_3P$ [O–Biph–Biph–O]. X. M. 146-148°, ^1H, ^{31}P +52.2.[533]

$(PhO)_3P$ [O–C$(CF_3)_2$–C$(CF_3)_2$–O]. X. M. 55°, ^{31}P +64.5.[534]

$(PhO)_3P$ [O–C(Me)(COMe)–C(Me)(COMe)–O]. IX. Not isolated, mixture, ^{31}P meso: +66.1, racem: +65.3.[534]

$(MeO)_3P$ [O–C(Me)(COMe)–C(=CH_2)–O]. IX. $B_{0.2}$ 62°, IR, ^1H, ^{31}P +51.4.[496]

$(MeO)_3P$ [O–C(Me)(COMe)–C(=NAr)–O]. IX. Ar = Ph: viscous oil, IR, ^1H, ^{31}P +57.8.[497] Ar = p-ClC_6H_4-: m. 120-121°, IR, ^1H, ^{31}P +58.2.[498] Ar = p-MeC_6H_4-: m. 110-111°, IR, ^1H, ^{31}P +57.3.[498]

$(EtO)_3P$ [1,3,2-dioxaphosphorinane]. VII. ^{31}P +72.[126]

$(EtO)_3P$ [4-Me-1,3,2-dioxaphosphorinane]. VII. ^{31}P +71.[126]

$(EtO)_3P$ [5,5-$(Me)_2$-1,3,2-dioxaphosphorinane]. VII. Not pure, IR, ^1H, ^{31}P +71.6.[123]

$(EtO)_2(MeO)P$ [4-Me-1,3,2-dioxaphosphorinane]. VII. ^{31}P +71.[126]

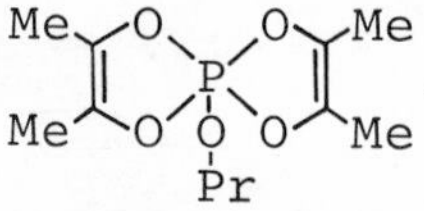

[Structure: Me/Me dioxaphosphole, O-Pr] . ^{31}P +50.3.[405]

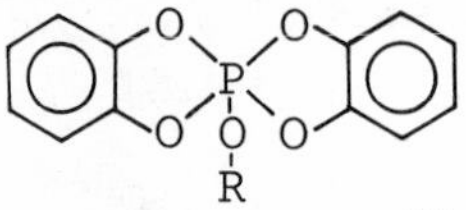

[Structure: bis(catechol) phosphorane, O-R] . R = Me: from corresponding fluoride and MeOH, $b_{1.2}$ 115°, n_D^{20} 1.5235, d_4^{20} 1.2780, IR, ^1H.[160] R = Ph: III. M. 110-112°, ^{31}P +29.8.[500]

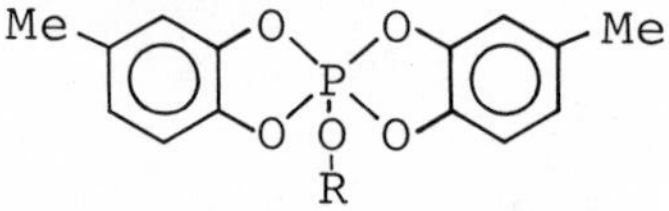

[Structure: bis(methylcatechol) phosphorane, O-R] . ? III. R = l-menthyl, from corresponding chloride and l-menthol, cryst., m. 65-70°.[347]

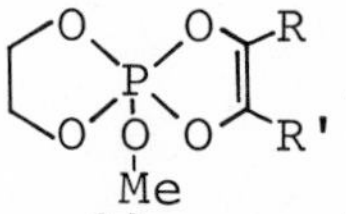

[Structure: O-Me, R, R'] . R, R' = Me: VIII. $B_{0.1}$ 73°, n_D^{25} 1.4622, IR, ^1H, ^{31}P +27.[518] R = Me, R' = Ph: VIII. $B_{0.2}$ 148-149°, n_D^{25} 1.5506, IR, ^1H, ^{31}P +28.[518] R, R' = Ph: VIII. M. 74-76°, IR, ^1H, ^{31}P +28.1.[518]

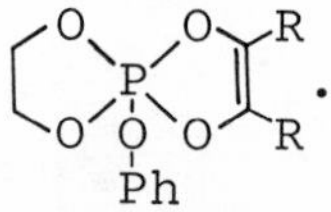

[Structure: O-Ph, R, R] . R = Me: VIII. ^{31}P +30.8.[518] R = Ph: VIII. M. 129-130°, IR, ^{31}P +32.2.[518]

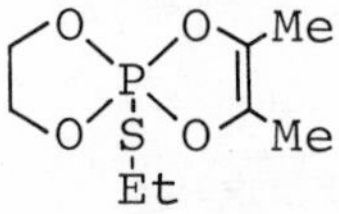

[Structure: S-Et, Me, Me] . VIII. $B_{0.05-0.07}$ 104-112°,[70] ^{31}P +6.7.[405]

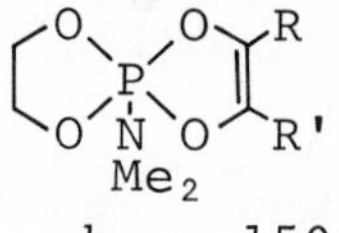

[Structure: NMe$_2$, R, R'] . VIII. R, R' = Me: $b_{0.05}$ 94-95°, n_D^{25} 1.4750, IR, ^1H, ^{31}P +26.1.[518] R, R' = Ph: m. 90-92°, IR, ^1H, ^{31}P +26.4.[518] R = Me, R' = Ph: $b_{0.5}$ 150-152°, n_D^{25} 1.5551, IR, ^1H, ^{31}P +26.9.[518]

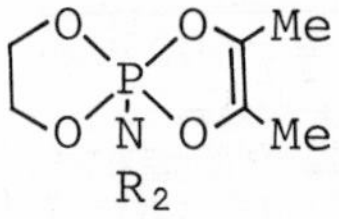

[Structure: NR$_2$, Me, Me] . R_2 = $-C_2H_4-$: VIII. $B_{0.03-0.05}$ 98-102°,[70] ^{31}P +20.2.[405] R = Et: ^{31}P +26.4.[405]

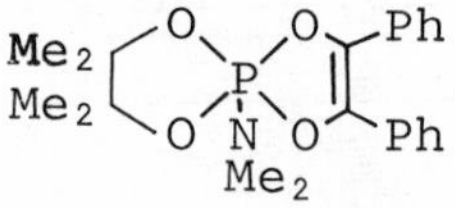

[Structure: Me$_2$, Me$_2$, NMe$_2$, Ph, Ph] . VIII. M. 156°, ^{31}P +35.5.[568]

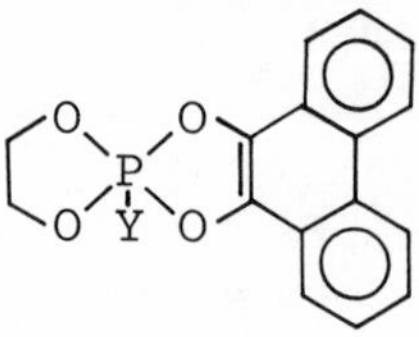

[Structure: phenanthrenediol phosphorane, Y] . VIII. Y = OMe: m. 61-63°, IR, ^1H, ^{31}P +23.0.[518] Y = OPh: m. 155-157°, IR, ^1H, ^{31}P +27.[518] Y = NMe$_2$: m. 157-159°, IR, ^1H, ^{31}P +21.3.[518]

[structure] . VIII. M. 136-137°, IR, ^{1}H, ^{31}P +40.7.[518]

[structure] . VIII. M. 94-96°, IR, ^{31}P +53.7.[518]

[structure] . ^{31}P +23.1.[405]

[structure] . VIII. M. 176-177°, IR, ^{1}H, the Me groups of the perhydro-1,3,2-dioxaphosphorine ring are not equivalent, ^{31}P +48.7.[518]

$(RO)_2(Et_2N)P$[ring] . VIII. R = Et: b_{10} 105-106°, n_D^{20} 1.4395, d_4^{20} 1.0178, IR, ^{1}H,[356] ^{31}P +45.[476] R = Pr: b_1 75-76°, n_D^{20} 1.4420, d_4^{20} 1.0019.[569] R = Bu: b_8 136.5°, n_D^{20} 1.4430, d_4^{20} 0.9688.[356]

$(PrO)_2(Et_2N)P$[ring] . VIII. $B_{0.007}$ 116-117°, n_D^{20} 1.5491, d_4^{20} 1.0883.[158]

$(PrO)_2(Et_2N)P$[ring] . VIII. $B_{0.03}$ 83-84°, n_D^{20} 1.4670.[569]

$(EtO)_2(Me_2N)P$[ring] = $(EtOOC)_2C(Et)$-O-P(O)(NMe$_2$)(OEt).[476] VIII, followed by rearrangement. $B_{0.5}$ 132°, $n_D^{17.5}$ 1.443, d_{17}^{17} 1.1324, IR.[88]

$(EtO)_2(Et_2N)P$[ring] . $B_{2.5}$ 158-160°, n_D^{20} 1.4435, d_4^{20} 1.0562.[475]

$(PrO)_2(Et_2N)P$[ring] . X. Water stable! $B_{0.013}$ 38-39°, n_D^{20} 1.3846, d_4^{20} 1.3756.[158]

$(EtO)_2(Et_2N)P$[ring] . X. R = Me: b_1 143-145°, n_D^{20} 1.4515, d_4^{20} 1.1803, ^{31}P +45.5.[475] R = Et: $b_{2.5}$ 158-160°, n_D^{20} 1.4451, d_4^{20} 1.2251.[475]

$(PrO)_2(Et_2N)P$[O–C(Me)(COOMe)–C(Me)(COOMe)–O]. X. B_1 150-155°, n_D^{20} 1.4489, d_4^{20} 1.2056.[475]

$(EtO)(Et_2N)_2P$[O–C(Me)=C(Me)–O]. VIII. $B_{2\cdot10^{-3}}$ 53-55°, n_D^{20} 1.4456, d_4^{20} 0.9872,[356] ^{31}P -16.6.[405]

$(Me_2N)_3P$[O–C(Ph)=C(Ph)–O] ⇌ $(Me_2N)_3P^{\oplus}$–O–C(Ph)=C(Ph)–$O^{\ominus}$. VIII. Yellow cryst., m. 105, IR;[88,89] m. 85-87°, IR, ^{1}H,[523] ^{31}P +25;[89] in solution mixture of covalent and orange dipolar forms: ^{31}P (hexane) +30.2, $^{31}P(CH_2Cl_2)$ +13.1![523]

$(Me_2N)_3P^{\oplus}$–O–(9,10-phenanthrenediyl)–$O^{\ominus}$. VIII. Yellow cryst., m. 100-101°, IR, ^{1}H, ^{31}P -38.5.[522]

$(Me_2N)_3P^{\oplus}$–O–C(Ph)=C(COPh)–$O^{\ominus}$. VIII. Orange prisms, m. 119-120°, IR, ^{1}H, ^{31}P -35.9.[522]

$(Me_2N)_3P^{\oplus}$–O–C(COOEt)=C(OEt)–$O^{\ominus}$. VIII. M. 111-113°, IR, ^{1}H, ^{31}P -38.2 to -38.8 in different solvents.[522]

$(RO)_3P$[O–C(Ph)=N–N(Ph)]. VIII. R = Me: stable against water! Yellow cryst., m. 78-80°, IR, ^{1}H, ^{31}P +62.[23] R = Ph: stable against water! M. 116-117°, IR, ^{31}P +71.6.[21]

$(PhO)_3P$[O–C(OR)=N–N(OCOR)]. VIII. R = Me: very hygroscopic, m. 93-95°, IR, ^{31}P +69.2.[21] R = Et: m. 80-80.5°, IR, μ 3.65 D.[193]

$(EtO)_3P$[N(Me)–CH_2CH_2–N(Me)] ?. VII. ^{31}P +56.[118]

$(EtO)_2(Et_2N)P$[N(Me)–CH_2CH_2–N(Me)] ?. VII. ^{31}P +49.[118]

[N(Me)–CH_2CH_2–N(Me)]P(Y)[O–C(Me)=C(Me)–O]. VIII. Y = –N(piperidino): ^{31}P +42.1.[523]

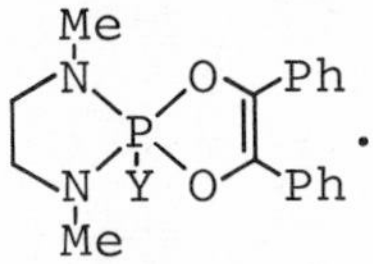

[structure]. VIII. Y = -OMe: gum, IR, ^{1}H, ^{31}P +38.6. Y = $-NMe_2$: m. 98-99°, IR, ^{1}H, ^{31}P +36.9. Y = -N(piperidino): m. 88-89°, IR, ^{1}H, ^{31}P +41.8.[523]

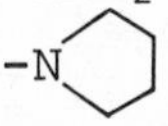

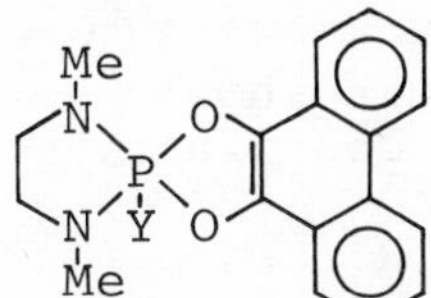

[structure]. VIII. Y = -OMe: m. 89-91°, ^{1}H, ^{31}P +33.5. Y = $-NMe_2$: m. 121-123°, IR, ^{1}H, ^{31}P +29.8. Y = -N(piperidino): m. 139-141°, IR, ^{1}H, ^{31}P +35.1.[522]

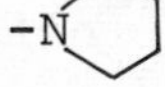

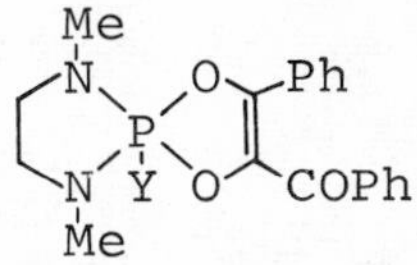

[structure]. VIII. Y = $-NMe_2$: yellow oil, ^{1}H, ^{31}P +31.9 to +33.8 in different solvents. Y = -N(piperidino): orange gum, ^{1}H, ^{31}P +37.5.[522]

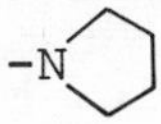

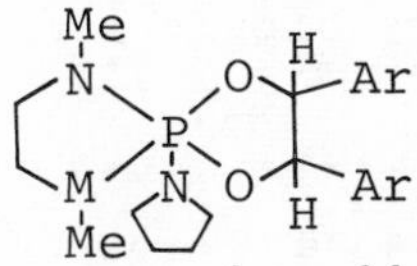

[structure]. X. Ar = o-$NO_2C_6H_4$-, cis: m. 136-137°, IR, ^{1}H, ^{31}P +42.4; trans: not isolated, ^{1}H. Ar = m-$NO_2C_6H_4$-, cis: m. 147-148°, IR, ^{1}H, ^{31}P +41.4; trans: not obtained. Ar = p-$NO_2C_6H_4$-, cis: m. 122-123°, IR, ^{1}H, ^{31}P +42.3, trans: not isolated, ^{1}H.[509]

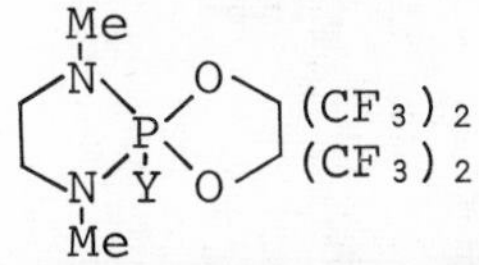

[structure]. X. Y = $-NMe_2$: m. 86-87°, $b_{0.1}$ 58-60°, IR, ^{1}H, ^{31}P +28.3. Y = -N(piperidino): m. 74-75°, IR, ^{31}P +30.9.[508]

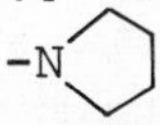

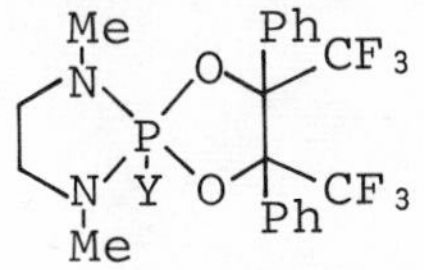

[structure]. X. Y = NMe_2, prob. trans: m. 121-122°, IR, ^{1}H, ^{31}P +34.9. Y = -N(piperidino), prob. trans: m. 105-107°, IR, ^{1}H, ^{31}P +40.5; cis, not isolated, ^{31}P +38.5.[508]

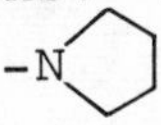

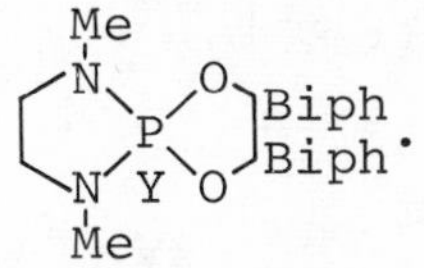

[structure]. Y = -N(piperidino): X. M. 140-142°, IR, ^{1}H, ^{31}P +40.3.[508]

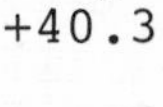

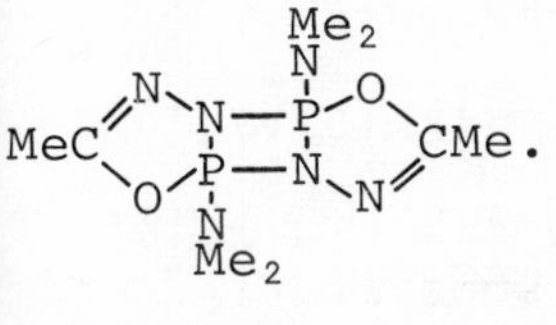

[structure]. III. Dec. 153°, ^{1}H, ^{31}P +40.5.[152]

D.2. Compounds of Type RPY_4

TYPE: RPX_4

$MePF_4$. XI.[343,576,580] XIII.[324] XIV.[369] XV.[631] B. 9°,[369] b. 10°,[576] b. 11.7°,[197] b. 11-12°,[324] b_{756} 12.5°,[343] m. -50°,[343] m. -52°,[197] log P_{mm} = -1448.9/T + 7.9765,[197] d_5^0 1.4538,[324,343] IR,[631] and Raman show C_{2v} trig. bipyr. with equat. Me,[133] normal coordinate analysis,[267] mass spect.,[73,197] ^{31}P +29.6,[580] ^{31}P +29.9,[445] ^1H,[446] ^{19}F,[426] heteronuclear INDOR spect.,[135a] elect. diff.[40] establishes C_{2v} trig. bipyr. with $P\text{-}F_{a,e}$ = 1.612, 1.543 Å, P-C = 1.780 Å, F_aPC 91.8°, F_ePC 122.2°.[40,229]

$ClCH_2PF_4$. XI.[591] B. 47°,[576] ^{31}P +43.7,[445,580] ^{19}F,[426,590] ^1H.[446]

$PhCH_2PF_4$. XI. B_{15} 53°,[576] ^{19}F, J_{FP} 1002 Hz.[426]

F_3CPF_4. XII. B. -35°, m. -117°,[398] b. -39°, m. -113°,[85] IR[398,426] and Raman show C_{2v} trig. bipyr. with equat. CF_3,[210] microwave spect. points to C_{3v} trig. bipyr. with apical CF_3,[102] ^{19}F, J_{PF} 1103 Hz[426] (400) ^{31}P +66.4.[445]

Cl_3CPF_4. XI.[343] XII.[443] XIII.[324] B. 68-70°,[324,343,443] d_4^{20} 1.7112,[343] d^{20} 1.7275,[324] log P_{mm} = -1643/T +7.668, IR, ^{19}F,[443] ^{31}P +67.[443,445]

$EtPF_4$. XI.[343,576,580] XIII.[324] XV.[631] B. 33-35°,[197,324,343,576] m. -70°,[197] log P_{mm} = -1522/T + 7.8408,[197] d_4^{20} 1.3074,[324,343] mass spect.,[73,197] ^{19}F, J_{FP} 987 Hz,[426 590] ^{31}P +30.2.[445,580]

$C_2F_5PF_4$. XI. B_{10} -78°,[590] ^{19}F, J_{FP} 1175 Hz.[425]

$H_2C{=}CHPF_4$. XV. Mol. wt., IR.[631]

$F_2C{=}CFPF_4$. XI. Reactive liquid, IR, ^{19}F, J_{FP} 927 Hz.[105]

$PhHC{=}CHPF_4$. XI. Cis product, b_{23} 97.5-98°,[576] ^{19}F, J_{FP} 940 Hz.[426]

$n\text{-}PrPF_4$. XV. Mol. wt., IR.[631]

$n\text{-}C_3F_7PF_4$. XII. B_{295} 0°,[590] ^{19}F, J_{FP} 1090 Hz.[444,590]

$i\text{-}PrPF_4$. XI. B_{757} 54-55°, d_4^0 1.2591.[343]

$AllylPF_4$. XI. B_{741} 48.5°.[343]

$n\text{-}BuPF_4$. XI. B. 85-86°,[576] ^{19}F, J_{FP} 990 Hz.[426]

(perfluorocyclobutyl: F_2 ring carbons, F on C bearing P)PF_4. From thermal rearrangement of (F-ring)PF_3, b. ∿50°, ^{19}F, J_{FP} 1057 Hz.[590]

$i\text{-}OctenylPF_4$. XI. B_3 60°,[616] b_{19} 43°, b. 156°,[574a] b_9 43°, $n_D^{24.8}$ 1.3931,[576] ^{19}F, J_{FP} 955 Hz.[426]

$PhPF_4$. XI.[343,576,585,591,616,681] XIII.[324,617,618] XV.[601] B. 134-135°[324,343,616-618] b_{60} 58°,[576] gas chrom.,[575] d_{20}^{20} 1.3839,[681] d_4^{15} 1.3888, n_D^{20} 1.4245,[324,343] mass spect.,[73] IR,[618] ^{19}F, J_{FP} 963 Hz,[426,590] ^{31}P +51.7.[445,580]

$C_6F_5PF_4$. XVIII. B. 121°, mass spect., ^{19}F, ^{31}P +52.2.[170]

$m\text{-}FC_6H_4PF_4$. XI. B_{90} 68°, ^{19}F, J_{FP} 964 Hz, substit.

effects.[483]
p-$FC_6H_4PF_4$. XI. B_{77} 68°, ^{19}F, J_{FP} 954 Hz, substit. effects.[483]
p-$ClC_6H_4PF_4$. XI. B_{760} 163-164°, d_{20}^{20} 1.5141.[681]
p-$MeC_6H_4PF_4$. XI. B_{760} 161-162°, d_{20}^{20} 1.3334.[681]
m-$CF_3C_6H_4PF_4$. XI. B_{40} 64°, ^{19}F, J_{FP} 965 Hz, ^{31}P +52.9.[590]
p-$CF_3C_6H_4PF_4$. XII. B. 140-141°, d_{20}^{24} 1.5584.[682]
$NO_2C_6H_4PF_4$. XI. Solid, IR, hydrolysis.[593]
RPF_4. XI.[574a] R, b. (mm): Me, 10°; Et, 32-33°; n-Bu, 85-86°; i-octenyl, 156°, 43°(19); Ph, 134.5-136°; m,p-MeC_6H_4-, 55°(8); m,p-iPrC_6H_4-, 64°(8); m,p-ClC_6H_4-, 49°(9); m,p-$MeOC_6H_4$-, 79-80°(10); PhHC=CH-, 97.5-98°(23); Bz, 53°(15).
$ArPF_4$. XI.[426] ^{19}F,[426] b.,[576] ^{31}P and ^{19}F,[590] all products are mixtures.

Ar	b. (mm)	J_{FP}	^{31}P
m,p-MeC_6H_4-	44° (4)	960	+49.7[445]
m,p-i-PrC_6H_4-	84-85.5° (18)	963,958	+50.8
m,p-ClC_6H_4-	49° (9)	960	—
m,p-$MeOC_6H_4$-	95-96° (20)	951	+51.2
2.5-$Me_2C_6H_3$-	56° (9)	981	+42.6

2-Thienyl-PF_4. XI. B_{30} 52°, ^{19}F, J_{FP} 929 Hz, ^{31}P +57.5.[590]
$PhPF_3Cl$. From $PhPF_3H$ and Cl_2,[305] ^{19}F shows 2 kinds of fluorines.[425] p-$ClC_6H_4PF_3Cl$ was similarly prepared.[305]
$CF_3PF_2Cl_2$. XVIII. Liquid, ^{19}F, ^{31}P +23.7.[444]
$CCl_3PF_2Cl_2$. XVIII. Solid, ^{19}F, ^{31}P +7.4.[443,445]
$C_3F_7PF_2Cl_2$. XVIII. Yellow liquid, ^{19}F,[444] ^{31}P +14.3.[444,445]
$CCl_3PF_2Br_2$. XVIII. Solid, ^{19}F, ^{31}P +26, +27.[443,445]
CF_3PFCl_3. XVIII. Liquid at -64°, ^{19}F(-50°) shows apical F and CF_3.[209]
$MePCl_4$. XVIII.[343,481] Dec. 132°,[41] dec. 198-199°,[559] white cryst., IR and Raman show ionic struct. for solid product[41,43] and C_{2v} trig. bipyr. with equat. Me in solution in nonionizing solvents,[43] cryst. adduct with $SbCl_5$.[43]
CF_3PCl_4. XVIII. B. 104°, m. -52.5°, log P_{mm} = -2106/T +8.187,[399] ^{19}F,[209] IR and Raman show C_{3v} trig. bipyr. with apical CF_3,[208] normal coordinate analysis also gives C_{3v} symmetry.[266]
$ClCH_2PCl_4$. XVIII. White cryst., dec. 102-104°, fumes in air,[688,690] is readily chlorinated to CCl_3PCl_4.[690]
Cl_2CHPCl_4. XVIII. Hygroscopic cryst.,[689] dec. before melting.[687]
Cl_3CPCl_4. XVIII.[481] Hygroscopic cryst., dec. 125-126°,[687,689] ^{31}P +19.[443]
$EtPCl_4$. XVII. Yellowish cryst. mass,[346] $[EtPCl_3]^+[AlCl_4]^-$: XVIII. M. ∿370°, electric conduct. in $MeNO_2$,[97] cubic cryst. m. 145-199°, needles m. 201-235°,[329] ^{31}P

-128.6.[262] $[EtPCl_3]^+[Al_2Cl_7]^-$: ^{31}P -128.6,[262] liquid.

$CH_3CHClPCl_4$. XVIII. White hygroscopic cryst., dec. easily.[688,690]

$CF_2ClCFHPCl_4$. XVIII. Liquid, IR.[105]

$MeCOOCHClCH_2PCl_4$: $\cdot PCl_5$. XVI. Cryst., full analysis but no further data.[391,392,394]

ClHC=$CHPCl_4$. XVIII. M. 42-46°.[629]

PhHC=$CHPCl_4$. XVI. Yellow needles. very moisture sensitive, m. 65-78°. $\cdot PCl_5 = -PCl_3]^+[PCl_6]^-$, colorless cryst.[162]

MeOHC=C(Ph)PCl_4: $\cdot PCl_5$. XVI. Colorless cryst., soluble in polar solvents, dec. 85-115°.[602]

Cl_3P=NC(Cl)=C(R)PCl_4: $\cdot PCl_5$. From RCH_2CN and PCl_5, no data.[607]

$PrPCl_4$. XVIII.[346] Hygroscopic cryst., dec. 126-137°, reacts with SO_2 to $PrPOCl_2$.[706]

i-$PrPCl_4$. XVIII. Yellowish cryst. mass.[346]

$AllylPCl_4$: $\cdot AlCl_3$. XVII. Gray cryst., dec. 270°, IR.[220]

i-$BuPCl_4$. XVIII. Yellowish cryst. mass.[346]

i-$AmPCl_4$. XVIII. Yellowish cryst. mass.[346]

t-$BuPCl_4$. XVIII. Pale yellow solid.[43]

$(NC)_3CPCl_4$. XVII. Extremely hygroscopic red solid, full analysis.[630]

$C_4H_7Cl_2PCl_4$. XVIII. Hygroscopic cryst., dec. 107-110°, reacts with SO_2 to $-POCl_2$.[706]

Me_2C=$CHCHClCH_2PCl_4$. XVI. M. 109-109.5°.[346]

$RPCl_4$. XVIII. The substances with R = i-Pr, Bu, i-Bu, Am, i-Am decompose above 0° and were identified by reaction with formic acid to $RPOCl_2$.[706]

$RCCl_2PCl_4$: $\cdot PCl_5$. XIX. R, dec.: Me, 100°; Et, 115°; n-Pr, 140°.[163]

$RPCl_4$: $\cdot Al_nCl_{3n+1}$, n = 1,2. XVII. R, m.: Et, 116-117°; i-Pr, 76°; Cl_3C, 132°; cyclohexyl, 44-46°.[570]

$PhPCl_4$. XVIII.[292] White needles, m. 75-76°,[257] m. 74°, $b_{0.001}$93-95°,[175] ^{31}P +39.3,[375] ^{31}P +43.[117] $\cdot PCl_5$: XVI. Yellowish cryst., dec. 200-220°.[162]

m-$ClC_6H_4PCl_4$. XVIII. M. 46-47°.[683]

p-$ClC_6H_4PCl_4$. XVIII. Yellowish mass.[346]

p-$BrC_6H_4PCl_4$. XVIII. Yellow, m. 55°.[346]

o-$MeC_6H_4PCl_4$. XVIII. Yellow, m. 63-66°.[346]

m-$MeC_6H_4PCl_4$. XVIII. Yellow oil, m. ∿0°.[346]

p-$MeC_6H_4PCl_4$. XVIII. Yellow, m. 42°, 69-71°.[346] $\cdot PCl_5$: XVI. M. 158-160°.[162]

p-$PhCH_2C_6H_4PCl_4$. XVIII: Yellow, m. 80°.[346]

p-$EtC_6H_4PCl_4$. XVIII. M. 51°.[346]

p(?)-$PhCH_2CH_2C_6H_4PCl_4$. XVIII. M. 65°.[346]

$ClMe_2SiCH_2CH_2C_6H_4PCl_4$. XVIII. M. 56°.[471]

$Cl_2MeSiCH_2CH_2C_6H_4PCl_4$. XVIII. M. 58°.[471]

$Cl_3SiCH_2CH_2C_6H_4PCl_4$. XVIII. M. 64°.[471]

$ClMe_2SiCH_2C_6H_4PCl_4$. XVIII. M. 32°.[471]

$Cl_2MeSiCH_2C_6H_4PCl_4$. XVIII. M. 36°.[471]

$Cl_3SiCH_2C_6H_4PCl_4$. XVIII. M. 38°.[471]

p(?)-i-$PrC_6H_4PCl_4$. XVIII. M. 53-55°.[346]
p-$MeOC_6H_4PCl_4$. XVIII. Needles, m. 35-40°.[346]
p-$EtOC_6H_4PCl_4$. XVIII. Needles, m. ∿40°.[346]
p-$PhOC_6H_4PCl_4$. XVIII. Cryst. mass.[346]
p-$Me_2NC_6H_4PCl_4$. XVIII. Yellow solid.[346]
m-$NO_2C_6H_4PCl_4$. From PCl_5 complex with red P, yellowish cryst., m. 96-99°. ·PCl_5: XIX. M. 163-165°.[705]
p-$NO_2C_6H_4PCl_4$. From PCl_5 complex with red P, yellowish cryst., m. 105-106°. ·PCl_5: XIX. M. 155-156°.[705]
2,5-$Me_2C_6H_3PCl_4$. XVIII. M. 60°.[346]
2(5)-Me-5(2)-i-$PrC_6H_3PCl_4$. XVIII. Oil.[346]
2-Cl-4-Me-$C_6H_3PCl_4$. XVIII. Needles.[346]
2,4,5-$Me_3C_6H_2PCl_4$. XVIII. Greenish, m. 75°.[346]
2,4,6-$Me_3C_6H_2PCl_4$. XVIII. Yellow, m. 70°.[346]
$PhC_6H_4PCl_4$. XVIII. Solid isomer mixture.[346]
1-Naphthyl-PCl_4. XVIII. M. 143°.[346]

(2-Thienyl)-PCl_4. XVIII. Yellow solid.[346]

(2-Indenyl)-PCl_4. XVI. Yellow solid.[346]

$EtPCl_3Br$: ·$AlCl_3$. XVII. Yellow cryst., m. 80°.[220]
i-$PrPCl_3Br$: ·$AlCl_3$. XVII. Yellow cryst., m. 101°.[220]
$AllylPCl_3Br$: ·$AlCl_3$. XVII. Brown oil, dec. 170°, IR.[220]
$MePCl_3I$: ·$AlCl_3$. XVII. Yellow cryst., m. 94°.[220]
$EtPCl_3I$: ·$AlCl_3$. XVII. Yellow cryst., m. 137°.[220]
$AllylPCl_3I$: ·$AlCl_3$. XVII. Brown oil, dec. 47°, IR.[220]
$PhPCl_2Br_2$. XVIII. Orange, m. 208°, excess Br_2 yields $PhPCl_2Br_4$, m. 209°.[346]
p-$ClC_6H_4PCl_2Br_2$. XVIII. Red, m. 216°.[346]
p-$MeC_6H_4PCl_2Br_2$. XVIII. Red, m. 128-130°.[346]
p-$Me_2NC_6H_4PCl_2Br_2$. XVIII. Red solid.[346]
1-Naphthyl-PCl_2Br_2. XVIII. Orange, m. 114-116°.[346]
$MePBr_4$. XVIII. Hygroscopic white powder, dec. 200°.[402]
CF_3PBr_4. XVIII. Orange solid, m. 28.3-28.5°.[86]
$EtPBr_4$. XVIII. Yellow cryst., hygroscopic, m. 188-190°.[298]
$AllylPBr_4$: ·$AlBr_3$. XVII. Thick, light brown liquid, full analysis, IR.[220]
$CyclohexylPBr_4$. XVIII. White cryst., dec. 158°.[300]
$PhPBr_4$. XVIII. Red mass, with excess Br_2: $PhPBr_6$, red solid, subl. 110°.[346]
p-$MeC_6H_4PBr_4$. XVIII. Red, m. 160-161°.[346]
$EtPI_4$. XVIII. Red-brown cryst., soluble in benzene, MeCN, m. 112-116°.[298]
$CyclohexylPI_4$. XVIII. Red-brown cryst., dec. 157°.[169a,300] ·HgI_2: red cryst., dec. 116°.[300]
$ArPI_4$. XVIII. Reddish violet cryst., fume in air. Ar, m.: C_6H_5-, 90-91°; p-MeC_6H_4-, 91-92°; p-FC_6H_4-, 93-95°; p-ClC_6H_4-, 91-92°; p-BrC_6H_4-, 94-95°.[168a]

TYPE: $R(R'Z)PX_3$

$MePF_3OEt$. XX. Not isolated, $^{19}F(-50°)$, $J_{FPe,a}$ 990, 850 Hz.[146]
$EtPF_3OEt$. XX. Not isolated, $^{19}F(-50°)$, $J_{FPe,a}$ 1005, 875 Hz.[146]
$MePF_3SMe$. XX. B_{75} 47°, ^{31}P -2, at room temp. 2 kinds of ^{19}F, at -90° 3 kinds of ^{19}F.[456a]
$MePF_3SEt$. XX. B_{60} 40-42°, ^{31}P -1, at room temp. 2 kinds of ^{19}F, at -80° 3 kinds of ^{19}F.[456a]
$PhPF_3SMe$. XX. $B_{0.02}$ 47-50°, ^{31}P +18.9,[456a] at room temp. 2 kinds of ^{19}F.[454,456a]
$PhPF_3SEt$. XX. $B_{0.01}$ 59-60°, ^{31}P +18.3,[456a] at room temp. 2 kinds of ^{19}F, at -60° 3 kinds of ^{19}F.[454,456,456a]
$PhPF_3SPh$. XX. Liquid, dec. 100°, at room temp. 2 kinds of ^{19}F, $J_{FPe,a}$ 1066, 970 Hz,[454,456a] ^{31}P +23.6.[456a]
$MePF_3NHMe$. XX. B_{100} 35-36°, n_D^{20} 1.3585, d_4^{20} 1.2630, 2 kinds of ^{19}F.[371]
$MePF_3NHEt$. XX. B_{90} 52-54°, n_D^{20} 1.3610, d_4^{20} 1.1800.[312]
$MePF_3NHAllyl$. XX. B_{70} 65-67°, n_D^{20} 1.3950, d_4^{20} 1.2090,[312] b_{40} 47-49°, n_D^{20} 1.3935, d_4^{20} 1.2069, 2 kinds of ^{19}F.[371]
$MePF_3NHCH_2CHBrCH_2Br$. From bromination of allyl derivative, b_9 110-114°, n_D^{20} 1.4810, d_4^{20} 1.8980.[312]
$MePF_3NHBu$. XX. B_{50} 71-72°, n_D^{20} 1.3830, d_4^{20} 1.1291,[312] b_{70} 68-69°, n_D^{20} 1.3795, d_4^{20} 1.1275,[371] b_{15} 37-38°, n_D^{20} 1.3843, d_4^{20} 1.1210, IR, ^{31}P +36,[145] 3 kinds of ^{19}F.[145,371]
$MePF_3NHi\text{-}Bu$. XX. B_{70} 77-78°, n_D^{20} 1.3840, d_4^{20} 1.1220,[312] b_{45} 58-60°, n_D^{20} 1.3760, d_4^{20} 1.1262,[143,371] 3 kinds of ^{19}F.[371]
$MePF_3NHAm$. XX. B_{50} 91-92°, n_D^{20} 1.3940, d_4^{20} 1.1040.[312]
$MePF_3NHi\text{-}Am$. XX. B_{10} 52-53°, n_D^{20} 1.3857, d_4^{20} 1.1075, 3 kinds of ^{19}F.[371]
$MePF_3NHHex$. XX. B_6 60-62°, n_D^{20} 1.4200, d_4^{20} 1.2241, 3 kinds of ^{19}F.[371]
$MePF_3NHC_6H_4Me$. XX. B_3 75-77°, n_D^{20} 1.4683, d_4^{20} 1.2720, 3 kinds of ^{19}F.[371]
$EtPF_3NHHex$. XX. B_2 65-66°, n_D^{20} 1.4245, d_4^{20} 1.1655, 3 kinds of ^{19}F.[371]
$PhPF_3NHMe$. XX. $B_{0.25}$ 42°, n_D^{26} 1.4732,[589,593] b_{10} 93-94°, n_D^{20} 1.4625,[371] IR, ^1H,[593] 2 kinds of ^{19}F.[371,589,593]
$PhPF_3NHi\text{-}Bu$. XX. B_{10} 114-115°, n_D^{20} 1.4675, d_4^{20} 1.1821, 3 kinds of ^{19}F.[371]
$MePF_3NMe_2$. XI.[586] XX.[555] B_{760} 85°,[586] b_{140} 48°,[555] ^{31}P +37.2,[586] ^{31}P +37.7,[555] ^1H, 2 kinds of ^{19}F.[555,586]
$MePF_3NEt_2$. XI.[143] XX.[145,371] B_{15} 33-35°, n_D^{20} 1.3784, d_4^{20} 1.1187,[145] b_{15} 32-35°, n_D^{20} 1.3790, d_4^{20} 1.1195,[143,371] IR, ^{31}P +35,[145] 2 kinds of ^{19}F.[145,371]
$MePF_3NBu_2$. XX. B_3 58-60°, n_D^{20} 1.4060, d_4^{20} 1.1799, 2 kinds of ^{19}F.[371]

MePF$_3$N(piperidino). XX. B_{10} 73-74°, n_D^{20} 1.4197, d_4^{20} 1.2205, 2 kinds of ^{19}F.[371]

NCPF$_3$NMe$_2$. From Me$_2$NPF$_2$ and BrCN, b_{20} 35°, m. -44°, IR, mass spect., ^{1}H, 2 kinds of ^{19}F.[98]

EtPF$_3$NMe$_2$. XX. B_{54} 45-47°, ^{31}P +37.1, 2 kinds of ^{19}F,[590] mass spect.[73]

EtPF$_3$NEt$_2$. XX. B_{30} 66.5°, n_D^{25} 1.3869, IR,[593] ^{31}P +35.5, 2 kinds of ^{19}F,[371,425,593] mass spect.[73]

EtPF$_3$N(pyrrolinyl). XX. B_{15} 45°, ^{31}P +38.5, 2 kinds of ^{19}F,[590] mass spect.[73]

PhPF$_3$NMe$_2$. XX (XI[586]). $B_{0.25}$ 48°,[578] $b_{0.3}$ 42°,[586] $b_{0.4}$ 48-49°,[593] $b_{6.5}$ 88°,[579] n_D^{25} 1.4796, IR,[593] ^{1}H,[586,593] ^{31}P +53.6,[579,586] 2 kinds of ^{19}F.[371,425,578,579,586,593]

PhPF$_3$NEt$_2$. XX (XI[586]). B_2 78-79°, n_D^{22} 1.4690, d_{20}^{22} 1.1835,[305] $b_{0.03}$ 51°,[586] $b_{0.5}$ 70°,[589,590] $b_{1.35}$ 81°,[593] $n_D^{24.6}$ 1.4689,[589,593] IR, ^{1}H, ^{31}P +52,[586,593] 2 kinds of ^{19}F,[371,425,586,593] mass spect.[73]

PhPF$_3$N(i-Bu)$_2$. XX. B_2 107-108°, n_D^{21} 1.4690, d_{20}^{21} 1.0972.[305]

PhPF$_3$NEtPh. XX. B_3 141-143°, n_D^{21} 1.5250, d_{20}^{20} 1.2343.[305]

PhPF$_3$N(piperidino). XX. B_2 115-116°, n_D^{20} 1.4940, d_{20}^{23} 1.2442,[305] $b_{0.05}$ 69°, n_D^{25} 1.4928, IR, ^{1}H,[593] 2 kinds of ^{19}F.[371,589,593]

PhPF$_3$N(morpholino)O. XX. B_2 108-109°, n_D^{23} 1.4942, d_{20}^{22} 1.3225.[305]

PhPF$_3$N(pyrrolinyl). XX. $B_{0.35}$ 73-74°, n_D^{26} 1.5134,[578,589] ^{31}P +60,[578] 2 kinds of ^{19}F,[578] mass spect.[73]

MeC$_6$H$_4$PF$_3$NEt$_2$. XX. B_2 87-89°, n_D^{20} 1.4714, d_4^{20} 1.1753, 2 kinds of ^{19}F.[371]

p-ClC$_6$H$_4$PF$_3$NR$_2$. XX. Colorless fuming liquids.[305]

R(-R$_2$-)	b. (mm)	n_D^x	d_{20}^x
Et	96 to 97° (2)	1.4820 ^{21}	1.2740 ^{25}
i-Bu	138 to 140° (3)	1.4754 ^{25}	1.1663 ^{25}
-(CH$_2$)$_5$-	128 to 130° (2)	1.5031 ^{18}	1.3258 ^{18}
-(C$_2$H$_4$OC$_2$H$_4$)-	132 to 134° (2)	1.5070 ^{19}	1.4008 ^{19}

MePFCl$_2$NEt$_2$. XVIII. Unstable at room temp., n_D^{20} 1.5060, d_4^{20} 1.3140, ^{19}F, J_{FP} 1122 Hz.[150]

MePFBr$_2$NEt$_2$. XVIII. Unstable at room temp., n_D^{20} 1.5045, ^{31}P -80°!, ^{19}F, J_{FP} 1164 Hz. The large negative ^{31}P value and the solubility in polar solvents indicate [MePFBrNEt$_2$]$^+$Br$^-$.[150]

PhPFCl$_2$NEt$_2$. XVIII. Unstable at room temp., n_D^{20} 1.5720, d_4^{20} 1.3520, ^{19}F, J_{FP} 1150 Hz, ^{31}P -67!, very probably [PhPFClNEt$_2$]$^+$Cl$^-$.[150]

PhPFBr$_2$NEt$_2$. XVIII. Unstable at room temp., n_D^{20} 1.5650, d_4^{20} 1.5403, ^{31}P -57!, J_{PF} 1200 Hz, very probably [PhPFBrNEt$_2$]$^+$Br$^-$.[150]

$(NC)_3CPCl_3OEt$. From $ClC(CN)_3$, PCl_3, and Et_2O, orange cryst., m. 101°, full analysis, IR.[630]

TYPE: $R(R'Z)_2PX_2$

$MePF_2(OPh)_2$. XX. One kind of ^{19}F, J_{FP} 825 Hz.[454]
$PhPF_2(OPh)_2$. XX. One kind of ^{19}F, J_{FP} 829 Hz.[454]
$MePF_2(SEt)_2$. XX. B_5 74-76°, n_D^{20} 1.5050, d_4^{20} 1.1900,[311] ^{31}P +15.[146]
$MePF_2(S\text{-}i\text{-}Pr)_2$. XX. B_1 63-67°, n_D^{20} 1.4833, d_4^{20} 1.1210.[311]
$EtPF_2(SEt)_2$. XX. B_1 67°, n_D^{20} 1.5020, d_4^{20} 1.1469.[311]
$PhPF_2(SEt)_2$. XX. B_3 136°, n_D^{20} 1.5760, d_4^{20} 1.2060.[311]
$MePF_2(NHMe)(NHCyclohex)$. XX. B_9 65-67°, n_D^{20} 1.4490, ^{19}F.[137]
$MePF_2(NHEt)(NEt_2)$. XX. B_{20} 48-50°, n_D^{20} 1.4050, d_4^{20} 1.0655, ^{19}F.[137]
$MePF_2(NHEt)(N\text{-}i\text{-}Pr_2)$. XX. B_{37} 65-67°, n_D^{20} 1.4007, d_4^{20} 1.0395, ^{19}F.[137]
$MePF_2(N\text{-}i\text{-}Pr_2)_2$. XX. B_{15} 42-44°, n_D^{20} 1.3978, d_4^{20} 1.0246, ^{31}P +19.[137]
$(MeF_2PNMe)_2$. XX. B_{14} 60°, m. 42°, IR, ^1H, ^{19}F, ^{31}P +50.7.[590]
$(MeF_2PNPh)_2$. XX. M. 140°, ^{19}F.[590]
$(ClCH_2F_2PNMe)_2$. XX. M. 47-48°,[446] $b_{0.05}$ 68°, ^{31}P +56.3,[582] IR, ^{19}F,[590] ^1H.[446,590]
$(EtF_2PNMe)_2$. XX. B_1 42°, $n_D^{25.5}$ 1.4171,[588] IR,[590] ^1H,[446,590] ^{31}P +45.6,[582] ^{19}F, J_{FP} ∿910 Hz.[590]
$(PhF_2PNMe)_2$. XX. M. 162°,[582,588] IR, ^1H, ^{19}F,[582,590] ^1H,[446] ^{31}P +56.1,[590] x-ray struct. determination[106] establishes trig. bipyr. with $P\text{-}F_{a,e}$ 1.62, 1.57 Å; $P\text{-}N_{a,e}$ 1.78, 1.64 Å; P-C 1.79 Å; N_aPN_e 80.6°; $CPN_{e,a}$ 122, 97.9°; $CPF_{e,a}$ 109, 91.8°; $F_ePN_{e,a}$ 128, 91.7°; F_ePF_a 87°.
$(PhF_2PNPh)_2$. XX. M. 134-136°, ^{19}F.[590]
$(PhF_2PNEt)_2$. XX. M. 108-110°, ^{31}P +53.6, ^{19}F, J_{PF} ∿990 Hz.[590]
$(m\text{-}CF_3C_6H_4F_2PNMe)_2$. XX. M. 164-166°, IR, ^{19}F,[590] ^1H.[446,590]
$(2,5\text{-}Me_2C_6H_3F_2PNMe)_2$. XX. M. 173-174°, IR, ^1H, ^{19}F.[590]
$(PhF_2PNH)_n$. XX. Semisolid, IR.[581]
$(MeCl_2PNAr)_2$. XX. R = Ph; m. 95-96°, R = p-MeC_6H_4-: m. 102-103°, in refluxing benzene these products are monomeric. R = o-ClC_6H_4-: m. 58-60°, R = o-BrC_6H_4-: m. 25.5°, these products are monomeric after distillation and dimerize on standing. The substances with R = p-ClC_6H_4- and p-BrC_6H_4- also seem to dimerize on long standing.[695]
$(PhCl_2PNMe)_2$. XX. M. 222°,[275] ^{31}P +54.9.[375]
$(PhCl_2PNAlk)_2$. XX. Prisms, not monomerized in boiling benzene.[695a] Alk, m.: Me, 180-185°; Et, 166-168°; Pr, 161-163°; Bu, 142-145°; Bz, 170-172°.
$(PhCl_2PNAr)_2$. XX. R, m.: Ph, 119-120°, m-MeC_6H_4-, 82-84°,

p-MeC_6H_4-, 124-126°, p-ClC_6H_4-, 87-89°, p-$MeOC_6H_4$-, 121-123°, p-$EtOC_6H_4$-, 92-94°, all products are colorless prisms and monomeric in boiling benzene solution.[702,704]

TYPE: $R(R'Z)_3PX$

$(CF_3)_2C{=}CFPF(OEt)_3$. XXIIIA. $B_{0.02}$ 58-60°, m. 10-11°, n_D^{20} 1.3730, d_4^{20} 1.3341.[334]

F_2(F-cyclobutenyl)$PF(OEt)_3$. XXIIIA. $B_{0.03}$ 60-62°, m. 8-8.5°, n_D^{20} 1.4020, d_4^{20} 1.3293,[334] ^{19}F, ^{31}P +1.5.[590]

$MePF(SPh)_3$. XX. B_2 148-155°.[311]

(Me)(Me)-dioxaphospholene $P(Me)(NEt_2)(F)$. VIII. n_D^{20} 1.4389, d^{20} 1.1074, ^1H, ^{31}P +28, J_{PF} 785 Hz.[139a]

TYPE: $R(R'Z)_4P$

$PhP(OEt)_4$. See VII. ^{31}P +55, ^1H.[118]

$(MeO)_3P$(oxaphospholene). XXI. $B_{0.01}$ 21-22°, n_D^{20} 1.4485, d_4^{20} 1.1713, IR, μ 2.85 D, ^1H, ^{31}P +11.5.[29]

$(MeO)_3P$(Me-oxaphospholene). XXI. B_3 56-57°, IR, ^1H shows inhibited pseudorotation at -86°,[201] activation parameters for pseudorotation.[198]

$(EtO)_3P$(oxaphospholene). XXI. $B_{0.01}$ 48-50°, n_D^{20} 1.4352, d_4^{20} 1.0607, IR,[29] ^{31}P +30.1.[405]

$(MeO)_3P$(Me, COMe, H, Me-oxaphospholene). XXI. $B_{0.01}$ 90-92°, n_D^{20} 1.4820, d_0^{20} 1.1469, IR, ^{31}P +19.3, ^1H,[22] ^{31}P +8.2, UV.[20]

$(MeO)_3P$(Me, COOEt, H, Me-oxaphospholene). XXI. $B_{0.01}$ 90-93°, n_D^{20} 1.4670, d_0^{20} 1.1393, IR, Raman, ^{31}P +8.2, ^1H,[22] ^{31}P +19.3.[20]

$(EtO)_3P$(Me, COMe, H, Me-oxaphospholene). XXI. $B_{0.01}$ 84-89°, n_D^{20} 1.4730, d_0^{20} 1.0749, IR.[20,22]

$(EtO)_3P$(Me, COOEt, H, Me-oxaphospholene). XXI. B_1 95-97°, n_D^{20} 1.4617, d_0^{20} 1.0769, IR, Raman,[22] UV.[20]

$(MeO)_3P$(Me, COMe, H, Ph-oxaphospholene). XXI. M. 49-51°, IR, ^1H, ^{31}P +27.9,[514] ^1H shows 3 kinds of MeO at -67° and 1 kind at room temp.[200]

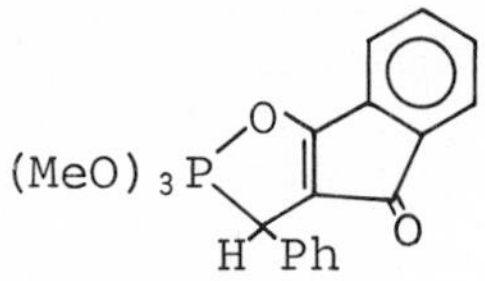

$(MeO)_3P$ (ring: O, H Ph, O) . XXI. Yellow cryst., dec. 50-55°, IR.[25] Corresponding $(EtO)_3$ derivative is unstable, IR.[25]

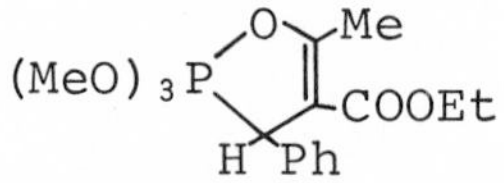

$(MeO)_3P$ (ring: O, Me, COOEt, H Ph) . XXI. M. 34-35°, IR, UV.[19]

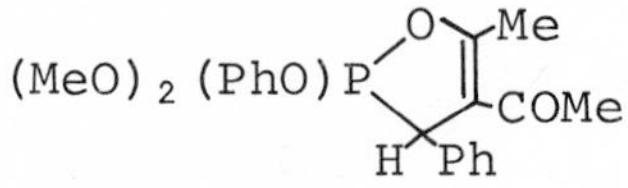

$(MeO)_2(PhO)P$ (ring: O, Me, COMe, H Ph) . XXI. M. 129-131°, IR, ^1H shows inhibited pseudorotation at -26°,[201] activation parameters.[198]

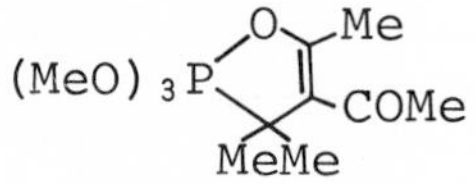

$(MeO)_3P$ (ring: O, Me, COMe, MeMe) . XXI. M. 58.5-61°, IR, ^1H shows inhibited pseudorotation at room temp.,[201] activation parameters.[198]

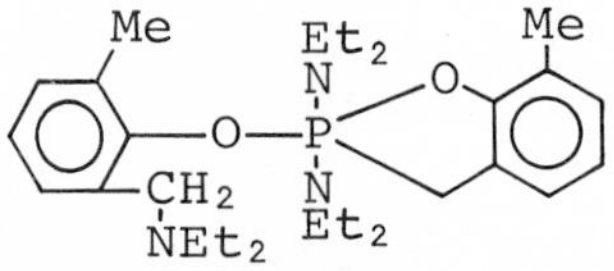

(Me, CH_2NEt_2-substituted phenyl)–O–P(NEt_2)$_2$ (ring: O, Me) . From phenol and $P(NEt)_3$, $b_{0.02}$ 170-172°, n_D^{20} 1.5345, d^{20} 1.0414, μ 3.16 D, NMR.[303]

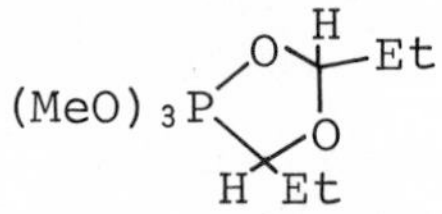

$(MeO)_3P$ (ring: O, H, Et, O, H Et) . XXII. Mixture of diastereomers, $b_{0.2}$ 50-51°, n_D^{25} 1.4313, IR, ^1H, ^{31}P +32.8, +34.2.[521]

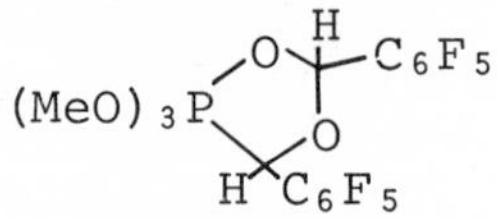

$(MeO)_3P$ (ring: O, H, C_6F_5, O, H C_6F_5) . XXII. Cis H/H: m. 83-86°, ^{31}P +34.1, ^1H at -135° shows fixed trig. bipyr. with 1 apical and 2 equat. MeO; trans H/H: only in mixture with cis H/H, ^1H, ^{31}P +38.4.[530]

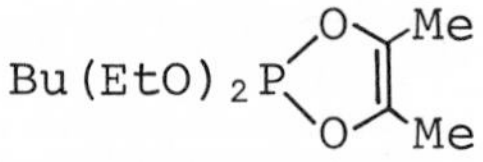

$Bu(EtO)_2P$ (ring: O, Me, O, Me) . VIII. $B_{0.15}$ 62-63°, n_D^{25} 1.4409,[70] ^{31}P +24.4.[405]

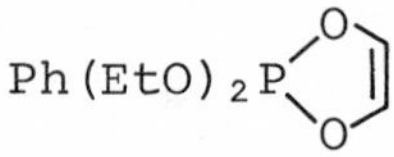

$Ph(EtO)_2P$ (ring: O, O) . ^{31}P +49.[405]

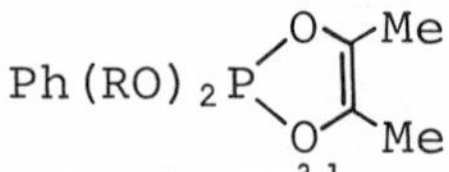

$Ph(RO)_2P$ (ring: O, Me, O, Me) . VIII. R = Et: $b_{0.2}$ 110-111°, n_D^{25} 1.5022.[70] R = Bu: ^{31}P +36.1.[534] R = Ph: ^{31}P +42.4.[534]

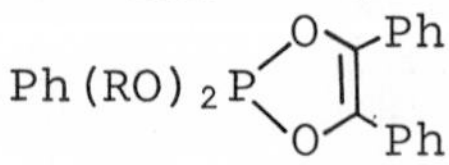

$Ph(RO)_2P$ (ring: O, Ph, O, Ph) . VIII. R = Bu: ^{31}P +36.6. R = Ph: ^{31}P +43.2.[534]

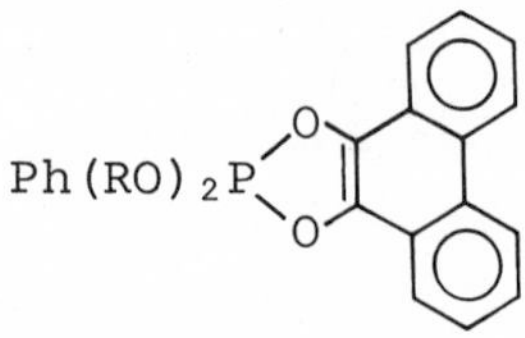

$Ph(RO)_2P$ (phenanthrenediol ring: O, O) . VIII. R = Et: ^{31}P +16.3 to +18.0.[405] R = Bu: m. 52°, ^{31}P +29.6; R = Ph: m. 142°, ^{31}P +38.1.[534]

H, O, COOBu; $Et(EtO)_2P$; O, COOBu, H. X. Viscous liquid, b_1 153-158°, n_D^{20} 1.4460, d_4^{20} 1.1148.[474]

H, O, COOBu; $Et(PrO)_2P$; O, COOBu, H. X. B_1 165-170°, n_D^{20} 1.4470, d_4^{20} 1.0773.[474]

H, O, C_6H_4X; $Ph(MeO)_2P$; O, C_6H_4X, H. X. ^1H shows for cis H/H 1 kind of MeO and for trans H/H 2 kinds of MeO.[488]

H, O, C_6H_4X; $Ph(MeO)_2P$; O, COMe, Me. IX. ^1H shows for cis and trans H/Me 2 kinds of MeO.[488]

O, $(CF_3)_2$; $Ph(RO)_2P$; O, $(CF_3)_2$. X. R = Me: ^1H shows 1 kind of MeO at all temp.[488] R = Bu: $b_{0.1}$ 99°, ^{31}P +33.4. R = Ph: ^{31}P +40.1.[534]

CF_3, O, CF_3; $Ph(MeO)_2P$; O, COMe, Me. IX. ^1H shows 2 kinds of MeO at all temp.[488]

Me, O, COMe; $Ph(RO)_2P$; O, COMe, Me. R = Me: IX. ^1H shows one kind of MeO for cis and 2 kinds of MeO for trans product.[488] R = Ph: X. Not isolated, probably cis product, ^{31}P +48.[534]

Ph, O, Ph, O, Ph; P; Ph, O, O, Ph. XXI. M. 216-218°.[574]

Me; O, O; P; O, O; R, R, R, R. XXI. R = H: m. 72-74°, $b_{0.1}$ 110°, ^1H (Me) 1.85,[660] ^{31}P 1.85.[661a] R = Cl: m. 118-120°, ^1H (Me) 2.05.[660]

Ph; O, O; P; O, O; R, R, R, R. XXI. R = H: m. 111°, ^1H,[660] m. 181°, ^{31}P +9.0.[661a] R = Cl: m. 136°, ^1H.[660]

Ph; O, O, Me; P; O, O, Me. XXI. B 120-160°, n_D^{25} 1.5179.[70]

R, H, O, Y, O, Me; P; R', H, O. XXI. Thick colorless substances, easily hydrolyzed, ^1H, IR.[647]

R	R'	Y	$b_{0.1}$	n_D^{20}	d_4^{20}
H	H	OMe	85°	1.4710	1.1999
Me	H	OEt	86°	1.4705	1.1879
Me	H	OMe	88°	1.4720	1.1819
Me	Me	OMe	90°	1.4675	1.1500
Me	H	NEt_2	92°	1.4757	1.0711

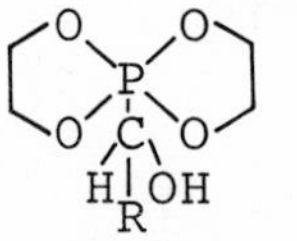

. XXIIIB. R = CCl_3: m. 109°, IR, ^{31}P +18. R = Ph: m. 105°, IR, ^{1}H ^{31}P +18.[91]

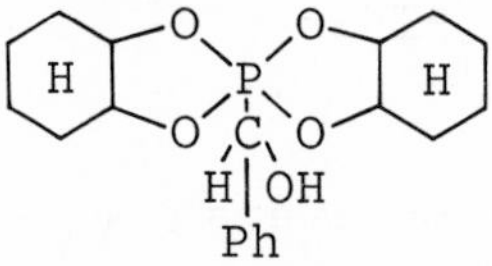

. XIIIB. M. 130°, IR, ^{31}P +20, ^{1}H shows 3 doublets for the benzylic protons; this speaks for 3 isomers.[91]

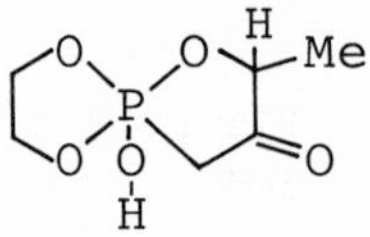

. ?? XXI. White cryst., full analysis, IR,[646] from 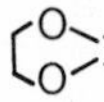 PCl and biacetyl.

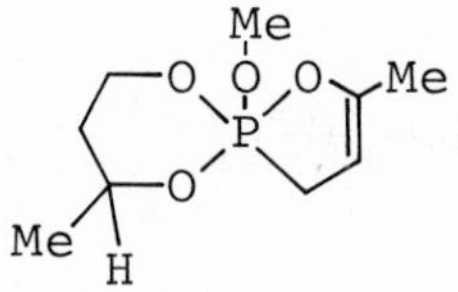

. XXI. $B_{0.1}$ 92°, n_D^{20} 1.4724, d_4^{20} 1.1509.[647]

D.3. Compounds of Type R_2PY_3

TYPE: R_2PX_3

Me_2PF_3. XXIV.[343,576] XXV.[324,577,591] XIV.[595] B_{745} 60-61°, n_D^{20} 1.3230, d_4^{20} 1.2155,[324,343] mass spect.,[73,595] ^{31}P -8.0,[445] ^{1}H,[446] 2 kinds of ^{19}F,[426,590] IR and Raman show C_{2v} trig. bipyr.,[134] elect. diff. shows trig. bipyr. with $PF_{e,a}$ 1.553, 1.643 Å, CP 1.798 Å, F_ePF_a 88.9°, F_ePC 118°,[40] heteronuclear INDOR spect.[135a]
$(CF_3)_2PF_3$. XXIV. B. -5°, m. -74°,[398] b. -4.7°, m. -76.2°,[85] IR,[398,426] ^{19}F,[426] ^{31}P +50.9.[445]
$MeEtPF_3$. XXIV.[343] B_{760} 81-82°, n_D^{20} 1.3470, d_4^{20} 1.1875,[343] ^{31}P +13.4,[445] $J_{PFe,a}$ 955, 813 Hz.
Et_2PF_3. XXIV.[343,576] XXV.[577] B_{763} 104-105°, n_D^{20} 1.3650, d_4^{20} 1.1486,[343] b_{100} 50°,[576] IR, 2 kinds of ^{19}F,[426] ^{31}P -6.2.[445,580]
$(C_2F_5)_2PF_3$. ^{19}F, J_{FP} 1245 Hz.[425]
$(F_2C{=}CF)_2PF_3$. XXIV. Reactive liquid, IR, ^{19}F shows at -60° 2 kinds of ^{19}F; this indicates trig. bipyr. with equat. alkenyl groups.[105]
$(C_3F_7)_2PF_3$. XXIV. Only 1 kind of ^{19}F!, J_{FP} 1172 Hz.[444,590]

Bu_2PF_3. XXV.[591] B_{10} 71°,[577] IR, 2 kinds of ^{19}F.[426]

(cyclopentyl)PF_3. XXIV.[576] XXV.[577] B_{90} 61-62°,[576,577] IR,[577] ^{31}P -29.6,[580] ^{31}P -29.8,[445] at -100° 2 kinds of ^{19}F,[426] attacks glass.[576]

(F_2-cyclopentyl, F_2 F_2 F_2)PF_3. XXIV. B. 49°, 1 kind of ^{19}F.[590]

(cyclohexyl)PF_3. XXV. B_{40} 64-65°, IR,[577] 2 kinds of ^{19}F.[426,590]

$MePhPF_3$. XXV. B_9 64°, $n_D^{26.5}$ 1.4646,[577] ^{31}P +13.1,[580] ^{31}P +13.4,[445] 2 kinds of ^{19}F.[426,590]
$PhPF_3(CHOHC_2H_5)$. XXIIIB. $B_{1.5}$ 58-60°.[305a]
p-$MeC_6H_4PF_3(CHOHC_2H_5)$. XXIIIB. B_1 52-54°.[305a]
p-$MeC_6H_4PF_3(CHOHCH_3)$. XXIIIB. B_2 62-64°.[305a]
Ph_2PF_3. XXIV.[576,591] XXV.[337,618] B_4 135°, b_3 116-117°,[337] b_2 106-107°,[618] $b_{0.4}$ 92-93°, n_D^{23} 1.5425, n_D^{20} 1.5410,[576] mass spect.,[73] ^{31}P +34.8,[445,580] IR,[426] 2 kinds of ^{19}F.[426,590]
$(C_6F_5)_2PF_3$. XXIV. B_1 120°, m. 55°, ^{31}P +28.8, 1 kind of ^{19}F.[170]
$(CF_3)_2PFCl_2$. XXVIII.[5,444] B. 63°, log P_{mm} = -1704/T +7.950, IR.[5]
Me_2PCl_3. XXIX. M. 183°, dec., IR and Raman show $[Me_2\overset{+}{P}Cl_2]Cl^-$,[41] ^{31}P -124.[48] $\cdot AlCl_3$: XXVII. M. 196-198°.[306]
$(CF_3)_2PCl_3$. XVIII. B_{355} 82°, b_{360} 107°,[154] b_{86} 44-45°,[626] m. -26 to -25°, no electric conduct.,[153] ^{19}F,[209,453] IR and Raman show trig. bipyr. (D_{3h}).[211]
$(Cl_3C)_2PCl_3$. XXIX.[559] XXX.[178,351] Colorless prisms, m. 192-193°,[351] dec. 187-188°;[559] white cryst., stable against air and water, IR, dec. 193-194°.[178]
$MeEtPCl_3$: $\cdot AlCl_3$. XXVII. Cryst., no data.[306]
$Me(CHF_2CF_2)PCl_3$. XXVIII. B_{18} 72°, from R_2PH and Cl_2.[84]
MeAllylPCl_3: $\cdot AlCl_3$. XXVII. Dark thick oil, dec. 125°, IR.[220]
$Me(C_3F_7)PCl_3$. XXVIII. B_{15} 64-66°, fumes in air.[212]
$F_3C(CF_3CH_2CH_2)PCl_3$. XXVIII. B_6 39-40.5°, b_{754} 158°, n_D^{20} 1.4150, fumes in air.[212]
$Cl_3CPhPCl_3$. XXVIII. Cryst., no data. $\cdot PCl_5$; XXIX. M. 202°, dec.[684]
Et_2PCl_3. XXIX. Dec. 65°, extremely sensitive against moisture, conduct. in DMF, nonsoluble in nonpolar solvents, soluble in polar solvents = $[Et_2PCl_2]^+Cl^-$,[353,354] m. 134-136°,[67] IR also shows ionic structure.[41] $\cdot AlCl_3$: XXVII. M. 155-160°.[306] $\cdot Al_nCl_{3n}$: XVIII. M. 76°.[570]
$(PhHC{=}CH)_2PCl_3$. XXVI. M. 72-76°, is obtained in the form of an adduct with HCl (m. 82°), which loses the

HCl on heating 3 hr to 100° i.V.[167a]
PhHC=CH⟩PCl_3 / PhClC=CH⟩. XXVI. M. 78-83°, sensitive to moisture.[166]
$(PhClC=CH)_2PCl_3$. XXVI. Dec. 108°.[167a]
$(EtOHC=CH)_2PCl_3$. XXVI. Colorless prisms, m. 70-75°.[164]
$EtAllylPCl_3$: $\cdot AlCl_3$. XXVII. Liquid, dec. 197°, IR.[220]
$(PhHC=CH)PhPCl_3$. XXVI. Viscous oil, no data, with SO_2 $>P(=O)Cl$.[167]
$(PhClC=CH)PhPCl_3$. XXVI. Red oil, no data, with SO_2 or H_2O $>P(=O)Cl(OH)$.[167]
$(EtOHC=CH)PhPCl_3$. XXVI. White hygroscopic cryst., full analysis.[12]
Pr_2PCl_3. XXIX. Dec. > 55°, moisture sensitive, conduct. in DMF shows $[Pr_2PCl_2]^+Cl^-$.[353,354]
$(C_3F_7)_2PCl_3$. XXVIII. B. 184 ± 2°, log P_{mm} = -2094/T +7.46.[157]
$(PhCH=CHCH=CH)_2PCl_3$. XXVI. Yellow cryst., dec. 130-133°.[165]
$(PhCH=CHCH=CH)PhPCl_3$. XXVI. Reddish liquid.[165]
$(PhCH=CH)(PhCH=CHCH=CH)PCl_3$. XXVI. Reddish liquid.[165]
3,4-Me₂-cyclic ring PCl_3 (Me, Me substituted ring). XXXII. M. 108-110°.[28]
Ph_2PCl_3. XXVIII.[117,224] XXIX.[260,277,608] White needles, dec. 194-200°,[224] m. 150°,[277] ^{31}P ($PhNO_2$) +26,[117] ^{31}P (polar solvents) -73.[48] $\cdot AlCl_3$ = $[Ph_2PCl_2]^+[AlCl_4]^-$, ^{31}P -92,[117] $[Ph_2PCl_2]^+[PCl_6]^-$.[375]
$(C_6F_5)_2PCl_3$. XXVIII. M. 110°, dec.,[6] mass spect.[6,414a]
$Ph(p\text{-}MeC_6H_4)PCl_3$. XXVIII. Yellow solid.[346]
$Ph(2,4,5\text{-}Me_3C_6H_2)PCl_3$. XXVIII. Yellow solid.[346]
$Ph(p\text{-}BrC_6H_4)PCl_3$. XXVIII. Solid.[346]
$Ph(p\text{-}RC_6H_4)PCl_3$: $\cdot PCl_5$. XXIX. Cryst. substances,[608] R, m.: H, 178-180°; Cl, 146-150°; Br, 156-159°; NO_2, 168-172°; MeO, 102-105°; Me, 124-127°.
$(p\text{-}MeC_6H_4)_2PCl_3$. XXVIII. Yellow solid.[346]
$RR'PCl_2X$: $\cdot AlCl_3$. XXVII. Light yellow to light brown cryst. or solids.[220] R, R', X, m.: Me, Me, I, 120°; Me, Et, Br, 52°; Me, i-Pr, Br, 93°; Me, allyl, Br, thick yellow oil, dec. 128°, IR; Me, Et, I, 91°; Me, allyl, I, 87° (dec., IR); Et, allyl, Br, liquid, dec. 108°, IR; Et, allyl, I, 70° (dec., IR).
$(CF_3)_2PBr_3$. XXVIII. White solid, m. 6-9°.[86]
Et_2PBr_3: $\cdot Br_2$. XXIX. Orange-red needles, m. 103.5°, nonsoluble in nonpolar solvents, soluble in polar solvents, conduct. in DMF shows $[Et_2PBr_2]^+Br_3^-$,[353,354] IR also shows ionic struct.[41]
Pr_2PBr_3: $\cdot Br_2$. XXIX. Orange-red needles, m. 87.5°, solubilities as with Et_2 derivative, conduct. in DMF shows ionic struct.[353,354]

$Cyclohex_2PBr_3$. XXVIII. White cryst., dec. 288°.[300]

(R, R'-substituted cyclopentenyl ring)PBr_3. XXXII. White cryst.,[27] R, R', m. : Me, Me, 151-153°; H, Me, 114-117°, H, H, 125-127°.

$Cyclohex_2PI_3$. XXVIII. Yellow cryst., dec. 248°. $\cdot HgI_2$, yellow cryst., dec. 194°.[300]

Ph_2PI_3. XXVIII. Hygroscopic solid, nonsoluble in benzene.[264]

TYPE: $R_2(R'Z)PX_2$

Ph_2PF_2OPh. XXXI. One kind of ^{19}F, J_{FP} 797 Hz.[454]

Ph_2PF_2SMe. XXXI. B_2 148-150°, ^{31}P +39.0, at -80° 2 kinds of ^{19}F, at room temp. 1 kind of ^{19}F.[454,456a]

Ph_2PF_2SEt. XXXI. $B_{0.1}$ 160-163°, ^{31}P +39.2, at -85° 2 kinds of ^{19}F, at room temp. 1 kind of ^{19}F.[454,456a]

Ph_2PF_2SPh. XXXI. Dec. 140°, 1 kind of ^{19}F at room temp., J_{FP} 796 Hz.[454,456a]

$Me_2PF_2NHAllyl$. XXXI. B_{20} 45-46°, n_D^{20} 1.4170, d_4^{20} 1.0590.[312]

$Me_2PF_2NMe_2$. Mass spect.[73]

$Ph_2PF_2NMe_2$. XXXI. B_1 128°, ^{31}P +54.0, ^{19}F, J_{FP} 709 Hz,[590] mass spect.[73]

$Et_2PCl_2(OCHClCH_3)$. XXXI. White hygroscopic cryst., fume in air, IR, analysis.[633]

$(MePCl_2)_nY$. XXXV.[194] Y, n, m.: -S-, 2, 140° (dec.); $-S_2-$, 2, 120° (dec.); $-Ti(OEt)_3$, 1, b_1 130°; $>Ti(OEt)_2$, 2, 120-130° (dec.); -Ti-, 4, 250-255° (dec.); -Sn-, 4, 112-115° (dec.).

$(Ph_2PCl_2)_4Ti$. XXXV. Dec. ∿500°.[194]

$(Cl_3C)_2PCl_2NH_2$. From $(Cl_3C)_2P{\lesssim}^{NH}_{Cl}$ and HCl, cryst., dec. 145-146°.[352]

$Ph_2PCl_2NH_2$. XXIX. From $Ph_2P{\lesssim}^{O}_{NH_2}$ and PCl_5, colorless needles, IR, ^{31}P -51.0 shows ionic struct.: $[Ph_2P{\lesssim}^{NH_2}_{Cl}]^+Cl^-$.[47,48]

Ph_2PCl_2NHMe. XXIX. Hygroscopic powder, conduct. in $PhNO_2$, ^{31}P -66 shows ionic struct.: $Ph_2P{\lesssim}^{NHMe}_{Cl}]^+Cl^-$.[48]

Ph_2PCl_2NHPh. XXIX. M. 133-137°, conduct. in $PhNO_2$, ^{31}P -55, ionic struct.[48]

$(CF_3)_2PCl_2N(CF_3)_2$. XXVIII. White cyrst. at -50°, IR, ^{19}F shows trig. bipyr. with apical F_3C groups.[5]

$Me_2PF_2PMe_2$. Intermediate in disproport. of Me_2PF to Me_2PF_3 and Me_2PPMe_2, ^{19}F, J_{FP} 560 Hz.[595]

TYPE: $R_2(R'Z)_2PX$

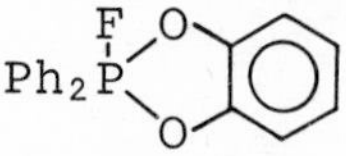
. XXXI. ^{19}F, J_{FP} 797 Hz.[131a]

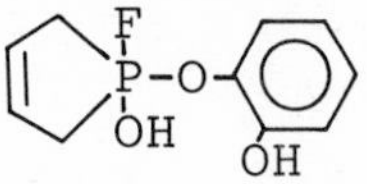
. ? From corresponding spiro compound and H_2O, $b_{0.5}$ 84°, m. 81°, n_D^{20} 1.5230, d_4^{20} 1.2990, IR, ^1H.[160]

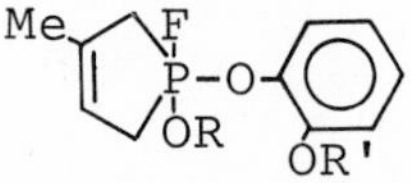
?. R, R' = H; from spiro compound and H_2O, $b_{0.9}$ 117°, m. 76°, n_D^{20} 1.5195, d_4^{20} 1.2460, IR, ^1H. R = Me, R' = H: b_1 119°, n_D^{20} 1.5220, d_4^{20} 1.2260, ^1H. R, R' = Me: $b_{0.9}$ 109°, n_D^{20} 1.5245, d_4^{20} 1.2101, ^1H.[160]

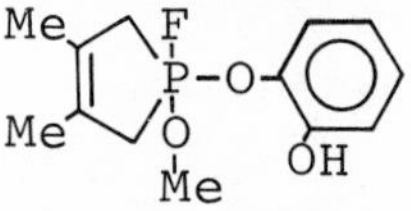
?. From spiro compound with MeOH, $b_{0.7}$ 112°, n_D^{20} 1.5240, d_4^{20} 1.2092, ^1H.[160]

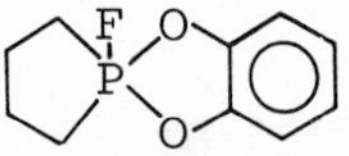
. XXXI. ^{19}F, J_{FP} 829 Hz.[131a]

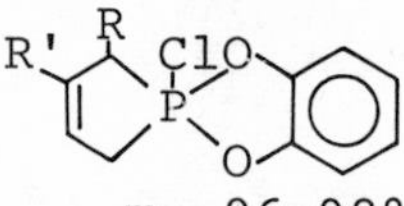
. XXXII. R, R' = H: m. 86-88°. R = Me, R' = H: m. 98-100°. R = H, R' = Me: m. 96-98°.[545]

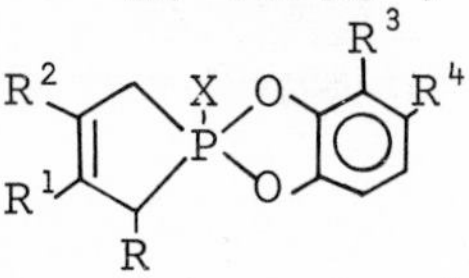
. XXXII. IR, ^1H.[543]

R	R^1	R^2	R^3	R^4	X	m.
H	H	H	H	H	Br	67-68°
H	H	Me	H	H	Br	60-62°
H	Me	Me	H	H	Br	63-65°
Me	H	H	H	H	Br	62-65°
H	H	Cl	H	H	Br	103-105°
H	H	H	H	H	F	82-83°
H	H	Me	H	H	F	86-87°
H	Me	Me	H	H	F	75-76°
Me	H	H	H	H	F	88-89°
H	H	Cl	H	H	F	79-80°
H	Me	Me	Me	H	F	66-67°
H	Me	Me	H	Me	F	64-65°

$(Ph_2ClPNMe)_2$. From Ph_2PCl and MeN_3,[275] ^{31}P +22.3.[275,379]
$(Ph_2FPNH)_n$. Viscous oil, IR.[581]

TYPE: $R_2(R'Z)_3P$

$(CF_3)_2P[ON(CF_3)_2]_3$. From $(CF_3)_2PH$ + $ON(CF_3)_2$, colorless liqu., ^{19}F, mass spect.[5a]
$Ph_2P(OEt)_3$. VII. ^{31}P +41, ^1H at -60° shows 2 kinds of EtO signals,[118] at room temp. only 1 EtO signal.

$Ph_2P(OEt)_2NEt_2$. VII. ^{31}P +47 (probably).[118]

$Ph(MeO)_2P$ (ring: O, C(Me)=C(COMe), CH_2). XXXIII. M. 102-105°, IR, ^1H shows inhibited pseudorotation at < -63°,[201] activation parameters.[198]

$Ph(RO)_2P$ (ring: O, C(Me)=C(COMe), CH(Ph)). XXXIII. R = Me, 2 inseparable stereoisomers at P, m. 105-107°, IR, ^{31}P +16.3, +13.3, ^1H at room temp and -20° shows trig. bipyr. with the C groups in equat. positions; at > +52° pseudorotation, at 125° ring opening.[199,528] R = Et: 2 isomers, m. 63-66°, ^{31}P +17, +15, ^1H at room temp. and at -20° shows stable trig. bipyr., at > +70° pseudorotation, at +110° ring opening.[528] R = Ph: 2 isomers, m. 132-134°, ^{31}P +30, +24.8, IR, ^1H like above, pseudorotation at > +51°.[528]

$Ph(MeO)_2P$ (ring: O, C(Me)=C(COMe), CMeMe). XXXIII. M. 76-81°, IR, ^1H shows inhibited pseudorotation at < +38°,[201] activation parameters.[198]

$Ph(MeO)_2P$ (ring: O, C(Me)=C(COOMe), CMeMe). XXXIII. M. 77-78.5°, IR, ^1H shows inhibited pseudorotation at < +39°,[201] activation parameters.[198]

$Ph(MeO)_2P$ (ring: O, CH(C_6F_5), O, CH(C_6F_5)). XXII. Cis H/H, ^{31}P +28.6, ^1H shows inhibited pseudorotation < -42°; trans H/H, ^{31}P +34, ^1H shows no inhibition of pseudorotation.[488]

$Ph_2(PhO)P$ (ring: O, C(R)=C(R), O). VIII. R = Me: ^{31}P +27.7. R = Ph: ^{31}P +27.0, m. 90°.[534]

$Ph(PhO)P$ (ring: O, phenanthrene-9,10-diyl, O). VIII. M. 147°, ^{31}P +16.3.[534]

$Ph_2(EtO)P$ (ring: O, CH($C_6H_4NO_2$-p), CH($C_6H_4NO_2$-p), O). X. One pure isomer of unknown struct., m. 164-167°, IR, ^1H shows inhibited pseudorotation at -75°,[201] activation parameters.[198]

$Ph_2(PhO)P$ (ring: O, $C(CF_3)_2$, $C(CF_3)_2$, O). X. M. 111°, ^{31}P +19.0.[534]

(P-OH phospholene spiro catechol structure). From corresponding bromide or chloride and H_2O, m. 125-126°,[545] IR, ^1H, pH (pot.) 5.36.[543] OMe derivative: XXXII.[542] From OH product with CH_2N_2,[543] m. 112-114°,[543] b_2 100-101°, n_D^{20} 1.5530, d^{20} 1.2531, IR, ^1H, solidifies on standing.[542] OEt derivative: m. 129-130°.[543,545]

. XXXII. B_2 120-121°, n_D^{20} 1.5500, d^{20} 1.2063, solidifies on standing, IR, ^1H.[542]

. From fluoride and MeOH, $b_{0.8}$ 115°, n_D^{20} 1.5232, d_4^{20} 1.1969, ^1H.[160]

. From $(Et_2N)_3P$ and corresponding phenol, $b_{0.02}$ 168-180°, m. 99.5-101°,[303] $b_{0.025}$ 150-160°, m. 99-100°.[303a]

. From Me_2PCl_3 and $MeCONHNH_2$, see III. Dec. 180°, $subl_{0.1}$ 130°, ^1H, ^{31}P +42.5.[152]

D.4. Compounds of Type R_3PY_2

TYPE: R_3PX_2

Me_3PF_2. XXXIV.[590] XXXVI.[134] B_{25} 0°, m. -31°,[590] b. 76°,[134] ^{31}P +15.8, ^1H; IR and Raman show D_{3h} trig. bipyr. with axial F,[134] ^{19}F,[134,426] heteronuclear INDOR spect.,[135a] mass spect.[73]

$(CF_3)_3PF_2$. XXXIV. B. 20°, m. -102°, IR,[398] ^{19}F,[426] ^{31}P +59.8.[445]

$(Et_2NCH_2)_3PF_2$. XXXIV. Not isolated, ^{19}F, J_{FP} 688 Hz.[415]

Me_2PF_2-CH_2CH_2-PF_2Me_2. XXXVI. B_5 80°, m. 47.1-48.4°.[577]

Et_3PF_2. XXXIV. B_{20} 53°, $n_D^{25.5}$ 1.4061, IR,[577] ^{19}F,[426,590] ^{31}P +13.2.[590]

Bu_3PF_2. XXXIV.[413,415] XXXVI.[577,591] $B_{0.4}$ 71-72°, n_D^{25} 1.4320, d_4^{25} 0.9398,[577] $b_{0.6}$ 76-77°, n_D^{25} 1.4336,[413] $b_{0.155}$ 63°, n_D^{24} 1.4332,[415] ^{31}P +15.4,[445,580] ^{19}F,[415,426,590] IR,[426,577] mass spect.[73]

Me_2PhPF_2. XXXVI. $B_{1.4}$ 64°, ^{19}F, ^{31}P +27.4.[590]

Bu_2PhPF_2. XXXVI. $B_{0.3}$ 89°, $n_D^{24.4}$ 1.5010,[577,587] $b_{0.08}$ 80°,[587] IR.[577]

$MePh_2PF_2$. XXXVI. $B_{0.15}$ 95°, ^{19}F, ^{31}P +43.2,[590] mass spect.[73]

Ph_2PF_2-CH_2-PF_2Ph_2. XXXIV. M. 122-124°, IR, ^{19}F, ^1H, mass spect.[73]

Ph_2PF_2COPh. XXXV. From Ph_2PF and PhCOF, not isolated, ^{31}P +61, J_{PF} 730 Hz.[79a]

Me–(ring)–P(Ph)(F)(F) . XXXVI. B_5 100-120°,[587] $b_{0.4}$ 89°, n_D^{25} 1.5139,[577] ^{19}F,[426,590] ^{31}P +2.3.[590]

Ph_3PF_2. XXXIV.[172,335,412,415] XXXVI.[618] Prisms, not pure, m. 134-162°, $subl_{10^{-4}}$ 160-170°,[335] m. 136-140°,[618] white needles, m. 144-146°,[412] m. 157°,[537] m. 160-162°,[172] IR, ^{19}F,[172,415,426] ^{31}P +58.1,[535,537] mass spect.[73]

$(C_6F_5)_3PF_2$. XXXVI. M. 159°, ^{19}F.[170]

Me_3PFCl. From Me_3PCl_2 and PhCOF, subl. 100-110°, IR of solid shows ionic struct. $[Me_3PF]^+Cl^-$, in gas phase probably trig. bipyr.[41]

Me_3PCl_2. XXXVI. M. 267-268°, IR and Raman of the solid show $[Me_3PCl]^+Cl^-$.[202] $\cdot AlCl_3$: XXXV. White cryst., m. 138-140°.[308]

$(CF_3)_3PCl_2$. XXXIV. M. 20.5°, b_{368} 71°, explodes at 94°,[57] m. 24-24.5°, b (extrapol.) 107°, IR, electric conduct. in MeCN.[153]

$(CCl_3)_3PCl_3$. XXXVI.[192] XXXIV.[687] Prisms, dec. 190°,[192,687] forms with $H_2O \rightarrow (Cl_3C)_3PCl(OH)$, needles, dec. 203°.[192,687]

Me_2PCl_2R. XXXV. Hygroscopic cryst., R, m.: -CN, 167-168° (dec.); -COOEt, 152-154°; $-CH_2COCH_3$, 144-145° (dec.); $-CH_2CH_2SO_2F$, 74-76°.[195]

$(Me_2PCl_2)_4C$. XXXV. Dec. 200°.[194]

$Me_2PCl_2-CH_2CH_2-PCl_2Me_2$. XXXV. M. 172-174°.[194]

Me_2EtPCl_2: $\cdot AlCl_3$. XXXV. M. 98-103°.[308]

$MeEt_2PCl_2$: $\cdot AlCl_3$. XXXV. Very hygroscopic.[308]

$Me(ClCH_2CH_2)_2PCl_2$. XXXVI. M. 110-112°, ^1H, extremely unstable in air.[615]

Et_2PCl_3. XXXIV.[619] XXXVI.[346] Dec. 240-250°,[346] m. 242-247°, ^1H,[619] is cytostatic on mice tumors.[611] $\cdot AlCl_3$: XXXV. Very hygroscopic.[308]

$Et_2PCl_2(C(Me)(H)(OCH_2CH_2Cl))$. XXXV. B_{25} 64-68° (redissociation), n_D^{20} 1.5108, d_4^{20} 1.2503, fumes in air.[636]

$Et_2PCl_2(C(Me)(H)(OBu))$. XXXV. B_{11} 44-45° (redissociation), n_D^{20} 1.5018, d_4^{20} 1.1018, fumes in air.[636]

PR_3PCl_2. XXXIV. White cryst., m. 141-148°.[619]

Me_2BuPCl_2. XXXV. M. 94-96°.[194]

Bu_3PCl_2. XXXIV.[619] M. 134-137°,[179,313] after subl. i.Vac. m. 147-148°,[313] IR,[665] ^1H,[313,665] ^{31}P (MeCN, $PhNO_2$) -106, -104 shows $[Bu_3PCl]^+Cl^-$.[665]

$Cyclohex_3PCl_2$. XXXIV. M. 172-175°,[295] $\cdot SbCl_5$: cryst., dec. 252°, conduct. measurements show ionic struct.: $[R_3PCl]^+SbCl_6^-$.[299]

Oct_3PCl_2. XXXIV. White cryst., m. 88-90°.[619]

$MeCl_2P$(ring). XXXVII. ^1H, ^{31}P -112 shows ionic struct.[479]

Me_2PhPCl_2. XXXVI. M. 174-176°.[235]
Et_2PhPCl_2. XXXIV. Oil?, d^{13} 1.216.[346]
$(PhCH_2)Ph_2PCl_2$. XXXV. M. 187°.[346]
Ph_3PCl_2. XXXIV.[66,117,299,346,562] XXXVI.[18,346] M. 176°?,[346] sint. > 100°,[18] ^{31}P ($PhNO_2$) +11 (equilibrium value?),[117] ^{31}P (MeCN) -62 shows $[Ph_3PCl]^+Cl^-$, ^{31}P ($PhNO_2$) -8 shows more covalent species, IR,[665] conduct. in MeCN indicates also ionization in the above sense[230] and not $[Ph_3PCl]^+[Ph_3PCl_3]^-$ as mentioned earlier;[66] conduct. in $PhNO_2$ shows about 25% ionization,[230] cryoscopic mol. wt. determinations in HCX_3 indicate a dimer $[Ph_3PCl_2]_2$, IR.[30] $Ph_3PCl_2 \cdot HCX_3$: cryst., IR, analysis;[30] $\cdot PCl_5$: m. 175°, high conduct. in MeCN and DMF shows $[Ph_3PCl]^+PCl_6^-$;[562] $\cdot SbCl_5$: cryst., dec. 164°, conduct. in $PhNO_2$ shows ionic struct.;[299] $\cdot MoOCl_3$: yellow hygroscopic crysts., m. 121°, conduct. in MeCN shows $[Ph_3PCl_3]^+[MoOCl_4]^-$; $\cdot WOCl_4$: green hygroscopic crysts., conduct. in MeCN shows ionic struct.[388]
$(C_6F_5)_3PCl_2$. XXXVI. M. 209-234° (after subl.), IR, ^{19}F, ^{31}P +104.7,[155] mass spect.[155,414a]
$(p\text{-}MeC_6H_4)_3PCl_2$. XXXIV. Solid.[346]
$(2,4,5\text{-}Me_3C_6H_2)_3PCl_2$. XXXIV. Cryst.[346]
$(1\text{-Naphthyl})_3PCl_2$. XXXIV. Isolated at $HCCl_3$ complex, m. 160°.[346]
R_3PCl_2. XXXIV. The following compounds are given without data:[17] Ph_3, $(p\text{-}MeC_6H_4)_3$, $(p\text{-}MeOC_6H_4)_3$, Et-Ph_2, Et_2Ph, Bu_3, Et_3, Me_3.
$Me_2PClBr(CH_2COOEt)$. XXXV. Hygroscopic cryst., m. 98.5-100°.[195]
$(PhCH_2)Ph_2PClBr$. XXXV. M. 171°.[346]
Me_3PBr_2. XXXIV. Yellow cryst., m. 277-278°, IR and Raman of solid show $[Me_3PBr]^+Br^-$.[202]
$(CF_3)_3PBr_2$. XXXIV. Colorless liquid, reactions.[86]
Et_3PBr_2. XXXIV. White cryst., dec. 253°,[299] cytostatic action on mice tumors.[611,612]
Pr_3PBr_2. XXXIV. White cryst., dec. 166°, conduct. shows $[Pr_3PBr]^+Br^-$,[299] cytostatic action on mice tumors.[611,612]
Bu_3PBr_2. XXXIV. ^{31}P (MeCN, $PhNO_2$) -105 shows $[Bu_3PBr]^+Br^-$,[665] cytostatic action on mice tumors.[611,612]
$(i\text{-Am})_3PBr_2$. XXXIV. White plates, dec. 174°.[299]
$Cyclohex_3PBr_2$. XXXIV. M. 260-265°.[295] $\cdot HgBr_2$: cryst., dec. 245°. $\cdot SbBr_3$: conduct. shows $[Cyclohex_3PBr]^+[SbBr_4]^-$, green cryst., dec. 208°.[299]
$MePrPhPBr_2$. XXXIV. The racemization of optically active MePrPhP on reaction with Br_2 indicates this pentacovalent intermediate.[284]

$CF_3Ph_2PBr_2$. XXXIV. Orange oil, stable against air and H_2O, reacts with NaOH to HCF_3.[56]

Ph_3PBr_2. XXXIV. Large, colorless hygroscopic cryst., m. 260-270°,[96] conduct. shows $[Ph_3PBr]^+Br^-$.[66] $\cdot Br_2$: orange cryst., m. 117°, UV, conduct. indicates ionic struct.: $[Ph_3PBr]^+Br_3^-$. $\cdot$BrI: orange cryst., m. 101°, UV, conduct. shows $[Ph_3PBr]^+IBr_2^-$. $\cdot I_2$: red needles, m. 79°, UV, conduct. shows $[Ph_3PBr]^+I_2Br^-$.[66]

(1-Naphthyl)$_3PBr_2$. XXXIV. Red solid.[346]

Pr_2(p-EtC_6H_4)PBr_2, Bu_2(p-EtC_6H_4)PBr_2, Pr_2(p-$MeOC_6H_4$)PBr_2, Am_2(p-$MeOC_6H_4$)PBr_2. Obtained as oils from R_3P and Br_2, no data given.[346]

Ph_3PBrI: $\cdot I_2$. XXXIV. Red cryst., m. 127°, UV, conduct. shows $[Ph_3PBr]^+I_3^-$.[66]

Me_3PI_2. XXXIV. Yellow cryst., m. 279-280°, IR and Raman of solid show $[Me_3PI]^+I^-$.[202]

Et_3PI_2. XXXIV. Yellow cryst., dec. 140°.[299]

Me(t-Bu)$_2PI_2$. XXXV. Yellow cryst., dec. 198-202°.[297]

Bu_3PI_2. XXXIV. Yellow needles, dec. 178°, conduct. shows $[Bu_3PI]^+I^-$,[299] almost colorless solid, m. 202-204°, from MeCN: m. 196-198°, ^1H.[313]

(i-Am)$_3PI_2$. XXXIV. Yellow cryst., stable toward moisture, dec. 150°.[299]

Ph_3C(Cyclohex)$_2PI_2$. XXXIV. Yellow cryst., dec. 223-226°.[301]

Cyclohex$_3PI_2$. XXXIV. Yellow needles, m. 174-175°,[295] conduct. shows $[R_3PI]^+I^-$. $\cdot HgI_2$: yellow rods, dec. 205°, conduct. shows $[R_3PI]^+[HgI_3]^-$.[299]

$MePrPhPI_2$. XXXIV. The racemization of optically active MePrPhP on reaction with I_2 indicates this pentacovalent intermediate.[284]

Cyclohex$_2PhPI_2$. XXXIV. M. 214-215°. $\cdot I_2$: m. 171-172°.[169]

$CF_3Ph_2PI_2$. XXXIV. Brown oil, stable against air and H_2O, with NaOH fluoroform is formed.[56]

Ph_3PJ_2. XXXIV. Yellow cryst., dec. 148°, conduct. shows ionic struct.[66,299] $\cdot I_2$: purple needles, m. 132°, UV, conduct. proves ionic struct.[66]

TYPE: $R_3(R'Z)_2P$

$Bu_3P(OEt)_2$?. VII. Not isolated ^{31}P +38.[119]

$MePh_2P(OEt)_2$. VII. ^1H, ^{31}P +47.[118]

$Ph_3P(OEt)_2$. VII. Not isolated, ^{31}P +55,[119] +54.[118]

Ph_2(RO)P[ring: O–C(Me)=C(COMe)–C(H)(Ph)]. XXXIII. R = Me: m. 122-124°, IR ^{31}P +25.7, ^1H shows trig. bipyr. with apical oxygen, ring opening at ∿127°. R = Et: m. 169-173°, IR, ^{31}P +26.3, ^1H, ring opening at ∿108°. R = Ph: m. 182-187°, IR, ^{31}P +40.1, ^1H ring opening at ∿160°.[528]

$R_2R'P$(dioxaphenanthrene structure). VIII. IR, some of these products show solvent-dependent ^{31}P chemical shifts; these suggest equilibria with open dipolar forms.[537]

R	R'	m.	^{31}P (benzene)	^{31}P (CH_2Cl_2)
Me	Me	Yellow	+3.0	-1.3 (^1H)
Me	Ph	95-96°	+9.6	+2.4
Et	Ph	150-151°	-0.9	-10.8
Ph	Et	125-126°	+9.3	+7.3 (no equil.)
Ph	Ph	165-166°	+15.6	+15.5 (no equil.)

$R_2R'P[OC(CF_3)_2C(CF_3)_2O]$. X. IR, ^1H, ^{19}F, all are cryst. substances.[537]

R	R'	m.	^{31}P (benzene)	^{31}P (CH_2Cl_2)
Me	Me	15°	+3.2	+3.2 (-3.2?[536])
Et	Et	Not isol.	—	-11.7
Bu	Bu	Not isol.	—	-7.3
Me	Ph	51-54°	—	+10.9
Et	Ph	44-47°	+1.1	-1.1
Ph	Et	69-73°	+6.1	+4.1
Ph	Ph	105-106°	+22.2	+21.6 (+21.0 $PhNO_2$)
Ph	Ph	118-119°[71]	—	+21.6 ($CDCl_3$)[71]
Ph	Ph	105-106°[621]	—	+18.6 (DMF)[621]

(Ph-phospholene-P spiro catechol structure). XXXVII. M. 127-129°, IR, ^1H.[541]

(Me-substituted phospholene, R, catechol structure). XXXVII. R = Me: m. 67°, ^1H. R = Ph: m. 127°, ^1H.[661]

(Ph-phospholene-P spiro dioxaphospholane structure). XXXVII. M. 60-62°, IR, ^1H.[541]

$Me_2[(CF_3)_2HCO]P[OC(CF_3)_2CH_2]$. From thermal rearrangement of $Me_3P[OC(CF_3)_2C(CF_3)_2O]$, m. 45°, IR, ^1H, ^{19}F, ^{31}P +23.7,[536] stable trig. bipyr.

$R,R'[(CF_3)_2HCO]P[OC(CF_3)_2CH_2]$. From thermal rearrangement of appropriate dioxaphospholane.[488] R = Ph, R' = Me: ^{31}P +31.4. R, R' = Ph: ^{31}P +35.4.

$R,R'[(CF_3)_2HCO]P[OC(CF_3)_2CH(Me)]$. From thermal rearrangement of $RR'EtP[OC(CF_3)_2C(CF_3)_2O]$, all these products have stable trig. bipyr. configurations with the oxygens in apical positions.[536] R, R' = Et, $b_{0.1}$

51-52°, IR, ^{19}F, ^{31}P +15.7. R = Et, R' = Ph: mixture, $b_{0.05}$ 74-76°, ^{19}F, cis Me/Et: ^{31}P +21.0, trans Me/Et: ^{31}P +30.3. R, R' = Ph: m. 70°, IR, ^{19}F, ^{31}P +32.1.

$Bu_2[(CF_3)_2HCO]P$(oxaphosphetane: O, $(CF_3)_2$, H, Pr). From thermal rearrangement of appropriate dioxapholane, ^{31}P +17.2.[488]

$Bu_2[CF_3PhHCO]P$(oxaphosphetane: O, CF_3, Ph, H, Pr). From CF_3COPh and Bu_3P, cis H/Ph: ^{31}P +21.0; trans H/Ph: ^{31}P +23.1.[488]

D.5. Compounds of Type R_4PY

TYPE: $R_4(R'Z)P$

Ph_4POPh. XXXIX. M. 138°, rather stable in air.[548]

Ph_3P(O, $(CF_3)_2$, PPh_3). XLI. Dec. 157-158°, ^1H, ^{19}F, ^{31}P +54, -7.3,[71,72] x-ray: distorted trig. bipyr.[95a]

Ph_3P(oxaphospholane). XL. M. 116-117°, IR, ^1H,[227] mass spect.[228]

Ph_3P(O, Ph, H). XLII. Prisms, m. 143-144°, IR, ^1H, ^{31}P +55.2.[679]

Ph_3P(O, COPh, Ph). XXXIX. M. 216-218°, IR, UV, ^1H, ^{31}P +49.5, mass spect.[593a]

Ph_3P(O, Ph, H, Me_2). XLII. Prisms, dec. 163-164°, ^1H, ^{31}P +49.2.[679]

Ph_3P(O, Ph, H, cyclopropane). XLII. M. 155°, ^{31}P +47.1.[63]

Ph_3P(O, H, H, H, cyclopropane). XLII. M. 165°, ^{31}P +49.6.[63]

Bu_3P(O, N, Ar). XL. Ar = Ph: liquid, mass spect. Ar = p-BrC_6H_4-: m. 70°.[184]

Ph_3P(O, N, Me). XL. M. 116° (dec.), IR, UV, ^1H.[184]

Ph_3P(O, N, Ar). Ar = Ph: XLII.[291] XL.[184,406] Rhombohedra, m. 129-130° (methanol),[184,291] needles, dec. 131° (MeCN),[406] IR, ^1H,[184,291,406] UV,[184] ^{31}P +37.[291] Ar = p-BrC_6H_4-: XL. M. 157°, IR, UV, ^1H.[184] Ar = p-$NO_2C_6H_4$-: XL. Yellow needles, dec. 149°, IR.[406]

Ar = 2,4,6-$Me_3C_6H_2$-: XLII. Prisms, dec. 110-111°, ^{1}H, ^{31}P +39.4.[291] Ar = β-Naphthyl-: XL. M. 142°, IR, UV, ^{1}H.[184]

Ph_3P(O–N=C(Ph)–C(H)(R)). R = Me: XL.[184] XLII.[64] M. 100°,[184] m. 92-93°,[64] ^{1}H,[64,184] UV, ^{1}H.[184] R = Pr: XL. M. 96°, IR, UV, ^{1}H.[184] R = Me: p-ClC_6H_4- for Ph: XLII. M. 107°, ^{1}H.[64]

Ph_3P(O–N=cyclohexane-fused, H). XL. M. 102°, IR, UV, ^{1}H.[184]

Ph_3P(O–N=C(Ph)–C(R)(Me)). XLII. R = Me: m. 122-123°, ^{1}H. R = Et: m. 65-75° (dec.).[64]

Ph_3P(O–N=C(Ph)–cyclopropylidene). XLII.[63,64] M. 186-187°, ^{1}H, ^{31}P +34.2, mass spect.[64]

Ph_3P(O–N=C(COOEt)–C(R)(Me)). XLII. R = Me, ·MeOH: m. 88-90° (dec.), ^{1}H, ^{31}P +21. R = Et: m. 68°.[64]

Ph_3P(O–N(R)–CH(Ph)–CH_2). XLII. R = Me: m. 134-135°, ^{31}P +60.3. R = Ph: m. 135-136°, IR, ^{1}H, ^{31}P +58.6.[290]

Ph_3P(O–N(tetrahydroisoquinoline)–C(H)(R)). XLII. R = H: m. 146-147°, ^{1}H, ^{31}P +56.4. R = Ph: dec. 150-152°, ^{1}H, ^{31}P +50.9.[290]

Ph_3P(O–N(Ph)–CH(Ph)–CH(Ph)). XLII. Dec. 146-147°, ^{1}H, ^{31}P +57.9.[290]

Ph_3P(O–N(Ph)–CH(Ph)–CMe_2). XLII. Dec. 142-143°, ^{1}H, ^{31}P +46.3.[290]

Bu_3P(O–C(Me)=CH–CH_2CH_2) + Bu_3P(O–C(=CH_2)–$CH_2CH_2CH_2$), 93%/7%. XL. $B_{0.06}$ 136-138°, n_D^{20} 1.4745, d_4^{20} 0.911, IR, ^{1}H, mol. wt. by mass spect.[110]

D.6. Compounds of Type R_5P

Ph_5P. XLIII.[675,676] XLIV.[572,671,673] Crystallizes from cyclohexane, dec. 124°,[673,675,676] μ 1.34 D,[674] R_P

15.36,[362] IR interpreted in terms of tetrag. pyr.,[113] IR shows trig. bipyr.,[203,(42)] ^{31}P, x-ray struct. determination establishes distorted trig. bipyr.:[657] P-$C_{a,e}$ 1.987, 1.850 Å, C_aPC_e 86.3-92.7°, C_ePC_e 117.8-122.6°, C_aPC_a 176.9°.

$Ph_4P^{14}C_6H_5$: ·1/2 cyclohexane. XLIII. Dec. 124°, 1674 counts/min.[553]

$Ph_4PC_6D_5$: ·1/2 cyclohexane. XLIII. Dec. 123°, 15,492 γ D.[553]

$Ph_4PC_6H_4$pMe: ·1/2 cyclohexane. XLIII. Colorless needles, dec. 123°.[670]

(p-$ClC_6H_4)_5$P. XLIV. Dec. 129.5°, ^{1}H.[238a]

(p-$MeC_6H_4)_5$P. XLIV. Dec. 124-125°,[238a,572] ^{1}H shows only 1 Me signal down to -60° (in Py/CS_2).[239]

(p-$PhC_6H_4)_5$P. XLIV. Dec. 130°.[571,572]

Ph_4PCPh_3. XLIII. M. 121°, becomes violet in light.[670]

Ph_4P-C≡CR. XLIII. R = Me, OMe, p-$MeOC_6H_4$: all these products decompose at temp. > -20°, IR; conduct. proves covalent struct.[440]

Ph_3P(2,2'-biphenylylene). XLIV. M. 155.5-156.5°,[673] thermolysis yields diphenyl-2-o-terphenylylphosphine.[674]

RP(2,2'-biphenylylene)$_2$. R = H: XLIII. Cryst., dec. 95-100°, IR, ν (H-P) = 2097 cm^{-1}, ^{1}H(P) 9.33, J_{HP} 482 Hz, reacts in solution spontaneously to the violet bis-2,2'-biphenylylene-phosphoranyl radical with a_P = 17.9 gauss,[240,246] ^{31}P +112.[246] R = D: XLIII. Dec. ~100°, ^{31}P +112, J_{PD} 75 Hz.[246] R = Me: XLIII. M. 217-218°, ^{1}H, ^{31}P +97.[237] R = Bz: XLIII. M. 173-174° (dec.).[247] R = Bu: XLIII.[237] XLV.[246,572] M. 177-178°,[237,572] m. 178-180°, ^{31}P +90.[246]

ArP(2,2'-biphenylylene)$_2$. Ar = Ph: XLIII.[237] XLIV.[673,674] M. 201-202°,[673,674] m. 201°, ^{31}P +85.[237] Ar = p-MeC_6H_4-: XLV. M. 185-186°.[572] Ar = p-$MeOC_6H_4$-: XLV. m. 198.5-200°.[572] Ar = p-$Me_2NC_6H_4$-: XLV. Needles, m. 208-208.5°, ·MeI: m. 157-158°; ·$MeBPh_4$: m. 188.5-189°; ·Me-camphor sulfonate, m. 178-181°, $[\alpha]^{25}_{578}$ +12°.[674] Ar = 2-biphenylyl: XLIII.[237] XLVI.[237,241] M. 244-245°,[237] ^{31}P +85.7,[246] stereochemistry.[241] Ar = 4-methyl-2-biphenylyl: XLIII. XLVI. M. 233°, ^{1}H.[242] Ar = 4'-methyl-2-biphenylyl: XLIII. XLVI. M. 188-189°, ^{1}H.[242] Ar = 2'-Br-2-biphenylyl: XLVI. Yellowish cryst., m. 220°, reacts with BuLi to hexacovalent phosphate.[252,668] Ar =

2'-(2,2'-biphenylylene-phosphino)-2-biphenylyl: from dimerization of the bis-2,2'-biphenylylene-phosphoranyl radical, m. 220-222°, ^{31}P +85.5, +19.5,[246] mass spect. shows $M^{2+}:M^{+}$ = 6.4:1.[253] Ar = 8-Me_2N-1-naphthyl: XLIII. M. 226°,[252,668] ^{1}H, ^{31}P +85.[250]

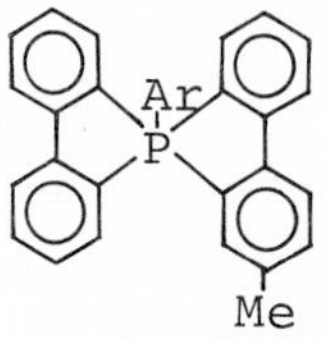

. Ar = 2-biphenylyl: XLIII. XLVI. Rac.: hexagonal prisms, m. 204-206°, ^{1}H, broad signal for Me group at room temp.,[242] at -55° ^{1}H shows four different Me signals, indication of 4 diastereomeric trig. bipyr. species.[248a] Optically active product: conical, hexagonal prisms, m. 208-209°, $[\alpha]^{24}_{578}$ ±94°, can undergo spontaneous racemate separation, ^{1}H shows broad Me signal.[242]

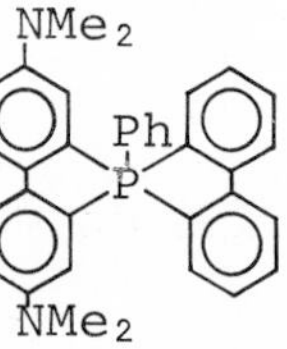

. XLIV. Yellow cryst., m. 212.5-213°,[673] m. 216-216.5°.[674]

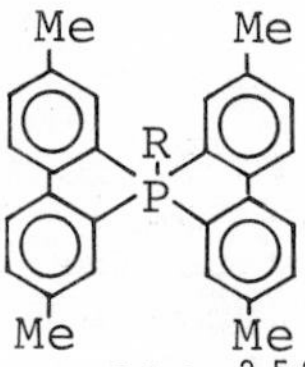

. XLIII. R = H: cryst., dec. 127-130°, IR, ν (H-P) = 2080 cm^{-1}, $^{1}H(P)$ 9.38, J_{HP} 483 Hz, reacts in solution spontaneously to violet phosphoranyl radical, ^{31}P +110.[252,668] The temperature dependence of the ^{1}H-NMR spectra of the following products shows that at lower temperatures relatively stable trig. bipyr. conformations are present, but that at higher temperatures rapid pseudorotations occur.[239,245]

R	m.	Number of Signals at High Temp.	Number of Signals at Low Temp.
Me	209-210°	1 (+35°)	2 (-60°)
Et	186-189°	1 (+35°)	2 (-60°)
Bz	146-148°	1 (+35°)	2 (-60°)
Ph	219-220°	1 (+35°)	1 (-60°)
2-i-PrC_6H_4	—	1 (+130°)	4 (+33°)[658]
2-Biphenylyl	215-216°	1 (+35°)	4 (-30°)
4,4'-Me_2-2-biphenylyl	220-222°[243]	3 (+35°), 1 is broad[248a]	—
2-Naphthyl	210-212°	1 (+35°)	2 (-54°)
1-Naphthyl	200-203°	1 (+120°)	2 (+35°)
8-MeO-1-naphthyl	237°[668]	4 (+35°)	Coalescence ∿160°[250]

R	m.	Number of Signals at High Temp.	Low Temp.
8-Me_2N-1-naphthyl(IR)	213°[668]	3 (+35°)	No coalescence up to 120°[250]
9-Anthryl	164° (dec.)	2 (+35°)	Dec. at higher temp.

δ ^{31}P for the 8-MeO-1-naphthyl derivative lies at +78 ppm.[250]

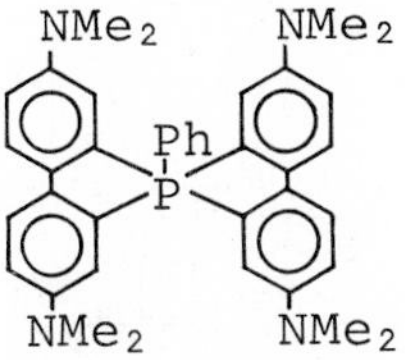

. XLIV. M. 217.5-218°, yellow cryst.[674]

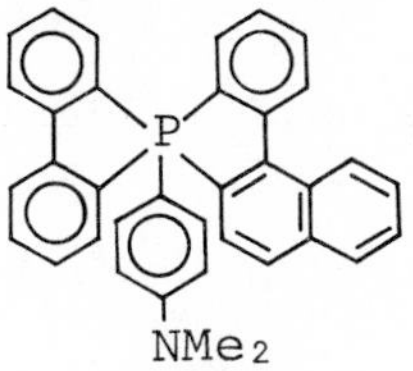

. XLIV. Light yellow, m. 167.5-168°, ·MeI: m. 186-188°; ·$MeBPh_4$: m. 186.5-188°; ·Me-camphor sulfonate, m. 205-206.5°, $[\alpha]^{25}_{578}$ 24.4.[674]

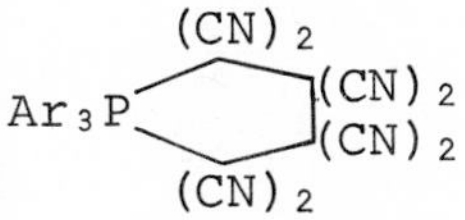

??. From Ar_3P and $(CN)_2C=C(CN)_2$. Ar = Ph: m. 168.5-170°, IR, UV, ^{31}P -22!. Ar = o-MeC_6H_4-: m. 235-237°. Ar = p-MeC_6H_4-: m. 205-206°.[556]

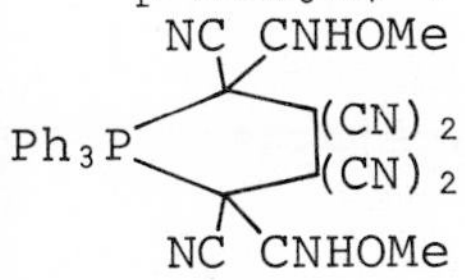

??. From above and MeOH, m. 215-220° (dec.), IR.[556]

D.7. Compounds of Type HR_nPY_{4-n}

HPF_4. From PF_5 and Me_3SnH,[632] from H_3PO_3 and HF,[74] b. -35.9°, m. -98°,[632] b. -39°, m. -100°,[273] b. -41 to -38°, m. -90 to -88°,[74] log P_{mm} = -1729/T +11.011 (solid), = -1152/T +7.737 (liquid),[632] log P_{mm} = -1158.9/T +7.8290,[273] IR, ^1H,[273,632] ^{19}F shows rapid pseudorotation from -90 to +30°, ^{31}P +53.6 (liquid), mass spect.;[632] μ = 1.32 D, microwave spect. shows C_{2v} trig. bipyr. with P-H 1.36 Å, P-$F_{a,e}$ 1.594, 1.55 Å, $HPF_{a,e}$ 90 ± 4, 124 ± 2°.[462]

DPF_4. From PF_5 and Me_3SnD,[632] ^{19}F, IR.[632]

H_2PF_3. From PF_5 and Me_3SnH,[632] from H_3PO_2 and HF,[74] b. +0.9°, m. -47°,[632] b. +3.8°, m. -52°,[273] b. +1 to +3°, m. -52 to -51°,[74] log P_{mm} = -2216/T +11.521 (solid), = -1561/T +8.579 (liquid),[632] log P_{mm} = -1473/T + 8.1991,[273] IR, ^1H,[273,632] at -46° 2 kinds of ^{19}F, at

-23° coalescence, ^{31}P +24.1,[632] IR and Raman show for the gas phase a C_{2v} trig. bipyr., but for the condensed states an octahedral dimer with 2 fluorine bridges,[202a] the same is concluded from temperature-dependent NMR.[273]

D_2PF_3. From PF_5 and Me_3SnD, ^{19}F, IR, ^{31}P +24.1.[632]

Me_2NPF_3H. From Me_2NPF_4 and R_3SnH, impure product with ν (H-P) 2455 cm^{-1}.[197]

$MePF_3H$. XLVII.[197] XLVIII.[145,594] B_{760} 34-36°, d_4^{20} 1.2832,[145] b. 37°,[594] b. 34.6°, log P_{mm} = -1602.0/T +8.0866,[197] IR, ^1H,[145,197] at -60° 2 kinds of ^{19}F, at room temp. equilibration,[197,594] ^1H 7.1,[197] ^{31}P (-50°) +7.7,[594] ^{31}P +10,[145] mass spect.[145,197]

$EtPF_3H$. XLVII.[197] XLVIII.[145] B_{760} 63-65°, d_4^{20} 1.2587,[145] b. 53.4°, log P_{mm} = -1685/T + 8.0425,[197] IR,[145,197] mass spect., ^1H 7.4, at -60° 2 kinds of ^{19}F, at room temp. equilibration.[197]

$BuPF_3H$. B_{60} 49-51°, d_4^{20} 1.1574.[145]

$PhPF_3H$. XLVIII. B_{20} 57-58°,[145,304] n_D^{20} 1.4705, d_4^{20} 1.3046, IR, ^{31}P +33,[145] 2 kinds of ^{19}F.[145,425]

p-$ClC_6H_4PF_3H$. XLVIII. B_{35} 90-91°.[304]

p-$MeC_6H_4PF_3H$. XLVIII. B_{40} 90-91°,[304] b_{20} 74-76°, n_D^{20} 1.4670, d_4^{20} 1.2750, IR, 2 kinds of ^{19}F.[145]

$MePF_2(OEt)H$. XLIX. B. 84-86°, n_D^{20} 1.3670, d_4^{20} 1.1304, IR, ^{31}P +20.5, ^{19}F.[136]

$MePF_2(O$-i-$Pr)H$. XLIX. B. 32-34°, n_D^{20} 1.3642, d_4^{20} 1.0919, IR, ^{19}F.[136]

$MePF_2(O$-i-$Bu)H$. XLIX. B_{100} 57-58°, n_D^{20} 1.3573, IR, ^{19}F, ^{31}P +22.[136,142]

$PhPF_2(OEt)H$. XLIX. L. B_1 42-45°, n_D^{20} 1.4876, d_4^{20} 1.1774, IR, ^{19}F.[136]

$PhPF_2(O$-i-$Pr)H$. XLIX. L. B_1 47-50°, n_D^{20} 1.4925, d_4^{20} 1.1980, IR, ^{19}F.[136]

$MeC_6H_4PF_2(OEt)H$. XLIX. L. B_4 98-100°, n_D^{20} 1.5005, d_4^{20} 1.1945, IR, ^{19}F.[136]

$MeC_6H_4PF_2(O$-i-$Pr)H$. XLIX. L. B_3 92-94°, n_D^{20} 1.5021, d_4^{20} 1.1984, IR, ^{19}F.[136]

$MePF_2(NHAllyl)H$. XLIX. B_{20} 31-32°, n_D^{20} 1.4132, d_4^{20} 1.1395, ^{31}P +42.[141-143]

$MePF_2(NH$-i-$Bu)H$. XLIX. B_{50} 55-57°, n_D^{20} 1.3750, d_4^{20} 0.9961, ^{31}P +44.5.[141,143]

$MePF_2(NHCyclohex)H$. XLIX. B_{10} 66-67°, n_D^{20} 1.4155.[141,143]

$MePF_2H$-$(NHCH_2CH_2NH)$-PF_2MeH. XLIX. B_3 78-82°, n_D^{20} 1.4355, d_4^{20} 1.3204, ^{19}F, ^{31}P +44.[139]

$EtPF_2H$-$(NHCH_2CH_2NH)$-PF_2EtH. XLIX. n_D^{20} 1.4412.[139]

Me_2PF_2H. XLVIII. B_{760} 66°,[595] b. 64°,[594] log P_{mm} = -1909/T +7.92,[595] IR, ^1H, ^{19}F, ^{31}P +31.7,[594,595] mass spect.[595]

Bu_2PF_2H. XXXIV. NMR and IR confirm presence of PF bonds![183]

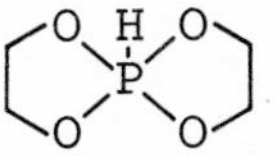

[structure]. LI. B_{12} 134-136°,[131] b_9 86°, m. 50°,[441] $b_{0.5}$ 70°, m. 50°,[87] IR, ^1H 7.1, ^{31}P +26.7.[92]

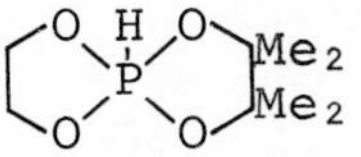

[structure, Me$_2$, Me$_2$]. LI. Mixture with 10% unsubstituted and 20% tetrasubstituted product, $b_{0.05}$ 52-53°, ^{31}P +30.4,[568] ^1H shows pseudorotation at temp. > -70°.[285]

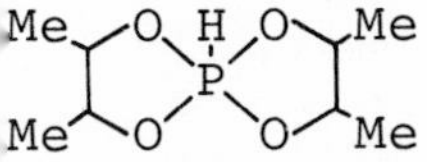

[structure, Me, Me, Me, Me]. LI. Cis product, from meso-2,3-butanediol, m. 82-85°,[441] is incorrectly formulated as ring-open substance.

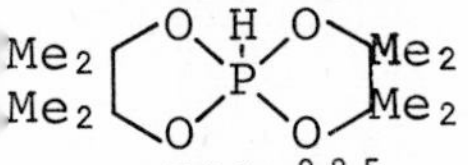

[structure, Me$_2$, Me$_2$, Me$_2$, Me$_2$]. LI. M. 88-89°,[441] m. 90°, IR, ^{31}P +40,[568] ^1H shows pseudorotation even at -70°.[285]

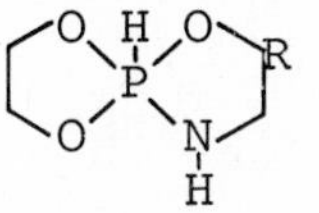

[structure, R]. LI. R = H: formulated as open species, m. 55-56°,[328] $b_{0.15}$ 73-75°,[87] IR[409,568] [ν(H-P) 2377 cm^{-1}], and ^{31}P +36.4[568] indicate pentacovalent struct., J_{PH} 780 Hz.[92,568] R = Me: mixture of two diastereomers, $b_{0.03}$ 72°, IR, ^{31}P +39.5,[90] IR,[409] ^1H 6.66, 6.72,[90,565] on heating intermolecular ligand reorganizations occur.[566] R = Ph: mixture with the 2 symmetrical products, ^{31}P +40,[566] mixture of 2 diastereomers, ^{31}P +40.8, ^1H 7.08, 7.02.[565]

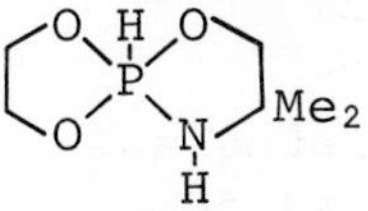

[structure, Me$_2$]. LI. M. 69-70°, ^{31}P +37,[566] IR,[409,565] ^1H 7.13, ^{31}P +41.2.[565]

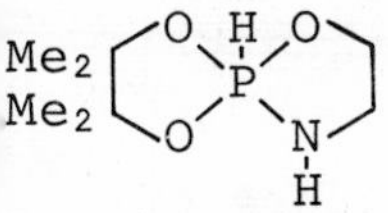

[structure, Me$_2$, Me$_2$]. LI. $B_{0.05}$ 80°, n_D^{20} 1.474, d_4^{20} 1.137, IR, ^{31}P +47.8.[568]

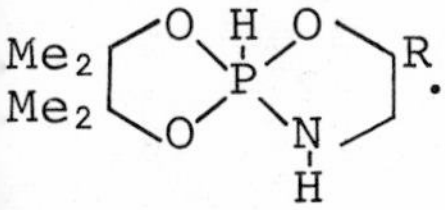

[structure, Me$_2$, Me$_2$, R]. LI. R = Me: 2 diastereomers, $b_{0.05}$ 70°, ^1H: 2 H-(P) doublets, ^{31}P +49.5,[90] IR.[90,409] R = Ph: 2 diastereomers, m. 102°, ^1H: 2 H-(P) signals, ^{31}P +49.5,[90] IR.[90,409]

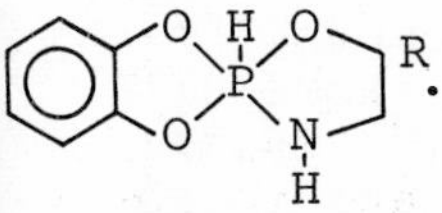

[structure, R]. LI. R = H: not isolated, ^{31}P +35.[566] R = Me: cryst., ^{31}P +35,[565,566] IR, ^1H shows 2 diastereomers, 7.79, 7.74.[565] R = Ph: m. 107°, ^{31}P +36,[565,566] IR, ^1H: 2 diastereomers 7, 90, 7.87.[565]

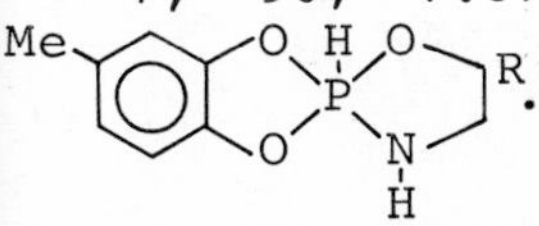

[structure, Me, R]. LI. R = Me: m. 115-116°,[566] IR, ^{31}P +37.6, ^1H shows only 2 of the 4 possible isomers.[565] R = Ph: m. 101-102°,[566] IR, ^{31}P +35.2, ^1H shows only 2 isomers.[565]

(structure: R-substituted benzo, O, H, O / P / O, N-H, Me_2). LI. R = H: m. 75°, ^{31}P +37.7,[566] IR, ^{1}H, ^{31}P +37.[565] R = Me: m. 96°,[566] ^{31}P +37.7,[565,566] IR, ^{1}H.[565]

(structure: i-Pr benzo, O, H, O / P / O, N-H, Me). LI. M. 99-100°,[566] ^{31}P +36.5, IR, ^{1}H shows 4 isomers.[565]

(structure: i-Pr benzo, O, H, O / P / O, N-H, Me_2). LI. $B_{0.4}$ 110-115°,[566] IR, ^{31}P +37.4, ^{1}H shows 2 isomers.[565]

(structure: benzo, O, H, O / P / N-H, O). LI. IR,[409,567] ^{1}H 7.76, J_{PH} 825 Hz, ^{31}P +37.7.[567]

(structure: O, H, O / P / O, N-Me). LI. Mixture with 85% ring-open form, IR, ^{1}H 7.08, ^{31}P +44.[567]

(structure: benzo, O, H, O / P / O, N-R). LI. Not isolated, mixtures with ring-open forms.[566] R = Me: ^{31}P +99.5, J_{PH} 807 Hz, ^{31}P (open form) -130. R = Et: ^{31}P +100, J_{PH} 802 Hz, ^{31}P (open form) -131. R = Bz: ^{31}P +100, J_{PH} 802 Hz, ^{31}P (open form) -132.

(structure: O, H, O / P / N-H, N-H). LI. M. 110-111°,[558] m. 116° (formulated as ring-open species),[87] IR,[206,409] Raman and NMR show pentacovalent struct. with ν (H-P) 2350 cm^{-1},[206] ^{31}P +54,[566] ^{31}P +53.6.[405]

(structure: Me, O, H, O / P / N-H, N-H, Me). LI. From D,L-2-aminopropanol, mixture of 3 diastereomers, m. 50°, IR, ^{1}H: 3 doublets, ^{31}P +56,[90] IR.[409] From L(+)-2-aminopropanol, mixture of 2 diastereomers, m. 79°, IR, $[\alpha]^{25}_{578}$ +36°, ^{1}H: 2 H-(P) doublets, ^{31}P +56.[90]

(structure: R, O, H, O / P / N-H, N-H, R). LI. R = Me: ^{1}H shows 3 diastereomers,[567] IR,[409] ^{31}P +54.6.[566] R = Ph: ^{1}H shows 3 diastereomers, m. 112°, ^{31}P +54.6,[90] IR,[90,409] mixture with corresponding trioxyamino- and tetraoxyphosphorane: ^{31}P +55.[566]

(structure: Me_2, O, H, O / P / N-H, N-H, Me_2). LI. M. 101-102°,[558,568] IR,[409] ^{1}H, ^{31}P +58.[568]

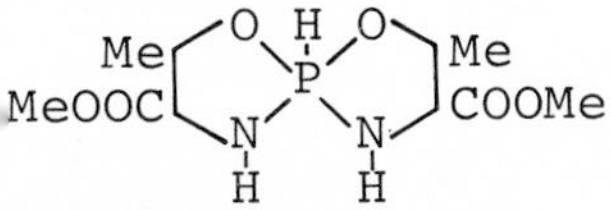

LI. From L-threonine, ^{1}H 7.13, J_{HP} 756 Hz.[566]

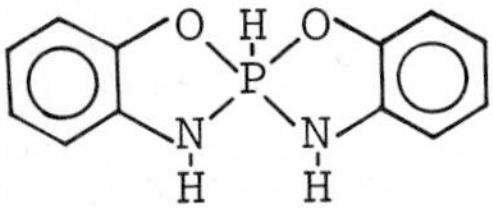

LI. ^{31}P +47.5, ^{1}H 8.62,[567] IR.[409,567]

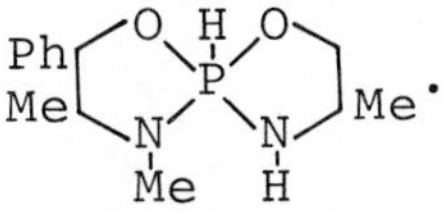

LI. From an oxazaphospholidine derivative of L-ephedrine and L-alaninol, mixture of 2 diastereomers, IR, ^{1}H, ^{31}P +64.[168]

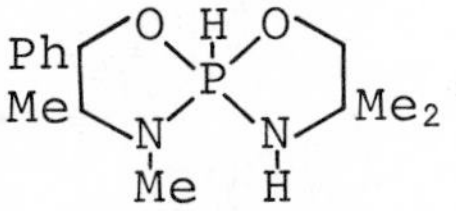

LI. 2 diastereomers, IR, ^{1}H, ^{31}P +64.5.[168]

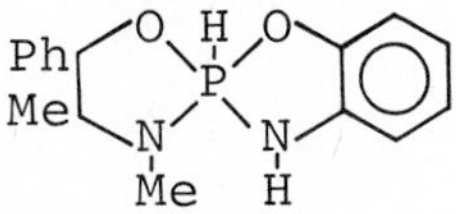

LI. 2 diastereomers, IR, ^{1}H, ^{31}P +60.[168]

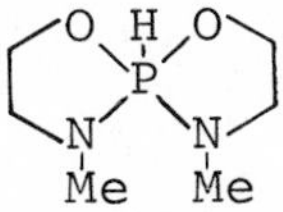

LI. Mixture with 90% ring-open form, $b_{0.005}$ 60°, n_D^{20} 1.4755, d_4^{20} 1.1310, IR, ^{1}H, ^{31}P +63.[568]

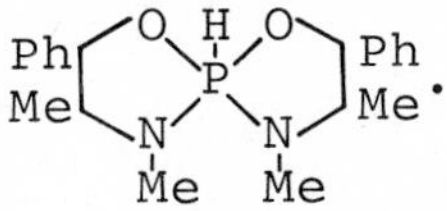

LI. From (-)- and (+)-ephedrine, respectively, and $P(NMe_2)_3$, $[\phi]_D^{20}$ -573 and +589°, IR, ^{1}H, only 1 of the 2 possible diastereomers is formed in each case, ^{31}P +72,[168] ^{31}P +71, ^{1}H.[566] When the oxazaphospholidine derivative of (-)- or (+)-ephedrine is reacted with (+)- or (-)-ephedrine, respectively, a racemate consisting of 2 enantiomeric trig. bipyr. species is obtained, IR, ^{1}H, ^{31}P +71.[168]

D.8. Compounds with Hexacovalent Phosphorus

$F_5\overset{\ominus}{P}-\overset{\oplus}{N}{\equiv}CMe$. LVIII. ^{31}P +146, J_{PF} 787 Hz,[662] ^{19}F.[623]

$F_5\overset{\ominus}{P}-\overset{\oplus}{P}R_3$. LVIII. On the basis of IR and ^{1}H NMR data the following products are believed to contain a P-P bond:[80] PR_3 = Me_3P (^{1}H), Pr_3P, Me_2PNMe_2 (^{1}H), $MeP(NMe_2)_2$, $P(NMe_2)_3$ (IR). When heated, the aminophosphine adducts decompose to Me_2NPF_4 and other products.

$F_5\overset{\ominus}{P}-\overset{\oplus}{B}$. LVIII. The following adducts show in the ^{19}F-NMR 2 kinds of fluorine positions; this is indicative of octahedral struct.[423] B, m.: Me_3N, 148°[623] ($b_{0.15}$ 170°); pyridine, 179-182°; $Me_2C{=}NOH$, dec.; Me_2NCHO,

118-120°; Me_2NCHS, 190° (dec.); Me_2SO, dec. The adducts with tetrahydrofurane, m. 55°, $b_{0.15}$ 116-118°, and tetrahydropyrane, m. 45-46°, dissociate at b.p., show only 1 ^{19}F signal that is probably due to rapid equilibration.

$F_2Cl_3P \cdot Py$. LVIII. White solid, 1 ^{19}F signal in $PhNO_2$, J_{FP} 983 Hz.[271,272]

$FCl_4P \cdot Py$. LVIII. Dec. ~63°, 1 ^{19}F signal in $PhNO_2$, J_{FP} 1049 Hz.[271,272]

$[Cl_5\overset{\ominus}{P}OPh][(PhO)_4\overset{\oplus}{P}]$. IV. Cryst., hydrolysis reactions.[564]

$Cl_5\overset{\ominus}{P}\text{-O-}\overset{\oplus}{P}R_3$. LVIII. R = Ph: m. 135-145°, ^{31}P +297, -67.2. R = Bu: oil, ^{31}P +296, -103.5.[68]

$Cl_5\overset{\ominus}{P}\text{-}\overset{\oplus}{N}C_5H_5$ (pyridine). LVIII. Solid, subl.,[44a] IR,[45] ^{31}P (solid) +296, ^{31}P (MeCN) +310,[662] ^{31}P (pyridin) +234.[375]

$F_4\overset{\ominus}{P}$(O–C(Me)=CH–C(Me)=O)$^{\oplus}$ (cyclic acetylacetonate). LIV. M. 85°, IR, UV, ^1H, 2 kinds of ^{19}F.[83]

$[Cl_4PB_2]^{\oplus}[SbCl_6]^{\ominus}$. LIV. Not isolated. B = pyridine: tetrahydrofuran, tetrahydrothiophene: IR, Raman, in solution.[44]

$Cl_4\overset{\ominus}{P}$(N(Me))$_2\overset{\oplus}{C}Cl$ (cyclic). LIV. M. 129°, IR, ^1H, ^{31}P +202.[378] X-ray struct. determination shows octahedral struct.[708] with P-Cl 2.07-2.15 Å, P-N 1.85, 1.91 Å, NPN 70.7°, Cl_aPCl_e 88.4-95.5°, Cl_aPCl_a 178.5°.

$[Cl_4\overset{\ominus}{P}(\overset{\oplus}{N})_2$-phenanthroline$]Cl^{\ominus}$. LIV. Solid, IR, Raman, conduct., cyroscopic mol. wt.[654] $SbCl_6^-$ salt: yellow solid, IR, Raman, relatively stable.[44]

$[Cl_3\overset{\ominus}{P}(OPh)_3][(PhO)_4\overset{\oplus}{P}]$. IV. Cryst., hydrolysis reactions.[564]

$[Cl_2\overset{\ominus}{P}(OPh)_4][(PhO)_4\overset{\oplus}{P}]$. IV. Cryst., hydrolysis reactions.[564]

ClC$\overset{\oplus}{}$(N(Me))$_2$$\overset{\ominus}{P}Cl_2$(N(Me))$_2$CO (cyclic). LIV. Not isolated, ^{31}P +183.[50]

ClC$\overset{\oplus}{}$(N(Me))$_2$$\overset{\ominus}{P}Cl_2$(N(Me))$_2$P(N(Me))$_2$CO (cyclic). LIV. Not isolated, ^{31}P +199, +70.[50]

$[Br_2P(OPh)_4]^{\ominus}[(PhO)_4P]^{\oplus}$. IV. Cryst., hydrolysis, alcoholysis reactions.[564]

$[(C_7H_4O_2^{\oplus})_3P]\,2I^{\ominus}$. LIV. Red cryst., no data.[428]

$[(C_6H_4O_2)_3P]^{\ominus}M^{\oplus}$. LIV. M = Li: not isolated, ^{31}P (DMF) +82. M = Na: not isolated, ^{31}P (EtOH) +86. M = Me_4N: cryst., dec. > 350°, IR, ^{31}P (DMF) +82. M = bis-2,2'-biphenylylene-ammonium, colorless plates, dec. 324°, IR, ^{31}P (DMF) +82.[251] M = Et_3NH: LV. $Subl_{0.05}$ 280°, IR, UV, mass spect., ^{1}H shows A_2B_2 system for aromatic protons.[1] X-ray: Octahedron.[1a]

$(C_6H_4(O)(NH))_2P^{\ominus}(C_6H_4(O)(NH_2)^{\oplus})$?. LV. M. 232-238°, UV, IR, ^{1}H, ^{31}P (acetone) +46.3, this indicates pentacovalent struct., but IR of the solid suggests hexacoordination!, mass spect.; acetylation yields the following pentacoordinated product: $(C_6H_4O_2)_2P\text{-}O\text{-}C_6H_4\text{-}NHCOMe$, m. 218-220°, IR, ^{1}H, ^{31}P +46.6, mass spect.[2]

$[(C_6H_4O_2)_2P^{\ominus}(OPh)_2]Li^{\oplus}$. LVI. Not isolated, ^{31}P +107.[251]

$[(C_6H_4O_2)_2P^{\ominus}(C_{12}H_8)]Li^{\ominus}$. LVII. Not isolated, ^{31}P +106.[251]

$[(C_6H_4O_2)P^{\ominus}(C_{12}H_8)_2]M^{\oplus}$. LVI. M = Li: not isolated, ^{31}P +147. M = bis-2,2'-biphenylylenephosphonium, yellow plates, dec. 285-295°, IR, ^{31}P +147, -24.[251]

$[(C_{10}H_6O_2)P^{\ominus}(C_{12}H_8)_2]M^{\oplus}$. LVI. M = Li: not isolated, ^{31}P +168. M = bis-2,2'-biphenylylenephosphonium, yellow cryst., dec. 270-275°.[251]

$[MePF_5]^{\ominus}H^{\oplus}$. LII. 2 kinds of ^{19}F establish octahedral C_{4v} struct.[424]

$[MePF_5]^{\ominus}M^{\oplus}$. LII. M = Me_2NH_2: $subl_{0.1}$ 90°, ^1H, ^{31}P +124.9, ^{19}F.[555] M = Et_3NH: b_2 124-127°, n_D^{25} 1.3940, d_4^{25} 1.2486, ^{19}F, ^{31}P +125.[147] M = $MePF(NMe_2)_2$: cryst., ^{31}P +126.4, +126.8, ^{19}F,[555,586] ^1H.[555] All these compounds show 2 kinds of F positions.

$[CF_3PF_5]^{\ominus}M^{\oplus}$. LII. M = Cs: white solid, resistant against and soluble in water!, dec. > 200°, IR, ^{19}F (2 kinds).[94a] M = NO.[589] M = $SnMe_3$: IR.[94a] M = Ph_4As: mixture with $(CF_3)_2PF_4^{\ominus}$ salt, IR, ^{19}F.[314]

$[EtPF_5]^{\ominus}M^{\oplus}$. LII. M = Et_2NH_2: $b_{0.1}$ 120°.[589] M = (ring: CH_2CH_2 bridged N(Me)–P(F)(Et)–N(Me)): cryst., subl., 2 kinds of ^{19}F, ^{31}P +126.5.[592] M = NO.[589]

$[PrPF_5]^{\ominus}NO^{\oplus}$. LII. No data.[589]

$[PhPF_5]^{\ominus}M^{\oplus}$. LII. M = H: ^{19}F.[424] M = NO: orange solid, stable below 0°, ^{19}F not reproducible, reactions.[593] (589) M = Me_2NH_2: m. 50°, $b_{0.15}$ 128-129°, IR, ^{19}F.[593] (589) M = Et_2NH_2: solid, ^{19}F, ^{31}P +136.[593] M = NH_2: solid, ^{19}F, ^{31}P +135.[593] M = $PhPF(NMe_2)_2$: cryst., conduct. in MeCN,[579] ^{19}F, ^{31}P +136.[555,579] M = (ring: CH_2CH_2 bridged NMe–P(F)(Ph)–NMe): cryst., ^1H, ^{19}F, ^{31}P +138.[592] M = (MeN/MeN ring)P(Ph)–N(Me)–CH_2CH_2–N(Me)–P(Ph)(NMe/NMe ring): m. 178-179°, IR, ^1H, ^{19}F, ^{31}P +136.[592] Again, all these salts have 2 different ^{19}F signals.

$[p\text{-}MeC_6H_4PF_5]^{\ominus}NO^{\oplus}$. LII. Without data.[589]

$PhPF_4 \cdot B$. LVIII. B = Me_2NCHO: pyridine, cryst., no data.[424]

$[(CF_3)_2PF_4]^{\ominus}M^{\oplus}$. LII. M = Cs: stable white solid, dec. 200°, IR, trans configuration, ^{19}F,[94a] ^{19}F (right interpretation).[72a] M = Ag: from $[(CF_3)_2PO_2]Ag$ and SF_4, dec. at ~120° to R_2PF_3.[94a] M = Ph_4As: mixture with CF_3PF_5 salt, IR, ^{19}F.[314]

$[(CF_3)_3PF_3]^{\ominus}M^{\oplus}$. LII. M = Cs: stable white solid, IR, ^{19}F,[94a] dec. > 200°. M = Me_3Sn: IR.[94a]

$[(CF_3)_4PF_2]^{\ominus}SnMe_3^{\oplus}$. LII. IR.[94a]

$[(2,2'\text{-biphenylylene})_3P]^{\ominus}M^{\oplus}$. LVII. M = Li: ·X acetone, white cryst., dec. 245-255°, ^{31}P +181. M = Na: ·2THF, white cryst., dec. 266-268°.[237] M = K: white cryst., dec. > 300°.[241] M = Ph_4P: white cryst., dec. 267-269°. M = Ph_4As: white cryst., dec. 246-247°. M = bis-2,2'-biphenylylenephosphonium: yellow cyrst., dec. 254-256°, ^{31}P +186.6, -26.5. M = bis-2,2'-biphenylylenearsonium, yellow cryst., dec. 243-245°.[237] Optically active salts: M = methyl-burcinium, $[\alpha]^{24.5}_{578}$ -1250°, m. 234-236°; $[\alpha]^{24.5}_{578}$ +950°, m. 230-234°. M = K: $[\alpha]^{24.5}_{578}$ ±1930 ± 20°, m. 295-298° (dec.). UV and circular dichroism spectra of (-) isomer show the $M(C_3)$ configuration.[249a] M = bis-2,2'-biphenylylenephosphonium: yellow cryst., m. 247-250 (dec.), $[\alpha]^{24.5}_{578}$ ±1265°;[238,241] stereochemistry of acidic cleavage.[241] [^{32}P-phosphonium][phosphate] complex.[244]

$[(2,2'\text{-biphenylylene})_2P(4\text{-methyl-2,2'-biphenylylene})]^{\ominus}M^{\oplus}$. LVII. M = K: white cryst., dec. 300°. M = bis-2,2'-biphenylylene-phosphonium: yellow cryst., dec. 270-272°. Optically active species: M = methylbrucinium, $[\alpha]^{24}_{578}$ -1200°, +986°; M = K, $[\alpha]^{24}_{578}$ ±1870°, yields optically active phosphorane when cleaved with acid.[242]

$[(4,4'\text{-dimethyl-2,2'-biphenylylene})_3P]^{\ominus}M^{\oplus}$. LVII. M = K: white cryst., dec. > 300°. M = bis(4,4'-dimethyl-2,2'-biphenylylene)-phosphonium: yellow cryst., dec. 252-255°, ^1H. Optically active species: M = methylbrucinium, m = 260-262°, $[\alpha]^{24}_{578}$ -1170°, +744°, in an asymmetric transformation of the second order the amorph (+)-diastereomer forms the crystalline (-)-diastereomer on warming in methanol; M = K, $[\alpha]^{24}_{578}$ -1450°, racemizes on warming in methanol; M = bis(4,4'-dimethyl-2,2'-biphenylylene)phosphonium, $[\alpha]^{24}_{578}$ -1070°.[243]

$[HPF_5]^{\ominus}$ K,$Me_2NH_2^{\oplus}$. LIII. ^1H, 2 kinds of ^{19}F prove octa-hedral struct.[448]

$[MePF_4H]^{\ominus}[Et_3NH]^{\oplus}$. LII. Insoluble, oily layer, readily hydrolyzed, n_D^{20} 1.4090, d_4^{20} 1.1735.[144]

$[CF_3PF_4H]^{\ominus}M^{\oplus}$. M = Me_2NH_2: LII. IR, ^1H, ^{19}F shows octahedron with 4 equivalent fluorines, J_{FP} 858 Hz.[94] M = K: LIII.[447]

$[EtPPF_4H]^{\ominus}[Et_3NH]^{\oplus}$. LII. Insoluble oily layer, readily hydrolyzed, n_D^{20} 1.4102, d_4^{20} 1.0828.[144]

$[p\text{-}MeC_6H_4PF_4H]^{\ominus}[Et_3NH]^{\oplus}$. LII. Insoluble oily layer, readily hydrolyzed, n_D^{20} 1.4665, d_4^{20} 1.1878.[144]

$[(CF_3)_2PF_3H]^{\ominus}K^{\oplus}$. LIII. ^1H, ^{19}F shows octahedron with 2 kinds of F and trans CF_3 groups.[447]

(received July 27, 1970).

REFERENCES

1. Allcock, H. R., J. Am. Chem. Soc., 85, 4050 (1963), 86, 2591 (1964).
1a. Allcock, H. R., and E. C. Bissell, Chem. Commun., 1972, 676.
2. Allcock, H. R., and R. L. Kugel, J. Am. Chem. Soc., 91, 5452 (1969); Chem. Commun., 1968, 1606.
3. Almenningen, A., B. Andersen, and E. E. Astrup, Acta Chem. Scand., 23, 2179 (1969).
4. Anderson, A. G., Jr., and F. J. Freenor, J. Am. Chem. Soc., 86, 5037 (1964).
5. Ang, H. G., J. Inorg. Nucl. Chem., 31, 3311 (1969).
5a. Ang, H. G., Chem. Commun., 1968, 1320.
6. Ang, H. G., and J. M. Miller, Chem. Ind. (London), 1966, 944.
7. Ang, H. G., and R. Schmutzler, J. Chem. Soc., A, 1969, 702.
8. Anisimov, K. N., Izv. Akad. Nauk SSSR, Ser. Khim., 1954, 803; C.A., 49, 13074e (1955).
9. Anisimov, K. N., and N. E. Kolobova, Izv. Akad. Nauk SSSR, Ser. Khim., 1956, 923; C.A., 51, 4933f (1957).
10. Anisimov, K. N., and N. E. Kolobova, Izv. Akad. Nauk SSSR, Ser. Khim., 1956, 927; C.A., 51, 4934a (1957).
11. Anisimov, K. N., and N. E. Kolobova, Izv. Akad. Nauk SSSR, Ser. Khim., 1962, 442; C.A. 57, 13790f (1962).
12. Anisimov, K. N., and N. E. Kolobova, Izv. Akad. Nauk SSSR, Ser. Khim., 1962, 444; C.A., 57, 12529c (1962).
13. Anisimov, K. N., N. E. Kolobova, and A. N. Nesmeyanov, Izv. Akad. Nauk SSSR, Ser. Khim., 1954, 796; C.A., 49, 13074a (1955).
14. Anisimov, K. N., N. E. Kolobova, and A. N. Nesmeyanov, Izv. Akad. Nauk SSSR, Ser. Khim., 1954, 799; C.A., 49, 13074c (1955).
15. Anisimov, K. N., G. M. Kunitskaya, and N. A. Slovokhotova, Izv. Akad. Nauk SSSR, Ser. Khim., 1961, 64;

C.A., 55, 18562e (1961).
16. Anisimov, K. N., and A. N. Nesmeyanov, Izv. Akad. Nauk SSSR, Ser. Khim., 1954, 610; C.A., 49, 11540i (1955).
17. Appel, R., B. Blaser, and G. Siegmund, Z. Anorg. Allgem. Chem., 363, 177 (1968).
18. Appel, R., and W. Heinzelmann, Ger. Pat. 1,192,205 (1965); C.A., 63, 8405g (1965).
19. Arbuzov, B. A., E. N. Dianova, and V. S. Vinogradova, Izv. Akad. Nauk SSSR, 1969, 1109; C.A., 71, 50070j (1969).
20. Arbuzov, B. A., N. A. Polezhaeva, and V. S. Vinogradova, Izv. Akad. Nauk SSSR, Ser. Khim., 1967, 2281; C.A., 68, 49684u (1968).
21. Arbuzov, B. A., N. A. Polezhaeva, and V. S. Vinogradova, Izv. Akad. Nauk SSSR, Ser. Khim., 1968, 2525; C.A., 70, 87682r (1969).
22. Arbuzov, B. A., N. A. Polezhaeva, V. S. Vinogradova, and Yu. Yu. Samitov, Dokl. Akad. Nauk SSSR, 173, 93 (1967); C.A. 67, 54221z (1967).
23. Arbuzov, B. A., N. A. Polezhaeva, V. S. Vinogradova, and Yu. Yu. Samitov, Izv. Akad. Nauk SSSR, Ser. Khim., 1967, 1605; C.A., 68, 95766v (1968).
24. Arbuzov, B. A., and L. A. Shapshinskaya, Izv. Akad. Nauk SSSR, Ser. Khim., 1962, 65; C.A., 57, 13791b (1962).
25. Arbuzov, B. A., T. D. Sorokina, N. P. Bogonostseva, and V. S. Vinogradova, Dokl. Akad. Nauk SSSR, 171, 605 (1966); C.A., 67, 32501p (1967).
26. Arbuzov, B. A., V. S. Vinogradova, and O. D. Zolova, Izv. Akad. Nauk SSSR, Ser. Khim., 1968, 2290; C.A., 70, 29004k (1969).
27. Arbuzov, B. A., and A. O. Vizel, Dokl. Akad. Nauk SSSR, Ser. Khim., 158, 1105 (1964); C.A., 62, 2791 (1965).
28. Arbuzov, B. A., O. A. Vizel, Yu. Yu. Samitov, and K. M. Ivanovskaya, Dokl. Akad. Nauk SSSR, 159, 582 (1964); C.A., 62, 6505e (1965).
29. Arbuzov, B. A., O. D. Zolova, V. S. Vinogradova, and Yu. Yu. Samitov, Dokl. Akad. Nauk SSSR, 173, 335 (1967); C.A., 67, 43886u (1967).
30. Arzoumanidis, G. G., Chem. Commun., 1969, 217.
31. Atherton, F. R., Chem. Soc. (London) Spec. Publ., 8, 77 (1957).
32. Bacon, W. E., U.S. Pat. 3,305,570 (1967); C.A., 67, 11584c (1967); Fr. Pat. 1,451,986 (1966); C.A., 66, 65627s (1967).
33. Balaban, A. T., D. Farcasiu, and R. Banica, Rev. Roumaine Chim., 11, 1205 (1966); C.A., 66, 64740m (1967).
34. Balon, W. J., U.S. Pat. 2,853,518 (1958); C.A., 53, 5202b (1959).

35. Banks, R. E., R. N. Haszeldine, and R. Hatton, Tetrahedron Letters, 1967, 3993.
36. Barnikow, G., and Th. Gabrio, Z. Chem., 8, 142 (1968).
37. Bayer, H. O., S. N. Lewis, and V. H. Unger, U.S. Pat. 3,468,946 (1969); C.A., 72, 12067b (1970).
38. Bartell, L. S., Inorg. Chem., 5, 1635 (1966).
39. Bartell, L. S., J. Chem. Educ., 45, 754 (1968).
40. Bartell, L. S., and K. W. Hansen, Inorg. Chem., 4, 1777 (1965).
41. Baumgärtner, R., W. Sawodny, and J. Goubeau, Z. Anorg. Allgem. Chem., 333, 171 (1964).
42. Beattie, I. R., T. R. Gibson, and G. A. Ozin, J. Chem. Soc., A, 1968, 2776.
43. Beattie, I. R., K. Livingston, and T. Gilson, J. Chem. Soc., A, 1968, 1.
44. Beattie, I. R., K. Livingston, and M. Webster, J. Chem. Soc., 1965, 7421.
44a. Beattie, I. R., and M. Webster, J. Chem. Soc., 1961, 1730.
45. Beattie, I. R., and M. Webster, J. Chem. Soc., 1963, 38.
46. Becke-Goehring, M., Fortschr. Chem. Forsch., 10, 207 (1968).
47. Becke-Goehring, M., and W. Haubold, Z. Anorg. Allgem. Chem., 338, 305 (1965).
48. Becke-Goehring, M., and W. Haubold, Z. Anorg. Allgem. Chem., 372, 273 (1970).
49. Becke-Goehring, M., L. Leichner, and B. Scharf, Z. Anorg. Allgem. Chem., 343, 154 (1966).
50. Becke-Goehring, M., and H. Schwind, Z. Anorg. Allgem. Chem., 372, 285 (1970).
51. Becke-Goehring, M., H. J. Wald, and H. Weber, Naturwissenschaften, 55, 491 (1968).
52. Becke-Goehring, M., and H. J. Wald, Z. Anorg. Allgem. Chem., 371, 88 (1969).
53. Becke-Goehring, M., and W. Weber, Z. Anorg. Allgem. Chem., 339, 281 (1965).
54. Becke-Goehring, M., and H. Weber, Z. Anorg. Allgem. Chem., 365, 185 (1969).
55. Becke-Goehring, M., and M. R. Wolf, Naturwissenschaften, 55, 543 (1968).
55a. Becke-Goehring, M., and M. R. Wolf, Z. Anorg. Allgem. Chem., 373, 245 (1970).
56. Beg, M. A. A., and H. C. Clark, Can. J. Chem., 40, 283 (1962).
57. Bennett, F. W., H. J. Emeleus, and R. N. Haszeldine, J. Chem. Soc., 1953, 1565.
58. Bentrude, W. G., J. Am. Chem. Soc., 87, 4026 (1965).
59. Bentrude, W. G., Chem. Commun., 1967, 174.
60. Bentrude, W. G., and K. R. Darnall, Tetrahedron Letters, 1967, 2511.

61. Bentrude, W. G., and K. R. Darnall, Chem. Commun., 1969, 862.
62. Berlin, K. D., and D. M. Hellwege, Topics in Phosphorus Chemistry, Vol. 6, Interscience, New York, 1969, p. 83.
62a. Berry, R. S., J. Chem. Phys., 32, 933 (1960).
63. Bestmann, H. J., Th. Denzel, R. Kunstmann, and J. Lengyel, Tetrahedron Letters, 1968, 2895.
64. Bestmann, H. J., and R. Kunstmann, Chem. Ber., 102, 1816 (1969).
65. Bestmann, H. J., and L. Mott, Ann., 693, 132 (1966).
66. Beveridge, A. B., G. S. Harris, and F. Inglis, J. Chem. Soc., A, 1966, 520.
67. Bilbo, A. J., Z. Naturforsch., 15b, 330 (1960).
68. Binder, H., and E. Fluck, Z. Anorg. Allgem. Chem., 365, 166 (1969).
69. Birum, G. H., and J. L. Dever, 134th National Meeting of the American Chemical Society, Chicago, September 1958, Abstract P101.
70. Birum, G. H., and J. L. Dever, U.S. Pat. 2,961,455 (1960); C.A., 55, 8292g (1961).
71. Birum, G. H., and C. N. Matthews, J. Org. Chem., 32, 3554 (1967); U.S. Pat. 3,488,408 (1970); C.A., 72, 67097s (1970).
72. Birum, G. H., and C. N. Matthews, Chem. Commun., 1967, 137.
72a. Bishop, E. O., P. R. Carey, J. F. Nixon, and J. R. Swain, J. Chem. Soc., A, 1970, 1074.
73. Blazer, T. A., R. Schmutzler, and I. K. Gregor, Z. Naturforsch., 24b, 1081 (1969).
74. Blaser, B., and K. H. Worms, Z. Anorg. Allgem. Chem., 361, 15 (1968); Angew. Chem., 73, 76 (1961).
75. Bodnarchuk, N. D., V. I. Shevchenko, and A. V. Kirsanov, Zh. Obshch. Khim., 35, 713 (1965); C.A. 63, 5549b (1965).
76. Borowitz, I. J., and M. Anschel, Tetrahedron Letters, 1967, 1517.
77. Borowitz, I. J., M. Anschel, and S. Firstenberg, J. Org. Chem., 32, 1723 (1967).
78. Boyd, D. B., J. Am. Chem. Soc., 91, 1200 (1969).
79. Brouns, H. G., U.S. Pat. 3,345,340 (1967); C.A., 67, 117514v (1967).
79a. Brown, C., M. Murray, and R. Schmutzler, J. Chem. Soc., C, 1970, 878.
79b. Brown, D. H., K. D. Crosbie, J. I. Darragh, D. S. Ross, and D. W. A. Sharp, J. Chem. Soc., A, 1970, 914.
80. Brown, D. H., K. D. Crosbie, G. W. Fraser, and D. W. A. Sharp, J. Chem. Soc., A, 1969, 551.
81. Brown, D. H., K. D. Crosbie, G. W. Fraser, and D. W. A. Sharp, J. Chem. Soc., A, 1969, 872.
82. Brown, D. H., G. W. Fraser, and D. W. A. Sharp, J.

Chem. Soc., A, 1966, 171 (Chem. Ind., 1964, 367).
83. Brown, N. M. D., and P. Bladon, Chem. Commun., 1966, 304.
84. Bruker, A. B., Kh. R. Raver, and L. Z. Soborovskii, Probl. Organ. Sinteza, Akad. Nauk SSSR, Ser. Obshch. i Tekhn. Khim., 1965, 285; C.A., 64, 6681c (1966).
85. Burg, A. B., G. Brendel, A. P. Caron, G. L. Juvinall, W. Mahler, K. Mödritzer, and P. J. Slota, W. A. D. C. Rept. 56-82, Part III, April 1958.
86. Burg, A. B., and J. E. Griffiths, J. Am. Chem. Soc., 82, 3514 (1960).
87. Burgada, R., Ann. Chim., 8, 347 (1963).
88. Burgada, R., Compt. Rend., 258, 4789 (1964).
89. Burgada, R., Bull. Soc. Chim. France, 1967, 347.
90. Burgada, R., M. Bon, and F. Mathis, Compt. Rend., 265, 1499 (1967).
91. Burgada, R., and H. Germa, Compt. Rend., 267, 270 (1968).
92. Burgada, R., D. Houalla, and R. Wolf, Compt. Rend., 264, 356 (1967).
93. Burgada, R., F. Mathis, and M. Bon, Compt. Rend., 264, 625 (1967).
94. Cavell, R. G., and J. F. Nixon, Proc. Chem. Soc., 1964, 229.
94a. Chan, S. S., and C. J. Willis, Can. J. Chem., 46, 1237 (1968).
95. Chapman, A. C., W. S. Holmes, N. L. Paddock, and H. T. Searle, J. Chem. Soc., 1961, 1825.
95a. Chioccola, G., and J. J. Daly, J. Chem. Soc., A, 1968, 568.
96. Chopard, P. A., and R. F. Hudson, J. Chem. Soc., B, 1966, 1089.
97. Clay, J. P., J. Org. Chem., 16, 892 (1951).
98. Clune, J. E., and K. Cohn, Inorg. Chem., 7, 2067 (1968).
99. Coates, H., and P. R. Carter, Brit. Pat. 734,187 (1955); C.A., 50, 7123a (1956); U.S. Pat. 2,853,515 (1958); C.A., 53, 7988g (1959).
100. Coe, D. G., S. R. Landauer, and H. N. Rydon, J. Chem. Soc., 1954, 2281.
101. Coe, D. G., H. N. Rydon, and B. L. Tonge, J. Chem. Soc., 1957, 323.
102. Cohen, E. A., and C. D. Cornwell, Inorg. Chem., 7, 398 (1968).
103. Cohn, K., and R. W. Parry, Inorg. Chem., 7, 46 (1968).
104. Cottrell, W. R. T., and R. A. N. Morris, Chem. Commun., 1968, 409.
105. Cowley, A. H., and M. W. Taylor, J. Am. Chem. Soc., 91, 1934 (1969).
106. Cox, J. W., and E. R. Corey, Chem. Commun., 1967, 123.
106a. Craig, D. P., A. Maccoll, R. S. Nyholm, L. E. Orgel, and L. E. Sutton, J. Chem. Soc., 1954, 332; D. P.

Craig and E. A. Magnusson, J. Chem. Soc., 1956, 4895.
107. Craig, W. G., and W. A. Higgins, U.S. Pat. 2,727,073 (1955); C.A., 50, 9445f (1956).
108. Crosbie, K. D., G. W. Fraser, and D. W. A. Sharp, Chem. Ind., 1968, 423.
109. Daniel, H., and J. Paetsch, Chem. Ber., 101, 1451 (1968).
110. Davydova, L. P., L. M. Kaboshina, E. A. Obol'nikova, I. M. Kustanovich, and G. I. Samokhvalov, Zh. Obshch. Khim., 38, 2091 (1968); C.A., 70, 87907t (1969).
111. DeBruin, K. E., and K. Mislow, J. Am. Chem. Soc., 91, 7393 (1969).
112. DeBruin, K. E., K. Naumann, G. Zon, and K. Mislow, J. Am. Chem. Soc., 91, 7031 (1969).
113. Degani, C., M. Halmann, I. Laulicht, and S. Pinchas, Spectrochim. Acta, 20, 1289 (1964).
114. Demitras, G. C., R. A. Kent, and A. G. MacDiarmid, Chem. Ind., 1964, 1712.
115. Demitras, G. C., and A. G. MacDiarmid, Inorg. Chem., 6, 1903 (1967).
116. Denney, D. B., and N. G. Adin, Tetrahedron Letters, 1966, 2569.
117. Denney, D. B., D. Z. Denney, and B. C. Chang, J. Am. Chem. Soc., 90, 6332 (1968).
118. Denney, D. B., D. Z. Denney, B. C. Chang, and K. L. Marsi, J. Am. Chem. Soc., 91, 5243 (1969).
119. Denney, D. B., D. Z. Denney, and L. A. Wilson, Tetrahedron Letters, 1968, 85.
120. Denney, D. B., and J. Giacin, Tetrahedron Letters, 1964, 1747.
121. Denney, D. B., W. F. Goodyear, and G. Goldstein, J. Am. Chem. Soc., 82, 1393 (1960).
122. Denney, D. B., W. F. Goodyear, and B. Goldstein, J. Am. Chem. Soc., 83, 1726 (1961).
123. Denney, D. B., and S. T. D. Gough, J. Am. Chem. Soc., 87, 138 (1965).
124. Denney, D. B., and M. A. Greenbaum, J. Am. Chem. Soc., 79, 979 (1957).
125. Denney, D. B., and F. J. Gross, J. Org. Chem., 32, 3710 (1967).
126. Denney, D. B., and D. H. Jones, J. Am. Chem. Soc., 91, 5821 (1969).
127. Denney, D. B., and H. M. Relles, J. Am. Chem. Soc., 86, 3897 (1964).
128. Denney, D. B., H. M. Relles, and A. K. Tsolis, J. Am. Chem. Soc., 86, 4487 (1964).
129. Denney, D. B., and L. Saferstein, J. Am. Chem. Soc., 88, 1839 (1966).
130. Dimitrieva, L. E., S. Z. Ivin, and K. V. Karavanov, Khim. Org. Soedin. Fosfora, Akad. Nauk SSSR, Ser.

Obshch. Tekh. i Khim., 1967, 155; C.A., 68, 114709x (1968).
131. Dimroth, K., and R. Ploch, Chem. Ber., 90, 801 (1957).
131a. Doak, G. O., and R. Schmutzler, Chem. Commun., 1970, 476.
132. Downs, A. J., Chem. Commun., 1967, 628.
133. Downs, A. J., and R. Schmutzler, Spectrochim. Acta, 21, 1927 (1965).
134. Downs, A. J., and R. Schmutzler, Spektrochim. Acta, 23A, 681 (1967).
135. Drach, B. S., and I. N. Zhmurova, Zh. Obshch. Khim., 37, 892 (1967); C.A., 68, 2533d (1968).
135a. Dreeskamp, H., C. Schumann, and R. Schmutzler, Chem. Commun., 1970, 671.
136. Drozd, G. I., S. Z. Ivin, V. N. Kulakova, and V. V. Sheluchenko, Zh. Obshch. Khim., 38, 576 (1968); C.A., 69, 43973m (1968).
137. Drozd, G. I., S. Z. Ivin, M. A. Landau, and V. V. Sheluchenko, Zh. Obshch. Khim., 38, 1654 (1968); C.A., 69, 87094e (1968).
138. Drozd, G. I., S. Z. Ivin, and V. V. Sheluchenko, Zh. Vses. Khim. Obshch., 12 (4), 474 (1967); C.A., 67, 108705f (1967).
139. Drozd, G. I., S. Z. Ivin, and V. V. Sheluchenko, Zh. Obshch. Khim., 38, 1655 (1968); C.A., 69, 96831v (1968).
139a. Drozd, G. I., S. Z. Ivin, and V. V. Sheluchenko, Zh. Obshch. Khim., 38, 1906 (1968); C.A. 69, 96825w (1968).
140. Drozd, G. I., S. Z. Ivin, V. V. Sheluchenko, and M. A. Landau, Zh. Obshch. Khim., 38, 1653 (1968); C.A., 70, 11754v (1969).
141. Drozd, G. I., S. Z. Ivin, V. V. Sheluchenko, and B. I. Tetel'baum, Zh. Obshch. Khim., 37, 957 (1967); C.A., 67, 108712f (1967).
142. Drozd, G. I., S. Z. Ivin, V. V. Sheluchenko, and B. I. Tetel'baum, Radiospektrosk. Kvantovokhim. Metody Strukt. Issled., 1967, 204; C.A., 70, 47538s (1969).
143. Drozd, G. I., S. Z. Ivin, V. V. Sheluchenko, B. I. Tetel'baum, G. M. Luganskii, and A. D. Varshavskii, Zh. Obshch. Khim., 37, 1631 (1967); C.A., 68, 78358w (1968).
144. Drozd, G. L., S. Z. Ivin, V. V. Sheluchenko, B. I. Tetel'baum, G. M. Luganskii, and A. V. Varshavskii, Zh. Obshch. Khim., 37, 1343 (1967); C.A., 67, 108707h (1967).
145. Drozd, G. I., S. Z. Ivin, V. V. Sheluchenko, B. I. Tetel'baum, and A. D. Varshavskii, Zh. Obshch. Khim., 38, 567 (1968); C.A., 69, 43972k (1968).
146. Drozd, G. I., S. Z. Ivin, and M. A. Sokalskii, Zh. Obshch. Khim., 39, 1177 (1969); C.A., 71, 50072m (1969).

147. Drozd, G. I., S. Z. Ivin, and A. D. Varshavskii, Zh. Obshch. Khim., 39, 1178 (1969); C.A., 71, 61496d (1969).
148. Drozd, G. I., M. A. Sokalskii, M. A. Landau, and S. Z. Ivin, Zh. Obshch. Khim., 39, 1888 (1969); C.A., 71, 123415n (1969).
149. Drozd, G. I., M. A. Sokalskii, V. V. Sheluchenko, M. A. Landau, and S. Z. Ivin, Zh. Obshch. Khim., 39, 935 (1969); C.A., 71, 50089x (1969).
150. Drozd, G. I., M. A. Sokalskii, V. V. Sheluchenko, M. A. Landau, and S. Z. Ivin, Zh. Obshch. Khim., 39, 936 (1969); C.A., 71, 50088w (1969).
151. Dunitz, J. D., and V. Prelog, Angew. Chem., 80, 700 (1968); Intern. Ed., 7, 725 (1968).
152. Ebeling, J., and A. Schmidpeter, Angew. Chem., 81, 707 (1969); Intern. Ed., 8, 674 (1969).
153. Emeleus, H. J., and G. S. Harris, J. Chem. Soc., 1959, 1494.
154. Emeleus, H. J., R. N. Haszeldine, and R. C. Paul, J. Chem. Soc., 1955, 563.
155. Emeleus, H. J., and J. M. Miller, J. Inorg. Nucl. Chem., 28, 662 (1966).
156. Emeleus, H. J., and T. Onak, J. Chem. Soc., A, 1966, 1291.
157. Emeleus, H. J., and J. D. Smith, J. Chem. Soc., 1959, 375.
158. Evdakov, V. P., L. I. Mizrak, and L. Yu. Sandalova, Zh. Obshch. Khim., 37, 1818 (1967); C.A., 68, 59507c (1968).
159. Evtikhov, Zh. L., N. A. Razumova, and A. A. Petrov, Dokl. Akad. Nauk SSSR, 177, 108 (1967); C.A., 69, 36218r (1968).
160. Evtikhov, Zh. L., N. A. Razumova, and A. A. Petrov, Dokl. Akad. Nauk SSSR, 181, 1385 (1968); C.A., 70, 20172f (1969).
161. Farkas, J., M. Menyhart, R. Bognar, and H. Gross, Ber., 98, 1419 (1965).
162. Fedorova, G. K., and A. V. Kirsanov, Zh. Obshch. Khim., 30, 4044 (1960); C.A., 55, 23401h (1961).
163. Fedorova, G. K., and A. V. Kirsanov, Zh. Obshch. Khim., 31, 594 (1961); C.A., 55, 27020a (1961).
164. Fedorova, G. K., and A. V. Kirsanov, Zh. Obshch. Khim., 35, 1483 (1965); C.A., 63, 14901a (1965).
165. Fedorova, G. K., and Ya. P. Shaturskii, Zh. Obshch. Khim., 36, 1262 (1966); C.A., 65, 15419b (1966).
166. Fedorova, G. K., Ya. P. Shaturskii, and A. V. Kirsanov, Zh. Obshch. Khim., 35, 1984 (1965); C.A., 64, 6682f (1966).
167. Fedorova, G. K., Ya. P. Shaturskii, and A. V. Kirsanov, Probl. Organ. Sinteza, Akad. Nauk SSSR, Ser. Obshch. i Tekhn. Khim., 1965, 258; C.A., 64, 8228a (1966).

167a. Fedorova, G. K., Ya. P. Shaturskii, and A. V. Kirsanov, Prob. Organ. Sinteza, Akad. Nauk SSSR, Ser. Obshch. i Tekhn. Khim., 1965, 263; C.A., 64, 8228d (1966).
168. Ferekh, J., J. F. Brazier, A. Munoz, and R. Wolf, Compt. Rend., 270, 865 (1970).
168a. Feshchenko, N. G., T. V. Kovaleva, and A. V. Kirsanov, Zh. Obshch. Khim., 39, 2184 (1969); C.A., 72, 43796t (1970).
169. Feshchenko, N. G., T. V. Kovaleva, and A. V. Kirsanov, Zh. Obshch. Khim., 39, 2188 (1969); C.A., 72, 43798v (1970).
169a. Feshchenko, N. G., E. A. Mel'nichuk, and A. V. Kirsanov, Zh. Obshch. Khim., 39, 2139 (1969); C.A., 72, 21743w (1970).
170. Fild, M., and R. Schmutzler, J. Chem. Soc., A, 1969, 840.
171. Finch, A., P. J. Gardner, A. Hameed, and K. K. Sen Gupta, Chem. Commun., 1969, 854.
172. Firth, W. C., S. Frank, M. Garber, and V. P. Wystrach, Inorg. Chem., 4, 765 (1965).
172a. Fluck, E., Z. Anorg. Allgem. Chem., 320, 64 (1963).
173. Fluck, E., Topics in Phosphorus Chemistry, Vol. 4, Interscience, New York, 1967, p. 396.
174. Fluck, E., H. Gross, H. Binder, and J. Gloede, Z. Naturforsch., 21b, 1125 (1966).
175. Fluck, E., and R. M. Reinisch, Chem. Ber., 96, 3085 (1963).
176. Forsman, J. P., and D. Lipkin, J. Am. Chem. Soc., 75, 3145 (1953).
177. Fowell, P. A., and C. T. Mortimer, Chem. Ind., 1960, 444.
178. Frank, A. W., Can. J. Chem., 46, 3573 (1968).
179. Frazier, S. E., R. P. Nielsen, and H. H. Sisler, Inorg. Chem., 3, 292 (1964).
180. Gallagher, M. J., and I. D. Jenkins, Topics in Stereochemistry, Vol. 3, Interscience, New York, 1968, p. 61.
181. Gambaryan, N. P., Yu. A. Cheburkov, and I. L. Knunyants, Izv. Akad. Nauk SSSR, Ser. Khim., 1964, 1526; C.A., 64, 14082a (1966).
182. Gambaryan, N. P., E. M. Rokhlin, Yu. V. Zeifman, and I. L. Knunyants, Izv. Akad. Nauk SSSR, Ser. Khim., 1965, 749; C.A., 63, 2914b (1965).
183. Garber, M., and W. C. Firth, Jr., U.S. Pat. 3,268,580 (1966); C.A., 65, 15425h (1966).
184. Gaudiano, G., R. Mondelli, P. P. Ponti, C. Ticozzi, and A. Umani-Ronchi, J. Org. Chem., 33, 4431 (1968).
184a. Germa, H., M. Willson, and R. Burgada, Compt. Rend., 270, 1426, 1474 (1970).
185. Gielen, M., M. DeClercq, and J. Nasielski, J. Organometal. Chem., 18, 217 (1969).

186. Gielen, M., C. Depasse-Delit, and J. Nasielski, Bull. Soc. Chim. Belges, 78, 357 (1969).
187. Gielen, M., G. Mayence, and J. Topart, J. Organometal. Chem., 18, 1 (1969).
188. Gielen, M., and J. Nasielski, Bull. Soc. Chim. Belges, 78, 339 (1969).
189. Gielen, M., and J. Topart, J. Organometal. Chem., 18, 7 (1969).
190. Gillespie, R. J., J. Chem. Educ., 47, 18 (1970); Angew. Chem., 79, 885 (1967); Intern. Ed., 6, 819 (1967).
191. Gillespie, R. J., Inorg. Chem., 5, 1634 (1966).
192. Ginsburg, V. A., A. Ia. Iakubovich, Zh. Obshch. Khim., 28, 728 (1958); C.A., 52, 1791i (1958).
193. Ginsburg, V. A., M. N. Vasil'eva, S. S. Dubov, and A. Ya. Yakubovich, Zh. Obshch. Khim., 30, 2854 (1960); C.A., 55, 17477b (1961).
194. Gladshtein, B. M., V. G. Noskov, and L. Z. Soborovskii, Zh. Obshch. Khim., 37, 2513 (1967); C.A., 69, 27493z (1968).
195. Gladshtein, B. M., and V. M. Zimin, Zh. Obshch. Khim., 37, 2055 (1967); C.A., 68, 49687x (1968).
196. Gloede, J., and H. Gross, Chem. Ber., 100, 1770 (1967).
197. Goodrich, R. A., and P. M. Treichel, Inorg. Chem., 7, 694 (1968).
198. Gorenstein, D., J. Am. Chem. Soc., 92, 644 (1970).
199. Gorenstein, D., and F. H. Westheimer, Proc. Natl. Acad. Sci. U.S., 58, 1747 (1967).
200. Gorenstein, D., and F. H. Westheimer, J. Am. Chem. Soc., 89, 2762 (1967).
201. Gorenstein, D., and F. H. Westheimer, J. Am. Chem. Soc., 92, 634 (1970).
202. Goubeau, J., and R. Baumgärtner, Z. Elektrochem., 64, 598 (1960).
202a. Goubeau, J., R. Baumgärtner, and H. Weiss, Z. Anorg. Allgem. Chem., 348, 286 (1966).
203. Goubeau, J., and G. Wenzel, Z. Physik. Chem. (Frankfurt), 45, 31 (1965).
204. Gough, S. T. D., and S. Trippett, J. Chem. Soc., 1961, 4263.
204a. Gough, S. T. D., and S. Trippett, Proc. Chem. Soc., 1961, 302.
204b. Gozman, I. P., Zh. Obshch. Khim., 39, 1954 (1969). C.A. 72, 31140m (1970).
205. Grafstein, D., R. Dudak, and M. S. Cohen, W. A. D. C. Rept., 57-45, Part III, March 1959.
206. Grechkin, N. P., R. R. Shagidullin, and L. N. Grishina, Dokl. Akad. Nauk SSSR, 161, 115 (1965); C.A., 62, 14478a (1965). See also Annual Session of Kazan Affiliate Academy of Sciences USSR, February 1960.
207. Greenbaum, M. A., D. B. Denney, and A. K. Hoffmann, J. Am. Chem. Soc., 78, 2563 (1956).

208. Griffiths, J. E., J. Chem. Phys., 41, 3510 (1964).
209. Griffiths, J. E., Inorg. Chim. Acta, 1, 127 (1967).
210. Griffiths, J. E., J. Chem. Phys., 49, 1307 (1968).
211. Griffiths, J. E., and A. L. Beach, J. Chem. Phys., 44, 2686 (1966).
212. Grinblat, M. P., A. L. Kelbanskii, and V. N. Prons, Zh. Obshch. Khim., 39, 172 (1969); C.A., 70, 106612m (1969).
213. Gross, H., and J. Gloede, Chem. Ber., 96, 1387 (1963).
214. Gross, H., and J. Gloede, Z. Chem., 5, 178 (1965).
214a. Gross, H., and J. Gloede, Ger. Pat. 1,443,074 (1969); C.A., 72, 54358c (1970).
215. Gross, H., J. Gloede, E. Eichhorn, C. Richter, and D. Krake, Ger. (East) Pat. 50,605 (1966); C.A., 67, 32468h (1967).
216. Gross, H., and U. Karsch, J. Prakt. Chem. [4], 29 315 (1965).
217. Gross, H., S. Katzwinkel, and J. Gloede, Chem. Ber., 99, 2631 (1966).
218. Gross, H., A. Rieche, and E. Höft, Chem. Ber., 94, 544 (1961).
219. Gruzdev, V. G., S. Z. Ivin, and K. V. Karavanov, Zh. Obshch. Khim., 35, 1027 (1965); C.A., 63, 9979f (1965).
220. Gruzdev, V. G., K. V. Karavanov, S. Z. Ivin, I. S. Mazel', and V. V. Tarasov, Zh. Obshch. Khim., 37, 450 (1967); C.A., 67, 43870j (1967).
221. Gutman, V., C. Kemenater, and K. Utvary, Monatsh. Chem., 96, 836 (1965).
222. Gutmann, V., K. Utvary, and M. Bermann, Monatsh. Chem., 97, 1745 (1966).
223. Haasemann, P., Dissertation, Stuttgart, 1963.
224. Haber, C. P., D. L. Herring, and E. A. Lawton, J. Am. Chem. Soc., 80, 2116 (1958).
225. Hamilton, W. C., S. J. LaPlaca, and F. Ramirez, J. Chem. Soc., 87, 127 (1965).
226. Hamilton, W. C., S. J. LaPlaca, F. Ramirez, and C. P. Smith, J. Am. Chem. Soc., 89, 2268 (1967).
227. Hands, A. R., and A. J. H. Mercer, J. Chem. Soc., C, 1967, 1099.
228. Hands, A. R., and A. J. H. Mercer, J. Chem. Soc., C, 1968, 2448.
229. Hansen, K. W., U.S. At. Energy Comm. IS-T-4 (1965).
230. Harris, G. S., and M. F. Ali, Tetrahedron Letters, 1968, 37.
231. Harris, G. S., and D. S. Payne, J. Chem. Soc., 1956, 3038.
232. Harris, J. J., U.S. Pat. 3,304,160 (1967); C.A., 66, 85852q (1967).
233. Harris, J. J., and B. Rudner, J. Org. Chem., 33,

1392 (1968).
234. Harris, R. K., and G. M. Woodmann, Mol. Phys., 10, 437 (1966).
235. Harwood, H. J., and K. A. Pollart, U.S. Pat. 3,082,256 (1963); C.A., 59, 10121e (1963).
236. Hasserodt, U., K. Hunger, and F. Korte, Tetrahedron, 19, 1563 (1963).
237. Hellwinkel, D., Chem. Ber., 98, 576 (1965).
238. Hellwinkel, D., Angew. Chem., 77, 378 (1965); Intern. Ed., 4, 356 (1965).
238a. Hellwinkel, D., Habilitationsschrift, Heidelberg, 1966.
239. Hellwinkel, D., Angew. Chem., 78, 749 (1966); Intern. Ed., 5, 725 (1966).
240. Hellwinkel, D., Angew. Chem., 78, 985 (1966); Intern. Ed., 5, 968 (1966).
241. Hellwinkel, D., Chem. Ber., 99, 3628 (1966).
242. Hellwinkel, D., Chem. Ber., 99, 3642 (1966).
243. Hellwinkel, D., Chem. Ber., 99, 3660 (1966).
244. Hellwinkel, D., Chem. Ber., 99, 3668 (1966).
244a. Hellwinkel, D., Ger. Pat. 1,235,913 (1967); C.A., 67, 64537w (1967).
245. Hellwinkel, D., Chimia (Aarau), 22, 488 (1968).
246. Hellwinkel, D., Chem. Ber., 102, 528 (1969).
247. Hellwinkel, D., Chem. Ber., 102, 548 (1969).
248. Hellwinkel, D., Colloques Intern. du C.N.R.S., 182, 177 (1970).
248a. Hellwinkel, D., unpublished results, 1966.
249. Hellwinkel, D., and W. Lindner, unpublished results.
249a. Hellwinkel, D., and S. F. Mason, J. Chem. Soc., B, 1970, 640.
250. Hellwinkel, D., and H. J. Wilfinger, Tetrahedron Letters, 1969, 3423.
251. Hellwinkel, D., and H. J. Wilfinger, Chem. Ber., 103, 1056 (1970).
252. Hellwinkel, D., and H. J. Wilfinger, unpublished results.
253. Hellwinkel, D., and C. Wünsche, Chem. Commun., 1969, 1412.
254. Hendrickson, J. B., J. Am. Chem. Soc., 83, 2018 (1961).
254a. Hendrickson, J. B., C. Hall, R. Rees, and J. F. Templeton, J. Org. Chem., 30, 3312 (1965).
255. Hendrickson, J. B., R. E. Spenger, and J. J. Sims, Tetrahedron Letters, 1961, 477.
256. Henning, H. G., Z. Chem., 5, 103 (1965).
256a. Henson, P. D., K. Naumann, and K. Mislow, J. Am. Chem. Soc., 91, 5646 (1969).
257. Herring, D. L., and C. M. Douglas, Inorg. Chem., 3, 428 (1964).
258. Hess, H., and D. Forst., Z. Anorg. Allgem. Chem.,

342, 240 (1966).
259. Higgins, W. A., G. R. Normann, and W. G. Craig, U.S. Pat. 2,779,787 (1957); C.A., 51, 8135a (1957).
260. Higgins, W. A., P. W. Vogel, and W. G. Craig, J. Am. Chem. Soc., 77, 1864 (1955).
261. Hoard, L. G., and R. A. Jacobson, J. Chem. Soc., A, 1966, 1203.
262. Hoffmann, F. W., T. C. Simmons, and L. J. Glunz, III, J. Am. Chem. Soc., 79, 3570 (1957).
262a. Hoffmann, H., and H. J. Dier, Tetrahedron Letters, 1962, 583.
263. Hoffmann, H., and H. Förster, Tetrahedron Letters, 1963, 1547.
264. Hoffmann, H., R. Grünewald, and L. Horner, Chem. Ber., 93, 861 (1960).
265. Hoffmann, H., L. Horner, H. G. Wippel, and D. Michael, Chem. Ber., 95, 523 (1962).
266. Holmes, R. R., J. Chem. Phys., 46, 3724 (1967).
267. Holmes, R. R., J. Chem. Phys., 46, 3730 (1967).
268. Holmes, R. R., and R. M. Deiters, Inorg. Chem., 7, 2229 (1968).
269. Holmes, R. R., and R. M. Dieters, J. Am. Chem. Soc., 90, 5021 (1968).
270. Holmes, R. R., R. M. Deiters, and J. A. Golen, Inorg. Chem., 8, 2612 (1969).
271. Holmes, R. R., and W. P. Gallagher, Inorg. Chem., 2, 433 (1963).
272. Holmes, R. R., W. P. Gallagher, and R. P. Carter, Inorg. Chem., 2, 437 (1963).
273. Holmes, R. R., and R. N. Storey, Inorg. Chem., 5, 2146 (1966).
274. Hormuth, P. B., and M. Becke-Goehring, Z. Anorg. Allgem. Chem., 372, 280 (1970).
275. Hormuth, P. B., and H. P. Latscha, Z. Anorg. Allgem. Chem., 365, 26 (1969).
276. Horner, L., P. Beck, and H. Hoffmann, Chem. Ber., 92, 2088 (1959).
277. Horner, L., P. Beck, and V. G. Toscano, Chem. Ber., 94, 2122 (1961).
278. Horner, L., H. Hoffmann, and P. Beck, Chem. Ber., 91, 1583 (1958).
279. Horner, L., and W. Jurgeleit, Ann., 591, 138 (1955).
280. Horner, L., and K. Klüpfel, Ann., 591, 69 (1955).
281. Horner, L., and A. Mentrup, Ann., 646, 65 (1961).
282. Horner, L., and H. Oedinger, Ann., 627, 142 (1959).
283. Horner, L., H. Oediger, and H. Hoffmann, Ann., 626, 26 (1959).
284. Horner, L., and H. Winkler, Tetrahedron Letters, 1964, 455.
285. Houalla, D., R. Wolf, D. Gagnaire, and J. B. Robert, Chem. Commun., 1969, 443.

285a. Houalla, D., R. Marty, and R. Wolf, Z. Naturforsch., 25b, 451 (1970).
286. Hudson, R. F., Structure and Mechanism in Organophosphorus Chemistry, Academic Press, New York, 1965; Angew. Chem., 79, 756 (1967); Intern. Ed., 6, 749 (1967).
287. Hudson, R. F., and G. Salvadori, Helv. Chim. Acta., 49, 96 (1966).
288. Hughes, A. N., and Chit Srivanavit, J. Heterocycl. Chem., 7, 1 (1970).
289. Hughes, A. N., and Siriporn Uaboonkul, Tetrahedron, 24, 3437 (1968).
290. Huisgen, R., and J. Wulff, Chem. Ber., 102, 746 (1969); Angew. Chem., 79, 472 (1967); Intern. Ed., 6, 457 (1967).
291. Huisgen, R., and J. Wulff, Chem. Ber., 102, 1833 (1969); Tetrahedron Letters, 1967, 917.
292. Humiec, F. S., and I. I. Bezman, J. Am. Chem. Soc., 83, 2210 (1961).
293. Hunger, K., U. Hasserodt, and F. Korte, Tetrahedron, 20, 1593 (1964).
294. Hunger, K., and F. Korte, Tetrahedron Letters, 1964, 2855.
295. Issleib, K., and A. Brack, Z. Anorg. Allgem. Chem., 277, 258 (1954).
296. Issleib, K., and W. Gründler, Theor. Chim. Acta (Berlin), 8, 70 (1967).
297. Issleib, K., and M. Hoffmann, Chem. Ber., 99, 1320 (1966).
298. Issleib, K., and B. Mitcherling, Z. Naturforsch., 15b, 267 (1960).
299. Issleib, K., and W. Seidel, Z. Anorg. Allgem. Chem., 288, 201 (1956).
300. Issleib, K., and W. Seidel, Z. Anorg. Allgem. Chem., 303, 155 (1960).
301. Issleib, K., and B. Walther, Chem. Ber., 97, 3424 (1964).
302. Issleib, K., and W. Wenschuh, Chem. Ber., 97, 715 (1964).
303. Ivanov, B. E., A. B. Ageeva, S. V. Pasmanyuk, R. R. Shagidullin, S. G. Salikhov, and E. I. Loginova, Izv. Akad. Nauk SSSR, 1969, 1757; C.A., 72, 3531w (1970).
303a. Ivanov, B. E., A. B. Ageeva, and Yu. Yu. Samitov, Dokl. Akad. Nauk SSSR, 174, 846 (1967); C.A., 68, 39736y (1968).
304. Ivanova, Zh. M., and A. V. Kirsanov, Zh. Obshch. Khim., 31, 3991 (1961); C.A., 57, 8605h (1962).
305. Ivanova, Zh. M., and A. V. Kirsanov, Zh. Obshch. Khim., 32, 2592 (1962); C.A., 58, 9122e (1963).
305a. Ivin, S. Z., and G. I. Drozd, U.S.S.R. Pat. 176,897 (1966); C.A., 64, 12723 (1966).

306. Ivin, S. Z., and K. V. Karavanov, Zh. Obshch. Khim., 28, 2958 (1958); C.A., 53, 9035be (1959).
307. Ivin, S. Z., K. V. Karavanov, and G. I. Drozd, U.S.S.R. Pat. 172,795 (1966); C.A., 64, 756d (1966).
308. Ivin, S. Z., K. V. Karavanov, and V. V. Lysenko, Zh. Obshch. Khim., 34, 852 (1964); C.A., 60, 15902g (1964).
309. Ivin, S. Z., K. V. Karavanov, and V. V. Lysenko, Khim., Org. Soedin. Fosfora, Akad. Nauk SSSR, Ser. Obshch. Tekh. i Khim., 1967, 152; C.A., 68, 114708w (1968).
310. Ivin, S. Z., K. V. Karavanov, and V. V. Lysenko, Zh. Obshch. Khim., 37, 492 (1967); C.A., 67, 43871k (1967).
311. Ivin, S. Z., K. V. Karavanov, V. V. Lysenko, and I. D. Shelakova, Zh. Obshch. Khim., 37, 1341 (1967); C.A., 67, 116920n (1967).
312. Ivin, S. Z., K. V. Karavanov, V. V. Lysenko, and T. N. Sosina, Zh. Obshch. Khim., 36, 1246 (1966); C.A., 65, 20158a (1966).
313. Jain, S. R., and H. H. Sisler, Inorg. Chem., 7, 2204 (1968).
314. Jander, J., D. Börner, and U. Engelhardt, Ann., 726, 19 (1969).
315. Johnson, A. W., and J. C. Tebby, J. Chem. Soc., 1961, 2126.
316. Jones, R. A., and A. R. Katritzky, Angew. Chem., 74, 60 (1962).
317. Kamai, Gil'm., and V. A. Kukhtin, Dokl. Akad. Nauk SSSR, 109, 91 (1956), and 112, 868 (1957); C.A., 51, 1827g, 13742f (1957).
318. Kamai, Gil'm., and V. A. Kukhtin, Zh. Obshch. Khim., 27, 2372 (1957); C.A., 52, 7127d (1958).
319. Kamai, Gil'm., and V. A. Kukhtin, Zh. Obshch. Khim., 27, 2376 (1957); C.A., 52, 7127h (1958).
320. Kamai, Gil'm., and V. A. Kukhtin, Zh. Obshch. Khim., 31, 1735 (1961); C.A., 55, 25732g (1961).
321. Kamai, Gil'm., V. A. Tsivunin, and S. Kh. Nurtdinov, Zh. Obshch. Khim., 35, 1817 (1965); C.A., 64, 3587e (1966).
322. Karavanov, K. V., L. E. Dimitrieva, and S. Z. Ivin, Zh. Obshch. Khim., 38, 1547 (1968); C.A., 69, 96823v (1968).
323. Karavanov, K. V., and S. Z. Ivin, Zh. Obshch. Khim., 35, 78 (1965); C.A., 62, 13175f (1965).
324. Karavanov, K. V., S. Z. Ivin, and V. G. Gruzdev, Khim. Org. Soedin. Fosfora Akad. Nauk SSSR, Otd. Obshch. i Tekh. Khim., 1967, 157; C.A., 68, 114700n (1968).
325. Karavanov, K. V., S. Z. Ivin, and V. V. Lysenko, Zh. Obshch. Khim., 35, 737 (1965); C.A., 63, 4327e (1965).
325a. Katz, T. J., and E. W. Turnblom, J. Am. Chem. Soc.,

92, 6701 (1970).

326. Kemmit, R. D. W., K. C. Moss, D. R. Russell, and D. W. A. Sharp, Second International Symposium on Fluorine Chemistry, Estes Park, Colo., July 17-20, 1962, Abstract, p. 264.

327. Khairullin, V. K., A. N. Pudovik, and N. I. Kharitonova, Zh. Obshch. Khim., 39, 608 (1969); C.A., 71, 39082u (1969).

328. Khaskin, A. N., P. M. Zavlin, and B. I. Ionin, Zh. Obshch. Khim., 39, 191 (1969); C.A., 70, 105878r (1969).

328a. Kimball, G. E., J. Chem. Phys., 8, 188 (1940).

329. Kinnear, A. M., and E. A. Perren, J. Chem. Soc., 1952, 3437.

330. Kirby, A. J., Tetrahedron, 22, 3001 (1966).

331. Kirillova, K. M., and V. A. Kukhtin, Zh. Obshch. Khim., 32, 2338 (1962); C.A., 58, 9128c (1963).

332. Kirillova, K. M., and V. A. Kukhtin, Zh. Obshch. Khim., 35, 544 (1965); C.A., 63, 523c (1965).

333. Kirsanov, O. V., M. D. Bodnarchuk, and V. I. Shevchenko, Dopovidi Akad. Nauk, Ukr. RSR, 1963 (2), 221; C.A., 59, 12666d (1963).

334. Knunyants, I. L., V. V. Tyuleneva, E. Ya. Pervova, and R. N. Sterlin, Izv. Akad. Nauk SSSR, Ser. Khim., 1964, 1797; C.A., 62, 2791d (1965).

335. Kobayashi, Y., and S. Akashi, Chem. Pharm. Bull. (Tokyo), 16, 1009 (1968); C.A., 69, 86240n (1968).

336. Kobayashi, Y., and S. Akashi, Japan Pat. 6,800,686 (1968); C.A., 69, 35392f (1968).

337. Kobayashi, Y., C. A. Akashi, and K. Morinaga, Chem. Pharm. Bull. (Tokyo), 16, 1784 (1968); C.A., 70, 3168s (1969).

338. Kochetkov, N. V., L. I. Kudryashov, and A. I. Usov, Dokl. Akad. Nauk SSSR, 133, 1094 (1960); C.A., 55, 27066i (1961).

339. Kolditz, L., and K. Lehmann, Z. Chem., 7, 356 (1967).

340. Kolditz, L., K. Lehmann, W. Wieker, and A. R. Grimmer, Z. Anorg. Allgem. Chem., 360, 259 (1968); Angew. Chem., 78, 451 (1966).

341. Kolditz, L., and W. Rehak, Z. Anorg. Allgem. Chem., 342, 32 (1966).

342. Komkov, I. P., S. Z. Ivin, and K. V. Karavanov, Zh. Obshch. Khim., 28, 2960 (1958); C.A., 53, 9035e (1959).

343. Komkov, I. P., S. Z. Ivin, K. V. Karavanov, and L. E. Smirnov, Zh. Obshch. Khim., 32, 301 (1962); C.A., 57, 16649a (1962).

344. Komkov, L. P., K. V. Karavanov, and S. Z. Ivin, Zh. Obshch. Khim., 28, 2963 (1958); C.A., 53, 9035h (1959).

345. Komkov, I. P., K. V. Karavanov, and L. E. Smirnov, Metody Polucheniya Khim. Reaktivov i Preparotov,

No. 12, 23 (1965); C.A., 65, 15416d (1966).
346. Kosolapoff, G. M., Organophosphorus Compounds, Wiley, New York, 1950, Chapter 4.
347. Kosolapoff, G. M., Organophosphorus Compounds, Wiley, 1950, Chapter 11.
348. Kosolapoff, G. M., and J. F. McCullough, J. Am. Chem. Soc., 73, 855 (1951).
349. Kolotilo, M. V., and G. I. Derkach, Zh. Obshch. Khim., 39, 463 (1969); C.A., 70, 115261y (1969).
350. Kozlov, E. S., and B. S. Drach, Zh. Obshch. Khim., 36, 760 (1966); C.A., 65, 8742g (1966).
351. Kozlov, E. S., and S. N. Gaidamaka, Zh. Obshch. Khim., 39, 933 (1969); C.A., 71, 50085t (1969).
352. Kozlov, E. S., S. N. Gaidamaka, and A. V. Kirsanov, Zh. Obshch. Khim., 39, 1648 (1969); C.A., 71, 91593a (1969).
353. Kuchen, W., H. Buchwald, K. Strolenberg, and J. Metten, Ann., 652, 28 (1962).
354. Kuchen, W., and K. Strolenberg, Angew. Chem., 74, 27 (1962).
355. Kukhtin, V. A., Dokl. Akad. Nauk SSSR, 121, 466 (1958); C.A., 53, 1105a (1959).
356. Kukhtin, V. A., and I. P. Gozman, Dokl. Akad. Nauk SSSR, 158, 157 (1964); C.A., 61, 14514b (1964).
357. Kukhtin, V. A., and Gil'm. Kamai, Zh. Obshch. Khim., 28, 1196 (1958); C.A., 52, 19909d (1958).
358. Kukhtin, V. A., Gil'm. Kamai, and L. A. Sinchenko, Dokl. Akad. Nauk SSSR, 118, 505 (1958); Engl. Transl., 1958, 73; C.A., 52, 10956i (1958).
359. Kukhtin, V. A., and L. A. Khismatullina, Zh. Obshch. Khim., 29, 3276 (1959); C.A., 54, 14103h (1960).
360. Kukhtin, V. A., and K. M. Kirillova, Dokl. Akad. Nauk SSSR, Ser. Khim., 140, 835 (1961); C.A., 56, 4607g (1962).
361. Kukhtin, V. A., and K. M. Kirillova, Zh. Obshch. Khim., 31, 2226 (1961); C.A., 56, 3507e (1962).
362. Kukhtin, V. A., and K. M. Kirillova, Zh. Obshch. Khim., 32, 2797 (1962); C.A., 58, 8489d (1963).
363. Kukhtin, V. A., K. M. Kirillova, and R. R. Shagidullin, Zh. Obshch. Khim., 32, 649 (1962); C.A., 58, 12404b (1963).
364. Kukhtin, V. A., K. M. Kirillova, R. R. Shagidullin, Yu. Yu. Samitov, N. A. Lyazina, and N. F. Rakova, Zh. Obshch. Khim., 32, 2039 (1962); C.A., 58, 4543a (1963).
365. Kukhtin, V. A., and K. M. Orekhova, Zh. Obshch. Khim., 28, 2790 (1958); C.A., 53, 7972a (1959).
366. Kukhtin, V. A., and K. M. Orekhova, Zh. Obshch. Khim., 30, 1208 (1960); C.A., 55, 358h (1961).
367. Kukhtin, V. A., and K. M. Orekhova, Zh. Obshch. Khim., 30, 1526 (1960); C.A., 55, 1567b (1961).

368. Kukhtin, V. A., T. N. Voskobeva, and K. M. Kirillova, Zh. Obshch. Khim., 32, 2333 (1962); C.A., 58, 9127g (1963).
369. Kulakova, V. N., Yu. M. Zinov'ev, and L. Z. Soborovskii, Zh. Obshch. Khim., 29, 3957 (1959); C.A., 54, 20846e (1960).
370. Laib, H., and H. Burger, Ger. Pat. 1,244,396 (1967); C.A., 67, 82800f (1967).
371. Landau, M. A., V. V. Sheluchenko, G. I. Drozd, S. S. Dubov, and S. Z. Ivin, Zh. Strukt. Khim., 1967, 1097; C.A., 69, 35214z (1968).
372. Larionova, M. A., A. L. Klebanskii, and V. A. Bartashev, Zh. Obshch. Khim., 33, 265 (1963); C.A., 59, 656h (1963).
373. Latscha, H. P., Z. Anorg. Allgem. Chem., 346, 166 (1966).
374. Latscha, H. P., Z. Anorg. Allgem. Chem., 355, 73 (1967).
375. Latscha, H. P., Z. Naturforsch., 23b, 139 (1968).
376. Latscha, H. P., and P. B. Hormuth, Z. Anorg. Allgem. Chem., 359, 78 (1968).
377. Latscha, H. P., and P. B. Hormuth, Z. Anorg. Allgem. Chem., 359, 81 (1968).
378. Latscha, H. P., and P. B. Hormuth, Angew. Chem., 80, 281 (1968); Intern. Ed., 7, 299 (1968).
379. Latscha, H. P., P. B. Hormuth, and H. Vollmer, Z. Naturforsch., 24b, 1237 (1969).
380. Latscha, H. P., and W. Klein, Angew. Chem., 81, 291 (1969); Intern. Ed., 8, 278 (1969).
381. Lauterbur, P. C., and F. Ramirez, J. Am. Chem. Soc., 90, 6722 (1968).
381a. Lecher, H. Z., and R. A. Greenwood, U.S. Pat. 2,870,204 (1959); C.A., 53, 11306b (1959).
382. Letcher, J. H., and J. R. Van Wazer, J. Chem. Phys., 45, 2926 (1966).
383. Letcher, J. H., and J. R. Van Wazer, Topics in Phosphorus Chemistry, Vol. 5, Interscience, New York, 1967, p. 75.
384. Levy, D., and R. Stevenson, Tetrahedron Letters, 1965, 341.
385. Levy, D., and R. Stevenson, J. Org. Chem., 30, 3469 (1965).
386. Levy, D., and R. Stevenson, J. Org. Chem., 32, 1265 (1967).
387. Lincoln, B. H., and G. D. Byrkit, U.S. Pat. 2,460,301 (1949); C.A., 43, 2765b (1949).
388. Lindner, E., R. Lehner, and H. Scheer, Chem. Ber., 100, 1331 (1967).
389. Lindner, W., Dissertation, Heidelberg, 1970.
389a. Linnett, J. W., and C. E. Mellish, Trans. Faraday Soc., 50, 655 (1954).

390. Lutsenko, I. F., and M. Kirilov, Dokl. Akad. Nauk SSSR, 128, 89 (1959); C.A. 54, 1288i (1960).
391. Lutsenko, I. F., and M. Kirilov, Dokl. Akad. Nauk SSSR, 132, 842 (1960); C.A., 54, 20842a (1960).
392. Lutsenko, I. F., and M. Kirilov, Zh. Obshch. Khim., 31, 3594 (1961); C.A., 57, 8606c (1962).
393. Lutsenko, I. F., M. Kirilov, and G. A. Ovchinnikova, Zh. Obshch. Khim., 31, 2028 (1961); C.A., 55, 27021f (1961).
394. Lutsenko, I. F., M. Kirilov, and G. B. Postnikova, Zh. Obshch. Khim., 31, 2034 (1961); C.A., 55, 27021i (1961).
395. Lysenko, V. V., I. D. Shelakova, K. V. Karavanov, and S. I. Ivin, Zh. Obshch. Khim., 36, 1507 (1966); C.A., 66, 11005m (1967).
396. Mahler, W., Second International Symposium on Fluorine Chemistry, Estes Park, Colo., July 17-20, 1962, Abstract, p. 441.
397. Mahler, W., J. Am. Chem. Soc., 84, 4600 (1962).
398. Mahler, W., Inorg. Chem., 2, 230 (1963).
399. Mahler, W., and A. B. Burg, J. Am. Chem. Soc., 80, 6161 (1958).
400. Mahler, W., and E. L. Muetterties, J. Chem. Phys., 33, 636 (1960).
401. Maier, L., Helv. Chim. Acta, 46, 2026 (1963).
402. Maier, L., Helv. Chim. Acta, 46, 2667 (1963).
403. Malatesta, P., and B. D'Atri, Farmaco (Pavia), Ed. Sci., 8, 398 (1953); C.A., 48, 9312f (1954).
404. Mark, V., J. Am. Chem. Soc., 85, 1884 (1963).
405. Mark, V., C. H. Dungan, M. M. Crutchfield, and J. R. Van Wazer, Topics in Phosphorus Chemistry, Vol. 5, Interscience, New York, 1967, p. 227.
405a. Marsmann, J., R. Van Vazer, and J. B. Robert, J. Chem. Soc., A, 1970, 1566.
406. Masaki, M., K. Fukui, and M. Ohta, J. Org. Chem., 32, 3564 (1967).
407. Mathey, F., Compt. Rend., 269, 1066 (1969).
408. Mathey, F., and G. Mavel, Compt. Rend., 268, 1902 (1969).
409. Mathis, R., R. Burgada, and M. Sanchez, Spectrochim. Acta, 25A, 1201 (1969).
410. McCormack, W. B., U.S. Pats. 2,663,737 (1953) and 2,663,736 (1953); C.A., 49, 7601a (1955).
411. McCormack, W. B., U.S. Pat. 2,663,738 (1953); C.A., 49, 7602d (1955).
412. Middleton, W. J., E. G. Howard, and W. H. Sharkey, J. Org. Chem., 30, 1375 (1965).
413. Middleton, W. J., and W. H. Sharkay, J. Org. Chem., 30, 1384 (1965).
414. Miller, B., Topics in Phosphorus Chemistry, Vol. 2, Interscience, New York, 1965, p. 133.

414a. Miller, J. M., J. Chem. Soc., A, 1967, 828.
415. Mitsch, R. A., J. Am. Chem. Soc., 89, 6297 (1967).
416. M. & T. Chemicals, Inc., Neth. Appl. 6,501,268 (1965); corresp. Brit. Pat. 1,029,924 (1966); C.A., 64, 2127g (1966).
417. Mizrakh, L. I., L. Yu. Sandalova, and V. P. Evdakov, Dokl. Akad. Nauk SSSR, 171, 1116 (1966); C.A., 66, 85732a (1967).
418. Moschel, W., H. Jonas, W. Thraum, and H. Knopf, Ger. Pat. 1,032,746 (1958); C.A., 54, 22497g (1960).
419. Muetterties, E. L., J. Org. Chem., 6, 635 (1967).
420. Muetterties, E. L., J. Am. Chem. Soc., 90, 5097 (1968).
421. Muetterties, E. L., J. Am. Chem. Soc., 91, 1636 (1969).
422. Muetterties, E. L., J. Am. Chem. Soc., 91, 4115 (1969).
423. Muetterties, E. L., T. A. Bither, M. W. Farlow, and D. D. Coffman, J. Inorg. Nucl. Chem., 16, 52 (1960).
424. Muetterties, E. L., and W. Mahler, Inorg. Chem., 4, 119 (1965).
425. Muetterties, E. L., W. Mahler, K. J. Packer, and R. Schmutzler, Inorg. Chem., 3, 1298 (1964).
426. Muetterties, E. L., W. Mahler, and R. Schmutzler, Inorg. Chem., 2, 613 (1963).
427. Muetterties, E. L., and R. A. Schunn, Quart. Rev., 1966, 245.
428. Muetterties, E. L., and C. M. Wright, J. Am. Chem. Soc., 86, 5132 (1964).
429. Mukaiyama, T., and T. Kumamoto, Bull. Chem. Soc. Japan, 39, 879 (1966).
430. Mukaiyama, T., J. Kuwajima, and K. Ohno, Bull. Chem. Soc. Japan, 38, 1954 (1965).
431. Mukaiyama, T., H. Nambu, and T. Kumamoto, J. Org. Chem., 29, 2243 (1964).
432. Mukaiyama, T., H. Nambu, and M. Okamoto, J. Org. Chem., 27, 3651 (1962).
433. Muroi, M., Y. Inouye, and M. Ohno, Bull Chem. Soc. Japan, 42, 2948 (1969).
434. Murray, M., and R. Schmutzler, Z. Chem., 8, 241 (1968).
435. Murray, M., and R. Schmutzler, Chem. Ind. (London), 1968, 1730.
436. Murray, R. W., and M. L. Kaplan, J. Am. Chem. Soc., 90, 537 (1968).
437. Murray, R. W., and M. L. Kaplan, J. Am. Chem. Soc., 91, 5358 (1969), and 90, 4161 (1968).
438. Musher, J. I., Angew. Chem., 81, 68 (1969); Intern. Ed., 8, 54 (1969).
439. Mustafa, A., M. M. Sidky, and F. M. Soliman, Tetrahedron, 22, 393 (1966).

440. Nast, R., and K. Käb, Ann., 706, 75 (1967).
441. Nesterov, L. V., R. A. Sabirova, N. E. Krepysheva, and R. I. Mutalapova, Dokl. Akad. Nauk SSSR, 148, 1085 (1963); C.A., 59, 3757a (1963).
442. Nifant'ev, E. E., and L. M. Matveeva, Zh. Obshch. Khim., 39, 1555 (1969); C.A., 71, 101220t (1969).
443. Nixon, J. F., J. Inorg. Nucl. Chem., 27, 1281 (1965); Chem. Ind. (London), 1963, 1555.
444. Nixon, J. F., J. Inorg. Nucl. Chem., 31, 1615 (1969).
445. Nixon, J. F., and R. Schmutzler, Spectrochim. Acta, 20, 1835 (1964).
446. Nixon, J. F., and R. Schmutzler, Spectrochim. Acta, 22, 565 (1966).
447. Nixon, J. F., and J. R. Swain, Chem. Commun., 1968, 997.
448. Nixon, J. F., and J. R. Swain, Inorg. Nucl. Chem. Letters, 5, 295 (1969).
449. Norris, W. P., and H. B. Jonassen, J. Org. Chem., 27, 1449 (1962).
450. Nöth, H., and H. J. Vetter, Chem. Ber., 94, 1505 (1961).
451. Nöth, H., and H. J. Vetter, Chem. Ber., 96, 1109 (1963).
452. Nöth, H., and H. J. Vetter, Chem. Ber., 96, 1816 (1963).
453. Packer, K. J., J. Chem. Soc., 1963, 960.
454. Peake, S. C., and R. Schmutzler, Chem. Commun., 1968, 665.
455. Peake, S. C., and R. Schmutzler, Chem. Ind., 1968, 1482.
456. Peake, S. C., and R. Schmutzler, Chem. Commun., 1968, 1662.
456a. Peake, S. C., and R. Schmutzler, J. Chem. Soc., A, 1970, 1049.
457. Pearson, R. G., J. Am. Chem. Soc., 91, 4947 (1969).
458. Perry, B. J., J. B. Reesor, and J. L. Ferron, Can. J. Chem., 41, 2299 (1963); Nature, 188, 227 (1960).
459. Petragnani, N., and M. De Moura Campos, Tetrahedron, 21, 13 (1965).
460. Petrov, K. A., G. A. Sokolsky, and B. M. Polees, Zh. Obshch. Khim., 26, 3381 (1956); C.A., 51, 9473i (1957).
461. Peiffer, G., E. Gaydou, and A. Guillemonat, Compt. Rend., 268, 529 (1969).
462. Pierce, S. B., and C. D. Cornwell, J. Chem. Phys., 48, 2118 (1968).
463. Pinchuk, A. M., I. M. Kosinskaya, and V. I. Shevchenko, Zh. Obshch. Khim., 39, 583 (1969); C.A., 71, 61300k (1969).
464. Pinkus, A. G., S. Y. Ma, and H. C. Custard, Jr., J. Am. Chem. Soc., 83, 3917 (1961).

465. Pinkus, A. G., and L. Y. C. Meng, J. Org. Chem., 31, 1038 (1966).
466. Pinkus, A. G., and P. G. Waldrep, J. Org. Chem., 24, 1012 (1959).
467. Pinkus, A. G., and P. G. Waldrep, Chem. Ind., 1962, 302.
468. Pinkus, A. G., and P. G. Waldrep, J. Org. Chem., 31, 575 (1966).
469. Pinkus, A. G., P. G. Waldrep, and S. Y. Ma, J. Heterocycl. Chem., 2, 357 (1965).
470. Ponomarenko, F. I., S. Z. Ivin, and K. V. Karavanov, Zh. Obshch. Khim., 39, 382 (1969); C.A., 71, 39087z (1969).
471. Ponomarev, V. V., S. A. Golubtsov, K. A. Andrianov, and G. N. Kondrashova, Izv. Akad. Nauk SSSR, Ser. Khim., 1969, 1743; C.A., 72, 3536b (1970).
472. Poshkus, A. C., J. E. Herweh, and L. F. Hass, J. Am. Chem. Soc., 80, 5022 (1958).
473. Poshkus, A. C., and J. E. Herweh, J. Org. Chem., 29, 2567 (1964).
474. Pudovik, A. N., I. V. Gur'yanova, and S. P. Perevezentseva, Zh. Obshch. Khim., 39, 1532 (1969); C.A., 71, 112336m (1969).
475. Pudovik, A. N., I. V. Gur'yanova, S. P. Perevezentseva, and S. A. Terent'eva, Zh. Obshch. Khim., 39, 337 (1969); C.A., 70, 115071m (1969).
476. Pudovik, A. N., I. V. Gur'yanova, S. P. Perevezentseva, and T. V. Zykova, Zh. Obshch. Khim., 37, 1317 (1967); C.A., 68, 22013n (1968).
477. Pudovik, A. N., and J. V. Konovalova, Zh. Obshch. Khim., 35, 1591 (1965); C.A., 63, 17887b (1965).
478. Quin, L. D., 1,4-Cycloaddition Reactions, Organic Chemistry, Vol. 8, Academic Press, New York, 1967, p. 47.
479. Quin, L. D., J. P. Gratz, and T. P. Barket, J. Org. Chem., 33, 1034 (1968).
480. Quin, L. D., and D. A. Mathewes, J. Org. Chem., 29, 836 (1964).
481. Quin, L. D., and C. H. Rolston, J. Org. Chem., 23, 1693 (1958).
482. Rabinowitz, R., and R. Marcus, J. Am. Chem. Soc., 84, 1312 (1962).
483. Rakshys, J. W., R. W. Taft, and W. A. Sheppard, J. Am. Chem. Soc., 90, 5236 (1968).
484. Ramirez, F., Pure Appl. Chem., 9, 337 (1964).
485. Ramirez, F., Bull. Soc. Chim. France, 1966, 2443.
486. Ramirez, F., Accounts Chem. Res., 1, 168 (1968).
487. Ramirez, F., Trans. N.Y. Acad. Sci., 30, 410 (1968).
488. Ramirez, F., Colloques Intern. du C.N.R.S., Bull. Soc. Chim. France, 1970, 3497, 182, 67 (1970).
489. Ramirez, F., S. B. Bhatia, A. J. Bigler, and C. P. Smith, J. Org. Chem., 33, 1192 (1968).

490. Ramirez, F., S. B. Bhatia, R. B. Mitra, Z. Hamlet, and N. B. Desai, J. Am. Chem. Soc., 86, 4394 (1964).
491. Ramirez, F., S. B. Bhatia, A. V. Patwardhan, E. H. Chen, and C. P. Smith, J. Org. Chem., 33, 20 (1968).
492. Ramirez, F., S. B. Bhatia, A. V. Patwardhan, and C. P. Smith, J. Org. Chem., 32, 2194 (1967).
493. Ramirez, F., S. B. Bhatia, A. V. Patwardhan, and C. P. Smith, J. Org. Chem., 32, 3547 (1967).
494. Ramirez, F., S. B. Bhatia, and C. P. Smith, J. Org. Chem., 31, 4105 (1966).
495. Ramirez, F., S. B. Bhatia, and C. P. Smith, Tetrahedron, 23, 2067 (1967).
496. Ramirez, F., S. B. Bhatia, and C. P. Smith, J. Am. Chem. Soc., 89, 3026 (1967).
497. Ramirez, F., S. B. Bhatia, and C. P. Smith, J. Am. Chem. Soc., 89, 3030 (1967).
498. Ramirez, F., S. B. Bhatia, C. D. Telefus, and C. P. Smith, Tetrahedron, 25, 771 (1969).
499. Ramirez, F., J. A. Bigler, and C. P. Smith, J. Am. Chem. Soc., 90, 3507 (1968).
500. Ramirez, F., A. J. Bigler, and C. P. Smith, Tetrahedron, 24, 5041 (1968).
501. Ramirez, F., and N. B. Desai, J. Am. Chem. Soc., 82, 2652 (1960).
502. Ramirez, F., and N. B. Desai, J. Am. Chem. Soc., 85, 3252 (1963).
503. Ramirez, F., N. B. Desai, and N. McKelvie, J. Am. Chem. Soc., 84, 1745 (1962).
504. Ramirez, F., N. B. Desai, and R. B. Mitra, J. Am. Chem. Soc., 83, 492 (1961).
505. Ramirez, F., N. B. Desai, and N. Ramanathan, Tetrahedron Letters, 1963, 323.
506. Ramirez, F., N. B. Desai, and N. Ramanathan, J. Am. Chem. Soc., 85, 1874 (1963).
507. Ramirez, F., S. L. Glaser, A. J. Bigler, and J. F. Pilot, J. Am. Chem. Soc., 91, 496 (1969).
508. Ramirez, F., A. S. Gulati, and C. P. Smith, J. Am. Chem. Soc., 89, 6283 (1967).
509. Ramirez, F., A. S. Gulati, and C. P. Smith, J. Org. Chem., 33, 13 (1968).
510. Ramirez, F., H. J. Kugler, A. V. Patwardhan, and C. P. Smith, J. Org. Chem., 33, 1185 (1968).
511. Ramirez, F., H. J. Kugler, and C. P. Smith, Tetrahedron, 24, 1931 (1968); Tetrahedron Letters, 1965, 261.
512. Ramirez, F., H. J. Kugler, and C. P. Smith, Tetrahedron, 24, 3153 (1968).
513. Ramirez, F., and G. V. Loewengart, J. Am. Chem. Soc., 91, 2293 (1969).
514. Ramirez, F., O. P. Madan, and S. R. Heller, J. Am. Chem. Soc., 87, 731 (1965).

515. Ramirez, F., O. P. Madan, and C. P. Smith, J. Am. Chem. Soc., 87, 670 (1965).
516. Ramirez, F., R. B. Mitra, and N. B. Desai, J. Am. Chem. Soc., 82, 2651 (1960).
517. Ramirez, F., R. B. Mitra, and N. B. Desai, J. Am. Chem. Soc., 82, 5763 (1960).
518. Ramirez, F., M. Nagabhushanam, and C. P. Smith, Tetrahedron, 24, 1785 (1968).
519. Ramirez, F., A. V. Patwardhan, N. B. Desai, and S. R. Heller, J. Am. Chem. Soc., 87, 549 (1965).
520. Ramirez, F., A. V. Patwardhan, N. B. Desai, N. Ramanathan, and C. V. Greco, J. Am. Chem. Soc., 85, 3056 (1963).
521. Ramirez, F., A. V. Patwardhan, and S. R. Heller, J. Am. Chem. Soc., 86, 514 (1964).
522. Ramirez, F., A. V. Patwardhan, H. J. Kugler, and C. P. Smith, J. Am. Chem. Soc., 89, 6276 (1967), and 87, 4973 (1965).
523. Ramirez, F., A. V. Patwardhan, H. J. Kugler, and C. P. Smith, Tetrahedron, 24, 2275 (1968); Tetrahedron Letters, 1966, 3053.
524. Ramirez, F., A. V. Patwardhan, N. Ramanathan, N. B. Desai, C. V. Greco, and S. R. Heller, J. Am. Chem. Soc., 87, 543 (1965).
525. Ramirez, F., A. V. Patwardhan, and C. P. Smith, J. Org. Chem., 30, 2575 (1965).
526. Ramirez, F., A. V. Patwardhan, and C. P. Smith, J. Org. Chem., 31, 474 (1966).
527. Ramirez, F., A. V. Patwardhan, and C. P. Smith, J. Org. Chem., 31, 3159 (1966).
528. Ramirez, F., J. F. Pilot, O. P. Madan, and C. P. Smith, J. Am. Chem. Soc., 90, 1275 (1968).
529. Ramirez, F., J. F. Pilot, and C. P. Smith, Tetrahedron, 24, 3735 (1968).
530. Ramirez, F., J. F. Pilot, C. P. Smith, S. B. Bhatia, and A. S. Gulati, J. Org. Chem., 34, 3385 (1969).
531. Ramirez, F., and N. Ramanathan, J. Org. Chem., 26, 3041 (1961).
532. Ramirez, F., N. Ramanathan, and N. B. Desai, J. Am. Chem. Soc., 85, 3465 (1963), and 84, 1317 (1962).
533. Ramirez, F., and C. P. Smith, Chem. Commun., 1967, 662.
534. Ramirez, F., C. P. Smith, A. S. Gulati, and A. V. Patwardhan, Tetrahedron Letters, 1966, 2151.
535. Ramirez, F., C. P. Smith, and S. Meyerson, Tetrahedron Letters, 1966, 3651.
536. Ramirez, F., C. P. Smith, and J. F. Pilot, J. Am. Chem. Soc., 90, 6726 (1968).
537. Ramirez, F., C. P. Smith, J. F. Pilot, and A. S. Gulati, J. Org. Chem., 33, 3787 (1968).
538. Ramirez, F., K. Tasaka, N. B. Desai, and C. P. Smith, J. Org. Chem., 33, 25 (1968).

539. Ramirez, F., K. Tasaka, N. B. Desai, and C. P. Smith, J. Am. Chem. Soc., 90, 751 (1968).
540. Ramirez, F., and C. D. Telefus, J. Org. Chem., 34, 376 (1969).
540a. Ramirez, F., and I. Ugi, Adv. Phys. Org. Chem., 9, 26 (1971); edited by V. Gold, Academic Press.
541. Razumova, N. A., F. V. Bagrov, and A. A. Petrov, Zh. Obshch. Khim., 39, 2368 (1969); C.A., 72, 43795s (1970).
542. Razumova, N. A., F. V. Bagrov, and A. A. Petrov, Zh. Obshch. Khim., 39, 2369 (1969); C.A., 72, 31920r (1970).
543. Razumova, N. A., Zh. L. Evtikhov, A. Kh. Voznesenskaya, and A. A. Petrov, Zh. Obshch. Khim., 39, 176 (1969); C.A., 70, 106608q (1969).
544. Razumova, N. A., Zh. L. Evtikhov, and A. A. Petrov, Zh. Obshch. Khim., 39, 1419 (1969); C.A., 71, 80443a (1969).
545. Razumova, N., and A. A. Petrov, Zh. Obshch. Khim., 33, 3858 (1963); C.A., 60, 10711h (1964).
546. Razuvaev, G. A., and N. A. Osanova, Dokl. Akad. Nauk SSSR, 104, 552 (1955); C.A., 50, 11268c (1956).
547. Razuvaev, G. A., and N. A. Osanova, Zh. Obshch. Khim., 26, 2531 (1956); C.A., 51, 1875b (1957).
548. Razuvaev, G. A., N. A. Osanova, and I. K. Grigor'eva, Izv. Akad. Nauk SSSR, Ser. Khim., 1969, 2234; C.A., 72, 31952c (1970).
549. Razuvaev, G. A., N. A. Osanova, and I. A. Shlyapnikova, Zh. Obshch. Khim., 27, 1466 (1957); C.A., 52, 3715b (1958).
550. Razuvaev, G. A., and G. G. Petukhov, Tr. po Khim. i Khim. Tekhnol., 4 (1), 150 (1961); C.A., 55, 27169f (1961).
551. Razuvaev, G. A., G. G. Petukhov, and N. A. Osanova, Dokl. Akad. Nauk SSSR, 104, 733 (1955); C.A., 50, 11268f (1956).
552. Razuvaev, G. A., G. G. Petukhov, and N. A. Osanova, Zh. Obshch. Khim., 29, 2980 (1959); C.A., 54, 12030e (1960).
553. Razuvaev, G. A., G. G. Petukhov, and N. A. Osanova, Zh. Obshch. Khim., 31, 2350 (1961); C.A., 56, 3506i (1962).
554. Razuvaev, G. A., G. G. Petukhov, M. A. Shubenko, and V. A. Voitovich, Ukr. Khim. Zh., 22, 45 (1956); C.A., 50, 13783c (1956).
555. Reddy, G. S., and R. Schmutzler, Inorg. Chem., 5, 164 (1966).
556. Reddy, G. S., and C. D. Weis, J. Org. Chem., 28, 1822 (1963).
557. Reesor, J. B., B. J. Perry, and E. Sherlock, Can. J. Chem., 38, 1416 (1960).
558. Reetz, T., and J. F. Powers, U.S. Pat. 3,172,903;

C.A., 63, 2981f (1965).
559. Reinhardt, H., D. Bianchi, and D. Mölle, Chem. Ber., 90, 1656 (1957).
560. Rokhlin, E. M., Yu. V. Zeifman, Y. A. Cheburkov, N. D. Gambaryan, and I. L. Knunyants, Dokl. Akad. Nauk SSSR, 161, 1356 (1965); C.A., 63, 4153g (1965).
561. Rozinov, V. G., and E. F. Grechkin, Zh. Obshch. Khim., 39, 934 (1969); C.A., 71, 50098z (1969).
562. Rozinov, V. G., E. F. Grechkin, and A. V. Kalabina, Zh. Obshch. Khim., 39, 712 (1969); C.A., 71, 50122c (1969).
563. Rozinov, V. G., A. L. Taskina, and E. F. Grechkin, Zh. Obshch. Khim., 39, 1647 (1969); C.A., 71, 91585z (1969).
563a. Rundle, R. E., Survey Progr. Chem., 1, 81 (1963).
564. Rydon, H. N., and B. L. Tonge, J. Chem. Soc., 1956, 3043.
564a. Rydon, H. N., Chem. Soc. (London) Spec. Publ., 8, 61 (1957).
565. Sanchez, M., L. Beslier, J. Roussel, and R. Wolf, Bull. Soc. Chim. France, 1969, 3053.
566. Sanchez, M., L. Beslier, and R. Wolf, Bull. Soc. Chim. France, 1969, 2778.
567. Sanchez, M., J. F. Brazier, D. Houalla, and R. Wolf, Bull. Soc. Chim. France, 1967, 3930.
568. Sanchez, M., R. Wolf, R. Burgada, and F. Mathis, Bull. Soc. Chim. France, 1968, 773.
569. Sandalova, L. Yu., L. I. Mizrakh, and V. P. Evdakov, Zh. Obshch. Khim., 36, 1451 (1966); C.A., 66, 10627r (1967).
570. Schliebs, R., Ger. Pat. 1,165,596 (1964); C.A., 60, 15913e (1964).
571. Schlosser, M., and T. Kadibelban, personal communication.
572. Schlosser, M., T. Kadibelban, and G. Steinhoff, Angew. Chem., 78, 1018 (1966); Intern. Ed., 5, 968 (1966), Am. 743, 25 (1971).
573. Schmidpeter, A., C. Weigand, and E. Hafner-Roll, Z. Naturforsch., 24b, 799 (1969).
574. Schmidt, U., J. Boie, C. Osterroth, R. Schröer, and H. F. Grützmacher, Chem. Ber., 101, 1381 (1968); Schmidt, U., and C. Osterroth, Angew. Chem., 77, 455 (1965); Intern. Ed., 4, 437 (1965).
574a. Schmutzler, R., Chem. Ind., 1962, 1868.
575. Schmutzler, R., J. Inorg. Nucl. Chem., 25, 335 (1963).
576. Schmutzler, R., Inorg. Chem., 3, 410 (1964).
577. Schmutzler, R., Inorg. Chem., 3, 421 (1964).
578. Schmutzler, R., Angew. Chem., 76, 893 (1964); Intern. Ed., 3, 753 (1964).
579. Schmutzler, R., J. Am. Chem. Soc., 86, 4501 (1964).
580. Schmutzler, R., J. Chem. Soc., 1964, 4551.

581. Schmutzler, R., Z. Naturforsch., 19b, 1101 (1964).
582. Schmutzler, R., Chem. Commun., 1965, 19.
583. Schmutzler, R., Advan. Fluorine Chem., 5, 31 (1965).
584. Schmutzler, R., Angew. Chem., 77, 530 (1965); Intern. Ed., 4, 496 (1965).
585. Schmutzler, R., Chem. Ber., 98, 552 (1965).
586. Schmutzler, R., J. Chem. Soc., 1965, 5630.
587. Schmutzler, R., U.S. Pat. 3,246,032 (1966); C.A., 64, 19678f (1966).
588. Schmutzler, R., U.S. Pat. 3,287,406 (1966); C.A., 66, 85857v (1966).
589. Schmutzler, R., U.S. Pat, 3,300,503 (1967); C.A., 66, 65631p (1967).
590. Schmutzler, R., Halogen Chemistry, Vol. 2, Academic Press, New York, 1967, p. 31.
591. Schmutzler, R., Inorg. Syn., 9, 63 (1967).
592. Schmutzler, R., Inorg. Chem., 7, 1327 (1968).
592a. Schmutzler, R., Z. Chem., 8, 241 (1968).
593. Schmutzler, R., and G. S. Reddy, Inorg. Chem., 4, 191 (1965).
593a. Schweizer, E. E., W. S. Creasy, J. C. Liehr, M. E. Jenkins, and D. L. Dalrymple, J. Org. Chem., 35, 601 (1970).
594. Seel, F., W. Gombler, and K. H. Rudolph, Z. Naturforsch., 23b, 387 (1968).
595. Seel, F., and K. Rudolph, Z. Anorg. Allgem. Chem., 363, 233 (1968).
596. Seel, F., K. Rudolph, and R. Budenz, Z. Anorg. Allgem. Chem., 341, 196 (1965).
597. Seyferth, D., J. Am. Chem. Soc., 80, 1336 (1968).
598. Seyferth, D., J. Fogel, and J. K. Heeren, J. Am. Chem. Soc., 88, 2207 (1966), and 86, 307 (1964).
599. Seyferth, D., W. B. Hughes, and J. K. Heeren, J. Am. Chem. Soc., 87, 2847 (1965), and 84, 1764 (1962).
600. Seyferth, D., W. B. Hughes, and J. K. Heeren, J. Am. Chem. Soc., 87, 3467 (1965).
601. Sharp, D. W. A., and J. M. Winfield, J. Chem. Soc., 1965, 2278.
602. Shaturskii, Ya. P., Yu. S. Grushin, G. K. Fedorova, and A. V. Kirsanov, Zh. Obshch. Khim., 39, 1467 (1969); C.A., 71, 113048f (1969).
603. Shaw, M. A., J. C. Tebby, J. Ronayne, and D. H. Williams, J. Chem. Soc., C, 1967, 944.
604. Shaw, M. A., J. C. Tebby, R. S. Ward, and D. H. Williams, J. Chem. Soc., C, 1968, 1609.
605. Shevchenko, V. I., M. El Dik, and A. M. Pinchuk, Zh. Obshch. Khim., 39, 1514 (1969); C.A., 71, 113057h (1969).
606. Shevchenko, V. I., and Zh. V. Merkulova, Zh. Obshch. Khim., 29, 1005 (1959); C.A., 54, 1371i (1960).
607. Shevchenko, V. I., E. E. Nizhnikova, N. D. Bodnarchuk,

and P. P. Konuta, Zh. Obshch. Khim., 37, 1358 (1967); C.A., 68, 39055a (1968).
608. Shevchenko, V. I., A. M. Pinchuk, and N. Ya. Kozlova, Zh. Obshch. Khim., 34, 3955 (1964); C.A., 62, 9170b (1965).
609. Shevchenko, V. I., V. T. Stratienko, and A. M. Pinchuk, Zh. Obshch. Khim., 30, 1566 (1960); C.A., 55, 1490g (1961).
610. Shokol, V. A., V. F. Gamaleya, and G. I. Derkach, Zh. Obshch. Khim., 38, 1081 (1968); C.A., 69, 59340p (1968). Also Zh. Obshch. Khim., 39, 856 (1969), and 39, 1703 (1969); C.A., 71, 61497e, 124593f (1969).
611. Siering, H., Naturwissenschaften, 46, 85, 405 (1959).
612. Siering, H., Tohoku J. Exptl. Med., 72, 104 (1960); C.A., 55, 4757c (1961).
613. Siering, H., Arzneimittel-Forsch., 10, 229 (1960); C.A., 54, 16653e (1960).
614. Siering, H., Arzneimittel-Forsch., 10, 836 (1960); C.A., 55, 5772d (1961).
615. Smith, D. C., and A. H. Soloway, J. Med. Chem., 11, 1060 (1968).
616. Smith, W. C., U.S. Pat. 2,904,588 (1959); C.A., 54, 2254d (1960).
617. Smith, W. C., U.S. Pat. 2,950,306 (1960); C.A., 55, 2569i (1961).
618. Smith, W. C., J. Am. Chem. Soc., 82, 6176 (1960).
619. Spangenberg, S. F., and H. H. Sisler, Inorg. Chem., 8, 1006 (1969).
620. Spratley, R. D., W. C. Hamilton, and J. Ladell, J. Am. Chem. Soc., 89, 2272 (1967).
621. Stockel, R. F., Tetrahedron Letters, 1966, 2833.
621a. Summers, J. C. and H. H. Sisler, Inorg. Chem., 9, 862 (1970).
622. Swank, D., C. N. Caughlan, F. Ramirez, O. P. Madan, and C. P. Smith, J. Am. Chem. Soc., 89, 6503 (1967).
623. Tebbe, F. N., and E. L. Muetteries, Inorg. Chem., 6, 129 (1967).
624. Tebbe, F. N., and E. L. Muetteries, Inorg. Chem., 7, 172 (1968).
625. Tee, O. S., J. Am. Chem. Soc., 91, 7144 (1969).
626. Tesi, G., and C. M. Douglas, J. Am. Chem. Soc., 84, 549 (1962).
627. Theilacker, W., and F. Bolsing, Angew. Chem., 71, 672 (1959).
628. Thompson, Q. E., J. Am. Chem. Soc., 83, 845 (1961).
629. Titov, A. I., M. V. Sizova, and P. O. Gitel, Dokl. Akad. Nauk SSSR, 159, 385 (1964); C.A., 62, 6509h (1965).
630. Toy, A. D. F., and H. J. Emeleus, J. Inorg. Nucl. Chem., 29, 269 (1967).
631. Treichel, P. M., and R. A. Goodrich, Inorg. Chem., 4, 1424 (1965); Chem. Eng. News, 1965 (Sept. 27), 53.

632. Treichel, P. M., R. A. Goodrich, and S. B. Pierce, J. Am. Chem. Soc., 89, 2017 (1967).
632a. Trippett, S., J. Chem. Soc., 1962, 4731.
633. Tsivunin, V. S., Gil'm. Kamai, and R. Sh. Khisamutdinova, Dokl. Akad. Nauk SSSR, 164, 594 (1965); C.A., 63, 18143g (1965).
634. Tsivunin, V. S., Gil'm. Kamai, and R. Sh. Khisamutdinova, Zh. Obshch. Khim., 35, 1815 (1965); C.A., 64, 2122f (1966).
635. Tsivunin, V. S., Gil'm. Kamai, and G. K. Makeeva, Dokl. Akad. Nauk SSSR, 135, 1157 (1960); C.A., 55, 12271d (1961).
636. Tsivunin, V. S., Gil'm. Kamai, R. R. Shagidullin, and R. Sh. Khisamutdinova, Zh. Obshch. Khim., 35, 1234 (1965); C.A., 63, 11607g (1965).
637. Tsivunin, V. S., Gil'm. Kamai, R. R. Shagidullin, and R. Sh. Khisamutdinova, Zh. Obshch. Khim., 35, 1811 (1965); C.A., 64, 3588g (1966).
638. Tsivunin, V. S., Gil'm. Kamai, and D. B. Sultanova, Zh. Obshch. Khim., 33, 2149 (1963); C.A., 59, 12839a (1963).
639. Tsivunin, V. S., G. Kh. Kamai, R. Sh. Khisamutdinova, and E. M. Smirnov, Zh. Obshch. Khim., 35, 1231 (1965); C.A., 63, 11608b (1965).
640. Tsivunin, V. S., G. Kh. Kamai, and V. V. Kormachev, Zh. Obshch. Khim., 35, 1819 (1965); C.A., 64, 2122d (1966).
641. Tsivunin, V. S., G. Kh. Kamai, and V. V. Kormachev, Zh. Obshch. Khim., 36, 271 (1966); C.A., 64, 15917g (1966).
642. Tung, C. C., and A. J. Speziale, J. Org. Chem., 28, 1521 (1963).
642a. Turnblom, E. W., and T. J. Katz, J. Am. Chem. Soc., 93, 4065 (1971).
643. Ulrich, H., and A. A. R. Sayigh, Angew. Chem., 76, 647 (1964); Intern. Ed. 3, 585 (1964).
644. Umani-Ronchi, A., M. Acampora, G. Gaudiano, and A. Selva, Chim. Ind. (Milan), 49, 388 (1967); C.A., 67, 64279p (1967).
645. Utvary, K., V. Gutmann, and C. Kemenater, Monatsh. Chem., 96, 1751 (1965).
646. Voznesenskaya, A. Kh., and N. A. Razumova, Zh. Obshch. Khim., 38, 1553 (1968); C.A., 69, 96824v (1968).
647. Voznesenskaya, A. Kh., N. A. Razumova, and A. A. Petrov, Zh. Obshch. Khim., 39, 1033 (1969); C.A., 71, 61486a (1969).
648. Waite, N. E., and J. C. Tebby, J. Chem. Soc., C, 1970, 386.
649. Waite, N. E., J. C. Tebby, R. S. Ward, and D. H. Williams, J. Chem. Soc., C, 1969, 1100.
650. Walling, C., and R. Rabinowitz, J. Am. Chem. Soc., 81, 1243 (1959).

651. Walsh, E. N., T. M. Beck, and W. H. Woodstock, J. Am. Chem. Soc., 77, 929 (1955).
652. Wasserman, E., R. W. Murray, M. L. Kaplan, and W. A. Yager, J. Am. Chem. Soc., 90, 4160 (1968).
653. Webster, M., Chem. Rev., 66, 87 (1966).
654. Webster, M., and M. J. Deveney, J. Chem. Soc., A, 1968, 2166.
655. Weiss, J., and G. Hartmann, Z. Anorg. Allgem. Chem., 351, 152 (1967).
656. Westheimer, F. H., Accounts Chem. Res., 1, 70 (1968).
657. Wheatly, P. J., and G. Wittig, Proc. Chem. Soc., 1962, 251; Wheatly, P. J., J. Chem. Soc., 1964, 2206.
658. Whitesides, G. M., and M. Bunting, J. Am. Chem. Soc., 89, 6801 (1967).
659. Whitesides, G. M., and H. Lee Mitchell, J. Am. Chem. Soc., 91, 5384 (1969).
660. Wieber, M., and W. R. Hoos, Tetrahedron Letters, 1968, 5333.
661. Wieber, M., and W. R. Hoos, Tetrahedron Letters, 1969, 4693.
661a. Wieber, M., and W. R. Hoos, Monatsh. Chem., 101, 776 (1970).
662. Wieker, W., and A. R. Grimmer, Z. Chem., 7, 434 (1967).
663. Wiley, G. A., R. L. Hershkowitz, B. M. Rein, and B. C. Chung, J. Am. Chem. Soc., 86, 964 (1964).
664. Wiley, G. A., B. M. Rein, and R. L. Hershkowitz, Tetrahedron Letters, 1964, 2509.
665. Wiley, G. A., and W. R. Stine, Tetrahedron Letters, 1967, 2321.
666. Wiley, P. W., and H. E. Simmons, J. Org. Chem., 29, 1876 (1964).
667. Wiley, R. H., and C. H. Jarboe, J. Am. Chem. Soc., 73, 4996 (1951).
668. Wilfinger, H. J., Dissertation, Heidelberg, 1970.
669. Wittig, G., Bull. Soc. Chim. France, 1966, 1162.
670. Wittig, G., and G. Geissler, Ann., 580, 44 (1953).
671. Wittig, G., and D. Hellwinkel, Angew. Chem., 74, 76 (1962).
672. Wittig, G., and E. Kochendörfer, Angew. Chem., 70, 506 (1958).
673. Wittig, G., and E. Kochendörfer, Chem. Ber., 97, 741 (1964).
674. Wittig, G., and A. Maercker, Chem. Ber., 97, 747 (1964).
675. Wittig, G., and M. Rieber, Naturwissenschaften, 35, 345 (1948).
676. Wittig, G., and M. Rieber, Ann., 562, 187 (1949).
677. Woodstock, W. H., U.S. Pat. 2,495,799 (1950); C.A., 44, 3517h (1950).
678. Woodstock, W. H., U.S. Pat. 2,471,472 (1949); C.A.,

43, 7499c (1949).
679. Wulff, J., and R. Huisgen, Chem. Ber., 102, 1841 (1969).
680. Wunsch, G., K. Wintersberger, and H. Geierhaas, Z. Anorg. Allgem. Chem., 369, 33 (1969).
681. Yagupol'skii, L. M., and Zh. M. Ivanova, Zh. Obshch. Khim., 29, 3766 (1959); C.A., 54, 19553d (1960).
682. Yagupol'skii, L. M., and Zh. M. Ivanova, Zh. Obshch. Khim., 30, 4026 (1960); C.A., 55, 22196e (1961).
683. Yagupol'skii, L. M., and P. A. Yufa, Zh. Obshch. Khim., 28, 2853 (1958); C.A., 53, 9109h (1959).
684. Yagupol'skii, L. M., and P. A. Yufa, Zh. Obshch. Khim., 30, 1294 (1960); C.A., 55, 431f (1961).
685. Yagupsky, M. P., Inorg. Chem., 6, 1770 (1967).
686. Yaku, F., and I. Yamashita, Osaka Kogyo Gijutsu Shikensho Kiho, 18 (2), 117 (1967); C.A., 68, 3237x (1968).
687. Yakubovich, A. Ya. and V. A. Ginsburg, Dokl. Akad. Nauk SSSR, 82, 273 (1962); C.A., 47, 2685f (1953).
688. Yakubovich, A. Ya., and V. A. Ginsburg, Zh. Obshch. Khim., 22, 1534 (1952); C.A., 47, 9255a,e (1952).
689. Yakubovich, A. Ya., and V. A. Ginsburg, Zh. Obshch. Khim., 24, 1465 (1954); C.A., 49, 10834h, 10835d (1955).
690. Yakubovich, A. Ya., V. A. Ginsburg, and S. P. Makarov, Dokl. Akad. Nauk SSSR, 71, 303 (1950); C.A., 44, 8320a (1950).
691. Yamashita, I., and A. Masaki, Kogyo Kagaku Zasshi, 70, 2031 (1967); C.A., 69, 10322h (1968). Also Kogyo Kagaku Zasshi, 71, 915 (1968); C.A., 69, 67790a (1968), and Kogyo Kagaku Zasshi, 71, 1061 (1968); C.A., 69, 97213g (1968).
692. Yoshioka, H., and S. Horie, Japan Pat. 11,823 ('62) (1959); C.A., 59, 10125d (1963). Also Japan Pat. 11,824 ('62) (1959); C.A., 59, 10125d (1963).
693. Zamojski, A., Chem. Ind., 1963, 117.
694. Zbiral, E., and L. Berner-Fenz, Monatsh. Chem., 98, 666 (1967).
694a. Zemann, J., Z. Anorg. Allgem. Chem., 324, 241 (1963).
695. Zhmurova, I. N., and I. Yu. Dolgushina, Khim. Org. Soedin. Fosfora, Akad. Nauk SSSR, Ser. Obshch. Tekh. Khim., 1967, 195; C.A., 69, 76781a (1968).
695a. Zhmurova, I. N., L. Yu. Dolgushina, and A. V. Kirsanov, Zh. Obshch. Khim., 37, 1797 (1967); C.A., 68, 22007p (1968).
696. Zhmurova, I. N., and B. S. Drach, Zh. Obshch. Khim., 34, 1441 (1964); C.A., 61, 5499f (1964).
697. Zhmurova, I. N., and B. S. Drach, Zh. Obshch. Khim., 34, 3055 (1964); C.A., 62, 14480d (1965).
698. Zhmurova, I. N., B. S. Drach, and A. V. Kirsanov, Zh. Obshch. Khim., 35, 344 (1965); C.A., 62, 13030b (1965).

699. Zhmurova, I. N., and A. V. Kirsanov, Zh. Obshch. Khim., 29, 1687 (1959); C.A., 54, 8689a (1960).
700. Zhmurova, I. N., and A. V. Kirsanov, Zh. Obshch. Khim., 30, 3044 (1960); C.A., 55, 17551c (1961).
701. Zhmurova, I. N., and A. V. Kirsanov, Zh. Obshch. Khim., 30, 4048 (1960); C.A., 55, 22197a (1961).
702. Zhmurova, I. N., and A. V. Kirsanov, Zh. Obshch. Khim., 31, 3685 (1961); C.A., 57, 13795i (1962).
703. Zhmurova, I. N., and A. V. Kirsanov, Zh. Obshch. Khim., 32, 2576 (1962); C.A., 58, 7848e (1963).
704. Zhmurova, I. N., A. A. Kisilenko, and A. V. Kirsanov, Zh. Obshch. Khim., 32, 2580 (1962); C.A., 58, 10877f (1963).
705. Zhmurova, I. N., and I. Yu. Voitsekhovskaya, Zh. Obshch. Khim., 34, 1171 (1964); C.A., 61, 1889c (1964).
706. Zhmurova, I. N., and I. Yu. Voitsekhovskaya, Zh. Obshch. Khim., 35, 2197 (1965); C.A., 64, 11244d (1966).
707. Zhmurova, I. N., I. Yu. Voitsekhovskaya, and A. V. Kirsanov, Zh. Obshch. Khim., 31, 3741 (1961); C.A., 57, 9702 (1962).
708. Ziegler, M. L., and J. Weiss, Angew. Chem., 81, 430 (1969), Intern. Ed. 8, 455 (1969).
709. Zinovyev, Yu. M., V. N. Kulakova, and L. Z. Soborovsky, Zh. Obshch. Khim., 27, 151 (1957); C.A., 51, 12815e (1957).

Chapter 6. Tertiary Phosphine Oxides

HUGH R. HAYS and DONALD J. PETERSON

The Procter & Gamble Company, Miami Valley Laboratories, Cincinnati, Ohio

This chapter is a review of the methods of preparation, the chemistry, and the properties of tertiary phosphine oxides. Primary emphasis has been placed on the different synthetic methods, their advantages and their limitations. By far the most widely used and versatile methods of preparing tertiary phosphine oxides are the oxidation of tertiary phosphines, the decomposition of phosphonium hydroxides, and the reactions of organometallic reagents with phosphorus esters or halides. These methods constitute the first three segments of Section A on preparation. Roman numerals have been assigned to the various methods and are used in the list of compounds to indicate the method of preparation for a specific tertiary phosphine oxide.

Emphasis has also been placed on the literature appearing in the two decades since G. M. Kosolapoff's book, Organophosphorus Compounds,[480] was published. During this period knowledge of the chemistry of tertiary phosphine oxides has grown immensely. The extent of this growth is illustrated by an increase in the number of known tertiary phosphine oxides from 88 to about 1000 and a similar tenfold increase in the number of publications relating to these compounds. Noteworthy in this period are the extensive reviews of tertiary phosphine oxides by K. D. Berlin and G. B. Butler[63] and by K. Sasse.[710] In addition, a number of important secondary reports or books on aspects of organophosphorus chemistry that include the chemistry of tertiary phosphine oxides have appeared in the past two decades.[62,68,169,170,193,320,400,470,559,565,662,676,786,813a,823] Additional references to these reports are made in the appropriate sections of this chapter.

We wish to express our appreciation to the Management of The Procter and Gamble Company Research Division for cooperation and assistance in making this chapter possible. In addition, we are deeply indebted to Miss Jane Goedl for her extensive help in compiling this chapter and to Dr. R. G. Laughlin for his many discussions and constructive comments.

A. METHODS OF PREPARATION

I. OXIDATION OF TERTIARY PHOSPHINES

One of the most important methods of preparing phosphine oxides involves the oxidation of the corresponding phosphines. A variety of oxidizing agents are available

$$R_3P + [O] \longrightarrow R_3PO$$

for this purpose, including oxygen, peroxy compounds, oxides of nitrogen, oxides of sulfur, and standard inorganic oxidants.

Phosphines vary extensively in ease of oxidation. The lower trialkylphosphines, some of which are spontaneously flammable, react vigorously with oxidants, and considerable care must be exercised in effecting the transformation of these compounds to phosphine oxides. The higher trialkylphosphines are somewhat less reactive toward oxidizing agents. The replacement of alkyl groups with aryl groups also decreases the ease of oxidation of tertiary phosphines to the extent that triphenylphosphine, e.g., is quite stable in air. However, even the triarylphosphines can be readily oxidized by other oxidants. The decreased sensitivity of arylphosphines toward oxidation is undoubtedly a reflection of a fundamental difference that exists between trialkyl- and triarylphosphines, i.e., the nonbonding pair of electrons on arylphosphine-phosphorus is less available for bond formation than the corresponding electrons of trialkylphosphine-phosphorus.

The role of the nonbonding pair of electrons of a tertiary phosphine in one oxidation reaction has been thoroughly elucidated.[292] Benzoyl peroxide-carbonyl-^{18}O was shown to react with triphenylphosphine to give triphenylphosphine oxide and benzoic anhydride. All of the ^{18}O in excess of the amount that occurs naturally was found in the anhydride and none in the phosphine oxide. This is consistent with a mechanism in which triphenylphosphine effected a nucleophilic attack on a peroxidic oxygen to give an ion pair that subsequently collapsed to yield the products isolated.[†]

$$Ph_3P: + PhC(=O^*)O-OC(=O^*)Ph \longrightarrow \left[Ph_3\overset{+}{P}\text{-}OC(=O^*)Ph \quad {}^{*/2}O \overset{-}{=\!=} C(O^{*/2})Ph \right] \longrightarrow$$

[†]The asterisk (*) denotes the original ^{18}O content of carbonyl groups, while */2 indicates that the ^{18}O is distributed equally between two oxygens of the product.

$$Ph_3PO + PhC(=O^*)—O^{*/2}—C(=O^{*/2})Ph$$

Ionic mechanisms are also favored for oxidations of tertiary phosphines with t-alkyl peresters,[217] hydroperoxides,[216,220] and hypohalites.[220]

The recent advent of optically active organophosphines has stimulated interest in determining the steric course of oxidations of phosphines. Oxidations with hydrogen peroxide occur with retention of configuration at phosphorus. Retention of configuration mechanisms occur in reactions of optically active phenylpropylmethylphosphine with t-butyl hydroperoxide in pentane or methylene chloride, and with t-butyl peracetate in benzene. Oxidations of the phosphine with t-butyl hypochlorite in pentane and in methanol resulted in racemization and in 70% inversion, respectively.[219]

The stereochemistry at phosphorus of the benzoyl peroxide-tertiary phosphine reaction has been elucidated.[390] In petroleum ether, methanol, and acetonitrile, deoxygenation of the peroxide with optically active methylphenylpropylphosphine resulted in 49, 78, and 88% racemization, respectively. It has been suggested that the racemization stems from the formation of a symmetrical trigonal-bipyramidal intermediate.[390]

$$ROOR + :P(Me)(Et)(Ph) \longrightarrow [RO{-}P^+(Me)(Et)(Ph)\ \ ^-OR] \longrightarrow O{=}P(Me)(Et)(Ph) + ROR$$

retention

$$[RO{-}P^+(Me)(Et)(Ph)\ \ ^-OR] \longrightarrow RO{-}P(Me)(Et)(Ph){-}OR \longrightarrow \text{Racemate} + ROR$$

Not all oxidations of phosphines proceed by ionic mechanisms. Thus oxidations with air[112] and di-t-butyl peroxide[112,812] have been shown to occur by radical paths. The peroxide reaction is of considerable significance

$$t\text{-BuOOBu-}t \longrightarrow 2t\text{-BuO}\cdot$$

$$t\text{-BuO}\cdot + Ph_3P \longrightarrow t\text{-BuO}\dot{P}Ph_3 \longrightarrow t\text{-Bu}\cdot + Ph_3PO$$

$$t\text{-BuO}\cdot + n\text{-Bu}_3\text{P} \begin{cases} \xrightarrow{80\%} t\text{-BuOPBu}_2\text{-}n + n\text{-Bu}\cdot \\ \xrightarrow{20\%} n\text{-Bu}_3\text{PO} + t\text{-Bu}\cdot \end{cases}$$

since a radical displacement at phosphorus is involved; this contrasts vividly with the lack of well-documented radical displacements at carbon.

1. AIR OXIDATION. The literature abounds with reports of air oxidations of tertiary phosphines. Only recently, however, has this reaction been subjected to thorough investigation.[112,249] The findings of these studies are of considerable significance since they clearly demonstrate that autoxidations of some trialkylphosphines do not yield phosphine oxides exclusively. Indeed, it was found that phosphinate esters accompany phosphine oxides as the major products of air oxidations, with phosphonates and phosphates being formed in lesser amounts. The autoxidation of tri-n-butylphosphine exemplifies this process. Autoxidation of tributylphosphine in hexane or acetone proceeded rapidly at

$$\text{Bu}_3\text{P} + \text{O}_2 \xrightarrow{\text{hexane}} \underset{(42\%)}{\text{Bu}_3\text{P=O}} + \underset{(49\%)}{\text{Bu}_2\overset{\overset{\text{O}}{\|}}{\text{P}}\text{OBu}} + \underset{(6\%)}{\text{Bu}\overset{\overset{\text{O}}{\|}}{\text{P}}(\text{OBu})_2} + \underset{(3\%)}{(\text{BuO})_3\text{P=O}}$$

room temperature and is presumably oxygen diffusion controlled. An examination of the oxidation of this phosphine in various solvents and under differing conditions revealed that the medium, in particular, influenced the product distribution. Increasing the solvent polarity (hexane → aqueous ethanol) increased the proportion of tributylphosphine oxide relative to the other products. Aromatic solvents were found to inhibit the autoxidation of tributylphosphine. Small amounts of diphenylamine, hydroquinone, diphenyl disulfide, thiophenol, sodium hydroxide, sodium ethoxide, and triphenylphosphine also effectively inhibited the air oxidation of trialkylphosphines. Triphenylphosphine is resistant to oxidation under similar conditions. The arylphosphine was partially oxidized, however, when a free-radical source [2,2'-azobis(2-methylpropionitrile)] was added to the reaction vessel.

This and other information suggested the following radical mechanism for autoxidations of trialkylphosphines:[112]

$$\text{Initiation} \longrightarrow R\cdot\,(RO\cdot,\ RO_2\cdot)$$

$$R\cdot + O_2 \longrightarrow RO_2\cdot$$

$$RO_2\cdot + R_3P \longrightarrow RO\cdot + R_3P{=}O$$

$$RO\cdot + R_3P \longrightarrow R\cdot + R_3P{=}O$$

$$RO\cdot + R_3P \longrightarrow R\cdot + R_2POR$$

$$RO_2\cdot + R_2POR \longrightarrow RO\cdot + R_2\overset{\overset{\displaystyle O}{\|}}{P}OR$$

$$RO\cdot + R_2POR \longrightarrow R\cdot + R_2\overset{\overset{\displaystyle O}{\|}}{P}OR$$

$$RO\cdot + R_2POR \longrightarrow R\cdot + RP(OR)_2$$

$$RO_2\cdot + RP(OR)_2 \longrightarrow RO\cdot + R\overset{\overset{\displaystyle O}{\|}}{P}(OR)_2$$

$$RO\cdot + RP(OR)_2 \longrightarrow R\cdot + R\overset{\overset{\displaystyle O}{\|}}{P}(OR)_2$$

$$RO\cdot + RP(OR)_2 \longrightarrow R\cdot + P(OR)_3$$

$$RO_2\cdot + P(OR)_3 \longrightarrow RO\cdot + (RO)_3P{=}O$$

Because of the complexity of autoxidation reactions of trialkylphosphines, the value of this process as a method for preparing phosphine oxides is questionable. The unwanted side reactions of autoxidations have apparently been suppressed, if not excluded, by employing metal catalysis.[734] For example, platinum and palladium tetra(triphenylphosphine) complexes catalyze the oxidation of triphenylphosphine and tributylphosphine to give the oxides.[77, 195, 779, 821] Triethylphosphine when coordinated with cobalt chloride autoxidizes to give triethylphosphine oxide exclusively.[734]

2. PEROXY COMPOUNDS. The most general method for oxidizing phosphines to phosphine oxides employs peroxy compounds as oxidants. Of this type of oxidant, aqueous

$$R_3P + R'OOR' \longrightarrow R_3PO + R'OR'$$

hydrogen peroxide is clearly the reagent of choice. The extensive use of hydrogen peroxide as an oxidizing agent for phosphines stems from several attributes of the reagent, which include availability, ease in handling, safety, the formation of water as a by-product, and potency as an oxidant. Some compounds typifying the wide variety of tertiary phosphines that have been oxidized with 1 to 30% aqueous hydrogen peroxide are triethylphosphine,[330] dimethyldodecylphosphine,[324] 1,3-bis(di-n-octylphosphino)-propane,[531] benzyldiphenylphosphine,[386] (p-dimethylaminophenyl)diphenylphosphine,[717] hexyldiphenylphosphine,[775] 2-hydroxyethyldiphenylphosphine,[415] cis- and trans-1,2-vinylenebis(diphenylphosphine),[6] and ethynyldiphenylphosphine.[151]

Interestingly, the oxidation of triphenylphosphine with hydrogen peroxide afforded an adduct having the formula $(Ph_3PO)_2 \cdot H_2O_2$, m. 131°.[167,780] The adduct is stable for several months at room temperature but decomposes on heating to give oxygen and triphenylphosphine oxide.

The results of a comprehensive study have demonstrated that several derivatives of hydrogen peroxide also react readily with tertiary phosphines to afford phosphine oxides.[382] Included in this list of peroxy compounds are alkyl hydroperoxides, unsaturated hydroperoxides, hydroxyalkyl hydroperoxides, dihydroxyalkyl hydroperoxides, peroxy acids, peroxy acid esters, diacyl peroxides, endo-peroxides, and ozonides. Triphenylphosphine and triethylphosphine were employed as model phosphines in this study.

Ozone oxidations of tributylphosphine and triphenylphosphine have been reported to give the corresponding oxides in 92 and 99% yields, respectively.[782] An ozone-triphenyl phosphite adduct similarly oxidized the trialkylphosphine.[782]

3. OXIDES OF NITROGEN. Several nitrogen oxides have been used for oxidizing phosphines to the corresponding phosphine oxides. Most important in this regard is nitric acid.[34,205,344,345,774] Because of its potency as a general oxidant, however, care must be exercised to avoid oxidizing any other functional group of the phosphine. Nitrosyl chloride (NOCl),[474] nitric oxide (NO),[1,359,516] and dinitrogen tetroxide (N_2O_4)[3,183] are similarly effective oxidants, even at low temperatures. Triethylphosphine oxide was formed from the reaction of triethylphosphine even with nitrous oxide (N_2O)[771] at 127°; nitrogen dioxide appeared more reactive in that it oxidized triphenylphosphine quantitatively at room temperature.[382]

Somewhat related to these oxidations with the simple oxides of nitrogen are formations of phosphine oxides as by-products from reactions of phosphines with hydroxylamine (NH_2OH),[554] a hydroxamide,[224] aliphatic[368] and aromatic[396] amine oxides, aldonitrones,[371] aromatic nitroso compounds,[98,120,371,383] azoxy compounds,[371] nitro compounds,[128,791] and arylnitrile oxides.[303] Because of the relatively exotic nature of the oxidants, however, these reactions are not generally useful for the preparation of phosphine oxides.

4. SULFUR OXIDES. Sulfur dioxide and related organic oxygen-containing sulfur compounds have been reported to effect oxidations of tertiary phosphines. Sulfur dioxide reacted vigorously at very low temperature with trimethylphosphine to give trimethylphosphine oxide and sulfur, while the corresponding reaction with triphenylphosphine afforded, in addition to the expected phosphine oxide, some triphenylphosphine sulfide.[762] Both phosphine oxide and sulfide were obtained from the reaction of tri-n-butylphosphine with sulfur dioxide at 50°.[250]

Of the organic oxygen-containing sulfur compounds that oxidize phosphines to the corresponding phosphine oxides, only reactions with dimethyl sulfoxide appear to constitute a useful preparative method.[24] Oxidations of both tributylphosphine and triphenylphosphine with this reagent require temperatures above 100° to achieve appreciable reaction rates.

$$R_3P + Me_2SO \longrightarrow R_3PO + Me_2S$$

The remaining miscellaneous sulfur oxidants are included in Table 1.

5. INORGANIC OXIDANTS. The standard inorganic oxidants, potassium permanganate,[229,270,411,413,415,775] chromic acid,[244,406,578,580] mercuric oxide,[196,300,579,735] ferric chloride,[196,587] potassium chlorate,[588] and potassium peroxymonosulfate,[382] provide additional ways of obtaining phosphine oxides from phosphines. Care must be exercised, however, to avoid unwanted oxidations of side chains of aromatic rings, functional groups, etc., when employing the stronger of these oxidants. Mercuric oxide and ferric chloride may find special utility in the oxidations of phosphines containing readily oxidizable substituents.

6. MISCELLANEOUS OXIDANTS. Phosphine oxides have been obtained as by-products from a wide variety of reactions that specifically utilize tertiary phosphines as reducing agents. Reactions of phosphines with oxygen-bearing organosulfur and nitrogen compounds of this type were discussed above. Additional reactions that fall into this category involve as oxidants trichloroacetamides,[768] "positive" halogen compounds, including secondary and tertiary α-haloketones,[91,790] α-halogenated aldehydes,[355] mono α-halogenated amides,[613] halogenated phenols,[357] cyclic car-

Table 1. Tertiary Phosphine Oxides as By-Products from Reactions of Organic and Inorganic Oxygen-Containing Compounds with Tertiary Phosphines

Phosphine	Oxygen Compound	Deoxygenated Compound	Ref.
Ph_3P	$4\text{-}MeC_6H_4S(O)SC_6H_4\text{-}4\text{-}Me$	$(4\text{-}MeC_6H_4S)_2$	143
$(4\text{-}ClC_6H_4)_3P$	$EtS(O)SEt$	EtSSEt	"
Et_3P	$PhSO_2SO_2Ph$	PhSSPh	357
Ph_3P	$PhSO_2H$	PhSH	388
Ph_3P	$PhSO_2Cl$	PhSH, PhSSPh	"
Ph_3P	$4\text{-}MeC_6H_4SO_2Cl$	$4\text{-}MeC_6H_4SH$ $(4\text{-}MeC_6H_4S)_2$	"
Ph_3P	Ethylene carbonate	$CH_2{=}CH_2$, CO_2	464
$(NCCH_2CH_2)_3P$	" "	"	"
Ph_3P	Phenylethylene carbonate	ØCH=CH_2, CO_2	"
Bu_3P	" "	"	"
Ph_3P	(4-Hydroxybutyl)ethylene carbonate	$CH_2{=}CH(CH_2)_3OH$	"
Bu_3P	cis-MeCH—CHMe (epoxide, O bridging)	81% trans-MeCH=CHMe 19% cis-MeCH=CHMe	92
Bu_3P	trans-MeCH—CHMe (epoxide, O bridging)	72% cis-MeCH=CHMe 28% trans-MeCH=CHMe	"
Ph_3P	$(HOCH_2CH_2S)_2$	CH_2—CH_2 (episulfide, S bridging)	287
$c\text{-}C_6H_{11}PPh_2$	"	"	"

bonates,[464] epoxides,[92] 2-hydroxyethyl disulfide,[287] N-bromoamides,[790] and bromoacetic acid.[222] Some of these reactions involve relatively stable intermediates that decompose to phosphine oxides only after treatment with a proton source. Table 1 lists the specific compounds involved in these reactions.

II. DECOMPOSITION OF PHOSPHONIUM HYDROXIDES

Tertiary phosphine oxides and hydrocarbons have long been recognized as the products of decomposition of phosphonium hydroxides.[130] The reaction is usually carried

$$R_4\overset{+}{P}\ \overset{-}{O}H \longrightarrow R_3PO + RH$$

out by distilling an aqueous solution of the phosphonium hydroxide prepared from the phosphonium halide and moist silver oxide[130,158,161,162,239,472,502,503,573,574,580,588,589,834] or by heating the phosphonium salt with 20 to 40% aqueous sodium hydroxide.[11,14,17,53,318,344,345,350,380,437,463,605,657,702,738,834] Tetraalkylphosphonium salts do not react readily with hot aqueous sodium hydroxide but do react smoothly upon heating with anhydrous sodium hydroxide.[329] The negative effect of water on the rate of reaction was demonstrated by the rapid decomposition of the anhydrous intermediate tetralkylphosphonium hydroxide at 25°.

The kinetics of the reaction is generally first order in phosphonium salt and second order in hydroxide ion,[14,82,351,496,627,761,842] although some reactions with overall second-order kinetics are known.[17,60] In the former case, methylethylphenylbenzylphosphonium hydroxide was shown to decompose stereospecifically with inversion to form optically active methylethylphenylphosphine oxide.[496,566] The mechanistic implications of these findings and supporting stereochemical evidence[367,391,392] have recently been reviewed.[565] The reaction is visualized to proceed as follows:

$$R_4\overset{+}{P}\,^{-}OH \rightleftharpoons R_4POH \xrightleftharpoons{^{-}OH} R_4P\text{-}O^{-} \xrightarrow{\text{slow}} R_3PO + R^{-} \longrightarrow RH$$

The phosphonium ion and a hydroxide ion rapidly form a pentacovalent intermediate that also reacts rapidly (and reversibly) with a second hydroxide ion to give a pentacovalent phosphorus oxyanion. The latter then proceeds to phosphine oxide in the rate-determining step with concurrent formation of a carbanion that is rapidly protonated to give the hydrocarbon.

In accordance with a mechanism that involves the extrusion of a carbanion in the rate-determining step, the phosphonium salts with different substituents generally lose the substituent corresponding to the most stable carbanion.[239,573,842] When sufficient differences in the stabilities of the incipient carbanions exist, the reaction constitutes an excellent synthesis of tertiary phosphine oxides. Thus the order of ease of displacement of groups from phosphonium salts is allyl, benzyl > phenyl > methyl > 2-phenylethyl > ethyl, higher alkyls. Representative examples of phosphonium salts containing these substituents and the yields of tertiary phosphine oxides obtained are shown in Table 2.

The usefulness of this method for the preparation of a wide variety of alkyl and substituted alkyldiphenylphosphine oxides, starting from alkyltriphenylphosphonium salts, is particularly noteworthy in view of the availability of

Table 2. Phosphonium Hydroxide Decompositions

Phosphonium Salt	Phosphine Oxide	% Yield	Ref.
Allylmethyldiphenyl	Methyldiphenyl	70	573
Benzyltriphenyl	Triphenyl	100	239
Dibenzylmethylphenyl	Benzylmethylphenyl	80	573
Diethylmethylphenyl	Diethylmethyl	79	573
Ethyl-tri-p-tolyl	Ethyldi-p-tolyl	100	605
Dodecyltrimethyl	Dodecyldimethyl	97	329
Tridodecylmethyl	Tridodecyl	87	329

triphenylphosphine and the ease with which it is quaternized.[12,76,333,352,420,521,589,605,732,739-741,783,828,832] The yields of tertiary phosphine oxides obtained by this procedure are generally excellent. Similarly, certain substituted triarylphosphine oxides have been prepared in 87 to 98% yields from substituted tetraarylphosphonium hydroxides.[380] Finally, in decompositions of phosphonium hydroxides that have four identical substituents, the yields of tertiary phosphine oxides generally appear to be essentially quantitative.[239,842]

In contrast to the reactions described above, reactions in which the carbanions displaced from phosphorus are structurally different, but have approximately the same stability, give mixtures of tertiary phosphine oxides.[239,329,380,842] The preparative value of this method is therefore diminished. For example, ethyltripropylphosphonium hydroxide gives a 77:23 mixture of ethyldipropylphosphine oxide and tripropylphosphine oxide. Similarly, phenyltri-p-tolylphosphonium hydroxide formed a mixture of the two possible phosphine oxides.[380]

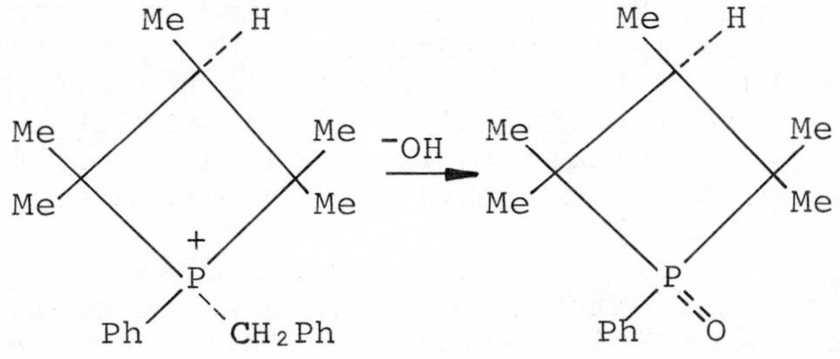

Certain four-, five-, and six-membered heterocyclic phosphine oxides can be prepared in good yield by decomposition of the properly substituted heterocyclic phosphonium salts.[68,171,234,235,657] Interestingly, alkaline hydrolysis of the benzylphosphetanium salt shown below proceeds with retention of configuration.[171,787] This is attributed to a preferred conformation in the trigonal-bipyramidal transition state in which an apical and an equatorial position are occupied by the four-membered ring. Accordingly the entering and leaving groups occupy the remaining apical position and an equatorial position, resulting in retention of configuration. A number of heterocyclic phosphonium salts have been shown to undergo ring cleavage upon reaction with aqueous sodium hydroxide.[11,21,60,547,830]

The reactions of tetrakis(hydroxymethyl) and related phosphonium salts with sodium hydroxide are unusual and may give different products, depending on the conditions. With excess base tetrakis(hydroxymethyl)phosphonium chloride yields tris(hydroxymethyl)phosphine oxide, hydrogen, and formaldehyde, but not methanol.[344,345] Methanol would be the expected product of this reaction if it proceeded analogously to the previously described phosphonium salts.

$$(HOCH_2)_4\overset{+}{P}\ Cl^- + NaOH \longrightarrow (HOCH_2)_3PO + H_2 + CH_2O$$

By way of contrast, with only 1 equivalent of sodium hydroxide, tetrakis(hydroxymethyl)phosphonium chloride gives tris-(hydroxymethyl)phosphine, which does not react further with sodium hydroxide to produce tris(hydroxymethyl)phosphine oxide.[286] Furthermore, only certain alkyltris(hydroxymethyl)phosphonium salts react with excess sodium hydroxide to give alkylbis(hydroxymethyl)phosphine oxides, hydrogen, and formaldehyde. Two of these salts are the cyclohexyl and isopropyl compounds, whereas the n-butyl-, methyl-, and phenyltris(hydroxymethyl)phosphonium salts with excess base form only substituted bis(hydroxymethyl)phosphines. Both dialkyl- and diarylbis(hydroxymethyl)phosphonium and, presumably, trisubstituted hydroxymethylphosphonium salts react with base to form only the tertiary phosphine and formaldehyde.

Chloromethylphosphonium salts react with sodium hydroxide to give anomalous products in addition to tertiary phosphine oxides.[333,334,733] For example, chloromethyltriphenylphosphonium chloride and sodium hydroxide, upon heating, form methyl chloride and a mixture of 50 to 70% triphenylphosphine oxide, 8% triphenylphosphine, 4% formaldehyde, and 10% of a rearrangement product, benzyldiphenylphosphine oxide. Phenyl-tris-(chloromethyl)phosphonium chloride and sodium hydroxide gave 51% 2-chloroethylchloromethylphenylphosphine oxide, resulting from rearrangement of a chloromethyl group, and 21% phenyl-bis-(chloromethyl)phosphine.

In addition to the limitations described above, certain phosphonium hydroxides with aromatic or electronegative substituents in the position beta to the phosphorus atom undergo elimination reactions to form tertiary phosphines

$$R_3\overset{+}{P}\text{-}CH_2CH_2X\ ^{-}OH \longrightarrow R_3P + CH_2{=}CHX + H_2O$$

and olefins.[12,239,340] Thus 2-phenylethyltriethylphosphonium hydroxide decomposes to a mixture of ethylbenzene, triethylphosphine, the oxide, and styrene. Decomposition of 2,2-diphenylethyltributylphosphonium hydroxide yields predominantly diphenylethylene and tributylphosphine. 2-Cyanoethylphosphonium salts react with strong bases in a similar manner to form tertiary phosphines in 32 to 61% yields.[289,352] The latter reaction has been applied to the synthesis of a number of unsymmetrically substituted tertiary phosphines.

Certain β-substituted vinylphosphonium salts react with aqueous base to give phosphine oxides resulting from phenyl migration from phosphorus to the α-carbon atom.[107,843] A vinylphosphonium hydroxide is believed to be a common intermediate in several diverse reactions that lead to

$$Ph_3P^{+}\text{-}CH{=}CHX \xrightarrow{OH^{-}} Ph_2\overset{O}{\overset{\|}{P}}\text{-}\underset{Ph}{\underset{|}{CH}}\text{-}CH_2X$$

$$X = CH{=}CH_2,\ Ph,\ PhCO,\ RCO$$

phenyl migration from phosphorus to the α-carbon.[698]

Phosphonium salts other than phosphonium hydroxides can also decompose to tertiary phosphine oxides. For example, certain phosphonium alkoxides, upon heating in refluxing butanol or to above 150°, lead to the tertiary phosphine oxide, ether, and hydrocarbon.[288,329,340,567] This decomposition requires more stringent conditions than

$$R_4\overset{+}{P}\ ^{-}OR \xrightarrow{\Delta} R_3PO + R_2O + RH$$

that of the analogous hydroxide, proceeds by a different mechanism, and can sometimes lead to elimination or displacement products. Alkyltriphenylphosphonium alkoxides, for example, give triphenylphosphine and alkylphenyl ethers upon heating to 280°.[232,784] Olefin formation is observed to some extent and becomes predominant when secondary and tertiary alkyltriphenylphosphonium alkoxides are decomposed. Other examples of phosphonium salts that give phosphine oxides upon heating at high temperatures are the nitrate, sulfate, bicarbonate, acetate, benzoate, and butyrate.[158,221,551,573] Methylethylphenylbenzylphosphonium acetate requires temperatures over 200° for pyrolysis. Accordingly, these reactions are of little practical value for the synthesis of tertiary phosphine oxides.

Tertiary phosphine oxides are also formed by a number of different reactions in which phosphonium salts are either precursors or intermediates. Representative examples are the highly useful Wittig olefin synthesis,[433,832] the hydrolysis of phosphinimines or aminophosphonium salts,[27,683,751,760,771] the hydrolysis of phosphorus ylids,[433] and the hydrolysis of alkylthiophosphonium salts.[314] These reactions, like the decomposition of phosphonium alkoxides, are generally not used to prepare tertiary phosphine oxides.

Within the limitations described, the synthesis of tertiary phosphine oxides from phosphonium salts and sodium hydroxide appears simple to carry out and efficient in terms of yields of the tertiary phosphine oxides. However, dependent on the structure of the desired phosphine oxide, other syntheses involving organometallic reagents or oxidation of the phosphine, or both combined, may give equally good yields in a more direct manner.

III. FROM PHOSPHORUS ESTERS OR HALIDES WITH ORGANOMETALLIC REAGENTS

The reactions of phosphorus halides or esters with organometallic reagents constitute a very important route to tertiary phosphine oxides, many of which have been prepared in good yields by this method. Nevertheless, much work remains to be done in this area before many of these reactions can be well understood. This is due in part to the fact that the actual nature of organometallic reagents is complex and remains under investigation today. The reactivity of an organometallic reagent, e.g., is a function of its molecular nature and depends greatly on the organic ligand, the metal, the halide in the case of Grignard reagents, and the solvent. The second important obstacle to understanding these reactions is that significant differences in the reactivity of phosphorus compounds exist with respect to a given organometallic reagent. These differences in reactivity are due to factors such as the group being displaced from phosphorus, the size of the groups attached to phosphorus, and the electronic effects of the substituents on phosphorus. Thus the temptation to generalize with regard to the reactions of phosphorus esters or halides with organometallic reagents must be tempered by the realization that results may apply only to a given set of experimental conditions.

1. PHOSPHORUS OXYHALIDES. Phosphorus oxychloride reacts exothermically with excess alkyl and aryl Grignard

reagents in diethyl ether to give low to good yields of symmetrical trialkyl- and triarylphosphine oxides, respectively.[83,148,149,153,205,301,462,597,649,712] Representative tertiary phosphine oxides prepared by this method

$$POCl_3 + 3RMgX \longrightarrow R_3PO + 3MgXCl$$

and their yields are trimethyl (52%),[123] trioctyl (42 to 46%),[820] triphenyl(65%),[301,712] tris(trimethylsilylmethyl) (51%),[139] tris(p-trimethylsilylphenyl) (30%),[255] tris-(phenylethynyl) (77%),[153] tri-2-pyrryl (42%), and tri-2-thienylphosphine oxide (42%).[297] When insufficient Grignard reagent is used the yields of tertiary phosphine oxides are lower because of formation of the disubstituted phosphinic acid. For example, inverse addition of an equivalent amount of arylmagnesium bromide to phosphorus oxychloride in ether gives predominantly the diarylphosphinic acid plus small amounts of the triarylphosphine oxide.[481] Similarly, sec-alkyl Grignard reagents can lead to low yields of the tertiary phosphine oxide; the disubstituted phosphinic acid is the predominant product.[639] Addition of cyclohexylmagnesium chloride to phosphorus oxychloride in diethyl ether, e.g., gave a 74% yield of dicyclohexylphosphinic acid.[406] Likewise, no tertiary phosphine oxide could be isolated from t-butylmagnesium chloride and phosphorus oxychloride in diethyl ether.[191] In view of these limitations and the availability or ease of preparation of the starting materials, this synthesis is best suited for the preparation of symmetrical trialkylphosphine oxides in which the α-carbon atom is not substituted and of triarylphosphine oxides.

Unsymmetrical tertiary phosphine oxides of the type RR'_2PO and RR'R"PO can be prepared in fair to excellent yields from the appropriately substituted phosphonyl dichloride[605] or phosphinyl chloride[93] and Grignard reagent. Phenylbis(p-tolyl)phosphine oxide was prepared in 82%

$$RP(O)Cl_2 + R'MgX \longrightarrow RR'_2PO$$

$$R_2P(O)Cl + R'MgX \longrightarrow R_2R'PO$$

$$R'R''P(O)Cl + RMgX \longrightarrow RR'R''PO$$

yield from phenylphosphonyl dichloride and p-tolylmagnesium bromide.[605] Similarly, a number of alkylenediphosphine dioxides have been prepared in yields of 39 to 75% from alkylenediphosphonyl tetrachlorides with alkyl and aryl Grignard reagents.[485,696] By the same procedure, bis-

$$\mathrm{RCH(\overset{O}{\overset{\|}{P}}Cl_2)_2 + R'MgX \longrightarrow RCH(\overset{O}{\overset{\|}{P}}R'_2)_2}$$

$$\mathrm{R = H, Pr \qquad R' = C_6H_{13}, 2\text{-}EtBu, Ph}$$

$$\mathrm{(CH_2)_n(\overset{O}{\overset{\|}{P}}Cl_2)_2 + RMgX \longrightarrow (CH_2)_n(\overset{O}{\overset{\|}{P}}R_2)_2}$$

(dimethylphosphinyl)methane was isolated in only 10% yield, partly because of its tenacity for water. As might be anticipated from the studies of phosphorus oxychloride, the reaction of t-butylmagnesium chlorides with alkylphosphonyl dichlorides is complicated. However, several alkyldi-t-butylphosphine oxides have been prepared in low yields by this procedure.[110] Diallylphenylphosphine oxide has recently been prepared from phenylphosphonyl dichloride and allylmagnesium bromide in 80% yield,[75] in contrast to an earlier report.[65] Methyl- and ethyldibutylphosphine oxides[487] and several alkyldiarylphosphine oxides[84,379,607] have been prepared in 44 to 81% yields from reactions of the dialkyl- and diarylphosphinyl chlorides, respectively, with Grignard reagents. Alkylenediphosphine dioxides containing four, five, and six bridging methylenes can also be obtained in good yields (36 to 79%) from α,ω-alkylene di-Grignard reagents and dialkyl- or diarylphosphinyl chlorides.[486,606]

$$\mathrm{(CH_2)_n(MgX)_2 + R_2\overset{O}{\overset{\|}{P}}Cl \longrightarrow R_2\overset{O}{\overset{\|}{P}}(CH_2)_n\overset{O}{\overset{\|}{P}}R_2}$$

$$\mathrm{n = 4, 5, 6 \qquad R = Me, Et, Bu, Ph}$$

These reactions constitute a good overall synthesis of unsymmetrical tertiary phosphine oxides with the exception that the phosphonyl and phosphinyl chlorides may require lengthy preparations. Accordingly the direct preparation of unsymmetrical phosphine oxides described in Section A.III.2 may be more convenient for type RR'_2PO.

Symmetrical triarylphosphine oxides have been obtained in 33 to 75% yields from diethyl phosphorochloridate and aryl Grignard reagents that do not possess ortho substituents.[125,208] In this procedure the diethyl phosphoro-

$$\mathrm{(EtO)_2\overset{O}{\overset{\|}{P}}Cl + 3PhMgBr \longrightarrow Ph_3PO}$$

chloridate is added to 3 equivalents of the Grignard reagent at 0° and then allowed to warm to 15 to 20°, at which point a vigorous, exothermic, and heterogeneous reaction occurs to rapidly form the phosphine oxide in good yields. Several factors are of interest with regard to this reaction and are important in determining the product ratio. The first is that ethoxy groups appear to be displaced by the Grignard reagent much more easily from diethyl phosphorochloridate than from diethyl phenylphosphonate.[482] Even inverse addition of only 1 equivalent of phenylmagnesium bromide to diethyl phosphorochloridate in ether either at 0°, then warming, or at 25° gave significant amounts of triphenylphosphine oxide (19 to 28%) along with diethyl phenylphosphonate (19 to 38%) and ethyl diphenylphosphinate (0.7 to 4%).[326] Slower rates of addition of the phenylmagnesium bromide resulted in higher yields of diethyl phenylphosphonate. This may account for the greater yields (55 to 60%) of diethyl phenylphosphonate reported earlier.[93,125,228] Also of significance is that, after the initial exothermic reaction, added phenylmagnesium bromide did not appear to react or to change significantly the product ratio described above. Finally, magnesium halides have been shown to retard the diethyl phenylphosphonate-phenylmagnesium bromide reaction.[328]

These findings suggest that the initial vigorous and highly exothermic displacement of chloride by Grignard reagent in a concentrated phase supplies the driving force for displacement of the ethoxy groups. In agreement with this explanation is the finding that diethyl phenylphosphonate is produced in yields of 97 to 98%, with only 1 to 2% of the other products, regardless of the mode of mixing, in tetrahydrofuran. In this solvent the reaction is homogeneous, using concentrations comparable to those in ether. Accordingly, substituted triarylphosphine oxides can be prepared in good yields provided the substituents are not ortho and the original conditions are closely followed.

2. PHOSPHORUS ESTERS. Tertiary phosphine oxides have been obtained from triaryl phosphates and Grignard reagents at high temperatures, although the yields are generally low.[272,273] As an exception, triphenylphosphine oxide has reportedly been prepared in 95% yield from triphenyl phosphate and phenylmagnesium bromide in ether-benzene solvent at 60° after 1 hr.[72] In contrast, triethyl phosphate and phenylmagnesium bromide in refluxing tetrahydrofuran gave only 5% triphenylphosphine oxide after 7 hrs.[326] Trimethyl phosphate and phenylmagnesium bromide in ether-benzene at 60° for 1 hr. yielded predominantly toluene from methylation of the Grignard reagent plus about 10% of triphenylphosphine oxide.[72]

Unsymmetrical tertiary phosphine oxides of type

$RP(O)R'_2$ have been prepared in good to excellent yields from the appropriate Grignard reagent and phosphonate.[64,65,483,498,605] Dialkylarylphosphine oxides were prepared from

$$RP(O)(OR')_2 + R''MgX \longrightarrow RP(O)R''_2$$

diphenyl phenylphosphonate and the alkyl Grignard reagent in 78 to 90% yields.[70] These reactions were carried out with an excess of Grignard reagent in ether-benzene at 60° for 4 hrs. Unsymmetrical trialkylphosphine oxides are similarly prepared in good yield from diphenyl alkylphosphonates and the alkyl Grignard reagent in refluxing THF for 6 to 12 hrs.[498] Both of these methods produce phenol, which reportedly can be difficult to separate from certain phosphine oxides. Triphenylphosphine oxide is produced by this route in about 70% yield from diethyl phenylphosphonate and phenylmagnesium bromide (1:2 reactant ratio) in THF at 68° after 6 hrs.[328] The by-product ethanol is easily separated in the aqueous phase. In general, the starting phosphonates must be prepared in the laboratory. Accordingly, the synthesis of phosphine oxides of type $RP(O)R'_2$ described below appears to be more direct.

Diethyl phosphonate reacts exothermically in ether or THF with both alkyl and aryl Grignard reagents to give magnesium salts of disubstituted phosphinous acids. Direct addition of alkyl halides or sulfonates with refluxing produces the unsymmetrical phosphine oxides in good to excellent yields.[226,327,484,695,756] Trialkyl- (RR'_2PO) and

$$(EtO)_2P(O)H + 3RMgX \longrightarrow RH + R_2POMgX \xrightarrow{R'X} R'R_2PO$$

alkyldiarylphosphine oxides were obtained in 25 to 80% yields by this method. Diethyl ether is not a suitable solvent for this preparation of alkyldimethyl- and alkyldiethylphosphine oxides,[327,657] presumably because of the heterogeneity of the reaction mixture in diethyl ether. When tetrahydrofuran was used as the reaction solvent, 65 to 89% yields of these phosphine oxides were obtained. The simplicity and the availability or ease of preparation in situ of the required phosphinite salt make this perhaps the most useful synthesis of phosphine oxides of type RR'_2PO. In addition to the compounds mentioned above, bridged diphosphine dioxides,[327,484] β-hydroxyalkyl,[327] alkoxymethyl,[226,327] alkylthiomethyl,[327] α-hydroxylbenzyl,[595] and allyl[226] tertiary phosphine oxides have been prepared in equally good yields.

Unsymmetrical phosphine oxides of types $R_2\overset{O}{\overset{\|}{P}}R'$ and $RR'R''\overset{O}{\overset{\|}{P}}$ can also be prepared from the appropriate phosphinate and Grignard reagent, although there are limitations. For example, ethyl diphenylphosphinate is 95% converted to

$$R'_2\overset{O}{\overset{\|}{P}}OR'' + RMgX \longrightarrow RR'_2PO$$

$$R'R''\overset{O}{\overset{\|}{P}}R + RMgX \longrightarrow RR'R''PO$$

triphenylphosphine oxide after refluxing for 3 hr in THF.[325,326] In contrast, ethyl diethylphosphinate is only 25% converted to diethylphenylphosphine oxide after refluxing in THF for 6 hr. In addition, the rate of reaction of alkyl diphenylphosphinates with alkyl Grignard reagents in ether-benzene has been shown to be susceptible to steric inhibition when bulky alkyl groups are present in either the Grignard reagent or the phosphinate ester.[71] Nevertheless a number of optically active phosphine oxides have recently been prepared from menthyl-disubstituted phosphinates and Grignard reagents in good yields.[476,477,507]

3. ORGANOLITHIUM REAGENTS. The synthesis of tertiary phosphine oxides by the reaction of organolithium reagents with phosphorus halides and esters has received relatively scant attention. Phosphorus oxychloride and aryllithiums react exothermically in ether to give the triarylphosphine oxides in 18 to 65%.[297,590,591,700] In one case a low

$$POCl_3 + 3PhLi \xrightarrow{Et_2O} Ph_3PO$$

yield of the diarylphosphinic acid was isolated along with the tertiary phosphine oxide.[700] Pentaphenylphosphole oxide has been obtained in 70% yield in a similar manner from phenylphosphonyl dichloride and the dilithiated tetraphenylbutadiene.[499] Similarly, cyclohexyldiphenylphosphine

$$\text{Ph(Li)C=C(Ph)-C(Ph)=C(Li)Ph} + Ph\overset{O}{\overset{\|}{P}}Cl_2 \longrightarrow \text{pentaphenylphosphole oxide (}C_4Ph_4P(O)Ph\text{)}$$

oxide was obtained in 58% yield from cyclohexylphosphonyl dichloride and phenyllithium.[742]

Attempts to prepare trimethyl-, tripropyl-, and tricyclohexylphosphine oxides from phosphorus oxychloride and the alkyllithium gave only a low yield of trimethylphosphine oxide.[742] These results were attributed to the difficulty in isolating the anhydrous trimethyl- and tripropylphosphine oxides. Another possible factor may be side reactions and by-products arising from the relatively facile metallation of alkylphosphine oxides by alkyllithium reagents. The fact that aryl groups are easily displaced from triphenylphosphine oxide by alkyllithiums[747,748] may also be a complicating factor in the synthesis of mixed arylalkylphosphine oxides from arylphosphorus esters and alkyllithium reagents.

Triethyl phosphate and phenyllithium form triphenylphosphine oxide in 85% yield under much milder conditions than are required with the relatively unreactive phenylmagnesium bromide.[824] In the same study tri-p-tolylphosphine oxide was prepared in 80% yield from diethyl phosphorochloridate and p-tolyllithium. As in the case of

$$(EtO)_2P(=O)Cl + 3CH_3\text{-}C_6H_4\text{-}Li \longrightarrow (CH_3\text{-}C_6H_4)_3PO$$

phenylmagnesium bromide, trimethyl phosphate and organolithium compounds can produce methylation of the organometallic.[271]

Aryllithium reagents also react readily with diethylphosphonate to give diarylphosphinyllithium compounds. The

$$(EtO)_2P(=O)H + PhLi \longrightarrow Ph_2POLi \xrightarrow{RX} RP(=O)Ph_2$$

latter are readily alkylated to form unsymmetrical tertiary phosphine oxides, although the direct synthesis of such compounds by the use of these reagents does not appear to have been investigated.

IV. MICHAELIS-ARBUZOV REARRANGEMENTS

A method of preparation that generally gives high yields of tertiary phosphine oxides involves the reaction of a phosphinite ester, R_2POR', with an alkyl halide. This reaction is only one specific example of a very general transformation of P(III) esters to P(IV) phosphoryl compounds that was discovered by Michaelis in 1898 and extensively investigated by Arbuzov.[320]

Michaelis-Arbuzov reactions of phosphinite esters proceed by way of intermediate alkylated phosphine oxides. Evidence for the participation of alkoxyphosphonium inter-

$$R_2POR' + R''X \longrightarrow R_2\overset{+}{P}(OR')(R'')\ X^- \longrightarrow R_2\overset{O}{\overset{\|}{P}}R'' + R'X$$

mediates in the Michaelis-Arbuzov reaction of phosphonite esters was obtained by the isolation of crystalline $Et_2Me\overset{+}{P}OEtI^-$, m. 56 to 58°.[685] A similar compound, $Et_3\overset{+}{P}OEtI^-$, m. 79 to 81°, was also obtained from a reaction of the phosphorus ester with ethyl iodide. Storage of the products over phosphorus pentoxide at 18 to 21° resulted in decomposition within 15 to 25 days to give the corresponding phosphine oxides.

Decompositions of the alkoxyphosphonium intermediates probably result from nucleophilic attack of X^- on R' to form phosphoryl groups that provide the driving force for the P(III) → P(IV) transformations. In accord with this mechanism of decomposition of $R_2R'\overset{+}{P}OR'X^-$ is the finding[266] that in the decomposition of the analogous alkoxyphosphonate, $[Me(C_6H_{13})\overset{*}{C}HO]_3\overset{+}{P}EtI^-$, inversion of configuration of the asymmetric carbon occurred. The importance of the nucleophilicity of the counterion in effecting the decom-

$$\underset{(+)}{Me\overset{*}{C}H(OH)C_6H_{13}} \longrightarrow [Me(C_6H_{13})\overset{*}{C}HO]_3P$$

$$[Me(C_6H_{13})\overset{*}{C}HO]_3P + EtI \longrightarrow \underset{(-)}{C_6H_{13}(I)\overset{*}{C}HMe} + Et\overset{O}{\overset{\|}{P}}[O\overset{*}{C}H(Me)C_6H_{13}]_2$$

position reaction is further exemplified by the stability of alkoxyphosphonium nitrates,[847] tetrafluoroborates,[215] and hexachloroantimonates.[214,847] The anions of these salts, of course, are essentially nonnucleophilic.

As expected from a consideration of the mechanism, the outcome of the final step of the Michaelis-Arbuzov reaction can be greatly influenced by the nature of R'. Examples of reaction are known for R' being methyl, primary, secondary, and tertiary alkyl. Since phosphinite esters bearing tertiary alkyl substituents (R') undergo the P(III) → P(IV) conversion,[300,641] it has been suggested[300,321] that carbonium processes can also occur in the decomposition step of the reaction.

$$MeEtPOCMe_2Bu + MeI \xrightarrow{100°} Me_2Et\overset{+}{P}OCMe_2Bu\ I^-$$

$$\downarrow \text{ionization}$$

$$Me_2EtP{=}O + {}^+CMe_2Bu\ I^-$$

Aryl groups as R' substituents, in accord with the known reluctance of unactivated aryl compounds to undergo nucleophilic substitution reactions, inhibit phosphine oxide formation. This is demonstrated by the finding[31] that phenyl diphenylphosphinite reacted with methyl iodide to give solid $MePh_2\overset{+}{P}(OPh)I^-$, which was thermally quite stable. Pyrolysis of the phenoxyphosphonium iodide at 300° afforded iodobenzene, an unidentified solid, and dimethyldiphenylphosphonium iodide.[31] Other phenoxyphosphonium salts have been converted to phosphine oxides by treatment with water[584] or aqueous base[706] or by heating.[706]

$$\geqslant\overset{+}{P}OAr\ X^- \xrightarrow[\text{heat or } H_2O]{\text{base or}} \geqslant P{=}O + ArOH + HX$$

Phosphinite esters, in general, undergo the Michaelis-Arbuzov reaction readily. For example, the reaction of ethyl diethylphosphinite with ethyl iodide was found to occur within 5 hr at 45°, while addition of methyl iodide to the ester at room temperature resulted in a very vigorous formation of diethylmethylphosphine oxide.[684] An aryl-substituted phosphinite ester, methyl diphenylphosphinite, appeared to be somewhat less reactive than the alkyl analogs in that the addition of methyl iodide gave a rapid but controllable reaction.[32]

A wide variety of alkylating agents have been used in the Michaelis-Arbuzov phosphinite ester reaction. Included in this list are primary and secondary alkyl chlorides, bromides, and iodides, α-haloketones, α-haloesters, α-heteroatom-substituted alkyl halides, carbon tetrachloride, acetyl chloride, benzal chloride, ethyl chloroformate, triphenylmethyl bromide, and cycloalkylvinyl halides (see Table 3). The diversity of alkylating agents capable of effecting the P(III) → P(IV) transformation of phosphinite esters suggests that the quaternization reactions occur by a variety of mechanisms contained only within the confines of the limiting extremes of S_N1 and S_N2 processes. There appears to be no experimental evidence pertaining to reactions of phosphinite esters with aryl halides that proceed by nucleophilic displacement mechanisms. However, it is reasonable to expect, by analogy with other phosphorus ester reactions,[320] that reactions of these reagents would occur only with considerable difficulty. Attention is called to the fact that the Michaelis-Arbuzov phosphinite ester reaction frequently generates compounds (R'X) capable of alkylating the starting esters. When this occurs, mixed phosphine oxides result which complicate work-up procedures.

It has long been known that some phosphinite esters undergo "self-isomerization" to the corresponding phosphine oxide. Isomerization can be effected by heating and is

$$R_2P\text{-}OR' \longrightarrow R_2\overset{\overset{\displaystyle O}{\|}}{P}\text{-}R'$$

catalyzed by impurities in the reaction mixtures or by additions of small quantities of iodine or R'X.[32] Indeed, because of the ease of the isomerization process, it frequently has been difficult to isolate phosphinite esters. Early attempts[32] to prepare methyl- and benzyldiphenylphosphinite esters led only to the corresponding phosphine oxides.

Notably reactive to thermal isomerization are allyl phosphinites[36,67,337a,659,713] and 2-alkynyl phosphinites.[89,544,545,745,746] The mechanisms of rearrangements of these esters have been studied, and it has been concluded, from extensive experimental evidence, that both occur by concerted intramolecular processes. These rearrangements therefore differ from the classical Michaelis-

$$>P\text{:}\;\cdots\;CR'_2{=}C\text{—}CR_2\text{—}O\text{—}P \xrightarrow{\Delta} >\overset{\overset{\displaystyle O}{\|}}{P}\text{-}CR'_2\text{-}C{=}CRR$$

$$>P\text{:}\;\cdots\;HC{\equiv}C\text{—}CH_2\text{—}O\text{—}P \xrightarrow{\Delta} >\overset{\overset{\displaystyle O}{\|}}{P}\text{-}CH{=}C{=}CH_2$$

Arbuzov reactions in that discrete phosphonium intermediates are probably not involved. However, this difference is a matter of degree and not of kind.

A unique free-radical method of isomerizing phosphinite esters to phosphine oxides was recently disclosed.[199,200] It was found that dimethylamino radicals, generated by the irradiation of tetramethyltetrazen, initiated a chain reaction that converted methyl diphenylphosphinite to methyldiphenylphosphine oxide in moderate yield.[199] Methyl radicals and isopropyl radicals, generated by photolysis of

$$Me_2N\cdot + Ph_2POMe \longrightarrow Ph_2P(=O)NMe_2 + Me\cdot \quad \text{(initiation)}$$

$$Me\cdot + Ph_2POMe \longrightarrow Ph_2P(=O)Me + Me\cdot$$

azoalkanes, reacted with methyl diphenylphosphinite to give methyl- and isopropyldiphenylphosphine oxide in 80 and 33% yields, respectively.[200] Benzyl radicals[200] did not react with the phosphorus ester, while t-butyl radicals,[200] like the dimethylamino radicals,[199] merely initiated a P(III) → P(IV) Arbuzov transformation of the ester to methyldiphenylphosphine oxide. Phenyl radicals with various alkyl diphenylphosphinite esters afforded moderate yields of triphenylphosphine oxide.[200]

The synthetic utility of the Michaelis-Arbuzov reaction for phosphine oxide preparations depends on the availability of the requisite phosphinite esters. The value of this method has increased with the finding[529,535] that these esters can be readily obtained by the following series of reactions:

$$R_2NPCl_2 + R'_3Al \longrightarrow R_2NPR'_2 + AlCl_3$$

$$R_2NPR'_2 + R''OH \longrightarrow R''OPR'_2 + R_2NH$$

$$R''OPR'_2 \xrightarrow{R''Br} R''P(=O)R'_2$$

Unsaturated alcohols, functionally substituted alcohols, and α,ω-alkanediols afford high overall yields of phosphine oxides.

$$2R'_2PNR_2 + HO(CH_2)_nOH \longrightarrow R'_2PO(CH_2)_nOPR'_2 + 2R_2NH$$

$$R'_2PO(CH_2)_nOPR'_2 \xrightarrow{Br(CH_2)_nBr} R'_2P(=O)(CH_2)_nP(=O)R'_2$$

The Michaelis-Arbuzov phosphinite reaction has recently been used to considerable advantage in the synthesis of several novel compounds containing two, three, and four phosphorus atoms.[535,537,563,570,571,664,672] The yields of products from these reactions were good to excellent.

$$R_2POR' + (ClCH_2)_nP(=O)R''_{3-n} \longrightarrow (R_2P(=O)CH_2)_nP(=O)R''_{3-n}$$

Table 3 includes some phosphine oxides prepared by the Michaelis-Arbuzov phosphinite ester reaction. These examples were chosen merely to show the scope of the method and to place it into perspective with other synthetic routes to phosphine oxides, and the tabulation should not be regarded as comprehensive.

V. FROM PRIMARY AND SECONDARY PHOSPHINE OXIDES

Aliphatic and functionally substituted tertiary phosphine oxides can be prepared by the addition of primary and secondary phosphine oxides to unsaturated compounds under a variety of conditions. Certain primary phosphine oxides, e.g., react with acrylonitrile under basic conditions to give alkylbis(cyanoethyl)phosphine oxides in 33 to 55% yields.[114,115] In addition, certain α,α'-dihydroxytertiary phosphine oxides can be prepared by the acid-

$$RP(=O)H_2 + CH_2{=}CHCN \xrightarrow{NaOEt} RP(=O)(CH_2CH_2CN)_2$$

catalyzed addition of primary phosphine oxides to aldehydes.[113-115] Finally, trioctylphosphine oxide has been

$$RP(=O)H_2 + RCHO \xrightarrow{H^+} RP(=O)(CH(OH)R)_2$$

prepared by the free-radical addition of octylphosphine oxide to 1-octene. In almost all cases, however, by-products arising from either disproportionation of the primary phosphine oxide or incomplete conversion to the tertiary phosphine oxides were observed.[114] These observations, plus the fact that primary phosphines undergo comparable addition reactions to form tertiary phosphines that are easily oxidized to tertiary phosphine oxides, limits the synthetic utility of primary phosphine oxides.

Secondary phosphine oxides in general undergo addition

Table 3. Phosphine Oxide Preparations by Michaelis-Arbuzov Reactions

Phosphinite Ester	Other Reactant	Phosphine Oxide	% Yield	Ref.
Et_2POEt	MeI	$Et_2P(O)Me$	—	685
$Ph_2POC_6H_{13}$	$C_6H_{13}Br$	$Ph_2P(O)C_6H_{13}$	—	775
Ph_2POEt	Me_3SiCH_2Cl	$Ph_2P(O)CH_2SiMe_3$	94.5	118
Ph_2POMe	N-(bromomethyl)phthalimide (NCH_2Br)	N-phthalimido-$CH_2P(O)Ph_2$ ($NCH_2P(O)Ph_2$)	—	653
Et_3POBu	$BrCH_2CO_2Et$	$Et_2P(O)CH_2CO_2Et$	—	36
Ph_2POBu-i	$PhC(O)CH_2Br$	$Ph_2P(O)CH_2C(O)Ph$	60	35
$Ph_2POCH_2CH_2Cl$	—	$Ph_2P(O)CH_2CH_2Cl$	89	295
Me_2POEt	$Br(CH_2)_2Br$	$Me_2P(O)CH_2CH_2Br$	—	774a
$Et_2PO(CH_2)_{12}OPEt_2$	"	$Et_2P(O)(CH_2)_{12}P(O)Et_2$	71	535
Me_2POBu	$(ClCH_2)_3PO$	$[Me_2P(O)CH_2]_3PO$	93.5	537
Ph_2POMe	cyclohexylidene$=CHCH_2Br$	$Ph_2P(O)CH_2CH=$cyclohexylidene	—	713
$Ph_2P(O)CH_2CH(Ph)OPPh_2 \cdot HCl$	—	$Ph_2P(O)CH_2CH(Ph)P(O)Ph_2$	60	788

$Ph_2POCH_2C{\equiv}CH$	—	$Ph_2P(O)CH{=}C{=}CH_2$	60	89
Ph_2POMe	Ph_3CBr	$Ph_2P(O)CPh_3$	100	32
Ph_2POMe	CCl_4	$Ph_2P(O)CCl_3$	60-90	466
Ph_2POMe	$CH_3C(O)Cl$	$Ph_2P(O)C(O)Me$	54	32
Ph_2POEt	1,2-dichloro-hexafluorocyclopentene (Cl, Cl, F_6)	1,2-bis($Ph_2P(O)$)-hexafluorocyclopentene ($Ph_2P(O)$, $P(O)Ph_2$, F_6)	75	252
$(CF_3)_2POBu\text{-}t$	CH_3I	$(CF_3)_2P(O)Bu\text{-}t$	78	300
Ph_2POEt	$ClCH_2P(O)(OBu)_2$	$Ph_2P(O)CH_2P(O)(OBu)_2$	60	444
$Ph_2POCHMe_2$	$PhCHCl_2$	$Ph_2P(O)CH(Cl)Ph$	—	336
Ph_2POMe	$EtOC(O)Cl$	$Ph_2P(O)CO_2Et$	—	32
Ph_2POEt	benzyne	Ph_3PO	17	295
phosphetane (Me, H, Me_2, Me_2, P:, MeO)	—	phosphetane oxide (Me, H, Me_2, Me_2, P, O, Me)	—	763a

reactions to give tertiary phosphine oxides in much the same manner as do primary phosphine oxides. Thus a number of α-hydroxy tertiary phosphine oxides have been obtained in good to excellent yields by the netural- or base-catalyzed addition of secondary phosphine oxides to aldehydes and ketones.[595,678,679] In relation to this reaction, the

$$R_2P(O)H + R'_2CO \xrightarrow{NaOEt} R_2P(O)C(OH)R'_2$$

halomagnesium salts of two secondary phosphine oxides, i.e., the precursors to secondary phosphine oxides from diethyl phosphonate and Grignard reagents, were also found to add to benzaldehyde to give moderate yields of tertiary α-hydroxybenzylphosphine oxides. The generality of this method does not appear to have been fully investigated.

Secondary phosphine oxides have been metallated to give alkali-metal or halomagnesium salts of disubstituted phosphinites, which were then alkylated to form tertiary phosphine oxides.[226,346,365,640,679,706] Yields of the tertiary

$$R_2P(O)H \longrightarrow R_2POM \xrightarrow{R'X} R'R_2PO$$

phosphine oxides are good to excellent. However, the reaction of diethyl phosphonate with Grignard reagents, followed by alkylation, is more direct and does not involve isolation of the secondary phosphine oxide.

Secondary phosphine oxides have been added to α,β-unsaturated nitriles, esters, ketones, and amides under basic conditions to give a variety of functionally substituted tertiary phosphine oxides.[490,492,592,594,678-680]

$$R_2P(O)H + CH_2{=}CHCN \xrightarrow{NaOH} R_2P(O)CH_2CH_2CN$$

Yields of the tertiary phosphine oxides are usually quite good, ranging from 26 to 92%. Additions of secondary phosphine oxides to unactivated olefins have been accomplished by using a free-radical initiator. In this manner n-octylbis(2-cyanoethyl)phosphine oxide[679] and dimethyldodecylphosphine oxide[327] were obtained in yields of 48 and 76%, respectively. Although secondary phosphine oxides are relatively more stable and less susceptible to oxidation than primary phosphine oxides, their use in synthesis is also limited, partly because of the ease with which either the secondary phosphines or the halomagnesium salt of the secondary phosphine oxides, prepared from diethyl phosphonate and Grignard reagents, are converted to tertiary phosphine oxides.

VI. HYDROLYSIS OF DIHALOPHOSPHORANES

Dihalophosphoranes are highly reactive compounds that are readily converted to tertiary phosphine oxides by several oxygen-containing compounds.[583,588] Because of their reactivity dihalides are seldom isolated but serve chiefly as intermediates to tertiary phosphine oxides.

Dihalophosphoranes are commonly prepared by direct halogenation of tertiary phosphines; e.g., bromine may be added to the tertiary phosphine in an inert solvent with cooling. The inert solvent may be chloroform, methylene

$$R_3P + Br_2 \longrightarrow R_3PBr_2 \xrightarrow{H_2O} R_3PO$$

chloride, ethylene chloride, or benzene. The intermediate dibromide is then hydrolyzed with water[85,163,225,248,577,580,582,773,834] or, more commonly, with aqueous sodium hydroxide[25,34,184,556,580,583,835] or alcohol.[248,419] The resulting phosphine oxide can then be dried, depending on its properties, by heating at 100° under vacuum, by the use of suitable drying agents such as molecular sieves, or by distillation. Reported yields of aliphatic and aromatic tertiary phosphine oxides are usually 40 to 77%.

Modifications of the procedure described above include the use of an aqueous reaction medium in place of the inert solvent[225,260,581,589] and the use of chlorine,[583] cyanogen bromide,[773] or metal halides[147,834] in place of bromine. Aqueous aniline has been substituted for aqueous sodium hydroxide or water to prepare tris(trichloromethyl)phosphine oxides from the dihalide.[274,429] In addition, the oxygen-containing compounds phosphorus pentoxide,[480] sulfur dioxide,[710] and carbon dioxide[389] can be used to convert the dihalides to the tertiary phosphine oxide under anhydrous conditions. Oxalic acid and tris(trifluoromethyl)-phosphine dichloride react to give tris(trifluoromethyl)-phosphine oxide in 70% yield.[630] An attempt to hydrolyze

$$(F_3C)_3PCl_2 + (CO_2H)_2 \longrightarrow (F_3C)_3PO + CO + CO_2 + 2HCl$$

the latter dihalide with water resulted in bis(trifluoromethyl)phosphinic acid from C-P bond cleavage.

From a synthetic standpoint the oxidation of tertiary phosphines to their oxides via the dihalides is limited, because of the ease and simplicity with which tertiary phosphines are oxidized with such reagents as hydrogen peroxide. The method is also limited to phosphines that do not contain substituents capable of reacting readily with halogen. For example, tri-o-anisylphosphine forms, upon bromination and hydrolysis, a tribrominated tri-o-anisylphosphine oxide and not tri-o-anisylphosphine oxide.[540]

Other syntheses of tertiary phosphine oxides via the dihalides involve the alkylation of phosphorus halides or other phosphorus compounds followed by hydrolysis. Typical examples of phosphine oxides prepared in this manner are given in Table 4 along with the starting phosphorus compound, the alkylating agent, and the yields of phosphine oxides.

Table 4. Phosphine Oxides from Dihalophosphoranes

Phosphorus Compound	Alkylating Agent	Phosphine Oxide	% Yield	Ref.
Ph_2PCl	$PhCH_2Cl$	$PhCH_2Ph_2PO$	—	225
Ph_2PCl	MeI	$MePh_2PO$	74	356
Ph_2PI_3	"	"	77	25
Ph_2PI_3	$PhCH_2Cl$	$PhCH_2Ph_2PO$	74	"
PCl_5	CH_2N_2	$(ClCH_2)_3PO$	—	430
$PhPCl_4$	"	$Ph(ClCH_2)_2PO$	31	428
$3\text{-}ClC_6H_4PCl_4$	"	$3\text{-}ClC_6H_4(ClCH_2)_2PO$	≈100	"
PI_3	$PhCH_2Cl$	$(PhCH_2)_3PO$	90	184, 356, 504
P_4	RX	R_3PO	—	141, 184
PH_4I	$PhCHCl_2$ or PhCHO	$(PhCH_2)_3PO$	—	248, 514

Noteworthy with respect to the dihalide method of preparing phosphine oxides is the recent synthesis of several heterocyclic phospholene oxides from substituted dichlorophosphines and dienes.[560-564] The synthesis appears to be

$$RPX_2 + \text{(diene)} \longrightarrow \text{RPX}_2\text{(phospholene)} \xrightarrow{\text{aq. NaOH}} \text{RP(O)(phospholene)}$$

general.[132,133,165,169,223,290,291,399,662,664] A variety of substituted aryl- and alkylphospholene oxides are formed in 7 to 82% yields. Isoprene, chloroprene, and a number of other dienes react in place of butadiene but at different rates, depending on both the diene and the dichlorophosphine. The synthesis[662] and the chemistry of these phosphine oxides have been recently reviewed along with those of other heterocyclic phosphorus compounds.[69]

VII. FROM PHOSPHINYLALKYLMETAL COMPOUNDS

One of the most significant advances in phosphine oxide synthesis of the last decade has been the advent[377] of phosphinyl-substituted organometallic compounds, $>\overset{O}{\overset{\|}{P}}-\overset{|}{\underset{|}{C}}-M$, where M = Li, Na, K, MgX. These organometallic compounds, which are readily available from reactions of the corresponding phosphine oxides with strong bases such as n-butyllithium, lithium amide, and potassium-t-butoxide, or from reactions of triphenylphosphine oxide with alkyllithium and alkylmagnesium compounds, undergo reactions characteristic of moderately reactive organometallic compounds to give functionally substituted phosphine oxides. Typical of this new class of reactive intermediates is (diphenylphosphinyl)phenylmethyllithium, $Ph_2\overset{O}{\overset{\|}{P}}CHLiPh$, which upon carbonation led to a 65% yield of (α-carboxybenzyl)-diphenylphosphine oxide, while reaction with benzaldehyde afforded (β-hydroxy-α,β-diphenylethyl)diphenylphosphine oxide.[382] Additional organometallic-like reactions of the

$$Ph_2\overset{O}{\overset{\|}{P}}CH_2Ph + RLi \longrightarrow Ph_2\overset{O}{\overset{\|}{P}}CH(Ph)Li + RH$$

$$Ph_2\overset{O}{\overset{\|}{P}}CH(Ph)Li \begin{cases} \xrightarrow{CO_2} \xrightarrow{H^+} Ph_2\overset{O}{\overset{\|}{P}}CH(Ph)CO_2H \\ \xrightarrow{PhCHO} \xrightarrow{H^+} Ph_2\overset{O}{\overset{\|}{P}}CH(Ph)C(OH)(H)Ph \end{cases}$$

metallated phosphine oxide are its acylation with benzoyl chloride[284] and methyl esters[349] to give the expected β-ketophosphine oxides, and aralkylations with benzyl bromide and m-chlorobenzyl bromide to yield (1,2-diphenylethyl)diphenylphosphine oxide[8] and (1-phenyl, 2-m-chlorophenylethyl)diphenylphosphine oxide,[788] respectively. Oxidation with elemental oxygen yields (α-hydroxybenzyl)diphenylphosphine oxide;[374] reaction with styrene oxide, the γ-hydroxy compound.[376] Some question exists as to whether the product is a primary or secondary alcohol.[434]

$$Ph_2P(O)CH(Ph)Li \xrightarrow{PhC(O)Cl} Ph_2P(O)CH(Ph)C(O)Ph$$

$$Ph_2P(O)CH(Ph)Li \xrightarrow{PhCH_2Br} Ph_2P(O)CH(Ph)CH_2Ph$$

$$Ph_2P(O)CH(Ph)Li \xrightarrow{\sim \frac{1}{2} O_2} Ph_2P(O)CH(OH)Ph$$

$$Ph_2P(O)CH(Ph)Li \xrightarrow{PhCH(O)CH_2} Ph_2P(O)CH(Ph)CH_2CH(OH)Ph \quad ?$$

A series of functionally substituted aliphatic organophosphine oxides has been prepared similarly.[695] (Di-n-hexylphosphinyl)methyllithium, $(n\text{-}C_6H_{13})_2P(O)CH_2Li$, on treatment with carbon dioxide, aldehydes, and ketones, afforded the expected products in moderate yields. Of special interest from this study was the synthesis of some symmetrical and unsymmetrical bis(phosphinyl)methanes. In a similar manner ethylene-bis(dibutylphosphine) dioxide has been

$$R_2P(O)CH_2Li + R'_2P(O)Cl \longrightarrow R_2P(O)CH_2P(O)R'_2 \quad (25 \text{ to } 35\%)$$

synthesized from the reaction of (dibutylphosphinyl)methyllithium with (chloromethyl)dibutylphosphine oxide.[483]

With the discovery of the phosphinylalkylmetal compounds several novel heteroatom-substituted phosphine oxides have become available.[747] Considerable potential value attaches to this overall method of preparation of

$$Ph_3P{=}O + RCH_2M \quad Ph_2P(O)CHRM + PhH$$

$$Ph_2P(O)CHRM \xrightarrow{Ph_3SiCl} Ph_2P(O)CHRSiPh_3 \quad (65\%)$$

$$Ph_2P(O)CHRM \xrightarrow{Ph_2PCl} \xrightarrow{S} Ph_2P(O)CHRP(S)Ph_2 \quad (25\%)$$

$$Ph_2P(O)CHRM \xrightarrow{Ph_3SnCl} Ph_2P(O)CHRSnPh_3 \quad (81\%)$$

substituted phosphine oxides since it utilizes readily available organometallic compounds and triphenylphosphine oxide as starting materials.

Other functionally substituted phosphine oxides that have been prepared using phosphinylalkylmetal compounds are diethyl(2-hydroxy-1-methyl-2,2-diphenylethyl)phosphine oxide,[330] diphenyl(2-hydroxy-1-cyclopropyl-2,2-diphenylethyl)phosphine oxide,[520] and a series of diethylphosphinyl-substituted acetic acid esters.[686]

$$Et_2P(O)CH(Me)Na + Ph_2C{=}O \longrightarrow \xrightarrow{H^+} Et_2P(O)CH(Me)C(OH)Ph_2 \quad (60\%)$$

$$Ph_2P(O)CH(Li)\triangleleft + Ph_2C{=}O \longrightarrow \xrightarrow{H^+} Ph_2P(O)CH(\triangleleft)C(OH)Ph_2 \quad (80\%)$$

$$Et_2P(O)CH_2CO_2Et + K \longrightarrow \xrightarrow{RX} Et_2P(O)CHRCO_2Et \quad (64\ to\ 71\%)$$

The syntheses of functionally substituted organophosphine oxides using phosphinylalkylmetal compounds need not necessarily be carried out in two individual steps, i.e., prior formation of the metallated phosphine oxides and subsequent functionalization. When the more selective alkoxides and amides are used as bases, phosphinylalkylmetal compounds can be generated in the presence of certain reactive functionalizing reagents. This factor has been used to advantage in the preparation of (α-nitrobenzyl)diphenylphosphine oxide,[373] diphenyldesylphosphine oxide,[382] and a phthalic ester-phenyldibenzylphosphine oxide condensation product.[382]

$$Ph_2P(O)CH_2Ph + t\text{-}BuOK \xrightarrow{EtONO_2} Ph_2P(O)CH(NO_2)Ph \quad (62\%)$$

$$Ph_2P(O)CH_2Ph + t\text{-}BuOK \xrightarrow{PhC(O)OEt} Ph_2P(O)CH(Ph)C(O)Ph \quad (50\%)$$

Some interesting methods of preparing unsaturated phosphine oxides that employ phosphinyl-substituted organometallic compounds have recently been disclosed. Bis(diphenylphosphinyl)methylpotassium[445] and (diphenylphosphino)diphenylphosphinylmethylpotassium[276] react with aldehydes to give "Wittig-like" products.

Several synthetically useful phosphinyl-substituted

$$(Ph_2P(O))_2CHK + RCHO \longrightarrow Ph_2P(O)CH{=}CHR + Ph_2P(O)K$$

$$Ph_2P(O)CHKPPh_2 + RCHO \longrightarrow Ph_2PCH{=}CHR + Ph_2P(O)K$$

carbanions, $R_2P(O)\overset{-}{C}\overset{+}{R}N_2$, have been described.[491,688] The carbanions, as expected, are highly stable but possess sufficient reactivity to afford chloromethylphosphine oxides, phosphazenes, and phosphinyl-substituted pyrazolines upon treatment with hydrochloric acid, tertiary phosphines, and unsaturated carbonyl compounds.[491] Photolysis of $Ph_2P(O)\overset{-}{C}(Ph)\overset{+}{N}_2$ in aqueous dioxane gave, via a carbene

$$R_2P(O)\overset{-}{C}H\overset{+}{N}_2 \xrightarrow{HCl} R_2P(O)CH_2Cl + N_2$$

$$R_2P(O)\overset{-}{C}H\overset{+}{N}_2 \xrightarrow{Ph_3P} R_2P(O)CH{=}N{-}\overset{-}{N}{-}\overset{+}{P}Ph_3$$

$$R_2P(O)\overset{-}{C}H\overset{+}{N}_2 \xrightarrow{MeC(O)CH{=}CH_2} Ph_2P(O)\text{-pyrazoline-}C(O)Me$$

intermediate, α-hydroxybenzyldiphenylphosphine oxide and (diphenylmethyl)phenylphosphinic acid.[688]

VIII. FROM ELEMENTAL PHOSPHORUS AND PHOSPHINE

Over the years a number of investigations have been initiated for the purpose of developing methods for preparing organophosphorus compounds directly from elemental phosphorus. Impetus for these efforts stemmed from the ready availability of phosphorus, either white or red, and the desire to circumvent the existing multistep preparative methods required for most organophosphorus syntheses. Many of these reactions involving elemental phosphorus have been shown to give phosphine oxides as final products. However, only a few of the reactions appear to constitute valuable synthetic routes to phosphine oxides since, in a number of instances, the yields are low because of the formation of other organophosphorus by-products.

Reactions of white phosphorus with sodium hydroxide and sodium ethoxide were described in 1904 as giving rise to dark red solutions which, upon treatment with methyl iodide, afforded some trimethylphosphine oxide along with methylphosphine, methylphosphonic acid, and dimethylphosphinic acid.[42] Recently this route to phosphine oxides has been more fully developed by using an electron-deficient alkylating agent, acrylonitrile or acrylamide, in place of methyl iodide.[676,677] Reaction of white phosphorus with aqueous potassium hydroxide in the presence of acrylonitrile gave tris(2-cyanoethyl)phosphine oxide in 53% yield (based on phosphorus used). An analogous reaction with acrylamide in aqueous ethanol resulted in the formation of a

$$P_4 + 2KOH + 4H_2O + 9CH_2{=}CH\text{-}CN \xrightarrow{CH_3CN} 3O{=}P(CH_2CH_2CN)_3 + K_2HPO_3$$

74% yield of tris(2-carbamoylethyl)phosphine oxide. Ethyl acrylate as an alkylating agent, however, led to only very low yields of the corresponding phosphine oxide.[677] The initial step of this complex reaction was thought to involve nucleophilic attack of hydroxide ion on tetrahedral white phosphorus to give a phosphide ion that subsequently underwent a Michael addition to the unsaturated alkylating agent present.[676]

$$P_4 + HO^- \longrightarrow {}^-P_3P\text{-}OH \xrightarrow{CH_2{=}CHX} HO\text{-}P_3P\text{-}CH_2\text{-}\overset{-}{C}HX \xrightarrow{H_2O} HO\text{-}P_3PCH_2CH_2X \xrightarrow{etc.} O{=}P(CH_2CH_2X)_3$$

White phosphorus also reacts with carbon nucleophiles to give complex phosphide intermediates capable of being alkylated. Reaction of the complex phosphide obtained from phosphorus and phenyllithium with benzaldehyde gave a 12% yield of bis(α-hydroxybenzyl)phenylphosphine oxide.[682]

$$PhLi + P_4 \longrightarrow [\text{Organophosphide}] \xrightarrow[(2)\ H_2O]{(1)\ PhCHO} (Ph\overset{OH}{\overset{|}{C}H})_2\overset{O}{\overset{\|}{P}}Ph$$

Historically, some of the earliest preparations of phosphine oxides resulted indirectly from alkylations of phosphorus by alkyl halides.[141,184,348,556] Thus alkyl iodides, upon reaction with white or red phosphorus at elevated temperatures, gave what appeared to be mixtures of dihalophosphorane and phosphonium salt. Treatment of the mixtures with ethanol and then potassium hydroxide resulted in the formation of trialkylphosphine oxide.

$$P_4 + RI \xrightarrow[22\text{ hr}]{180°} R_3PI_2 + R_4PI \xrightarrow[(2)\ KOH]{(1)\ EtOH} R_3P{=}O \quad (25 \text{ to } 52\%)$$

Interest in phosphorus alkylation reactions has recently been revived by several groups of workers, the result being considerable improvement in these reactions as preparative routes to phosphine oxides.

Red phosphorus, alkyl iodides (C_1 to C_4), and a trace of iodine, when heated in an autoclave at 200° for 8 hr, followed by treatment with aqueous sodium sulfite or successive treatments with nitric acid and aqueous base, gave trimethylphosphine oxide (54%), triethylphosphine oxide (25%), and tributylphosphine oxide (66%).[504] Alkylations of phosphorus with higher alkyl iodides did not have to be run in an autoclave.[242] Isoamyl iodide, e.g., when refluxed with red phosphorus and small amounts of iodide for 100 hr, gave tri-i-amylphosphine oxide in 83% yield. It has been suggested[504] that these reactions proceeded by the initial formation of P_2I_4, which subsequently reacted readily with RI to give $R_6P_2I_x$ and R_3PI_y. Indeed, it was shown that P_2I_4 reacted with alkylating agents to form good

$$P_{red} + I_2 \longrightarrow P_2I_4 \xrightarrow{RI} R_6P_2I_x + R_3PI_y \xrightarrow{H_2O} R_3P{=}O$$

yields (25 to 75%) of tertiary phosphine oxides under similar conditions. A reaction of benzyl iodide with P_2I_4 occurred exothermically at 110° and resulted in the formation of tribenzylphosphine oxide in greater than 90% yield.[504] A similar reaction of diphosphorus tetraiodide with benzyl chloride resulted in a 37% yield of phosphine oxide.[356]

Several substituted tribenzylphosphine oxides have been obtained in yields of 67 to 86% by similar reactions of red phosphorus, substituted benzyl chloride, and iodine at 120 to 130°.[844] In a further variation, phosphine oxides were prepared from white phosphorus and benzyl halides with the aid of metal catalysis.[142]

In a modification of this reaction, higher alcohols ($> C_5$) were substituted for the alkyl halides as the alkylating agents. The yields of phosphine oxides were reported to be in excess of 80% from these reactions.[241]

$$ROH + P_{red} + I_2 \xrightarrow{\sim 200°} \xrightarrow{NaOH} R_3P{=}O$$

In another systematic study,[533] reactions of white phosphorus with methyl chloride at temperatures ranging from 200 to 310°, and under pressure, were found to give tetramethylphosphonium chloride in yields varying from 60 to 93%. The desired phosphonium salt was accompanied by methylphosphinous chloride (4 to 23%) and phosphorus trichloride (11 to 47%). Again, phosphine oxides were formed by alkaline decompositions of the phosphonium salts.

$$2P_4 + 4MeCl \longrightarrow PCl_3 + Me_4\overset{+}{P}Cl^- \xrightarrow{base} Me_3P{=}O$$

An interesting alkylation of white phosphorus has been achieved using N-hydroxymethyldialkylamines.[523,534] Tertiary phosphine oxides comprise up to ca. 45% of the reaction products. The yields of phosphine oxides were shown to be dependent on mole ratios of reactants, variations in solvent, variations in temperature, and, in

$$R_2NH + CH_2{=}O \longrightarrow R_2NCH_2OH \xrightarrow[H_2O]{P_4}$$

$$(R_2NCH_2)_3P{=}O + (R_2NCH_2)_2\overset{O}{\overset{\|}{P}}OH + R_2NCH_2\overset{O}{\overset{\|}{P}}(OH)_2$$

particular, on the pH of the reaction mixtures.[534] Thus, at a pH of 10 or above, the reaction occurred very rapidly, prompting proposal of a mechanism that involved initial alkoxide attack on phosphorus. This reaction, as suggested,

$$R_2NCH_2OH + B \longrightarrow R_2NCH_2O^- + BH^+$$

$$P_4 + {}^-O\text{-}CH_2NR_2 \longrightarrow {}^-P_3\text{—}P\text{-}OCH_2NR_2$$

is reminiscent of the mechanism proposed for hydroxide attack on phosphorus.[676] A direct attack of hydroxymethyldialkylamine on P_4 was favored for the reaction at a pH of 7 or below.

The reaction of phosphorus with metals,[100,129,503] first investigated more than 100 years ago,[129] has recently been shown[638] to occur with sodium at 100°, and sodium-potassium alloy at 25°, to give phosphides characterized as M_3P. These phosphides reacted readily with methyl halides in glyme solvents to produce varying mixtures of tetramethylphosphonium and trimethylphosphonium halides. Treatments of the reaction mixtures with sodium hydroxide gave trimethylphosphine oxide in overall yields up to ca. 60%. A significant improvement in the reaction of white phosphorus with sodium was subsequently achieved[635] by

$$Na + P \longrightarrow Na_3P \xrightarrow{MeX} Me_4PX \xrightarrow{NaOH} Me_3P{=}O$$

utilizing various aromatic hydrocarbons as "electron transport agents," which facilitated the reaction to the extent

$$Na + C_{10}H_8 \longrightarrow [C_{10}H_8^{\cdot -}]\,Na^+ \xrightarrow{P_4} Na_3P + C_{10}H_8$$

that phosphide formation was rapid at room temperatures.

In a modification of this process, trialkylphosphine oxides were obtained from a "Wurtz-type" reaction involving phosphorus, sodium, and alkyl halide at elevated temperatures in organic solvents.[804]

Alkali-metal phosphides, M_3P, apparently have also been obtained from exhaustive metallation[421] of phosphine and metal reduction[364] of phosphorus trichloride. These reactions, therefore, provide additional preparative routes

$$PH_3 + 3PhLi \longrightarrow PLi_3 + 3PhH$$

$$PCl_3 + 6M \longrightarrow PM_3 + 3MCl$$

to phosphine oxides, starting with compounds available from elemental phosphorus.

$$P \xrightarrow{Cl_2} PCl_3$$

$$P \xrightarrow{^-OH} PH_3 + \text{"Phosphorus acids"}$$ [801]

Phosphine is also the starting material for the preparation of tertiary phosphine oxides directly from aromatic aldehydes.[113,231,514] Strong acid is necessary to effect the reactions, which represent interesting examples of the transfer of oxygen from carbon to phosphorus.

$$PH_3 + X\text{-}C_6H_4\text{-}CHO \xrightarrow[Et_2O]{HCl} (X\text{-}C_6H_4\text{-}\overset{OH}{\overset{|}{C}}H)_2\overset{O}{\overset{\|}{P}}CH_2\text{-}C_6H_4\text{-}X \sim 60\%$$

$$X = H, CH_3, Cl$$

IX. MISCELLANEOUS PREPARATIONS

A number of cyclohexylphosphine oxides have been prepared by the photophosphonylation of cyclohexane in good yields.[616] Diethylchlorophosphine, upon photolysis for 10 days in cyclohexane, gave an 80% yield of diethylcyclohexylphosphine oxide. It is anticipated that this reaction

$$C_6H_{12} + R_2PCl \xrightarrow[O_2]{h\nu} R_2\overset{O}{\overset{\|}{P}}\text{-}C_6H_{11} + HCl$$

will give mixtures of positional isomers when linear hydrocarbons are used in place of cyclohexane.

Disubstituted α-hydroxyalkylphosphines undergo a thermal rearrangement to tertiary phosphine oxides.[111,334,783] This rearrangement occurs upon heating the phosphine for several

$$R_2PC(OH)R_2^1 \xrightarrow{\Delta} R_2\overset{O}{\overset{\|}{P}}CHR_2^1$$

hours or in refluxing toluene with toluenesulfonic acid as catalyst. The yields of the tertiary phosphine oxides from these reactions are only fair (20 to 57%), presumably because of a competing decomposition of the hydroxyalkylphosphine to the disubstituted phosphine and the aldehyde.

Diphenylchlorophosphine, which is commercially available, reacts with certain aldehydes in the presence of acid to form diphenyl-α-hydroxy-substituted phosphine oxides.[550] The yields of the tertiary phosphine oxides derived from

$$Ph_2Cl + RCHO \xrightarrow{H^+} R_2\overset{O}{\overset{\|}{P}}CHOHR$$

formaldehyde and benzaldehyde were 63% and 85%, respectively. No tertiary phosphine oxide could be isolated from the reaction of butyraldehyde with diphenylchlorophosphine.

Tetraphenyldiphosphine reacts with carbonyl compounds to give several novel types of phosphine oxides.[201,202]

For example, tetraphenyldiphosphine and benzoic acid at 180° give α-(diphenylphosphinyloxy)benzyldiphenylphosphine oxide. Aliphatic carboxylic acids and tetraphenyldiphosphine at 180° initially give 1-hydroxy-1,1-bis(diphenyl-

$$Ph_2PPPh_2 + PhCO_2H \xrightarrow{\Delta} PhCH\begin{cases} P(O)Ph_2 \\ OP(O)Ph_2 \end{cases}$$

phosphinyl)alkanes. With further heating at 180° the diphosphine dioxide slowly rearranges to the 1-diphenyl-

$$Ph_2PPPh_2 + RCO_2H \xrightarrow{\Delta} RC(OH)\begin{cases} P(O)Ph_2 \\ P(O)Ph_2 \end{cases}$$

phosphinyloxyalkyldiphenylphosphine oxide. This rearrangement of the diphosphine dioxide is favored by R groups that stabilize adjacent carbanions.

$$RC(OH)\begin{cases} P(O)Ph_2 \\ P(O)Ph_2 \end{cases} \xrightarrow{\Delta} RCH\begin{cases} P(O)Ph_2 \\ OP(O)Ph_2 \end{cases}$$

Triferrocenylphosphine oxide appears to be a unique tertiary phosphine oxide, which is isolated from the reaction of ferrocene, aluminum chloride, and phosphorus trichloride[619,764] or N,N-diethylphosphoramidous dichloride.[766] Triferrocenylphosphine is formed initially and

$$RH + PCl_3 \xrightarrow[(2)\ H_2O]{(1)\ AlCl_3} R_3PO + R_3P + R_2P(O)OH + RP(O)(OH)_2$$

presumably undergoes air oxidation to the phosphine oxide. By comparison, air oxidation of the triferrocenylphosphine appears much easier than oxidation of triphenylphosphine. The triferrocenylphosphine oxide appears unusual in that it forms hydrates and hydrohalide complexes via interactions with the cyclopentadienyl rings and not the phosphoryl oxygen atom. Such interactions are quite surprising in view of the nature of the involvement of the phosphoryl group in the hydrate and hydrohalides of triphenylphosphine oxide.

Triphenylphosphine, phenylacetylene, and water react upon heating to form 1,2-diphenylethyldiphenylphosphine oxide.[22,23,698] This tertiary phosphine oxide is also

$$Ph_3P + PhC{\equiv}CH + H_2O \longrightarrow Ph_2P(O)CH(Ph)CH_2Ph$$

formed by the reaction of triphenylmethylenephosphorane with benzaldehyde[788,789] and by the reaction of triphenylphosphine with styrene oxide.[788] Evidence suggests that all of these reactions may proceed by a common intermediate believed to be a vinylphosphonium hydroxide.

In addition to electrophilic and nucleophilic aromatic substitution reactions arylphosphine oxides undergo many other synthetically useful chemical transformations. The aryl groups of some phosphine oxides have been hydrogenated, using platinum[471] or Raney nickel[471,693] catalyst, to give the corresponding cyclohexyl compounds. Oxidation of ring substituent methyl groups with potassium permanganate has been employed extensively for preparing phosphinylarylcarboxylic acids and their derivatives.[479,605,725] A claim

$$>P(O)-C_6H_4-CH_3 \xrightarrow[\text{or } O_2]{KMnO_4} >P(O)-C_6H_4-CO_2H$$

has been made for this same oxidation using molecular oxygen.[243] The methyl groups of phosphinylanisoles have been removed by acid- and aluminum chloride-catalyzed hydrolysis to afford high yields of the corresponding phenols.[463,497,620,743] Ring substituent nitro groups undergo reduction to

$$>P(O)-C_6H_4-OMe \xrightarrow{H^+,\ H_2O} >P(O)-C_6H_4-OH$$

amino groups when treated with stannous chloride-acid[837] or iron-acid.[693] Amino substituents in turn have been converted to hydroxy groups via the diazonium salts.[693]

A number of reactions have been reported for substitutions at carbon atoms of phosphine oxides alpha to the phosphoryl group. Of particular importance are those involving formation and substitutions of α-haloalkylphosphine oxides. α-Chloroalkyl derivatives result from reactions of α-hydroxyalkylphosphine oxides with thionyl chloride,[647] phosphorus pentachloride,[537] and triphenylphosphine dichloride.[537] Halogens have also been introduced directly into the α-position of phosphine oxides, using t-butyl hypochlorite,[373] N-bromosuccinimide,[373] and chlorine.[837]

$$>\overset{O}{\overset{\|}{P}}CH_2OH + \text{"Chlorinating agent"} \longrightarrow >\overset{O}{\overset{\|}{P}}CH_2Cl$$

Unsymmetrical chloromethylphosphine oxides can also be prepared by the reaction of disubstituted chlorophosphines with formaldehyde.[451,809] Exchange of chloride for iodide

$$R_2PCl + CH_2O \xrightarrow{\Delta\ 220°} R_2\overset{O}{\overset{\|}{P}}CH_2Cl$$

has been effected for tris(chloromethyl)phosphine oxide, using sodium iodide.[26] The iodo compound was subsequently employed for preparations of several ester derivatives.[26]

$$\overset{O}{\overset{\|}{P}}(CH_2Cl)_3 \xrightarrow{NaI} \overset{O}{\overset{\|}{P}}(CH_2I)_3 \xrightarrow{RCO_2M} \overset{O}{\overset{\|}{P}}(CH_2O\overset{O}{\overset{\|}{C}}R)_3$$

Furthermore, chloride displacement from α-chloroalkylphosphine oxides was realized from reactions with sodium ethoxide[452] and ammonia.[491] α-Amino-substituted phosphine oxides have also been obtained directly from a reaction of a secondary amine with a phosphinylmethylol,[642] and a Raney nickel[373] reduction of the corresponding nitro compound. Treatment of the α-aminoalkyl compounds with nitrous acid resulted in high yields of the very stable diazo compounds.[373,491]

$$>\overset{O}{\overset{\|}{P}}CH(R)NH_2 + HNO_2 \longrightarrow >\overset{O^-}{\overset{\|}{P}}C(R)\overset{+}{N_2}$$

Halogen atoms were introduced into both the α- and the β-positions when diphenylvinylphosphine oxide was treated with chlorine and bromine.[449] The dihalo compounds cleanly dehydrohalogenated in the presence of triethylamine to give

$$Ph_2\overset{O}{\overset{\|}{P}}CH{=}CH_2 + X_2 \longrightarrow Ph_2\overset{O}{\overset{\|}{P}}CHXCH_2X \quad (76 \text{ to } 92\%)$$

excellent yields of α-halovinyldiphenylphosphine oxide.[449,466]

Dehydrohalogenation also occurred in the reaction of β-

$$Ph_2P(O)CHXCH_2X + Et_3N \longrightarrow Ph_2P(O)CX{=}CH_2 \quad (88\text{ to }90\%)$$

chloropropyldiphenylphosphine oxide with triethylamine.[819]

Bromination of β-phenylethyldiphenylphosphine oxide with N-bromosuccinimide gave only a product with bromine in the position beta to the phosphoryl group.[5]

$$Ph_2P(O)CH_2CH_2Ph + NBS \longrightarrow Ph_2P(O)CH_2CH(Br)Ph$$

Interestingly, an α-phosphinyl-substituted ketone, 1-diphenylphosphinyl-1,2-dibenzoylethane, has been reported to undergo a spontaneous dehydration reaction to give a furan.[322] The dehydration probably occurs because of the acidifying influence of the phosphinyl group on the α-hydrogen, which would favor intermediate enol formation.

$$Ph_2P(O)CH(COPh)CH_2COPh \rightleftharpoons Ph_2P(O)C(=C(OH)Ph)CH_2COPh \xrightarrow{-H_2O} \text{3-}Ph_2P(O)\text{-2,5-diphenylfuran}$$

X. POLYMERIZATION OF UNSATURATED PHOSPHINE OXIDES

Polymers of diphenylvinylphosphine oxide have been prepared by free-radical and ionic polymerizations of the monomer. The nature and the properties of the polymer depend on the conditions and the mode of polymerization. For example, low-molecular-weight polymers were obtained when diphenylvinylphosphine oxide was heated with di-t-butyl peroxide at 140° for 24 hr,[66] or in solution with azobis(isobutyronitrile) at 75° for several days.[66,667,669] Irradiation of molten diphenylvinylphosphine oxide with x-rays produced polymers with molecular weights of about 10,000[19] and 30,000,[792] depending on the conditions. Infrared spectral evidence has been obtained that suggests phenyl groups are involved in the chain transfer processes. This may account in part for the differences in properties of the polymers obtained by the radical and the ionic processes.[20] In the latter case, Grignard reagents effectively catalyze the polymerization of diphenylvinylphosphine oxide to polymers with molecular weights up to 10,000. In contrast to the relative order of reactivity in the radical process, the reactivity of diphenylvinylphosphine oxide in anionic copolymerizations is between that of styrene and that of methyl methacrylate.

The polymerization of metal halide complexes of diphenylvinylphosphine oxide, dibutylvinylphosphine oxide, and diphenyl-trans-1-propenylphosphine oxide has been investigated.[818] In general, polymerizations of the complexes under free-radical conditions are sluggish, as in the case of the unsaturated phosphine oxides. However, copolymers of the complexes with a number of vinyl monomers can be readily prepared provided the monomeric complex is kept below 40% of the monomer composition. Intramolecular crosslinking between the two unsaturated phosphine oxides in the 2:1 phosphine oxide:metal halide complexes was not observed.

Poly tertiary phosphine oxides have been prepared by free-radical polymerizations of a number of substituted diallylphosphine oxides.[64,65] The free-radical polymerizations are usually quite slow and yield low-molecular-weight polymers. Diallylphenylphosphine oxide, e.g., after heating with benzoyl peroxide at 75° for 18 days gave a 31% yield of polyphosphine oxide with an intrinsic viscosity of 0.026. Monomer:reactant ratios measured for the copolymerization of diallylphenylphosphine oxide with n-lauryl methacrylate indicate that the latter is much more reactive.[75] Polymers of the substituted diallylphosphine oxides are believed to be linear, as a result of a cyclic ("ladder") polymerization mechanism.

Diphenyl-p-styrylphosphine oxide polymerizes quite readily to high-molecular-weight polymers, using free-radical initiation.[668] Furthermore, copolymerization studies have shown that diphenyl-p-styrylphosphine oxide is more reactive than either styrene or methyl methacrylate. Thus the p-diphenylphosphinyl substituent activates styrene to radical addition by inductively withdrawing electrons. Because steric effects are minimal in this case, the failure of vinylphosphine oxides to homopolymerize to high-molecular-weight polymers is believed to be due to steric inhibition of the propagation step.

A variety of high-molecular-weight polyamides and polyesters that contain either aryl or alkylphosphinyl groups in the carbon backbone have been made.[258-260,478,604,634,644,755,805] Representative examples of these polyphosphine oxides are shown below. Aside from slightly lower softening temperatures and increased solubility in a wider variety of solvents, the introduction of the substituted phosphinyl

$$[NH(CH_2)_3\overset{O}{\overset{\|}{\underset{R}{\underset{|}{P}}}}(CH_2)_3NH-\overset{O}{\overset{\|}{C}}(CH_2)_4\overset{O}{\overset{\|}{C}}]_n \qquad [NH(CH_2)_6NH\overset{O}{\overset{\|}{C}}(CH_2)_2\overset{O}{\overset{\|}{\underset{R}{\underset{|}{P}}}}(CH_2)_2\overset{O}{\overset{\|}{C}}]_n$$

$$[O-\overset{O}{\overset{\|}{C}}-C_6H_4-\underset{R}{\overset{O}{\overset{\|}{P}}}-C_6H_4-\overset{O}{\overset{\|}{C}}O(CH_2)_2]_n \qquad [NH(CH_2)_6NH\overset{O}{\overset{\|}{C}}-C_6H_4-\underset{R}{\overset{O}{\overset{\|}{P}}}-C_6H_4-\overset{O}{\overset{\|}{C}}]_n$$

R = Ph, CH_3

groups into the polymer backbone does not appear to greatly alter the properties of polyesters and polyamides.

B. CHEMISTRY

B.1. Phosphoryl Bond

Tertiary phosphine oxides are compounds represented by the formula R_3PO, in which the substituents are linked to the phosphoryl group by C-to-P bonds. The characteristic properties of tertiary phosphine oxides, such as their high thermal stability and weak basicity, depend, of course, on the nature of both the phosphoryl group and its substituents. However, in considering phosphine oxides, particular attention must be focused on the phosphoryl bond, since a knowledge of this linkage is the major prerequisite to understanding the chemical behavior and physical properties of phosphine oxides.

The phosphoryl bond energy is of considerable magnitude and supplies the primary driving force for several reactions in which tertiary phosphine oxides are formed. Notable in this respect are the facile conversions of tertiary phosphines, alkyl phosphinite esters, and quaternary phosphonium hydroxides to phosphine oxides by oxidation, alkyl halide-catalyzed rearrangement, and cleavage reactions, respectively.

$$R_3P + [O] \longrightarrow R_3PO$$

$$R_2POR' + R'X \longrightarrow R_2R'PO + R'X$$

$$R_4\overset{+}{P}\,^{-}OH \longrightarrow R_3PO + RH$$

The nature of the bonding of the phosphoryl groups of phosphine oxides, as well as the phosphoryl groups of related organophosphorus esters, has been the subject of numerous investigations and discussions during the past two decades. Although a clear understanding of the type of bonding involved in these groups has not resulted, the fundamental concepts needed for the evolution of a satisfactory solution to the problem appear to have been established.

The valence electrons of phosphorus are contained in the M-shell. In the ground state the phosphorus atom has the electronic configuration $3s^2 3p_x^1 3p_y^1 3p_z^1$. The K- and L-shells are filled, and the electrons contained therein presumably do not enter into bonding. The 3p-orbitals are of higher energy than the 3s-orbital, which is, in addition, filled, and therefore the electrons of the former orbitals enter into bonding first. For example, in accord with valence bond theory in its simplest form, phosphine approaches p^3-bonding quite closely, as indicated by its 93.3° P-H bond angles.[269,631] The replacement of the hydrogens of phosphine by methyl groups increases the bond angles to 98.6° in trimethylphosphine,[522] suggesting the participation of the 3s-orbital in bond formation. Finally, in forming a fourth bond, as with oxygen, hybridization to the sp^3-state is essentially complete.[815] In an elementary sense, the P-O linkage of a phosphine oxide can be regarded

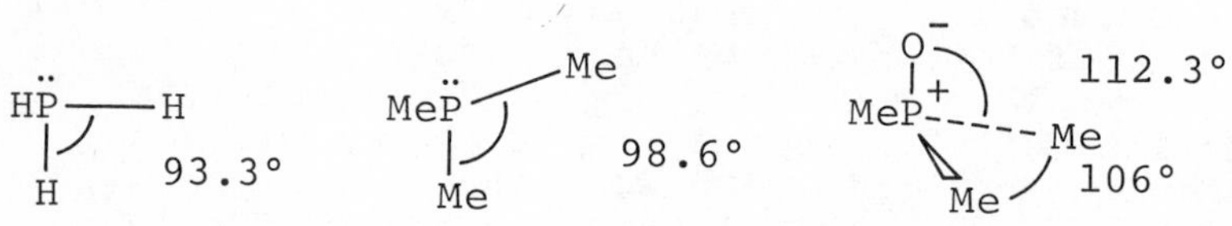

HP̈—H, H; 93.3°   MeP̈—Me, Me; 98.6° O^-, MeP^+, Me, Me; 112.3°, 106°

as an ionic bond superimposed on a σ-bond. Bonding of this type is reminiscent of an N-O linkage of an amine oxide. However, since phosphorus is a second-row element and, unlike nitrogen, has low-lying vacant 3d-orbitals thought by some to be capable of entering into bond formation,[185,599] the phosphoryl bond may possess multiple-bond character.

The five 3d-orbitals are of higher energy than the 3p-orbitals and in the isolated atom are degenerate. This degeneracy is lost in a ligand field, and some of the d-orbitals possess the correct symmetry for participation in π-bonding. Ignoring other factors for the moment, then, it appears that some of the electron density on the negatively charged oxygen of a phosphoryl group would be delocalized into the appropriate d-orbital of phosphorus. The parameters of phosphoryl bonding are therefore established, i.e., a phosphoryl bond represented as a single bond [form (A)] at one extreme and a double bond [form (B)] at the other extreme.

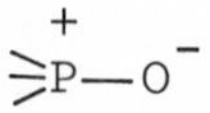

$$\geqslant\overset{+}{P}-O^- \qquad\qquad \geqslant P{=}O$$

(A) 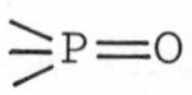   (B)

The question that surrounds d-orbital participation in

π-bond formation, and consequently, the significance of form (B), appears to stem from the uncertainty in determining the "diffuseness" of d-orbitals. In the phosphorus atom the d-orbitals are thought to be too diffuse for significant bond formation owing to an interrelated combination of effective inner-shell electron screening of the d-orbitals and the distance of the d-orbitals from the nuclear charge. However, in a classical paper[185] dealing with d-orbital bonding, it was concluded that, in an actual molecule, the ligands attached to the central atom contract the d-orbitals, relative to the orbitals of the free atom, to the extent that overlap with adjacent centers of high electron density becomes likely (Fig. 1). The more electronegative the ligands, the greater is the degree of

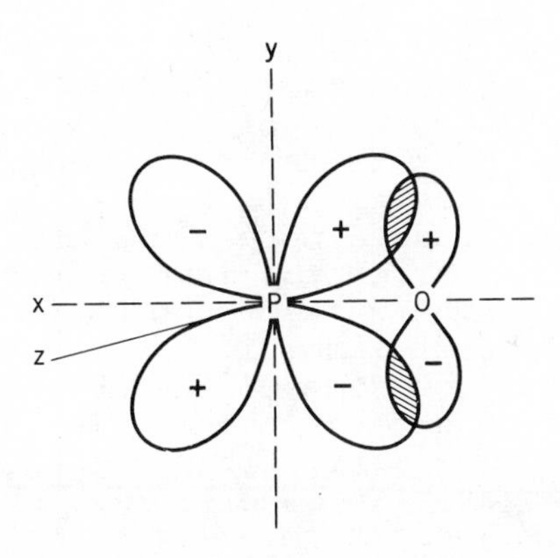

Fig. 1. Overlap of a p_y-orbital with a d_{xy}-orbital.

d-orbital contraction. The order of ligand contraction of d-orbitals has been calculated to be F > Cl > C > H, hydrogen being by far the least effective.[598] It has also been argued that greater overlap between a d-orbital and a compact p-orbital will occur if the atom possessing the vacant d-orbitals carries a formal charge.[185,426] In considering the participation of d-orbitals in bonding, however, it must be remembered that, for a d_π-p_π bond, overlap is presumably better when the d-orbital involved is considerably more diffuse than the p-orbital; i.e., overlap tends to be at a maximum when the lobes of the d-orbital are nearly above and below the orbital of the second atom.[185] Because maximum overlap occurs in an internuclear region closer to the more electronegative oxygen atom, an asymmetric bond results and the phosphoryl bond is highly polar.

Further insight into the nature of the phosphoryl bond

has been gained from a nuclear quadrupole resonance study.[517] The results of this investigation led to the conclusion that the phosphoryl group of phosphorus oxytrichloride has triple-bond characteristics because of overlap of two nonbonding p-orbitals of oxygen with the appropriate two 3d-orbitals of phosphorus. Phosphine oxides having the appropriate symmetry should therefore also possess phos-

$$\geqslant P \equiv O$$

phoryl groups with triple-bond character, as suggested[152] earlier for tripropyl- and tributylphosphine oxides.

A similar picture of overlap between oxygen and positively charged phosphorus has been proposed for phosphorus oxyanions.[800] Indeed, x-ray fluorescence data from this study have been viewed as direct evidence for the participation of 3d-orbitals in π-bonds.

Some of the disagreement that exists regarding the extent of conjugative participation of phosphorus undoubtedly arises because of the diversity of methods used to determine the significance of this factor. For example, spectral evidence frequently indicates a lack of phosphorus participation, whereas chemical data more often than not are rationalized in terms of phosphorus involvement in the total "electronic picture." The work of Kabachnik is of particular value in this context since it has resulted in a discussion of the validity of some of the methods. In brief summary, it has been concluded[441] that, in general, the chemical probe is more sensitive than optical methods to small energy differences. In other words, even though the participation of phosphorus may result in only small energy changes in a system, the effect is magnified greatly in a chemical sense, e.g., by a marked increase in reaction rate, whereas an optical method that reveals only strong interactions is insensitive to the small energy change.

In spite of some uncertainty, the concept of $d\pi$-$p\pi$ bonding in phosphoryl groups of phosphine oxides appears to be accepted by most workers in the field. The main justification for using the d-orbital model is that it provides a common basis for rationalizing many of the chemical and physical properties of phosphine oxides that are not explained satisfactorily by less subtle models. Much of the discussion in the rest of this chapter has been written with this viewpoint in mind.

B.2. Basicity of the Phosphoryl Group

Tertiary phosphine oxides are much weaker bases than tertiary amine oxides and tertiary arsine oxides but significantly stronger bases than analogous carbonyl compounds.[304] The pK_b of dimethylphenylphosphine oxide, e.g., is -2.4 versus 4.7 for trimethylamine oxide and -6.5 for acetophenone. The reported value of the pK_b for trimethylphosphine oxide is 0,[623] but this is probably too high.[304] Trimethylarsine oxide and trimethylstibine oxide are 10^6 to 10^8 times more basic than either phenyldimethylphosphine oxide or methyldiphenylphosphine oxide. These results are consistent with the idea that there is considerable p_π-d_π overlap in the semipolar phosphoryl bond of tertiary phosphine oxides.[304,307]

Tertiary phosphine oxides interact with protic acids[628,825-827] to give either hydrogen-bonded complexes[90,106,236,285,307,310,519] or hydroxyphosphonium salts.[306,519] The

$$R_3PO + HA \rightleftharpoons R_3PO \text{ ---- } HA \rightleftharpoons R_3\overset{+}{P}OH\ A^-$$

nature of the product is dependent on the basicity of the phosphine oxide, the size of the substituents, the strength of the acid, and the solvent. Tertiary phosphine oxides and weak acids such as chloroform[15,310,622] water,[309,624] hydrogen peroxide,[167 780] alcohols,[519] phenols,[15,308,432,519] acetic acid,[519] trihaloacetic acids,[306,519] and tertiary ammonium salts[236,267,268,600] form hydrogen-bonded complexes. Recent evidence suggests that the crystalline triphenylphosphine oxide complexes with water,[309] with hydrogen peroxide,[167,780] and with phenols[671] are surprisingly stable 2:1 rather than 1:1 complexes. Conflicting evidence for a 1:1 complex[624] and a 1:2 complex[612] of water with trioctylphosphine oxide has been reported. Strong acids such as hydrogen bromide[306] and sulfuric acid[277,519] react with triphenylphosphine oxide to form triphenylhydroxyphosphonium salts. The phosphorus NMR spectrum of triphenylhydroxyphosphonium sulfate[519] is nearly identical to that of triphenylethoxyphosphonium tetrafluoroborate.[215] Therefore, proton transfer is essentially complete in the case of triphenylphosphine oxide and sulfuric acid.

Recent evidence indicates that protonated tertiary phosphine oxides are involved in the racemization of optically active tertiary phosphine oxides.[222a,390] Optically active methylpropylphenylphosphine oxide, for example, is quite stereochemically stable and does not exchange oxygen in refluxing water or in refluxing 2N sodium hydroxide. In refluxing concentrated hydrochloric acid, however, methylpropylphenylphosphine oxide is rapidly racemized and undergoes oxygen exchange. The rate of racemization of methylpropylphenylphosphine oxide is much slower than that of the analogous arsine oxide, perhaps because the tertiary phosphine oxide is a much weaker base and is less easily protonated than the arsine oxide.

Tertiary phosphine oxides also form complexes with Lewis acids such as halogens,[285,845,846] sulfur trioxide,[277] sulfur dioxide,[123] boron trichloride,[632] boron trifluoride,[123] and silicon tetrahalides.[54] In most cases the adducts are 1:1, although higher complexes are observed for arylphosphine oxides and the halogen molecules. Silicon tetrahalides form 1:2 and 1:4 adducts with triphenyl- and trimethylphosphine oxide.

Tertiary phosphine oxides coordinate strongly with certain metal atoms of salts and organometallics. The complexes that are formed are generally crystalline and thermally stable, and vary in composition from 1:1 to 5:1 in the ratio of phosphine oxide to metal atom. The composition of the complex is dependent on the metal, the ligands, and the phosphine oxide. Trimethylphosphine oxide, e.g., forms distillable 1:1 complexes with trimethyl compounds of aluminum, gallium, and indium[728] and tris(trimethylsiloxy) compounds of aluminum, gallium, and iron.[729] Triphenylphosphine oxide forms 2:1 complexes with certain salts of Sn(IV), Ti(IV), Fe(III),[753] Cd(II), Zn(II),[50,174] Co(II),[179] Ni(II),[512] Cu(II), Sb(IV), Sb(V),[512] V(IV),[394] and Mo(V).[393] The 3:1 complexes of triphenylphosphine oxide with salts of Y(III) and La(III) are thermally stable at 320°.[181] Tertiary phosphine oxides, in general, form $M(L)_4^{n+}$ complexes with the perchlorates of Fe(II), Fe(III),[174] V(IV),[394] Y(III), La(III),[181] Cr(III), Ce(III), Mg(II), Mn(II), Co(II), Ni(II),[459] Cu(II),[105] Zn(II),[49,104,173,414,459] and Li(I).[157,312,410] Cationic complexes of the type ML_5^{2+} have recently been reported for manganese, cobalt, and nickel and for trimethylphosphine oxide.[105] Noteworthy is the fact that many lithium and sodium salts form $M(L)_5^+$ complexes[312] with triphenylphosphine oxide.

Upon complexation of tertiary phosphine oxides with metal salts the phosphoryl stretching frequency is decreased. This is attributed to an approximate 20% decrease in the phosphoryl bond force constant and a 20 to 30% decrease in the phosphoryl bond order.[174] The magnitude of this change in the phosphoryl stretching frequency of triphenylphosphine oxide has been used as a measure of the strengths of a variety of Lewis acids.[253]

Tertiary phosphine oxides with multiple phosphoryl groups are good multidentate ligands and form very stable crystalline complexes with metal salts.[609,610,612,629,814] Noteworthy in this respect is the extensive use of alkylenediphosphine dioxides as extractants for metal salts such as uranium and thorium nitrates.

Tertiary phosphine oxides are weak bases and require powerful alkylating agents to form alkoxyphosphonium salts.[215] Nevertheless, the phosphoryl group of tertiary phosphine oxides is capable of acting as a nucleophile and of entering into several reactions. The isomerization of

$$R_3PO + Et_3O^+BF_4^- \longrightarrow R_3\overset{+}{P}OEt \quad BF_4^-$$

epoxides to carbonyl compounds[81] and the conversion of isocyanates to carbodiimides[602] are examples of reactions that are catalyzed by tertiary phosphine oxides. An alkoxyphosphonium zwitterion resulting from nucleophilic

$$RCH\overset{O}{—}CHR \xrightarrow{R_3PO} R\overset{O}{\overset{\|}{C}}CH_2R$$

$$2RNCO \xrightarrow{R_3PO} RNCNR + CO_2$$

attack of the phosphoryl oxygen on the epoxide ring is believed to be involved in the epoxide isomerization. In

$$R_3PO + RCH\overset{O}{—}CHR \longrightarrow R_3\overset{+}{P}OCHRCHRO^-$$

the carbodiimide formation the phosphine oxide catalyst is believed to be converted to a phosphinimine, which then

$$R_3PO + RNCO \longrightarrow \begin{matrix} R_3P-O \\ | \quad | \\ R-N-C=O \end{matrix} \longrightarrow CO_2 + R_3P=NR$$

$$R_3P=NR + RNCO \longrightarrow RNCNR + R_3PO$$

reacts with isocyanate to form the carbodiimide and regenerate the tertiary phosphine oxide. Cyclic phosphine oxides are especially effective as catalysts in this reaction.

B.3. Phosphinyl Substituent Effects

As would be expected from a consideration of the nature of the polar phosphoryl group, in which the phosphorus atom bears some positive charge and is capable of accepting electrons by a conjugative mechanism, phosphinyl groups, $R_2P(O)-$, as substituents greatly influence the "electronic makeup" of molecules to which they are attached. The existence of positive charge on the phosphorus atom is vividly reflected in the ability of the phosphinyl group to inductively withdraw electrons (-I); this has been evaluated quantitatively by several methods. Most straightforward in this respect is the finding that diphenylphosphinylacetic acid is approximately one pK unit more acidic than the parent acid.[553] By application of a linear free-energy correlation, an approximate substituent constant (σ^*) of +0.60 for the diphenylphosphinyl group was determined. A further conversion of this value to a polar substituent constant of +1.68 was effected by mathematically removing the methylene group, which attenuates the inductive effect of the substituent. The polarizing influence of the diphenylphosphinyl substituent has been put into perspective with the effects of the more common substituents by drawing attention to its near equivalence as a -I group ($\sigma^* = 0.52$) with the carbomethoxy moiety.[553] Of additional interest is a comparison of the substituent constant value of the diphenylphosphinyl group with that of the triphenylphosphonio substituent, $Ph_3\overset{+}{P}-$ ($\sigma^* = 4.70$).[553] The latter is a particularly strong polarizing group because of its full positive charge, whereas the charge on the phosphorus atom of the $Ph_2\overset{O}{\overset{\|}{P}}-$ group is reduced because of Coulombic interaction and back bonding (d_π-p_π bond) of the oxyanion with the charged phosphorus atom.

Quantitative measures of the degree of the interaction of the diphenylphosphinyl group with aromatic systems have also been obtained. In these compounds, in addition to the inductive effect, conjugative interaction of the aromatic ring with the phosphorus atom is possible. In order to determine the resonance contribution to the overall electron-withdrawing power of the diphenylphosphinyl group, σ_m and σ_p values were ascertained, using meta- and para-substituted diphenylphosphinyl phenols and benzoic acids.[603,797,797a] The resulting data, as summarized in Table 5, are consistent with moderate resonance interactions of the phosphinyl groups with the aromatic rings. The predominant effect, however, results from inductive electron withdrawal, as shown by the magnitude of σ_m relative to σ_p. The para-substituent value for the diphenylphosphinyl group in benzoic acid approximates the value (+0.52) determined by ^{1}H-NMR spectral analysis.[689] This method of determination was based on a linear relationship that exists between π-electron density, as determined by the ^{13}C chemical shifts of para carbons in aromatic phosphorus compounds, and Hammett substituent constants.

In further studies[438,670] ^{19}F-NMR spectroscopy was used as a probe for determining the electronic interactions of aromatic rings with diphenylphosphinyl- and bis(4-fluorophenyl)phosphinyl substituents. The ^{19}F shielding in fluorobenzenes is very sensitive to intramolecular perturbations by ring substituents,[778] such as phosphinyl groups, and precise linear relationships from the shielding parameters enable the separation of the total electronic

Table 5. $Ph_2P(O)$-Substituent Constants

$Ph_2P(O)C_6H_4X$	pK	σ_m	σ_p	Ref.
m-CO_2H	5.01	0.422	—	603
	5.11*[a]	0.38	—	797a
p-CO_2H	4.79	—	0.553	603
	4.88*[a]	—	0.53	797a
m-OH	10.20	0.434	—	603
p-OH	9.48	—	0.680	603
	9.38*	—	0.68	797a

[a]Aqueous ethanol 1:1.

effect of the substituent into inductive and resonance effects. The substituent constant values so obtained[438]

$$F_{(m,p)}C_6H_4P(O)Ar_2$$

for the diphenylphosphinyl group were σ_I = +0.30 and σ_R = 0.12 (in CCl_4 solvent). The substitution of fluorophenyl for phenyl groups on the substituent led to an increase in the inductive effect to 0.45 and no change in σ_R. The increase in σ_I was thought to result from a magnification of the net positive charge on phosphorus because of the greater electron-withdrawing power of the entire group, i.e.,

$$(FC_6H_4)_2P(O)- > Ph_2P(O)-$$

Similarly, in the other direction, the replacement of phenyl by methyl groups in a phosphoryl substituent predictably should result in a decreased σ_I value. In an independent study,[658,726] σ_I = 0.62 was found for the dimethylphosphinyl group in alcohol. However, since the shielding parameters of the phosphorus-substituted fluorobenzenes are affected quite dramatically by solvent, a comparison of the two values may not be valid.

In brief review, the information gained from the studies previously described in this section sheds light on a question that has been of major concern for some time; i.e., how and to what extent does positively charged phosphorus interact with adjacent centers possessing delocalizable electrons? A partial answer to this question has been adequately summarized[438] as follows:

"(1) $R_3\overset{+}{P}\text{-}(R_2\overset{O^-}{\overset{|}{P}}{}^{+}\text{-})$ groups exert both inductive and conjugative withdrawal effects on aryl rings; (2) in the general relationship $\sigma_p = a\sigma_I + b\sigma_R$ as applied to an $X\text{-}C_6H_4P^+$ system, a is larger than b, i.e., there is relatively little conjugative interaction between phenyl and tetravalent phosphorus; (3) substantial changes in the σ_I values of tetravalent phosphorus-containing groups create only minor changes in the σ_R values of the same groups; (4) substantial conjugative interaction of tetravalent phosphorus with an adjacent carbanion or oxyanion rules out simultaneous substantial conjugative interaction between the phosphorus and an attached aryl group."

Substituent constants determined by other spectral methods for phosphinyl groups are in reasonable agreement with those just discussed.[720-722,724]

Information regarding the effects of phosphinyl substituents has also been obtained from kinetic studies involving the rates of cleavages of substituted benzyltrimethylsilanes by aqueous methanolic sodium hydroxide.[94] These reactions occur by way of benzyl carbanion intermediates (not necessarily "free" carbanions)[18] and are markedly facilitated by electron-withdrawing substituents.

$$X\text{-}C_6H_4\text{-}CH_2SiMe_3 + {}^-OH \longrightarrow \left[X\text{-}C_6H_4\text{-}CH_2^-\right] + HOSiMe_3$$

$$\left[X\text{-}C_6H_4\text{-}CH_2^-\right] \xrightarrow{H_2O} X\text{-}C_6H_4\text{-}Me + HO^-$$

For this reaction, the order of activation by the substituents (X) was established as follows:

$$p\text{-}NO_2\text{-} > p\text{-}Me_3\overset{+}{P}\text{-} > p\text{-}Ph_2\overset{O}{\overset{\|}{P}}\text{-} > p\text{-}Me_2\overset{O}{\overset{\|}{P}}\text{-}$$

$$> m\text{-}Ph_2\overset{O}{\overset{\|}{P}}\text{-} > m\text{-}Me_2\overset{O}{\overset{\|}{P}}\text{-} > m\text{-}CF_3\text{-}$$

Exalted σ^- constants were determined for the para substituents, while normal σ constants were realized for the meta groups (see Table 6). The difference in substituent

Table 6. Substituent Constants from Base Cleavage of $XC_6H_4CH_2SiMe_3$

X	σ	σ^-
$p\text{-}\overset{+}{P}Me_3$	—	1.14
$p\text{-}\overset{O}{\overset{\|}{P}}Ph_2$	—	0.845
$p\text{-}\overset{O}{\overset{\|}{P}}Me_2$	—	0.725
$m\text{-}\overset{O}{\overset{\|}{P}}Ph_2$	0.485	—
$m\text{-}\overset{O}{\overset{\|}{P}}Me_2$	0.43	—

constants for the meta and para isomers is in accord with $d_\pi\text{-}p_\pi$ bonding between the ring and the para substituents, which stabilizes the forming benzyl carbanion.

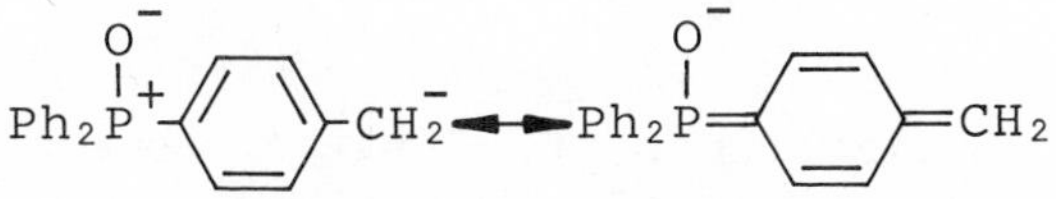

$$Ph_2\overset{O^-}{\overset{|}{P^+}}\text{-}C_6H_4\text{-}\overset{-}{C}H_2 \longleftrightarrow Ph_2\overset{O^-}{\overset{|}{P}}{=}C_6H_4{=}CH_2$$

These data also show that the diphenylphosphinyl group is more strongly electron-withdrawing than the dimethylphosphinyl moiety. This suggests that the -I effect of the phenyl groups is more significant than the resonance interaction between the phenyl groups and the phosphorus atom and hence is in agreement with the findings of the spectral studies discussed previously.

In a closely related study,[95] the rates of cleavage of some substituted phenyltrimethylsilanes by acetic-sulfuric acid-water were determined. In this electrophilic aromatic substitution reaction, the order of decreasing deactivation by the groups was established as follows:

$$p\text{-}Ph_2\overset{O}{\overset{\|}{P}}\text{-} > p\text{-}Me_2\overset{O}{\overset{\|}{P}}\text{-} > p\text{-}Me_3\overset{+}{P}\text{-} > p\text{-}NO_2\text{-} > p\text{-}Me_3\overset{+}{N}\text{-} > m\text{-}Ph_2\overset{O}{\overset{\|}{P}}\text{-}$$

The greater deactivation caused by the para-phosphinyl

$$X\text{-}C_6H_4\text{-}SiMe_3 + H^+Y^- \longrightarrow X\text{-}C_6H_4(H)(SiMe_3)^+\ Y^- \longrightarrow X\text{-}C_6H_5 + Me_3SiY$$

substituents, relative to the para-phosphonio group, has been rationalized as being due to protonation of the phosphoryl oxygens in strongly acidic media to give $\overset{+}{>}\overset{OH}{P}\text{-}^+$, which withdrew electrons more effectively than $R_3\overset{+}{P}$-.

Another indication of the influence of phosphinyl groups that are presumably protonated as substituents comes from nitrations of arylphosphine oxides. In this electrophilic substitution reaction, methyldiphenylphosphine oxide,[102,693] triphenylphosphine oxide,[148,769] tris(3-nitro-4-methylphenyl)phosphine oxide,[769] tris(4-chlorophenyl)phosphine oxide,[769] tris(4-dimethylaminophenyl)-phosphine oxide,[769] and bis(chloromethyl)phenylphosphine oxide[837] gave only products corresponding to substitution meta to the phosphinyl group.

$$X\text{-}C_6H_4\text{-}\overset{O}{\overset{\|}{P}}< + HNO_3/H_2SO_4 \longrightarrow X\text{-}C_6H_3(NO_2)\text{-}\overset{O}{\overset{\|}{P}}< \longrightarrow X\text{-}C_6H_2(NO_2)_2\text{-}\overset{O}{\overset{\|}{P}}<$$

$$X = H,\ CH_3,\ Cl,\ (CH_3)_2N$$

In an inverse sense, then, phosphinyl substituents should enhance the facility of nucleophilic displacements of ortho and para leaving groups, since such reactions involve rate-determining nucleophilic additions to the aromatic ring. Reactions of tris(3-nitro-4-chlorophenyl)-phosphine oxide with aniline, methylamine, and phenol in the presence of base, or with sodium methoxide, to give chloride displacement are in agreement with phosphoryl activation of the ring.[769] It is difficult to assess the extent of phosphoryl group activation in these reactions, however, since the nitro group is also an activating substituent.

$$N^- + Cl\text{-}C_6H_3(NO_2)\text{-}\overset{O^-}{\underset{}{P^+}}\!< \longrightarrow (Cl)(N)C_6H_3(NO_2){=}\overset{O^-}{P}\!< \longrightarrow N\text{-}C_6H_3(NO_2)\text{-}\overset{O^-}{P^+}\!< + Cl^-$$

$$N^- = Ph\bar{N}H,\ Me\bar{N}H,\ PhO^-,\ MeO^-$$

B.4. Addition Reactions of Vinylphosphine Oxides

Phosphinyl groups as substituents are also known to significantly alter the behavior of other unsaturated systems. Most studied in this regard are the vinylphosphine oxides, which undergo Michael-type addition reactions with various nucleophiles with a facility that exceeds that of the unsubstituted parent olefin. Since Michael addition reactions usually occur only with conjugated olefins under

$$>\overset{O}{\overset{\|}{P}}CH{=}CH_2 + N^- \longrightarrow >\overset{O}{\overset{\|}{P}}\text{-}\bar{C}H\text{-}CH_2N \xrightarrow{NH} >\overset{O}{\overset{\|}{P}}CH_2CH_2N + N^-$$

normal conditions, it seems reasonable to conclude that the phosphinyl group activates the alkene linkage of the vinylphosphine oxides to nucleophilic attack. In view of the preceding discussion of carbanion stabilization by phosphorus substituents, the degree of activation of the alkene linkage would be expected to parallel the degree of positive charge on phosphorus. One would anticipate, therefore, that diarylvinylphosphine oxides would be more reactive than dialkylvinylphosphine oxides. Experimental evidence is in accord with this concept.[441,797a] Since vinylphosphorus compounds undergo Michael addition reactions less readily than α,β-unsaturated carbonyl compounds and are less readily polymerized, it has been logically concluded[441] that phosphoryl groups possess a weaker polarizing effect than carbonyl groups.

A variety of nucleophiles have been added to vinylphosphine oxides. Included in the list of nucleophiles are piperidine,[454,456] ethanol,[441] ethylamine,[449] secondary phosphites,[449] malonic ester carbanion,[449] secondary phosphine oxides,[441,464] Grignards,[20,264] organolithium compounds,[20] and electrons.[46] Whereas the first six nucleophiles gave monomeric adducts, the organometallic compounds

$$Bu_2\overset{O}{\overset{\|}{P}}CH{=}CH_2 + C_5H_{10}NH \longrightarrow Bu_2\overset{O}{\overset{\|}{P}}CH_2CH_2NC_5H_{10}$$

$$Ph_2\overset{O}{\overset{\|}{P}}H + Ph_2\overset{O}{\overset{\|}{P}}CH{=}CH_2 \xrightarrow{base} Ph_2\overset{O}{\overset{\|}{P}}CH_2CH_2\overset{O}{\overset{\|}{P}}Ph_2$$

have been used to effect anionic polymerizations of the vinylphosphine oxides[20,264,818] (see Section A.X. on phosphorus polymers). The simplest nucleophile, the electron, resulted in a reductive coupling of the phosphorus Michael

$$>\overset{O}{\overset{\|}{P}}CH{=}CH_2 + RM \longrightarrow >\overset{O}{\overset{\|}{P}}\text{-}\underset{M}{\underset{|}{C}}HCH_2R \xrightarrow{>\overset{O}{\overset{\|}{P}}CH=CH_2}$$

$$>\overset{O}{\overset{\|}{P}}\text{-}CHCH_2R \text{ (with } CH_2\text{-}CH(M)\overset{O}{\overset{\|}{P}}{=} \text{ on CH)} \xrightarrow{\text{etc.}} \text{Polymer}$$

acceptor to give some 1,4-tetramethylenebis(diphenylphosphine) oxide.[46]

$$Ph_2\overset{O}{\overset{\|}{P}}CH{=}CH_2 \xrightarrow[Et_4N^+X^-\text{(aq.)}]{e^-} Ph_2\overset{O}{\overset{\|}{P}}(CH_2)_4\overset{O}{\overset{\|}{P}}Ph_2 + H_2$$

B.5. Formation and Reactions of Phosphinylalkylmetal Compounds

Hydrogens attached to α-carbon atoms of phosphine oxides are rendered acidic by the phosphoryl group. Because of the acidity of these hydrogens, phosphine oxides readily react with strong bases such as organometallic compounds, alkali-metal amides, and alkali-metal alkoxides to give the corresponding phosphinyl-substituted carbanions. The activating influence of the phosphoryl group on the α-

$$R_2\overset{O}{\overset{\|}{P}}\text{-}\overset{/}{\underset{\backslash}{C}}\text{-}H + B^-M^+ \longrightarrow R_2\overset{O}{\overset{\|}{P}}\text{-}\overset{/}{\underset{\backslash}{C}}^- \; M^+ + BH$$

hydrogens results from an interrelated combination of inductive (-I)(Coulombic) and resonance effects. In attempting to evaluate the contributions of these factors, it is helpful to consider the two major resonance forms of the phosphoryl group, i.e., the limiting extreme of a P-O double bond (A) and a dipolar single bond (B). The phosphorus atom in form (B) bears a formal positive charge, and it follows that a phosphoryl group having some of this character would be effective in inductively enhancing the acidity of α-protons of phosphine oxides and stabilizing the resulting phosphinyl carbanions. An indication of the

$$>\overset{O}{\overset{\|}{P}}- \longleftrightarrow >\overset{O^-}{\overset{|}{P}}{}^{+}-$$

(A) (B)

ability of the phosphoryl group to inductively withdraw

$$>\overset{O^-}{\overset{|}{P}}{}^{+} \leftarrow \overset{|}{\underset{|}{C}} \leftarrow H \qquad >\overset{O^-}{\overset{|}{P}}{}^{+} \leftarrow \overset{-}{\underset{|}{C}} -$$

electrons is found, e.g., in a report[553] that ascribes a polar (inductive) substituent constant (σ^*) value of +1.68 for the diphenylphosphinyl moiety, $Ph_2\overset{O}{\overset{\|}{P}}-$. This value demonstrates that the diphenylphosphinyl substituent is a moderately strong electron acceptor and possesses an inductive effect comparable to the effects of phenyl and carboalkoxy groups.[553]

The type of bonding in phosphorus-stabilized carbanions is open to question. However, a reasonable rationale for resonance in these systems invokes the overlap of the appropriate empty d-orbitals of phosphorus with the orbital on carbon which contains the nonbonding pair of electrons (Fig. 2). Presumably, but by no means necessarily, the carbanionic carbon atom would be sp^2-hybridized for maximum overlap.

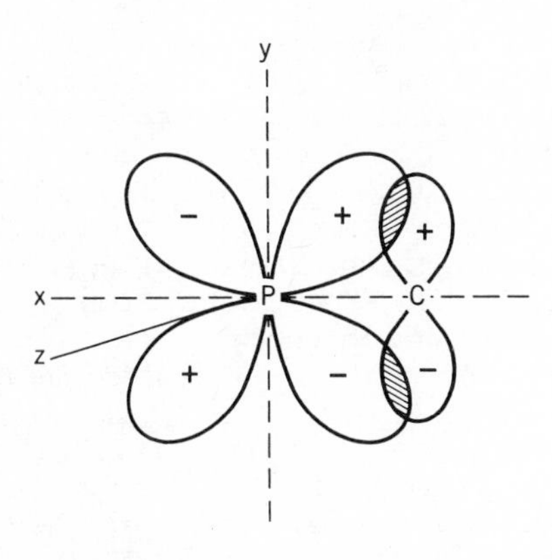

Fig. 2. The π overlap of a d_{xy}-orbital with a p_y-orbital

The effect of the phosphoryl group on the configuration of a carbanion to which it is attached has been investigated by determining the stereochemical course of the base-catalyzed hydrogen-deuterium exchange of diphenyl-2-octylphosphine oxide.[186,187] The ratios of rate constants for isotopic exchange (k_e) and for racemization (k_α) in this system were found to vary only from 1 to 3 in widely differing base-solvent combinations. Since the exchange reaction

$$B^- + Ph_2P(O)\text{-}CH(CH_3)(C_6H_{13}\text{-}n) \xrightarrow{-BH} Ph_2P(O)\text{-}C^-(CH_3)(C_6H_{13}\text{-}n) \xrightarrow[\text{retention}]{BD} Ph_2P(O)\text{-}CD(CH_3)(C_6H_{13}\text{-}n)$$

$$\rightleftharpoons \quad {}^-C(CH_3)(C_6H_{13}\text{-}n)\text{—}P(O)Ph_2 \xrightarrow[\text{inversion}]{BD} D\text{-}C(CH_3)(C_6H_{13}\text{-}n)\text{—}P(O)Ph_2$$

approaches 100% racemization as $k_e/k_\alpha \rightarrow 1$, it is apparent that the carbanionic carbon of the optically active phosphine oxide readily assumes an sp^2-hybridization state. Unfortunately, these results do not lead to a clear understanding of the stereochemical influence of the phosphoryl moiety on an adjacent carbanion, and several interpretations of these findings have been advanced.[186,187] It would appear reasonable, however, to conclude that the exchange data are not in disagreement with the idea that the carbanion undergoes sp^3- to sp^2-hybridization to maximize resonance stabilization by the d-orbitals of phosphorus.

Theoreticians have argued that d-orbital interactions in multiple-bond systems of the $3d_\pi$-$2p_\pi$ type will be of greatest importance in cases in which the electron-accepting atom bears some positive charge. The positive charge tends to contract the relatively diffuse 3d-orbitals, making overlap with compact 2p-orbitals more favorable. Contributing canonical form (B) fulfills this requirement. A higher degree of d-orbital resonance would be expected to occur, therefore, in phosphinyl carbanions than in the nonpositively charged phosphinocarbanions,[636,637] $R_2PCH_2^-M^+$. If this reasoning is followed through to the other extreme, phosphinyl carbanions should be somewhat less stable than the corresponding phosphoniocarbanions, $R_3\overset{+}{P}\text{-}\overset{-}{C}H_2$, because of the full positive charge of the phosphorus atom in the latter. Evidence in support of this proposal is found in an interesting report which discloses that phosphonium salts are converted more rapidly than phosphine oxides to the corresponding phosphorus-substituted carbanions, and that, once formed, the phosphinyl carbanions are more reactive (less stable) than the phosphoniocarbanions.[385]

Further verification of the ability of phosphinyl groups to activate hydrogens attached to adjacent carbon atoms comes from the finding that methanolic potassium hydroxide is a sufficiently strong base to effect extensive prototropic isomerization of diallylphenylphosphine oxide.[204] After 72 hr of heating at reflux, the reaction mixture appeared to be composed predominantly of the trans-propenyl isomer.

$$PhP(O)(CH_2CH{=}CH_2)_2 + KOH \xrightarrow{MeOH} PhP(O)(CH{=}CHMe)_2 \quad (trans)$$

The hydrogen atoms in the methylene bridge of bis(diphenylphosphinyl)methane, $Ph_2P(O)CH_2P(O)Ph_2$, are quite acidic.[744] The rate of isotopic exchange at the methylene position with a base as weak as aniline was determined to be 3.7×10^{-5} sec^{-1} at room temperature ($t_{1/2} \sim 5$ hr). Reaction with potassium in xylene gives bis(diphenylphosphinyl)methylpotassium.[445] The analogous sodium salt was prepared with sodium hydride.

$$(Ph_2P(O))_2CH_2 + K \longrightarrow (Ph_2P(O))_2CHK + \tfrac{1}{2}H_2$$

With regard to the preparation of phosphinylalkylmetal compounds, the finding[747,748] that triphenylphosphine oxide reacts with alkyllithium and alkylmagnesium compounds to give the corresponding diphenylphosphinyl carbanions is of special value because of the ready availability of starting materials. These reactions have been shown to proceed in two discrete steps: a rapid displacement of a phenyl group by the organometallic at phosphorus, followed by a slower metallation step. The question of whether the initial exchange reaction proceeded via a pentacovalent

$$Ph_3PO + RR'CHM \longrightarrow Ph_2P(O)CHRR' + PhM$$

$$Ph_2P(O)CHRR' + PhM \longrightarrow Ph_2P(O)CRR'M + PhH$$

phosphorus intermediate or by a direct S_N2 displacement at phosphorus was not answered.

The reaction of phenyllithium with triphenylphosphine oxide results in a mono- and a dimetallation of the meta position of the phosphine oxide.[829] The metallation reaction is accompanied by aryl ligand exchange.

Phosphinylalkylmetal compounds are chemically quite reactive and have recently found considerable utility in organic synthesis (see also Section A.VII on the synthesis of phosphine oxides using metallated phosphine oxides). Perhaps the most useful application of these reagents is in the conversion of carbonyl compounds to the corresponding methylene or substituted-methylene extended olefins.[313,370,436] This reaction, discovered[377] in 1958, is reminiscent of another carbonyl olefination process, the extensively used Wittig reaction, which involves a phosphonio-

$$>P(=O)CRR'M + >C{=}O \longrightarrow [>P(=O)CRR'C(O^-){-}]M^+ \longrightarrow$$

$$>C{=}CRR' + >P(=O){-}OM$$

carbanion. The two olefination reactions probably proceed by similar mechanisms and occur because of a common driving force, i.e., the formation of strong P-O bonds.

The influences of the alkali metal, mole ratio of reactants, temperature, and solvent on the carbonyl olefination reaction have received attention.[375] Of particular importance is the effect of the base. For example, no 1,1-diphenylprop-1-ene was obtained from an ethyldiphenylphosphine oxide-benzophenone mixture when interacted with sodium methoxide, sodium ethoxide, methylmagnesium bromide, sodium, or potassium. In contrast, the desired reaction was realized when potassium t-butoxide, potassium amide, phenylpotassium, and sodium amide were used as bases for deprotonation. Although these differences probably reflect the differing reactivities of bases, the cationic counterions of bases have also been shown to independently influence the olefination reaction.[382] Thus (diphenylphosphinyl)phenylmethyllithium reacts with benzaldehyde to give an alkoxide sufficiently stable to be converted by acid to (1,2-diphenyl-2-hydroxyethyl)diphenylphosphine oxide. Conversion of the 2-hydroxyalkylphosphine oxide to the potassium alkoxide using potassium t-butoxide resulted in the

$$Ph_2P(=O)CH_2Ph + PhLi \longrightarrow Ph_2P(=O)CHLiPh \xrightarrow{PhCHO}$$

$$Ph_2P(O)CH(Ph)CH(OLi)Ph \xrightarrow{H^+} Ph_2P(O)CH(Ph)CH(OH)Ph$$

formation of stilbene in high yield.

$$Ph_2P(O)CH(Ph)CH(OH)Ph + KOC_4H_9\text{-}t \longrightarrow$$

$$[(Ph)_2P(O)CH(Ph)CH(OK)Ph] \longrightarrow PhCH{=}CHPh + (Ph)_2P(O)OK$$

Some of the stereochemical consequences of olefination reactions that employ metallated phosphine oxides have been investigated. The reaction of metallated optically active methylphenylbenzylphosphine oxide with benzal aniline was shown to occur with retention of configuration at phosphorus.[374] A cis-elimination process was favored for this reaction and the related ones with carbonyl compounds.[384]

$$>P(O)\text{-}C^{-}< \; + \; >C{=}O \longrightarrow \left[>P(O)\text{-}C(\text{-})\text{-}C(\text{-})\text{-}O^{-}\right] \longrightarrow >P(O)O^{-} \; + \; >C{=}C<$$

Upon treatment with benzaldehyde, (diphenylphosphinyl)-phenylmethyllithium, $Ph_2P(O)CH(Ph)Li$, in ether afforded a mixture of alcohols comprised predominantly (90% yield) of one diastereomer melting at 193° with a small amount (5% yield) of another diastereomer melting at 188°. The higher- and lower-melting compounds were thought to be threo and erythro, respectively, based on the products resulting from base-catalyzed decompositions of these compounds. That is, assuming a cis elimination of lithium diphenylphosphinate, trans- and cis-stilbene would form from the threo and erythro isomers, respectively. The overall stereochemistry of this system has been questioned.[435]

$$\underset{\text{(threo)}}{Ph_2P(O)CH(Ph)\text{-}CH(OLi)Ph} \xrightarrow{Et_2O} Ph_2P(O)OLi + \text{trans-}PhCH{=}CHPh$$

$$Ph_2P(O)-CH(Ph)-CH(OLi)(Ph)\ (erythro) \xrightarrow{Et_2O} Ph_2P(O)OLi + HC(Ph){=}C(Ph)H$$

The stereochemistry of the reaction of the analogous sodium compound, $Ph_2P(O)CH(Ph)Na$, with benzaldehyde has been elucidated in solvents other than ether.[59] A high degree of stereoselectivity occurred in this system also, as the ratios of cis- and trans-stilbenes were 3:97 and 5:95 in N,N-dimethylformamide and cyclohexane, respectively. The same reaction, but with propionaldehyde, gave 15:84 and 1:99 cis-trans ratios of olefins. Finally, (diphenylphosphinyl)ethylmethyllithium, $Ph_2P(O)CH(Et)Li$, with benzaldehyde in dimethylformamide and cyclohexane yielded cis-trans isomer ratios of 4:96 and 24:76 of 1-phenylbutene. For purposes of stereochemical comparison, the analogous olefination reactions were effected using $Ph_3\overset{+}{P}\overset{-}{C}HPh$. Solvent polarities were found to have a greater influence on the cis-trans ratios of olefins resulting from ylid reactions than from the metallated phosphine oxides. Also, the phosphorus ylid tended to form larger amounts of the cis isomer.

The phosphine oxide method of converting carbonyl compounds to methylene-extended olefins has an additional feature in common with the Wittig reaction. Thus it has been shown that the intermediate alkoxide, $>P(O)-C-C-O^-$, like the analogous betaine of the Wittig reaction, $\geqslant\overset{+}{P}-C-C-O^-$, is capable of reverting to carbanion and carbonyl compound in addition to decomposing to olefin and oxygenated phosphorus compound.[384] To put retrograde reactions of this type into perspective, attention is called to the finding[785]

$$>P(O)-\overset{|}{\underset{|}{C}}-\overset{|}{\underset{|}{C}}-O^- \longrightarrow >P(O)-O^- + >C{=}C<$$

$$>P(O)-\overset{|}{\underset{|}{C}}-\overset{|}{\underset{|}{C}}-O^- \rightleftharpoons >P(O)-\overset{-}{C}< + O{=}C<$$

that an analogous trivalent phosphine also reverted to a phosphorus-substituted carbanion and carbonyl compound. This retrograde reaction is particularly interesting since the leaving group is only slightly stabilized relative to

$$Ph_2PCH(Ph)CH(Ph)O^- \rightleftharpoons Ph\overset{-}{C}HPPh_2 + PhCHO$$

the corresponding phosphoniocarbanion and phosphinyl carbanions.

Table 7 contains some selected examples of olefins obtained from reactions of phosphinylalkylmetal compounds with carbonyl compounds.

Bis(diphenylphosphinyl)methylpotassium has also been shown to effect carbonyl olefination reactions.[445] Reactions of this carbanion with benzaldehyde, formaldehyde, propionaldehyde, isobutyraldehyde, p-nitrobenzaldehyde, and cinnamaldehyde in xylene afforded the corresponding substituted diphenylvinylphosphine oxides in 25 to 49% yields. Similar results were obtained from reactions run in 1,2-dimethoxyethane, and in cases where the anion was generated from the parent hydrocarbon using potassium-t-butoxide in t-butanol.

$$(Ph_2\overset{O}{\overset{\|}{P}})_2CHK + RCHO \longrightarrow Ph_2\overset{O}{\overset{\|}{P}}CH{=}CHR + Ph_2\overset{O}{\overset{\|}{P}}OK$$

In addition to their reactions with carbonyl compounds to form methylene-extended olefins and with numerous substrates to form functionally substituted phosphine oxides, phosphinylalkylmetal compounds have been found to effect other useful chemical transformations. Reactions of benzyldiphenylphosphine oxide with potassium-t-butoxide in the presence of p-nitrosodimethylaniline and styrene oxide afforded a Schiff base[382] and 1,2-diphenylcyclopropane,[376] respectively. It should be noted that the analogous lithium carbanion with styrene oxide stopped at the inter-

$$Ph_2\overset{O}{\overset{\|}{P}}CH_2Ph + t\text{-}BuOK \xrightarrow{(CH_3)_2N\text{-}C_6H_4\text{-}N{=}O} (CH_3)_2N\text{-}C_6H_4\text{-}N{=}CHPh + Ph_2\overset{O}{\overset{\|}{P}}OK$$

$$Ph_2\overset{O}{\overset{\|}{P}}CH_2Ph + t\text{-}BuOK \xrightarrow{PhCH\overset{O}{—}CH_2} Ph\triangle Ph + Ph_2\overset{O}{\overset{\|}{P}}OK$$

mediate alkoxide stage.[376] It is apparent, then, that the stabilities of the O-alkali-metal γ-alkoxyalkylphosphine

Table 7. Carbonyl Olefination Reactions Using Metallated Phosphine Oxides

Phosphine Oxide	Carbonyl Compound	Base	Olefin (% Yield)		Ref.
$Ph_2P(O)Me$	Ph_2CO	t-BuOK	$Ph_2C{=}CH_2$	(61)	382
$Ph_2P(O)Me$	"	$NaNH_2$	"	(70)	377
$Ph_2P(O)Et$	Ph_2CO	t-BuOK	$Ph_2C{=}CHMe$	(51)	382
$Ph_2P(O)Et$	"	$NaNH_2$	"	(54)	377
$Ph_2P(O)CH_2Ph$	"	t-BuOK	$Ph_2C{=}CHPh$	(70)	382
$Ph_2P(O)CH_2Ph$	PhCOMe	"	Ph(Me)C=CHPh	(60)	"
$Ph_2P(O)CH_2Ph$	PhCHO	"	PhCH=CHPh	(70)	"
$Ph_2P(O)CH_2Ph$	cyclohexanone (ring)=O	"	cyclohexylidene (ring)=CHPh	(47)	"
$Ph_2P(O)CH_2Ph$	PhCH=CHCHO	"	PhCH=CHCH=CHPh	(46)	"
$Ph_2P(O)CH_2Ph$	PhCOCOPh	"	PhCH=C(Ph)C(Ph)=CHPh	(50)	"
$Ph_2P(O)CH_2Ph$	OCH–C_6H_4–CHO	"	PhCH=CH–C_6H_4–CH=CHPh	(57)	"
$Ph_2P(O)CH_2Ph_2$	4-pyridyl–CHO	"	4-pyridyl–CH=CHPh	(47)	"
$(CH_2)_2[P(O)Ph_2]_2$	Ph_2CO	"	$Ph_2C{=}CHCH{=}CPh_2$	(41)	375
$(CH_2)_6]P(O)Ph_2]_2$	"	"	$Ph_2C{=}CH(CH_2)_4CH{=}CPh_2$	(25-35)	"

oxides parallel those of the analogous beta series. (Diphenylphosphinyl)phenylmethyllithium, generated from the phosphine oxide using ethereal-lithium dimethylamide as base, gave, upon treatment with phenylazide and subsequent hydrolysis, a triazine.[353]

$$Ph_2\overset{O}{\overset{\|}{P}}CH(Ph)Li + PhN_3 \longrightarrow \xrightarrow{H^+} Ph_2\overset{O}{\overset{\|}{P}}CH(Ph){-}N{=}N{-}NHPh \quad (26\%)$$

By employing successive lithium and potassium metallated phosphine oxides, an interesting alkyne synthesis has been devised.[284]

$$Ph_2\overset{O}{\overset{\|}{P}}CH(Ph)Li + Ph\overset{O}{\overset{\|}{C}}Cl \longrightarrow Ph_2\overset{O}{\overset{\|}{P}}CH(Ph)\overset{O}{\overset{\|}{C}}Ph \xrightarrow{t\text{-}BuOK}$$

$$Ph_2\overset{O}{\overset{\|}{P}}{-}\bar{C}(Ph){-}C(Ph){=}O \; K^+ \longrightarrow \left[\begin{array}{l} \;\;\; O^- \\ \;\;\; | \\ Ph_2P{-}CPh \\ \;\;\; | \;\;\; \| \\ \;\;\; O{-}C{-}Ph \end{array} \right] K^+ \longrightarrow Ph_2\overset{O}{\overset{\|}{P}}{-}O^-K^+ + PhC{\equiv}CPh$$

Phosphinylalkylmetal compounds reveal additional organometallic-like character in their reactions with oxygen. Thus, in a manner reminiscent of the oxidation of the classical types of organolithium and organomagnesium compounds,[813] (diphenylphosphinyl)phenylmethyllithium has been shown to react with oxygen to give a metal alkoxide.[374] The reaction presumably proceeds via a peroxy intermediate. Since O-metallated α-hydroxyphosphine oxide salts are capable of undergoing decomposition to phosphinylmetal and carbonyl compounds,[365] it is not surprising that olefins also can result from these oxidation reactions when less than stoichiometric amounts of oxygen are employed.

$$>\overset{O}{\overset{\|}{P}}{-}\overset{|}{C}{-}O^-M^+ \longrightarrow >\overset{O}{\overset{\|}{P}}M + >C{=}O \xrightarrow{>\overset{O}{\overset{\|}{P}}{-}\bar{C}<\;M^+}$$

$$>C{=}C< \quad + \quad >\overset{O}{\overset{\|}{P}}{-}OM$$

Three miscellaneous points of interest with regard to the acidity of phosphine oxides and the formation of phosphinylalkylmetal compounds may be mentioned.

1. The finding[749] that bis(methoxymethyl)methylphos-

phine oxide possesses properties very similar to those of dimethyl sulfoxide as a solvent for carrying out anionic exchange reactions. To cite one specific example, $PhCD_3$ underwent H-D exchange when treated with potassium-t-butoxide in this phosphine oxide solvent at a rate comparable to that realized in dimethyl sulfoxide.

2. The formation of olefins from high-temperature reactions of aldehydes with diphenylphosphinylsodium.[366] Presumably, α-metallated phosphine oxides are involved in these reactions as outlined below.

$$Ph_2P(=O)Na + PhCHO \longrightarrow Ph_2P(=O)\text{-}CH(ONa)Ph \xrightarrow{Ph_2P(=O)Na} Ph_2P(=O)CH(Na)Ph + Ph_2P(=O)ONa$$

3. The preparation[491,688] of some highly stabilized phosphinylcarbanions, $R_2P(=O^-)CH\overset{+}{N}_2$.

B.6. Reductions of the Phosphoryl Group

One area of organophosphine oxide chemistry which has received considerable attention in recent years is that pertaining to reductions of the phosphoryl groups of these compounds to form tertiary phosphines. Lithium aluminum hydride,[331,408] aluminum hydride,[408] calcium aluminum hydride,[331] organoboron compounds,[489] trichlorosilane,[189,257,323,360,362] phenylsilane,[256,713] diphenylsilane,[256] methylpolysiloxane $[(\text{-O-}Si(H)(CH_3)\text{-})_n]$,[256] and hexachlorodisilane[617] have been used as reagents for effecting these reductions (see also Chapter 1).

Reactions of lithium aluminum hydride with trialkylphosphine oxides afford the corresponding phosphines in moderate to high yields.[331] By way of contrast, reactions of the hydride with aromatic phosphine oxides appear to be solvent sensitive, since substantial quantities of diphenylphosphine resulted from lithium aluminum hydride reductions of triphenylphosphine oxide in dioxane and tetrahydrofuran.[331,408] It was subsequently revealed that lithium diphenylphosphide was the precursor of the diphenylphos-

$$Ph_3PO + LiAlH_4 \xrightarrow[\text{or THF}]{\text{dioxane}} Ph_2PLi + Ph_3P$$
$$Ph_2PLi \longrightarrow Ph_2PH$$

phine.[408] The cleavage reaction was not of significance, however, when the reduction was carried out in aliphatic ethers as solvents.[372,408]

Phenyl group cleavage was accompanied by disproportionation with an attempt was made to reduce triphenylphosphine oxide with calcium hydride.[256] Aryl group cleavages also result from reactions of triphenylphosphine oxide

$$Ph_3PO + CaH_2 \xrightarrow{350°} \xrightarrow{H^+} Ph_3P + Ph_2P(O)OH + PhH$$

with sodium hydride.[365] In this reaction the resulting sodium diphenylphosphinite maintains its chemical integrity and can be used to considerable advantage in the preparation of phosphine oxides of type $Ph_2P(O)R'$ (see Section A.III).

$$Ph_3PO + NaH \longrightarrow Ph_2PONa + PhH$$

Diphenylphosphine has similarly been obtained from a reaction of diphenyl(1-phenylallyl)phosphine oxide with lithium aluminum hydride.[256] These cleavage reactions are

$$Ph_2\overset{O}{\overset{\|}{P}}CH(Ph)CH{=}CH_2 + LiAlH_4 \longrightarrow \xrightarrow{H^+} Ph_2PH$$

reminiscent of the reductions of unsymmetrical phosphonium salts effected by lithium aluminum hydride, which result in cleavage of the group best able to accommodate a negative charge.[44] For example, benzylethylmethylphenylphosphonium iodide gave only ethylmethylphenylphosphine on treatment with the hydride, demonstrating the ease of cleavage of the benzylphosphorus linkage in relation to

$$PhCH_2\overset{+}{P}(Me)(Et)Ph\,I^- + LiAlH_4 \longrightarrow Ph(Et)(Me)P + PhMe$$

ruptures of the other bonds. Hydride cleavages of groups from phosphine oxides undoubtedly follow an analogous course, i.e., $PhCH_2\text{-} \sim CH_2{=}CHCH_2\text{-} > Ph\text{-} \gg$ alkyl, in accord with a transition state for these reactions which involves a leaving group bearing some negative charge.

$$>\overset{O^-}{\overset{|}{P^+}}-R + H^- \longrightarrow \left[>\overset{O^-}{\overset{|}{\underset{H}{\underset{|}{P}}}}-R \right] \longrightarrow >\overset{O^-}{\overset{|}{P^+}}-H + R^-$$

Reductions of optically active phosphine oxides with lithium aluminum hydride result in complete loss of stereochemical integrity at phosphorus.[137,337] It has been demonstrated that the phosphine oxides undergo rapid stereomutation before reduction.[137] For example, optically active methylphenyl-n-propylphosphine oxide racemized completely before more than 10% of the reduction reaction had occurred. Epimerization of the diastereomeric cis- or trans-1-phenyl-2,2,3,4,4-pentamethylphosphetane 1-oxide was also effected by lithium aluminum hydride.[337] The changes in configuration at phosphorus were rationalized in terms of reversible additions of lithium aluminum hydride to the phosphine oxides to give pentacoordinate phosphorus intermediates. The latter then undergo pseudorotation, followed by dissociation.[337]

From a mechanistic standpoint, the most informative developments in the chemistry of phosphine oxide-phosphoryl groups have resulted from studies of reactions of phosphine oxides with various silicon compounds. Initially, only silanes containing Si-H bonds were employed for these reductions,[189,256,257,362,713] but more recently it has been shown[617] that perchlorosilanes also are very efficient in effecting these transformations:

$$R_3PO + \geq SiH \longrightarrow R_3P + [\geq SiOH]$$

$$R_3PO + Cl_3SiSiCl_3 \longrightarrow R_3P + Cl_3SiOSiCl_3 + SiCl_4 + \text{"Other siloxanes"}$$

Reduction of optically active phosphine oxides with trichlorosilane alone or with pyridine or dimethylaniline present gave phosphines with net retention of configuration and in high optical purity.[362] On the other hand, the same

$$Cl_3SiH + O{=}P(R)(R')(R'') \longrightarrow :P(R)(R')(R'') + [Cl_3SiOH]$$

reaction, when conducted in the presence of the more basic triethylamine, afforded phosphines with inverted configuration.[362] The retention reaction, in the absence of base, was rationalized in terms of a complexation-intramolecular hydride transfer mechanism.[362] It has been pointed out[617] that, if this reaction proceeds by way of a phosphorus pentacovalent intermediate, and if hydride attack and leaving group departure follow the same stereochemical course,

$$Cl_3SiH + O{=}P(R)(R')(R'') \longrightarrow O(Cl_3\bar{S}i{-}H){-}\overset{+}{P}(R)(R')(R'') \longrightarrow$$

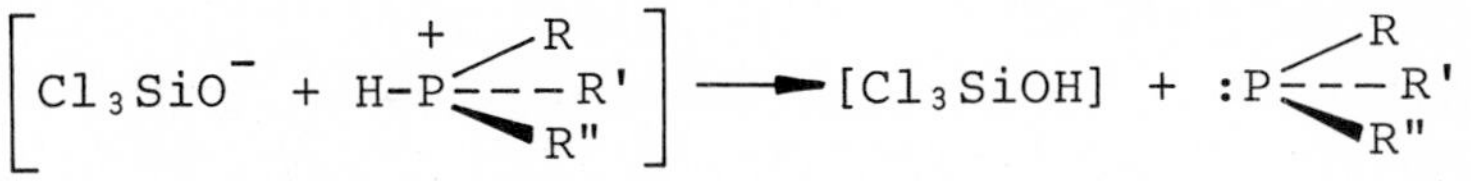

$$\left[Cl_3SiO^- + H\text{-}\overset{+}{P}RR'R''\right] \longrightarrow [Cl_3SiOH] + :PRR'R''$$

i.e., equatorial or apical, pseudorotation[73] must intervene. The following equation exemplifies the apical stereochemical transformation:

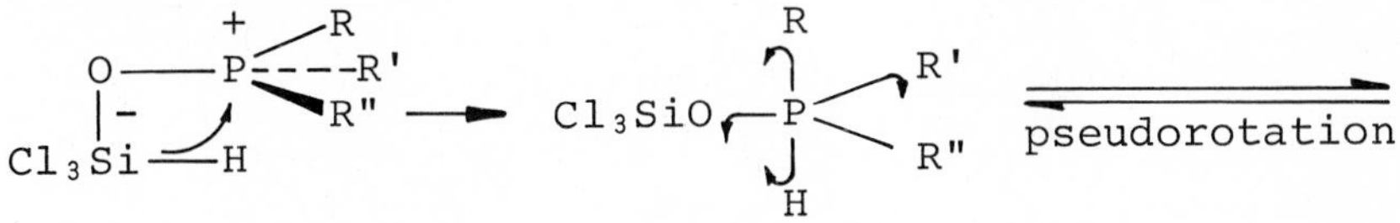

$$Cl_3Si(H)\text{-}O^-\text{-}\overset{+}{P}RR'R'' \longrightarrow Cl_3SiO\text{-}P(R)(H)R'R'' \xrightleftharpoons{\text{pseudorotation}}$$

Apical attack

$$Cl_3SiO\text{-}P(R)(H)R'R'' \xrightarrow{\text{apical departure}} RR'H\overset{+}{P}R'' + [Cl_3SiO^-]$$

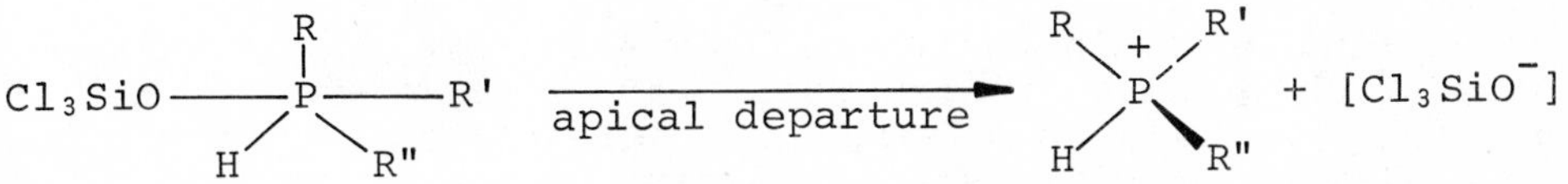

To account for the inversion of configuration which was realized in the presence of triethylamine, it was proposed that the chlorosilane-phosphine oxide complex was reduced by a hydride transfer from a second 1:1 complex of triethylamine and trichlorosilane.[362] A subsequent study[617]

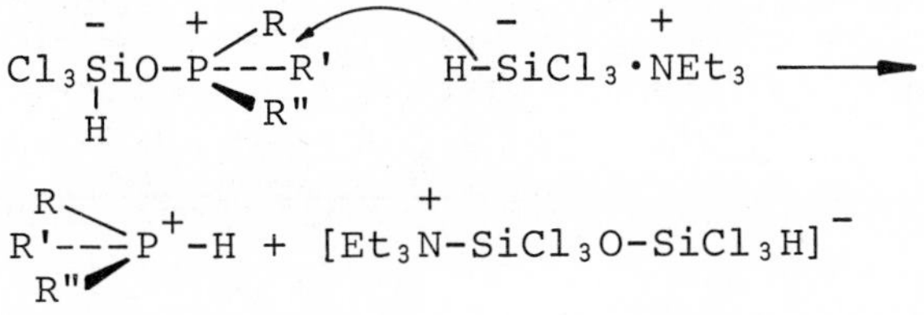

$$Cl_3\overset{-}{Si}(H)O\text{-}\overset{+}{P}RR'R'' \quad H\text{-}\overset{-}{Si}Cl_3\cdot\overset{+}{N}Et_3 \longrightarrow$$

$$RR'R''P^+\text{-}H + [Et_3\overset{+}{N}\text{-}SiCl_3O\text{-}SiCl_3H]^-$$

demonstrated, however, that the amine-silane complexes as such were irrelevant to the stereochemical outcome of the reaction. It was alternatively proposed that strongly basic amines (pK_b < ca. 5), such as triethylamine, react with trichlorosilane to give polysilanes, e.g., $Cl_3SiSiCl_2H$ and $Cl_3SiSiCl_2SiCl_3$, which function as the active species in the reduction reactions. In accord with this proposal, it was established that hexachlorodisilane and octachlorotrisilane were capable of reducing phosphine oxides, and, furthermore, that reductions with these reagents proceeded with inversion of configuration and with high stereospecificity. The following reaction scheme was advanced to rationalize the observed stoichiometry and stereochemistry of the reductions:

$$RR'R''PO + Si_2Cl_6 \longrightarrow RR'R''\overset{+}{P}OSiCl_3 + {}^{-}SiCl_3 \longrightarrow$$

$$Cl_3Si\overset{+}{P}R''R'R + {}^{-}OSiCl_3$$

$$Cl_3SiO^{-} + Cl_3Si\overset{+}{P}R''R'R \longrightarrow Cl_3SiOSiCl_3 + :PR''R'R$$

$$Cl_3SiO^{-} \longrightarrow [OSiCl_2] + Cl^{-}$$

$$[OSiCl_2] + Cl^{-} \longrightarrow SiCl_4 + :PR''R'R$$

One aspect of this reaction which detracts somewhat from its usefulness stems from the finding[617] that, althought the reduction process was highly stereospecific, the starting phosphine oxide and the resulting phosphine underwent chemically induced racemization quite readily under the very mild conditions of the reaction.[617] Fortunately, this problem is not too serious since the reduction reaction occurred very rapidly (it was complete within a few minutes in most instances) and long contact times could be avoided.

Of particular interest with regard to the reduction of phosphine oxides by organosilicon compounds are the findings that trichlorosilane in the presence of triethylamine[189,323] and hexachlorodisilane[213,214] reduces phosphetane oxides with complete retention of configuration at phosphorus. The following mechanism was advanced to rationalize the stereochemistry of the hexachlorodisilane reaction:

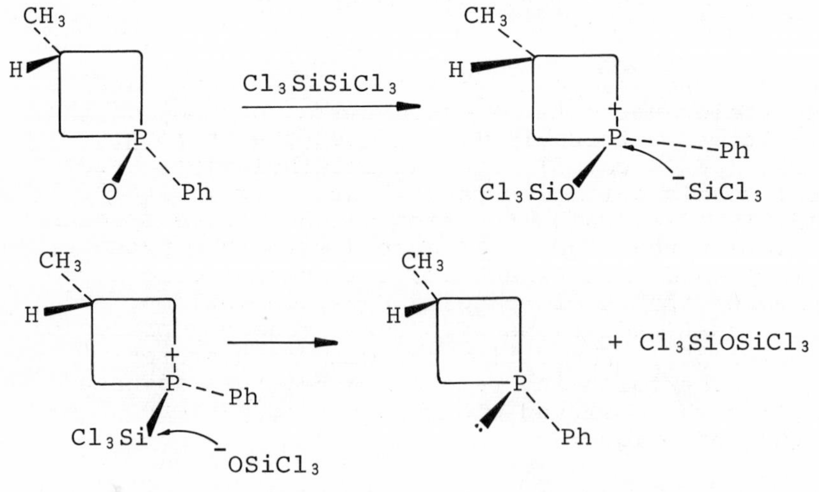

CH3, H, P, O, Ph $\xrightarrow{Cl_3SiSiCl_3}$ CH3, H, $\overset{+}{P}$, Ph, Cl_3SiO, ${}^{-}SiCl_3$

CH3, H, $\overset{+}{P}$, Ph, Cl_3Si, ${}^{-}OSiCl_3$ $\longrightarrow$ CH3, H, P, Ph $+ Cl_3SiOSiCl_3$

The rapid reduction of phosphetane oxides, relative to the reduction of unbranched acyclic analogs, such as methylphenyl-n-propylphosphine oxide, was regarded as evidence for the direct attack of the trichlorosilane anion on phosphorus. It has been reasoned that the rate acceleration results from relief of angle strain in going from the phosphine oxide to the pentacovalent phosphorus intermediate; i.e., the 90° C-P-C bond angle of the ring becomes essentially strain free in the trigonal-bipyramidal intermediate.

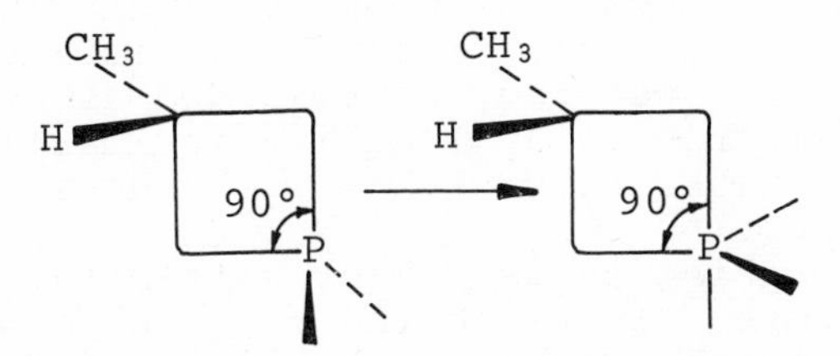

It was further shown[214] in support of the proposed mechanism that attack of an external nucleophile on tetracoordinate phosphorus can occur stereospecifically with retention of configuration. Specifically, the base-catalyzed hydrolysis of cis- and trans-1-ethoxy-1-phenyl-2,2,3,4,4-pentamethylphosphetanium hexachloroantimonate using ^{18}O-enriched water afforded the corresponding phosphetane oxides with ^{18}O-enriched phosphoryl groups and with retention of configuration at phosphorus.

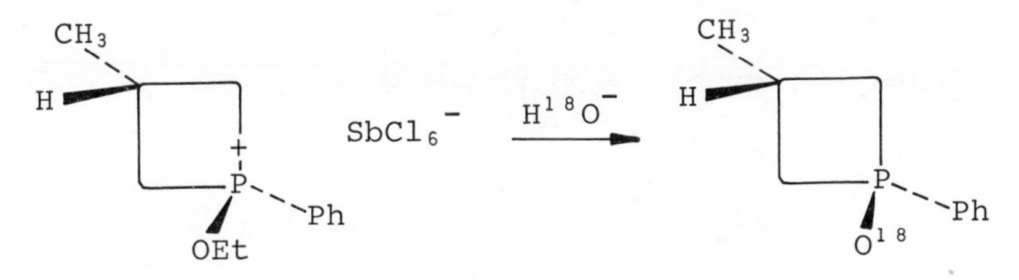

B.7. Metal and Electrochemical Reactions

Historically, reactions of phosphine oxides with metals go back more than 100 years, when an attempt was made to reduce triethylphosphine oxide with sodium.[131] Little reaction was thought to occur with sodium and other metals[163] below a temperature at which the phosphine oxide entirely decomposed. More recently,[364] sodium has been shown to react with tributylphosphine oxide at an elevated temperature (> 240°) to give a 50% yield of dibutylphosphine.

$$Bu_3P{=}O + Na \xrightarrow{> 240°} Bu_2PH$$

Reactions of sodium and potassium with triethylphosphine oxide in tetrahydrofuran as solvent have been reported[322,330] to give "phosphyls" which are supposedly analogous to the metal ketyls, $>\dot{C}-O^-M^+$. Spectral evidence supporting the formation of a radical was found in

$$Et_3P{=}O + M\cdot \xrightarrow{THF} Et_3\dot{P}{-}O^-M^+$$

the EPR spectrum of the product of a tributylphosphine oxide-potassium reaction; it exhibited a single line with a poorly defined superfine structure consisting of twelve components.[441] The spectrum was regarded as evidence for the (extra) d-electron being localized primarily on phosphorus with only weak interactions with the six α-protons of the butyl groups.

Aromatic phosphine oxides react with metals in solvents in a number of ways. In refluxing toluene, a phenyl group is cleaved from triphenylphosphine oxide by sodium to give sodium diphenylphosphinite.[363] The phenylsodium generated metallates toluene; the benzylsodium, in turn, attacks

$$Ph_3P{=}O + Na \xrightarrow{toluene} Ph_2PONa + PhNa$$

additional phosphine oxide with displacement of phenylsodium and formation of metallated benzyldiphenylphosphine oxide.

$$Ph_3P{=}O + PhCH_2Na \longrightarrow Ph_2P(O)CH_2Ph + PhNa \longrightarrow Ph_2P(O)CH(Ph)Na + PhH$$

In ether solvents aromatic phosphine oxides readily react with alkali metals to afford radical anions. Extensive electron delocalization into aromatic groups appears to occur in these systems, as shown by the splitting

$$>P(O)Ph + M \xrightarrow{THF} PhP^{\cdot}{<}{-}O^-M^+$$

patterns of the EPR spectra of solutions of the compounds.[441]

The biphenyl radical anion has been identified by EPR spectroscopy as a product of reactions of triphenylphosphine oxide with lithium or sodium in 1,2-dimethoxyethane.[346]

With potassium, a different paramagnetic species was formed initially; this slowly disappeared and was replaced with potassium biphenylide. Some chemical characterization of the products of the triphenylphosphine oxide-sodium reaction was achieved by derivatization with various alkylating agents. Thus from benzyl chloride, methyl iodide, ethyl bromide, and 1,4-tetramethylene dibromide there were obtained benzyldiphenyl-, methyldiphenyl-, and ethyldiphenylphosphine oxides and 1,4-tetramethylene-bis(diphenylphosphine oxide) in 62 to 77% yields. The predominant phosphorus-containing product of the triphenylphosphine oxide-sodium reaction was, therefore, sodium diphenylphosphinite.

$$Ph_3PO + Na \xrightarrow{CH_3OCH_2CH_2OCH_3} Ph_2PONa \xrightarrow{RX} Ph_2\overset{O}{\overset{\|}{P}}R$$

The products resulting from reactions of triphenylphosphine oxide with 2 moles of lithium or potassium in 1,2-dimethoxyethane were somewhat more complex.[346] Treatment of these reaction mixtures with benzyl chloride afforded a product thought to be the meta-substituted triphenylphosphine oxide.[346]

$$PhCH_2-C_6H_4-\overset{O}{\overset{\|}{P}}Ph_2 \quad (meta)$$

A reaction of benzyldiphenylphosphine oxide with sodium in 1,2-dimethoxymethane resulted in the formation of (diphenylphosphinyl)phenylmethylsodium.[346] The metallated phosphine oxide accounted for less than half of the total phosphorus, and it is likely that cleavage reactions occurred in this reaction also.

$$Ph_2\overset{O}{\overset{\|}{P}}CH_2Ph + Na \xrightarrow{CH_3OCH_2CH_2OCH_3} Ph_2\overset{O}{\overset{\|}{P}}CH(Ph)Na$$

Aryl group cleavage prevailed in the reaction of ethyldiphenylphosphine oxide with sodium.[346]

$$Ph_2\overset{O}{\overset{\|}{P}}Et + Na \longrightarrow \xrightarrow{\emptyset CH_2Cl} Ph(Et)\overset{O}{\overset{\|}{P}}CH_2Ph \quad (39\%)$$

The role of the metals in the reactions just discussed is to supply electrons to the acceptor phosphine oxide molecules. Recently, the electron transfer process has been effected electrochemically. Thus phenyldibenzylphosphine oxide has been reduced to cyclohexadienyldibenzylphosphine oxide by electrolysis in methanol, using tetramethylammonium bromide as the electrolyte.[387] Interestingly, cleavage reactions occurred to only a minor extent (∿ 5%). The controlled-potential reduction of triphenylphosphine oxide has received attention,[708] and it has been concluded that reduction on the polarographic and voltammetric time scale proceeded by a one-electron reduction to a radical anion.

Diphenylvinylphosphine oxide was found to reduce polarographically at an $-E_{1/2}$ of 2.29V (s.c.e.).[46] Electrolysis of a 19% solution of the phosphine oxide in aqueous tetraethylammonium p-toluenesulfonate gave a low yield of the hydrodimer, 1,4-tetramethylene-bis(diphenylphosphine oxide), and hydrogen.

B.8. Thermal Stability

Tertiary phosphine oxides as a class are generally quite thermally stable in relation to other phosphorus compounds,[251,254] amine oxides,[166] and sulfoxides.[469a] Trimethyl- and triphenylphosphine oxides, e.g., are thermally stable up to almost 700°.[45] Other aliphatic tertiary phosphine oxides that contain β-hydrogens are somewhat less stable but still decompose at temperatures several hundred degrees above the decomposition point of amine oxides. For example, dimethylethylphosphine oxide decomposes above 330° to ethylene and dimethylphosphine oxide, presumably by a cyclic concerted mechanism. Under these conditions

$$CH_3CH_2\overset{O}{\overset{\|}{P}}(CH_3)_2 \xrightarrow{>330°} CH_2{=}CH_2 + [(CH_3)_2\overset{O}{\overset{\|}{P}}H]$$

the dimethylphosphine oxide rapidly disproportionates to dimethylphosphine and dimethylphosphinic acid.[45,327] Decomposition of tertiary phosphine oxides in this manner appears general for phosphine oxides with β-hydrogen atoms. Furthermore, the decomposition appears to be favored slightly by electron-donating substituents.

Tertiary phosphine oxides containing α-hydroxyl groups are less thermally stable than aliphatic tertiary phosphine oxides. They undergo thermal elimination to form secondary phosphine oxides and carbonyl compounds rather than olefins.[595] This decomposition occurs at 100° or slightly

$$R_2\overset{O}{\overset{\|}{P}}-\overset{OH}{\overset{|}{C}}R'R'' \longrightarrow R_2\overset{O}{\overset{\|}{P}}H + R'R''CO$$

above when either R' or R" or both R' and R" are substituents other than hydrogen. In cases in which R' and R" are both hydrogens, the hydroxymethylphosphine oxides appear stable to about 200° or above.

A study of several silicon-substituted organophosphorus compounds indicates that they thermally decompose at 200 to 300°.[106,138] Tris-(trimethylsilylmethyl)phosphine oxide, e.g., decomposes slowly at temperatures above 200° to hexamethyldisiloxane and unidentified phosphorus compounds.

B.9. Chemical Degradation

Most tertiary phosphine oxides are relatively resistent to chemical cleavage of the C-P bonds. Triphenylphosphine oxide, e.g., reacts very slowly with refluxing alcoholic sodium hydroxide to give benzene and sodium metaphosphate.[16] Tri(2-pyrryl)phosphine oxide and tri(2-thienyl)phosphine oxide[298,552] are also degraded with hot alcoholic sodium hydroxide in the same manner. In the latter case, however, di(2-thienyl)phosphinic acid is also formed in varying amounts and is presumably a precursor to the metaphosphate.

Fusion of phenyl- and benzyl-containing tertiary phosphine oxides with powdered sodium hydroxide at 200 to 300° gives hydrocarbons and the sodium-disubstituted phosphinate.[234,235,318,377,378,501] Yields of most of the sodium

$$R_3PO + NaOH \xrightarrow{200\text{-}300°} R_2P(=O)ONa + RH$$

phosphinates ranged from 80% to quantitative. As is the case in the decomposition of phosphonium hydroxides, the substituent that is displaced by hydroxide ion corresponds to the most stable carbanion. Thus the order of ease of displacement is benzyl > phenyl > alkyl, and naphthyl > phenyl > tolyl.

That the stability of the incipient carbanion is of utmost importance in determining the ease of displacement is probably best illustrated by the chemistry of tris(trifluoromethyl)phosphine oxide. This highly reactive phosphine oxide reacts readily with water at room temperature to give fluoroform and bis(trifluoromethyl)phosphinic acid.[630] Similarly, dimethylamine displaces fluoroform

$$(CF_3)_3PO + H_2O \longrightarrow CF_3H + (CF_3)_2P(=O)OH$$

from the phosphine oxide to give the phosphinamide in quantitative yield.[124] Accordingly, with respect to chemical cleavage of a C-P bond, tris(trifluoromethyl)phosphine

$$(CF_3)_3PO + Me_2NH \longrightarrow F_3CH + (CF_3)_2P(=O)NMe_2$$

oxide presently appears to be the most reactive phosphine oxide known.

Tertiary phosphine oxides can also be degraded by heating to high temperatures with metals or with metal hydrides. These reactions are discussed in Sections B.7 and B.6, respectively.

Phosphine oxides that contain α-hydroxyalkyl substituents are also chemically labile. They are degraded by acid and, even more readily, by base.[595] For example, 2-hydroxy-iso-propyldiphenylphosphine oxide, upon warming with 4N sodium hydroxide, forms hydrogen, sodium diphenylphosphinate, and acetone.[135] It was demonstrated that the first step in this reaction is the formation of acetone and

$$Ph_2P(=O)\text{-}C(OH)Me_2 + NaOH \longrightarrow H_2 + Ph_2P(=O)ONa + Me_2CO$$

diphenylphosphine oxide. The latter was then shown to react with alcoholic sodium hydroxide to give hydrogen and sodium diphenylphosphinate. No 2-propanol, which would result from initial attack at phosphorus by hydroxide ion, was produced.

$$Ph_2P(=O)COHMe_2 + NaOH \longrightarrow Ph_2P(=O)H + Me_2CO$$

$$Ph_2P(=O)H + NaOH \longrightarrow H_2 \quad Ph_2P(=O)ONa$$

C. PHYSICAL PROPERTIES

C.1. Properties of the Phosphoryl Bond

The bond distances within trimethylphosphine oxide have been determined by electron diffraction methods.[815] The P-O bond length was found to be 1.48 Å, which is considerably shorter than the predicted single-bond distance of 1.76 Å.[626] Arguments[192,802] correlating bond lengths with π-bond orders have been advanced for phosphorus compounds such as PO_4^{-3}, and accordingly, using similar reasoning, it has been concluded that the phorphoryl group of trimethylphosphine oxide has considerable π-bond character. It is interesting to contrast the shortened phosphoryl bond length with the corresponding amine oxide bond, which has a length approximating that calculated for the single bond.[401]

Considerable caution must be exercised in relating bond lengths to π-bond order. This is best exemplified by the very similar bond lengths found for phosphorus oxytrifluoride (1.45 Å), phosphorus oxytrichloride (1.45 Å), and trimethylphosphine oxide (1.48 Å), whereas it has been suggested that the bond orders of these compounds vary from a triple[801] to a double[701] bond for POF_3 to a single bond[801] for the phosphine oxide. By way of contrast to the latter, a π-bond order of two has been assigned to the phosphoryl group of trimethylphosphine oxide on the basis of ^{13}C-^{1}H coupling data.[305]

In a similar vein, experimentally determined bond energies higher than the energies expected from calculations of the "theoretical values" are frequently regarded as evidence for d_π-p_π bonding in compounds containing second-row elements.[802] Again it is instructive to draw a comparison between the phosphine oxide P-O bond and the dipolar N-O linkage of amine oxides. The phosphoryl bond dissociation energies that have been determined for phosphine oxides lie between 128 and 139 kcal/mole,[56,152,155,317,608] whereas, by way of contrast, the amine oxide bond energy is approximately 50 to 70 kcal/mole. Some, but not all, of the difference in bond energy that exists between

$$R_3PO(g) \longrightarrow O(g) + R_3P(g) \quad D(P{=}O) = 128 \text{ to } 138 \text{ kcal/mole}$$

these two moieties can be accounted for by the greater ionization potential of R_3N as compared to R_3P, and by the interelectronic repulsion of the valence electrons in bonds of first-row elements.[402] It has been concluded,[402] therefore, that part of the high bond energy results from d_π-p_π bonding in the phosphoryl groups of phosphine oxides.

Phosphoryl bond energies have also been contrasted to bond energies determined for P-O "single" bonds,[317,599] such as the P-O bonds of trialkyl phosphites, which have bond energies of approximately 92 kcal/mole. The major difficulty in this type of comparison is that even the P-O bond of trialkyl phosphites possesses some multiple-bond character. However, the energy difference (36 to 46 kcal/mole) that exists between the phosphoryl bonds and P-O "single" bonds is believed to give a rough indication of the π-contribution in the former.[599] Finally, it has been noted that the π-contribution is far less significant in phosphoryl groups than in carbonyl groups (80 to 90 kcal/mole).[599]

As done before for bond lengths and energies, the bond moments of phosphoryl groups of phosphine oxides are contrasted with those of the N-O linkages of amine oxides. On the basis of bond moments, it has been estimated that the charge distributions in the P-O and N-O bonds are as follows:[79]

$$\overset{+0.4}{\geqslant P}\text{———}\overset{-0.4}{O} \qquad \overset{+0.7}{\geqslant N}\text{———}\overset{-0.7}{O}$$

The low value for the phosphoryl group suggests that π-bonding results in a reduction of charge since, in the absence of this effect, the phosphoryl group should be more polar than the N-O bond because of the lower electronegativity of phosphorus.

The dipole moment of triphenylphosphine oxide (4.44 D) and its apparent P-O moment (2.95 D) are somewhat greater than the corresponding values for the alkylphosphine oxides, where the moments vary from 4.35 to 4.37 D and the P-O moments range from 2.86 to 2.89 D.[194]

C.2. Water Solubility

Low-molecular-weight tertiary phosphine oxides are highly water soluble and very hygroscopic. Whereas alkyldimethylphosphine oxides with alkyl chains at or above C_8 are readily extracted from water with chloroform, the lower analogs are not extracted from water to an appreciable extent. Therefore special techniques are required for the isolation of compounds such as trimethylphosphine oxide[123] and 2-hydroxybutyldimethylphosphine oxide[327] from aqueous solution. Furthermore, once the low-molecular-weight phosphine oxides are dry, special precautions are required to prevent rapid water pickup from the atmosphere.

Higher unsymmetrical tertiary phosphine oxides such as dimethyldodecylphosphine oxide are surface active and possess unusual solubility and phase properties.[338,339,523a] Figure 3 shows the phase diagram of dimethyldodecylphosphine oxide in water. Most noteworthy are the relatively large isotropic solution region and region C, in which two isotropic liquid layers exist. These layers are immiscible and are referred to as conjugate solutions. This region does not exist in the diagram of the dimethyldodecyalamine oxide-water system.[518] Regions A and B are the middle and the neat phases, respectively, and are typical anisotropic liquid crystalline phases. The phase diagram of dimethyldecylphosphine oxide is interesting in that region C is entirely surrounded by isotropic solution. In contrast the dimethyloctylphosphine oxide-water diagram does not show a region C.

C.3. Proton NMR Spectroscopy

^{1}H-NMR spectroscopy complements ^{31}P-NMR spectroscopy as a

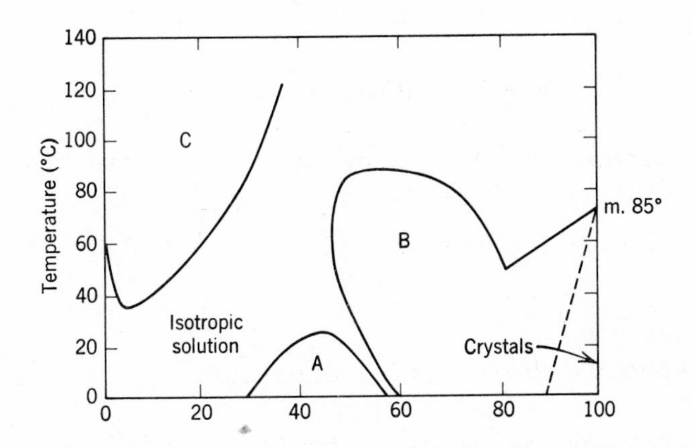

Fig. 3. Phase diagram of dimethyldodecylphosphine oxide in water.

technique for studying the structures of tertiary phosphine oxides. Like hydrogen, ^{31}P nuclei (100% of natural abundance) possess a nuclear spin (I) of 1/2 and, when coupled to hydrogen atoms in the phosphine oxide molecule, give rise to readily distinguished splitting patterns in the proton spectrum. Frequently, the two types of NMR spectra in conjunction provide unambiguous proof of structure of the compound in question.

The chemical shifts of the protons of the alkyl groups in organophosphine oxides depend mainly on their proximity to the phosphorus atom. Protons attached to the α-carbon atom fall within the approximate chemical shift range of 1.5 to 2 ppm downfield from tetramethylsilane.[559] The β-hydrogens of alkylphosphine oxides lie closer to tetramethylsilane, e.g., 1.1 ppm for triethylphosphine oxide.[335] The resonances of protons ortho to the phosphoryl groups of some para-substituted triarylphosphine oxides were determined to be shifted downfield relative to the protons ortho to electron donor substituents.[293] The magnitude of this internal shift (25 to 40 Hz) was regarded as evidence for the electron-accepting properties of the

$Y = HO, (CH_3)_2N, CH_3O$

phosphoryl group.[293] A similar conclusion resulted from a study of the ^{1}H-NMR spectra of $(C_4H_3O)_3P{=}O$,[461] and $>P(O)C{\equiv}CH$.[227]

Proton-phosphorus coupling constants for phosphine oxides vary from approximately 1 to 5 Hz for $>P(O)$–(*meta*-H-phenyl), 5 to 15 Hz for $>P(O)$-C-H, and 10 to 15 Hz for $>P(O)$–(*ortho*-H-phenyl).[293,296,559] The analogous β-coupling constants in alkylphosphine oxides are relatively large, e.g., 18 Hz for triethylphosphine oxide. Long-range ^{31}P-^1H spin-spin couplings have been determined and discussed for other unsaturated systems.[296,559] The 'H-NMR spectra of $>P(O)CH_2P(O)<$, $>P(O)CH{=}CHP(O)<$, $>P(O)CH_2CH_2P(O)<$, etc., have been reported.[108,144]

The ^{19}F-NMR spectrum of tris(p-fluorophenyl)phosphine oxide has been recorded at 94.1 and 56.4 MHz.[294] The six coupling constants and the three chemical shifts for this compound fall within the ranges observed for the analogous triarylphosphine oxides and para-substituted fluorobenzenes.

Proton NMR has also been used as a tool for studying the hydrogen-bond interactions of phenol with phosphine oxides[233] and the stereochemistry of allenicdiphenylphosphine oxides,[745] and for evaluating the basicity of phosphine oxides.[304]

C.4. Phosphorus NMR Spectroscopy

The theory and the application of phosphorus NMR spectroscopy are the subjects of a recent comprehensive review.[193] This review contains a compilation of approximately 3250

^{31}P-NMR chemical shifts and coupling constants, including data on 61 tertiary phosphine oxides.

Without a doubt ^{31}P-NMR spectroscopy is an extremely useful and popular tool for the study of phosphorus compounds and their chemistry. Its applications include structural determination, qualitative and quantitative analysis of pure compounds and of mixtures, determination of reaction kinetics, equilibrium studies, and in certain cases the detection and study of inter- and intramolecular interactions. Examples of the latter are hydrogen bonding and metal-salt complex formation with phosphoryl compounds.

The chemical shifts of tertiary phosphine oxides generally occur within the range of 0 to 55 ppm downfield from the 85% phosphoric acid reference. Representative tertiary phosphine oxides and their chemical shifts are given in Table 8.

Table 8. ^{31}P-NMR Chemical Shifts

R_3PO	Chemical Shift	Ref.
Me_3PO	-36.2	193
$Me_2(C_{12}H_{25})PO$	-42.4	601
$(C_{12}H_{25})_3PO$	-53.0	324
Et_3PO	-48.3	614
$(C_6H_5)_3PO$	-24.9	193
$(CF_3)_3PO$	-2.3	625
$(ClCH_2)_3PO$	-38.1	601
$(MeO_2CCH_2)_3PO$	-33.7	524

Detailed studies of phosphoryl compounds indicate that chemical shifts are dependent on several factors. These are primarily the combined inductive effects of the substituents, the extent of 3d-orbital involvement in π-bonding, and the geometrical symmetry of the molecule. The quantitate importance of these factors in determining chemical shifts has been described.[500]

Particularly noteworthy is the sensitivity of the phosphorus chemical shift of tertiary phosphine oxides to the solvent.[519] The chemical shift of triphenylphosphine oxide, e.g., varied over 35 ppm in a study involving eleven solvents. Proceeding from dioxane, the chemical shift of triphenylphosphine oxide changed from -24.8 ppm to -32.6 ppm in methanol to -37.3 ppm in formic acid to -59.8 ppm in concentrated sulfuric acid. These results are interpreted in terms of the equilibrium involving the phosphine oxide, a hydrogen-bonded intermediate, and a protonated ionic species. That triphenylphosphine oxide is either

$$Ph_3PO + HA \rightleftharpoons Ph_3PO\text{---}HA \rightleftharpoons Ph_3\overset{+}{P}OH,\ A^-$$

completely protonated or nearly so in sulfuric acid is suggested by close agreement with the chemical shift of $Ph_3\overset{+}{P}OEt\ \overset{-}{BF_4}$ (-62 ppm).[215]

Phosphorus NMR spectroscopy is also a useful and sensitive tool for studying the interactions of tertiary phosphine oxides and certain other phosphoryl compounds with other molecules.[325] Triphenylphosphine oxide and phenylmagnesium bromide in tetrahydrofuran have been shown to form a relatively strong and rapidly equilibrating 1:1 complex. The ^{31}P chemical shift of the complex is 12.7 ppm downfield from that of triphenylphosphine oxide. Tri-

$$Ph_3PO + PhMgBr(THF)_2 \rightleftharpoons Ph_3PO{:}MgPhBr(THF)$$

butylphosphine oxide and phenylmagnesium bromide in tetrahydrofuran form a very stable 1:1 complex with a downfield change in chemical shift of 18.9 ppm. Downfield shifts of this magnitude are indicative of a substantial decrease in the electron density of the phosphoryl bond, resulting from coordination of the phosphoryl oxygen with the metal atom. This fact and related data suggest that ^{31}P chemical shifts may provide a more sensitive and useful method of determining and studying molecular interactions than changes in the phosphoryl absorption frequency in infrared spectroscopy.

C.5. Infrared Spectroscopy

Infrared spectroscopy constitutes a very useful tool for studying organophosphorus compounds.[58,170] Qualitatively, tertiary phosphine oxides are readily detected by an intense phosphoryl absorption in the 1100 to 1330 cm^{-1} range. More detailed information pertaining to the structures of phosphine oxides, such as the nature of the phosphoryl bond, i.e., bond order, polarity, and basicity, can also be obtained by infrared spectral studies.

The phosphoryl stretching frequency for simple aliphatic and aromatic tertiary phosphine oxides occurs in the 1150 to 1220 cm^{-1} range, and the absorption is intense. The exact frequency of the phosphoryl absorption depends on several factors. Thus the absorption frequency increases with increasing electronegativity of the substituents.[845] This is illustrated nicely in the following series of phosphine oxides: $(CF_3)_3PO$, 1327 cm^{-1};[124,630] $CH_3(CF_3)_2PO$, 1316 cm^{-1}; t-Bu$(CF_3)_2PO$, 1282 cm^{-1};[300] Ph_3PO, 1193 ±

13 cm^{-1};[13,198] R_3PO, 1170 ± 15 cm^{-1}.[168,197,324,498] That the substituents should affect the frequency of the phosphoryl absorption in this manner is in agreement with the results of broader infrared studies of a large number and variety of other phosphoryl compounds.[57,781]

The frequency and, in certain cases, the multiplicity of the phosphoryl absorption are also dependent on the phase of the sample.[265,283,596,845] For example, the phosphoryl absorption of solid triphenylphosphine oxide consists of two intense absorptions at 1188 cm^{-1} and 1193 cm^{-1} versus a single absorption at 1206 cm^{-1} in carbon disulfide solution. Tri-2-ethylhexylphosphine oxide, in contrast, absorbs at 1164 cm^{-1} and 1198 cm^{-1} in carbon disulfide. Although a number of possible explanations for doubling of the phosphoryl absorption have been given, the phenomenon does not appear to be well understood at this time.[58,170,845]

The effect of solvents on the frequency of phosphoryl absorption has been the subject of several investigations.[15,52,57,240,265,309,432] Protic solvents in general form hydrogen bonds with the phosphoryl oxygen atom which result in a shift of the phosphoryl absorption to lower frequencies. These shifts are approximately 20 to 80 cm^{-1}. The phosphoryl absorption of trioctylphosphine oxide, e.g., is shifted to lower frequency in methanol than in carbon tetrachloride by 37 cm^{-1}. Hydrogen-bonding shifts may be as small as 8 cm^{-1}, as in the case of tributylphosphine oxide in chloroform.[310] Shifts of 135 cm^{-1} in the case of triphenylphosphine oxide in nitric acid[306] and of 200 cm^{-1} in the triphenylphosphine oxide-hydrogen bromide complex probably result from protonation of the phosphoryl oxygen rather than from hydrogen bonding.

Substantial shifts of the phosphoryl absorption to lower frequencies also occur when hydrogen-bonding substituents are present in the molecule. The phosphoryl absorption of certain α-hydroxytrialkylphosphine oxides occurs as low as 1100 cm^{-1} in the solid state, as compared to 1170 ± 15 cm^{-1} for most trialkylphosphine oxides.[596] This and related evidence indicates that relatively strong hydrogen bonding exists between α-hydroxyl groups and the phosphoryl oxygen atom.

Infrared spectroscopy has also been used to study and characterize complexes of tertiary phosphine oxides with boron trichloride,[632] halogens,[285,845] and metal halides.[174,181,285,393,729,753,822,845] In general, complexation lowers the phosphoryl stretching frequency by 20 to 80 cm^{-1}. This is attributed to a reduction of the electron density in the phosphoryl bond, i.e., to a decrease in phosphoryl bond order and strength as a result of coordination between the phosphoryl oxygen and the metal atom or heteroatom.

Infrared spectroscopy is useful as a supplementary tool in identifying specific tertiary phosphine oxides, but its value is limited in this respect because many other compounds, including some organophosphorus compounds, also absorb in the same region. For example, the phosphorus-methyl stretching frequency appears at 750 cm^{-1} for trimethylphosphine oxide.[197] Nevertheless, phosphorus methyl groups in tertiary phosphine oxides can generally be characterized by a weak but detectable absorption at about 1420 cm^{-1}, a medium-intensity absorption at about 1300 cm^{-1}, and a medium absorption in the range of 930 to 840 cm^{-1}. For trimethyl- and alkyldimethylphosphine oxides, the 1300 cm^{-1} absorption appears as a closely spaced doublet. The third absorption also appears as a closely spaced doublet at 866 cm^{-1} and 872 cm^{-1} in trimethylphosphine oxide. In addition, alkyldimethylphosphine oxides absorb moderately to strongly at about 850, 870, 885, and 920 cm^{-1}.

Phosphorus[568] ethyl-containing compounds are reported to absorb at 1227 cm^{-1}. Weak to medium absorptions occur at 1240 and 1260 cm^{-1} in a number of spectra of alkyldiethylphosphine oxides.[324,498] These absorptions appear similar to the 1300 cm^{-1} P-CH_3 absorptions in trimethylphosphine oxide.

Arylphosphine oxides are characterized by medium or weak absorptions in the 1450 to 1425 cm^{-1} and 1010 to 990 cm^{-1} regions. As a specific example, these absorptions occur at 1437 cm^{-1} and 992 cm^{-1} for Ph_3PO.[493]

C.6. Ultraviolet Spectroscopy and Optical Rotatory Dispersion

Ultraviolet spectroscopy has been used sparingly in the study of tertiary phosphine oxides. The spectrum of triphenylphosphine oxide has been recorded and compared with the spectra of benzene, other phenylphosphoryl compounds, triphenylarsine oxide, and diphenyl sulfone.[425,427] The general appearance of the spectrum and, in particular, the vibrational structure of Ph_3PO appear essentially the same as those of benzene, with only a small bathochromic shift of the secondary band of triphenylphosphine oxide being observed. This shift was initially thought to be due to the inductive effect of the phosphoryl group but has since been attributed to a very weak p_π-d_π resonance interaction between the phenyl ring and the d-orbitals on phosphorus.[425]

The UV spectrum of triphenylphosphine oxide, unlike the spectra of aryl ketones, is unchanged upon going from hexane to ethanol solvent. It has been suggested, therefore, that the degree of zwitterionic character, $\equiv\overset{+}{P}-\overset{-}{O}$, versus the degree of double-bond character, $\equiv P=O$, in the phosphoryl bond does not change upon transition from the ground state to the excited state.

Evidence supporting the concept of p_{π}-d_{π} interactions between aryl rings and phosphorus d-orbitals has recently been reported.[278,297,299] In all these cases, the aryl ring contained a strong electron-donating group either in the ring or in the para position. The spectrum of tri-2-pyrrylphosphine oxide shows an intense absorption at 237.5 mμ which is quite different from that of pyrrole. Intense absorption in the 228 to 252 mμ region appears to be characteristic of pyrroles with substituents in the 2-position that are capable of conjugation with the aryl ring. For example, the λ_{max} of a pyrrole with a -CHO group in the 2-position occurs at 252 mμ (ε_{max} 5000). For p-dimethylaminophenyldiphenylphosphine oxide the analogous secondary absorption maximum occurs at 281.7 mμ and is intense (ε_{max} 2.5 x 10^4).[278]

Ultraviolet spectroscopy and visible spectroscopy have been utilized to investigate halogen complexes of phosphine oxides, sulfides, and selenides.[846] When complexes with 1:1 stoichiometries were formed, the phosphine oxides formed the least stable complexes with respect to dissociation.

The optical rotatory dispersions of several tertiary phosphine oxides have been reported.[48,705] The absolute configurations of ortho-, meta-, and para-methoxyphenylmethylphenylphosphine oxides have been determined by intersystem matching of Cotton effects with the respective spectra of o-, m-, and p-methoxyphenyl-p-tolyl sulfoxide. This configurational correlation of spectra, however, appears to be dependent on the methoxy group in some undetermined manner.

C.7. Mass Spectrophotometry

Mass spectrometry is becoming increasingly important as a tool for determining the structures of tertiary phosphine oxides. In addition to determining the molecular weights of phosphine oxides, information about the behavior of these molecules after ionization in the mass spectrometer is slowly accumulating. From one of the most detailed studies[823] initiated to date, reactions induced by electron impact in triphenylphosphine oxide have been identified. Evidence was presented to support the view that the main decomposition pathways occurred with the formation of bridged phosphafluorenyl ions.

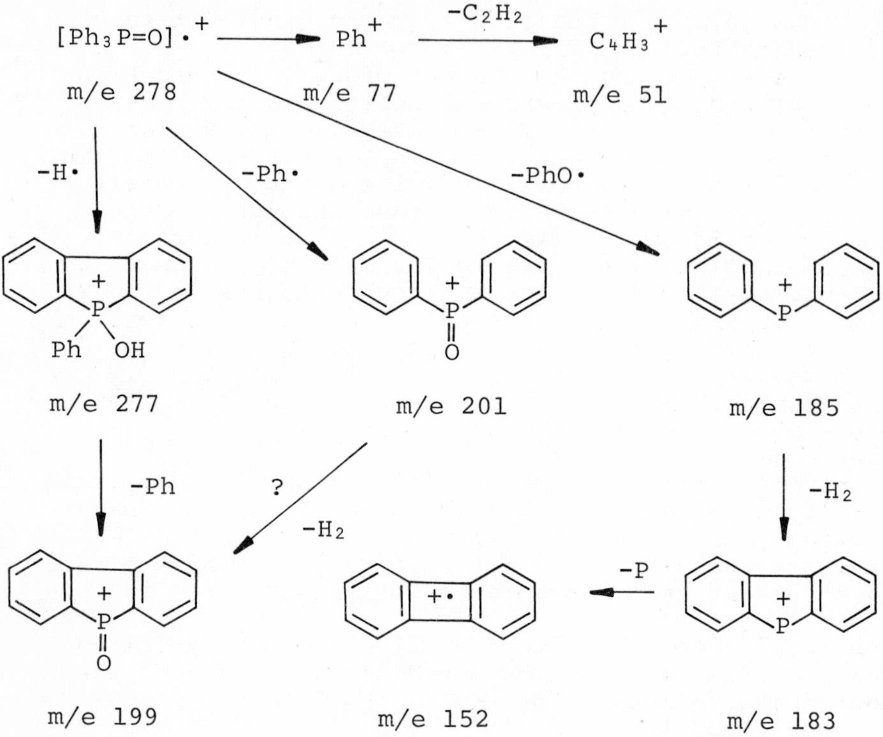

$$[Ph_3P{=}O]^{\cdot +} \xrightarrow{} Ph^+ \xrightarrow{-C_2H_2} C_4H_3^+$$

m/e 278 m/e 77 m/e 51

-H· -Ph· -PhO·

m/e 277 m/e 201 m/e 185

-Ph ? $-H_2$ $-H_2$ -P

m/e 199 m/e 152 m/e 183

Mass spectral fragmentation of 1,2-diphenylethyldiphenylphosphine oxide resulted in only three significant ions.[23,117] The production of these fragment ions was believed to be due to simple P-C fission with hydrogen migration (m/e 202) and without hydrogen migration (m/e 201 and 180).

The fragmentation pattern of β-ketoalkyldiphenylphosphine oxides has been characterized.[592] For example, a

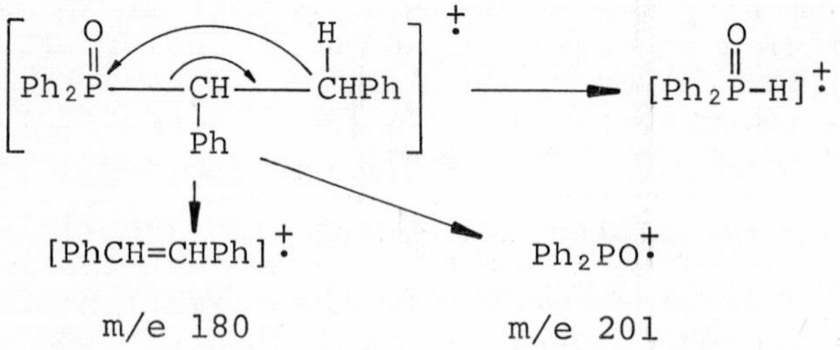

$$\left[Ph_2P(O){-}CH(Ph){-}CH(H)Ph\right]^{\ddagger} \longrightarrow [Ph_2P(O){-}H]^{\ddagger}$$

$[PhCH{=}CHPh]^{\ddagger}$ $Ph_2PO^{\ddagger}$

m/e 180 m/e 201

diphenylphosphine oxide adduct of a tetracyclone shows in its spectrum a strong M-140 peak which was thought to arise by the following pathway:

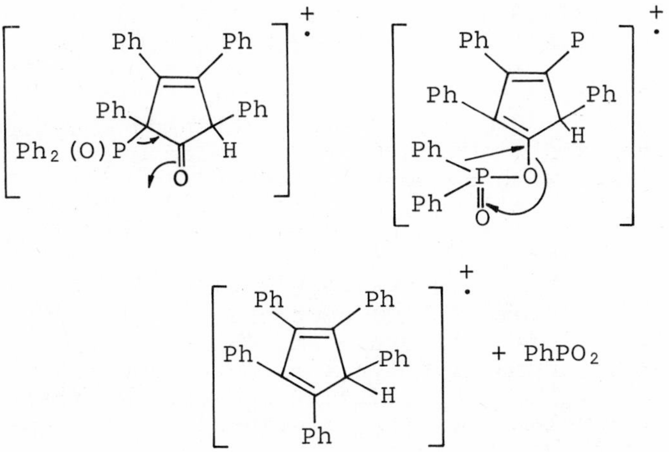

M-140

D. LIST OF TERTIARY PHOSPHINE OXIDES

The following compilation of tertiary phosphine oxides is divided into five types. They are the symmetrical phosphine oxides, R_3PO, the two types of unsymmetrical phosphine oxides, $R_2R'PO$ and $RR'R''PO$, the compounds containing multiple phosphine oxide groups, and heterocyclic phosphine oxides in which the phosphoryl group is in the ring. Within these five divisions individual phosphine oxides are listed according to the empirical formula in the same manner as in the formula index of Chemical Abstracts. Methods of preparation of specific phosphine oxides are indicated by a Roman numeral that refers to the appropriate method in Section A on synthesis.

D.1. R_3PO

C_3Cl_9OP

$(Cl_3C)_3PO$. VI. M. 53°.[839]
$(CF_3)_3PO$. I.[124] VI.[630] M. -89°,[124] b. 32°,[124] b. 23.6°,[630] ^{31}P -2.6.[625]
$(BrCH_2)_3PO$. ^{31}P -42.5.[193] M. 137°,[26] m. 128.5-129.5°.[153a]
$(ClCH_2)_3PO$. IX.[537] VI. M. 88-89°,[537] m. 78°,[344] m. 100.5°,[839] ^{31}P -39.2, ^1H. Hemihydrate, m. 88-89°.[344]
$(ICH_2)_3PO$. IX.[26] M. 223°.
$(H_2NCH_2)_3PO$. IX. M. 40°.[791a]

C_3H_9OP

Me_3PO. I.[130] II.[130,159,239] III.[649] M. 140-141°,[239]

m. 140°,[815] m. 137-138°,[130] b. 214-215°,[159] Raman,[197] IR,[41,197] ^{31}P -36.2,[193] elect. diff.[815] establishes P-O 1.48 Å,[626,815] P-C 1.81 Å,[626,815] C-P-C 106.0°,[626,815] and C-P-O 112.3°,[815] D(P=O) 139,[155,626] P-C stretching const. 3.0 md/Å,[626] P=O stretching const. 8.0 md/Å.[626] Complexes with Cl_3CCO_2H,[649] aminoborane,[593] I_2.[285] Complexes with compounds of Sb,[99,511,841] As,[511] Sc,[511] Hg,[511,841] Co,[104,175,179,403,626] Al,[728,729,730] Zn,[175,403,626,649] Sn,[403] Ga,[728,729] Fe,[649,694] Au,[649] Cr,[649] Pt,[649] Te,[729] In,[728] Cu,[279] U,[263] Si.[54]

$(HOCH_2)_3PO$. II.[286,344,345] M. 44-45°,[237,302,691] IR,[286] ^{31}P -45.4.[601] Complex with Cu compound.[279] Tribenzoate, m. 110°.[344,345] Trilaurate, m. 65.5-66.6°.[395]

$[(HO)_2P_\beta(O)CH_3]\ P_\alpha O$. From hydrolysis of corresponding alkyl esters.[537] ^{1}H, $^{31}P_\alpha$ -40.7, $^{31}P_\beta$ -16. Crystallized as cyclohexylamine salts, m. 218-226°. Tris-p-toluidinium salt, m. 203-204°.[571]

C_6H_9OP

$(H_2C{=}CH)_3PO$. I. M. 99-101°, UV.[817]

Et_3PO. I.[130,771] VI.[556] II.[160,184,239,340] III.[331] Hygroscopic needle, m. 50°,[130] m. 46°,[771] b. 238-240°,[771] b. 243°,[160] ^{31}P -48.3.[440,614] Complexes with compounds of Cr,[419,422,649] Sn,[87,661] Ni,[414] Co,[414,649] V,[405] Al,[730,731] I_2,[285] Au,[649] Pt,[649] Cu,[649] Hg,[649] U.[263]

$C_9H_{12}N_3OP$

$(NCCH_2CH_2)_3PO$. I.[681] VIII. M. 172-173°,[677,681] ^{31}P -46.[193]

$(CH_2{=}CHCH_2)_3PO$. M. 15-17°, $b_{0.5}$ 98°.[462]

$(\overline{CH_2CH_2CH})_3PO$. I. IR, ^{1}H.[218]

$(HO_2CCH_2CH_2)_3PO$. IX. M. 155-156°.[681]

$(MeO_2CCH_2)_3PO$. ^{31}P -34.4.[193]

$(MeCO_2CH_2)_3PO$. IX.[26] Oil.

$(H_2NCOCH_2CH_2)_3PO$. VIII. M. 206-209°.[677]

Pr_3PO. III.[273,649] M. 36°,[273] m. 38°,[649] b. 280-282°,[273] b. 260-265°.[649] Complexes with compounds of V,[405] Cr,[419] Hg.[649]

$(EtOCH_2)_3PO$. IX. B_1 109-111°.[690]

$(Me_2NCH_2)_3PO$. VIII. M. 154-157°, IR, ^{1}H, ^{31}P -48.5.[527,534]

$[(MeO)_2P_\beta(O)CH_2]_3P_\alpha O$. IV.[537] M. 169-171°. ^{31}P -31, ^{31}P -22.8.

$C_{12}H_9OPS_3$

$(\overline{S{-}CH{=}CHCH{=}C})_3PO$. III.[297] I.[406] M. 130-130.5°,[297] m. 128°,[406] UV,[297] ^{1}H.[461]

$(\overline{NH{-}CH{=}CHCH{=}C})_3PO$. III.[297] M. 136-137°,[297] UV,[297] ^{1}H.[461]

$(CH_2{=}CMeCH_2)_3PO$. I. III. M. 132°.[439]

Bu_3PO. I.[205] III.[205] B. 300°, ^{31}P -43.2, -45.8.[440,614,673] Complexes with compounds of U,[127,759,808]

Cr,[419,422] halogen,[846] Sn.[87]
(s-Bu)$_3$PO. III. B$_{0.3}$ 105-110°, n_D^{20} 1.4598, d^{25} 0.922.[756]
(i-Bu)$_3$PO. III.[639] I.[748] M. 89°,[639] m. 123-125°,[774] b$_4$ 119-120°.[639]
(PrOCH$_2$)$_3$PO. IX. b$_1$ 122-123°.[690]
(Me$_3$SiCH$_2$)$_3$PO. III.[139] M. 182°.

C$_{15}$H$_9$OP

(MeC≡CC≡C)$_3$PO. I.[319] M. 85° (dec.).
(N=CHCH=CHCH=C)$_3$PO. I. M. 209°. Picrate, m. 144-148°.[543]
(MeNCH=CHCH=C)$_3$PO. III.[297] M. 136-137.5°, UV,[297] ^{1}H.[461]
(EtO$_2$CCH$_2$CH$_2$)$_3$PO. VIII. B$_{0.2}$ 199-203°, n_D^{25} 1.4682.[677]
(Me$_2$CHO$_2$CCH$_2$)$_3$PO. ^{31}P -33.7.[193]
(PrCO$_2$CH$_2$)$_3$PO. IX.[26] Impure oil.
[CH$_2$(CH$_2$)$_3$NCH$_2$]$_3$PO. VIII. M. 150.5-153°, IR, ^{1}H, ^{31}P -48.4 (EtOH).[527,534]
[CH$_2$CH$_2$O(CH$_2$)$_2$NCH$_2$]$_3$PO. VIII. M. 160-161.3°, IR, ^{31}P -50.6 (H$_2$O).[527,534]
Am$_3$PO. I. M. 59°.[542]
i-Am$_3$PO. I. M. 60-65°.[347]
(Et$_2$NCH$_2$)$_3$PO. VIII. M. 73.4-74.2°, IR, ^{1}H, ^{31}P -43.2 (C$_6$H$_6$).[527,534]
[(EtO)$_2$P$_\beta$(O)CH$_2$]$_3$PO. IV.[537] M. 168-170°, m. 167.5-168°,[571] ^{31}P$_\alpha$ -34, ^{31}P$_\beta$ -21.

C$_{18}$F$_{15}$OP

(C$_6$F$_5$)$_3$PO. VI. IX. IR, ^{31}P +8.2 (CHCl$_3$), ^{19}F, mass spect.[230]
(3,5-(O$_2$N)$_2$-4-ClC$_6$H$_2$)$_3$PO. IX. M. 343-345°.[769]
(4-Cl-3-O$_2$NC$_6$H$_3$)$_3$PO. IX. M. 187-189°.[769]
(3,5-(NO$_2$)$_2$C$_6$H$_3$)$_3$PO. IX. M. 365-367°.[769]
(2-ClC$_6$H$_4$)$_3$PO. VI. Hemihydrate, m. 226-236°.[540]
(3-ClC$_6$H$_4$)$_3$PO. VI. IX.[540] III.[148,540] M. 135°.
(4-ClC$_6$H$_4$)$_3$PO. I.[143] III.[481] M. 171.5-172.0,[481] m. 177-178°,[143] m. 175°,[540] ^{31}P -23, -22.7.[245,440,803]
(4-ClSO$_2$C$_6$H$_4$)$_3$PO. IX. M. 210-212°.[769]
(3-FC$_6$H$_4$)$_3$PO. I.[727] M. 104°.
(4-FC$_6$H$_4$)$_3$PO. I. M. 121-123°, ^{19}F.[726]
(3-O$_2$NC$_6$H$_4$)$_3$PO. IX. M. 242°.[148,588]
[3,5-(NO$_2$)$_2$-4-H$_2$NC$_6$H$_2$]$_3$PO. IX. M. 360° (dec.).[769]
(4-NaO$_3$SC$_6$H$_4$)$_3$PO. IX.[769]
Ph$_3$PO. III.[272,273,481] VI.[270,737] M. 152-153°,[481] IR,[198] UV,[427] ^{31}P -25 ± 2,[193,601,673] mass spect.,[823] μ 4.44 D,[194] 4.40 D,[246] 4.49 D,[40] P=O moment 2.99 D,[40] D(O=PPh) 128.4 ± 5.5,[56] ΔH of formation -15.6,[317] molar Kerr const. (-828 x 10^{12}),[40] conformation,[40] tri-p-deuterio.[457] Complexes with SO$_3$,[277,511] Br$_2$,[770,846] NO$_2$,[770] H$_2$O$_2$,[397,780] lithium halides,[312] HNO$_3$,[306] CCl$_3$CO$_2$H,[306,649] HCl,[306,649] HBr,[306] hydroquinone

compounds,[671] I_2,[285] $ClCH_2CO_2H$,[694] 2,4,6-$(O_2N)_3C_6H_2OH$,[694] $C_6H_5CO_2H$.[694] Complexes with compounds of Ni,[50,174,176,281,694] Tl,[177,211,212,253] Mo,[86,150,343,465,508,752] W,[86,103,150,508] Eu,[181,575] Au,[157,649] V,[342,394,404,538] Cr,[55,253,419,422,716] B,[253,282,632] Co,[50,174,175,414,649] Zn,[50,174,175,210,253,675] Pu,[652] Mg,[253] Cd,[174,253,649,674,675] Hg,[253,511] Al,[253,510] Ga,[253] In,[2,145,253,703] Si,[54,253] As,[253,736] Sb,[253,511] Bi,[253] Fe,[174,253,513,649,753] P,[253] U,[119,127,209,247,316,510] Cu,[50,74,174,279,694] Ti,[510,753] Sn,[87,154,753,822] Pa,[109] Nb,[109,510] Ta,[109,510] Re,[178,515,703,709] Pb,[156,398] Th,[538,763] Zr,[510,538] La,[181,182] Mn,[174,280] Nd,[180,181] Sm,[180,181] Gd,[181] Dy,[181] Ho,[181] Er,[181] Tm,[181] Lu,[181] Y,[181] Ce,[181] Pr,[181] Sc.[188]

$(2\text{-}HOC_6H_4)_3PO$. IX.[463] M. 214.5-216°.

$(3\text{-}HOC_6H_4)_3PO$. I.[497] M. 270-272°.

$(4\text{-}HOC_6H_4)_3PO$. I. IX. M. 273-275°.[743]

$[3,5\text{-}(HO)_2C_6H_3]_3PO$. IX.[497] Hydrate, m. 365-378°. Acetate, m. 231-233°.

$(3\text{-}H_2NC_6H_4)_3PO$. IX. M. 258.[148,588]

$(4\text{-}H_2NSO_2C_6H_4)_3PO$. IX. M. 154-155°.[769]

$[\overline{CH_2(CH_2)_4CH}]_3PO$. I.[406] IX. M. 156.5°,[616] m. 155-157°,[406] ^{31}P -50.[193] Complexes with SO_3,[277] HI.[419] Complexes with compounds of V,[405] Cr,[419] Hg.[419]

$C_{18}H_{36}N_3OP$

$[\overline{CH_2(CH_2)_4N}CH_2]_3PO$. VIII. M. 119-120°, IR, ^1H, ^{31}P -51 (EtOH).[534]

$(C_6H_{13})_3PO$. Crystals,[423] m. 34-35°,[696] $b_{0.8}$ 146-150.[696] Complexes with compounds of Cr,[422] Ni,[414] Co.[414]

$(3\text{-}NCC_6H_4)_3PO$. IX. M. 198-200°.[725]

$(4\text{-}NCC_6H_4)_3PO$. IX. M. 215-219°.[725]

$C_{21}H_{15}N_6O_{13}P$

$[4\text{-}Me\text{-}3,5\text{-}(NO_2)_2C_6H_2]_3PO$. IX. M. 264-266°.[769]

$(3\text{-}HO_2CC_6H_4)_3PO$. I. M. 335-337°.[777]

$(4\text{-}HO_2CC_6H_4)_3PO$. IX.[605] M. 323-330°. Methyl ester, m. 123-125°.

$(4\text{-}BrC_6H_4CH_2)_3PO$. VIII. M. 182-183°.[844]

$[2\text{-}MeO\text{-}3(?)\text{-}BrC_6H_3]_3PO$. VI. M. 151-152°.[540]

$(4\text{-}ClC_6H_4CH_2)_3PO$. III. M. 179-180°.[699]

$(3\text{-}H_2NCOC_6H_4)_3PO$. IX. M. 288-289°.[725]

$(4\text{-}H_2NCOC_6H_4)_3PO$. IX.[725] M. 272-275°.

$(4\text{-}Me\text{-}3\text{-}O_2NC_6H_3)_3PO$. IX. M. 153°,[580] 152-153°.[769]

$(4\text{-}O_2NC_6H_4CH_2)_3PO$. IX. M. 273,[146] m. 100°(?).[159]

$(3\text{-}NO_2\text{-}4\text{-}MeOC_6H_3)_3PO$. IX. M. 188-190°.[769]

$(PhCH_2)_3PO$. I.[501] VI.[248] II.[501] III.[649] IV.[706] M. 213°,[248] m. 214°,[649] m. 210-213°.[706] Hydrochloride, m. 169°; chloroaurate, m. 222.5°.[159]
$(2\text{-}MeC_6H_4)_3PO$. VI. IX. Hemihydrate, m. 153°.[540]
$(3\text{-}MeC_6H_4)_3PO$. VI. M. 111°.[540]
$(4\text{-}MeC_6H_4)_3PO$. VI. M. 145°,[580] ^1H.[296]
$(2\text{-}MeOC_6H_4)_3PO$. I. M. 215-217°.[833]
$(3\text{-}MeOC_6H_4)_3PO$. I. M. 151-152°,[540] m. 150-152°.[497]
$(4\text{-}MeOC_6H_4)_3PO$. I.[743] IX.[47] M. 143-144°,[743] m. 142-143°.[47]
$(PhOCH_2)_3PO$. IX. M. 155°.[690]
$(4\text{-}Me\text{-}3(?)\text{-}H_2NC_6H_3)_3PO$. IX. M. 235°.[580]
$[(HO)(Ph)P_\beta(O)CH_2]_3P_\alpha O$. From hydrolysis of corresponding alkyl ester.[537] M. 90° (sintered, clear at 138°), $^{31}P_\alpha$ -35, $^{31}P_\beta$ -29.8, ^1H.
$[\overline{CH_2(CH_2)_4C}HCH_2]_3PO$. IX. M. 170-171°.[473]
$(AmCO_2CH_2)_3PO$. IX.[26] M. 29-30°.
$(C_7H_{15})_3PO$. I.[423]
$[(i\text{-}PrO)_2P_\beta(O)CH_2]_3P_\alpha O$. IV.[537] M. 85-87°, $^{31}P_\alpha$ -28.9, $^{31}P_\beta$ -18.6.

$C_{24}H_{15}OP$

$(PhC{\equiv}C)_3PO$. III. M. 126°.[153]
$(PhCCl{=}CH)_3PO$. IX. M. 216-217°.[750]
$(3\text{-}Indolyl)_3PO$. III. M. 138-140°.[597]
$(PhCH{=}CH)_3PO$. IX. M. 236-237°.[750]
$(MeO_2CC_6H_4)_3PO$. IX.[725] M. 118-120°.
$(PhO_2CCH_2)_3PO$. ^{31}P -36.0.[193]
$(2\text{-}MeCO_2C_6H_4)_3PO$. IX. M. 197-199°.[463]
$[3,5\text{-}(O_2N)_2\text{-}4\text{-}Me_2NC_6H_2]_3PO$. IX. Ignition point 226°.[769]
$(2,4\text{-}Me_2C_6H_3)_3PO$. VI.[580]
$(2,5\text{-}Me_2C_6H_3)_3PO$. VI. M. 173°.[580]
$(PhCH_2CH_2)_3PO$. III. M. 150-152°.[83]
$(2\text{-}MeC_6H_4CH_2)_3PO$. VIII. M. 143-144°.[844]
$(3\text{-}MeC_6H_4CH_2)_3PO$. VIII. M. 188-189°.[844]
$(4\text{-}MeC_6H_4CH_2)_3PO$. VIII. M. 168-169°.[844]
$[PhCH(OMe)]_3PO$. I. M. 137-142°.[231]
$[3,5\text{-}(MeO)_2C_6H_3]_3PO$. I.[497] M. 139-141°.
$(3\text{-}Me_2NC_6H_4)_3PO$. IX. M. 149-152°.[588]
$(4\text{-}Me_2NC_6H_4)_3PO$. I.[700] II.[717] IX. M. 290°,[475] m. 262°,[97] m. 305-306°,[700] m. 288-290°, IR,[717] UV.[723] Hydrate, m. 321°.[97]
$(C_8H_{17})_3PO$. I. M. 51-51.5°,[83] b_1 201-202°,[83] ^{31}P -42.0.[121] Solvent effects on PO IR absorption.[240] Halogen complexes.[846]

Complexes with compounds of U,[121,341,675,811] Cd,[674,675] Zn,[675,811] Cu.[811]
$[Me(CH_2)_3CHEtCH_2]_3PO$. III. B_{2-3} 205-213°.[83]
$(4\text{-}MeC_6H_4C{\equiv}C)_3PO$. I. M. 177°.[319a]
(2-Methyl-3-indolyl)$_3$PO. III. M. 170°.[597]
$(4\text{-}EtC_6H_4CH_2)_3PO$. VIII. M. 140-142°.[844]
$(PhCH_2CH_2CH_2)_3PO$. III. M. 84-85°.[83]
$(2,4,5\text{-}Me_3C_6H_2)$ PO. VI. M. 222°.[580]
$[4\text{-}ClC_6H_4Si(CH_3)_2CH_2]_3PO$. III. M. 121.5°.[138]
$(4\text{-}Me_3SiC_6H_4)_3PO$. III. M. 259°.[139,255]
$(C_7H_{15}CO_2CH_2)_3PO$. IX.[26] M. 38-39°.
$(C_9H_{19})_3PO$. III. B_4 235°, m. 35-36°.[841a]
$(Me_3CCH_2CHMeCH_2CH_2)_3PO$. III. $B_{0.1}$ 185-190°.[83]
$(Bu_2NCH_2)_3PO$. VIII. M. -1 to -0.5°, IR, ^1H, ^{31}P -43.5.[527,534]
$[(BuO)_2P_\beta(O)CH_2]_3P_\alpha O$. IV.[537] M. 109-111°, $^{31}P_\alpha$ -29.7, $^{31}P_\beta$ -19.8.

$C_{30}H_{21}OP$

$(1\text{-}C_{10}H_7)_3PO$. III.[590,712] VI.[25] M. 341-342°,[25] m. 342-344°.[190]
$(2\text{-}C_{10}H_7)_3PO$. III. M. 248-249°.[125]
(Ferrocenyl)$_3$PO. IX. Yellow powder, infusible, IR,[764] UV.[621] Trimethyl ester dihydrate of trisulfonic acid, m. 118-120°.[618]
$(4\text{-}PrC_6H_4CH_2)_3PO$. VIII. M. 160-162°.[844]
$(4\text{-}i\text{-}PrC_6H_4CH_2)_3PO$. VIII. M. 190-192°.[844]
$(4\text{-}Et_2NC_6H_4)_3PO$. IX. M. 239°.[475]
$(4\text{-}MeC_6H_4SiMe_2CH_2)_3PO$. III. M. 111.5°.[139]
$(C_{10}H_{21})_3PO$. III. M. 40-42°, b_{3-4} 278-283°.[83]
$[Ph(BuO)P_\beta(O)CH_2]_3P_\alpha O$. IV.[537] Oil, $^{31}P_\alpha$ -29.0, $^{31}P_\beta$ -34.7.
$(3\text{-}NO_2\text{-}4\text{-}PhOC_6H_3)_3PO$. IX. M. 248-250°.[769]
$(3,5\text{-}(NO_2)_2\text{-}4PhNHC_6H_2]_3PO$. IX. M. 318° (dec.).[769]
$(3\text{-}NO_2\text{-}4\text{-}PhNHC_6H_3)_3PO$. IX. M. 258-260°.[769]
$(2\text{-}PhC_6H_4)_3PO$. VI. M. 184-185°.[835]
$(4\text{-}PhC_6H_4)_3PO$. I.[718] II. VI.[834] M. 244-248°,[718] m. 233-234°.[834]
$(C_{12}H_{25})_3PO$. I.[324] III. M. 44-48°, $b_{0.1}$ 235-240°,[83] m. 59-59.5°, IR, ^{31}P -53.[324]

$C_{39}H_{72}N_3OP$

$\{[\overline{CH_2(CH_2)_4C}H]_2NCH_2\}_3PO$. VIII. M. 214-215°, IR, ^1H.[527 543]

$C_{39}H_{75}O_7P$

$(C_{11}H_{23}CO_2CH_2)_3PO$. IX.[26] M. 66-67°.
(9-Phenanthryl)$_3$PO. III. M. 354-356°.[590 591]
$(C_{15}H_{31}CO_2CH_2)_3PO$. IX.[26] M. 71-72°.
$[(C_8H_{17})_2NCH_2]_3PO$. VIII. M. 35-37°.[527]
$[(BuEtCHCH_2O)_2P_\beta(O)CH_2]_3P_\alpha O$. IV.[537] Oil, $^{31}P_\alpha$ -29.3,

$^{31}P_{\beta}$ -21.
[Benz(a)anthr-7-yl]$_3$PO. III. M. 191-193°.[590,591]
$(C_{17}H_{35}CO_2CH_2)_3PO$. IX.[26] M. 75-76°.
$(C_{23}H_{47}CO_2CH_2)_3PO$. IX.[26] M. 93-94°.
$[(C_{12}H_{25})_2NCH_2]_3PO$. VIII. M. 77-79°.[527]

D.2. $R_2R'PO$

$C_3H_3F_6OP$

$(CF_3)_2MePO$. I. IV. M. -27 to -26.9°, b. 129.7°, IR.[300]
$Me_2(NC)PO$. VI. B. 128-130°.[275]
$(ClCH_2)_2MePO$. II.[455] III.[526] IX. M. 45-47°, ^{31}P -41.6, ^1H.[537]

C_3H_8ClOP

$Me_2(ClCH_2)PO$. III. M. 69-72°.[469] ^{31}P -42.[601]
$[(HO)_2P(O)CH_2]_2MePO$. From hydrolysis of corresponding alkyl ester. Viscous oil, ^1H. 4-Toluidinium salt, m. 181-183°.[571]
$(ClH_2C)_2EtPO$. III.[536] M. 42-44°, ^{31}P -45.7, ^1H.
Me_2EtPO. II. M. 73-75°, b. 223-225°.[239]
$Me_2(H_2C{=}C{=}CH)PO$. IV.[545] VI.[546] M. 58-60°,[545,546] ^{31}P -41.0.[545]
$Me_2(MeCOCH_2)PO$. VI. B_1 149-151°, n_D^{20} 1.4931, d^{20} 1.1584.[275]
$Me_2(EtO_2C)PO$. VI. M. 54-55°, b_1 143-146°.[275]
Et_2MePO. IV.[651] II.[573] IX. B_{10} 103-106°.[334]
$(MeOCH_2)_2MePO$. IX. M. 10.7°, $b_{3.5}$ 102-103°, n_D^{20} 1.4591, d^{20} 1.1044, ^1H, μ 3.6 D.[749]

$C_6H_9F_6OP$

$(CF_3)_2(t\text{-}Bu)PO$. I. IV.[300] B. 177°, IR.
$Et_2(H_2C{=}CH)PO$. IX.[447] B_2 62.3°, m. 35-36°.
$Me_2(EtCH{=}CH)PO$. IX. $B_{0.1}$ 78-82°, IR, ^1H, ^{31}P.[327]
$Et_2(MeCO)PO$. IX. B. 143-144°, n_D^{20} 1.4807, d^{20} 1.0857, IR.[795]
$Me_2(EtO_2CCH_2)PO$. VI. B_1 147-149°.[275]
$Et_2(ClCH_2CH_2)PO$. IX.[447] B_1 100-101°, m. 33-34°, n_D^{20} 1.4855, d^{20} 1.1154.
$Me_2t\text{-}BuPO$. ^{31}P -38.1.[601]
$Me_2(EtCHOHCH_2)PO$. III. $B_{0.01}$ 104-108°, IR, ^1H, ^{31}P -41 (C_6H_6).[327]
$Et_2(MeOCH_2)PO$. VI. B_{17} 128-129°, n_D^{20} 1.4571, d^{20} 1.0152.[794]
$(MeC{\equiv}C)_2MePO$. n_D^{20} 1.566, d^{20} 1.066, μ 4.980.[88]
$(NCCH_2CH_2)_2CH_2(OH)PO$. V.[679] M. 99-101°.
$Et_2(\overline{OCH{=}NN{=}C}CH_2)PO$. IX. M. 111-112°.[655]
$(HO_2CCH_2CH_2)_2MePO$. From hydrolysis of corresponding

nitrile.[634] M. 161-163°.
$Et_2(H_2C{=}CHCH_2)PO$. IV.[687] B_2 78-80°,[659] n_D^{20} 1.4766,[687] d^{20} 0.9738,[687] d^{20} 0.9665.[659]
Et_2PrPO. II. M. 37°, b. 245-247°.[239]
$Et_2(EtOCH_2)PO$. VI.[794] B_{11} 121-122°, d^{20} 0.9972, n_D^{20} 1.4560.
$(EtOCH_2)_2MePO$. From reaction of tri(chloromethyl)phosphine with sodium ethoxide.[452] B_2 97-99°, n_D^{20} 1.4491, d^{20} 1.0275.
$(H_2NCH_2CH_2CH_2)_2MePO$. I.[634] $B_{0.2}$ 165-166°, n_D^{25} 1.5098.

$C_8H_3Cl_7NO_3P$

$(Cl_3C)_2(3\text{-}O_2N,4\text{-}ClC_6H_4)PO$. IX.[838] M. 172-173°.
$(Cl_3C)_2(3\text{-}O_2NC_6H_4)PO$. IX.[838] M. 158-159°.
$(Cl_3C)_2(4\text{-}ClC_6H_4)PO$. IX.[838] M. 148-149°.
$(Cl_3C)_2C_6H_5PO$. IX.[838] M. 133-134°.
$Me_2C_6F_5PO$. I. M. 56-58°.[244]
$(ClCH_2)_2(3\text{-}O_2NC_6H_4)PO$. IX. M. 155-156°.[837]
$(ClCH_2)_2(3\text{-}ClC_6H_4)PO$. IX. M. 129-130°.[837]
$(ClCH_2)_2(4\text{-}ClC_6H_4)PO$. VI.[261] M. 150-151°.
$(ClH_2C)_2PhPO$. III.[526,536] M. 141-143°, m. 142°,[333] ^{31}P -33.9,[536] ^1H.[526,536]
$(ICH_2)_2PhPO$. IX. M. 172-173°.[837]
$Me_2(4\text{-}BrC_6H_4)PO$. III.[557] M. 137-138°.
$(ClCH_2)_2(3\text{-}H_2NC_6H_4)PO$. IX. M. 110-111°.[837]
$(NCCH_2CH_2)_2Cl_3CCH(OH)PO$. V.[679] M. 159-160° (dec.).
$Me_2(3\text{-}FC_6H_4)PO$. I.[727] M. 83°.
$Me_2(4\text{-}FC_6H_4)PO$. I. M. 104-108°, ^{19}F.[726]
Me_2PhPO. I. IV. M. 100°, b. 300-308°,[573,576] IR, ^1H, pK_b.[304] Complexes with compounds of Pt, Pd,[431] and Hg.[361,573]

$C_8H_{11}O_3P$

$(HOH_2C)_2PhPO$. I.[334] V.[645] Oil,[645] m. 102.5-105°.[334]
$Pr_2(H_2C{=}CH)PO$. I.[442] IX.[334] M. 36-37°,[334] b_1 77-78°.[334]
$Et_2(EtO_2CCH_2)PO$. IV.[36] B_3 113-113.5, d^{20} 1.0733, n_D^{20} 1.4638.
$Et_2(MeO_2CCH_2CH_2)PO$. V.[453] M. 28-30°, n_D^{20} 1.4710, d^{20} 1.0872 (supercooled).
Et_2BuPO. IV.[36,687] B_1 85.5-86°,[36] n_D^{20} 1.4589,[687] d^{20} 0.9290.[687]
$Et_2(EtSCHMe)PO$. VI. B_{10} 145-147°, n_D^{20} 1.5049, d^{20} 1.0469.[794]
$Et_2(PrOCH_2)PO$. VI. $B_{11.5}$ 136-137°, n_D^{20} 1.4547, d^{20} 0.9693.[794]
$Et_2[Me(EtO)HC]PO$. VI.[796] B_2 141-142°, n_D^{20} 1.4797, d^{20} 1.1081, IR.

$C_9H_{11}Cl_2OPS$

$(ClCH_2)_2PhSCH_2PO$. IX. M. 113°.[690]
$Me_2(4\text{-}HO_2CC_6H_4)PO$. IX. M. 240°, b_{15} 360°. Chloroplatinate,

m. 234°. Ammonium salt, dec. 212°. Anilide, m. 235°.[578]
$Me_2(4\text{-}ClC_6H_4CH_2)PO$. III. M. 130-132°.[699]
$Me_2(4\text{-}Me\text{-}3\ (?)\ O_2NC_6H_3)PO$. IX. M. 175°. Mercuric chloride complex, m. 127°.[578]
$Me_2(4\text{-}MeC_6H_4)PO$. I.[196,578] M. 95°.[196]
$Me_2(PhCH_2)PO$. II. M. 58-60°, b. 303-308°. Mercuric chloride complex, m. 115°.[573]
$(NCCH_2CH_2)_2H_2NCOCH_2CH_2PO$. V.[769] M. 161-163°.
$Me_2[Ph(HO)P(O)CH_2]PO$. IV. M. 143-145°, Zn salt, m. 229-233°; Cr salt, m. 365-367°.[469]
$(NCCH_2CH_2)_2Me_2C(OH)PO$. V.[127] M. 124-125°.
$[H_2C{=}C(Me)CH_2]_2MePO$. III.[64] B_2 135-141°, IR.
$Et_2[EtO(Me)C{=}CH]PO$. IV.[776] $B_{0.3}$ 107°.
$Bu_2(ClCH_2)PO$. IX.[647] B_2 122-125°, n_D^{20} 1.4793, d^{20} 1.0940.
$(t\text{-}Bu)_2MePO$. III. $B_{0.02}$ 57°, IR, ^1H.[110]
Bu_2MePO. III.[487] B_2 181°, m. 35°.
$Et_2(BuOCH_2)PO$. By reaction of $Et_2P(O)Na$ with $ClCH_2OBu$.[453] B_2 104-105°, n_D^{20} 1.4532, d^{20} 0.9571.
$Bu_2(H_2O_3PCH_2)PO$. IX. M. 217-129°.[450]

$C_{10}H_{10}F_5OP$

$Et_2C_6H_5PO$. I. $B_{0.4}$ 151°.[244]
$(H_2C{=}CH)_2PhPO$. I.[442] B_2 130-130.5°, m. 50-51°.
$Et_2(4\text{-}BrC_6H_4)PO$. III.[557] M. 113-115°.
$(NCCH_2CH_2)_2MeO_2CCH_2CH_2PO$. V.[679] M. 79-81°.
$Me_2(2,5\text{-}Me_2C_6H_3)PO$. I. M. 94-95°.[424]
Et_2PhPO. II.[573,576] IV.[651] VI. M. 55-56°,[576] m. 55-57°,[651] ^{31}P -42.4.[673] Complex with compounds of U,[263] Sn.[87]
$[HO(Me)HC]_2PhPO$. V.[645] $B_{0.5}$ 100-103°.
$Me_2(4\text{-}Me_2NC_6H_4)PO$. I. M. 62°,[587] UV.[723]
$(NCCH_2CH_2)_2Me_2CHCH(OH)PO$. V.[679] M. 77-79°.
$[H_2C{=}C(Me)CH_2]_2EtPO$. III.[64] $B_{0.3}$ 115-118°.
$Bu_2(Cl_3C)CH(OH)PO$. V.[678] M. 132-133°.
$(i\text{-}Bu)_2Cl_3CCH(OH)PO$. V.[678] M. 175° (dec.).
$(ClCH_2)_2(C_8H_{17})PO$. III.[536] M. 56-59°, ^{31}P -45.4, ^1H.
$Bu_2(H_2C{=}CH)PO$. I.[442] $B_{1.5}$ 103.5-104°, m. 37.5-38°, Raman, UV.[654] Complexes with compounds of Zn,[818] Hg,[818] Sb,[818] Sn,[818] Ti,[818] Fe,[818] U.[818]
$(i\text{-}Bu)_2(CH_2{=}CH)PO$. IX. M. 18°, b. 122-123°.[669]

$C_{10}H_{21}OP$

$Et_2[\overline{CH_2(CH_2)_4CH}]PO$. IX. M. 70°.[616]
$Et_2[H_2C{=}C(OBu)]PO$. VI.[795] B_2 122-123°, n_D^{20} 1.4696, d^{20} 0.9872.

$Bu_2(MeCO)PO$. IX. B_{8-10} 224-247°.[412]
Bu_2EtPO. III. $B_{0.15}$ 104°, $B_{0.14}$ 95°, n_D^{30} 1.4635, d_4^{30} 0.9218,[487] ^{31}P -45.7.[193]
$(t\text{-}Bu)_2EtPO$. III. $B_{0.1}$ 73°, IR, ^1H.[110]
$Et_2(C_6H_{13})PO$. III.[487] B_{1-3} 125°, n_D^{30} 1.4580, d_4^{30} 0.9066.
$Et_2[Me(BuO)HC]PO$. VI.[796] B_{13} 124-125°, n_D^{20} 1.4572, d^{20} 0.9741, IR.
$(HOCH_2CH_2CH_2)_2BuPO$. I.[38] $B_{0.8}$ 196-197°, n_D^{20} 1.4930, d^{20} 1.0761.
$Et_2(Et_2NCH_2CH_2)PO$. I. B_5 135°.[118,417]
$(H_2NCH_2CH_2CH_2)_2BuPO$. IX.[38] $B_{0.9}$ 148.5°, n_D^{20} 1.5030, d^{20} 1.0165.

$C_{11}H_{15}O_3P$

$Et_2(4\text{-}HO_2CC_6H_4)PO$. IX. Anilide, m. 198°.[578]
$Et_2(PhCH_2)PO$. VI.[793] $B_{0.5}$ 148-149°, n_D^{20} 1.5337, d^{20} 1.0582.
$Et_2(4\text{-}MeC_6H_4)PO$. I. M. 74°. Mercuric chloride complex, m. 135°. Trinitro derivative.[578]
$Et_2[Ph(OH)HC]PO$. V.[453] M. 87-88.5°.
$Me_2(4\text{-}Me_3SiC_6H_4)PO$. III. M. 114°.[93]
$[(EtO)_2P_\beta(O)CH_2]_2MeP_\alpha O$. IV. M. 107°,[536] m. 99-101°,[571] $^{31}P_\alpha$ -34.5, $^{31}P_\beta$ -21.3.

$C_{12}H_{13}N_2OP$

$(NCCH_2CH_2)_2PhPO$. I. M. 199-202°.[681]
$(CH_3CH{=}CH)_2PhPO$. I. III. $B_{0.6}$ 130-140°.[204]
$(CH_2{=}CHCH_2)_2PhPO$. III.[65,75] M. 48-50°,[65] 52°,[204] 54°, b_2 135-137°,[75] IR.[65] Polymers.[65,75]
$(HO_2CCH_2CH_2)_2PhPO$. From hydrolysis of corresponding nitrile.[137] M. 203.5-205°. Complex with compound of Cu.[263]
$Pr_2(4\text{-}BrC_6H_4)PO$. III.[557] M. 60-62°.
$(NCCH_2CH_2)_2MeCOCH_2C(Me)_2PO$. V.[679] M. 111-112°.
$(Pr)_2PhPO$. III. M. 43.5°, $b_{0.4}$ 120°, IR.[65]
$(i\text{-}Pr)_2PhPO$. III.[485] M. 49-50°,[485] m. 43-44°,[70] ^1H.[757]
$Et_2[PhS(Me)HC]PO$. VI.[794] B_1 162-162.5°, d^{20} 1.1134, n_D^{20} 1.5635.
$(HOCH_2CH_2CH_2)_2PhPO$. I. M. 77-78°.[39]
$Et_2(4\text{-}Me_2NC_6H_4)PO$. I. M. 65°.[587]
$(H_2NCH_2CH_2CH_2)_2PhPO$. I.[634] IX.[39] $B_{0.45}$ 290-302°,[634] n_D^{25} 1.5690, d^{20} 1.1208.[39]
$(NCCH_2CH_2CH_2)_2BuPO$. IX.[38] $B_{2.5}$ 213-125°, n_D^{20} 1.4888, d^{20} 1.0485.
$Me_2(3\text{-}Me_3SiCH_2C_6H_4)PO$. III.[94] $B_{0.07}$ 132°, m. 36-38°.
$Me_2(4\text{-}Me_3SiCH_2C_6H_4)PO$. III.[94] M. 134°.
$Am_2(H_2C{=}CH)PO$. I.[442] $B_{1.5}$ 123-123.5°, m. 37.5-38°.
$(I\text{-}Bu)_2(MeO_2CCH_2CH_2)PO$. V.[453] B_2 173-174°, m. 42.5-44°.
$(s\text{-}Bu)_2BuPO$. III. $B_{0.3}$ 100°, n_D^{20} 1.4663, d^{25} 0.913.[756]

$(t\text{-}Bu)_2BuPO$. III. $B_{0.025}$ 73°, IR, ^1H.[110]
$[(EtO)_2P_\beta(O)CH_2]_2EtP_\alpha O$. IV.[536] M. 45-52°. $^{31}P_\alpha$ -38.6, $^{31}P_\beta$ -21.1.

$C_{13}H_3F_{10}OP$

$(C_6F_5)_2MePO$. I. M. 131-132°.[244]
$(4\text{-}ClC_6H_4)_2(Cl_3C)PO$. M. 161-161.5°.[468]
$(4\text{-}BrC_6H_4)_2(ClCH_2)PO$. IX. M. 105-107°.[840]
$(3\text{-}BrC_6H_4)_2(ClCH_2)PO$. IX. $B_{0.01}$ 197-202°.[840]
$(4\text{-}ClC_6H_4)_2(ClCH_2)PO$. IX. M. 98-100°.[840]
$(3\text{-}ClC_6H_4)_2(ClCH_2)PO$. IX. B_2 238-242°.[840]
$Ph_2(Cl_3C)PO$. IV.[466] M. 138.5-139.5°.
$Ph_2(CD_3)PO$. II.[732] M. 113-114°, IR, ^1H.
$Ph_2(F_3C)PO$. I. M. 167-169°.[509]
$(3\text{-}FC_6H_4)_2MePO$. I.[727] M. 70°.
$(4\text{-}FC_6H_4)_2MePO$. I.[658] M. 113-114°.
$Ph_2(N_2CH)PO$. IX.[491] M. 56°.
$(3\text{-}O_2NC_6H_4)_2MePO$. IX.[693] M. 190°.
$Ph_2(BrMgH_2C)PO$. VII.[747]
$Ph_2(ClCH_2)PO$. From reaction of Ph_2PCl with formaldehyde. IX. B_2 146-147°, m. 122-124° (clear at 135-137°), m. 135-136°,[451] ^{31}P -30.4.[601]
$Ph_2(LiH_2C)PO$. VII.[695]
$Ph_2(NaH_2C)PO$. VII.[377]
$(NCCH_2CH_2)_2(4\text{-}ClC_6H_4NHCO)PO$. V.[679] M. 145°.
$(NCCH_2CH_2)_2$2,4-$Cl_2C_6H_3CH_2PO$. V.[679] M. 136-137°.
Ph_2MePO. I.[585] II.[573,584,588] III.[272] IV.[28,29,33] M. 110°,[585] m. 111-112°,[584] m. 111.5-112°,[528] IR, ^1H, pK_b.[304]

$C_{13}H_{13}O_2P$

$Ph_2(HOCH_2)PO$. I. M. 192-193°,[642] m. 137-139°,[334] m. 138-140°.[783]
$(2\text{-}HOC_6H_4)_2MePO$. IX. M. 204.5-206°.[463]
$(3\text{-}HOC_6H_4)_2MePO$. IX.[693] M. 194°.
$Ph_2(H_2NCH_2)PO$. IX.[491] From hydrolysis of corresponding phthalimide.[653] M. 102-103°,[653] m. 101°.[491]
$(HO_2CCH_2CH_2)_2(2,4\text{-}Cl_2C_6H_3CH_2PO)$. IX.[679] M. 170-172°.
$(3\text{-}H_2NC_6H_4)_2MePO$. IX.[693] M. 48°.
$(4\text{-}H_2NC_6H_4)_2MePO$. M. 145-147°.[569]
$(NCCH_2CH_2)_2PhCH(OH)PO$. V.[679] M. 113-115°, ^{31}P -46.[193]
$(i\text{-}Pr)_2PhCH_2PO$. III. $B_{0.3}$ 115°.[756]
$(i\text{-}Pr)_2[Ph(OH)HC]PO$. V.[453] M. 112.5-113.5°.
$[CH_2(CH_2)_4CH]_2(HOCH_2)PO$. I. M. 156-157°.[334]
$[CH_2CHOH(CH_2)_3CH]_2MePO$. IX. M. 55-60°.[693]
$[HO_2C(CH_2)_2]_2(C_7H_{15})PO$. ^{31}P -52.[193]
$(C_6H_{13})_2(LiH_2C)PO$. VII.[695]
$(C_6H_{13})_2MePO$. III. M. 27-28°, $b_{0.8}$ 116-118°.[696]
$Me_2(C_{10}H_{21}SCH_2)PO$. III. B_3 170-179°.[327]

$Me_2(C_{10}H_{21}OCH_2)PO$. III. B_1 152-154°.[327]
$Bu_2(Et_2NCH_2)PO$. IX. $B_{0.001}$ 98°, n_D^{20} 1.4659, IR.[809]

$C_{14}H_5F_{10}OP$

$(C_6F_5)_2EtPO$. I. M. 121°.[244]
$Ph_2(ClCH{=}CCl)PO$. IX. M. 121-123°.[450]
$Ph_2[H_2C{=}C(Br)]PO$. IX.[449] M. 68-70°.
$Ph_2[H_2C{=}C(Cl)]PO$. IX.[449] M. 76-77°.
$Ph_2(ClCH_2Cl_2C)PO$. IX. M. 104-106°.[450]
$Ph_2(Cl_3CCHOH)PO$. VI. M. 169.5-170.5°.[550]
$Ph_2[BrCH_2CH(Br)]PO$. IX.[449] M. 139-141°.
$(4\text{-}BrC_6H_4)_2EtPO$. III.[495] B_5 210-260°, m. 126.5-128.5°.
$Ph_2(ClCH_2CHCl)PO$. IX.[449] M. 126-127°.
$(4\text{-}ClC_6H_4)_2EtPO$. III.[495] B_8 230-260°, m. 98.5-99°.
$Ph_2(H_2C{=}CH)PO$. I.[669] IV.[448] VII.[445] M. 116-117°,[448] m. 116.5-118°,[669] IR.[66] Polymers.[20,669] Complexes with compounds of Zn,[818] Hg,[818] Sb,[818] Sn,[818] Ti,[818] Fe,[818] U.[818]
$Ph_2(MeCO)PO$. IV. M. 186-188°.[33]
$Ph_2(HO_2CCH_2)PO$. II.[420] VII.[695,747] M. 145-146°,[695] m. 142-144°.[420]
$Ph_2[BrMg(Me)CH]PO$. VII.[747]

$C_{14}H_{14}ClOP$

$Ph_2(ClCH_2CH_2)PO$. II.[334] IV. M. 125-127°,[449] m. 125-126°.[334]
$Ph_2[Li(Me)CH]PO$. VII.[747]
$Ph_2[Na(Me)CH]PO$. VII.[377]
Ph_2EtPO. I.[585] II.[378] M. 121°,[378] ^{31}P -35.[193] Complex with compounds of U,[263] Sn.[87]
$Me_2(3\text{-}PhC_6H_4)PO$. II. M. 92-93°, IR, ^1H, UV.[21]
$Ph_2(MeSCH_2)PO$. IX. M. 134-135°, ^1H.[10]
$Ph_2(MeOCH_2)PO$. III.[226] IX. M. 116.7°,[783] m. 113.5-114°,[831] ^1H.[10]
$Ph_2(HOCH_2CH_2)PO$. I. M. 111-114°.[415]
$(BrCH_2CMeBrCH_2)_2(4\text{-}BrC_6H_4)PO$. VI. M. 152°.[439]
$(BrCH_2CMeBrCH_2)_2PhPO$. VI. M. 105.[439]
$(CH_2{=}CMeCH_2)_2PhPO$. III. M. 67-69°, b_3 156-161°, polymer.[65]
$(MeO_2CCH_2CH_2)_2PhPO$. I. M. 71-71.5°.[39]
$Et_2(3\text{-}O_2NC_6H_4CH{=}NNHCOCH_2CH_2)PO$. IX. M. 171-172°.[655]
$(NCCH_2CH_2)_2[EtO_2CCH_2(EtO_2C)CH]PO$. V.[679] M. 72-73°.
$Bu_2(4\text{-}BrC_6H_4)PO$. III.[557] M. 69-70°.

$C_{14}H_{23}OP$

Bu_2PhPO. ^{31}P -41.5.[193]
$(NCCH_2CH_2)_2(C_8H_{17})PO$. V. M. 64-66°,[679] m. 127-129°.[115]
$[\overline{CH_2(CH_2)_4C}H]_2EtPO$. I. B_4 170-173°.[415]
$(MeCO_2CH_2CH_2CH_2)_2BuPO$. Acetylation of corresponding

alcohol.[38] $B_{2.5}$ 190-192°, n_D^{20} 1.4700, d^{20} 1.0816.
$(ClCH_2)_2C_{12}H_{25})PO$. III.[536] M. 57-59°, ^1H.
$(H_2NCOCH_2CH_2)_2(C_8H_{17})PO$. V. M. 149-151°.[114]
$(C_6H_{13})_2(HO_2CCH_2)PO$. VII.[695]
$(i\text{-}Pr)_2(C_8H_{17})PO$. III. $B_{0.3}$ 125°, n_D^{20} 1.4631, d^{20} 0.889.[756]
$(i\text{-}Pr)_2(2\text{-}EtC_6H_{12})PO$. III. $B_{0.3}$ 105°, n_D^{20} 1.4650, d^{25} 0.902.[756]
$Me_2(C_{12}H_{25})PO$. III. M. 84-85°, IR, ^1H, ^{31}P -48.[498]
$(C_6H_{13})_2EtPO$. III.[487] $B_{1.4}$ 161°, n_D^{30} 1.4583, d_4^{30} 0.8865.
$Me_2(C_{10}H_{21}CHOHCH_2)PO$. III. M. 76-76.5°.[327]
$[(HO)_2P_\beta(O)CH_2]_2C_{12}H_{25}PO$. From hydrolysis of corresponding alkyl ester.[536] $^{31}P_\alpha$ -45.8, $^{31}P_\beta$ -16.3. Disodium salt, m. 405-410°. Tri- and tetrasodium salt, m. >460°.

$C_{15}H_{10}ClF_4OP$

$Ph_2(F_2C{=}CClCF_2)PO$. IV. M. 80-82°, IR, ^1H, ^{19}F.[80]
$(3\text{-}CF_3C_6H_4)_2(ClCH_2)PO$. IX. 103-105°.[840]
$(4\text{-}CF_3C_6H_4)_2(ClCH_2)PO$. IX. 97-99°.[840]
$(4\text{-}ClCOC_6H_4)_2MePO$. From reaction of $SoCl_2$ with corresponding acid. M. 64-65° (dec.).[644]
$(4\text{-}ClCOC_6H_4)_2MePO$. IX.[445] M. 110-112°.
$Ph_2(MeCOCN_2)PO$. VII. M. 83-85°.[688]
$Ph_2(OCH{=}NN{=}CCH_2)PO$. IX. M. 150-152°.[655]
$Ph_2(H_2C{=}C{=}CH)PO$. IV.[89] M. 96-99°,[89] ^1H.[758]
$(3\text{-}HO_2CC_6H_4)_2MePO$. IX.[259,445] M. 282-283°.
$(4\text{-}HO_2CC_6H_4)_2MePO$. IX.[445,605,644] M. 285°,[605] m. 295.[644] Methyl ester, m. 225.5°.[605]
$Ph_2[ClCH{=}C(Me)]PO$. IX.[446] M. 127-128°.
$(4\text{-}MeC_6H_4)_2(Cl_3C)PO$. IV. M. 142-142.5°.[467]
$Ph_2(ClCH_2ClCMe)PO$. IX. M. 146-148°.[446]
$(4\text{-}ClC_6H_4CH_2)_2MePO$. III. M. 145-147°.[699]

$C_{15}H_{15}N_2OP$

$(PhCH_2)_2(N_2CH)PO$. IX.[491] M. 99°.
cis-$Ph_2(MeCH{=}CH)PO$. I.[819] M. 113-116°, IR, ^1H.
trans-$Ph_2(MeCH{=}CH)PO$. IX.[819] M. 124-125°, IR, ^1H. Complexes with compounds of Zn,[818] Hg,[818] Sb,[818] Sn,[818] Ti,[818] Fe,[818] U.[818]
$Ph_2(H_2C{=}CMe)PO$. I.[819] IX. M. 126-128°,[446] IR, ^1H.[819]
$Ph_2(CH_2{=}CHCH_2)PO$. III. M. 108-109°,[226] m. 94-94.5°.[659] Copolymer.[75]
$Ph_2(CH_2CH_2CH)PO$. II. M. 132-133°, IR, ^1H.[739]
$Ph_2(MeCOCH_2)PO$. I.[711] IV.[35] M. 127-128°, m. 127-129°,[711] IR, UV. 2,4-DNP, m. 197.5-198°.[35]
$Ph_2(HO_2CCHMe)PO$. VII.[748] M. 138-140°.
$Ph_2[BrMg(Et)CH]PO$. VII.[747]
$(3\text{-}MeC_6H_4)_2(ClCH_2)PO$. IX. M. 104-105°.[840]

$(4\text{-}MeC_6H_4)_2(ClCH_2)PO$. IX. M. 104-105°.[840]
$(3\text{-}MeOC_6H_4)_2(ClCH_2)PO$. IX. B_1 234-236°.[840]
$(4\text{-}MeOC_6H_4)_2(ClCH_2)PO$. IX. $B_{0.01}$ 178-184°.[840]
$(PhOCH_2)_2ClCH_2PO$. IX. M. 131°.[690]

$C_{15}H_{17}OP$

$(PhCH_2)_2MePO$. I.[43] M. 133.5-134°.
Ph_2PrPO. IX. M. 100-101°.[783] Complex with compound of Sn.[87]
$Ph_2(i\text{-}Pr)PO$. IV.[28,29] IX.[159,446] M. 145-146°,[28,29] m. 144-145°,[446] ^{31}P -37.[193]
$(2\text{-}MeC_6H_4)_2MePO$. III.[259] M. 120-121°.
$(3\text{-}MeC_6H_4)_2MePO$. III.[259] M. 78°.
$(4\text{-}MeC_6H_4)_2MePO$. II.[580,605] III.[443] M. 145.5-146.5°,[443] m. 143°,[580] m. 146-147°.[605]
$Ph_2(MeOCH_2CH_2)PO$. II. M. 50-52°.[311]
$Ph_2(Me_2COH)PO$. Reaction with NaOH.[135]
$(4\text{-}HOC_6H_4CH_2)_2MePO$. IX. M. 260-263°.[699]
$(2\text{-}MeOC_6H_4)_2MePO$. II. M. 150°.[463]
$Ph_2(Me_2NCH_2)PO$. III.[126] M. 189-190°,[126] m. 185-187°.[10] Hydrobromide salt, m. 237-238° (dec.).
$(PhCH_2)_2(H_2NCH_2)PO$. IX.[491] M. 112°.
Et_2(1-naphthylmethyl)PO. VI.[793] M. 85°, $b_{0.005}$ 173°.

$C_{15}H_{25}OP$

$Bu_2(PhCH_2)PO$. From reaction of BuP(O)K with $PhCH_2Cl$.[706] M. 64-65°.
$[CH_2(CH_2)_3CH]_2(Et_2NCO)PO$. III. $B_{0.3}$ 150°, n_D^{20} 1.5023, d^{25} 1.056.[756]
$(C_6H_{13})_2(MeCOCH_2)PO$. VII. $B_{0.1}$ 137-144°.[695]
$(C_6H_{13})_2(MeO_2CCH_2)PO$. VII. $B_{0.3}$ 168°,[695] ^{31}P -45.[193]
$Bu_2[CH_2(CH_2)_4N\text{-}CH_2CH_2]PO$. From Michael addition of piperidine to dibutylvinylphosphine oxide.[454] $B_{1.5}$ 158.5-159°, m. 34-35°.
$(Am)_2(Et_2NCO)PO$. III. $B_{0.3}$ 145-150°, n_D^{20} 1.4685, d^{25} 0.953.[756]
$(2\text{-}Am)_2(Et_2NCO)PO$. III. $B_{0.3}$ 145-150°, n_D^{20} 1.4752, d^{25} 0.963.[756]
$(C_6H_{13})_2(MeCHOHCH_2)PO$. VII. $B_{0.2}$ 165-170°.[695]

$C_{16}H_{15}N_2O_2P$

$Ph_2(OCH{=}NN{=}CCH_2CH_2)PO$. IX. M. 80-82°.[655]
$Ph_2(EtO_2CCN_2)PO$. IX. M. 104°.[648]
$Ph_2(CH_2{=}C{=}CMe)PO$. ^1H.[758]
$(4\text{-}HO_2CC_6H_4)_2EtPO$. IX.[605] M. 265°. Methyl ester, m. 162°.
$(3\text{-}O_2NC_6H_4)_2BuPO$. III. M. 124-125.5°.[607]
$Ph_2(CH_2{=}CMeCH_2)PO$. III. M. 137-138°.[226]
$Ph_2(CH_2{=}CHCH_2CH_2)PO$. M. 102-103°, IR, ^1H.[741]

$Ph_2(MeCH=CHCH_2)PO$. M. 84-86°, IR.[741]
$Ph_2(EtCH=CH)PO$. VII. M. 148-149°.[445]
$Ph_2(CH_2CH_2CHCH_2)PO$. M. 131-133°,[741] m. 134-136°.[520]
$Ph_2[CH_2(CH_2)_2CH]PO$. M. 173-174°, IR, ^1H.[741]
$Ph_2(H_2C=CHCHMe)PO$. IV.[715] M. 90-91°, IR.
$Ph_2[H_2C=CHOCH(Me)]PO$. VI.[795] B_7 219-220°, n_D^{20} 1.5783, d^{20} 1.1144.
$Ph_2(EtO_2CCH_2)PO$. IV. M. 186-188°.[33]
$Ph_2[K(Pr)CH]PO$. VII.[182]
$Ph_2[Li(Pr)CH]PO$. VII.[747]
Ph_2BuPO. I.[494] III.[607] IX.[783] M. 89.5°,[607] m. 93-94°,[783] m. 95°.[494]
$Ph_2(i\text{-}Bu)PO$. IV. M. 137.5-138°,[28,29] ^{31}P -21.[193]
$(4\text{-}MeC_6H_4)_2EtPO$. II.[605] M. 77.5°.
$Ph_2(EtSCHMe)PO$. VI. M. 118°, $b_{0.5}$ 210-211°.[794]
$(PhCH_2)_2(MeCHOH)PO$. V. M. 153.7-154°.[595]
$Ph_2(EtOCH_2CH_2)PO$. II.[311] IX.[449] M. 69-70°,[311] m. 69-71°.[449]
$Ph_2(EtNHCH_2CH_2)PO$. IX.[449] M. 72-74°.
$Ph_2(Me_2NCH_2CH_2)PO$. III.[126] M. 111.5-112.5°.

$C_{16}H_{21}OPSi$

$Ph_2(Me_3SiCH_2)PO$. IV. M. 107.5-108.5°.[118]
$(MeCOCHMeCH_2)_2PhPO$. I. M. 54.2°, $b_{1.5}$ 191.5°, n_D^{20} 1.5178, d^{20} 1.1664.[39]
$(MeCO_2CH_2CH_2CH_2)_2PhPO$. IX. B_1 223-224°, n_D^{20} 1.5178, d^{20} 1.1617.[39]
$(NCCH_2CH_2)_2(Bu_2Me\overset{+}{N}CH_2)PO\cdot I^-$. V.[679] M. 129-131°.
$[CH_2(CH_2)_4CH]_2BuPO$. III. $B_{0.3}$ 145°.[756]
$Me_2(C_{14}H_{29})PO$. From hydrogen peroxide oxidation of corresponding phosphine sulfide.[532] III.[498] M. 89-90°,[498] m. 84-86°,[532] ^{31}P -37.2.[532]
$Et_2(C_{12}H_{25})PO$. From hydrogen peroxide oxidation of corresponding phosphine sulfide.[532] III.[498] M. 47-49°, m. 48-48.5°,[498] $b_{0.2}$ 170-172°, ^{31}P -56.1.

$C_{17}H_6KO_3P$

$Ph_2[MeO_2CCH=CHCH(K)]PO$. VII.[381]
$Ph_2(CH=CH\text{-}CH=CH\text{-}NCH_2)PO$. III.[126] M. 154-156°. Hydrobromide salt, m. 240° (dec.).
$Ph_2[N=NCH_2CH(COMe)CH]PO$. IX.[491] M. 196°.
$Ph_2(Me_2C=C=CH)PO$. IV. M. 70-73°,[89] IR,[89] ^1H.[758]
$(2\text{-}MeCO_2C_6H_4)_2MePO$. IX. M. 201-203°.[463]
$(4\text{-}MeO_2CC_6H_4)_2MePO$. From esterification of corresponding acid.[644] M. 225°.
$(PhCH_2)_2(NCCH_2CH_2)PO$. V. M. 109-110°.[594]
$(4\text{-}ClC_6H_4CHOH)_2(i\text{-}Pr)PO$. V. M. 161-162°.[113]

$Ph_2(Me_2C{=}CHCH_2)PO$. IV.[715] M. 124-125°.
$Ph_2(i\text{-}PrCH{=}CH)PO$. VII. M. 148-150°.[445]
$Ph_2(CH_2{=}CHCMe_2)PO$. IV. M. 105-106°, IR, ^1H.[713]
$Ph_2[O(CH_2)_3CHCH_2]PO$. II. M. 122-124°, IR, ^1H.[740]
$Ph_2[EtOCH{=}C(Me)]PO$. IX.[446] M. 146-147°.
$(PhCH_2)_2HO_2CCH_2CH_2PO$. From hydrolysis of corresponding nitrile.[594] M. 180.6-181.4°.
$(PhCH_2)_2H_2NCOCH_2CH_2PO$. V.[594] M. 195-195.6°.
$Ph_2(CH_2CH_2OCH_2CH_2NCH_2)PO$. III.[126] M. 163-165°.

$C_{17}H_{21}OP$

$Ph_2(Et_2CH)PO$. ^{31}P -38.[193]
$Ph_2(i\text{-}Am)PO$. II. M. 96-97°.[588]
$(PhCH_2)_2(EtCHOH)PO$. V. M. 165.8-166.4°.[595]
$Ph_2(BuOCH_2)PO$. II. M. 60-61°.[828]
$(PhCH_2)_2(Me_2COH)PO$. V. M. 144.5-145.5°.[595]
$(PhCHOH)_2(i\text{-}Pr)PO$. V. M. 162-163°.[115]
$Ph_2(Et_2NCH_2)PO$. IX.[642] M. 147-151°.
$(C_6H_{13})_2(Et_2NCO)PO$. III. $B_{0.3}$ 160°, n_D^{20} 1.4682, d^{25} 0.939.[756]
$(C_6H_{13})_2(PrCHOHCH_2)PO$. VII. $B_{0.25}$ 175°.[695]
$Bu_2[(BuO)_2P(O)CH_2]PO$. IV. M. 43-45°, $b_{0.0004}$ 180-190°.[450]

$C_{18}H_5F_{10}OP$

$(C_6F_5)_2PhPO$. I. M. 103°.[244]
$Ph_2C_6F_5PO$. I. M. 123°,[244] m. 122-123°.[122]
$(4\text{-}BrC_6H_4)_2PhPO$. III.[495] B_4 240-280°.
$(Ph_2(3,4\text{-}Cl_2C_6H_3)PO$. VII. M. 170-171°.[374]
$(4\text{-}ClC_6H_4)_2PhPO$. III.[495] B_3 240-280°, m. 104.5-105°.
$(3\text{-}FC_6H_4)_2PhPO$. I.[727] M. 112°.
$Ph_2(4\text{-}BrC_6H_4)PO$. I. M. 133-134°, IR, UV, μ.[278]
$Ph_2(4\text{-}ClC_6H_4)PO$. I. M. 143-144°, IR, UV, μ.[278]
$Ph_2(3\text{-}FC_6H_4)PO$. I.[727] M. 135°.
$Ph_2(4\text{-}FC_6H_4)PO$. I. M. 133-135°, ^{19}F.[726]
$Ph_2(3\text{-}HOC_6H_4)PO$. II.[380] IX.[497,603] M. 158-160°,[380] m. 186-187°, m. 185-186°.[603]
$Ph_2(4\text{-}HOC_6H_4)PO$. IX.[603,743] M. 243-244°,[743] m. 241.5-242.8°.[603]
$(4\text{-}HOC_6H_4)_2PhPO$. IX.[743] M. 233-234°.
$Ph_2(CH{=}CHCH{=}N\text{-}CH{=}CCH_2)PO$. I. IX. M. 220-221°, IR, mass spect.[202]
$Ph_2(CH{=}CH\text{-}N{=}CH\text{-}CH{=}CCH_2)PO$. From reaction of diphenylphosphinylsodium with 4-pyridyl methyl chloride.[555] M. 220-221°. Hydrochloride salt, m. 223-224°.
$Ph_2(N{=}CH\text{-}CH{=}CH\text{-}CH{=}CCH_2)PO$. From reaction of diphenylphosphinylsodium with 2-pyridyl methyl chloride.[555] M. 133-134°. Hydrochloride salt, m. 186-188°. Complexes with compounds of Co, Ni.[798]
$Ph_2(3\text{-}H_2NC_6H_4)PO$. II.[380] M. 160-161°.

$Ph_2(4\text{-}H_2NC_6H_4)PO$. II. M. 235°.[380]

$C_{18}H_{16}NOP$

$Ph_2(\overline{CH=CHCH=N\text{-}CH=C}\text{-}CHOH)PO$. I. M. 198-199°, IR.[202]
$Ph_2[(CH_2=CHOCOCH_2)CHClCH_2]PO$. ^{31}P -30.2.[193]
$Ph_2(\overline{CH=CH\text{-}CH=CH\text{-}N}CH_2CH_2)PO$. III.[126] M. 83-85°. Sulfuric acid salt, m. 206-208°.
$(4\text{-}O_2NC_6H_4NHCO)_2(i\text{-}Bu)PO$. V.[116] M. 200-205°.
$Ph_2(PrCH=C=CH)PO$. IV.[89] M. 43-46°.
$Ph_2(EtMeC=C=CH)PO$. IV.[89] M. 69-71°.
$(PhCH_2)_2HO_2CCH_2(HO_2C)CHPO$. From hydrolysis of ethyl ester.[594] M. 218-218.5°.
$(4\text{-}MeO_2CC_6H_4)_2EtPO$. III. M. 162°.[605]
$(PhCH_2)_2(NCCHMeCH_2)PO$. V. M. 121.5-122°.[594]
$(PhCH_2)_2(NCCH_2CHMe)PO$. V. M. 93.5-94.1°.[594]
$Ph_2[\overline{CH_2(CH_2)_4CH}]PO$. I.[287] III.[378] IX.[616] M. 165°,[365, 616] m. 164-165°,[287] m. 165-166°,[378] ^{31}P -33.[193]
$Ph_2[CH_2=C(OBu)]PO$. VI. M. 191.5-192°.[795]
$Ph_2[\overline{CH_2(CH_2)_3CHOHCH}]PO$. I. M. 153-155° (dec.).[418]

$C_{18}H_{21}O_3P$

$(PhCH_2)_2(MeO_2CCH_2CH_2)PO$. V. M. 122.4-123.0°.[594]
$(PhCH_2)_2(HO_2CCHMeCH_2)PO$. V. M. 142.5-143.0°.[594]
$(PhCH_2)_2(HO_2CCH_2CHMe)PO$. V. M. 151-151.8°.[594]
$Ph_2[\overline{CH_2(CH_2)_2NCH_2}]PO$. III.[126] Hydrobromide salt, m. 245-247°.
$Ph_2(\overline{CH_2CH_2OCH_2CH_2NCH_2CH_2})PO$. III.[126] M. 120-122°. Hydrochloride salt, m. 220° (dec.).
$Ph_2[\overline{CH_2CH_2N(Me)CH_2CH_2NCH_2}]PO$. III.[126] M. 165-166°.
$Ph_2(C_6H_{13})PO$. I.[775] M. 61-62°.
$(PhCH_2)_2(EtMeCOH)PO$. V. M. 153.5-154.0°.[595]
$Ph_2(t\text{-}BuOCH_2CH_2)PO$. II. M. 90.5-92°.[311]
$(PhCH_2)_2[i\text{-}Pr(HO)CH]PO$. V.[595] M. 175-175.6°.
$(PhCH_2)_2(PrCHOH)PO$. V. M. 169.1-169.8°.[595]
$Ph_2(Et_2COHCH_2)PO$. VII. M. 114-116°.[695]
$Et_2(PhCHOHPhCOH)PO$. I. M. 94-95°.[409]
$(PhCHOH)_2(i\text{-}Bu)PO$. V. M. 155-156°.[114]
$(H_2C=CHCH_2O_2CCH_2CH_2)_2PhPO$. I.[806] $B_{0.016}$ 205-206°, n_D^{20} 1.5291.

$C_{18}H_{23}O_5P$

$(H_2C=CHCH_2O_2CCH_2CH_2)$ PhPO. I.[806] $B_{0.016}$ 205-206°, n_D^{20} 1.5291, 1.1557.
$Ph_2(Et_2NCH_2CH_2)PO$. I.[118]
$Ph_2[(EtO)_2P(O)CH_2CH_2]PO$. IX.[449] M. 108-109°.
$Ph_2[(Me_3SiO)(Et)CH]PO$. ^{31}P -28.[193]
$(EtO_2CCHMeCH_2)_2PhPO$. I. $B_{1.5}$ 161-168°, n_D^{20} 1.5079, d^{20} 1.1276.[39]

$(PrO_2CCH_2CH_2)_2PhPO$. I. B_1 187-198°, n_D^{20} 1.5006, d^{20} 1.0863.[39]
$(C_6H_{13})_2PhPO$. I.[238] M. 55°.
$[\overline{CH_2(CH_2)_3C}H]_2C_8H_{17}PO$. III. $B_{0.3}$ 170°.[756]
$[\overline{CH_2(CH_2)_4C}H]_2(Et_2NCH_2CH_2)PO$. I.[118]
$Et_2(C_{14}H_{29})PO$. III. M. 56-57°.[498]
$Me_2(C_{16}H_{33})PO$. III. M. 93-94°.[498]
$(C_6H_{13})_2(Et_2COHCH_2)PO$. VII. $B_{0.1}$ 158°.[695]
$[(EtO)_2P_\beta(O)CH_2]_2C_8H_{17}P_\alpha O$. IV.[536] M. 42-45°. $^{31}P_\alpha$ -37.2, $^{31}P_\beta$ -20.8.

$C_{19}H_{14}F_4NOP$

$Ph_2(2\text{-}MeHNC_6F_4)PO$. IX. M. 119-121°.[122]
$Ph_2(4\text{-}MeHNC_6F_4)PO$. IX. M. 175-177°, ^1H, ^{19}F.[122]
$Ph_2(3\text{-}NCC_6H_4)PO$. From reaction of corresponding amide with phosphorus oxytrichloride.[725] M. 110-111°.
$Ph_2(4\text{-}NCC_6H_4)PO$. I. M. 141-143°.[718]
$Ph_2(PhCN_2)PO$. VII. M. 156°,[688] m. 160°.[648]
$Ph_2(4\text{-}HCOC_6H_4)PO$. I.[718] M. 103-108°.
$Ph_2(4\text{-}HO_2CC_6H_4)PO$. I.[718] IX.[603] M. 273-274°,[718] m. 263-265°.[603]
$Ph_2(3\text{-}HO_2CC_6H_4)PO$. IX. M. 234-236°,[603] m. 232°.[270]
$Ph_2[PhCH(Br)]PO$. IX.[373] M. 204-206°.
$Ph_2[PhCH(Cl)]PO$. IX.[373] IV.[336] M. 231°,[373] m. 198°.[336]
$Ph_2(4\text{-}ClC_6H_4CHOH)PO$. IX. M. 188°.[550]
$Ph_2[K(Ph)CH]PO$. VII.[381]
$Ph_2[Li(Ph)CH]PO$. VII.[374]
$Ph_2(3\text{-}H_2NCOC_6H_4)PO$. IX.[725] M. 215-217°.
$Ph_2(4\text{-}H_2NCOC_6H_4)PO$. IX.[725] M. 190-191°.

$C_{19}H_{16}NO_3P$

$Ph_2[O_2NCH(Ph)]PO$. VII.[373] M. 183-184°.
$Ph_2(4\text{-}O_2NC_6H_4CHOH)PO$. IX. M. 191.5-193°.[550]
$Ph_2(2\text{-}MeC_6H_4)PO$. ^1H.[296]
$Ph_2(PhCH_2)PO$. II.[584] III.[226] IV.[28,29,33] VI.[584] IX. M. 195-196°,[783] m. 196°.[584] Trinitro derivative, m. 206°.[225]
$Ph_2(3\text{-}MeC_6H_4)PO$. I.[603] II.[380] M. 123-124°,[380] m. 126.6-127.5°,[603] ^1H.[296]
$Ph_2(4\text{-}MeC_6H_4)PO$. II.[380] III.[378] VI.[225] M. 129-130°,[225,378,380] ^1H.[296]
$Ph_2(3\text{-}MeOC_6H_4)PO$. I.[497] III.[380] M. 112-113°,[380] m. 110-111°.[497]
$Ph_2(4\text{-}MeOC_6H_4)PO$. I.[603] II.[278] III.[380] M. 105-108°,[278] m. 115-116°,[603] m. 112-113°,[380] m. 117-118°,[743] IR, UV, μ.[278]
$Ph_2(PhCHOH)PO$. V.[595] IX. M. 177-179.5°,[783] m. 178-179.5°,[595] ^{31}P -30.[193]
$Ph_2(4\text{-}HOCH_2C_6H_4)PO$. I.[718] M. 189-191°.

$Ph_2[H_2NCH(Ph)]PO$. From Raney-Nickel reduction of corresponding nitro compound.[373] M. 132° (dec.).

$C_{19}H_{18}O_3P_2$

$Ph_2[Ph(HO)P(O)CH_2]PO$. IX. M. 130-131°. Zn salt, m. 265-273°; Cr salt, m. 200-220°.[469]
$(PhCH_2)_2(\overline{OCH{=}CHCH{=}C}CHOH)PO$. V. M. 170.8-171.7°.[595]
$(4\text{-}EtO_2CC_6H_4)_2MePO$. From esterification of corresponding acid.[644] M. 147-148°.
$(4\text{-}ClC_6H_4CHOH)_2(2\text{-}Am)PO$. V. M. 146-148°.[113]
$(4\text{-}ClC_6H_4CHOH)_2(3\text{-}Am)PO$. V. M. 136-138°.[113]
$(PhCH_2)_2[\overline{CH_2(CH_2)_3C}OH]PO$. V. M. 184.3-185.0°.[595]
$Ph_2[\overline{CH_2(CH_2)_3CHOHCH}CH_2]PO$. VII. M. 186-187°.[376]
$(PhCH_2)_2(EtO_2CCH_2CH_2)PO$. V. M. 93-93.6°.[594]
$(PhCHOH)_2[\overline{CH_2(CH_2)_3C}H]PO$. V. M. 145-147°.[113]
$Ph_2[\overline{CH_2(CH_2)_4N}CH_2CH_2]PO$. III.[126] M. 111°, hydrochloride salt, m. 248-250° (dec.).
$Et_2[Ph_2C(OH)CH(Me)]PO$. VII.[330] M. 162°.
$(PhCHOH)_2(2\text{-}Am)PO$. V. M. 139-140°.[113]
$(PhCHOH)_2(3\text{-}Am)PO$. V. M. 143-144°.[113]

$C_{19}H_{29}O_2P$

$(C_6H_{11})_2(PhCHOH)PO$. ^{31}P -53.[193]
$(C_8H_{17})_2(NCCH_2CH_2)PO$. V. M. 51-52°,[680] m. 53.4-54.2°.[594]
$(C_8H_{17})_2(HO_2CCH_2CH_2)PO$. V. M. 81.5-82.1°.[594]
$(C_8H_{17})_2(H_2NCOCH_2CH_2)PO$. V. M. 132.6-133.0°.[594]
$(C_8H_{17})_2(Me_2COH)PO$. V. M. 54.5-56.0°.[595]
$(Am)_2[(BuO)_2P(O)CH_2]PO$. III. $B_{0.0001}$ 165-180°, n_D^{20} 1.4598, d^{25} 0.993.[756]
$(4\text{-}ClCOC_6H_4)_2PhPO$. IX. M. 128-129°,[755] m. 127-128°.[445]
$(4\text{-}NCC_6H_4)_2PhPO$. From reaction of corresponding amide with phosphorus oxytrichloride.[725] M. 173-175°.
$(3\text{-}NCC_6H_4)_2PhPO$. From reaction of corresponding amide with phosphorus oxytrichloride.[725] M. 154-155°.
$(4\text{-}OCNC_6H_4)_2PhPO$. IX. M. >295°.[755]
$(4\text{-}N_3COC_6H_4)_2PhPO$. IX. M. 123-123.5°.[755]

$C_{20}H_{15}N_2O_2P$

$Ph_2(PhCOCN_2)PO$. VII. M. 155°,[688] m. 139°.[648]
$(3\text{-}HO_2CC_6H_4)_2PhPO$. IX.[725] 269-273°.
$(4\text{-}HO_2CC_6H_4)_2PhPO$. IX. M. 307-310°,[605] m. 340-342°, m. 334-336°.[634] Methyl ester, m. 165-166°.[605]
$Ph_2(4\text{-}BrC_6H_4COCH_2)PO$. Complexes with compounds of Cu, Sn.[754]
$Ph_2(4\text{-}NCC_6H_4CH_2)PO$. IX. M. 195-196°, IR, ^1H.[202]
$Ph_2(4\text{-}NCC_6H_4CHOH)PO$. I. M. 207-208°, IR.[202]
$Ph_2(4\text{-}O_2NC_6H_4CH{=}CH)PO$. VII. M. 162-163°.[445]
$Ph_2(4\text{-}O_2NC_6H_4COCH_2)PO$. Complexes with compounds of Cu, Sn.[754]

$(PhNHCO)_2PhPO$. V. M. 169-170°.[114]
$(3\text{-}H_2NCOC_6H_4)_2PhPO$. IX.[725] M. 158-162°.
$Ph_2(H_2C{=}CPh)PO$. I.[714] M. 114-115°, IR.
trans-$Ph_2(PhCH{=}CH)PO$. I.[5] M. 168-169°, m. 165-167°,[354] IR, ^1H.[5]
$Ph_2(4\text{-}CH_2{=}CHC_6H_4)PO$. I. M. 98-101.5°, UV, polymers.[666]
$Ph_2(PhCOCH_2)PO$. IV. VII.[695] M. 140.0-140.5°,[754] m. 139-140°,[695] IR, UV.[35] Complexes with compounds of Cu, Sn.[754]

$C_{20}H_{17}O_2P$

$Ph_2(4\text{-}MeCOC_6H_4)PO$. I.[719] M. 119-120°.
$Ph_2(HO_2CCHPh)PO$. III. M. 135-137°,[84] m. 136°.[381]
$Ph_2(3\text{-}MeCO_2C_6H_4)PO$. IX.[497] M. 158-159°.
$Ph_2(4\text{-}MeO_2CC_6H_4)PO$. I.[718] M. 113-116°.
$Ph_2[PhCH(Br)CH_2]PO$. IX.[5] M. 147-148°, IR, ^1H.
$(4\text{-}MeC_6H_4)_2(4\text{-}ClC_6H_4)PO$. VI. M. 130°.[580]
$Ph_2[4\text{-}MeOC_6H_4CH(Cl)]PO$. IX.[550] M. 181-182°.
$(2\text{-}MeC_6H_4)_2PhPO$. III. $B_{0.2}$ 192-194°.[70]
$(3\text{-}MeC_6H_4)_2PhPO$. VI.[315] M. 108.5-109°.
$(4\text{-}MeC_6H_4)_2PhPO$. II.[380] M. 81-83°.[380]
$Ph_2(PhMeCH)PO$. ^{31}P -31.[193]
$Ph_2(PhCH_2CH_2)PO$. I.[5] II. IX. M. 103-104°,[354] m. 102-103°, IR, ^1H.[5]
$(PhCH_2)_2PhPO$. I. M. 174°.[573]
$Ph_2[Ph(MeO)CH]PO$. ^{31}P -28.[193]
$Ph_2[4\text{-}MeC_6H_4CH(OH)]PO$. IX.[550] M. 152.5-155°.
$Ph_2(4\text{-}MeOC_6H_4CH_2)PO$. IX. M. 212-213°, IR, ^1H.[201]

$C_{20}H_{19}O_2P$

$Ph_2(PhCHOHCH_2)PO$. VII. M. 141-142°.[695]
$(4\text{-}MeOC_6H_4)_2PhPO$. I.[743] M. 96-97°.
$Ph_2(4\text{-}MeOC_6H_4CHOH)PO$. IX. M. 160-162°.[550]
$(PhCHOH)_2PhPO$. V. M. 187-188°,[114] $b_{0.001}$ 147-150°.[645]
$(PhCH_2)_2(CH{=}CHCH{=}CHN{=}C\text{-}CH_2)PO$. From reaction of dibenzylphosphinylsodium with 2-pyridyl methyl chloride.[555] M. 221-222°.
$Ph_2(4\text{-}Me_2NC_6H_4)PO$. I.[587,717] M. 183.5°,[587] m. 181-183°,[603] m. 186-189°, IR,[278,717] UV, μ.[278]
$Ph_2[H_2C\text{-}(CH_2)_4\text{-}C{=}C{=}CH]PO$. IV.[89] M. 107-109°, IR.
$Ph_2[MeO_2CCH_2(MeO_2C)C{=}CHCH_2]PO$. IX. M. 130-132°.[402a]
$(4\text{-}ClC_6H_4CHOH)_2[CH_2(CH_2)_4CH]PO$. V. M. 176-178°.[113]
$Ph_2[H_2C\text{-}(CH_2)_4\text{-}C{=}CHCH_2]PO$. IV.[715] M. 166-167°, IR.
$(PhCH_2)_2[CH_2(CH_2)_4COH]PO$. V. M. 170.0-170.6°.[595]
$(PhCH_2)_2(EtO_2CCHMeCH_2)PO$. V. M. 92.7-93.1°.[594]
$(PhCH_2)_2(EtO_2CCH_2CHMe)PO$. V. M. 138.5-139°.[594]
$(PhCHOH)_2[CH_2(CH_2)_4CH]PO$. V. M. 167-168°.[113]

$C_{20}H_{26}KOP$

$Ph_2[C_6H_{13}CK(Me)]PO$. VII.[187]
$Ph_2[C_6H_{13}CH(Me)]PO$. VII.[187] M. 94-96°, $[\alpha]^{26}_{546}$ -14.7.
$Ph_2(C_8H_{17})PO$. I.[775] M. 64-65°.
$[H_2C{=}CHCH_2O_2CCH(Me)CH_2]_2PhPO$. I.[806] $B_{0.014}$ 173-174°, n_D^{20} 1.5189, d^{20} 1.1196.
$Ph_2[Me(EtO)_2Si(CH_2)_3]PO$. IV. B_2 200-205°.[118]
$(C_6H_{13})_2(PhCH{=}CH)PO$. IX. $B_{0.15}$ 205-210°.[695]
$(C_6H_{13})_2(PhCOCH_2)PO$. VII. $B_{0.05}$ 182-187°.[695]
$(C_7H_{15})_2PhPO$. I.[238] M. 38-40°, b_5 215-217°.
$(C_6H_{13})_2(PhCHOHCH_2)PO$. VII. $B_{0.15}$ 220°.[695]
$(C_8H_{17})_2[(HO_2C)CH_2(HO_2C)CH]PO$. V. M. 97.5-99.0°.[594]
$Me_2(C_{18}H_{37})PO$. III. M. 94.5-96°.[498]
$Et_2(C_{16}H_{33})PO$. III. M. 62-63°.[498]

$C_{21}H_{16}NO_3P$

Ph_2(N-Phthalimidylmethyl)PO. IV.[653] M. 206-207°.
$Ph_2(H_2C{=}C{=}CPh)PO$. IV.[714] M. 125-126°, IR, ^1H.
$Ph_2[\overline{CH{=}CHN{=}CHCH{=}C}(NCCH_2CH_2)CH]PO$. VII.[555] M. 199-200°.
$Ph_2(\overline{CH_2CH_2C}Ph)PO$. IX. M. 131-132°.[376]
$Ph_2(PhCH{=}CHCH_2)PO$. IV.[715] M. 181-182°, IR.
$Ph_2(H_2C{=}CHCHPh)PO$. IV.[715] M. 193-193.5°, IR.
$Ph_2(2\text{-}MeCH{=}CHC_6H_4)PO$. II. M. 140-142°.[547]
$Ph_2[MeO_2CCH(Ph)]PO$. IX. M. 175°,[373] m. 205-207°.[84]
$Ph_2(4\text{-}MeOC_6H_4COCH_2)PO$. Complex with compound of Sn.[754]
$Ph_2(ClCH_2CH_2CHPh)PO$. IX. M. 208-209°.[376]
$(PhCH_2)_2(2\text{-}ClC_6H_4CHOH)PO$. V. M. 173.8-174.5°.[595]
$Ph_2(4\text{-}Me_2NCOC_6H_4)PO$. I.[718] M. 195-196°.
$Ph_2[(\overline{CH{=}CHN{=}CH\text{-}CH{=}C})(HO_2CCH_2CH_2)CH]PO$. From hydrolysis of corresponding nitrile.[555] M. 200°.
$(PhCH_2)_2[2\text{-}O_2NC_6H_4CH(OH)]PO$. V.[595] M. 187.1-187.5°.
$(PhCH_2)_2[3\text{-}O_2NC_6H_4CH(OH)]PO$. V.[595] M. 176.7-177.0°.
$(PhCH_2)_2[4\text{-}O_2NC_6H_4CH(OH)]PO$. V.[595] M. 186.5-187.0°.
$Ph_2(PhCH_2OCH_2CH_2)PO$. II. M. 82-82.5°.[311]
$(PhCH_2)_2(PhCHOH)PO$. V. M. 163.2-164.0°, ^{31}P -41.[111,440]

$C_{21}H_{21}O_2P$

$Ph_2(HOCH_2CH_2CHPh)PO$. VII. M. 235-237°.[376]
$Ph_2[Ph(CH_2)_2CHOH]PO$. I. M. 187-188°, IR.[202]
$Ph_2(HOCH_2CHPhCH_2)PO$. VII. M. 142-144°.[376]
$(PhCH_2)_2(2\text{-}HOC_6H_4CHOH)PO$. V. M. 133.7-134.2°.[595]
$Ph_2(4\text{-}Me_2NC_6H_4CH_2)PO$. IX. M. 212-213°, IR.[201]
$Ph_2[(MeO)_2P(O)CPh(OH)]PO$. V. M. 120-121°.[201]
$Ph_2(PhMe_2SiCH_2)PO$. IV. M. 95-96°, b_2 185-202°.[118]
$Ph_2(4\text{-}Me_3SiC_6H_4)PO$. III. M. 93°.[93]
$Ph_2(3\text{-}Me_3SiC_6H_4)PO$. III. $B_{0.1}$ 212-215°.[96]
$(PhCHOH)_2(2\text{-}C_7H_{15})PO$. V. M. 139-141°.[113]
$(PhCHOH)_2(4\text{-}C_7H_{15})PO$. V. M. 141-142°.[113]

$Ph_2[Et(EtO)_2Si(CH_2)_3]PO$. IV. B_2 205-207°.[118]
$Ph_2[(EtO)_3Si(CH_2)_3]PO$. IV. M. 53-56°, b_1 206-208°.[118]
$(NCCH_2CH_2)_2(C_{12}H_{25}O_2CCH_2CH_2)PO$. V.[679] M. 65-67°.
$(C_6H_{13})_2(PhMeC(OH)CH_2)PO$. VII. M. 83-84°, $b_{0.18}$ 207°.[695]
$(C_6H_{13})_2[(BuO)_2P(O)CH_2]PO$. III. $B_{0.0001}$ 190-195°, n_D^{20} 1.4617, d^{25} 0.970.[756]

$C_{21}H_{46}O_4P_2$

$(C_5H_{11})_2[(2-C_5H_{11}O)_2P(O)CH_2]PO$. III. $B_{0.0001}$ 190-195°, n_D^{20} 1.4598, d^{25} 0.970.[756]

$C_{22}H_{19}Cl_2OP$

$(PhC(Cl)=CH_2)_2PhPO$. IX. M. 148-149°.[750]
Ph_2(1-naphthyl)PO. I.[848] M. 178-179°.
Ph_2(ferrocenyl)PO. IX. Orange needles, m. 163-165°, IR.[765]
$Ph_2(PhCH=CHCH=CH)PO$. VII. M. 175-176°.[445]
$(PhCH=CH)_2PhPO$. IX. M. 198-199°,[750] m. 189-190°.[571]
$(4-MeO_2CC_6H_4)_2PhPO$. IX.[605,725] M. 160-167°,[725] m. 167-166°.[605]
$Ph_2(3-O_2NC_6H_4CH=NNHCOCH_2CH_2)PO$. M. 134-136°.[655]
$(PhCH_2)_2[3,4(-OCH_2O-)C_6H_3CHOH]PO$. V. M. 195.5-196.3°.[595]
$Ph_2[MeCH=CHCH(NHPh)]PO$. IX. M. 60-61°.[660]
$Ph_2(Me_2NCOCHPh)PO$. I. M. 270-272°.[84]
$(PhCH_2CH_2)_2PhPO$. III. M. 90-91°.[70]
$Ph_2(4-i-PrC_6H_4CH_2)PO$. IX. M. 189-191°.[201]
$Ph_2(4-i-PrC_6H_4CHOH)PO$. I. M. 168-169°, IR.[202]
$(PhCHOH)_2(PhMeCH)PO$. V. M. 157-158°.[113]
$(4-ClC_6H_4)_2\{EtMeC=C=C[EtMeC(OH)]\}PO$. IV.[89] IR.
$(2-Me-5-pyridylCH_2CH_2)_2PhPO$. I. M. 84-85°.[39]
$(4-Me_2NC_6H_4)_2PhPO$. I. M. 203.5-204.5°, IR,[717] UV.[723]
$Ph_2(3-Me_3SiCH_2C_6H_4)PO$. III. $B_{0.1}$ 215-218°.[96]
$Ph_2(4-Me_3SiCH_2C_6H_4)PO$. III. M. 151-152°.[96]
$(4-O_2NC_6H_4NHCO)_2(C_8H_{17})PO$. V. M. 113-115°,[114] ^{31}P -15.[193]
$Ph_2[Me_2C=CH(CH_2)_2CMe=CHCH_2]PO$. I. M. 113-114°.[713]
$Ph_2[Me_2C=CH(CH_2)_2CMe(CH=CH_2)]PO$. IV. M. 68-69°, IR.[713]
$(PhNHCO)_2(C_8H_{17})PO$. V. M. 105-107°,[114] ^{31}P -23.[193]
$(PhCHOH)_2(C_8H_{17})PO$. V. M. 127-129°,[115] ^{31}P -50.[193]
$(BuO_2CCHMeCH_2)_2PhPO$. I. B_1 209-210°, n_D^{20} 1.4987, d^{20} 1.0680.[39]
$(C_8H_{17})_2PhPO$. I.[238] M. 42-43°, b_5 245-247°.
$(C_8H_{17})_2[\overline{CH_2(CH_2)_4C}OH]PO$. V. M. 67.5-68.1°.[595]
$Et_2(C_{18}H_{37})PO$. III. M. 65-67°.[498]
$[(EtO)_2P_\beta(O)CH_2]_2C_{12}H_{25}PO$. IV.[536] M. 48-54°, $^{31}P_\alpha$ -36.9, $^{31}P_\beta$ -21.1.

$C_{23}H_{19}OP$

$Ph_2(1\text{-}C_{10}H_7CH_2)PO$. VII. M. 157.9°.[374]
$Ph_2(PhCH{=}CHCH{=}CHCH_2)PO$. IV.[715] M. 221-222°.
$(4\text{-}MeC_6H_4)_2(3\text{-}O_2NC_6H_4CH{=}NNHCOCH_2)PO$. M. 90-92°.[655]
$Ph_2[PhCO(CH_2)_4]PO$. III.[76] M. 150-150.5°.
$(PhCH_2)_2(HO_2CCH_2CHPh)PO$. V. M. 211-212°.[594]
$(PhCH_2)_2[3,4\text{-}O(CH_2)_2O\text{-}C_6H_4CHOH]PO$. V. M. 195.5-196.3°.[595]
$(PhCH_2)_2(4\text{-}Me_2NC_6H_4CHOH)PO$. V. M. 153.6-154.2°.[595]
$(4\text{-}BuO_2CC_6H_4)_2MePO$. From esterification of corresponding acid.[644] M. 122°.
$(C_8H_{17})_2(2\text{-}ClC_6H_4CHOH)PO$. V. M. 97.5-98.1°.[595]
$(C_8H_{17})_2(2\text{-}O_2NC_6H_4CHOH)PO$. V. M. 122.2-122.6°.[595]
$(C_8H_{17})_2(PhCHOH)PO$. V. M. 62.5-64.0°.[595]

$C_{24}H_{19}Cl_2OP$

$(PhCCl{=}CH)_2(PhCH{=}CH)PO$. IX. M. 159-160°.[750]
$Ph_2(4\text{-}PhC_6H_4)PO$. I. M. 143-144°.[9]
$Ph_2[\overline{CH{=}CHCH{=}CH\text{-}N{=}C}\text{-}CH_2)(\overline{CH{=}CHCH{=}CH\text{-}N{=}C})CH]PO$. VII.[555] M. 143-143.5°. Hydrochloride salt, m. 201-202°.
$(PhCH_2)_2(MeCOCH_2CHPh)PO$. V. M. 171.6-172.1°.[594]
$(PhCH_2)_2(MeO_2CCH_2CHPh)PO$. V. M. 178.2-178.6°.[594]
$Ph_2[MeCH{=}CHCH(4\text{-}MeC_6H_4NMe)]PO$. IX. M. 75-76°.[660]
$Ph_2(C_{12}H_{25})PO$. II.[378] M. 65-66°. Complex with compound of Sn.[87]
$(C_9H_{19})_2PhPO$. I.[238] M. 48-49°.
$(4\text{-}PhC_6H_4)_2MePO$. II. M. 223-224°.[834]
$Ph_2(Ph_2CH)PO$. III.[443] VII.[374] M. 303-304°,[443] m. 303°.[374]
$Ph_2(PhNHCHPh)PO$. I.[404] IX.[353] M. 242-243°,[404] m. 243-245°.[353]
$Ph_2[CH_2{=}CHCH(C_{10}H_7NH)]PO$. IX. M. 129-130°.[660]
$Ph_2(PhNHN{=}NCHPh)PO$. IX. M. 160-163° (dec.).[353]
$Ph_2[Ph_2P(S)CH_2]PO$. VII. M. 213-214°.[747]
$(PhCH_2)_2EtO_2CCH_2CH(Ph)PO$. V.[594] M. 138.5-139°.
$Ph_2[PhCH{=}CHCH(NEt_2)]PO$. IX. M. 161°.[660]
$[\overline{CH_2(CH_2)_4CH}]_2(PhNHCHPh)PO$. IX.[404]
$(C_{12}H_{25})_2MePO$. I. M. 72.5-73.5°, IR, ^1H, ^{31}P -53.[324]
$(C_6H_{13})_2[(C_6H_{13}O)_2P(O)CH_2]PO$. III. $B_{0.0001}$ 195-210°, n_D^{20} 1.4554, d^{25} 0.952.[756]

$C_{26}H_{19}OP$

$Ph_2(2\text{-anthryl})PO$. I.[848] M. 206-207°.
$Ph_2(1\text{-anthryl})PO$. I.[848] M. 235°.
$(1\text{-}C_{10}H_7)_2PhPO$. III. M. 234-235°.[70]
$Ph_2(PhCH{=}CPh)PO$. IX. M. 159-161°, IR, UV.[22]
$Ph_2[PhCOCH(Ph)]PO$. VII.[381] M. 229-230.5°.
$Ph_2(PhCO_2CHPh)PO$. IX. M. 188-189°, IR.[202]
$Ph_2(PhCHBrCHPh)PO$. IX. M. 182-183°, IR.[22]
$(Ferrocenyl)_2(4\text{-}ClC_6H_4)PO$. IX. M. 162-164°.[767]

$Ph_2[3\text{-}O_2NC_6H_4CH_2CH(Ph)]PO$. II.[788] M. 203-204°.
(Ferrocenyl)$_2$PhPO. IX. Orange crystals, m. 239-241°, IR,[765] UV.[621]
$Ph_2(PhCH_2CHPh)PO$. II.[788] IX.[836] M. 233-234°, IR, UV, ^1H.[22]
$Ph_2(PhCHOHCHPh)PO$. V. M. 196°,[381] m. threo 193°, m. erythro 188°.[384]
$Ph_2(Ph_2COHCH_2)PO$. VII. M. 192-193°.[695]
$Ph_2[(PhNH)(4\text{-}MeC_6H_4)CH]PO$. IX. M. 230-232°.[353]
$Ph_2[(PhNH)PhCHCH_2]PO$. IX. M. 204-205°, IR.[836]
$Ph_2[Ph_2P(O)OCHMe]PO$. IX. $B_{0.2}$ 280-290°, IR, ^1H, mass spect.[202]
$(PhCH_2)_2[(MeO_2C)_2CHCHPh]PO$. V. M. 139°.[490]
$[\overline{CH_2(CH_2)_4CH}]_2(Ph_2CHCH_2)PO$. I. M. 205°.[704]

$C_{26}H_{35}O_2P$

$[\overline{CH_2(CH_2)_4CH}]_2[PhCH_2C(OH)Ph]PO$. I. M. 100-103°.[409]
$(C_6H_{13})_2(Ph_2COHCH_2)PO$. VII. M. 82-83°.[695]
$[(HO)(Ph)P(O)CH_2]_2C_{12}H_{25}PO$. From hydrolysis of corresponding alkyl ester.[536] Viscous oil, ^1H.
$(C_8H_{17})_2(MeCOCH_2CHPh)PO$. V. M. 62.3-63.0°.[594]
$(C_{10}H_{21})_2PhPO$. I.[238] M. 55-56°.

$C_{27}H_{20}ClO_2P$

$Ph_2(4\text{-}ClC_6H_4COCH{=}CPh)PO$. IX. M. 151°.[164]
$Ph_2(4\text{-}ClC_6H_4COCHBrCHPh)PO$. IX. M. 187°.[164]
$Ph_2(Ph_2C{=}C{=}CH)PO$. ^1H.[758]
$Ph_2(PhCOCH{=}CPh)PO$. IX. M. 143°.[164]
$Ph_2(PhCOCHBrCHPh)PO$. IX. M. 187°.[164]
$Ph_2(4\text{-}ClC_6H_4COCH_2CHPh)PO$. IX. M. 225-226°.[164]
(Ferrocenyl)$_2(4\text{-}NCC_6H_4)$PO. IX.[767]
$(4\text{-}PhC_6H_4)_2(CH_2{=}CHCH_2)PO$. II. M. 192-193°.[834]
$Ph_2(PhCH{=}CPhCH_2)PO$. VII. M. 159-162°.[374]
$Ph_2(PhCOCH_2CHPh)PO$. IX. M. 235-237°,[354] m. 227°.[164]
Ph_2[1-phenyl-2-(3,4-methylenedioxyphenyl)ethyl]PO. II.[788] M. 204-205°.
(Ferrocenyl)$_2(4\text{-}MeC_6H_4)$PO. IX. M. 151-153°.[767]
(Ferrocenyl)$_2(4\text{-}MeOC_6H_4)$PO. IX. M. 111-113°.[767]
$Ph_2(MeOCHPhCHPh)PO$. IX. M. 263-264°, IR, ^1H.[22]
$Ph_2(Ph_2COHCHMe)PO$. VII. M. 260-262°.[747]
$Ph_2(HOCH_2CHPhCHPh)PO$. VII. M. 271-272°.[376]
$Ph_2[(PhNH)PhCHCHMe]PO$. IX. M. 228°.[74a]

$C_{28}H_{21}O_2P$

$Ph_2[\overline{CH{=}C(Ph)OC(Ph){=}C}]PO$. From dehydration of 1-diphenylphosphinyl-1,2-dibenzoylethane.[322] M. 193-194°.
Phenylene-1,2-$[COCH(Ph)]_2PhPO$. VII.[381] M. 228-231°.
$Ph_2[(PhCOCH_2)(PhCO)CH]PO$. Unstable, dehydrates during recrystallization to give 2,5-diphenyl-3-diphenylphosphinylfuran.[322] ^{31}P -27.5.[673]
$Ph_2[2,4,6\text{-}(O_2N)_3C_6H_2NHN{=}C(Me)CH_2CH(Ph)]PO$. V.[594] M. 181-182°.

trans-$Ph_2[4\text{-}MeC_6H_4CH{=}C(Ph)CH_2]PO$. II.[714] M. 186-187°, IR, UV, ^1H, mass spect.
cis-$Ph_2[4\text{-}MeC_6H_4CH{=}C(Ph)CH_2]PO$. II.[714] M. 160-161°, IR, UV, ^1H, mass spect.

$C_{28}H_{26}NOP$

$Ph_2[PhCH{=}CHCH(4\text{-}MeC_6H_4NH)]PO$. IX. M. 186°.[660]
$Ph_2[(PhNH)PhCHCMe_2]PO$. IX. M. 217-219°.[836]
$Ph_2[(4\text{-}MeC_6H_4NH)PhCHCHMe]PO$. IX. M. 205°.[74a]
$Ph_2(C_{16}H_{33})PO$. I.[494] M. 80°.
$Ph_2[(PhNH)(1\text{-}C_{10}H_7)CH]PO$. IX. M. 205-210° (dec.).[355]
trans-$Ph_2[4\text{-}MeC_6H_4CH{=}C(Ph)CH_2CH_2]PO$. III.[714] M. 163-164°, IR, UV, ^1H, mass spect.
cis-$Ph_2[4\text{-}MeC_6H_4CH{=}C(Ph)CH_2CH_2]PO$. II.[714] M. 169-170°, IR, UV.
$(PhCH_2)_2(PhCOCH_2CHPh)PO$. V. M. 205-206.5°.[594]
$(4\text{-}MeC_6H_4)_2[PhCH_2(4\text{-}MeC_6H_4)CH]PO$. IX. M. 218-219°, IR, ^1H.[22]
$Ph_2[(PhNH)PhCHCHPr]PO$. IX. M. 228°.[74a]
$Ph_2[Me_2CHCH_2CH(Ph_2PO_2)]PO$. IX. M. 143-144°.[202]
$Ph_2[(Ph_2PO_2)(3\text{-}pyridyl)CH]PO$. IX. M. 186-187°, IR.[201]
$Ph_2[PhCH{=}C(Ph)CH_2CH(CN)CH_2]PO$. II.[714] M. 118-119°, IR, UV, mass spect.
$Ph_2[(4\text{-}MeC_6H_4NH)PhCHCHPr]PO$. IX. M. 228°.[74a]
$Ph_2(C_{18}H_{37})PO$. II.[595] M. 77-78°.

$C_{31}H_{25}OP$

$Ph_2(Ph_3C)PO$. IV. M. 227.5-228°,[30] m. 227-229°.[33]
$Ph_2[PhCH{=}CHCH(1\text{-}C_{10}H_7NH)]PO$. IX. M. 106-107.5°.[660]
$Ph_2[(Ph_2PO_2)PhCH]PO$. IX. M. 190-191°, IR, mass spect.[201]
$Ph_2(Ph_3SiCH_2)PO$. VII. M. 149-152°, IR.[747]
$Ph_2(Ph_3SnCH_2)PO$. VII. M. 141-142°, IR, ^1H.[747]
$Ph_2[4\text{-}MeC_6H_4CH{=}C(Ph)CH_2CH(CN)CH_2]PO$. II.[714] M. 127-128°, IR, UV, mass spect.
$Ph_2[(Ph_2PO_2)(4\text{-}NCC_6H_4)CH]PO$. IX. M. 180-181°, IR.[201]
$Ph_2[(PhNH)PhCHCHPh]PO$. IX. M. 228-229°, IR.[836]
$Ph_2[(PhCH_2)(Ph_2PO_2)CH]PO$. IX. M. 83-84°, IR.[202]
$Ph_2(Ph_3SnCHMe)PO$. VII. M. 176-177°, IR.[747]
$Ph_2[(4\text{-}i\text{-}PrC_6H_4)(4\text{-}i\text{-}PrC_6H_4CO_2)CH]PO$. IX. M. 191-192°, IR, ^1H.[202]

$C_{33}H_{25}N_2O_5P$

$Ph_2[3,5\text{-}(O_2N)_2C_6H_3CO_2CHPhCHPh]PO$. V.[381] M. 243-244°.
$Ph_2(4\text{-}O_2NC_6H_4CO_2CHPhCHPh)PO$. V.[381] M. 266-267°.
$Ph_2(PhCO_2CHPhCHPh)PO$. V.[381] M. 253°.
$Ph_2(ClCHPhCHPhCHPh)PO$. IX. M. 234-236°.[376]
$Ph_2(HOCHPhCHPhCHPh)PO$. VII. M. 270-272°.[376]
$Ph_2[PhCH_2CH_2(Ph_2PO_2)CH]PO$. IX. M. 147-148°, IR, mass spect.[202]

$C_{34}H_{32}O_3P_2$

$Ph_2[(4\text{-}i\text{-}PrC_6H_4)(Ph_2PO_2)CH]PO$. IX. M. 181-182°, IR.[201]
$Ph_2(Ph_3SnCHPr)PO$. VII. M. 148-151°, IR.[747]
$[Ph(BuO)P_\beta(O)CH_2]_2C_{12}H_{25}P_\alpha O$. IV.[536] Oil. $^{31}P_\alpha$ -37.7, $^{31}P_\beta$ -34.9.
$(1\text{-}C_{10}H_7)_2(PhCOCH_2CHPh)PO$. IX. M. 220°.[190]
$Ph_2[(PhCH_2O)PhC{=}C(CH_2COPh)]PO$. IX. M. 179-180°, ^{31}P -28.6.[672,673]
$Ph_2[(Ph_2PO_2)(PhCO_2)CPh]PO$. IX. M. 207-208°, IR.[201]

D.3. RR'R"PO

$C_4H_{10}ClOP$

$MeEt(ClCH_2)PO$. ^{31}P -48.0.[601]
MeEtPrPO. III. $B_{0.1}$ 90°, ^1H.[507]
$Me(ClCH_2)PhPO$. III. M. 106-108°,[469] ^{31}P -47.2.[601]
MeEtAmPO. I.[44,45] M. 8-9°, $b_{0.4}$ 76-76.5°, n_D^{20} 1.4591.
$Me[H_2N(CH_2)_3][H_2N(CH_2)_4]PO$. I.[634] $B_{0.5}$ 186-188°, n^{25} 1.5060.

$C_9H_{10}Cl_3OP$

$EtPh(Cl_3C)PO$. IV.[466] M. 152.5-153.5°.
$(ClCH_2)(ClCH_2CH_2)PhPO$. II. M. 68-70°, IR.[333]
$MePh(H_2C{=}CH)PO$. I.[442] M. 78-79°, b_2 127-127.5°.
MeEtPhPO. II.[496] $[\alpha]_D^{25}$ +22.4 ± 1.0°,[496] m. 47-48°;[496] $[\alpha]_D^{25}$ 21.5°,[566] m. 56-58°;[566] $[\alpha]_D^{25}$ -22.8 ± 1.0°, m. 48-49°.
$Et(CH_2{=}CH)PhPO$. IX. M. 43-44°, b_1 112-113°.[447]
MePrPhPO. II. M. 41°,[391] b_{13} 180°.[573]
EtPrPhPO. II. M. 44-45°, b_{15} 184-185°.[573]
$Ph(NCCH_2CH_2)(NCCH_2CH_2CH_2)PO$. I. M. 73°, b_2 264°.[39]
$ClCH_2Ph(PhCH_2)PO$. II. M. 132-133°, IR,[333] m. 130-133°.[334]
$MePh(PhCH_2)PO$. II. M. 148-149°, b_{15} 235. d-Camphor sulfonate of the l-form, m. 94°. l-Camphor sulfonate of the d-form, m. 111°. $[\alpha]_D^{20}$ 78° (in HCl).[573]
$MePh(2\text{-}MeOC_6H_4)PO$. IR.[539]
$MePh(3\text{-}MeOC_6H_4)PO$. IR.[539]
$MePh(4\text{-}MeOC_6H_4)PO$. IR.[539]

$C_{15}H_{16}O_3P_2$

$MePh[Ph(HO)P(O)CH_2]PO$. IV. M. 206-208°. Zn salt, m. 135-160°; Cr salt, m. 210-218°.[469]
$EtPh(PhCH_2)PO$. I. II. M. 110-111°.[573]
$MePh(4\text{-}Me_2NC_6H_4)PO$. II. M. 146°.[587]
$Me_3C(Ph)(PhCH_2)PO$. From benzylation of t-butylphenylphosphine oxide.[358] M. 187-189°.
$MePh(2\text{-}PhC_6H_4)PO$. II. Complex with Co compound. IR, UV, ^1H.[21]
$(Ph(H_2C{=}CHCH_2O_2CCH_2CH_2)[H_2C{=}CHCH_2O_2CCH(Me)CH]PO$. I.[806]

$B_{0.5}$ 195-197°, n_D^{20} 1.5250, d^{20} 1.1313.
Ph(2-MeC_6H_4)(3-$CF_3C_6H_4$)PO. M. 93-94°.[134]
Me(4-$Me_2NC_6H_4$)(2-PhC_6H_4)PO. II. M. 196-197°.[830]
Ph(2-$MeOC_6H_4$)(2-$C_{10}H_7$)PO. III. M. 139-142°, $b_{0.1}$ 210-220°, ^1H.[507]
Ph(4-$MeOC_6H_4$)(2-$C_{10}H_7$)PO. III. $B_{0.5}$ 260°, ^1H.[507]
MePh[PhCH=CHCH=C(Ph)]PO. II.[60] M. 169-170°.
BuPh(4-$PhCO_2C_6H_4$)PO. I. M. 136°.[206]
MePh[4,4'-$(Me_2N)_2$biphenylyl-2]PO. II. M. 97-100°.[830]
Ph(4-$MeOC_6H_4$)[3-$ClC_6H_4CH_2CH$(Ph)]PO. II.[788] M. 185-186°.
Ph(4-$MeOC_6H_4$)[$PhCH_2CH$(Ph)]PO. II.[788]
MePh[PhCH=C(Ph)C(Ph)=C(Ph)]PO. II.[60] M. 204°.

D.4. Di-, Tri-, and Tetraphosphines

$C_5H_{14}O_2P_2$

$Me_2P(O)CH_2P(O)Me_2$. III.[696] VI.[275] M. 132-134°,[696] b. 160-161°, n_D^{20} 1.4500, d^{20} 1.1598.[275]
$Me_2P(O)(CH_2)_2P(O)Me_2$. III.[488] M. 232-233°.
$Me_2P(O)(CH_2)_3P(O)Me_2$. III.[488] II.[289] M. 211-212°,[488] $b_{0.2}$ 115°.[289]
$[Me_2P_\beta(O)CH_2]_2MeP_\alpha O$. IV. M. 194-196°, $^{31}P_\alpha$ -39.6, $^{31}P_\beta$ -42.4.[536]
$Me_2P(O)(CH_2)_4P(O)Me_2$. III.[486] M. 204-205°, $b_{0.35}$ 219-222°.
$Me_2P(O)CHEtCH_2P(O)Me_2$. III. $B_{0.5}$ 172-175°, IR, ^1H, ^{31}P -41.7, -48.3.[327]
$Me_2P(O)(CH_2)_5P(O)Me_2$. III.[486] M. 167-168°, $b_{1.3}$ 249-252°.
$[Me_2P_\beta(O)CH_2]_3P_\alpha O$. IV.[537] M. 204-207°, $^{31}P_\alpha$ -36.9, $^{31}P_\beta$ -43.1.
$Me_2P(O)(CH_2)_6P(O)Me_2$. III.[486] M. 179-180°, $b_{0.35}$ 217°.
$Et_2P(O)(CH_2)_2P(O)Et_2$. IX.[447] M. 122-123°.
$Et_2P(O)(CH_2)_4P(O)Et_2$. III.[486] M. 111-112°, $b_{0.25}$ 181-182°.
$EtPhP(O)CH_2P(O)EtPH$. IV. M. 194-196°.[450]
$Bu_2P(O)CH_2P(O)Bu_2$. III.[488] M. 172-174°. Complexes with compounds of Fe, Co, Ni.[814]

$C_{18}H_{40}O_2P_2$

$Bu_2P(O)(CH_2)_2P(O)Bu_2$. I.[646] III. M. 174-175°,[488] m. 168-169°,[646] $b_{0.8}$ 222-224°.[488]
$EtPhP(O)(CH_2)_4P(O)PhEt$. I.[411] M. 159-161°.
$Et_2P(O)(CH_2)_{12}P(O)Et_2$. IV.[535] M. 96-97.5°, ^{31}P -55.5, IR.
$Bu_2P(O)(CH_2)_4P(O)Bu_2$. III.[486] M. 116-118°, b_2 270-271°.
$EtPhP(O)CH_2P(O)Ph_2$. IV. M. 200-202°.[450]
$Bu_2P(O)(CH_2)_5P(O)Bu_2$. III. M. 106-107°, $b_{0.75}$ 259°,[486] ^{31}P -47.7.[193]
$Bu_2P(O)(CH_2)_6P(O)Bu_2$. III.[486,606] M. 110-111°,[486] m.

98.5-100°,[606] $b_{0.9}$ 278-281°.[486]
$Ph_2P(O)CH_2P(O)(CH_2NEt_2)_2$. I. M. 98-100°, IR.[530]
$Ph_2P(O)CH_2P(O)Ph(CH_2NEt_2)$. I. M. 183-184.5°.[530]
$Et(PhCH_2)P(O)(CH_2)_6P(O)(CH_2Ph)Et$. IV.[772] M. 157-158°.

$C_{25}H_{18}Br_4O_2P_2$

$(3-BrC_6H_4)_2P(O)CH_2P(O)(3-BrC_6H_4)_2$. IV. M. 189-191°.[840]
$(4-BrC_6H_4)_2P(O)CH_2P(O)(4-BrC_6H_4)_2$. IV. M. 286-287°.[840]
$(3-ClC_6H_4)_2P(O)CH_2P(O)(3-ClC_6H_4)_2$. IV. M. 194-196°.[840]
$(4-ClC_6H_4)_2P(O)CH_2P(O)(4-ClC_6H_4)_2$. IV. M. 276-277°.[840]
$Ph_2P(O)CN_2P(O)Ph_2$. IX. M. 154°.[648]
$[Ph_2P(O)]_2CHNa$. From reaction of sodium hydride with the phosphine oxide.[445]
$Ph_2P(O)CH_2P(O)Ph_2$. III.[696] IV. M. 181-182°,[450] m. 181-182°,[696] ^1H.[450] Complex with compound of U.[263]
$(C_6H_{13})_2P(O)CH_2P(O)(C_6H_{13})_2$. III.[696] VII.[695] M. 33-35°, $b_{0.2}$ 218-223°,[696] $b_{0.23}$ 225-235°.[695]
$(Et_2CHCH_2)_2P(O)CH_2P(O)(CH_2CHEt_2)_2$. III. M. 35-37°, $b_{1.2}$ 215-218°.[696]
$Ph_2P(O)CH{=}CHP(O)Ph_2$. I. M. trans isomer, 310-311°,[6] m. 312-313°,[450] IR, ^1H. M. cis isomer, 244-245°, IR, ^1H.
$Ph_2P(O)CHClCH_2P(O)Ph_2$. IV. M. 186-188°.[450]
$Ph_2P(O)(CH_2)_2P(O)Ph_2$. I.[416] III.[443,488] IV.[535] VI.[375] M. 266-268.5°,[535] m. 276-278°,[535] m. 276°,[375] m. 252-254°,[416] m. 269-270°,[443] ^{31}P -35.8,[535] IR,[535] ^1H.[144] Complexes with compounds of Sn,[707] Mo,[140,508] W,[140,508] U,[263] Cu.[558]
$Ph_2P(O)C(OH)MeP(O)Ph_2$. IX. M. 159-160°, IR, ^1H.[202]
$[Ph_2P(O)CH_2]_2PO_2H$. IV.[525] IX. M. 258-260°,[571] m. 246-250°.[525]

$C_{27}H_{26}O_2P_2$

$Ph_2P(O)(CH_2)_3P(O)Ph_2$. II. VI.[375] III.[488] IV.[535] M. 141-143°,[375] m. 142-143°,[485] $b_{0.1}$ 295-300°,[488] $b_{0.1}$ 280-288°, ^{31}P -34.4.[535]
$Ph_2P(O)(CH_2)_4P(O)Ph_2$. VI.[375] III.[486] II. M. 258-260°,[375] m. 154-155°,[486] m. 257°, $b_{1.6}$ 290-291°, ^{31}P -38.6.[193] Complexes with compounds of Sn.[707]
$[Ph_2P(O)CH_2]_2P(O)OEt$. IV. M. 211.5-212°.[571]
$[\overline{CH_2(CH_2)_4CH}]PhP(O)(CH_2)_4P(O)Ph[\overline{HC(CH_2)_4CH_2}]$. I.[411] M. 201-202°.
$(C_6H_{13})_2P(O)(CH_2)_4P(O)(C_6H_{13})_2$. III. M. 27-28°, $b_{0.2}$ 223°.[696]

$C_{29}H_{18}F_{12}O_2P_2$

$(3-CF_3C_6H_4)_2P(O)CH_2P(O)(3-CF_3C_6H_4)_2$. IV. M. 183-185°.[840]
$(4-CF_3C_6H_4)_2P(O)CH_2P(O)(4-CF_3C_6H_4)_2$. IV. M. 234-236°.[840]

$Ph_2P(O)C\overset{(CF_2)_3}{=\!=}CP(O)Ph_2$. IV.[252] M. 176-177°.
$Ph_2P(O)(CH_2)_5P(O)Ph_2$. VI.[375] III. M. 119-120°,[486] m.

124-126°,[375] $b_{0.27}$ 326-327°.[486] ^{31}P -31.8.[193]
$(3\text{-}MeC_6H_4)_2P(O)CH_2P(O)(3\text{-}MeC_6H_4)_2$. IV. M. 149-151°.[840]
$(4\text{-}MeC_6H_4)_2P(O)CH_2P(O)(4\text{-}MeC_6H_4)_2$. IV. M. 212-214°.[840]
$(3\text{-}MeOC_6H_4)_2P(O)CH_2P(O)3\text{-}MeOC_6H_4)_2$. IV. M. 150-151°.[840]
$(4\text{-}MeOC_6H_4)_2P(O)CH_2P(O)(4\text{-}MeOC_6H_4)_2$. IV. M. 197-200°.[840]
$Ph_2P(O)CH_2P(O)(C_8H_{17})_2$. VII. M. 96-97°.[695]
$(C_7H_{15})_2P(O)CH_2P(O)(C_7H_{15})_2$. I. M. 29-31°, $b_{0.02}$ 165-180°.[646]
1,4-$[Ph_2P(O)]_2C_6H_4$. I. VI.[848] M. 298-300°.
(3-Pyridyl)$CH[P(O)Ph_2]_2$. IX. M. 214-215°, IR.[201]
$Ph_2P(O)(CH_2)_6P(O)Ph_2$. III. M. 196-198°, $b_{0\ 2}$ 292-294°.[486]
$[Ph_2P(O)CH_2CH_2]_2NEt$. IX.[449] M. 190-191°.
$(C_7H_{15})_2P(O)(CH_2)_2P(O)(C_7H_{15})_2$. I. M. 33-35°, $b_{0.02}$ 210-220°.[646]
$PhCH[P(O)Ph_2]_2$. IV. M. 306-307°.[450]
$[Ph_2P(O)CH_2CH_2]_2C(CO_2H)_2$. From hydrolysis of corresponding ethyl ester.[449] M. 180° (dec.).
$Ph_2P(O)(CH_2)_7P(O)Ph_2$. III. M. 128-129°.[485]
$Pr_2C[P(O)(C_6H_{13})_2]_2$. III. M. 27-28°, $b_{0.13}$ 220-221°.[696]

$C_{32}H_{28}O_2P_2$

$Ph_2P(O)CH_2CH(Ph)P(O)Ph_2$. VII.[7] IV.[725] M. 277-278°,[7] m. 283-284°.[725]
1,2-$[Ph_2P(O)CH_2]_2C_6H_4$. II. M. 240°, IR.[11]
1,4-$[Ph_2P(O)CH_2]_2C_6H_4$. VI.[375] M. 322°.
$[Ph_2P_\beta(O)CH_2]_2PhP_\alpha O$. IV. M. 196-207°. $^{31}P_\alpha$ -27.6, $^{31}P_\beta$ -27.6.[536]
$Ph_2P(O)(CH_2)_8P(O)Ph_2$. III. M. 170.5-172°.[485]
$[Ph_2P(O)CH_2CH_2]_2NC_8H_{17}$. IX.[449] M. 158-159°.
$Ph_2P(O)CH_2(Ph)C[Ph_2P(O)]{=}CH_2$. V.[714] M. 195-196°, IR, mass spect.
$PhCH_2CH_2(HO)C[Ph_2P(O)]_2$. IX. M. 155-156°, IR, mass spect.[202]
$Ph_2P(O)(CH_2)_9P(O)Ph_2$. III. M. 123-125°.[485]
$(4\text{-}MeC_6H_4)_2P(O)(CH_2)_5P(O)(4\text{-}MeC_6H_4)_2$. III. M. 156-158°.[485]
Naphthalene-1,4-$[Ph_2P(O)]_2$. I.[848] M. 282-284°.
Naphthalene-1,5-$[Ph_2P(O)]_2$. I.[848] M. >330°.
Naphthalene-2,6-$[Ph_2P(O)]_2$. I.[848] M. 276-277°.
Naphthalene-2,7-$[Ph_2P(O)]_2$. I.[848] M. 121-122°.
4-$[Ph_2P(O)CH{=}CH]C_6H_4CH{=}CHP(O)Ph_2$. VII. M. 354-355°.[445]
$Ph_2P(O)(CH_2)_{10}P(O)Ph_2$. III. M. 181-182°.[485]
$(4\text{-}MeOC_6H_4)_2P(O)(CH_2)_6P(O)(4\text{-}MeOC_6H_4)_2$. III. M. 161-162°.[485]
$(C_8H_{17})_2P(O)(CH_2)_2P(O)(C_8H_{17})_2$. III. M. 152-152.5°.[483]

$C_{35}H_{38}O_6P_2$

$[Ph_2P(O)CH_2CH_2]_2C(CO_2Et)_2$. IX.[449] M. 184-185°.

$(C_8H_{17})_2P(O)(CH_2)_3P(O)(C_8H_{17})_2$. I. M. 145-150°, IR.[531]
$Ph_2P(O)(CH_2)_{12}P(O)Ph_2$. IV.[535] M. 165-165.5°, ^{31}P -35.3.
Anthracene-2,6-$[(O)PPh_2]_2$. I.[848] M. >350°.
$[Ph_2P(O)]_2CHCH_2P(O)Ph_2$. IV. M. 207-208°.[450]
$[Ph_2P_\beta(O)CH_2]_2C_{12}H_{25}P_\alpha O$. IV. M. 63-69°, $^{31}P_\alpha$ -39.5, $^{31}P_\beta$ -25.7,[536] ^1H.[526]
$[Ph_2P(O)CH_2]_3PO$. IV. M. 238-239°,[571] m. 223-224°, ^{31}P -24.4, -31.0.[526,537]
$[(C_8H_{17})_2P(O)CH_2]_3PO$. IV. M. 64-65°.[571]
$Ph_2P(O)C(Ph)_2C{\equiv}CC(Ph)_2(O)PPh_2$. I.[849] M. 348° (dec.), IR.
$(C_{12}H_{25})_2P(O)(CH_2)_4P(O)(C_{12}H_{25})_2$. I.[531] M. 91-91.8°, IR. Hydrate, m. 65-67°.
$[Ph_2P(O)CH_2]_4C$. I. M. 258-260°, IR.[229]
1,4-$[Ph_2P(O)CH(O_2PPh_2)]_2C_6H_4$. IX. M. 295-296°, IR.[201]

D.5. Cyclic Phosphine Oxides

$C_3H_6O_5P$

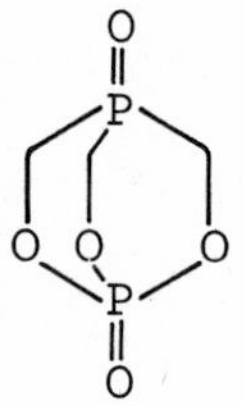

[bicyclic structure: O=P(CH2O)3P=O]. I. M. 210-213°, ^{31}P -6.4, 18.1.[172]

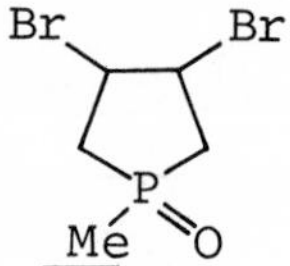

[structure: 3,4-dibromo-1-methylphospholane 1-oxide]. IX.[665] M. 145-146°.

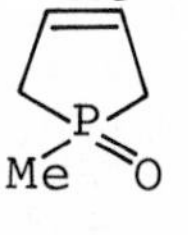

[structure: 1-methyl-3-phospholene 1-oxide]. VI.[665] $B_{0.16}$ 59°.

$C_7H_{13}OP$

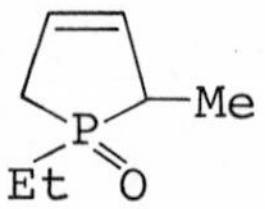

[structure: 1-ethyl-2-methyl-3-phospholene 1-oxide]. VI.[37] B_5 114-115°, n_D^{20} 1.5080.

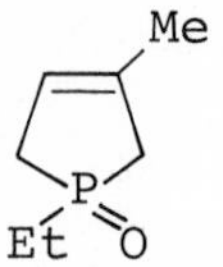

[structure: 1-ethyl-3-methyl-3-phospholene 1-oxide]. VI.[560,561] $B_{0.6}$ 116-117°, n_D^{23} 1.5049.

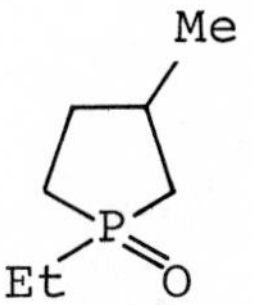

Me / Et / P=O (3-methyl-1-ethylphospholane oxide). VI.[563] $B_{0.6}$ 95-100°.

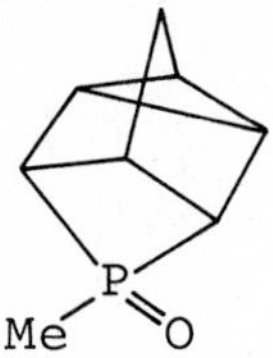

Me / P=O (polycyclic). VI.[290,291] M. 157°, ^{1}H, UV.[290]

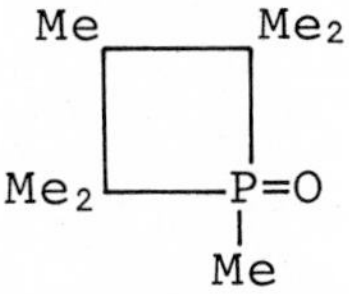

Me, Me_2, Me_2, P=O, Me (phosphetane). VI.[189] M. 170-171°, ^{1}H.

$C_{10}H_{10}BrOP$

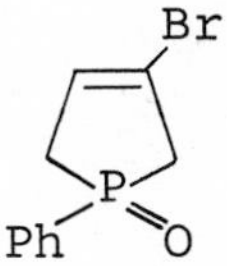

Br / Ph / P=O. VI.[560,561] $B_{0.5}$ 160-165°.

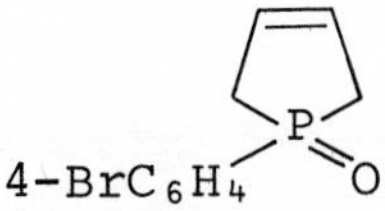

4-BrC_6H_4 / P=O. VI.[37] M. 66-71°, b_3 151-152°.

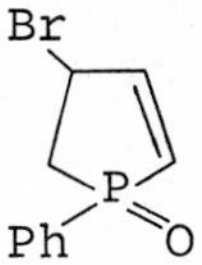

Br / Ph / P=O. VI.[223] M. 110°.

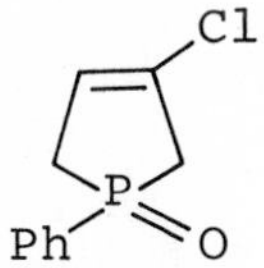

Cl / Ph / P=O. VI.[560,561] $B_{0.1}$ 158-164°.

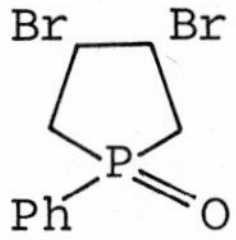

Br, Br / Ph / P=O. IX.[223] M. 139°.

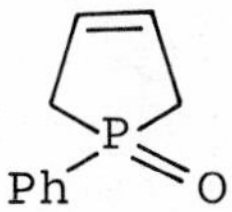

Ph / P=O. VI.[560,561] M. 67-75°, $b_{0.2}$ 153-155°.

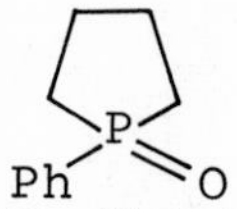

. I.[203] M. 56-57°, $b_{0.15}$ 99-100°.

$C_{11}H_9O_2P$

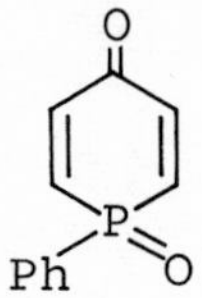

. From selenium dioxide dehydrogenation of corresponding saturated compound.[549] M. 130-131°.

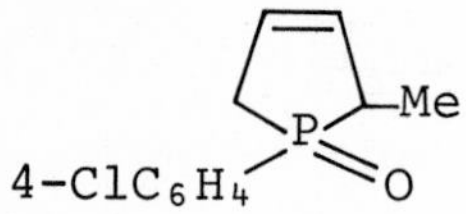

4-ClC_6H_4 . VI.[37] M. 94-96°, $b_{3.5}$ 165-166°.

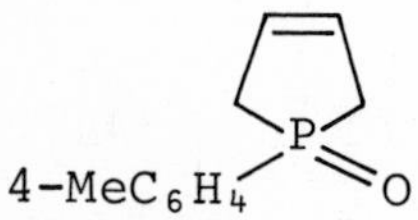

4-MeC_6H_4 . VI.[37] M. 60-67°, b_3 183-184°, n_D^{20} 1.5795.

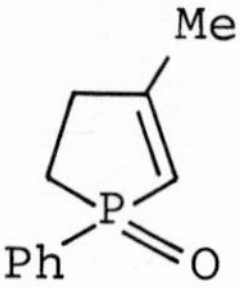

. IV.[807] M. 65-67°, $b_{0.6}$ 154-165°.

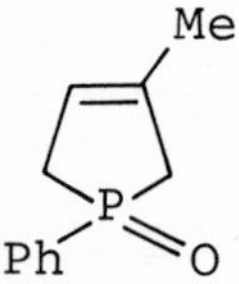

. VI.[37,561] M. 75-77°, $b_{2.5}$ 182-183°.

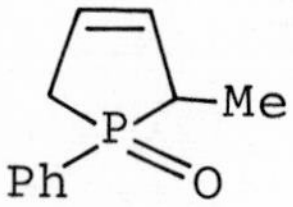

. VI.[37,560,561] $B_{3.5}$ 158-159°.

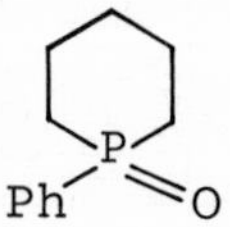

. I.[203] M. 130°, b_1 140°.

$C_{12}H_{15}OP$

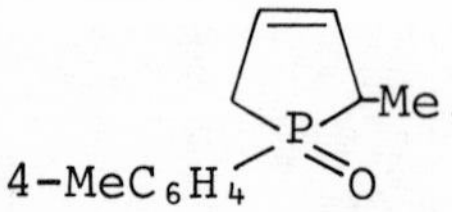

4-MeC_6H_4 Me. VI.[37] $B_{2.5}$ 171-172°, n_D^{20} 1.5702.

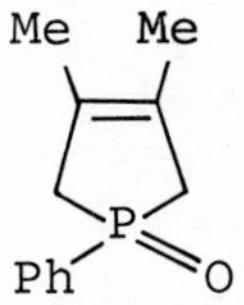

Me Me
Ph P=O (3,4-dimethyl-3-phospholene ring)

. VI.[560,561] $B_{0.3}$ 173-175°.

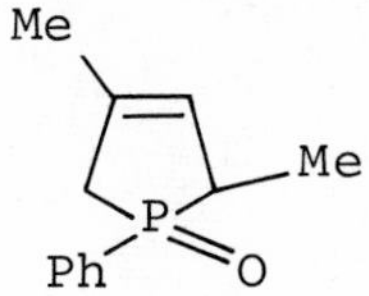

Me
Me
Ph P=O

. VI.[560,561] B_8 202-208°.

Me
4-MeOC$_6$H$_4$ P=O

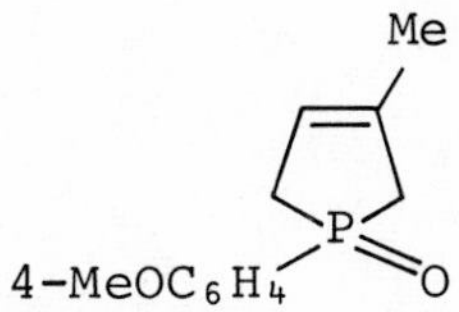

. VI.[560,561] $B_{0.7}$ 210-212°, n_D^{23} 1.5754.

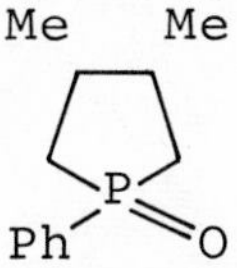

Me Me
Ph P=O

. VI.[563] $B_{0.7}$ 160-168°.

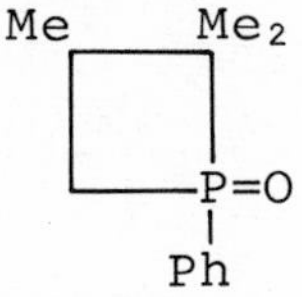

Me Me$_2$
P=O
Ph

. VI.[189] M. 81-83°, ^{1}H.

Me
4-MeOC$_6$H$_4$ P=O

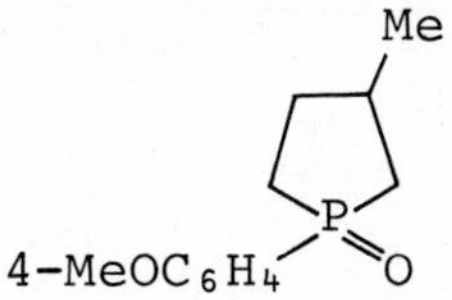

. VI.[563] $B_{0.3}$ 178-185°.

$C_{12}H_{25}O_3P$

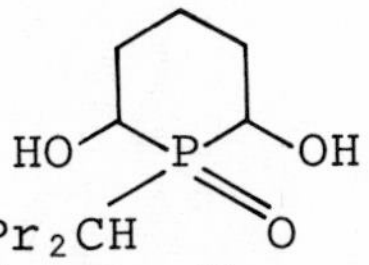

HO P OH
Pr$_2$CH O

. V.[114] M. 226-227°.

Me$_2$ Me$_2$
P=O
Ph

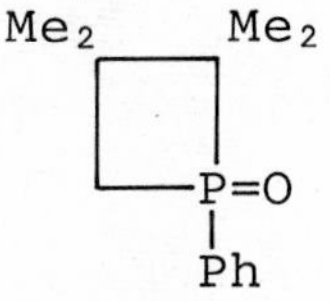

. VI.[189] M. 82.5-84°.

. I.[541] M. 98-100°.

. IX.[262] M. 124-126°. 2,4-Dinitrophenylhydrazone, m. 249-250°.

Me_2N $\overset{+}{N}Me_3I^-$

. IX.[223] M. 236°.

$C_{16}H_{15}OP$

Ph . VI.[560,561] M. 96-98°, $b_{0.9}$ 224-226°.

C_6H_5 C_6H_5 . VI.[560,561] M. 125°, $b_{0.2}$ 235-240°.

Et Et . I.[207] M. 234-235°.

$CH_2CH_2C{=}CMe_2$. VI.[560,561] $B_{0.2}$ 192-193°.

$C_{12}H_{25}$. II. M. 48°,[799] m. 62°.[657]

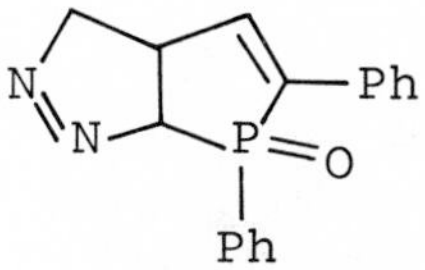

[Structure] . IX.[133] M. 190° (dec.).

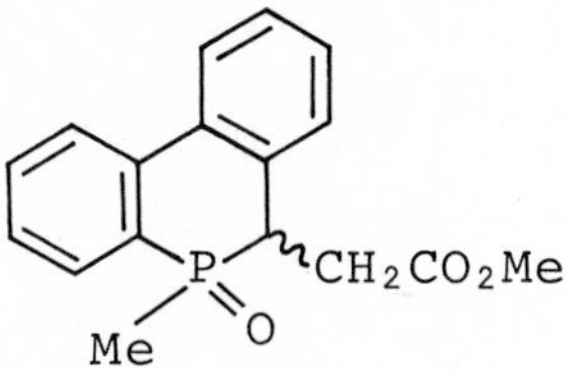

[Structure] . IX.[697] M. 145-148°.

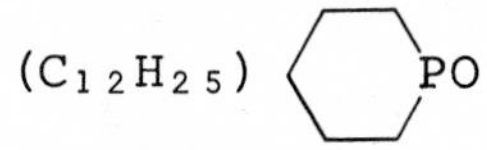

$(C_{12}H_{25})$ [Structure] PO . II. M. 78°,[657] m. 61°.[799]

$C_{18}H_{12}NO_3P$

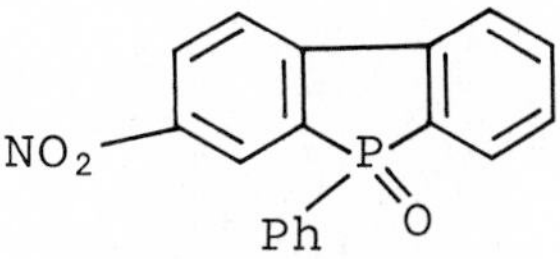

[Structure] . IX.[136,137] M. 203°.

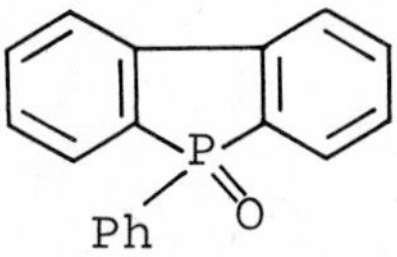

[Structure] . From cyclization of corresponding phosphinic acid.[136,137] M. 167-168°.

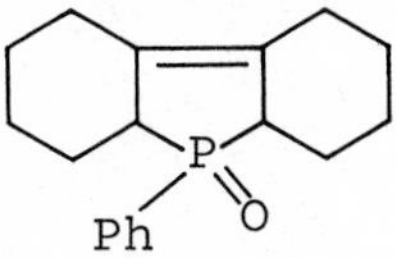

[Structure] . VI.[132,133,560,561] $B_{0.1}$ 200-220°.

$C_{19}H_{25}OP$

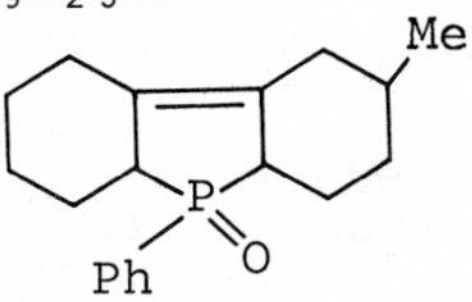

[Structure] . VI.[132.133] $B_{0.01}$ 196-206°.

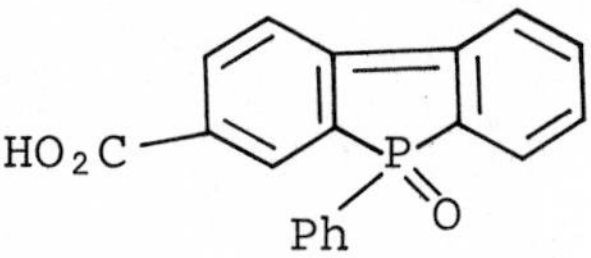

[Structure] . IX.[137] M. 323-326°.

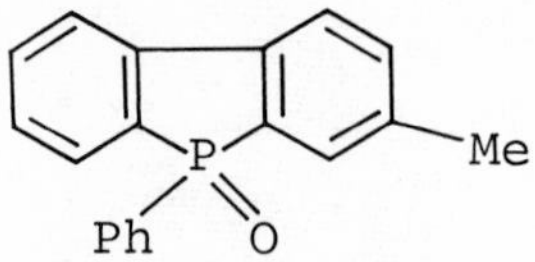

Ph, O, Me . IX.[137] M. 146-147°.

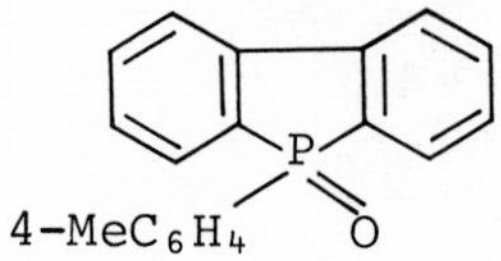

$4\text{-}MeC_6H_4$, O . VI.[505] M. 190-192°.

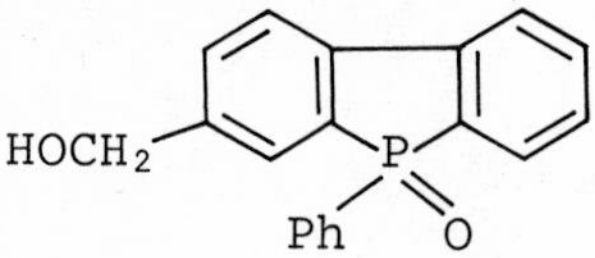

$HOCH_2$, Ph, O . IX.[137] M. 195-196°.

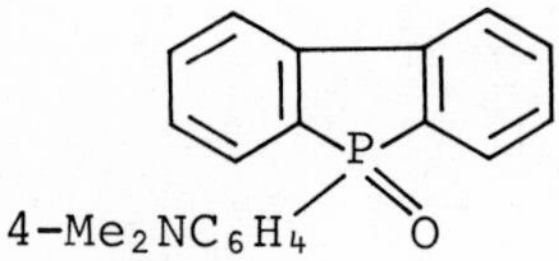

$4\text{-}Me_2NC_6H_4$, O . I.[830] M. 177-177.5°.

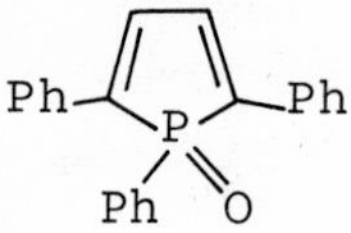

Ph, Ph, Ph, O . VI.[132] M. 237-239°.

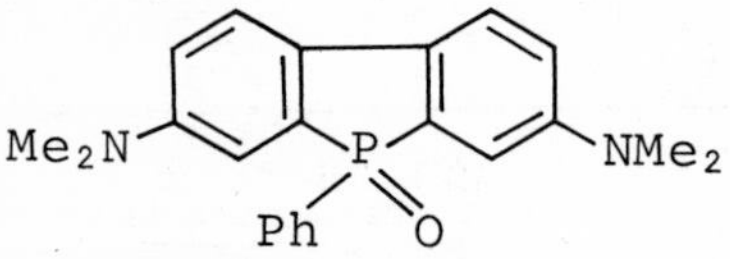

Me_2N, NMe_2, Ph, O . I.[830] M. 271-273°.

$C_{23}H_{17}O_2P$

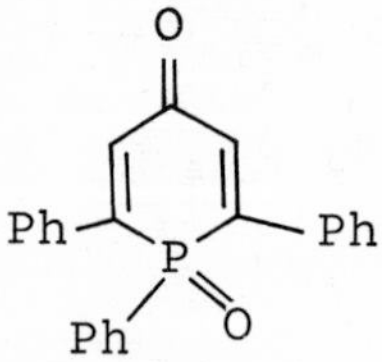

O, Ph, Ph, Ph, O . From selenium dioxide dehydrogenation of corresponding saturated compound.[549] M. 162°.

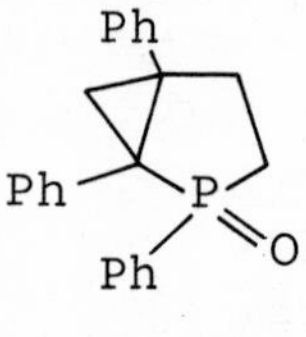

Ph, Ph, Ph, O . IX.[133] M. 220-222°.

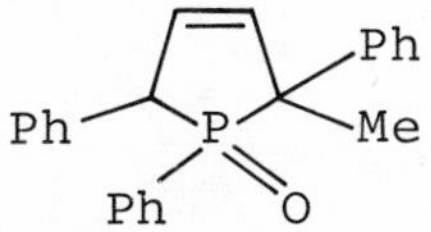

. VI.[133] M. 175.5-176.5°.

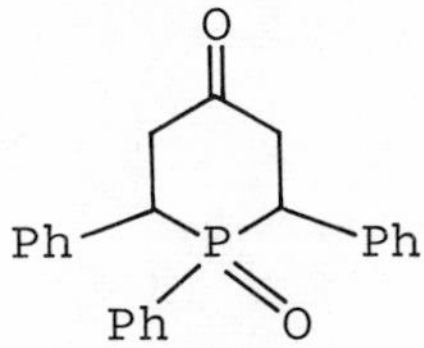

. IX.[549] M. 178°.

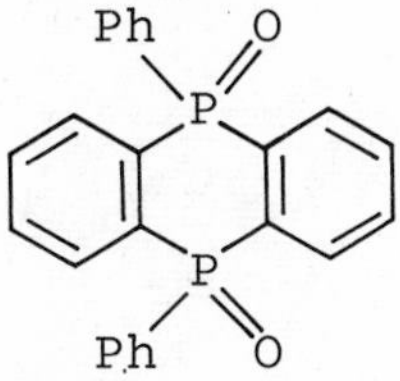

. I.[207] M. 276-278°.

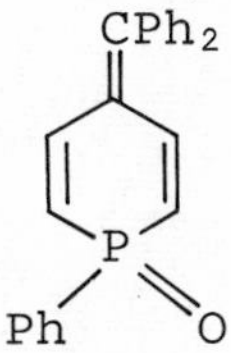

. From reaction of corresponding dienone with diphenyl ketene.[549] M. 136-138°.

$C_{25}H_{20}NOP$

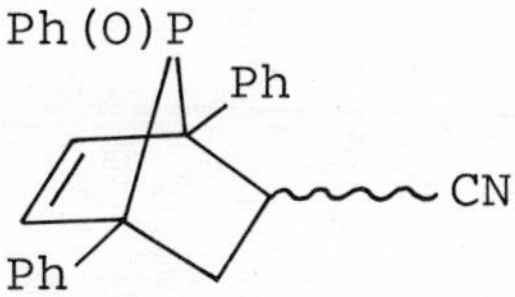

. IX.[133] M. 180°.

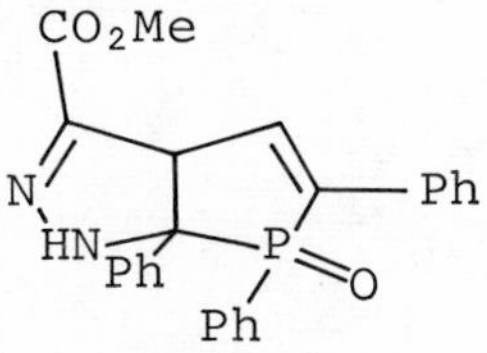

. IX.[133] M. 233-235° (dec.).

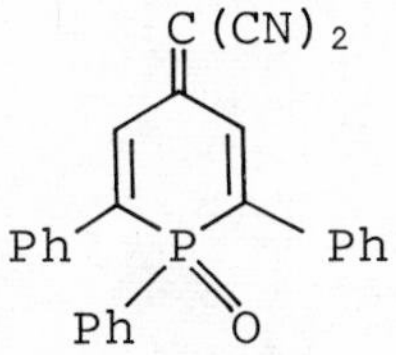

. From Knoevenagel condensation of dienone with malonitrile.[549] M. 259-260°.

Ph(O)P, Ph, Ph, O, O, O. IX.[133] M. 229-231°.

NCCCO$_2$Et, Ph, P, Ph, Ph, O. From Knoevenagel condensation of dienone with cyano ester.[549] M. 205-206°.

$C_{29}H_{23}OP$

Ph, H, Ph, P, Ph, Ph, O. I.[548] M. 156-158°.

Ph, Ph, P, Ph, Ph, O. IX.[133] M. 234-235° (dec.).

Ph, Ph, Ph, Ph, P, Me, O. I.[101] M. 240-241°.

Ph, Ph, OH, Ph, P, Ph, O. IX.[656] M. 257°, ^{31}P -30.

Ph, Ph, Ph, Ph, P, Ph, O. III.[499] M. 292-293°.

Ph, Ph, Ph, Ph, P, PhCH$_2$, O. VI.[101] M. 228-230°.

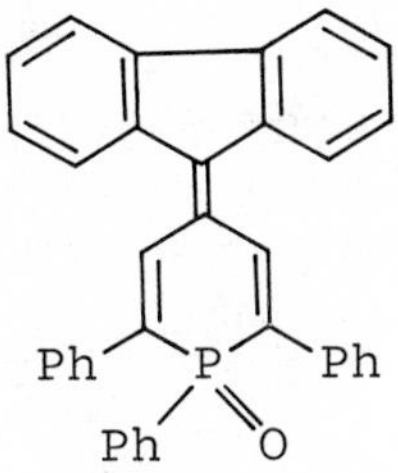

. From reaction of the dienone with fluorene.[549] M. 264-266°

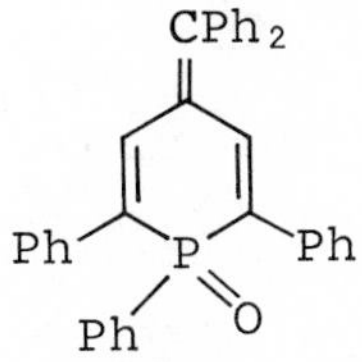

. From reaction of corresponding dienone with diphenyl ketone.[549] M. 319-321°.

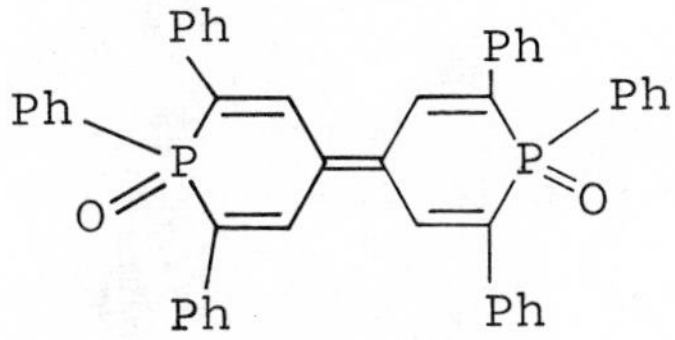

. From triethylphosphite condensation of corresponding dienone.[549] M. 425-428°.

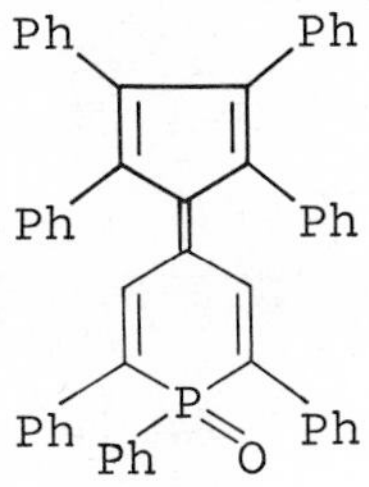

. From reaction of corresponding dienone with cyclopentadiene compound.[549] M. 261-263°.

(received July 7, 1970)

REFERENCES

1. Abraham, M. H., J. H. N. Garland, J. A. Hill, and L. F. Larkworthy, Chem. Ind. (London), 1962, 1615.
2. Adams, D. M., J. Chem. Soc., A, 1968, 162.
3. Addison, C. C., and J. C. Sheldon, J. Chem. Soc., 1956, 2705.
4. Aguiar, A. M., J. Beisler, and A. Mills, J. Org. Chem., 27, 1001 (1962).
5. Aguiar, A. M., and D. Daigle, J. Org. Chem., 30, 2826 (1965).
6. Aguiar, A. M., and D. Daigle, J. Am. Chem. Soc., 86,

2299 (1964).
7. Aguiar, A. M., and D. Daigle, J. Org. Chem., 30, 3527 (1965).
8. Aguiar, A. M., J. Giacin, and A. Mills, J. Org. Chem., 27, 674 (1962).
9. Aguiar, A. M., H. J. Greenberg, and K. E. Rubenstein, J. Org. Chem., 28, 2091 (1963).
10. Aguiar, A. M., K. C. Hansen, and J. T. Mague, J. Org. Chem., 32, 2383 (1967).
11. Aguiar, A. M., and M. G. Raghanan Nair, J. Org. Chem., 33, 579 (1968).
12. Aksnes, G., Acta Chem. Scand., 15, 438 (1961).
13. Aksnes, D., and G. Aksnes, Acta Chem. Scand., 17, 1262 (1963).
14. Aksnes, G. and L. J. Brudvik, Acta Chem. Scand., 17, 1616 (1963).
15. Aksnes, G., and T. Gramstad, Acta Chem. Scand., 14, 1485 (1960).
16. Aksnes, G., and J. Songstad, Acta Chem. Scand., 16, 507 (1962).
17. Aksnes, G., and J. Songstad, Acta Chem. Scand., 16, 1426 (1962).
18. Alexander, R., C. Eaborn, and T. G. Traylor, J. Organometal. Chem., 21, 65 (1970).
19. Allcock, H. R., J. Polymer Sci., Part A-2, 4087 (1964).
20. Allcock, H. R., and R. L. Kugel, J. Polymer Sci., Part A-1, 3627 (1963).
21. Allen, D. W., and I. T. Miller, J. Chem. Soc., C, 1967, 1869.
22. Allen, D. W., and J. C. Tebby, Tetrahedron, 23, 2795 (1967).
23. Allen, D. W., J. C. Tebby, and D. H. Williams, Tetrahedron Letters, 1965, 2361.
24. Amonoo-Neizer, E. G., S. K. Ray, R. A. Shaw, and B. C. Smith, J. Chem. Soc., 1965, 4296.
25. Anschutz, L., H. Kraft, and K. Schmidt, Ann. Chem., 542, 14 (1939).
26. Anteunis, M., M. Verzele, and G. Dacremont, Bull. Soc. Chim. Belges, 74, 622 (1965).
27. Appel, R., and A. Hauss, Z. Anorg. Allgem. Chem., 311, 290 (1961).
28. Arbuzov, A. E., Dissertation, Kazan, 1914.
29. Arbuzov, A. E., J. Russ. Phys. Chem. Soc., 42, 395 (1910); C.A., 5, 1397 (1911).
30. Arbuzov, A. E., and B. A. Arbuzov, J. Russ. Phys. Chem. Soc., 61, 217 (1929).
31. Arbuzov, A. E., G. Kamai, and L. V. Nesterov, Tr. Kazansk. Khim.-Tekhnol. Inst., 1951, 17; C.A., 51, 5720 (1957).
32. Arbuzov, A. E., and K. V. Nikonorov, Zh. Obshch. Khim., 18, 2008 (1948); C.A., 43, 3801i (1949).

33. Arbuzov, A. E., and K. V. Nikonorov, Zh. Obshch. Khim., 18, 2012 (1948); C.A., 43, 3802b (1949).
34. Arbuzov, B. A., J. Prakt. Chem., 131, 357 (1931).
35. Arbuzov, B. A., N. A. Polezhaeva, V. S. Vinogradova, and A. K. Shamsutdinova, Izv. Akad. Nauk SSSR, Ser. Khim., 1965, 669; C.A., 63, 2998 (1965).
36. Arbuzov, B. A., and N. I. Rizpolozhenskii, Dokl. Akad. Nauk SSSR, 89, 291 (1953); C.A., 48, 7540 (1954).
37. Arbuzov, B. A., and L. A. Shapshinskaya, Izv. Akad. Nauk SSSR, Otd. Khim. Nauk, 1962, 65; C.A., 57, 13791 (1962).
38. Arbuzov, B. A., and G. M. Vinokurova, Izv. Akad. Nauk SSSR, Otd. Khim. Nauk, 1963, 502; C.A., 59, 3947 (1963).
39. Arbuzov, B. A., G. M. Vinokurova, I. A. Aleksandrova, and S. G. Fattakhov, Nekotorye Vopr. Organ. Khim., Sb., 1964, 244; C.A., 65, 3899g (1966).
40. Aroney, M. J., R. J. W. LeFevre, and J. D. Saxby, J. Chem. Soc., 1964, 6180.
41. Arshad, M., A. Beg, and M. S. Siddiqui, Can. J. Chem., 43, 608 (1965).
42. Auger, V., Compt. Rend., 139, 639 (1904).
43. Bailey, W. J., and S. A. Buckler, J. Am. Chem. Soc., 79, 3567 (1957).
44. Bailey, W. J., S. A. Buckler, and F. Marktscheffel, J. Org. Chem., 25, 1996 (1960).
45. Bailey, W. J., W. M. Muir, and F. Marktscheffel, J. Org. Chem., 27, 4404 (1962).
46. Baizer, M. M., and J. D. Anderson, J. Org. Chem., 30, 3138 (1965).
47. Baliah, V., and P. Subbarayan, J. Org. Chem., 25, 1833 (1960).
48. Balzer, W. D., Tetrahedron Letters, 1968, 1189.
49. Bannister, E., and F. A. Cotton, J. Chem. Soc., 1960, 1878.
50. Bannister, E., and F. A. Cotton, J. Chem. Soc., 1960, 2276.
51. Bartecki, A., and D. Dembicka, Rocz. Chem., 39, 1783 (1965); C.A., 64, 18932d (1966).
52. Baker, H. T., and C. F. Baes, Jr., J. Inorg. Nucl. Chem., 24, 1277 (1962).
53. Batiste, M. A., and C. T. Sprouse, Jr., Tetrahedron Letters, 1969, 3165.
54. Beattie, I. R., and M. Webster, J. Chem. Soc., 1965, 3672.
55. Beck, W., and K. Lottes, Chem. Ber., 96, 1046 (1963).
56. Bedford, A. F., and C. T. Mortimer, J. Chem. Soc., 1960, 1622.
57. Bell, J. V., J. Heisler, H. Tannenbaum, and J. Goldenson, J. Am. Chem. Soc., 76, 5185 (1954).
58. Bellamy, L. J., The Infrared Spectra of Complex

Molecules, Wiley, New York, 1958, Chapter 18.
59. Bergel'son, L. D., V. A. Vaver, L. I. Barsukov, and M. M. Shemyakin, Izv. Akad. Nauk SSSR, Ser. Khim. 1966, 506; C.A., 65, 10615 (1966).
60. Bergesen, K., Acta Chem. Scand., 20, 899 (1966).
61. Bergman, E. D., and M. Droc, Israel J. Chem., 3, 239 (1966).
62. Berlin, K. D., T. H. Austin, M. Peterson, and M. Nagabhushanam, Topics in Phosphorus Chemistry, Vol. I, M. Grayson and E. J. Griffith, Eds., Interscience, New York, 1964, p. 17.
63. Berlin, K. D., and G. B. Butler, Chem. Rev., 60, 243 (1960).
64. Berlin, K. D., and G. B. Butler, J. Org. Chem., 25, 2006 (1960).
65. Berlin, K. D., and G. B. Butler, J. Am. Chem. Soc., 82, 2712 (1960).
66. Berlin, K. D., and G. B. Butler, J. Org. Chem., 26, 2537 (1961).
67. Berlin, K. D., and J. F. Calvert, Proc. Okla. Acad. Sci., 46, 78 (1966).
68. Berlin, K. D., and D. M. Hellwege, Topics in Phosphorus Chemistry, Vol VI, M. Grayson and E. J. Griffith, Eds., Interscience, New York, 1969, pp. 11, 28, 54, 65, 68, 112, 116, and 119.
69. Reference 68, p. 1.
70. Berlin, K. D., and M. Nagabhushanarm, Chem. Ind. (London), 1964, 974.
71. Berlin, K. D., and R. U. Pagilagan, J. Org. Chem., 32, 129 (1967).
72. Berlin, K. D., and M. E. Peterson, J. Org. Chem., 32, 125 (1967).
73. Berry, R. S., J. Chem. Phys., 32, 933 (1960).
74. Bertrand, J. A., Inorg. Chem., 6, 495 (1967).
74a. Bestmann, H. J., and F. Seng, Tetrahedron, 21, 1373 (1965).
75. Beynon, K. I., J. Polymer Sci., Part A-1, 3357 (1963).
76. Bieber, T. I., and E. H. Eisman, J. Org. Chem., 27, 678 (1962).
77. Birk, J. P., J. Halpern, and A. L. Pickard, J. Am. Chem. Soc., 90, 4491 (1968).
78. Birum, G. H., and C. N. Matthews, J. Am. Chem. Soc., 88, 4198 (1966).
79. Bissey, J. E., J. Chem. Educ., 44, 95 (1967).
80. Bissey, J. E., H. Goldwhite, and D. G. Rowsell, J. Org. Chem., 32, 1542 (1967).
81. Bissing, D. E., and A. J. Speziale, J. Am. Chem. Soc., 87, 1405 (1965).
82. Bladé-Font, A., C. A. Vender Werf, and W. E. McEwen, J. Am. Chem. Soc., 82, 2397 (1960).
83. Blake, C. A., K. B. Brown, and C. F. Coleman, U.S. At.

Energy Comm. ORNL-1964 (1955); C.A., 50, 15320i (1956).
84. Blicke, F. F., and S. Raines, J. Org. Chem., 29, 204 (1964).
85. Blount, B. K., J. Chem. Soc., 1891, 1931.
86. Bock, H., and H. T. Dieck, Z. Naturforsch., 21, 739 (1966).
87. Boehringer, C. H., Ger. Pat. 1,216,300 (1966).
88. Bogolyubov, G. M., K. S. Mingaleva, and A. A. Petrov, Zh. Obshch. Khim., 35, 1566 (1965); C.A., 63, 17860e (1965).
89. Boisselle, A. P., and N. A. Meinhardt, J. Org. Chem., 27, 1828 (1962).
90. Bokanov, A. I., B. A. Korolev, and B. I. Stepanov, Z. Obshch. Khim., 39, 321 (1969); C.A., 71, 12395m (1969).
91. Borowitz, I. J., K. C. Kirby, Jr., and R. Virkhaus, J. Org. Chem., 31, 4031 (1966).
92. Boskin, M. J., and D. B. Denny, Chem. Ind. (London), 1959, 330.
93. Bott, R. W., B. F. Dowden, and C. Eaborn, J. Organometal. Chem., 4, 291 (1965).
94. Bott, R. W., B. F. Dowden, and C. Eaborn, J. Chem. Soc., 1965, 4994.
95. Bott, R. W., B. F. Dowden, and C. Eaborn, J. Chem. Soc., 1965, 6306.
96. Bott, R. W., B. F. Dowden, and C. Eaborn, Intern. Symp. Organosilicon Chem., Sci. Commun. Prague, 1965, 290; C.A., 65, 10606 (1966).
97. Bourneuf, M., Bull. Soc. Chim. France, 33 (4), 1808 (1923).
98. Boyer, J. H., and S. E. Ellzey, Jr., J. Org. Chem., 26, 4684 (1961).
99. Branden, C. I., and I. Lindquist, Acta Chem. Scand., 17, 353 (1963).
100. Brauer, G., and E. Zintl, Z. Phys. Chem., B-37, 323 (1937).
101. Braye, E. H., W. Hübel, and I. Caplier, J. Am. Chem. Soc., 83, 4406 (1961).
102. Bride, M. H., W. A. W. Cummings, and W. Pickles, J. Appl. Chem., 11, 352 (1961).
103. Brisdon, B. J., Inorg. Chem., 6, 1791 (1967).
104. Brodie, A. M., S. H. Hunter, G. A. Rodley, and C. J. Wilkins, J. Chem. Soc., A, 1968, 2039.
105. Brodie, A. M., S. H. Hunter, G. A. Rodley, and C. J. Wilkins, Inorg. Chim. Acta, 1968, 195.
106. Brook, J. H. T., Trans Faraday Soc., 63, 2034 (1967).
107. Brophy, J. J., K. L. Freeman, and M. J. Gallagher, J. Chem. Soc., C, 1968, 2760.
108. Brophy, J. J., and M. J. Gallagher, Australian J. Chem., 20 (3), 503 (1967).

109. Brown, D., J. F. Easey, and J. G. H. du Preez, J. Chem. Soc., A, 1966, 258.
110. Brown, A. D., and G. M. Kosolapoff, J. Chem. Soc., C, 1968, 839.
111. Buckler, S. A., J. Am. Chem. Soc., 82, 4215 (1960).
112. Buckler, S. A., J. Am. Chem. Soc., 84, 3093 (1962).
113. Buckler, S. A., and M. Epstein, Tetrahedron, 18, 1211 (1962).
114. Buckler, S. A., and M. Epstein, Tetrahedron, 18, 1221 (1962).
115. Buckler, S. A., and M. Epstein, J. Am. Chem. Soc., 82, 2076 (1960).
116. Buckler, S. A., and M. Epstein, U.S. Pat. 3,052,719 (1962); C.A., 58, 551c (1963).
117. Budzikiewicz, H., C. Djerassi, and D. H. Williams, Mass Spectrometry of Organic Compounds, Holden-Day, San Francisco, Calif., 1967, p. 647.
118. Bugerenko, E. F., E. A. Chernyshev, and A. D. Petrov, Izv. Akad. Nauk SSSR, Ser. Khim., 1965, 286; C.A., 62, 14721h (1965).
119. Bullock, J. I., J. Inorg. Nucl. Chem., 29, 2257 (1967).
120. Bunyan, P. J., and J. I. G. Cadogan, J. Chem. Soc., 1963, 42.
121. Burdett, J. L., and L. L. Burger, Can. J. Chem., 44, 111 (1966).
122. Burdon, J., I. N. Rozhkov, and G. M. Perry, J. Chem. Soc., C, 1969, 2615.
123. Burg, A. B., and W. E. McKee, J. Am. Chem. Soc., 73, 4590 (1951).
124. Burg, A. B., and A. J. Sarkis, J. Am. Chem. Soc., 87, 238 (1965).
125. Burger, A., and N. D. Dawson, J. Org. Chem., 16, 1250 (1951).
126. Burger, A., and W. H. Shelver, J. Med. Pharm. Chem., 4, 225 (1961); C.A., 56, 1534d (1962).
127. Burger, L. L., J. Phys. Chem., 62, 590 (1958).
128. Cadogan, J. I. G., M. Cameron-Wood, R. K. Mackie, and R. J. G. Searle, J. Chem. Soc., 1965, 4831.
129. Cahours, A., Ann. Chem., 122, 329 (1862).
130. Cahours, A., and A. W. Hoffmann, Ann. Chem., 104, 1 (1857).
131. Cahours, A., and A. W. Hoffmann, Ann. Chem., 104, 20 (1857).
132. Campbell, I. G. M., R. C. Cookson, and M. G. Hocking, Chem. Ind. (London), 1962, 359.
133. Campbell, I. G. M., R. C. Cookson, M. B. Hocking, and A. N. Hughes, J. Chem. Soc., 1965, 2184.
134. Campbell, J. R., and R. E. Hatton, U.S Pat. 3,020,315 (1962); C.A., 56, 12948c (1962).
135. Campbell, I. G. M., and I. D. R. Stevens, Chem.

Commun., 1966, 505
136. Campbell, I. G. M., and J. K. Way, Proc. Chem. Soc., 1959, 231.
137. Campbell, I. G. M., and J. K. Way, J. Chem. Soc., 1961, 2133.
138. Canavan, A. E., and C. Eaborn, J. Chem. Soc., 1962, 592.
139. Canavan, A. E., and C. Eaborn, J. Chem. Soc., 1959, 3751.
140. Canziani, F., F. Zingales, and U. Sartorelli, Gazz. Chim. Ital., 94 (7), 841 (1964); C.A., 61, 12030a (1964).
141. Carius, L., Ann. Chem., 137, 117 (1866).
142. Carr, R. L. V., and C. F. Baranauckas, U.S. Pat. 3,316,293 (1967); C.A., 67, 43921b (1967).
143. Carson, J. F., and F. Wong, J. Org. Chem., 26, 1467 (1961).
144. Carty, A. J., and R. K. Harris, Chem. Commun., 1967, 234.
145. Carty, A. J., and D. G. Tuck, J. Chem. Soc., 1964, 6012.
146. Challenger, F., and A. T. Peters, J. Chem. Soc., 1929, 2610.
147. Challenger, F., and F. Pritchard, J. Chem. Soc., 1924, 864.
148. Challenger, F., and J. F. Wilkinson, J. Chem. Soc., 1924, 2675.
149. Chance, L. H., and J. D. Guthrie, J. Appl. Chem., 10, 395 (1960).
150. Chandrasegaran, L., and G. A. Rodley, Inorg. Chem., 4, 1360 (1965).
151. Charrier, C., M. P. Simonnin, W. Chodkiewicz, and P. Cadiot, Compt. Rend., 258 (4), 1537 (1964).
152. Chernick, C. L., and H. A. Skinner, J. Chem. Soc., 1956, 1401.
153. Chodkrewicz, W., P. Chadiot, and A. Willemart, Compt. Rend., 250, 866 (1960).
153a. Clampitt, R. B., G. H. Birum, and R. M. Anderson, U.S. Patent 3,306,937 (1967).
154. Clark, J. P., V. M. Langford, and C. J. Wilkins, J. Chem. Soc., A, 1967, 792.
155. Claydon, A. P., P. A. Fowell, and C. T. Mortimer, J. Chem. Soc., 1960, 3284.
156. Clees, V. H., and F. Huber, Z. Anorg. Allgem. Chem., 352, 200 (1967).
157. Collamati, I., Ric. Sci., 6, 363 (1964); C.A., 62, 15738a (1965).
158. Collie, N., J. Chem. Soc., 53, 636 (1888).
159. Collie, N., J. Chem. Soc., 55, 223 (1889).
160. Collie, N., J. Chem. Soc., 53, 714 (1888).
161. Collie, N., Phil. Mag., 24, 27 (1887).
162. Collie, N., and E. A. Letts, Phil. Mag., 22, 183 (1886).

163. Collie, J. N., and F. Reynolds, J. Chem. Soc., 1915, 367.
164. Conant, J. B., J. B. S. Braverman, and R. E. Hussey, J. Am. Chem. Soc., 45, 165 (1923).
165. Cookson, R. C., G. W. A. Fowles, and D. K. Jenkins, J. Chem. Soc., 1965, 6406.
166. Cope, A. C., N. A. LeBel, H. H. Lee, and W. R. Moore, J. Am. Chem. Soc., 79, 4720 (1957).
167. Copley, D. B., F. Fairbrother, J. R. Miller, and A. Thompson, Proc. Chem. Soc., 1964, 300.
168. Corbridge, D. E. C., J. Appl. Chem. (London), 6, 456 (1956).
169. Corbridge, D. E. C., Topics in Phosphorus Chemistry, Vol. III, M. Grayson and E. J. Griffith, Eds., Interscience, New York, 1966, p. 57.
170. Corbridge, D. E. C., Topics in Phosphorus Chemistry, Vol. VI, M. Grayson and E. J. Griffith, Eds., Interscience, New York, 1969, p. 235.
171. Corfield, J. R., J. R. Shutt, and S. Trippett, Chem. Commun., 1969, 789.
172. Coskran, K. J., and J. G. Verkade, Inorg. Chem., 4, 1655 (1965).
173. Cotton, F. A., and E. Bannister, J. Chem. Soc., 1960, 1873.
174. Cotton, F. A., R. D. Barnes, and E. Bannister, J. Chem. Soc., 1960, 2199.
175. Cotton, F. A., D. M. L. Goodgame, and R. H. Soderberg, Inorg. Chem., 2, 1162 (1963).
176. Cotton, F. A., and D. M. L. Goodgame, J. Am. Chem. Soc., 82, 5771 (1960).
177. Cotton, F. A., B. F. G. Johnson and R. M. Wing, Inorg. Chem., 4, 502 (1965).
178. Cotton, F. A., S. J. Lippard, and J. T. Mague, Inorg. Chem., 4, 508 (1965).
179. Cotton, F. A., and R. H. Soderberg, J. Am. Chem. Soc., 85, 2402 (1963).
180. Cousins, D. R., and F. A. Hart, Chem. Ind. (London), 1965, 1259.
181. Cousins, D. R., and F. A. Hart, J. Inorg. Nucl. Chem., 30, 3009 (1968).
182. Cousins, D. R., and F. A. Hart, J. Inorg. Nucl. Chem., 29, 1745 (1967).
183. Cox, J. R., Jr., and F. H. Westheimer, J. Am. Chem. Soc., 80, 5441 (1958).
184. Crafts, J. M., and R. Silva, J. Chem. Soc., 24, 629 (1871).
185. Craig, D. P., A. Maccoll, R. S. Nyholm, L. E. Orgel, and L. E. Sutton, J. Chem. Soc., 1954, 332.
186. Cram, D. J., and R. D. Partos, J. Am. Chem. Soc., 85, 1093 (1963).
-37. Cram, D. J., R. D. Partos, S. H. Pine, and H. Jager,

J. Am. Chem. Soc., 84, 1742 (1962).
188. Crawford, N. P., and G. A. Melson, J. Chem. Soc., A, 1969, 1049.
189. Cremer, S. E., and R. J. Charvat, J. Org. Chem., 32, 4066 (1967).
190. Crofts, P. C., I. M. Downie, and K. Williamson, J. Chem. Soc., 1964, 1240.
191. Crofts, P. C., and G. M. Kosolapoff, J. Am. Chem. Soc., 75, 3379 (1953).
192. Cruickshank, D. W. J., J. Chem. Soc., 1961, 5486.
193. Crutchfield, M. M., C. H. Dungan, J. H. Letcher, V. Mark, and J. R. Van Wazer, Topics in Phosphorus Chemistry, Vol. V, Interscience, M. Grayson and E. J. Griffith, Eds., New York, 1967, pp. 1-489.
194. Cumper, C. W. N., A. A. Foxton, J. Read, and A. I. Vogel, J. Chem. Soc., 1964, 430.
195. Curry, J. D., The Procter & Gamble Company, personal communication, 1970.
196. Czimatis, L., Chem. Ber., 15, 2014 (1882).
197. Daasch, L. W., and D. C. Smith, J. Chem. Phys., 19, 22 (1951).
198. Daasch, L. W., and D. C. Smith, Anal. Chem., 23, 853 (1951).
199. Davidson, R. S., Tetrahedron Letters, 1968, 3029.
200. Davidson, R. S., Tetrahedron, 25, 3383 (1969).
201. Davidson, R. S., R. A. Sheldon, and S. Trippett, J. Chem. Soc., C, 1967, 1547.
202. Davidson, R. S., R. A. Sheldon, and S. Trippett, J. Chem. Soc., C, 1968, 1700.
203. Davies, J. H., J. D. Downer, and P. Kirby, J. Chem. Soc., C, 1966, 245.
204. Davies, J. H., and P. Kirby, J. Chem. Soc., 1964, 3425.
205. Davies, W. C., and W. J. Jones, J. Chem. Soc., 1929, 33.
206. Davies, W. C., and F. G. Mann, J. Chem. Soc., 1944, 276.
207. Davis, M., and F. G. Mann, J. Chem. Soc., 1964, 3370.
308. Dawson, N. D., and A. Burger, J. Org. Chem., 18, 207 (1953).
209. Day, J. P., and L. M. Venanzi, J. Chem. Soc., A, 1966, 197.
210. Deacon, G. B., and J. H. S. Green, Chem. Ind. (London), 1965, 1031.
211. Deacon, G. B., J. H. S. Green, and W. Kynaston, J. Chem. Soc., A, 1967, 158.
212. Deacon, G. B., and R. S. Nyholm, J. Chem. Soc., 1965, 6107.
213. DeBruin, K. E., K. Naumann, G. Zon, and K. Mislow, J. Am. Chem. Soc., 91, 7031 (1969).
214. DeBruin, K. E., G. Zon, K. Naumann, and K. Mislow,

J. Am. Chem. Soc., 91, 7027 (1969).
215. Denney, D. B., D. Z. Denney, and L. A. Wilson, Tetrahedron Letters, 1968, 85.
216. Denney, D. B., W. F. Goodyear, and B. Goldstein, J. Am. Chem. Soc., 82, 1393 (1960).
217. Denney, D. B., W. F. Goodyear, and B. Goldstein, J. Am. Chem. Soc., 83, 1726 (1961).
218. Denney, D. B., and F. J. Gross, J. Org. Chem., 32, 2445 (1967).
219. Denney, D. B., and J. W. Hanifin, Jr., Tetrahedron Letters, 1963, 2178.
220. Denney, D. G., and J. W. Hanifin, Jr., Tetrahedron Letters, 1963, 58.
221. Denney, D. B., and H. A. Kindsgrab, J. Org. Chem., 28, 1133 (1963).
222. Denney, D. B., and L. C. Smith, J. Org. Chem., 27, 3404 (1962).
222a. Denney, D. B., A. K. Tsolis, and K. Mislow, J. Am. Chem. Soc., 86, 4486 (1964).
223. Donadio, R. E., Dissertation Abstr., 20, 495 (1959).
224. Döpp, D., Chem. Commun., 1968, 1284.
225. Dörken, C., Chem. Ber., 21, 1505 (1888).
226. Downie, I. M., and G. Morris, J. Chem. Soc., 1965, 5771.
227. Drenth, W., and D. Rosenberg, Rec. Trav. Chim., 86, 26 (1967).
228. Edmundson, R. S., and J. O. L. Wrigley, Tetrahedron, 23, 283 (1967).
229. Ellermann, J., and D. Schirmacher, Chem. Ber., 100, 2220 (1967).
230. Emeleus, H. F., and J. M. Miller, J. Inorg. Nucl. Chem., 28, 662 (1966).
231. Ettel, V., and J. Horak, Collection Czech. Chem. Commun., 26, 1949 (1961); C.A., 56, 5994d (1962).
232. Eyles, E. T., and S. Trippett, J. Chem. Soc., C, 1966, 67.
233. Eyman, D. P., and R. S. Drago, J. Am. Chem. Soc., 88, 1617 (1966).
234. Ezzell, B. R., and L. D. Freedman, J. Org. Chem., 34, 1777 (1969).
235. Ezzell, B. R., and L. D. Freedman, J. Org. Chem., 35, 241 (1970).
236. Ezzell, J. B., and W. R. Gilkerson, J. Am. Chem. Soc., 88, 3486 (1966).
237. Farbwerke Hoechst, DAS 1040549 (1957).
238. Fedorova, G. K., and G. A. Lanchuk, Zh. Obshch. Khim., 34 (2), 511 (1964); C.A., 60, 12048f (1964).
239. Fenton, G. W., and C. K. Ingold, J. Chem. Soc., 1929, 2342.
240. Ferraro, J. R., Appl. Spectry., 17, 12 (1963).
241. Feshchenko, N. G., and A. V. Kirsanov, Zh. Obshch.

Khim., 36 (3), 564 (1966); C.A., 65, 742 (1966).
242. Feshchenko, N. G., I. K. Mazepa, Zh. K. Gorbatenko, Yu. P. Makovetskii, Yu. P. Kukhar, and A. V. Kirsanov, Zh. Obshch. Khim., 39, (6), 1219 (1969); C.A., 71, 124582b (1969).
243. Fields, E. K., Ger. Pat. 1,122,521 (1962); C.A., 58, 551f (1962).
244. Fild, M., O. Glemser, and I. Hollenberg, Naturwissenschaften, 52 (21), 590 (1965).
245. Finegold, H., Ann. N. Y. Acad. Sci., 70, 875 (1958).
246. Fischer, E., I. Laulicht, and S. Pinchas, J. Phys. Chem., 66, 2708 (1962).
247. Fitzsimmons, B. W., P. Gans, B. Hayton, and B. C. Smith, J. Inorg. Nucl. Chem., 28, 915 (1966).
248. Fleissner, F., Chem. Ber., 13, 1665 (1880).
249. Floyd, M. B., and C. E. Boozer, J. Am. Chem. Soc., 85, 984 (1963).
250. Fluck, E., and H. Binder, Z. Anorg. Allgem. Chem., 354, 139 (1967).
251. Frank, A. W., J. Org. Chem., 24, 966 (1959).
252. Frank, A. W., J. Org. Chem., 30, 3663 (1965).
253. Frazer, M. J., W. Gerrard, and R. Twaits, J. Inorg. Nucl. Chem., 25, 637 (1963).
254. Freedman, L. D., and G. O. Doak, Chem. Rev., 57, 479 (1957).
255. Frisch, K. C., and H. Lyons, J. Am. Chem. Soc., 75, 4078 (1953).
256. Fritzsche, H., U. Hasserodt, and F. Korte, Chem. Ber., 97, 1988 (1964).
257. Fritzsche, H., U. Hasserodt, and F. Korte, Chem. Ber., 98, 171 (1965).
258. Frunze, T. M., V. V. Korshak, L. V. Kozlov, and V. V. Kurashev, Vysokomol. Soedin., 1, 677 (1959); C.A., 54, 19554b (1960).
259. Frunze, T. M., V. V. Korshak, and V. V. Kurashev, Vysokomol. Soedin., 1, 670 (1959); C.A., 54, 19553f (1960).
260. Frunze, T. M., V. V. Korshak, V. V. Kurashev, G. S. Kolesnikov, and B. A. Zhubanov, Izv. Akad. Nauk SSSR, Otd. Khim. Nauk, 1958, 783; C.A., 52, 20001 (1958).
261. Gallagher, M. J., and I. D. Jenkins, Advances in Stereochemistry, Vol. III, E. L. Eliel and N. L. Allinger, Eds., Interscience, New York, 1968, Chapter 1.
262. Gallagher, M. J., E. C. Kirby, and F. G. Mann, J. Chem. Soc., 1963, 4846.
263. Gans, P., and B. C. Smith, J. Chem. Soc., 1964, 4172.
264. Gebelein, C. G., and H. Howard, Jr., Third Delaware Region Meeting, Philadelphia, Feb. 25, 1960, Abstracts of Papers, p. 79.

265. Geddes, A. L., J. Phys. Chem., 58, 1062 (1954).
266. Gerrard, W., and W. J. Green, J. Chem. Soc., 1951, 2550.
267. Gilkerson, W. R., and J. B. Ezell, J. Am. Chem. Soc., 89, 808 (1967).
268. Gilkerson, W. R., and J. B. Ezell, J. Am. Chem. Soc., 87, 3812 (1965).
269. Gillespie, R. J., J. Am. Chem. Soc., 82, 5978 (1960).
270. Gilman, H., and G. E. Brown, J. Am. Chem. Soc., 67, 824 (1945).
271. Gilman, H., and B. J. Gaj, J. Am. Chem. Soc., 82, 6326 (1960).
272. Gilman, H., and J. Robinson, Rec. Trav. Chim., 48, 328 (1929).
273. Gilman, H., and C. C. Vernon, J. Am. Chem. Soc., 48, 1063 (1926).
274. Ginsburg, V. A., and A. Ya Jakubovich, Zh. Obshch. Khim., 28, 728 (1958); C.A., 52, 17091 (1958).
275. Gladshtein, B. M., and V. M. Zimin, Zh. Obshch. Khim., 37, 2055 (1967); C.A., 68, 49687 (1968).
276. Gloyna, D., and H. G. Henning, Angew. Chem., Intern. Ed., 5 (9), 847 (1966).
277. Goehring, M. B., and H. Thielemann, Z. Anorg. Allgem. Chem., 308, 33 (1961).
278. Goetz, H., F. Nerdel, and K. H. Wiechel, Ann. Chem., 665, 1 (1963).
279. Goodgame, D. M. L., and F. A. Cotton, J. Chem. Soc., 1961, 2298.
280. Goodgame, D. M. L., and F. A. Cotton, J. Chem. Soc., 1961, 3735.
281. Goodgame, D. M. L., M. Goodgame, and F. A. Cotton, J. Am. Chem. Soc., 83, 4161 (1961).
282. Goodgame, D. M. L., and M. A. Hitchman, Inorg. Chem., 4, 721 (1965).
283. Goubeau, Von J., and W. Berger, Z. Anorg. Allgem. Chem., 304, 147 (1960).
284. Gough, S. T. D., and S. Trippett, J. Chem. Soc., 1962, 2333.
285. Gramstad, T., and S. J. Snaprud, Acta Chem., Scand., 16, 999 (1962).
286. Grayson, M., J. Am. Chem. Soc., 85, 79 (1963).
287. Grayson, M., and C. E. Farley, Chem. Commun., 1967, 831.
288. Grayson, M., and P. T. Keough, J. Am. Chem. Soc., 82, 3919 (1960).
289. Grayson, M., P. T. Keough, and G. A. Johnson, J. Am. Chem. Soc., 81, 4803 (1959).
290. Green, M., Proc. Chem. Soc., 1963, 177.
291. Green, M., J. Chem. Soc., 1965, 541.
292. Greenbaum, M. A., D. B. Denny, and A. K. Hoffman, J. Am. Chem. Soc., 78, 2563 (1956).

293. Griffin, C. E., Tetrahedron, 20, 2399 (1964).
294. Griffin, C. E., J. J. Burke, F. E. Dickson, M. Gordon, H. H. Hsieh, R. Obrycki, and M. P. Williamson, J. Phys. Chem., 71 (3), 4558.
295. Griffin, C. E., and N. T. Castellucci, J. Org. Chem., 26, 629 (1961).
296. Griffin, C. E., and M. Gordon, J. Am. Chem. Soc., 89, 4427 (1967).
297. Griffin, C. E., R. P. Peller, K. R. Martin, and J. A. Peters, J. Org. Chem., 30, 97 (1965).
298. Griffin, C. E., R. P. Peller, and J. A. Peters, J. Org. Chem., 30, 91 (1965).
299. Griffin, C. E., and R. A. Polsky, J. Org. Chem., 26, 4772 (1961).
300. Griffiths, J. E., and A. B. Burg, J. Am. Chem. Soc., 84, 3442 (1962).
301. Grignard, V., and J. Savard, Compt. Rend., 192, 592 (1931).
302. Grinshtein, E. I., A. B. Bruker, and L. Z. Soborovskii, Dokl. Akad. Nauk SSSR, 139, 1359 (1961).
303. Gruenanger, P., and M. R. Langella, Atti Accad. Naz. Lincei, Rend., Sci. Fis. Mat. Nat., 36 (3), 387 (1964); C.A., 62, 3973b (1965).
304. Haake, P., R. D. Cook, and G. H. Hurst, J. Am. Chem. Soc., 89, 2650 (1967).
305. Haake, P., W. B. Miller, and D. A. Tyssee, J. Am. Chem. Soc., 86, 3577 (1964).
306. Hadzi, D., J. Chem. Soc., 1962, 5128.
307. Hadzi, D., C. Klofutar, and S. Oblak, J. Chem. Soc., A, 1968, 905.
308. Hadzi, D., H. Ratajczak, and L. Sobczyk, J. Chem. Soc., A, 1967, 48.
309. Halmann, M., and S. Pinchas, J. Chem. Soc., 1958, 3264.
310. Halpern, E., J. Bouck, H. Finegold, and J. Goldenson, J. Am. Chem. Soc., 77, 4472 (1955).
311. Hands, A. R., and A. J. H. Mercer, J. Chem. Soc., 1965, 6055.
312. Hands, A. R., and A. J. H. Mercer, J. Chem. Soc., A, 1968, 449.
313. Hantz, A., Stud. Cercet. Chim., 16 (9), 665 (1968).
314. Hantzsch, A., and H. Hibbert, Chem. Ber., 40, 1508 (1907).
315. Hart, F. A., and F. G. Mann, J. Chem. Soc., 1955, 4107.
316. Hart, F. A., and J. E. Newbery, J. Inorg. Nucl. Chem., 30, 318 (1968).
317. Hartley, S. B., W. S. Holmes, J. K. Jackques, M. F. Mole, and J. C. McCoubrey, Quart, Rev. (London), 17, 204 (1963).
318. Hartmann, H., C. Beermann, and H. Z. Czempik, Z.

Anorg. Allgem. Chem., 287, 261 (1956).
319. Hartmann, H., and H. Fratzscher, Naturwissenschaften, 51, 213 (1964).
319a. Hartmann, H., and A. Meixner, Naturwissenschaften, 50, 403 (1963).
320. Harvey, R. G., and E. R. DeSombre, Topics in Phosphorus Chemistry, Vol. I, M. Grayson and E. J. Griffith, Eds., Interscience, New York, 1964, pp. 57-101.
321. Reference 320, p. 63.
322. Harvey, R. G., and E. V. Jensen, Tetrahedron Letters, 1963, 1801.
323. Hawes, W., and S. Trippett, J. Chem. Soc., C, 1969, 1465.
324. Hays, H. R., J. Org. Chem., 31, 3817 (1966).
325. Hays, H. R., J. Am. Chem. Soc., 91, 2736 (1969).
326. Hays, H. R., J. Org. Chem., 36, 98 (1971), compound analyzed by vapor-phase chromatography.
327. Hays, H. R., J. Org. Chem., 33, 3690 (1968).
328. Hays, H. R., J. Org. Chem., 33, 4201 (1968).
329. Hays, H. R., and R. G. Laughlin, J. Org. Chem., 32, 1060 (1967).
330. Hein, F., and H. Hecker, Chem. Ber., 93, 1339 (1960).
331. Hein, Fr., K. Issleib, and H. Rabold, Z. Anorg. Allgem. Chem., 287, 208 (1956).
332. Hein, F., H. Plust, and H. Pohlemann, Z. Anorg. Allgem. Chem., 272, 25 (1953).
333. Hellmann, H., and J. Bader, Tetrahedron Letters, 1961, 724.
334. Hellmann, H., J. Bader, H. Birkner, and O. Schumacher, Ann. Chem., 659, 49 (1962).
335. Hendrickson, J. B., M. L. Maddox, J. J. Sims, and H. D. Kaesz, Tetrahedron, 20, 449 (1964).
336. Henning, H. G., and K. Forner, Z. Chem., 6 (8), 314 (1966); C.A., 65, 16995 (1966).
337. Henson, P. D., K. Naumann, and K. Mislow, J. Am. Chem. Soc., 91, 5645 (1969).
337a. Herriott, A. W., and K. Mislow, Tetrahedron Letters, 1968, 3013.
338. Herrmann, K. W., and L. Benjamin, J. Colloid Sci., 23, 478 (1967).
339. Herrmann, K. W., J. G. Brushmiller, and W. L. Courchene, J. Phys. Chem., 70, 2909 (1966).
340. Hey, L., and C. K. Ingold, J. Chem. Soc., 1933, 531.
341. Heyn, A. H. A., and Y. D. Soman, J. Inorg. Nucl. Chem., 26, 287 (1964).
342. Hieber, W., J. Peterhans, and E. Winter, Chem. Ber., 94, 2572 (1961).
343. Higginson, W. C. E., R. T. Leigh, and R. Nightingale, J. Chem. Soc., 1962, 435.

344. Hoffman, A. W., Ann. Chem. Suppl., 1, 1 (1861).
345. Hoffman, A., J. Am. Chem. Soc., 43, 1684 (1921).
346. Hoffmann, A. K., and A. G. Tesch, J. Am. Chem. Soc., 81, 5519 (1959).
347. Hofmann, A. W., Ber., 6, 292, 301, 303 (1873).
348. Hofmann, A. W., Ann. Chem. Suppl., 1, 1 (1861).
349. Hoffmann, H., Angew. Chem., 71, 379 (1959).
350. Hoffmann, H., Angew. Chem., 74, 881 (1962).
351. Hoffmann, H., Ann. Chem., 634, 1 (1960).
352. Hoffmann, H., Chem. Ber., 94, 1331 (1961).
353. Hoffmann, H., Chem. Ber., 95, 2563 (1962).
354. Hoffmann, H., and H. Diehr, Chem. Ber., 98, 363 (1965).
355. Hoffmann, H., and H. J. Diehr, Tetrahedron Letters, 1962, 583.
356. Hoffmann, H., R. Gunewald, and L. Horner, Chem. Ber., 93, 861 (1960).
357. Hoffmann, H., H. L. Horner, H. G. Wippel, and D. Michael, Chem. Ber., 95, 523 (1962).
358. Hoffmann, H., and P. Schellenberk, Chem. Ber., 99, 1134 (1966).
359. Halmann, M., and L. Kugel, J. Chem. Soc., 1962, 3272.
360. Horner, L., Helv. Chim. Acta, Alfred Werner Commemoration Volume, 1969, 114.
361. Reference 360, p. 93.
362. Horner, L., and W. D. Balzer, Tetrahedron Letters, 1965, 1157.
363. Horner, L., and P. Beck, Chem. Ber., 93, 1371 (1960).
364. Horner, L., P. Beck, and H. Hoffmann, Chem. Ber., 92, 2088 (1959).
365. Horner, L., P. Beck, and V. G. Toscano, Chem. Ber., 94, 1317 (1961).
366. Horner, L., P. Beck, and V. G. Toscano, Chem. Ber., 94, 1323 (1961).
367. Horner, L., H. Fuchs, H. Winkler, and A. Rapp, Tetrahedron Letters, 1963, 965.
368. Horner, L., and H. Hoffmann, Angew. Chem., 68, 473 (1956).
369. Horner, L., and H. Hoffmann, Newer Methods of Preparative Organic Chemistry, Vol. II, W. Foerst, Ed., Academic Press, New York, 1963, p. 163.
370. Reference 369, pp. 184-187.
371. Reference 369, p. 180.
372. Horner, L., H. Hoffmann, and P. Beck, Chem. Ber., 91, 1583 (1958).
373. Horner, L., H. Hoffmann, H. Ertel, and G. Klahre, Tetrahedron Letters, 1961, 9.
374. Horner, L., H. Hoffmann, G. Klarhe, V. G. Toscano, and H. Ertel, Chem. Ber., 94, 1987 (1961).
375. Horner, L., H. Hoffmann, W. Klink, H. Ertel, and V. G. Toscano, Chem. Ber., 95, 581 (1962).

376. Horner, L., H. Hoffmann, and V. G. Toscano, Chem. Ber., 95, 536 (1962).
377. Horner, L., H. Hoffmann, and H. G. Wippel, Chem. Ber., 91, 61 (1958).
378. Horner, L., H. Hoffmann, and H. G. Wippel, Chem. Ber., 91, 64 (1958).
379. Horner, L., H. Hoffmann, and H. G. Wippel, Ger. Pat. 1,044,813 (1957); Chem. Zentr., 12057 (1959); C.A., 55, 3521 (1961).
380. Horner, L., H. Hoffmann, H. G. Wippel, and G. Hassel, Chem. Ber., 91, 52 (1958).
381. Horner, L., H. Hoffmann, H. G. Wippel, and G. Klahre, Chem. Ber., 92, 2499 (1959).
382. Horner, L., and W. Jurgeleit, Ann. Chem., 591, 138 (1955).
383. Horner, L., and E. Jurgens, Chem. Ber., 90, 2184 (1957).
384. Horner, L., and W. Klink, Tetrahedron Letters, 1964, 2467.
385. Horner, L., W. Klink, and H. Hoffmann, Chem. Ber., 96, 3133 (1963).
386. Horner, L., and H. Moser, Chem. Ber., 99, 2789 (1966).
387. Horner, L., and H. Neumann, Chem. Ber., 102, 3953 (1969).
388. Horner, L., and H. Nickel, Ann. Chem., 597, 20 (1955).
389. Horner, L., H. Oediger, and H. Hoffmann, Ann. Chem., 626, 26 (1959).
390. Horner, L., and H. Winkler, Tetrahedron Letters, 1964, 3265, 3271, 3275.
391. Horner, L., and H. Winkler, Ann. Chem., 685, 1 (1965).
392. Horner, L., H. Winkler, A. Rapp, A. Mentrup, H. Hoffmann, and P. Beck, Tetrahedron Letters, 1961, 161.
393. Horner, S. M., and S. Y. Tyree, Inorg. Chem., 1, 122 (1962).
394. Horner, S. M., S. Y. Tyree, and D. L. Venezky, Inorg. Chem., 1, 844 (1962).
395. Houston, D. F., J. Am. Chem. Soc., 68, 914 (1946).
396. Howard, E., Jr., and W. F. Olszewski, J. Am. Chem. Soc., 81, 1483 (1959).
397. Howell, G. V., and R. L. Williams, J. Chem. Soc., A, 1968, 117.
309. Huber, F., and M. S. A. El-Meliby, Angew. Chem., Intern. Ed., 7, 946 (1968).
399. Hudson, R. F., Pure Appl. Chem., 9, 371 (1964).
400. Hudson, R. F., Structure and Mechanism in Organo-Phosphorus Chemistry, Vol. VI, A. T. Blomquist, Ed., Academic Press, New York, 1965.
401. Reference 400, p. 67.
402. Reference 400, p. 68.
402a. Hughes, A. N., and M. Davies, Chem. Ind. (London),

1969, 138.
403. Hunter, S. H., V. M. Langford, G. A. Rodley, and C. J. Wilkins, J. Chem. Soc., A, 1968, 305.
404. Issleib, K., and R. D. Bleck, Z. Anorg. Allgem. Chem., 336, 234 (1965).
405. Issleib, K., and G. Bohn, Z. Anorg. Allgem. Chem., 301, 188 (1959).
406. Issleib, K., and A. Brack, Z. Anorg. Allgem. Chem., 277, 258 (1954).
407. Issleib, K., and A. Brack, Z. Anorg. Allgem. Chem., 292, 245 (1957).
408. Issleib, K., and G. Grams, Z. Anorg. Allgem. Chem., 299, 58 (1959).
409. Issleib, K., and K. Jasche, Chem. Ber., 100, 412 (1967).
410. Issleib, K., and Kr. Krech, Z. Anorg. Allgem. Chem., 328, 69 (1964).
411. Issleib, K., and F. Krech, Chem. Ber., 94, 2656 (1961).
412. Issleib, K., and O. Low, Z. Anorg. Allgem. Chem., 346, 241 (1966).
413. Issleib, K., and D. W. Muller, Chem. Ber., 92, 3175 (1959).
414. Issleib, K., and B. Mitscherling, Z. Anorg. Allgem. Chem., 304, 73 (1960).
415. Issleib, K., and H. M. Mobius, Chem. Ber., 94, 102 (1961).
416. Issleib, K., and D. W. Muller, Chem. Ber., 92, 3175 (1959).
417. Isleib, K., and R. Rieschel, Chem. Ber., 98, 2086 (1965).
418. Issleib, K., and H. R. Roloff, Chem. Ber., 98, 2091 (1965).
419. Issleib, K., and W. Seidel, Z. Anorg. Allgem. Chem., 288, 201 (1956).
420. Issleib, K., and G. Thomas, Chem. Ber., 94, 2244 (1961).
421. Issleib, K., and A. Tzschach, Chem. Ber., 92, 1118 (1959).
422. Issleib, K., A. Tzschach, and H. O. Fröhlich, Z. Anorg. Allgem. Chem., 298, 164 (1959).
423. Jackson, I. K., W. C. Davies, and W. J. Jones, J. Chem. Soc., 1931, 2109.
424. Jackson, I. K., and W. J. Jones, J. Chem. Soc., 1931, 575.
425. Jaffe, H. H., J. Chem. Phys., 22, 1430 (1954).
426. Jaffe, H. H., J. Phys. Chem., 58, 185 (1954).
427. Jaffe, H. H., and L. D. Freedman, J. Am. Chem. Soc., 74, 1069 (1952).
428. Reference 837, p. 2855.
429. Reference 838a, p. 275.

430. Reference 838b, p. 1538.
431. Jenkins, J. M., and B. L. Shaw, J. Chem. Soc., A, 1966, 770.
432. Joesten, M. D., and R. S. Drago, J. Am. Chem. Soc., 84, 3817 (1962).
433. Johnson, A. W., Ylid Chemistry, Vol. VII, A. T. Blomquist, Ed., Academic Press, New York, 1966, Chapter 5.
434. Reference 433, p. 202.
435. Reference 433, p. 197.
436. Reference 433, pp. 193-203.
437. Johnson, A. W., Chem. Ind. (London), 504 (1964).
438. Johnson, A. W., and H. L. Jones, J. Am. Chem. Soc., 90, 5232 (1968).
439. Jones, W. J., W. C. Davies, S. T. Bowden, C. Edwards, V. E. Davis, and L. H. Thomas, J. Chem. Soc., 1947, 1446.
440. Jones, R. A. Y., and A. R. Katritzky, Angew. Chem., Intern. Ed., 1, 32 (1962).
441. Kabachnik, M. I., Tetrahedron, 20, 655 (1964).
442. Kabachnik, M. I., C. Yu Chang, and E. N. Tsvetkov, Dokl Akad. Nauk SSSR, 135, 603 (1960); C.A., 55, 12272a (1961).
443. Kabachnik, M. I., T. A. Mastryukova, and A. E. Shipov, Zh. Obshch. Khim., 35 (9), 1574 (1965).
444. Kabachnik, M. I., and T. Ya. Medved, Izv. Akad. Nauk SSSR, Otd. Khim. Nauk, 1962, 2103; C.A., 58, 9127f (1963).
445. Kabachnik, M. I., T. Ya. Medved, and E. I. Matrosov, Dokl. Akad. Nauk SSSR, 162, 339 (1965); C.A., 63, 5674b (1965).
446. Kabachnik, M. I., T. Ya. Medved, and Yu. M. Polikarpov, Izv. Akad. Nauk SSSR, Ser. Khim., 1966, 368; C.A., 64, 15917 (1966).
447. Kabachnik, M. I., T. Ya. Medved, and Yu. M. Polikarpov, Dokl. Akad. Nauk SSSR, 135, 849 (1960); C.A., 55, 14288 (1961).
448. Kabachnik, M. I., T. Ya. Medved, Yu. M. Polikarpov, and K. S. Yudina, Izv. Akad. Nauk SSSR, Otd. Khim. Nauk, 1961, 2029; C.A., 56, 11609d (1962).
449. Kabachnik, M. I., T. Ya. Medved, Yu. M. Polikarpov, and K. S. Yudina, Izv. Akad. Nauk SSSR, Otd. Khim. Nauk, 1962, 1584; C.A., 58, 5720c (1963).
450. Kabachnik, M. I., T. Ya. Medved, Yu. M. Polikarpov, and K. S. Yudina, Izv. Akad. Nauk SSSR, Ser. Khim., 1967, 591; C.A., 68, 39743 (1968).
451. Kabachnik, M. I., and E. S. Shepeleva, Izv. Akad. Nauk SSSR, Otd. Khim. Nauk, 1953, 862; C.A., 49, 843f (1955).
452. Kabachnik, M. I., and E. N. Tsvetkov, Dokl. Akad. Nauk SSSR, 143, 592 (1962); C.A., 4694d (1962).

453. Kabachnik, M. I., and E. N. Tsvetkov, Izv. Akad. Nauk SSSR, Ser. Khim., 1963, 1227; C.A., 59, 12839h (1963).
454. Kabachnik, M. I., E. N. Tsvetkov, and Chung-Yu Chang, Zh. Obshch. Khim., 32, 3340 (1962); C.A., 58, 9120c (1963).
455. Kabachnik, M. I., E. N. Tsvetkov, and R. A. Malevannaya, U.S.S.R. Pat. 170,972 (1965); C.A., 63, 9991 (1965).
456. Kabachnik, M. I., E. N. Tsvetkov, and C. C. Yu, Tetrahedron Letters, 5 (1962).
457. Kabachnik, M. I., V. V. Voevodskii, T. A. Mastryukova, S. P. Solodovnikov, and T. A. Melent'eva, Zh. Obshch. Khim., 34 (10), 3234 (1964); C.A., 62, 3906f (1965).
458. Kamai, G., Dokl. Akad. Nauk SSSR, 66, 389 (1949); C.A., 44, 127 (1950).
459. Karayannis, N. M., C. M. Mikulski, L. L. Pytlewski, and M. M. Labes, Inorg. Chem., 9, 582 (1970).
460. Katritzky, A. R., and B. Ternai, J. Chem. Soc., B, 1966, 631.
461. Kemp, R. H., W. A. Thomas, M. Gordon, and C. E. Griffin, J. Chem. Soc., B, 1969, 527.
462. Kennedy, J., E. S. Lane, and B. K. Robinson, J. Appl. Chem. (London), 8, 459 (1958).
463. Kennedy, J., E. S. Lane, and J. L. Willans, J. Chem. Soc., 1956, 4670.
464. Keough, P. T., and M. Grayson, J. Org. Chem., 27, 1817 (1962).
465. Kepert, D. L., and R. Mandyczewsky, J. Chem. Soc., A. 1968, 530.
466. Kharrasova, F. M., and G. Kamai, Zh. Obshch. Khim., 34 (7), 2195 (1964); C.A., 61, 10705f (1964).
467. Kharrasova, F. M., G. Kamai, and R. R. Shagidullin, Zh. Obshch. Khim., 36, 1987 (1966); C.A., 66, 65592 (1967).
468. Kharrasova, F. M., G. Kh. Kamai, R. B. Sultanova, and R. R. Shagidullin, Zh. Obshch. Khim., 37, 687 (1967); C.A., 67, 53437n (1967).
469. King, J. P., B. P. Block, and I. C. Popoff, Inorg. Chem., 4, 198 (1965).
469a. Kingsbury, C. A., and D. J. Cram, J. Am. Chem. Soc., 82, 1810 (1960).
470. Kirby, A. J., and S. G. Warren, The Organic Chemistry of Phosphorus, Vol. V, E. Eaborn and N. B. Chapman, Eds., Elsevier, New York, 1967.
471. Reference 470, p. 250.
472. Kirk, N., Ind. Eng. Chem., 51, 515 (1959).
473. Kirsanov, A. V., L. P. Zhuravleva, M. I. Zóla, G. A. Butova, and M. G. Suleimanova, Zh. Obshch. Khim., 37, 510 (1967); C.A., 67, 82253 (1967).
474. Klamann, D., and P. Weyerstahl, Angew. Chem., 75, 89 (1963).

475. Koenigs, E., and H. Friedrich, Ann., 509, 138 (1934).
476. Korpium, O., R. A. Lewis, J. Chickos, and K. Mislow, J. Am. Chem. Soc., 90, 4842 (1968).
477. Korpium, O., and K. Mislow, J. Am. Chem. Soc., 89, 4784 (1967).
478. Korshak, V. V., J. Polymer Sci., 31, 319 (1958); C.A., 53, 1816d (1959).
479. Korshak, V. V., T. M. Frunze, V. V. Kurashev, T. Ya. Medved, Yu. M. Polikarpov, Ch'ing-Mei Hu, and M. I. Kabachnik, Sintezy i Svoistva Monomerov, Akad. Nauk SSSR, Inst. Neftekhim. Sinteza, Sb. Rabot 12-oi [Dvenadtsatoi] Konf. po Vysokomolekul. Soedin 63 (1962); C.A., 62, 6507d (1965).
480. Kosolapoff, G. M., Organophosphours Compounds, Wiley, New York, 1950.
481. Kosolapoff, G. M., J. Am. Chem. Soc., 64, 2982 (1942).
482. Kosolapoff, G. M., J. Am. Chem. Soc., 72, 5508 (1950).
483. Kosolapoff, G. M., J. Chem. Soc., 1965, 6638.
484. Kosolapoff, G. M., and R. F. Struck, J. Chem. Soc., 1961, 2423.
485. Kosolapoff, G. M., and A. D. Brown, Jr., J. Chem. Soc., C, 1967, 1789.
486. Kosolapoff, G. M., and R. F. Struck, J. Chem. Soc., 1959, 3950.
487. Kosolapoff, G. M., and R. F. Struck, Proc. Chem. Soc., 1960, 351.
488. Kosolapoff, G. M., and R. F. Struck, J. Chem. Soc., 1961, 2423.
489. Köster, R., and Y. Morita, Angew. Chem., 77, 589 (1965).
490. Kreutzkamp, N., Naturwissenschaften, 48, 620 (1961).
491. Kreutzkamp, D. N., E. Schmidt-Samoa, and K. Herberg, Angew. Chem., Intern. Ed., 4, 1078 (1965).
492. Kreutzkamp, N., and K. Storck, Naturwissenshaften, 47, 497 (1960).
493. Kuchen, W., and H. Buchwald, Chem. Ber., 91, 2871 (1958).
494. Kuchen, W., and H. Buchwald, Chem. Ber., 92, 227 (1959).
495. Kumada, M., and K. Noda, Mem. Fac. Eng. Osaka City Univ., 4, 173 (1962); C.A., 59, 8782d (1963).
496. Kumli, K. F., W. E. McEwen, and C. A. Vander Werf, J. Am. Chem. Soc., 81, 3805 (1959).
497. Lamza, L., J. Prakt. Chem., 25, 294 (1964); C.A., 61, 16089b (1964).
498. Laughlin, R. G., J. Org. Chem., 30, 1322 (1965).
499. Leavitt, F. C., T. A. Manuel, F. Johnson, L. U. Matternas, and D. S. Lehman, J. Am. Chem. Soc., 82, 5099 (1960).
500. Letcher, J. H., and J. R. Van Wazer, J. Chem. Phys., 45, 2916 (1966).

501. Letts, E. A., and R. F. Blake, J. Chem. Soc., 58, 766 (1890).
502. Letts, E. A., and N. Collie, J. Chem. Soc., 42, 724 (1882).
503. Letts, E. A., and N. Collie, Proc. Roy. Soc. Edinburgh, 11, 46 (1881).
504. Levchenko, E. S., Yu. V. Piven, and A. V. Kirsanov, Zh. Obshch. Khim., 30, 1976 (1960); C.A., 55, 6418 (1961).
505. Levy, J. B., G. O. Doak, and L. D. Freedman, J. Org. Chem., 30, 660 (1965).
506. Lewis, R. A., O. Korpium, and K. Mislow, J. Am. Chem. Soc., 89, 4786 (1967).
507. Lewis, R. A., and K. Mislow, J. Am. Chem. Soc., 91, 7009 (1969).
508. Lewis, J., and R. Whyman, J. Chem. Soc., 1965, 6027.
509. Lindner, E., and H. Kranz, Z. Naturforsch., B, 22, 675 (1967); C.A., 67, 116925 (1967).
510. Lindner, E., R. Lehner, and H. Scheer, Chem. Ber., 100, 1331 (1967).
511. Lindqvist, I., and G. Olofsson, Acta Chem. Scand., 13, 1753 (1959).
512. Lindquist, I., and M. Zackrisson, Acta Chem. Scand., 14, 453 (1960).
513. Lipp, A., and W. Hieber, Chem. Ber., 92, 2085 (1959).
514. Litthauer, S., Chem. Ber., 22, 2144 (1889).
515. Lock, C. J. L., and G. Wilkinson, Chem. Ind. (London), 1962, 40.
516. Longhi, R., R. O. Ragsdale, and R. S. Drago, Inorg. Chem., 1, 768 (1962).
517. Lucken, E. A. C., and M. A. Whitehead, J. Chem. Soc., 1961, 2459.
518. Lutton, E. S., J. Am. Oil Chemists' Soc., 43, 28 (1966).
519. Maciel, G. E., and R. V. James, Inorg. Chem., 3, 1650 (1964).
520. Maercker, A., Angew. Chem., 79, 576 (1967).
521. Maercker, A., Angew. Chem., Intern. Ed., 6, 557 (1967).
522. Maier, L., Progr. Inorg. Chem., Vol. V., F. A. Cotton, Ed., Interscience, New York, 1936, p. 31.
523. Maier, L., Angew. Chem., Intern. Ed., 4 (6), 527 (1965).
523a. Maier, L., Chimia, 23, 323 (1969).
524. Maier, L., Angew. Chem., 77, 549 (1965).
525. Maier, L., Angew. Chem., Intern. Ed., 7, 384 (1968).
526. Maier, L., Angew. Chem., Intern. Ed., 7, 385 (1968).
527. Maier, L., Brit. Pat. 1,120,374 (1968); C.A., 69, 77492u (1968).
528. Maier, L., Chem. Ber., 94, 3043 (1961).
529. Maier, L., Helv. Chim. Acta, 47 (8), 2129 (1964).

530. Maier, L., Helv. Chim. Acta, 48, 1034 (1965).
531. Maier, L., Helv. Chim. Acta, 49, 842 (1966).
532. Maier, L., Helv. Chim. Acta, 49, 1249 (1966).
533. Maier, L., Helv. Chim. Acta, 49 (8), 2458 (1966).
534. Maier, L., Helv. Chim. Acta, 50, 1723 (1967).
535. Maier, L., Helv. Chim. Acta, 51, 405 (1968).
536. Maier, L., Helv. Chim. Acta, 52, 845 (1969).
537. Maier, L., R. Gredig, A. Hauser, and R. Battagello, Helv. Chim. Acta, 52, 858 (1969).
538. Majumdar, A. K., and R. G. Bhattacharyya, J. Inorg. Nucl. Chem., 29, 2359 (1967).
539. Mallion, K. B., F. G. Mann, B. P. Tong, and V. P. Wystrach, J. Chem. Soc., 1963, 1327.
540. Mann, F. G., and E. J. Chaplin, J. Chem. Soc., 1937, 527.
541. Mann, F. G., I. T. Millar, and H. R. Watson, J. Chem. Soc., 1958, 2516.
542. Mann, F. G., and D. Purdie, J. Chem. Soc., 1935, 1549.
543. Mann, F. G., and J. Watson, J. Org. Chem., 13, 502 (1948).
544. Mark, V., Mechanisms of Molecular Migrations, Vol. II, B. S. Thyagarajan, Ed., Interscience, New York, 1969, pp. 319-437.
545. Mark, V., Tetrahedron Letters, 1962, 281.
546. Mark, V., U.S. Pat. 3,197,497 (1965).
547. Märkl, G., Angew. Chem., 75, 168 (1963).
548. Märkl, G., F. Lieb, and A. Merz, Angew. Chem., Intern. Ed., 6, 87 (1967).
549. Märkl, G., and H. Olbrich, Angew. Chem., 78, 598 (1966).
550. Marmor, R. S., and D. Seyferth, J. Org. Chem., 34, 748 (1969).
551. Marsi, K. L., and G. D. Homer, J. Org. Chem., 28, 2150 (1963).
552. Martin, K. R., and C. E. Griffin, J. Heterocycl. Chem., 3, 92 (1966).
553. Martin, D. J., and C. E. Griffin, J. Org. Chem., 30, 4034 (1965).
554. Martz, M. D., and L. D. Quin, J. Org. Chem., 34, 3195 (1969).
555. Maruszewska-Wieczorkowska, E., and J. Michalski, Rocz. Chem., 38 (4), 625 (1964); C.A., 61, 10702h (1964).
556. Masson, O., and J. B. Kirkland, J. Chem. Soc., 55, 135 (1889).
557. Mastalerz, P., and Z. E. Golubski, Rocz. Chem., 39, (6), 951 (1965); C.A., 64, 3591b (1966).
558. Mathew, M., and G. J. Palenik, Can. J. Chem., 47, 1093 (1969).
559. Mavel, G., Progress in NMR Spectroscopy, Vol. I,

J. W. Emsley, J. Feeney, and L. H. Sutcliff, Eds., Pergamon Press, New York, 1966, Chapter 4, pp. 251-373.
560. McCormack, W. B., U.S. Pat. 2,663,736 (1953); C.A., 49, 7602c (1955).
561. McCormack, W. B., U.S. Pat. 2,663,737 (1953); C.A., 49, 7601a (1955).
562. McCormack, W. B., U.S. Pat. 2,663,738 (1953); C.A., 49, 7602d (1955).
563. McCormack, W. B., U.S. Pat. 2,663,739 (1953); C.A., 49, 7602f (1955).
564. McCormack, W. B., Organic Syntheses, Vol. 43, B. C. McKusick, Ed., Wiley, New York, 1963, p. 73.
565. McEwen, W. E., Topics in Phosphorus Chemistry, Vol. II, M. Grayson and E. J. Griffith, Eds., Interscience, New York, 1965, pp. 5-9.
566. McEwen, W. E., K. F. Kumli, A. Blade-Font, M. Zanger, and C. A. Vander Werf, J. Am. Chem. Soc., 86, 2378 (1964).
567. McEwen, W. E., C. Vander Werf, and C. B. Pariset, J. Am. Chem. Soc., 82, 5503 (1960).
568. McIvor, R. A., G. A. Grant, and C. E. Hubley, Can. J. Chem., 34, 1611 (1956).
569. Medved, T. Ya., T. M. Frunze, Ch'ing-Mei Hu, V. V. Kurashev, V. V. Korshak, and M. I. Kabachnik, Vysokomol. Soedin., 5 (9), 1309 (1963); C.A., 59, 15391g (1963).
570. Medved, T. Ya., Yu. M. Polikarpov, S. A. Pisareva, and M. I. Kabachnik, Izv. Akad. Nauk SSSR, Ser. Khim., 1968, 1417; C.A. 69, 87103 (1968).
571. Medved, T. Ya., Yu. M. Polikarpov, S. A. Pisareva, E. I. Matrosov, and M. I. Kabachnik, Izv. Akad. Nauk SSSR, Ser. Khim., 1968, 2062; C.A., 70, 37884 (1969).
572. Medved, T. Ya., Yu. M. Polikarpov, and K. S. Yudina, Izv. Akad. Nauk SSSR, Ser. Khim., 1967, 591; C.A., 68, 39743 (1968).
573. Meisenheimer, J., J. Casper, M. Höring, W. Lauter, L. Lichtenstadt, and W. Samuel, Ann. Chem., 449, 213 (1926).
574. Meisenheimer, J., and L. Lichtenstadt, Chem. Ber., 44, 356 (1911).
575. Melby, L. R., N. J. Rose, E. Abramson, and J. C. Caris, J. Am. Chem. Soc., 86, 5117 (1964).
576. Michaelis, A., Ann. Chem., 181, 265 (1876).
577. Michaelis, A., Ann. Chem., 181, 354 (1897).
578. Michaelis, A., Ann. Chem., 293, 193 (1897).
579. Michaelis, A., Ann. Chem., 293, 261 (1896).
580. Michaelis, A., Ann. Chem., 315, 43 (1901).
581. Michaelis, A., Ann. Chem., 315, 75 (1901).
582. Michaelis, A., and J. Ananoff, Chem. Ber., 8, 493 (1875).

583. Michaelis, A., and L. Gleichmann, Chem. Ber., 15, 801 (1882).
584. Michaelis, A., and W. LaCoste, Chem. Ber., 18, 2109 (1885).
585. Michaelis, A., and A. Link, Ann. Chem., 207, 193 (1881).
586. Michaelis, A., and H. von Sonden, Ann. Chem., 229, 295 (1885).
587. Michaelis, A., and A. Schenk, Ann. Chem., 260, 1 (1890).
588. Michaelis, A., and H. von Soden, Ann. Chem., 229, 299, (1885).
589. Michaelis, A., and H. von Soden, Ann. Chem., 229, 295, 315 (1885).
590. Mikhailov, B. M., and N. F. Kucherova, Zh. Obshch. Khim., 22, 792 (1952); C.A., 47, 5388 (1953).
591. Mikhailov, B. M., and N. F. Kucherova, Dokl. Akad. Nauk SSSR, 74, 501 (1950); C.A., 45, 3343c (1951).
592. Miller, J. A., Tetrahedron Letters, 1969, 4335.
593. Miller, N. E., Inorg. Chem., 8, 1693 (1969).
594. Miller, R. C., J. S. Bradley, and L. A. Hamilton, J. Am. Chem. Soc., 78, 5299 (1956).
595. Miller, R. C., C. D. Miller, W. Rogers, Jr., and L. A. Hamilton, J. Am. Chem. Soc., 79, 424 (1957).
596. Miller, R. C., C. D. Miller, and W. Rogers, Jr., J. Am. Chem. Soc., 80, 1562 (1958).
597. Mingoia, Q., Gazz. Chim. Ital., 62, 333 (1932); C.A., 26, 4813 (1932).
598. Mitchell, K. A. R., Can. J. Chem., 46, 3499 (1968).
599. Mitchell, K. A. R., Chem. Rev., 69, 157 (1969).
600. Mixon, A. L., and W. R. Gilkerson, J. Am. Chem. Soc., 89, 6410 (1967).
601. Moedritzer, K., L. Maier, and L. C. D. Groenweghe, J. Chem. Eng. Data, 7, 307 (1962).
602. Monagle, J. J., J. Org. Chem., 27, 3851 (1962).
603. Monagle, J. J., J. V. Mengenhauer, D. A. Jones, Jr., J. Org. Chem., 32, 2477 (1967).
604. Morgan, P. W., U.S. Pat. 2,646,420 (1953); C.A., 47, 10276g (1954).
605. Morgan, P. W., and B. C. Herr, J. Am. Chem. Soc., 74, 4526 (1952).
606. Morris, R. C., and J. L. Van Winkle, U.S. Pat. 2,642,461 (1953); C.A., 48, 8814a (1954).
607. Morrison, D. C., J. Am. Chem. Soc., 72, 4820 (1950).
608. Mortimer, C. T., Pure Appl. Chem., 2, 71 (1961).
609. Mrochek, J. E., and C. V. Banks, J. Inorg. Nucl. Chem., 27, 589 (1965).
610. Mrochek, J. E., J. W. O'Laughlin, and C. V. Banks, J. Inorg. Nucl. Chem., 27, 603 (1965).
611. Mrochek, M. E., J. W. O'Laughlin, H. Sakurai, and C. V. Banks, J. Inorg. Nucl. Chem., 25, 955 (1963).

612. Mrochek, J. E., J. J. Richard, and C. V. Banks, J. Inorg. Nucl. Chem., 27, 625 (1965).
613. Mukaiyama, T., O. Mitsunobu, and T. Obata, J. Org. Chem., 30, 101 (1965).
614. Muller, N., P. C. Lauterbur, and J. Goldenson, J. Am. Chem. Soc., 78, 3557 (1956).
615. Muller, N., and J. Goldenson, J. Am. Chem. Soc., 78, 5182 (1956).
616. Muller, E., and H. G. Padeken, Chem. Ber., 100, 521 (1967).
617. Naumann, K., G. Zon, and K. Mislow, J. Am. Chem. Soc., 91, 7012 (1969).
618. Nesmeyanov, A. N., D. N. Kursanov, V. D. Vil'chevskaya, N. S. Kochetkova, V. N. Setkina, and Yu. N. Novikov, Dokl. Akad. Nauk SSSR, 160, 1090 (1965); C.A., 62, 14725d (1965).
619. Nesmeyanov, A. N., V. D. Vil'chevskaya, N. S. Kochetkova, and N. P. Palitsyn, Izv. Akad. Nauk SSSR, Ser. Khim., 1963, 2051; C.A., 60, 5548g (1964).
620. Neunhoeffer, O., and L. Lamza, Chem. Ber., 94, 2519 (1961).
621. Neuse, E. W., J. Organometal. Chem., 7, 349 (1967).
622. Nishemura, S., C. H. Ke, and N. C. Li, J. Phys. Chem., 72, 1297 (1968).
623. Nylen, P., Z. Anorg. Allgem. Chem., 246, 227 (1941).
624. O'Laughlin, J. W., J. J. Richard, J. W. Ferguson, and C. V. Banks, Anal. Chem., 40, 146 (1968).
625. Packer, K. J., J. Chem. Soc., 1963, 960.
626. Paddock, N. L., Structure and Reactions in Phosphorus Chemistry, The Royal Institute of Chemistry, Lecture Series 1962, No. 2, London, 1962, p. 32.
627. Pagilagan, R. U., and W. E. McEwen, Chem. Commun., 1966, 652.
628. Parker, J. R., and C. V. Banks, Anal. Chem., 36, 2191 (1964).
629. Parker, J. R., and C. V. Banks, J. Inorg. Nucl. Chem., 27, 583 (1965).
630. Paul, R. C., J. Chem. Soc., 1955, 574.
631. Pauling, L., The Nature of the Chemical Bond and the Structure of Molecules and Crystals, 3rd Ed., Cornell University Press, Ithaca, N.Y., 1960, p. 110.
632. Peach, M. E., and T. C. Waddington, J. Chem. Soc., 1962, 3450.
633. Pebal, L., Ann. Chem., 120, 194 (1861); and 118, 22 (1861).
634. Pellon, J., and W. G. Carpenter, J. Polymer Sci., Part A-1, 863 (1963).
635. Peterson, D. J., U.S. Pat. 3,397,039 (1968).
636. Peterson, D. J., J. Organometal. Chem., 8, 199 (1967).
637. Peterson, D. J., and H. R. Hays, J. Org. Chem., 30, 1939 (1965).

638. Peterson, D. J., and T. J. Logan, J. Inorg. Nucl. Chem., 28, 53 (1966).
639. Petrov, K. A., N. K. Bliznyuk, and V. P. Korotkova, Zh. Obshch. Khim., 30, 2995 (1960); C.A., 55, 18561 (1961).
640. Petrov, K. A., N. K. Bliznyuk, and T. N. Lysenko, Zh. Obshch. Khim., 30, 1964 (1960); C.A., 55, 6362g (1961).
641. Petrov, K. A., E. E. Nifant'ev, L. V. Khorkhoyanu, and A. I. Trushkov, Zh. Obshch. Khim., 31, 3085 (1961); C.A., 57, 858i (1962).
642. Petrov, K. A., and V. A. Parshina, Zh. Obshch. Khim., 31, 3417 (1961); C.A., 57, 4692i (1962).
643. Petrov, K. A., V. A. Parshina, and G. L. Daruze, Zh. Obshch. Khim., 30, 3000 (1960); C.A., 55, 23399h (1961).
644. Petrov, K. A., and V. A. Parshina, Khim. Prom., 4, 686 (1959); C.A., 54, 8714a (1960).
645. Petrov, K. A., V. A. Parshina, and V. A. Gaidamak, Zh. Obshch. Khim., 31, 3411 (1961); C.A., 57, 4692e (1962).
646. Petrov, K. A., V. A. Parshina, and A. F. Manuilov, Zh. Obshch. Khim., 35, 1602 (1965); C.A., 63, 18145c (1965).
647. Petrov, K. A., V. A. Parshina, and A. F. Manuilov, Zh. Obshch. Khim., 35, 2062 (1965); C.A., 64, 6681h (1966).
648. Petzold, G., and H. G. Henning, Naturwissenschaften, 54, 469 (1967).
649. Pickard, R. H., and J. Kenyon, J. Chem. Soc., 89, 262 (1906).
650. Pirkle, W. H., S. D. Beare, and R. L. Muntz, J. Am. Chem. Soc., 91, 4575 (1969).
651. Plets, V. M., Dissertation, Kazan, 1938.
652. Plymale, D. L., J. Inorg. Nucl. Chem., 31, 236 (1969).
653. Popoff, I. C., L. K. Huber, B. P. Block, P. D. Morton, and R. P. Riordan, J. Org. Chem., 28, 2898 (1963).
654. Popov, E. M., E. N. Tsvetkov, Yung-Yu Chang, and T. Ya. Medved, Zh. Obshch. Khim., 32, 3255 (1962); C.A., 58, 5165c (1963).
655. Poznyak, R. L., and A. I. Razumov, Zh. Obshch. Khim., 37, 424 (1967); C.A., 67, 100209 (1967).
656. Price, C. C., T. Parasaran, and T. V. Lakshminarayan, J. Am. Chem. Soc., 88, 1034 (1966).
657. Priestly, H. M., J. Chem. Eng. Data, 1967, 618.
658. Prikoszovich, W., and H. Schindlbauer, Chem. Ber., 102, 2922 (1969).
659. Pudovik, A. N., I. M. Aladzheva, and L. V. Spirina, Zh. Obshch. Khim., 37, 700 (1967); C.A., 67, 21972 (1967).
660. Pudovik, A. N., and E. S. Batyeva, Izv. Akad. Nauk

SSSR, Ser., Khim., 1968, 2391; C.A., 71, 3438 (1968).
661. Pudovik, A. N., A. A. Muratova, and E. P. Semkina, Zh. Obshch. Khim., 33, 3350 (1963); C.A., 60, 4175d (1964).
662. Quin, L. D., 1,4-Cycloaddition Reactions: The Diels-Alder Reaction in Heterocyclic Syntheses, Vol. VIII, J. Hamer, Ed., Academic Press, New York, 1967, Chapter 3.
663. Reference 662, p. 47.
664. Quin, L. D., and H. G. Anderson, J. Org. Chem., 29, 1859 (1964).
665. Quin, L. D., J. A. Peters, C. E. Griffin, and M. Gordon, Tetrahedron Letters, 1964, 3689.
666. Rabinowitz, R., R. Marcus, and J. Pellon, J. Polymer Sci., Part A-2, 1245 (1964).
667. Rabinowitz, R., R. Marcus, and J. Pellon, J. Polymer Sci., Part A-2, 1233 (1964).
668. Rabinowitz, R., R. Marcus, and J. Pellon, J. Polymer Sci., Part A-2, 1241 (1964).
669. Rabinowitz, R., and J. Pellon, J. Org. Chem., 26, 4623 (1961).
670. Rakshys, J. W., R. W. Taft, and W. A. Sheppard, J. Am. Chem. Soc., 90, 5236 (1968).
671. Ramirez, F., and S. Dershowitz, J. Org. Chem., 24, 704 (1959).
672. Ramirez, F., O. P. Madan, and C. P. Smith, J. Am. Chem. Soc., 86, 5339 (1964).
673. Ramirez, F., O. P. Madan, and C. P. Smith, Tetrahedron, 22, 567 (1966).
674. Rao, G. N., and N. C. Li, J. Inorg. Nucl. Chem., 28, 2931 (1966).
675. Rao, G. N., and N. C. Li, Can. J. Chem., 44, 2775 (1966).
676. Rauhut, M. M., Topics in Phosphorus Chemistry, Vol. I, M. Grayson and E. J. Griffin, Eds., Interscience, New York, 1964, pp. 1-16.
677. Rauhut, M. M., R. Bernheimer, and A. M. Semsel, J. Org. Chem., 28, 478 (1963).
678. Rauhut, M. M., and H. A. Currier, J. Org. Chem., 26, 4626 (1961).
679. Rauhut, M. M., and H. A. Currier, J. Org. Chem., 26, 4628 (1961).
680. Rauhut, M. M., I. Hechenbleikner, H. A. Currier, and V. P. Wystrach, J. Am. Chem. Soc., 80, 6690 (1958).
681. Rauhut, M. M., I. Hechenbleikner, H. A. Currier, F. C. Schaefer, and V. P. Wystrach, J. Am. Chem. Soc., 81, 1103 (1959).
682. Rauhut, M. M., and A. M. Semsel, J. Org. Chem., 28, 473 (1963).
683. Rave, T. W., and H. R. Hays, J. Org. Chem., 31, 2894 (1966).

684. Razumov, A. I., Zh. Obshch. Khim., 29, 1635 (1959); C.A., 54, 8608b (1960).
685. Razumov, A. I., and N. N. Bankovskaya, Dokl. Akad. Nauk SSSR, 116, 241 (1957); C.A., 52, 6164i (1958).
686. Razumov, A. I., and N. G. Zabusova, Zh. Obshch. Khim., 32, 2688 (1962); C.A., 58, 9119b (1963).
687. Razumov, A. I., and N. G. Zabusova, Zh. Obshch. Khim., 32, 2691 (1962); C.A., 58, 9119 (1963).
688. Regitz, M., W. Anschütz, W. Bartz, and A. Liedhegener, Tetrahedron Letters, 1968, 3171.
689. Retcofsky, H. L., and C. E. Griffin, Tetrahedron Letters, 1966, 1975.
690. Reuter, M., Ger. Pat. 1,230,800 (1966); C.A., 66, 95192 (1967).
691. Reuter, M., and F. Jakob, Ger. Pat. 1,040,549 (1958); C.A., 55, 7289b (1961).
692. Richards, E. M., and J. C. Tebby, Chem. Commun., 1969, 495.
693. Rio, A., Fr. Pat. 1,321,431 (1963); C.A., 59, 10122c (1963).
694. Rio, A., Fr. Pat. 1,443,121 (1966).
695. Richard, J. J., and C. V. Banks, J. Org. Chem., 28, 123 (1963).
696. Richard, J. J., K. E. Burke, J. W. O'Laughlin, and C. V. Banks, J. Am. Chem. Soc., 83, 1722 (1961).
697. Richards, E. M., and J. C. Tebby, Chem. Commun., 1967, 957.
698. Richards, E. M., and J. C. Tebby, Chem. Commun., 1969, 494.
699. Ritt, P. E., and C. M. Kindley, U.S. Pat. 3,213,057 (1965).
700. Robins, R. K., and B. E. Christensen, J. Org. Chem., 16, 324 (1951).
701. Robinson, E. A., Can. J. Chem., 41, 3021 (1963).
702. Rothberg, I., and E. R. Thornton, J. Am. Chem. Soc., 86, 3296 (1964).
703. Rouschias, G., and G. Wilkinson, J. Chem. Soc., A, 1967, 993.
704. Rudolhh, F., Jenaer Jahrb, 1966, 221; C.A., 68, 114707 (1968).
705. Saeva, F. D., D. R. Rayner, and K. Mislow, J. Am. Chem. Soc., 90, 4176 (1968).
706. Sander, M., Chem. Ber., 93, 1220 (1960).
707. Sandhu, S. S., and S. S. Sandhu, J. Inorg. Nucl. Chem., 31, 1363 (1969).
708. Santhanam, K. S. V., and A. J. Bard, J. Am. Chem. Soc., 90, 1118 (1968).
709. Sartorelli, U., F. Canziani, and F. Zingales, Inorg. Chem., 5, 2233 (1966).
710. Sasse, K., Methoden Der Organischen Chemie, Organische Phosphorverbindungen, Vol. XII, Georg Thiem Verlag,

Stuttgart, 1963, Part 1, p. 127.
711. Saunders, M., and G. Burchman, Tetrahedron Letters, No. 1, 8 (1959).
712. Sauvage, R., Compt. Rend., 139, 674 (1904).
713. Savage, M. P., and S. Trippett, J. Chem. Soc., C, 1967, 1998.
714. Savage, M. P., and S. Trippett, J. Chem. Soc., C, 1968, 591.
715. Savage, M. P., and S. Trippett, J. Chem. Soc., C, 1966, 1842.
716. Scaife, D. E., Australian J. Chem., 20, 845 (1967).
717. Schiemenz, G. P., Chem. Ber., 98, 65 (1965).
718. Schiemenz, G. P., Chem. Ber., 99, 504 (1966).
719. Schiemenz, G. P., Chem. Ber., 99, 514 (1966).
720. Schiemenz, G. P., Angew. Chem., Intern. Ed., 7, 544 (1968).
721. Schiemenz, G. P., Angew. Chem., Intern. Ed., 7, 545 (1968).
722. Schiemenz, G. P., Angew. Chem., Intern. Ed., 6, 564 (1967).
723. Schiemenz, G. P., Tetrahedron Letters, 1964, 2729.
724. Schiemenz, G. P., Tetrahedron Letters, 1966, 3023.
725. Schiemenz, G. P., and H-U. Siebeneick, Chem. Ber., 102, 1883 (1969).
726. Schindlbauer, H., Chem. Ber., 100, 3432 (1967).
727. Schindlbauer, H., and W. Prikoszovich, Chem. Ber., 102, 2914 (1969).
728. Schindler, F., and H. Schmidbaur, Chem. Ber., 100, 3655 (1967).
729. Schindler, F., and H. Schmidbaur, Chem. Ber., 101, 1656 (1968).
730. Schindler, F., H. Schmidbaur, and G. Jonas, Chem. Ber., 98, 3345 (1965).
731. Schindler, F., H. Schmidbaur, and G. Jonas, Angew. Chem., 77, 170 (1965).
732. Schlosser, M., Chem. Ber., 97, 3219 (1964).
733. Schlosser, M., Angew. Chem., 74, 201 (1962).
734. Schmidt, D. D., and J. T. Yoke, 158th National Meeting of the American Chemical Society, New York, N.Y., Sept. 7-12, 1969, Inorg. Paper 203.
735. Schmidt, D. D., and J. T. Yoke, Inorg. Chem., 9, 1176 (1970).
736. Schmulbach, C. D., Inorg. Chem., 4, 1232 (1965).
737. Schönberg, A., and H. Krüll, Chem. Ber., 59, 1403 (1926).
738. Schönberg, A., and R. Michaelis, Chem. Ber., 69, 1080 (1936).
739. Schweizer, E. E., C. J. Berninger, and J. G. Thompson, J. Org. Chem., 33, 336 (1968).
740. Schweizer, E. E., W. S. Creasy, K. K. Light, and E. T. Shaffer, J. Org. Chem., 34, 212 (1969).

741. Schweizer, E. E., J. G. Thompson, and T. A. Ulrich, J. Org. Chem., 33, 3082 (1968).
742. Screttas, C., and A. F. Isbell, J. Org. Chem., 27, 2573 (1962).
743. Senear, A. E., W. Valient, and J. Wirth, J. Org. Chem., 25, 2001 (1960).
774. Setkina, V. N., T. Ya. Medved, and M. I. Kabachnik, Izv. Akad. Nauk SSSR, Ser. Khim., 1967, 1399; C.A., 68, 38643 (1968).
745. Sevin, A., and W. Chodkiewicz, Bull. Soc. Chim. France, 1969, 4016.
746. Sevin, A., and W. Chodkiewicz, Bull. Soc. Chim. France, 1969, 4023.
747. Seyferth, D., D. E. Welch, and J. K. Heeren, J. Am. Chem. Soc., 86, 1100 (1964).
748. Seyferth, D., D. E. Welch, and J. K. Heeren, J. Am. Chem. Soc., 85, 642 (1963).
749. Shapiro, I. O., E. N. Tsvetkov, A. I. Shatenshtein, and M. I. Kabachnik, Dokl. Akad. Nauk SSSR, 179, 888 (1968); C.A., 69, 77351 (1968).
750. Shaturskii, Ya. P., G. K. Fedorova, L. S. Moskalevskaya, Yu. S. Grushin, and A. V. Kirsanov, Zh. Obshch. Khim., 37, 2686 (1967); C.A. 69, 67490 (1968).
751. Shaw, R. A., B. W. Fitzsimmons, and B. C. Smith, Chem. Rev., 62, 247 (1962).
752. Sheldon, J. C., J. Chem. Soc., 1961, 750.
753. Sheldon, J. C., and S. Y. Tyree, J. Am. Chem. Soc., 80, 4775 (1958).
754. Shvets, A. A., O. A. Osipov, and E. L. Korol, Zh. Obshch. Khim., 38, 676 (1968); C.A., 69, 7923 (1968).
755. Shizunobu, H., and F. Isao, Kobunshi Kagaku, 25, 11 (1968); C.A., 70, 58346 (1969).
756. Siddall, T. H., III, and M. A. Davis, J. Chem. Eng. Data, 10, 303 (1965).
757. Siddall, T. H., III, and C. A. Prohaska, J. Am. Chem. Soc., 84, 3467 (1962).
758. Simonnin, M.-P., and B. Borecka, Bull. Soc. Chim. France, 1966, 3842.
759. Sinitsyna, S. M., and N. M. Sinitsyn, Dokl. Akad. Nauk SSSR, 164, 351 (1965); C.A., 64, 2997h (1966).
760. Sisler, H. H., H. S. Ahiya, and N. L. Smith, J. Org. Chem., 26, 1819 (1961).
761. Smalley, A. W., C. E. Sullivan, and W. E. McEwen, Chem. Commun., 1967, 5.
762. Smith, B. C., and G. H. Smith, J. Chem. Soc., 1965, 5516.
763. Smith, B. C., and M. A. Wassef, J. Chem. Soc., A, 1968, 1817.
763a. Smith, J. H., and S. Trippett, Chem. Commun., 1969, 855.
764. Sollott, G. P., and E. Howard, Jr., J. Org. Chem.,

27, 4034 (1962).
765. Sollott, G. P., H. E. Mertwoy, S. Portnoy, and J. L. Snead, J. Org. Chem., 28, 1090 (1963).
766. Sollott, G. P., and W. R. Peterson, Jr., J. Organometal. Chem., 4, 491 (1965).
767. Sollott, G. P., J. L. Snead, S. Portnoy, W. R. Peterson, Jr., and H. E. Mertwoy, U. S. Dept. Com. Office Tech. Serv. AD 611869, Vol. II, 441 (1965); C.A., 63, 18147b (1965).
768. Speziale, A. J., and R. C. Freeman, J. Am. Chem. Soc., 82, 903 (1960).
769. Stachlewska-Wroblowa, A., and K. Okon, Biul. Wojskowej Akad. Tech., 10, 14 (1961); C.A., 56, 14322g (1962).
770. Stachlewska-Wroblowa, A., and K. Okon, Bull. Acad. Polon. Sci., Ser. Sci. Chim., 9, 281 (1961); C.A., 60, 5545a (1964).
771. Staudinger, H., and E. Hauser, Helv. Chim. Acta, 4, 861 (1921).
772. Steininger, E., Chem. Ber., 96, 3184 (1963).
773. Steinkopf, W., and K. Buchheim, Chem. Ber., 54, 1024 (1921).
774. Stiles, A. R., F. F. Rust, and W. E. Vaughan, J. Am. Chem. Soc., 74, 3282 (1952).
774a. Struck, R. F., and Y. Fulmer, J. Med. Chem., 9, 414 (1966).
775. Stuebe, C., W. M. LeSuer, and G. R. Norman, J. Am. Chem. Soc., 77, 3526 (1955).
776. Sturtz, G., Bull. Soc. Chim. France, 1964, 2340.
777. Suga, K., and S. Watanabe, Koryo, No. 62, 44 (1961); C.A., 58, 551h (1963).
778. Taft, R. W., E. Price, I. R. Fox, I. C. Lewis, K. K. Andersen, and G. T. Davis, J. Am. Chem. Soc., 85, 3146 (1963).
779. Takahashi, S., K. Sonogashira, and N. Hagihara, Nippon Kagaku Zasshi, 87, 610 (1966); C.A., 65, 14485d (1966).
780. Temple, R. D., Y. Tsuno, and J. E. Leffler, J. Org. Chem., 28, 2495 (1963).
781. Thomas, L. C., and R. A. Chittenden, Spectrochim. Acta, 20, 467 (1964).
782. Thompson, Q. E., J. Am. Chem. Soc., 83, 845 (1961).
783. Trippett, S., J. Chem. Soc., 1961, 2813.
784. Trippett, S., Proc. Chem. Soc., 1963, 19.
785. Trippett, S., Pure Appl. Chem., 9, 255 (1964).
786. Trippett, S., R. S. Davidson, D. W. Hutchinson, R. Keat, R. A. Shaw, and J. C. Tebby, Organophosphorus Chemistry, Vol. I, The Chemical Society, England.
787. Trippett, S., and W. Hawes, Chem. Commun., 1968, 295.
788. Trippett, S., and B. J. Walker, J. Chem. Soc., C, 1966, 887.

789. Trippett, S., and B. J. Walker, Chem. Commun., 1965, 106.
790. Trippett, S., and D. M. Walker, J. Chem. Soc., 1960, 2976.
791. Trippett, S., B. J. Walker, and H. Hoffmann, J. Chem. Soc., 1965, 7140.
791a. Trostyanskaya, E. B., E. S. Venkova, and Yu. A. Mikhailin, Zh. Obshch. Khim., 37, 1655 (1967).
792. Tsetlin, B. L., T. Ya. Medved, Yu. G. Chikishev, Yu. M. Polikarpov, S. R. Rafikov, and M. I. Kabachnik, Vysokomol. Soedin., 3, 1117 (1961); C.A., 56, 2568h (1962).
793. Tsivunin, V. S., G. Kh. Kamai, and V. V. Kormachev, Zh. Obshch. Khim., 35, 1819 (1965); C.A., 64, 2122d (1966).
794. Tsivunin, V. S., G. Kh. Kamai, and V. V. Kormachev, Zh. Obshch. Khim., 35, 2190 (1965); C.A., 64, 11243e (1966).
795. Tsivunin, V. S., Gil'm Kamai, R. R. Shagidullin, and R. Sh. Khisamutdinova, Zh. Obshch. Khim., 35, 1811 (1965); C.A., 64, 3588g (1966).
796. Tsivunin, V. S., Gil'm Kamai, R. R. Shagidullin, and R. Sh. Khisamutdinova, Zh. Obshch. Khim., 35 (7), 1234 (1965).
797. Tsvetkov, E. N., D. I. Lobanov, and M. I. Kabachnik, Teor. Eksp. Khim., 1 (6), 729 (1965); C.A., 64, 12523f (1966).
797a. Tsvetkov, E. N., D. I. Lobanov, M. M. Makhamatkhanov, and M. I. Kabachnik, Tetrahedron, 25, 5623 (1969).
798. Uhlig, E., and M. Schafer, J. Inorg. Nucl. Chem., 30, 3109 (1968).
799. Unilever Ltd., Brit. Pat. 976,974 (1964); C.A., 62, 8014c (1965).
800. Urch, D. S., J. Chem. Soc., A, 1969, 3026.
801. Van Wazer, J. R., Phosphorus and Its Compounds. Vol. I; Chemistry, Interscience, New York, 1958, p. 536.
802. Van Wazer, J. R., J. Am. Chem. Soc., 78, 5709 (1956).
803. Van Wazer, J. R., C. F. Callis, J. N. Shoolery, and R. C. Jones, J. Am. Chem. Soc., 78, 5715 (1956).
804. Varshavskii, S. L., L. V. Kaabak, M. I. Kabachnik, and A. P. Tomilov, U.S.S.R. Pat. 222,382 (1968); C.A., 70, 4271g (1969).
805. Vinogradova, S. V., V. V. Korshak, G. S. Kolesnikov, and B. A. Zhubanov, Vysokomol. Soedin., 1, 357 (1959); C.A., 54, 7214g (1960).
806. Vinokurova, G. M., and S. G. Fattakhov, Zh. Obshch. Khim., 36, 67 (1966); C.A., 64, 14209g (1966).
807. Vizel, A. O., M. A. Zvereva, K. M. Ivanovskaya, I. A. Studentsova, V. G. Dunaev, and M. G. Berim, Dokl. Akad. Nauk SSSR, 160, 826 (1965); C.A., 62, 14721e (1965).
808. Voden, V. G., G. P. Nikitina, and M. F. Pushlenkov,

Radiokhimiya, 1, 121 (1959); C.A., 53, 19663c (1959).
809. Voskuil, W., and J. F. Arens, Rec. Trav. Chim., 81, 993 (1962).
810. Wagner, E. L., J. Am. Chem. Soc., 85, 161 (1963).
811. Walker, W. R., and N. C. Li, J. Inorg. Nucl. Chem., 27, 411 (1965).
812. Walling, C., O. H. Basedon, and E. S. Savas, J. Am. Chem. Soc., 82, 2181 (1960).
813. Walling, C., and S. A. Buckler, J. Am. Chem. Soc., 77, 6032 (1955).
813a. Walling, C., and M. S. Pearson, Topics in Phosphorus Chemistry, Vol. III, E. J. Griffith and M. Grayson, Eds., Interscience, New York, 1966, p. 1.
814. Walmsley, J. A., and S. Y. Tyree, Inorg. Chem., 2, 312 (1963).
815. Wang, H. K., Acta Chem. Scand., 19, 879 (1965).
816. Reference 712, p. 674.
817. Weiner, M. A., and G. Pasternack, J. Org. Chem., 32, 3707 (1967).
818. Welch, F. J., and H. J. Paxton, J. Polymer Sci., Part A-3, 3427 (1965).
819. Welch, F. J., and H. J. Paxton, J. Polymer Sci., Part A-3, 3439 (1965).
820. White, J. C., U. S. At. Energy Comm., ORNL 2161 (1956); C.A., 51, 4205c (1957).
821. Wilke, G., H. Schott, and P. Heimbach, Angew. Chem., Intern. Ed., 6, 92 (1967).
822. Wilkins, C. J., and H. M. Haendler, J. Chem. Soc., 1965, 3174.
823. Williams, D. H., R. S. Ward, and R. G. Cooks, J. Am. Chem. Soc., 90, 966 (1968).
824. Willans, J. L., Chem. Ind. (London), 1957, 235.
825. Wimer, D. C., Anal. Chem., 30, 77 (1958).
826. Wimer, D. C., Anal. Chem., 30, 2060 (1958).
827. Wimer, D. C., Anal. Chem., 34, 873 (1962).
828. Wittig, G., and W. Boll, Chem. Ber., 95, 2526 (1962).
829. Wittig, G., and H. J. Cristau, Bull. Soc. Chim. France, 1969, 1293.
830. Wittig, G., and A. Maercker, Chem. Ber., 97, 747 (1964).
831. Wittig, G., and M. Schlosser, Chem. Ber., 94, 1373 (1961).
832. Wittig, G., and V. Schöllkopf, Chem. Ber., 87, 1318 (1954).
833. Wittig, G., H.-D. Weigmann, and M. Schlosser, Chem. Ber., 94, 676 (1961).
834. Worrall, D. E., J. Am. Chem. Soc., 52, 2933 (1930).
835. Worrall, D. E., J. Am. Chem. Soc., 62, 2514 (1940).
836. Wulff, J., and R. Huisgen, Angew. Chem., Intern. Ed., 6, 457 (1967).
837. Yagupol'skii, L. M., and P. A. Yufa, Zh. Obshch.

Khim., 28, 2853 (1958); C.A., 53, 9109i (1959).
838. Yagupol'skii, L. M., and P. A. Yufa, Zh. Obshch. Khim., 30, 1294 (1960); C.A., 55, 431f (1961).
838a. Yakubovich, A. Ya., and V. A. Ginsburg, Dokl. Akad. Nauk SSSR, 82, 273 (1952); C.A., 47, 2685 (1953).
838b. Yakubovich, A. Ya., and V. A. Ginsburg, Zh. Obshch. Chim., 22, 1534 (1952); C.A., 47, 9254 (1953).
839. Yakubovich, A. Ya., V. A. Ginsburg, and S. P. Makarov, Dokl. Akad. Nauk SSSR, 71, 303 (1950); C.A., 44, 8320a (1950).
840. Yudina, K. S., T. Ya. Medved, and M. I. Kabachnik, Izv. Akad. Nauk SSSR, Ser. Khim., 1966, 1954; C.A., 66, 76096 (1967).
841. Zackrisson, M., and K. I. Alden, Acta Chem. Scand., 14, 994 (1960); C.A., 56, 8180f (1962).
841a. Zakharkin, L. I., O. Yu. Okhlobystin, and B. N. Strunin, Izv. Akad. Nauk SSSR, Otd. Khim. Nauk, 1962, 2002; C.A., 58, 9131h (1963).
842. Zanger, M., C. A. Vander Werf, and W. E. McEwen, J. Am. Chem. Soc., 81, 3806 (1959).
843. Zbiral, E., and L. Berner-Fenz, Tetrahedron, 24, 1363 (1968).
844. Zhuravleva, L. P., M. I. Zóla, M. G. Suleimanova, and A. V. Kirsanov, Zh. Obshch. Khim., 38, 342 (1968); C.A., 69, 27484 (1968).
845. Zingaro, R. A., and R. M. Hedges, J. Phys. Chem., 65, 1132 (1961).
846. Zingaro, R. A., R. E. McGlothlin, and E. A. Meyers, J. Phys. Chem., 66, 2579 (1962).
847. Zon, G., K. E. DeBruin, K. Naumann, and K. Mislow, J. Am. Chem. Soc., 91, 7023 (1969).
848. Zorn, H., H. Schindlbauer, and H. Hagen, Chem. Ber., 98, 2431 (1965).
849. Zweig, A., and A. K. Hoffmann, J. Am. Chem. Soc., 84, 3278 (1962).